Advances in Volcanology

An Official Book Series of the International Association of Volcanology and Chemistry of the Earth's Interior

Series editor

Karoly Nemeth, Palmerston North, New Zealand

Editor-in-chief

IAVCEI, Barcelona, Spain

More information about this series at http://www.springer.com/series/11157

Angelo Peccerillo

Cenozoic Volcanism in the Tyrrhenian Sea Region

Second Edition

Springer

Angelo Peccerillo
Perugia
Italy

Additional material to this book can be downloaded from http://extras.springer.com.

ISSN 2364-3277 ISSN 2364-3285 (electronic)
Advances in Volcanology
ISBN 978-3-319-42489-7 ISBN 978-3-319-42491-0 (eBook)
DOI 10.1007/978-3-319-42491-0

Library of Congress Control Number: 2016948611

Printed on acid-free paper

This Springer imprint is published by Springer Nature
The registered company is Springer International Publishing AG
The registered company address is: Gewerbestrasse 11, 6330 Cham, Switzerland

The original version of the book back matter was revised: Incorrect appendices have been corrected. The Erratum to the book backmatter is available at DOI 10.1007/978-3-319-42491-0_14

Preface

The Cenozoic volcanism in the Tyrrhenian Sea region shows a very wide range of petrological, geochemical and isotopic compositions, which cover almost entirely the field of igneous rock occurring worldwide. Volcanic activity took place contemporaneously with the opening of the Ligurian-Provençal and Tyrrhenian Sea basins, and with the formation of the Apennine-Maghrebian chain. Its occurrence in the hearth of the Mediterranean Sea and the spectacular eruptions of active volcanoes such as Etna and Vesuvio, attracted the attention of scholars since ancient times. Modern science has been interested to this volcanism because of its compositional complexity and the variety of volcanological processes taking place at the active and dormant centres. My book published by Springer in 2005 and entitled *Plio-Quaternary Volcanism in Italy. Petrology, Geochemistry, Geodynamics* summarised the state of knowledge of Tyrrhenian Sea volcanism at the time of publication. However, studies have been enormously expanding in the last 10 years, and huge amounts of data and ideas have been published on various aspects of this complex subject.

The present work is an extensively rewritten and expanded version of my 2005 book. Its objective is to update and integrate previous knowledge by reporting on the voluminous body of data and ideas published in the last 10 years and to elucidate some aspects of volcanism, which had not been adequately considered in the previous book. Therefore, information on volcanological and geophysical aspects of magmatism has been considerably increased, and some topics such as the Oligo-Miocene orogenic activity from Sardinia, Ligurian Sea and Provence are more fully treated. Moreover, a very short summary is provided of the Alpine stage magmatic activity that took place from Upper Cretaceous-Eocene to Miocene along the Alps and at some localities of the Adriatic and Pelagian blocks. Such magmatism shares many compositional characteristics with the Tyrrhenian-Apennine activity, but also exhibits significant differences. The comparison between the two magmatic suites is briefly discussed in Appendix 1, with the aim of stimulating further comparative studies that may provide a better insight into petrogenesis and geodynamics along the Africa-Europe converging boundaries.

The book is subdivided into thirteen chapters. The first one provides an introduction to the main petrological and geochemical characteristics and gives the rationale for subdivision of the magmatism into several distinct igneous provinces. The last chapter is a summary of the petrological, volcanological and structural characteristics of the magmatic provinces and of the most popular hypotheses that have been proposed to explain the relationship between geodynamics and volcanism. The other chapters are devoted to the volcanology, geochemistry, petrogenesis and geodynamic significance of single magmatic provinces into which the Tyrrhenian Sea magmatism has been subdivided.

The data discussed in the text have been generally taken from the most recent literature. They have been carefully checked, and some have been discarded because clearly incorrect. This, however, does not guarantee that the resulting files are free of errors. A few representative analyses for the single magmatic provinces have been reported in the tables attached to each chapter. More data can be found at the publisher website or can be obtained by the author on request (angelo.peccerillo@unipg.it; lathebiosas46@tiscali.it).

In most diagrams, especially those showing major and trace elements, a limited number of representative data have been plotted. Such a choice has been dictated by the need of avoiding excessive crowding, which would make diagrams difficult to read. It goes without saying that, in the choice of data points, care has been taken to select representative compositions, in order to preserve all necessary information.

Figures have been all drawn by the author, which explains the quality of diagrams and maps. Photos are partly by the author and partly by colleagues and friends. I thank Giampiero Poli, James T. Kirk, Federico Lucchi, Gianfilippo De Astis, Luciano Giannini and Giovanni Marano for providing some nice and informative images. The aerial photograph of Campi Flegrei has been downloaded from the NASA website.

For classification and nomenclature of volcanic rocks, the IUGS scheme of Le Maitre (2002) has been consistently adopted through the book. For rocks related to volcanic arcs, the classification scheme of Peccerillo and Taylor (1976) has been used. These classification schemes, together with a few notes on the classification of potassium-rich rocks, have been explained in Appendix 2. Rocks' names not reported in these schemes and eventually used through the book have been defined in footnotes or through the text. Notes on classification and nomenclature of igneous rocks are trivial for petrologists and geochemists, but may be useful for other potential readers who are not familiar with petrological-geochemical issues and jargon. Acronyms are unavoidable in modern petrology and geochemistry, but I have tried to keep them to a minimum.

Papers cited through the text are reported in the attached reference lists. These consist of about 1200 bibliographic entries, but are far from being comprehensive of the large quantity of papers and books published on magmatism and geodynamics of the Tyrrhenian Sea area. The bibliography on this subject is enormous and would deserve a book by itself. Therefore, many important contributions, especially in the field of mineralogy, physical volcanology, geophysics and geodynamics, have been omitted. While apologies are due to authors whose papers and books have not been cited, a somewhat more extended list of papers has been reported on the publisher website.

Several colleagues reviewed the various chapters and made important suggestions, corrections, and modifications, which significantly improved the style of writing and quality of science. It is obvious, however, that the responsibility for the contents of this book leans entirely on the author. I wish to express my gratitude to Michele Lustrino, Alba Santo, Laura Pinarelli, Gianfilippo De Astis and Marco Malusà for their help. I am also indebted to Donatella De Rita, Danilo Palladino, Paola Marianelli and Aidamaria Conte for providing precious information on Vulsini, Tolfa, Pontine Islands and Ischia volcanoes. Aidamaria Conte, Danilo Palladino and Attilio Giacobbe also allowed the use of their unpublished data on the Pontine Islands, Vulsini and Etna.

Special thanks are due to Simone Tarquini and to the Istituto Nazionale di Geofisica e Vulcanologia for kindly providing the 10-m resolution relief maps of volcanoes. The kindness and expertise of Simone were fundamental for map handling.

Professor Caterina Petrillo, Head of the Department of Physics and Geology of the University of Perugia, was extremely helpful for kindly allowing access to the library at the University of Perugia. Without the use of such a facility, writing this book would have been a mission impossible.

This effort is dedicated to my wife, my daughter and to the memory of my parents.

Perugia, Italy — Angelo Peccerillo

Contents

About the Author

Angelo Peccerillo graduated in Geology cum laude at the University of Florence, has been working as assistant professor, associate professor of Volcanology and full professor of Petrology at the Universities of Florence, Messina, Cosenza and Perugia. He has been visiting scientist at several national and international research centres (e.g. Australian National University, University of Göttingen, University of Paris XI at Orsay, etc.) where he carried out research activity on igneous rocks and volcanology. He participated in several national and international panels for the evaluation of research projects (NSF, ESF, CNR, MIUR, NATO, etc.) and scientific institutions (e.g. Padua and Siena Universities). He has been editor or member of the editorial board of several international and national journals such as the European Journal of Mineralogy, Lithos, Journal of Volcanology and Geothermal Research, Open Mineralogy Journal, Journal of Virtual Explorer, Journal of the Italian Geological Society, Acta Vulcanologica, Periodico di Mineralogia. He has been topic editor, for Igneous and Metamorphic Petrology, of the Encyclopedia of Life Support System (EOLSS) printed by UNESCO. His scientific research has been focused on petrology, geochemistry and geodynamic significance of igneous processes. For his scientific activity he has been awarded the Feltrinelli Medal by the Accademia Italiana dei Lincei, for the year 2006. He is author or co-author of some 200 scientific papers, mostly published on peer review international journals, several popular and didactic publications and scientific books, published with both national and international printers. He is Member of Academia Europaea and Honorary Member of the Geological Society of Italy.

Abstract

The Tyrrhenian Sea region is a tectonically complex area sited at the convergence between Europe and Africa. Because of its long-standing position along plate margins, a complex series of fundamental geological features developed in the area during the Cenozoic, including the Alps and Apennine fold-thrust belts, and several extensional basins. A wide variety of magma types erupted in the region, and their petrological, geochemical and isotopic compositions cover almost entirely the field of igneous rock occurring worldwide. Magmatism developed from Eocene to Present, exhibiting strong time-related compositional variations as a result of modification of geodynamic setting during Africa–Europe convergence. Most magmas originated within the upper mantle testifying to very heterogeneous sources and complex evolutionary processes during emplacement. Because of its position in the hearth of Europe and the extremely large variety of magma types and styles of eruptions, volcanism in the Tyrrhenian Sea region has attracted the attention of scholars since early ancient times and has been the subject of several thousand papers in the last few decades. The book is an up-to-date account of the geochemical data and of the petrogenetic hypotheses on such a complex geological issue.

The book is subdivided into 13 chapters plus Appendix 1 and Appendix 2. The first chapter is an introduction to the main petrological and geochemical characteristics of magmatism. Geochemical data are used to subdivide Cenozoic magmas occurring in the Tyrrhenian Sea region into two main families, respectively showing arc (orogenic magmas) and intraplate signatures (anorogenic magmas). Based on ages and compositional characteristics, several distinct magmatic provinces are recognised. The following chapters are devoted to the volcanology, geochemistry, petrogenesis, and geodynamic significance of single magmatic provinces into which the Tyrrhenian Sea magmatism has been subdivided. The last chapter is a summary of the petrological, volcanological and structural characteristics of the magmatic provinces and of the most popular hypotheses that have been proposed to explain the relationship between geodynamics and volcanism. Appendix 1 reports on a comparison between the Tyrrhenian Sea volcanics and the igneous rocks emplaced from Eocene to Miocene along the Alps and in nearby regions. Although the twofold orogenic and anorogenic subdivision is also recognised for Alpine rocks, many differences are highlighted to show that Alpine magmas were generated in distinct mantle sources than the Tyrrhenian Sea volcanism and such a variability might be related to the different geometry (e.g. dipping angle) and/or compositions of the subduction systems along Alps and Apennines. Appendix 2 is a short summary of classification and nomenclature of igneous rocks, with special emphasis on ultrapotassic compositions.

Keywords Petrology · Geochemistry · Volcanology · Igneous rocks · Geodynamics · Cenozoic · Tyrrhenian Sea · Apennines · Alps

1 Magmatism in the Tyrrhenian Sea Region: An Introductory Overview

Abstract

Cenozoic magmatism in the Tyrrhenian Sea region exhibits a wide range of major, trace element, and radiogenic isotope characteristics. Based on petrological and trace element signatures, especially LILE/HFSE ratios of mafic rocks, two main families of magmas are recognised respectively showing arc-type (orogenic) and intraplate (anorogenic) compositional features. Orogenic magmas have calcalkaline, shoshonitic and potassic alkaline affinities, and are spread over the entire Tyrrhenian region becoming younger eastward, from the Oligo-Miocene volcanic belt of Sardinia to the young to active calcalkaline to potassic alkaline volcanoes of the Italian peninsula and the southeastern Tyrrhenian Sea. Large volumes of Quaternary potassic and ultrapotassic rocks in central Italy represent the most important magma types of young orogenic activity. Anorogenic rocks include tholeiites to Na-alkaline compositions and occur in Sardinia, in the Tyrrhenian Sea basin and Sicily. No clear age polarity is observed for anorogenic rocks. Areal distribution, ages, and major, trace element and radiogenic isotope compositions allow recognising several distinct volcanic provinces. Except for a few rhyolites and granitoid rocks from Tuscany, the bulk of magmas is of mantle origin, testifying to extremely heterogeneous upper mantle sources. Geochronological, petrological and isotopic data suggest that multiple metasomatic events over compositionally variable pre-metasomatic mantle rocks are responsible for this heterogeneity.

Keywords

Tyrrhenian Sea · Magmatic provinces · Orogenic magmatism · Anorogenic magmatism · Mantle metasomatism · Apennines · Alps · Italy · Geodynamics

A. Peccerillo, *Cenozoic Volcanism in the Tyrrhenian Sea Region*,
Advances in Volcanology, DOI 10.1007/978-3-319-42491-0_1

1.1 Introduction

The Tyrrhenian Sea region is one of the most complex geological settings on Earth (e.g. Cavazza and Wezel 2003; Cavazza et al. 2004; Beltrando et al. 2010 and references therein). Such a complexity is basically the result of its position at the boundary between the converging plates of Europe and Africa, from Cretaceous to Present. Two main orogenic belts (Alps and Apennines), a number of pull-apart and back-arc basins and a widespread magmatic activity are the main effects of Late Cretaceous-Cenozoic geodynamic evolution in this region.

Magmatism developed almost continuously from Upper Cretaceous-Eocene to Present. Location of the main magmatic centres is shown in Fig. 1.1.

Late-Cretaceous to Miocene activity along the Alps and at few places of the Adria and Pelagian blocks is here referred to as the "Alpine stage" magmatic activity. Rocks in the Alps consist of several stocks and plutons (Valle del Cervo, Bregaglia-Bergel, Adamello, Vedrette di Ries-Riserferner, etc.), abundant dykes and a few lavas and volcanoclastic deposits, showing arc-type cal-calkaline to ultrapotassic compositions (e.g. Alagna et al. 2010; Lustrino et al. 2011). These are collectively referred to as the Periadriatic Magmatic Province. Magmatism developed in the Adriatic-Africa plate away from the Adria-Europe colliding margins (Pietre Nere, La Queglia, and southeastern Sicily) shows tholeiitic to Na-alkaline compositions and intraplate geochemical and isotopic signatures (e.g. Avanzinelli et al. 2012).

A distinct stage of magmatism developed mainly form Upper Eocene-Oligocene to Present in the Tyrrhenian Sea region, and was coeval with the opening of the Western Mediterran extensional basins and the formation of the Apennine-Maghrebian chain (Lustrino et al. 2011). This younger phase is here referred to as the "Apennine stage" magmatism.

Composition, origin and geodynamic significance of the Apennine-stage magmatic activity are the main focus of this book. Short information on the Alpine-stage magmatism is given in Appendix 1.

Magmas erupted during the Apennine stage activity show a wide range of compositions, from mafic to felsic and from subalkaline to ultra-alkaline. The largest compositional variations are observed among the youngest rocks, mainly erupted by the Plio-Quaternary volcanoes of central-southern Italy. Most studies carried out in the last three decades agree that such a variability testifies to magma origin in a strongly heterogeneous and anomalous upper mantle, whose composition is the result of continuous input of various types of crustal materials, accompanied by upwelling of asthenospheric or deep mantle rocks, in some areas (e.g. Peccerillo 2005; Peccerillo and Frezzotti 2015 and references therein). Intensive magma evolution at several volcanic centres further complicates the magmatic setting of the Tyrrhenian Sea area.

In this chapter, an introductory overview of the geochronology, petrology and geochemistry of the Upper Eocene-Oligocene to Present magmatism in the Tyrrhenian Sea region is given. The scope is that of describing the main compositional characteristics of the rocks, regional variations of magma compositions, major constraints for petrogenesis, and geodynamic setting in which magmatism developed.

1.2 Geochronology and Petrology

Petrological characteristics and ages for the main magmatic centres in the Tyrrhenian Sea region are summarised in Fig. 1.1 and Table 1.1. The oldest activity occurred in western Sardinia, Ligurian Sea, and Provence (e.g. Réhault et al. 2012), and successively shifted to Tuscany, the Tyrrhenian Sea basin, and to central-southern Italy.

Early volcanism in Sardinia is mainly Oligo-Miocene in age, with scanty Eocene occurrences (Older Sardinia volcanism). This is followed by Miocene to Quaternary activity in the same area (Younger Sardinia volcanism). Igneous rocks in Tuscany range from about 8.5 to 0.3 Ma, and generally young towards the east, from the Tuscan archipelago to the mainland. The age of this magmatism extends to about

Fig. 1.1 Distribution, petrochemical affinity and ages of Upper Eocene to Plio-Quaternary magmatic centres in the Tyrrhenian Sea region. White symbols refer to outcrops below the sea level. Main Alpine-stage magmatic centres are also reported (*yellow symbols*). Relief shaded map is from Tarquini et al. (2012)

Table 1.1 Petrological characteristics and ages of Cenozoic volcanic provinces in the Tyrrhenian Sea region

Magmatic Province	Main magmatic centers and ages (in Ma)	Petrology and geochemistry
Sardinia Oligo-Miocene (33–15 Ma; single dating at 38 Ma)	Bosa-Logudoro district (38–15), Arcuentu (~30–17), Sarroch (~24–22), Sulcis (~28–15). Volcanism extends to offshore western coast of Corsica, to the Ligurian Sea (20–7 Ma with a single dating at 43 Ma), and Provence (33–7 Ma)	- Dissected stratovolcanoes, ignimbrite sheets, subvolcanic bodies, etc. formed of dominant calcalkaline basalt to rhyolite and a few tholeiitic, shoshonitic and comenditic rocks. Variable isotope signatures (e.g. $^{87}Sr/^{86}Sr$ ~ 0.704–0.710), mostly resulting from crustal assimilation. High LILE/HFSE ratios, typical of subduction related magmas, with a very few occurrences in the Ligurian Sea and Provence, southeastern France, showing intraplate-type trace element signatures
Tuscany (~14 to 0.3 Ma)	***Acid intrusions***: Elba (8.5–6.5), Montecristo (7.1), Giglio (5), Campiglia-Gavorrano (5–4.3), Vercelli Seamount (7.2) Hidden intrusions in the Larderello area (<5) ***Acid volcanoes***: San Vincenzo (4.5), Roccastrada (2.5), Amiata (0.3), Older Cimini (1.3), Tolfa-Manziana-Cerite (3.5) ***Mafic centres***: Sisco (14.5), Capraia (7.6–4.6), Eastern Elba (5.8), Campiglia (4.3), Orciatico and Montecatini (4.1), Radicofani (1.3), Younger Cimini (1.3–0.9), Torre Alfina (0.8)	**- *Acid rocks***: Peraluminous to metaluminous monzogranite, granodiorite, granites, aplites, pegmatites (Elba, Montecristo, Giglio, etc.), lava flows and domes (e.g. San Vincenzo, Roccastrada), and a few ignimbrites (Cimini and Tolfa). Crustal-type geochemical and radiogenic isotope signatures suggesting crustal anatexis plus mixing with mafic-intermediate melts
		- *Mafic-intermediate rocks*: monogenetic extrusive and subvolcanic bodies sometimes containing mantle xenoliths, stratovolcano (Capraia), mafic enclaves in silicic rocks, showing calcalkaline to ultrapotassic (*lamproite*) compositions. High Mg#, Ni and Cr, indicating a mantle origin, coupled with crustal-like trace element and radiogenic isotope signatures (e.g. $^{87}Sr/^{86}Sr$ up to about 0.717)
Intra-Apennine (~0.8–0.25 Ma)	San Venanzo (0.26), Cupaello (0.64), Polino (0.25), Acquasparta (0.39), Oricola (0.53), Colle Fabbri (0.8), etc.	- Monogenic mafic to felsic pyroclastic centres and rare mafic lavas with an ultrapotassic melilititic (*kamafugite*) composition. Calcite-rich pyroclastic and subvolcanic rocks interpreted as carbonatites. Melilitolites of dubious origin at Colle Fabbri. Trace element and radiogenic isotope compositions similar to the Roman Province.
Roman (Latium) (>0.8–0.02 Ma)	Vulsini (>0.6–0.13), Vico (0.4–0.1), Sabatini (0.8–0.09), Colli Albani (0.6–0.02)	- Large volcanic complexes (Vulsini, Sabatini) and stratovolcanoes (Vico, Colli Albani) formed of potassic (trachybasalt to trachyte) and ultrapotassic (leucite tephrite to phonolite) pyroclastics and minor lavas. High enrichments in incompatible elements. $^{87}Sr/^{86}Sr$ around 0.709–0.711.

(continued)

Table 1.1 (continued)

Magmatic Province	Main magmatic centers and ages (in Ma)	Petrology and geochemistry
Ernici-Roccamonfina (0.6–0.15 Ma)	Ernici: Pofi, Ceccano, Patrica, etc. (0.63–0.27) Roccamonfina (0.55–0.15)	- Monogenetic mafic pyroclastic and lava centres at Ernici (Middle Latina Valley), and a stratovolcano at Roccamonfina. Calcalkaline and shoshonitic to potassic and ultrapotassic compositions, with variable incompatible element abundances and radiogenic isotope signatures, encompassing compositions of adjoining Roman and Campania provinces
Pontine Islands (4.2 to less than 0.13 Ma)	Ponza (4.2–1.0), Palmarola (1.64–1.54), Zannone, La Botte islet (1.2), Ventotene (0.8–0.13)	- Domes, cryptodomes, hyaloclastites and pyroclastic rocks with rhyolite, peralkaline rhyolite and trachyte compositions in the western islands of Ponza, Palmarola, Zannone, and La Botte, plus some submarine mafic-intermediate calcalkaline to potassic rocks. Basalt-trachybasalt, latite, trachyte and phonolite lava flows, pyroclastic layers and domes in the eastern islands of Ventotene and Santo Stefano, where trace element and radiogenic isotope compositions are similar to the potassic suites from Ernici-Roccamonfina
Campania-Stromboli (>0.2 Ma to Present)	Somma-Vesuvio (0.03–1944 AD), Campi Flegrei (0.2–1538 AD), Ischia (0.15–1302 AD), Procida (0.07–0.014). About 2 Ma-old volcanism buried beneath the Campanian Plain. Stromboli (0.2 to Present)	- Shoshonitic, potassic (trachybasalt to trachyte) and ultrapotassic (leucite tephrite to phonolite) rocks forming stratovolcanoes and multi-centre complexes. Lower incompatible element contents, LILE/HFSE and radiogenic Sr than ultrapotassic rocks from the Roman Province, and similar compositions as Stromboli. Mafic-intermediate calcalkaline composition for volcanism buried beneath the Campanian Plain. Stromboli consists of mafic-intermediate calcalkaline to leucite-bearing potassic volcanics
Vulture (0.75–0.14 Ma)	Vulture stratovolcano with poligenetic caldera and a few parasitic centres	- Na-K-rich lavas and pyroclastics with tephrite, foidite to phonolite compositions, containing haüyne as a main foid. Occurrence of haüynite lava at Melfi. Late erupted carbonatite lavas and pyroclastics. Coexisting arc-type and intraplate OIB-type trace element signatures
Western-Central Aeolian arc (>0.27 Ma to Present)	Alicudi (0.1–0.03), Filicudi (0.25–0.03), Salina (0.25–0.016), Lipari (0.27–1220 AD), Vulcano (0.13-1888-1890 AD), Panarea (0.15–0.009)	- Stratovolcanoes with calcalkaline compositions in the western-central sector of the arc, and dominant shoshonitic to potassic rocks in the central arc. Abundant rhyolitic rocks in the central islands (Lipari, Vulcano). Typical arc-type geochemical signatures; Sr-isotope ratios increase and Nd–Pb isotope ratios decrease toward the eastern Island of Stromboli, which shows isotope and trace element compositions similar to the Campania volcanoes

(continued)

Table 1.1 (continued)

Magmatic Province	Main magmatic centers and ages (in Ma)	Petrology and geochemistry
Sicily (7 Ma to Present)	Etna (0.5-Present), Iblei (7–1.5), Ustica (0.75–0.13), Pantelleria (0.3–0.01), Linosa (1–0.5), Sicily Channel seamounts (Miocene to 1891 AD)	- Stratovolcanoes, diatremes, small plateaux, etc. formed of tholeiitic to Na-alkaline rocks (nephelinite, basanite, hawaiite, trachyte, peralkaline trachyte and rhyolite). Several seamounts in the Sicily Channel. Typical intraplate compositions, with low LILE/HFSE ratios, low-moderate Sr-isotope and high Nd-isotope ratios, and variable Pb isotope signatures
Sardinia Miocene-Quaternary (12–0.1)	*Mio-Pliocene stage:* Isola del Toro (11.8), Capo Ferrato (~6.6–5), Guspini (6.4), Rio Girone (4.4). *Plio-Quaternary stage:* Montiferro (3.9–1.6), Orosei-Dorgali (3.6–2), Monte Arci (3.8–2.6), Logudoro (3.1–0.1), etc.	- *Miocene-Pliocene stage*: small outcrops of basanite-trachybasalt to trachyte in Southern Sardinia. Low LILE/HFSE ratios and moderately radiogenic Pb-isotope compositions. - *Plio-Quaternary stage*: Stratovolcanoes, basaltic plateaux and monogenetic centres composed of subalkaline to Na-alkaline rocks, sometimes with a K-affinity. Typical intraplate compositions with low LILE/HFSE ratios and low uranogenic Pb isotope ratios similar to EM1, a unique case in Europe
Tyrrhenian Sea Floor (12 Ma to Present)	Cornacya (12), Magnaghi (3), Marsili (1.1–0.1), Vavilov, Aceste, Anchise, Lametini, Palinuro, etc.	- Coexisting intraplate (MORB-type, Na-transitional and alkaline OIB-type) and arc-type (arc-tholeiitic, calcalkaline, potassic) rocks, the latter showing a decrease in age from west to the southeast

14.5 Ma if the ultrapotassic dyke from Sisco, NE Corsica, is considered. Rocks cropping out from Vulsini to the Ernici (Middle Latina Valley) and Roccamonfina cover a time span of about 0.8 to 0.02 Ma. Active volcanism starts in Campania (Vesuvio, Campi Flegrei and Ischia) and continues in the Aeolian Islands, in eastern Sicily (Etna), and along the Sicily Channel (Ferdinandea-Graham, Foerstner).

Magmatic rocks in the Tyrrhenian Sea region exhibit a very large range of petrological characteristics and cover almost all the different compositional fields on Total Alkali vs. Silica diagram (TAS; Le Maitre 2002), a scheme designated for classifying all the main volcanic rocks occurring on Earth (Fig. 1.2a). Rocks range from subalkaline to alkaline, and from mafic to silicic. In several areas, mafic rocks occur in subordinate amounts with respect to intermediate and silicic volcanics. Most of the evolved rocks represent daughter magmas derived from mafic parents by complex evolutionary processes. Notable exceptions are represented by some silicic rocks in Tuscany, which were formed by crustal anatexis or by mixing between crustal anatectic and mantle-derived melts (Poli and Peccerillo 2016 with references).

Although present in small amounts in most volcanic centres, mafic rocks (here defined as those with MgO > 4.0 wt%) are particularly interesting since they represent poorly to moderately evolved compositions, which may provide important information on primary (i.e. unmodified) mantle-derived melts, on the nature of the sources and on mechanisms of magma formation.

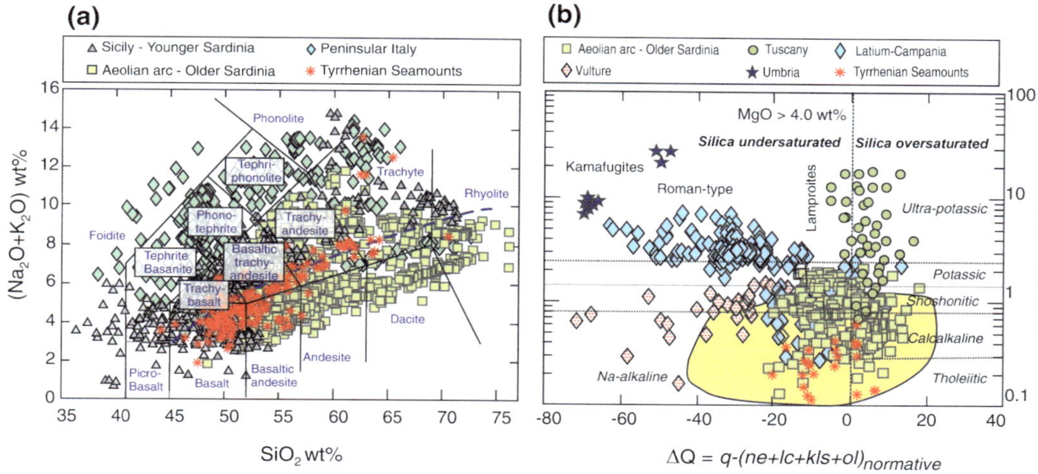

Fig. 1.2 **a** Total alkali vs. silica (TAS) classification diagram (Le Maitre 2002) for representative Cenozoic volcanic rocks from the Tyrrhenian Sea region. The *thick dashed line* is the divide between the subalkaline and the alkaline fields of Irvine and Baragar (1971). Older Sardinia and Younger Sardinia are Oligo-Miocene and Miocene-Quaternary volcanic stages, respectively. Analyses normalised to 100 % on a LOI-free basis; **b** ΔQ versus K_2O/Na_2O classification diagram for mafic volcanic rocks (MgO > 4.0 wt%) from the Tyrrhenian Sea region. ΔQ is the algebraic sum of normative quartz, minus undersaturated minerals (nepheline, leucite, kalsilite, and Mg-olivine). The yellow field encloses the Miocene to Quaternary rocks from Sardinia and Sicily. For further explanation see text and Appendix 2

A classification scheme, modified after Sahama (1974), shows the ΔQ versus K_2O/Na_2O (expressed as wt%) relationships for mafic rocks (Fig. 1.2b; Peccerillo 2002, 2003). ΔQ is the algebraic sum of normative quartz, minus undersaturated minerals (nepheline, leucite, kalsilite, and olivine). It quantifies the degree of silica saturation: saturated magmas have $\Delta Q \sim 0$, undersaturated magmas have $\Delta Q < 0$, whereas oversaturated magmas have $\Delta Q > 0$. The diagram shows that mafic rocks range from strongly undersaturated to oversaturated in silica. The largest compositional variability is observed among Plio-Quaternary mafic rocks, which cover the tholeiitic, calcalkaline, shoshonitic, and the K-, Na-alkaline and ultra-alkaline fields. Carbonatitic rocks also occur at some volcanoes, but are not plotted in the diagram. Among potassic and ultrapotassic rocks, various groups can be distinguished on the basis of silica saturation and K_2O/Na_2O ratios. These include lamproites, kamafugites, potassic series (or KS) and Roman-type ultrapotassic series (HKS). Lamproites are characterised by SiO_2 (~56–58 wt%) and CaO (~3–5 wt%) contents typical of intermediate-silicic magmas, but show high MgO (up to 10 wt%), and relative depletion in Al_2O_3 (~10–13 wt%) and Na_2O (~1.0–1.5 wt%). The latter two oxides also show low abundances in kamafugites, which, however, are enriched in CaO (~15 wt%), MgO (~10–15 wt%) and depleted in SiO_2 (~40–45 wt%). Because of their chemistry, both lamproites and kamafugites are plagioclase-free rocks. Roman-type rocks show higher Al_2O_3 and Na_2O and SiO_2 than kamafugites, with consequent occurrence of plagioclase in the phenocryst mineralogy (see Appendix 2; Foley et al. 1987; Peccerillo 1992; Conticelli et al. 2015).

1.3 Regional Distribution of Magma Types

Tholeiitic and Na-alkaline rocks occur in eastern Sicily at Etna and Iblei, in the Sicily Channel, in Sardinia (Miocene-Quaternary or Younger Sardinia volcanism) and in the Tyrrhenian Sea basin (e.g. Vavilov, Magnaghi, Prometeo seamounts). A few arc-tholeiites have ben found at some Aeolian seamounts. Calcalkaline and shoshonitic rocks are concentrated in the Aeolian arc and in Sardinia (Oligo-Miocene or Older Sardinia volcanism), with minor occurrences in the Naples area, in Central Italy (e.g. Ernici, Capraia, Radicofani), and in the Tyrrhenian Sea basin.

Potassic and ultrapotassic compositions are typical of Central Italy, although a few leucite-bearing potassic rocks also occur in the Aeolian arc (e.g. Conticelli et al. 2002; Peccerillo 2002; Peccerillo et al. 2013). Ultrapotassic rocks from Tuscany are nearly saturated to oversaturated in silica and are classified as lamproites. Older lamproites occur at Sisco (Corsica; about 14.5 Ma), and in the Western Alps, where ages are around 30 Ma (see Appendix 1). Strongly undersaturated ultrapotassic rocks (kamafugites) make up a few monogenetic centres in the internal zones of northern-central Apennines (Umbria, Abruzzi), but are also found as small outcrops at other volcanoes in Central Italy. Volcanism extending from Vulsini to Vesuvio is formed by dominant potassic and Roman-type ultrapotassic rocks. Mount Vulture volcano, east of Vesuvio, is composed of undersaturated alkaline rocks that are rich both in K_2O and Na_2O (e.g. De Fino et al. 1986).

1.4 Regional Variation of Trace Element and Sr-Nd–Pb-O Isotopic Compositions

Abundances and ratios of incompatible[1] elements are best illustrated by normalised diagrams (spider-diagrams), where concentrations of single elements in the rocks are divided by the abundances of the same elements in the primitive mantle. Spider-diagrams of representative mafic rocks are shown in Fig. 1.3. Samples from eastern Sicily, Sicily Channel, Younger Sardinia and some Tyrrhenian seamounts have a rather smooth pattern with a maximum at Ta and Nb, and negative anomalies of K and Rb for some samples (Fig. 1.3a). These features are typical of magmas erupted in intraplate tectonic settings or along diverging plate margins. Therefore, rocks showing these patterns have been referred to as "anorogenic". In contrast, the mafic rocks from the Aeolian arc, the Italian peninsula, Older Sardinia and some Tyrrhenian seamounts are more enriched in Large Ion Lithophile Elements (LILE: Rb, Cs, Ba, U, Th, K, Light REE, and Pb) and display depletion of the High Field Strength Elements (HFSE: Ta, Nb, Zr, Hf and Ti; Fig. 1.3b). Negative anomalies of HFSE and enrichments in LILE are typical of magmas erupted along converging plate margins (e.g. Gill 1981). Therefore, rocks with these characteristics have been defined as "arc-type" or "orogenic" (e.g. Peccerillo 2001, 2005; Carminati et al. 2010; Lustrino et al. 2011, 2013).

Mafic magmas erupted in different tectonic settings, especially anorogenic vs. arc-type volcanic rocks, can be well discriminated by their relative abundances of LILE and HFSE. A discriminant diagram, in which Th and Ta are normalised against a non-discriminant element (Yb), is shown in Fig. 1.4 (Pearce 1982). Mafic rocks

[1]Incompatible elements are not hosted by common igneous rock-forming minerals and prefer to go into the liquid during melting and magma crystallization.

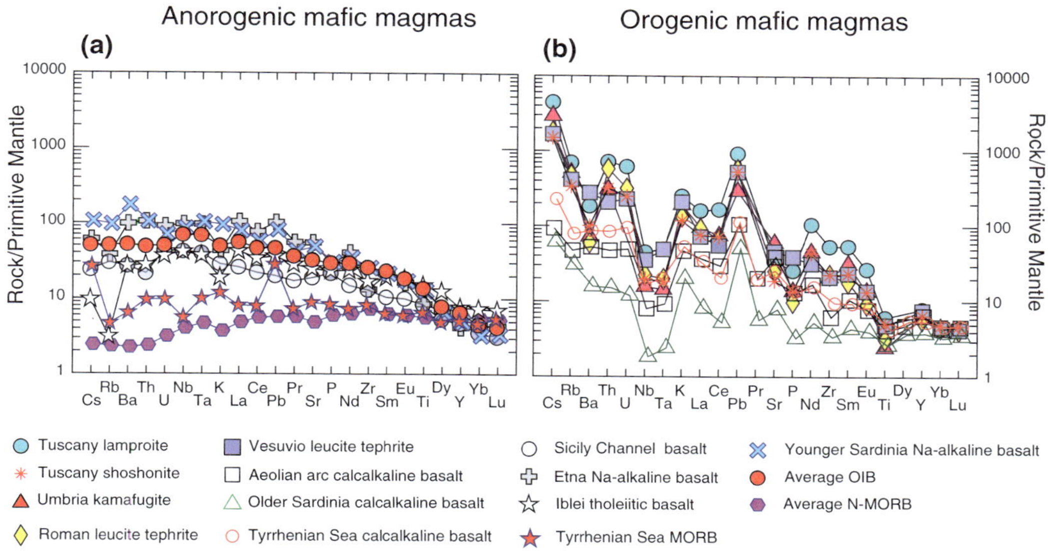

Fig. 1.3 Patterns of incompatible elements normalised against primitive mantle composition (Sun and McDonough 1989) for some mafic rocks from the Tyrrhenian Sea region. Patterns of average Ocean Island Basalts (OIB; Sun and McDonough 1989) and Normal Mid-Ocean Ridge Basalts (N-MORB, Gale et al. 2013) are shown for comparison. Older Sardinia and Younger Sardinia indicate Oligo-Miocene and Miocene-Quaternary volcanic activity, respectively

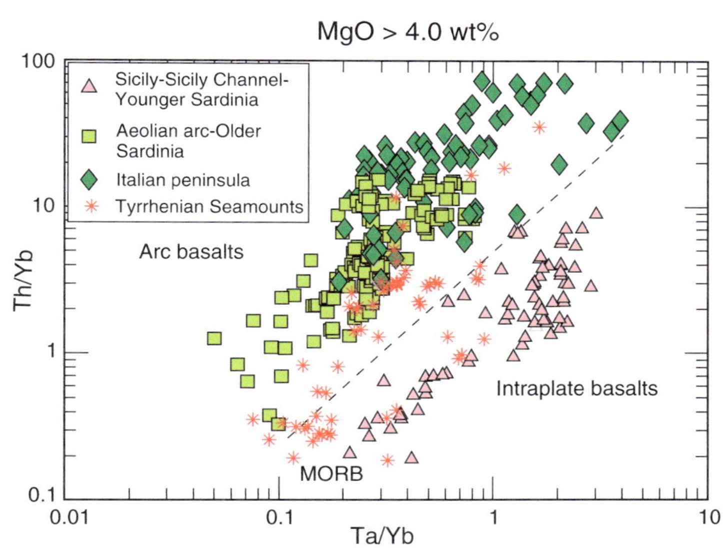

Fig. 1.4 Th/Yb vs. Ta/Yb discriminant diagram (Pearce 1982) for representative Cenozoic mafic rocks (MgO > 4.0 wt %) from the Tyrrhenian Sea region. Older Sardinia and Younger Sardinia indicate Oligo-Miocene and Miocene-Quaternary volcanic stages, respectively

from the Tyrrhenian Sea region span intraplate (anorogenic) and volcanic arc (orogenic) compositions. Rocks from Etna, Iblei, Ustica, Younger Sardinia, and Sicily Channel plot in the field of intraplate basalts, whereas the Aeolian arc, Central Italy, and Older Sardinia plot in the field of arc basalts. Rocks from the Tyrrhenian seamounts plot in both the fields.

Incompatible element and radiogenic isotope ratios furnish additional information, especially on rocks classified as orogenic. Plots of LILE/LILE, LILE/HFSE and $^{87}Sr/^{86}Sr$ ratios

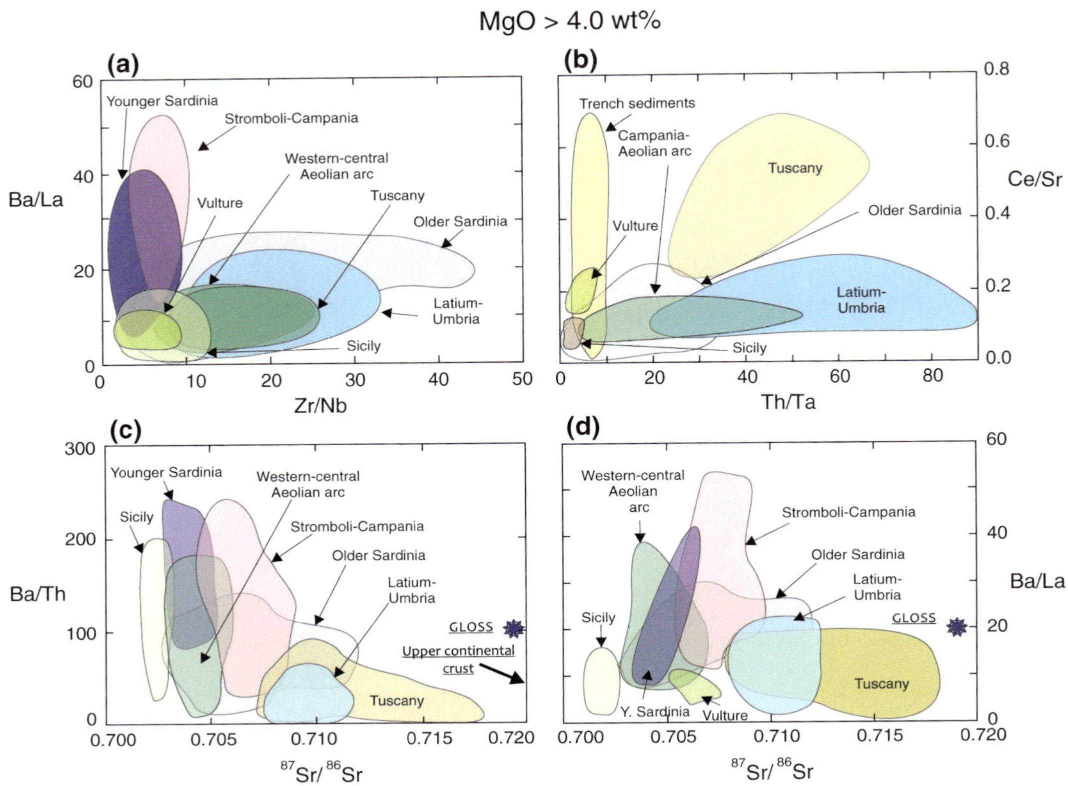

Fig. 1.5 Variation diagrams of incompatible trace element ratios and $^{87}Sr/^{86}Sr$ for Cenozoic mafic volcanic rocks (MgO > 4.0 wt%) from the Tyrrhenian Sea region. Older Sardinia and Younger Sardinia indicate Oligo-Miocene and Miocene-Quaternary volcanic activity, respectively. GLOSS composition (Global Subducted Sediments) is from Plank and Langmuir (1998)

show that mafic rocks from the Tyrrhenian Sea region define distinct trends that are closely related to the geographic position of volcanoes rather than to petrological characteristics of magmas (Fig. 1.5). For instance, Tuscany rocks, which include calcalkaline, shoshonitic and ultrapotassic compositions, plot along different trends than the equivalent association from Stromboli-Campania.

Radiogenic isotope signatures (Sr, Nd, Pb, Hf) show large variations (e.g. Conticelli et al. 2002, 2010; Gasperini et al. 2002; Bell et al. 2004, 2013 and references therein). However, various regions exhibit rather restricted and distinct isotopic signatures, which further highlight regionality of rock compositions (Fig. 1.6).

1.5 Magmatic Provinces in the Tyrrhenian Sea Region

The strong regionality for major, trace element and isotopic compositions of volcanism in the Tyrrhenian Sea region allows subdivision into several compositionally different magmatic provinces. For the purpose of this review, a "magmatic province" is defined as a restricted zone in which a significant amount of igneous activity has been going on over a relatively short period of time, for a few Ma or less. The rocks of each province show peculiar compositional characteristics, such as petrochemical affinity, geochemical signatures, or even a particular

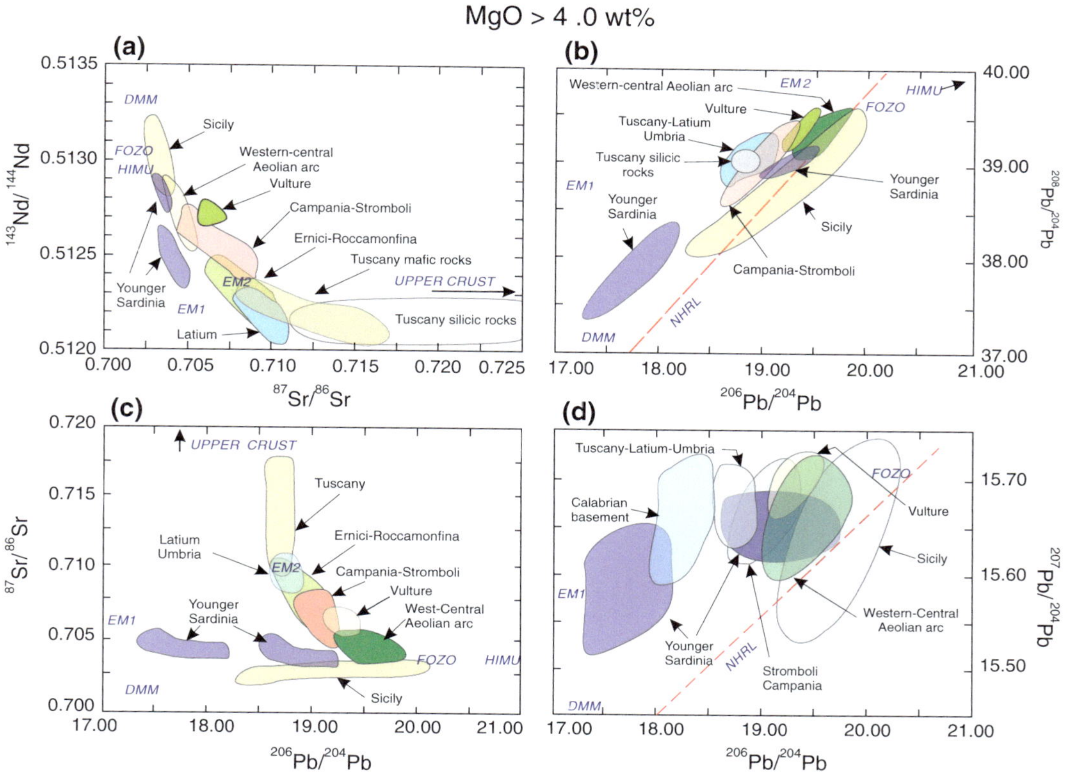

Fig. 1.6 Sr-Nd–Pb isotope variations for Cenozoic mafic volcanic rocks (MgO > 4.0 wt%) from the Tyrrhenian Sea region. The *dashed line* on Pb-isotope diagrams (NHRL) is the Northern Hemisphere Reference Line (Hart 1984). Some mantle end-member compositions are also indicated (FOZO = Focal Zone; HIMU = High-μ, i.e. high $^{238}U/^{204}Pb$; EM1 = Enriched Mantle 1; EM2 = Enriched Mantle 2; DMM = Depleted MORB Mantle; Zindler and Hart 1986; Hofmann 1997; Stracke et al. 2005). (EM1, EM2, HIMU, FOZO, and DMM compositions are deduced from isotopic measurements on intraoceanic volcanoes. They developed as a result of ageing of mantle rocks that had undergone very ancient melt extraction or contamination by different types of crustal materials, such as sediments, altered oceanic crust, or lower continental crust. These processes imparted variable Rb/Sr, Sm/Nd, Lu/Hf, Th/Pb and U/Pb to the mantle rocks, which developed distinct isotopic signatures with ageing)

association of magma types, which make them different from rock associations occurring in other zones. However, magmas of a given province might not be strictly cogenetic, i.e. do not necessarily derive from a single source or magma type. The underlying assumption of this definition is that magmas closely associated in space and time and showing distinctive compositional characteristics, are likely related to specific geological processes, which are somewhat different from those occurred in other areas. Based on the present definition, the Cenozoic igneous activity in the Tyrrhenian Sea region has been grouped into the following magmatic provinces (Fig. 1.7, Table 1.1; Peccerillo 2002, 2005):

- **The Tuscany Province**
  It extends from the Tuscan archipelago to the mainland of southern Tuscany and to the Tolfa-Manziana-Cerite area, about 80 km northwest of Rome. A potassic alkaline sill cropping out at Sisco, NE Corsica, is also included in the Tuscany Province, because of its ultrapotassic petrogenetic affinity similar to some Tuscany occurrences. Mafic to silicic magmas occur in this province. Silicic rocks are polygenetic: some are of crustal anatectic

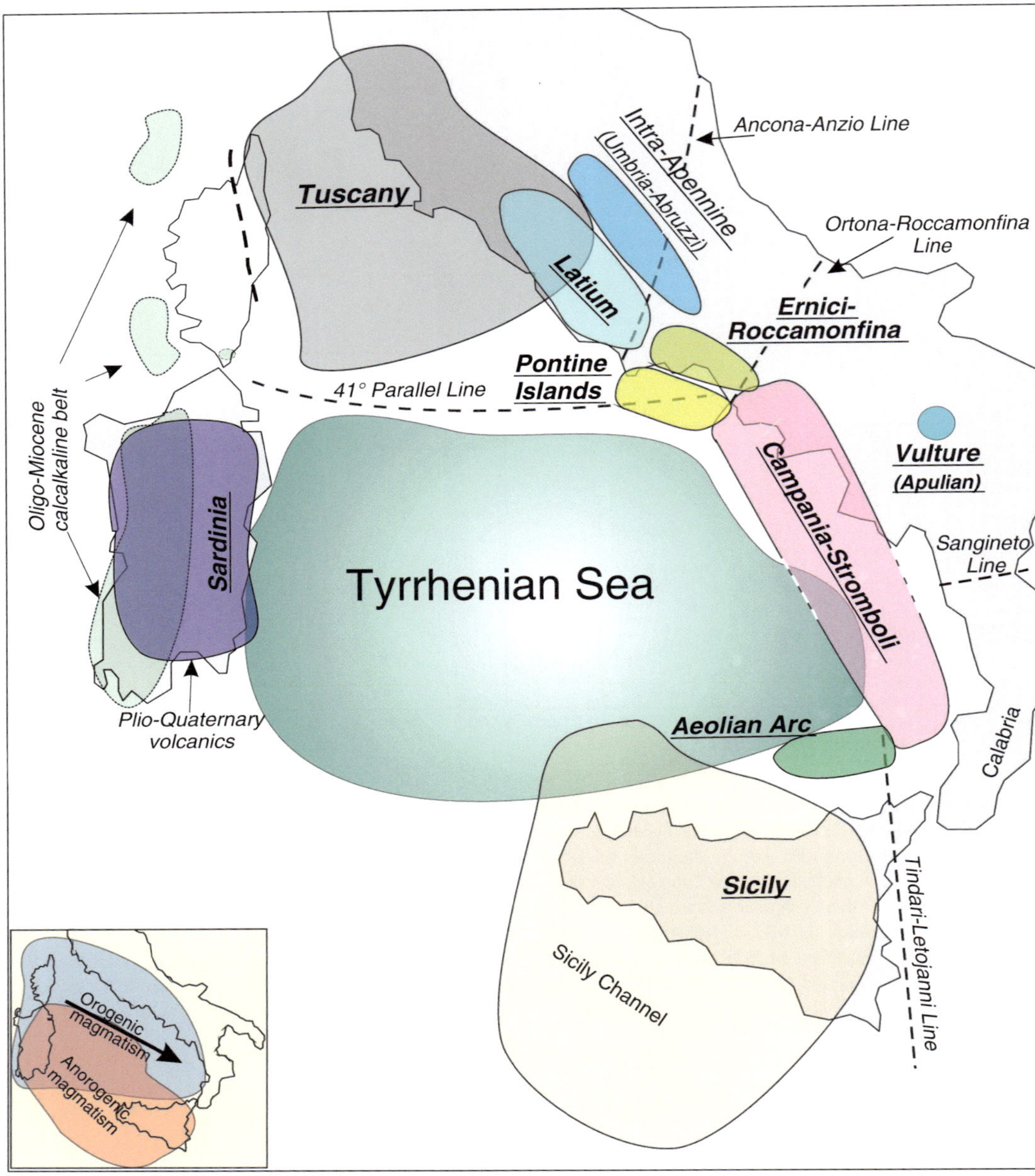

Fig. 1.7 Magmatic provinces in the Tyrrhenian Sea region, as identified from major, trace element and isotopic characteristics of mafic rocks. Some major tectonic lines across the Tyrrhenian Sea, the Apennines and Sicily (Ancona-Anzio, Ortona-Roccamonfina, 41° Parallel, Sangineto, Tindari-Letojanni) are also indicated. Inset: distribution of volcanism with orogenic (i.e. high LILE/HFSE ratios) and anorogenic (i.e. low LILE/HFSE ratios) compositions. The arrow indicates south-eastward migration of orogenic magmatism from Upper Eocene-Oligocene to Present. See text for further explanation

origin; others are mixtures of crustal magmas and minor amounts of mantle-derived melts; a few minor bodies represent daughter products derived from mafic-intermediate magmas by fractional crystallization or other evolutionary processes. Mafic rocks range from calcalkaline and shoshonitic to potassic and ultrapotassic. Some ultrapotassic rocks contain high-pressure ultramafic xenoliths and xenocrysts (e.g. Peccerillo et al. 1988; Conticelli and Peccerillo 1990). The mafic rocks from this province show compositions that are closer to typical crust-derived than to mantle-derived magmas (e.g. $^{87}Sr/^{86}Sr$ up to 0.717, $^{143}Nd/^{144}Nd$ ~ 0.5122; $^{206}Pb/^{204}Pb$ ~ 18.80; Poli et al. 1984; Conticelli and Peccerillo 1992; Conticelli et al. 2002; Poli and Peccerillo 2016 and references therein). A similar association of calcalkaline, shoshonitic and lamproitic mafic rocks occur in southeastern Spain and in the Western Alps (e.g. Peccerillo and Martinotti 2006).

- **The Intra-Apennine Province**
  This province comprises a number of small centres (e.g. San Venanzo, Cupaello, Polino, Oricola etc.) scattered through the internal zones of Apennines, from Umbria to Abruzzi. The best known occurrences consist of ultrapotassic kalsilite-pyroxene and olivine melilitite lavas (kamafugites) cropping out at San Venanzo and Cupaello. Many pyroclastic rocks contain abundant calcite and have been classified as carbonatites (Stoppa and Woolley 1997). Incompatible element ratios and isotopic signatures are similar to those of volcanic rocks from Latium. However, the petrological characteristics are different, with the intra-Apennine kamafugites displaying lower Al_2O_3 and Na_2O, and higher CaO, K_2O/Na_2O and degree of silica undersaturation.
- **The Roman Province or Latium Province**
  This is part of the belt of potassic and ultrapotassic rocks running from northern Latium to the Neapolitan area, which was defined as the Roman Comagmatic Region by Washington (1906). The Roman Province (or Latium Province) defined here only includes Vulsini, Vico, Sabatini and Colli Albani (Alban Hills) volcanoes. The newly defined Roman Province is limited at its southern end by an important tectonic line that crosses the Apennines and the Italian peninsula and is known as the Ancona-Anzio line, or Olevano-Antrodoco line (e.g. Castellarin et al. 1982; Turtù et al. 2013). Potassic rocks (KS) basically consist of trachybasalts, shoshonites, latites and trachytes; ultrapotassic rocks (HKS) are represented by leucite-tephrites, leucitites to leucite-phonolites. Evolved rocks largely prevail over mafic ones, and mainly occur as caldera-forming ignimbrite and fallout deposits. Isotope compositions are still close to crustal values, but are less extreme than in Tuscany (for example $^{87}Sr/^{86}Sr$ ratio is around 0.7090–0.7110; Hawkesworth and Vollmer 1979; Conticelli et al. 2002, 2010 and references therein).
- **The Ernici-Roccamonfina Province**
  The Ernici-Roccamonfina volcanoes erupted minor volumes of magmas, relative to the adjoining volcanic provinces. They consist of several monogenetic centres (Ernici or Middle Latina Valley) and the strotovolcano of Roccamonfina. Rock compositions range from calcalkaline to KS and Roman-type HKS. A few kamafugites have also been found (Frezzotti et al. 2007; Boari et al. 2009). The least potassic rocks display incompatible trace element and radiogenic isotope ratios that are close to those of the Neapolitan volcanoes (Vesuvio, Campi Flegrei, Ischia). On the contrary, ultrapotassic rocks resemble the Colle Albani and other Latium volcanoes. Therefore, the Ernici-Roccamonfina Province is considered as a sort of transition zone between the Roman and the Campania provinces (Peccerillo 2001; Peccerillo and Frezzotti 2015).
- **Pontine Islands**
  The volcanoes of this province developed along the eastern segment of the 41° Parallel line, a main tectonic feature across the Tyrrhenian Sea (e.g. Bruno et al. 2000). Compositions of rocks exposed at the surface are silicic (rhyolite, peralkaline rhyolite, trachyte) in the western islands of Ponza, Palmarola, Zannone and La Botte, but mafic-intermediate submarine calcalkaline to potassic rocks have been recently recovered (Conte et al. 2016). Basalt and trachybasalt to trachyte and phonolite crop out in the eastern islands of Ventotene and Santo Stefano. Ages are older in the western (4.2–1.0 Ma) than in the eastern islands (0.8 to <1.3 Ma).

Ventotene mafic magmas show geochemical affinity with the moderately potassic rocks from Ernici-Roccamonfina.

- **The Campania Province and Stromboli**
Somma-Vesuvio, Campi Flegrei (Phlegraean Fields) and Ischia are the largest and best known volcanoes in this province (see Giacomelli and Scandone 2007). All gave eruptions during the historical time and are presently quiescent. Magmas are potassic to ultrapotassic; calcalkaline rocks are also found by borehole drillings and among lithic ejecta. The mafic rocks have lower concentrations of several incompatible elements and lower Sr-isotope ratios than the equivalent rocks of the Roman Province (e.g. $^{87}Sr/^{86}Sr \sim 0.705$ to 0.708). Overall, trace element and radiogenic isotope ratios basically coincide with the rocks from Stromboli in the eastern Aeolian arc. This has led to the conclusion that Vesuvio and adjoining volcanoes do not represent the southern end of the Roman Province as suggested by Washington (1906) but rather the northern extension of the eastern Aeolian arc (Peccerillo 2001).
- **Mount Vulture or Apulian Province**
This volcano is located east of the southern Apennines and is composed of alkaline rocks that are enriched in both Na_2O and K_2O (De Fino et al. 1986). Haüyne is common in these rocks. Vulture is petrologically and geochemically different from any other Italian volcano. Such diversity was early recognised by Washington (1906) who established a separate magmatic province for this volcano (Apulian Region). Carbonatitic pyroclastics and lavas also erupted at Vulture, a unique case for orogenic settings (Stoppa and Woolley 1997; D'Orazio et al. 2007).
- **The Aeolian arc Province**
The Aeolian arc has been subdivided into a western, central and eastern sector. The western arc (Alicudi, Filicudi, Salina) consists of dominant calcalkaline mafic and intermediate rocks, with minor silicic volcanics. The central islands (Vulcano and Lipari) are dominated by calcalkaline to shoshonitic mafic to silicic rocks; mafic rocks from this sector show radiogenic isotope compositions similar to the western islands. The eastern arc (Panarea and Stromboli) consists of calcalkaline to potassic alkaline rocks. Stromboli shows geochemical and isotopic signatures akin to the Neapolitan volcanoes, as mentioned earlier. The Island of Panarea, located between Stromboli and Lipari, has intermediate characteristics between these two volcanoes (Calanchi et al. 2002).
- **The Sicily Province**
This includes several recent to active volcanoes occurring in eastern Sicily (Etna, Iblei), Sicily Channel (Pantelleria, Linosa and several seamounts) and in the southern Tyrrhenian Sea (Ustica and Prometeo lava field). Compositions range from tholeiitic to Na-alkaline, and from mafic to silicic, with abundant peralkaline rhyolites (pantellerites) at Pantelleria. All these volcanoes display intraplate geochemical signatures, although Etna and Ustica exhibit some petrological and geochemical features (e.g. relatively low TiO_2, high H_2O and volatile element contents) that recall arc-rock compositions.
- **The Sardinia Provinces**
Sardinia is the site of three stages of magmatism, showing distinct petrogenetic affinities and geochemical-isotopic signatures (e.g. Lustrino et al. 2013). The older volcanism (here simply indicated as Older Sardinia) developed during Oligo-Miocene, with very scarce Eocene activity. Compositions are dominated by arc-type calcalkaline basalt, andesite, dacite and rhyolite, with minor tholeiitic and shoshonitic volcanics. Calcalkaline to shoshonitic rocks occurring in the Corsica-Liguria basin and in Provence are considered as part of this magmatic stage. Two younger volcanic stages (Younger Sardinia) developed during Miocene-Pliocene in the southern Sardinia, and during Plio-Quaternary in the northern sector of the island. Both have OIB-type geochemical signatures, but differ strongly for volumes of

erupted products, petrogenetic affinity, and radiogenic isotope signatures.

- **The Tyrrhenian Sea basin**
  Several volcanoes with arc-type and anorogenic affinity coexist in the Tyrrhenian Sea basin. According to some authors (Savelli 1988; Locardi 1993; Argnani and Savelli 1999), the calcalkaline and shoshonitic seamounts developed along arc-shaped structures that become younger from west to southeast. Ponza and adjoining volcanoes may represent the northern end of Plio-Quaternary arcs developed from the Anchise seamount in the southern Tyrrhenian Sea to the Pontine Islands and northern Campania (see Chap. 6).

The magmatic provinces outlined above are often separated from each other by tectonic lines, such as the Tindari-Letojanni-Malta Escarpment fault, dividing western and eastern Aeolian arc, and the Ancona-Anzio (or Olevano-Antrodoco) line, separating the Roman and the Ernici-Roccamonfina provinces (Fig. 1.7). It has been demonstrated that these provinces are also characterised by different geophysical features of the crust-mantle system, such as thickness of the crust, mechanical characteristics of the lid, and depth of earthquake foci (Peccerillo and Panza 1999; Panza et al. 2007; Chiarabba et al. 2008; Neri et al. 2009).

1.6 Petrogenesis and Geodynamic Setting: A Preliminary Perspective

The extreme variability of petrological and geochemical characteristics of volcanism in the Tyrrhenian Sea region calls for complex genetic and evolutionary processes for magmas. These will be discussed in the following chapters. Here, the main issues relevant to petrogenesis and geodynamics will be introduced.

As stated earlier, two broad groups of rocks can be distinguished, respectively with orogenic (i.e. high LILE/HFSE) and anorogenic (i.e. low LILE/HFSE) geochemical signatures (Figs. 1.3 and 1.4). Orogenic rocks have variable ages, decreasing from Sardinia to the Southern Tyrrhenian Sea. Active orogenic volcanoes are associated with a zone of deep-focus earthquakes defining a steep Benioff zone that extends from the eastern Aeolian arc to Campania (e.g. Neri et al. 2009; Orecchio et al. 2014). Rocks with anorogenic geochemical affinity (i.e. low LILE/HFSE ratios) were erupted from Miocene to Present, and active volcanism occurs in eastern Sicily and Sicily Channel.

Sr-Nd–Pb isotopic data (Fig. 1.6) show that orogenic rocks define rather regular trends between upper continental crust and upper mantle compositions. This clearly highlights an interaction between mantle- and crust-derived components in the origin of magmatism. Such an interaction could have occurred either by introduction of upper crust into the mantle (*source contamination*) or/and by assimilation of crustal rocks during magma ascent to the surface (*magma contamination*). Many studies (e.g. Hawkesworth and Vollmer 1979; Peccerillo 2002; Conticelli 1998) demonstrated that, although magma contamination has played an important role in many volcanoes, it is unable to explain the large range of Sr-Nd–Pb isotopic compositions encountered throughout Italy and the Tyrrhenian Sea basin. Therefore, *mantle contamination* or *metasomatism* by upper crustal components has been widely accepted as an explanation for the geochemical and isotopic variability of the orogenic magmas. Timing and modalities of these processes, and compositions of crustal and mantle end-members are still open problems. Some authors favour a two-end member mixing between extreme mantle compositions, such as FOZO and crustal-like mantle rocks occurring beneath Tuscany and the Western Alps (e.g. Bell et al. 2013). Others suggest multiple mantle contamination events by different types of crustal material for various magmatic provinces (Peccerillo 1999). These issues will be discussed in detail in the following chapters.

In contrast, anorogenic rocks have poorly variable Sr-Nd isotopic ratios, indicating little or no role of upper crustal components in their origin. In contrast, they show large Pb-isotope variation (Fig. 1.6), which plot between various mantle end-member compositions, especially FOZO, DMM, and EM1. These variations suggest intense intra-mantle mixing processes in the source region of anorogenic magmas.

Therefore, the overall history of Cenozoic magmatism in the Tyrrhenian Sea region can be viewed as related to a complex interplay of several distinct processes, including interaction among various mantle components and mixing between these and different amounts and types of upper crustal materials. Additional complexity to the all picture was added by evolution processes, which took place during magmas ascent within the crust.

1.7 Summary and Conclusions

Eocene to Present magmatism in the Tyrrhenian Sea region exhibits a wide range of major, trace element, and isotopic characteristics. Single volcanic provinces, however, have more restricted variations. Since most of the volcanic rocks occurring in the Tyrrhenian Sea region are of ultimate mantle origin, this extreme magmatic diversity requires heterogeneous upper mantle sources. Geochronological, petrological and isotopic data suggest that multiple metasomatic events over compositionally variable pre-metasomatic mantle rocks are responsible for this heterogeneity.

Quaternary rocks occurring in central-southern Italy show the widest range of petrological and geochemical signatures. Moreover, large volumes of extremely enriched potassic and ultrapotassic rocks characterise this young volcanism. Compositions and age constraints, therefore, indicate that mantle metasomatic anomalies and heterogeneities increased with time, reaching a maximum during the latest stages of geodynamic evolution of the Tyrrhenian Sea region.

References

Alagna K, Peccerillo A, Martin S, Donati D (2010) Tertiary to present evolution of orogenic magmatism in Italy. J Virtual Explorer 36, paper 18 doi:10.3809/jvirtex.2010.00233

Argnani A, Savelli C (1999) Cenozoic volcanism and tectonics in the southern Tyrrhenian Sea: space-time distribution and geodynamic significance. Geodynamics 27:409–432

Avanzinelli R, Sapienza GT, Conticelli S (2012) The Cretaceous to Paleogene within-plate magmatism of Pachino-Capo Passero (southeastern Sicily) and Adria (La Queglia and Pietre Nere, southern Italy): geochemical and isotopic evidence against a plume-related origin of circum-Mediterranean magmas. Eur J Mineral 24:73–96

Bell K, Castorina F, Lavecchia G, Rosatelli G, Stoppa F (2004) Is there a mantle plume below Italy? EOS 85:541–547

Bell K, Lavecchia G, Rosatelli G (2013) Cenozoic Italian magmatism—isotope constraints for possible plume-related activity. J South Am Earth Sci 41:22–40

Beltrando M, Peccerillo A, Mattei M, Conticelli S, Doglioni C (eds) (2010) The Geology of Italy: tectonics and life along plate margins. J Virtual Explorer 36

Boari E, Avanzinelli R, Melluso L, Giordano G, Mattei M, Morra V, Conticelli S (2009) Isotope geochemistry (Sr- Nd-Pb) and petrogenesis of leucite-bearing volcanic rocks from "Colli Albani" volcano, Roman Magmatic Province, Central Italy: inferences on volcano evolution and magma genesis. Bull Volcanol 71:977–1005

Bruno PP, Di Fiore V, Ventura G (2000) Seismic study of the "41st Parallel" fault system offshore the Campanian-Latial continental margin, Italy. Tectonophysics 324:37–55

Calanchi N, Lucchi F, Pirazzoli PA, Romagnoli C, Tranne CA, Radtke U, Reyss JL, Rossi PL (2002) Late Quaternary relative sea-level changes and vertical movements at Lipari (Aeolian Islands). J Quater Sci 17:459–467

Carminati E, Lustrino M, Cuffaro M, Doglioni C (2010) Tectonics, magmatism and geodynamics of Italy: What we know and what we imagine. In: Beltrando M, Peccerillo A, Mattei M, Conticelli S, Doglioni C (eds) J Virtual Explorer 36, paper 9 doi:10.3809/jvirtex.2010.00226

Castellarin A, Colacicchi R, Praturlon A, Cantelli C (1982) The Jurassic-Lower Pliocene history of the Ancona-Anzio line (Central Italy). Mem Soc Geol It 24:243–260

Cavazza W, Wezel FC (2003) The Mediterranean region-a geological primer. Episodes 26:160–168

Cavazza W, Roure F, Spakman W, Stampfli GM, Ziegler PA (eds) (2004) The TRANSMED Atlas. The Mediterranean region from crust to mantle (with CD-ROM). Springer, 133 pp

Chiarabba C, De Gori P, Speranza F (2008) The southern Tyrrhenian subduction zone: deep geometry, magmatism and Plio-Pleistocene evolution. Earth Planet Sci Lett 268:408–423

Conte AM, Perinelli C, Bianchini G, Natali C, Martorelli E, Chiocci FL (2016) New insights on the petrology of submarine volcanics from the Western Pontine Archipelago (Tyrrhenian Sea, Italy). J Volcanol Geoth Res, in press

Conticelli S (1998) The effect of crustal contamination on ultrapotassic magmas with lamproitic affinity: mineralogical, geochemical and isotope data from the Torre Alfina lavas and xenoliths, Central Italy. Chem Geol 149:51–81

Conticelli S, Peccerillo A (1990) Petrological significance of high-pressure ultramafic xenoliths from ultrapotassic rocks of Central Italy. Lithos 24:305–322

Conticelli S, Peccerillo A (1992) Petrology and geochemistry of potassic and ultrapotassic volcanism in Central Italy: petrogenesis and inferences on the evolution of the mantle sources. Lithos 28:221–240

Conticelli S, D'Antonio M, Pinarelli L, Civetta L (2002) Source contamination and mantle heterogeneity in the genesis of Italian potassic and ultrapotassic volcanic rocks: Sr-Nd-Pb isotope data from Roman Province and Southern Tuscany. Mineral Petrol 74:189–222

Conticelli S, Laurenzi MA, Giordano G, Mattei M, Avanzinelli R, Melluso L, Tommasini S, Boari E, Cifelli F, Perini G (2010) Leucite-bearing (kamafugitic/leucititic) and -free (lamproitic) ultrapotassic rocks and associated shoshonites from Italy: constraints on petrogenesis and geodynamics. In: Beltrando M, Peccerillo A, Mattei M, Conticelli S, Doglioni C (eds) J Virtual Explorer 36, paper 20 doi:10.3809/jvirtex.2010.00251

Conticelli S, Avanzinelli R, Ammannati E, Casalini M (2015) The role of carbon from recycled sediments in the origin of ultrapotassic igneous rocks in the Central Mediterranean. Lithos 232:174–196

D'Orazio M, Innocenti F, Tonarini S, Doglioni C (2007) Carbonatites in a subduction system: The Pleistocene alvikites from Mt. Vulture (southern Italy). Lithos 98:313–334

De Fino M, La Volpe L, Peccerillo A, Piccarreta G, Poli G (1986) Petrogenesis of Monte Vulture volcano, Italy: inferences from mineral chemistry, major and trace element data. Contrib Mineral Petrol 92:135–145

Foley SF, Venturelli G, Green DH, Toscani L (1987) The ultrapotassic rocks: characteristics classification, and constraints for petrogenetic models. Earth Sci Rev 24:81–134

Frezzotti ML, De Astis G, Dallai L, Ghezzo C (2007) Coexisting calc-alkaline and ultrapotassic magmatism at Monti Ernici, Mid Latina Valley (Latium, central Italy). Eur J Mineral 19:479–497

Gale A, Dalton CA, Langmuir CH, Su Y, Schilling J-G (2013) The mean composition of ocean ridge basalts. Gechem Geophys Geosyst 14 doi:10.1029/2012GC004334

Gasperini D, Blichert Toft J, Bosch D, Del Moro A, Macera P, Albarède F (2002) Upwelling of deep mantle material through a plate window: evidence from the geochemistry of Italian basaltic volcanics. J Geophys Res 107(B12):2367. doi:10.1029/2001JB000418

Giacomelli L, Scandone R (2007) Vulcani d'Italia. Liguori, Naples, 475 pp

Gill JB (1981) Orogenic andesites and plate tectonics. Springer, Berlin, 358 pp

Hart SR (1984) A large-scale isotopic anomaly in the Southern Hemisphere mantle. Nature 309:753–757

Hawkesworth CJ, Vollmer R (1979) Crustal contamination vs. enriched mantle: $^{143}Nd/^{144}Nd$ and $^{87}Sr/^{86}Sr$ evidence from the Italian volcanics. Contrib Mineral Petrol 69:151–165

Hofmann AW (1997) Mantle geochemistry: the message from oceanic volcanism. Nature 385:219–229

Irvine TN, Baragar WRA (1971) A guide to chemical classification of common volcanic rocks. Can J Earth Sci 8:523–548

Le Maitre RW (ed) (2002) A classification of igneous rocks and glossary of terms. Cambridge University Press, Cambridge 252 pp

Locardi E (1993) Dynamics of deep structures in the Tyrrhenian-Apennines area and its relation to neotectonics. Il Quaternario 6:59–66

Lustrino M, Duggen S, Rosenberg CL (2011) The Central-Western Mediterranean: Anomalous igneous activity in an anomalous collisional tectonic setting. Earth Sci Rev 104:1–40

Lustrino M, Fedele L, Melluso L, Morra V, Ronga F, Geldmacher J, Duggen S, Agostini S, Cucciniello C, Franciosi L, Meisel T (2013) Origin and evolution of Cenozoic magmatism of Sardinia (Italy). A combined isotopic (Sr-Nd-Pb-O-Hf-Os) and petrological view. Lithos 180:138–158

Neri G, Orecchio B, Totaro C, Falcone G, Presti D (2009) Subduction beneath southern Italy close the ending: results from seismic tomography. Seism Res Lett 80:63–70

Orecchio B, Presti D, Totaro C, Neri G (2014) What earthquakes say concerning residual subduction and STEP dynamics in the Calabrian Arc region, south Italy. Geophys J Int 199:1929–1942

Panza GF, Peccerillo A, Aoudia A, Farina B (2007) Geophysical and petrological modeling of the structure and composition of the crust and upper mantle in complex geodynamic settings: The Tyrrhenian Sea and surroundings. Earth Sci Rev 80:1–46

Pearce JA (1982) Trace element characteristics of lavas from destructive plate boundaries. In: Thorpe RS (ed) Andesites: Orogenic andesites and related rocks. Wiley, Chichester, pp 525–548

Peccerillo A (1992) Potassic and ultrapotassic magmatism: compositional characteristics, genesis and geologic significance. Episodes 15:243–251

Peccerillo A (1999) Multiple mantle metasomatism in central-southern Italy: geochemical effects, timing and geodynamic implications. Geology 27:315–318

Peccerillo A (2001) Geochemical similarities between Vesuvius, Phlegraean Fields and Stromboli volcanoes:

petrogenetic, geodynamic and volcanological implications. Mineral Petrol 73:93–105
Peccerillo A (2002) Plio-Quaternary magmatism in central-southern Italy: a new classification scheme for volcanic provinces and its geodynamic implications. In: Barchi RM, Cirilli S, Minelli G (eds) Geological and geodynamic evolution of the Apennines. Boll Soc Geol It (Spec Vol 1):113–127
Peccerillo A (2003) Plio-Quaternary magmatism in Italy. Episodes 26:222–226
Peccerillo A (2005) Plio-Quaternary volcanism in Italy. Petrology, Geohemistry, Geodinamics. Springer, Berlin, 365 pp
Peccerillo A, Frezzotti ML (2015) Magmatism, mantle evolution and geodynamics at the converging plate margins of Italy. J Geol Soc London 172:407–427
Peccerillo A, Martinotti G (2006) The Western Mediterranean lamproitic magmatism: origin and geodynamic significance. Terra Nova 18:109–117
Peccerillo A, Panza GF (1999) Upper mantle domains beneath central-southern Italy: petrological, geochemical and geophysical constraints. Pure Appl Geophys 156:421–443
Peccerillo A, Poli G, Serri G (1988) Petrogenesis of orenditic and kamafugitic rocks from Central Italy. Can Mineral 26:45–65
Peccerillo A, De Astis G, Faraone D, Forni F, Frezzotti ML (2013) Compositional variations of magmas in the Aeolian arc: implications for petrogenesis and geodynamics. In: Lucchi F, Peccerillo A, Keller J, Tranne CA, Rossi PL (eds) The Aeolian Islands Volcanoes. Geol Soc London Memoirs 37:491–510
Plank T, Langmuir CH (1998) The chemical composition of subducting sediment and its consequences for the crust and mantle. Chem Geol 145:325–394
Poli G, Peccerillo A (2016) The Upper Miocene magmatism of the Elba Island: compositional characteristics, petrogenesis and implications for the origin of the Tuscany Magmatic Province. Mineral Petrol, in press
Poli G, Frey FA, Ferrara G (1984) Geochemical characteristics of the south Tuscany (Italy) volcanic province, constraints on lava petrogenesis. Chem Geol 43:203–221
Réhault J-P, Honthaas C, Guennoc P, Bellon H, Ruffet G, Cotton J, Sosson M, Maury RC (2012) Offshore Oligo-Miocene volcanic fields within the Corsica-Liguria Basin: magmatic diversity and slab evolution in the western Mediterranean Sea. J Geodyn 58:73–95
Sahama TG (1974) Potassium-rich alkaline rocks. In: Sorensen H (ed) The alkaline rocks. Wiley, London, pp 94–109
Savelli C (1988) Late Oligocene to recent episodes of magmatism in and around the Tyrrhenian Sea: implications for the processes of opening in a young inter-arc basin of intra-orogenic (Mediterranean) type. Tectonophysics 146:163–181
Stoppa F, Woolley AR (1997) The Italian carbonatites: field occurrence, petrology and regional significance. Mineral Petrol 59:43–67
Stracke A, Hofmann AW, Hart SR (2005) FOZO, HIMU, and the rest of the mantle zoo. Geochem Geophys Geosyst 6:Q05007
Sun SS, McDonough WF (1989) Chemical and isotopic systematics of oceanic basalts: implications for mantle composition and processes. In: Saunders AD, Norry MJ (eds) Magmatism in ocean basins. Geol Soc London (Spec Publ 42):313–345
Tarquini S, Vinci S, Favalli M, Doumaz F, Fornaciai A, Nannipieri L (2012) Release of a 10-m-resolution DEM for the Italian territory: comparison with global-coverage DEMs and anaglyph-mode exploration via the web. Comput Geosci 38:168–170
Turtù A, Satolli S, Maniscalco R, Calamita F, Speranza F (2013) Understanding progressive-arc- and strike-slip-related rotations in curve-shaped orogenic belts: The case of the Olevano-Antrodoco-Sibillini thrust (Northern Apennines, Italy). J Geophys Res, Solid Earth 118:1–15
Washington HS (1906) The Roman Comagmatic Region. Carnegie Inst Washington (Publ 57): 199 pp
Zindler A, Hart SR (1986) Chemical geodynamics. Ann Rev Earth Planet Sci 14:493–571

2 The Tuscany Province

Abstract

The Tuscany Magmatic Province consists of various intrusive and extrusive bodies ranging in composition from mafic to felsic and from calcalkaline to ultrapotassic lamproitic. Rock age ranges from about 8 to 0.3 Ma, and decreases eastward from the Tuscan Archipelago to the Southern Tuscany mainland. An isolated sill from Sisco (Alpine Corsica, 14.5 Ma) is also included in the Tuscany Province. Silicic magmas make up several granitoid bodies and minor lavas. They are polygenetic and have been formed by crustal melting, mixing between crustal anatectic and minor amounts of mafic melts, and fractional crystallisation or AFC of intermediate-mafic parents. Except for the Capraia Island stratovolcano, mafic-intermediate rocks mostly occur as small subvolcanic and extrusive bodies, and as mafic enclaves in the silicic rocks. Mafic magmas originated within the upper mantle but have striking crustal-like trace element and radiogenic isotope signatures, resembling closely some upper crustal rocks such as metapelites. The coexistence of both crustal and mantle signatures in these magmas reveals an origin from anomalous sources, consisting of a mélange of mantle and crustal rocks. Such a particular type of source in Tuscany probably formed during the Late Cretaceous to Eocene subduction of the European plate beneath the African margin. Partial melting of subduction mélange took place much later, during the Miocene to present opening of the northern Tyrrhenian Sea behind the west dipping and eastward retreating Adriatic subducting slab. Age polarity of Tuscany magmatism reflects the timing of backarc extension that migrated from west to the east.

Keywords

Tuscany magmatism · Crustal anatexis · Granitoids · Calcalkaline magmas · Lamproites · Adakite · Mantle contamination · Subduction mélange · Northern Apennines · Geodynamics

A. Peccerillo, *Cenozoic Volcanism in the Tyrrhenian Sea Region*, Advances in Volcanology, DOI 10.1007/978-3-319-42491-0_2

2.1 Introduction

The Tuscany Magmatic Province (Fig. 2.1) consists of several mafic to silicic intrusive and extrusive centres scattered through southern Tuscany, the Tuscan Archipelago and northern Latium. An older mafic ultrapotassic sill from Sisco (northeastern Alpine Corsica) is here included into the Tuscany Province, because of its lamproitic composition similar to some Tuscany occurrences.

The Tuscany igneous rocks form stocks, dikes, necks, lava flows and domes, and the large volcanoes of the Island of Capraia, Monte Amiata, and Monti Cimini. Pyroclastic rocks occur as ignimbritic deposits at Monti Cimini and Cerite area (e.g. Poli et al. 2003; Cimarelli and De Rita 2006a, b; LaBerge et al. 2006), but are scarce or absent in other Tuscany centres. Ages range from about 8.5 Ma for some rocks in the Elba island to about 0.3 Ma for Monte Amiata and show a tendency to decrease from west to east (Fig. 2.1). The Sisco rock is about 14.5 Ma old (Civetta et al. 1978).

Although volumetrically subordinate with respect to other volcanic provinces in Italy, the Tuscany magmatism exhibits an extreme compositional complexity, which calls for the interplay of several mechanisms of magma genesis and evolution, and various types of mantle and crustal sources. An overview of magmatic centres is given in Table 2.1. Representative whole rock analyses are given in Table 2.2.

2.2 Regional Geology

The geology of Tuscany consists of stacked tectonic thrust units, mostly verging eastwards, overlain by Mio-Pleistocene post-orogenic neo-

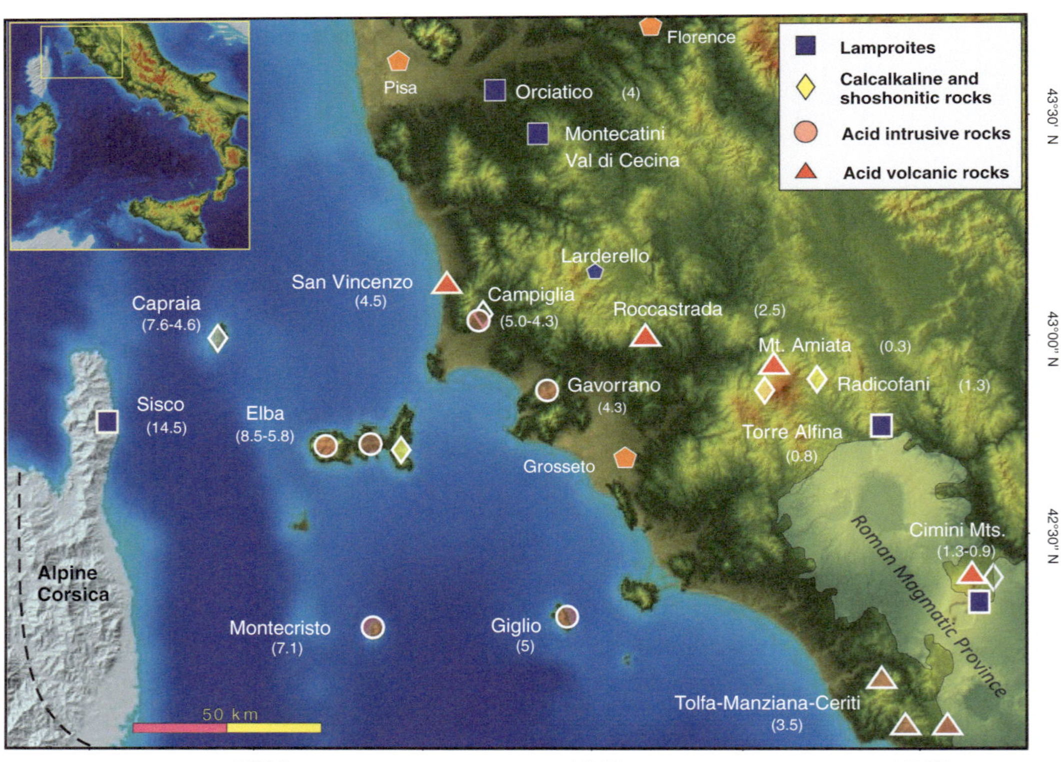

Fig. 2.1 Location of intrusive and extrusive centres of the Tuscany Magmatic Province. Numbers in parentheses indicate ages (in Ma). Relief shaded map from Tarquini et al. (2012)

Table 2.1 Petrological characteristics and ages (in Ma) of magmatic centres in the Tuscany Province

Magmatic centres	Age (Ma)	Occurrence	Petrology-geochemistry
Silicic volcanic centres			
San Vincenzo	4.5	– Lava flows and domes	– Calcalkaline peraluminous rhyolites containing latite enclaves. Variable Sr isotope ratios ($^{87}Sr/^{86}Sr$ ~ 0.713–0.725) resulting from mixing of crustal anatectic rhyolites and mafic magmas
Tolfa-Manziana-Cerite	3.5	– Dome complex and pyroclastic deposits strongly affected by secondary alteration	– Trachydacites to high-silica rhyolites, containing abundant mafic enclaves. Formed by crustal melting plus fractional crystallisation and mixing with mafic magmas
Roccastrada	2.5	– Lava flows and dome	– Calcalkaline peraluminous rhyolites with $^{87}Sr/^{86}Sr$ ~ 0.7182 to 0.7198, originated by crustal melting
Monte Amiata	0.3	– Central cone made of prevailing silicic lava flows and domes, with a few late mafic lavas	– Dominant trachydacites with minor olivine-latites and shoshonites. Origin by mixing of crustal anatectic melts and mafic Roman-type potassic alkaline magmas
Silicic plutonic centres			
Island of Elba	8.5–5.8	– Monzogranite stock with aplites and pegmatites (Mt. Capanne) and various laccoliths, dikes and sills. Late mafic dikes	– Dominant acid peraluminous monzogranites and granodiorites containing microgranular mafic enclaves, formed by crustal anatexis plus interaction with calcalkaline mantle-derived magmas, and fractional crystallisation. Late intrusion of mantle-derived shoshonitic dikes
Vercelli Seamount	7.2	– Seamount with summit at a depth of about 58 m bsl	– Syenogranitic composition, based on a single sample
Island of Montecristo	7.1	– Monzogranite stock cut by aplite and pegmatite veins and porphyritic dikes	– Peraluminous monzogranite stock cut by andesitic dikes. Origin by mixing of crustal anatectic and mafic calcalkaline magma, plus late intrusion of shoshonites
Island of Giglio	5	– Multiple intrusions of monzogranite, leucocratic monzogranite, and aplite-pegmatite	– Calcalkaline peraluminous rocks with variable $^{87}Sr/^{86}Sr$ (~ 0.7163–0.7203) increasing from monzogranites to leucogranites. Similar origin to other Tuscany granitoids
Campiglia	5	– Stock mostly hidden beneath surface. Younger felsic and mafic porphyritic dikes in the same area	– Intensely altered monzogranite to alkali-feldspar granite depleted in Fe by secondary processes
Gavorrano	4.3	– Intrusive body forming a 3 km NNW elongated outcrop	– Monzogranite to tourmaline-rich alkali feldspar granite. Same origin as other Tuscany granitoids
Hidden intrusions	5 - 0?	– Various bodies near Campiglia and Gavorrano, and in the Larderello area, the latter being still partially molten	– Granitoid rocks (monzogranites, syenogranites, granodiorite etc.) with $^{87}Sr/^{86}Sr$ ~ 0.7145–0.7222. Formed by crustal anatexis plus probable mixing with mafic melts
Mafic volcanic and hypabyssal centres			
Sisco	14.5	– Thin sill cutting through high-P Alpine metamorphic units	– Peralkaline high-silica lamproite originated in mantle sources contaminated by siliceous rocks

(continued)

Table 2.1 (continued)

Magmatic centres	Age (Ma)	Occurrence	Petrology-geochemistry
Island of Capraia	7.6-4.6	– Remnants of a large stratovolcano formed mainly of intermediate lavas. Final mafic lavas and scoriae	– Dominant high-K calcalkaline andesites and dacites, and final shoshonitic basalts, showing the least radiogenic Sr isotope composition in Tuscany ($^{87}Sr/^{86}Sr$ = 0.7073-0.7102). Occurrence of Sr-Ba- LREE rich adakite-type intermediate magmas. Formed by melting of sources less contaminated than at other Tuscany mafic centres
Campiglia	4.3	– N-S trending strongly altered mafic dikes	– Ultrapotassic rocks showing trace element and Sr isotopic compositions similar to Capraia calcalkaline rocks. Ultrapotassic composition probably related to secondary processes
Montecatini Val di Cecina	4.1	– Phologopite-rich plug permeated by abundant leucocratic veinlets	– High-silica lamproite with higher Sr isotope ratios than Sisco ($^{87}Sr/^{86}Sr$ = 0.7169). Formed by melting of anomalous lithospheric mantle, contaminated by crustal siliceous rocks. Veinlets generated by unmixing from the partially crystallised host intrusion
Orciatico	4.1	– Mafic hypabyssal body	– High-silica lamproite with composition and origin similar to Montecatini Val di Cecina
Monti Cimini	1.3-0.9	– Early felsic lava domes and flows, and ignimbrites followed by intermediate to mafic lavas	– Shoshonitic to ultrapotassic trachytes, trachydacites, latites and shoshonites showing variable $^{87}Sr/^{86}Sr$ ($\sim$0.7122–0.7157) in the mafic rocks. Melting of heterogeneous mantle sources and mixing with crustal anatectic melts
Radicofani	1.3	– Mafic neck and remnants of lava flows	– Basaltic andesite to shoshonite formed by mixing between high-silica lamproite and calcalkaline basalt magmas
Torre Alfina	0.8	– Mafic necks and lava flows containing crustal xenoliths and some high-P ultramafic nodules	– High-silica lamproite with composition and origin similar to Montecatini and Orciatico

autochthonous sediments (e.g. Abbate et al. 1970; Barchi et al. 2001; Carmignani et al. 2001). The main tectono-stratigraphic units include:

1. Paleozoic metamorphic rocks (quartz-metaconglomerates, quartzites, phyllites, metavolcanics and micaschists) that have been affected by Alpine metamorphism and are superimposed over a gneiss complex;
2. The Tuscan units consisting of Late Triassic to early Miocene sedimentary rocks (conglomerates, evaporites, limestones, marls and arenaceous flysch) overlying the metamorphic basement, and affected by Alpine metamorphism in some places (e.g. Alpi Apuane);
3. The Ligurian complex tectonically superimposed on the Tuscan units, consisting of middle to late Jurassic ophiolites and radiolarites plus Cretaceous to Eocene pelagic limestones and flysch sequences;
4. Oligo-Miocene clastic marine sediments resting unconformably over the Liguride sequences;
5. Neo-autochthonous Miocene to Pleistocene sediments (evaporites, lignite, fresh water limestones, conglomerates, clays, sands, etc.) infilling tectonic depressions.

These rock units were formed during a complex series of tectonic events, which include

Table 2.2 Compositions of representative igneous rocks from Tuscany

Magmatic centre	Roccastrada	San Vincenzo	Tolfa	Manziana	Mt. Amiata	Mt. Amiata	Mt. Cimini	Mt. Cimini
Rock type	Rhyolite	Rhyolite	Latite	Rhyolite	Trachydacite	Mafic enclave	Trachydacite	Latite
Data source	4,1,2	4,1,2	5	5	13	13	20	20
SiO_2 wt%	72.48	68.76	64.61	71.66	64.90	53.76	64.16	56.94
TiO_2	0.24	0.38	0.66	0.39	0.57	0.72	0.69	1.10
Al_2O_3	13.56	14.94	16.45	14.50	15.90	17.16	16.60	15.39
FeO_{total}	1.91	2.33	3.57	2.19	3.88	6.02	4.41	4.80
MnO	0.03	0.03	0.04	0.03	0.06	0.10	0.08	0.10
MgO	0.36	0.91	1.40	0.31	1.74	4.64	2.18	7.65
CaO	1.02	1.96	3.18	1.51	2.97	6.85	3.16	5.50
Na_2O	2.61	2.86	2.47	3.28	2.08	1.64	2.31	1.33
K_2O	4.86	4.63	5.23	4.99	5.2	5.27	5.18	5.77
P_2O_5	0.14	0.20	0.12	0.03	0.2	0.28	0.21	0.27
LOI	1.56	1.93	1.98	0.81	1.92	2.76	0.83	1.01
Sc ppm	5	6	–	–	(12)	(37)	10.7	(21)
V	12	36	–	–	59	141	57	103
Cr	10	28	–	–	31	121	32	389
Co	2	5	–	–	–	–	9	–
Ni	19	21	–	–	14	40	20	154
Rb	395	310	305	385	369	295	290	382
Sr	68	223	264	91	282	663	416	409
Y	33	24	28	35	27	32	31	23
Zr	113	149	275	214	265	288	262	429
Nb	14	13	19	10	17	14	19	24
Cs	(13.8)	–	–	–	26.01	18.09	(33)	(27)
Ba	172	400	567	231	343	781	764	756
La	30	43	51	71	71	74	95	93
Ce	68	91	97	133	142	146	155	193
Nd	29	43	36	50	57	68	55	85
Sm	6.8	8.9	6.7	9.4	10.28	12.19	12.8	15.9
Eu	0.6	1.3	1.2	0.82	1.16	2.18	1.69	2.32
Tb	1.1	1.0	–	–	1.02	1.15	0.99	1.10
Yb	3.1	1.8	2.4	3.4	2.7	2.61	2.40	1.96
Lu	0.3	0.2	0.5	0.63	0.41	0.39	0.50	0.34
Hf	3.8	4.1	–	–	7.49	7.21	7.0	10.9
Ta	(1.83)	(1.9)	–	–	1.7	1.12	1.8	1.9
Pb	45	50	56	75	57	52	69	49
Th	24	20	30	53	43	36	45	58
U	17	11	8.7	10	11.2	8.36	6.9	8.5
$^{87}Sr/^{86}Sr$	0.71799	0.71558	0.71383	0.71308	0.712574	0.71139	0.713902	0.715648
$^{143}Nd/^{144}Nd$	(0.51221)	(0.51225)	–	–	0.512096	0.51209	0.512139	0.512063
$^{206}Pb/^{204}Pb$	(18.68)	(18.71)	18.74	18.723	18.721	18.731	18.708	18.723

(continued)

Table 2.2 (continued)

Magmatic centre	Roccastrada	San Vincenzo	Tolfa	Manziana	Mt. Amiata	Mt. Amiata	Mt. Cimini	Mt. Cimini
Rock type	Rhyolite	Rhyolite	Latite	Rhyolite	Trachydacite	Mafic enclave	Trachydacite	Latite
$^{207}Pb/^{204}Pb$	(15.65)	(15.66)	15.70	15.663	15.681	15.678	15.671	15.689
$^{208}Pb/^{204}Pb$	(38.91)	(38.88)	38.79	38.859	39.003	39.002	38.959	39.023

Magmatic centre	Capraia	Capraia	Sisco	Montecatini Val di Cecina	Orciatico	Torre Alfina	Radicofani
Rock type	Shoshonitic basalt	Andesite	Lamproite	Lamproite	Lamproite	Lamproite	Shoshonite
Data source	14	14	12,15,17	12,15, 17	12,15, 17	18,17	18,10
SiO_2 wt%	50.65	61.90	58.50	56.86	57.79	55.50	54.30
TiO_2	1.65	0.69	2.27	1.37	1.51	1.36	0.91
Al_2O_3	15.50	15.30	10.84	12.61	11.79	13.40	17.21
FeO_{total}	9.18	4.85	3.15	5.76	5.13	5.78	6.07
MnO	0.13	0.07	0.06	0.1	0.08	0.10	0.11
MgO	6.41	3.48	6.63	7.15	8.23	9.36	7.77
CaO	7.92	5.48	3.12	3.47	3.46	4.70	7.59
Na_2O	2.83	3.14	1.02	1.2	1.31	1.18	2.02
K_2O	2.42	3.33	10.73	7.91	8.06	7.46	2.89
P_2O_5	0.48	0.18	0.67	0.92	0.85	0.54	0.25
LOI	1.13	1.36	2.09	2.43	1.55	0.58	0.84
Sc ppm	23	–	11	20.2	18.5	17	(23)
V	166	128	84	137	116	118	173
Cr	400	177	340	380	500	641	417
Co	30	11.9	19	27	28.1	36	29.4
Ni	69	29	230	140	280	349	124
Rb	115	116	318	768	612	453	168
Sr	399	733	640	408	577	726	339
Y	20	19.3	19	28	23.8	33	22.6
Zr	221	185	1040	491	749	674	216
Nb	15	11.9	55	30.1	39	31	11.7
Cs	4	9.2	2	14	25.2	30	9.46
Ba	540	711	1450	1370	1400	1293	579
La	29	58.5	183	79.8	148	98	41.4
Ce	68	109	367	206	352	294	92.6
Nd	51.9	43.6	146	133	193	127	43.6
Sm	9.6	8.01	19.1	23.5	26.9	20.7	7.7
Eu	2.09	1.76	3.5	4.02	4.32	3.51	1.75
Tb	1	0.77	1.09	1.31	1.27	1.2	0.81
Yb	2.3	1.84	1.59	2.19	1.85	2.38	1.97
Lu	0.35	0.28	0.21	0.3	0.25	0.39	0.32
Hf	5.6	5.4	32.1	13.4	21.4	15.5	5.48
Ta	1.2	0.9	3.96	2.17	2.93	2.3	0.95
Pb	12	61	12	19	30	57	29
Th	24	23.1	37.9	112	119	61	24.7
U	3.8	6.43	4.61	18.2	14.9	14.2	4.62

(continued)

Table 2.2 (continued)

Magmatic centre	Capraia	Capraia	Sisco	Montecatini Val di Cecina	Orciatico	Torre Alfina	Radicofani
Rock type	Shoshonitic basalt	Andesite	Lamproite	Lamproite	Lamproite	Lamproite	Shoshonite
$^{87}Sr/^{86}Sr$	0.708135	0.708719	0.712264	0.71672	0.71579	0.71579	0.71333
$^{143}Nd/^{144}Nd$	0.512254	0.512346	0.512149	0.512086	0.512094	0.51212	0.51218
$^{206}Pb/^{204}Pb$	18.652	18.735	18.808	18.624	18.697	18.677	18.711
$^{207}Pb/^{204}Pb$	15.663	15.702	15.696	15.638	15.698	15.678	15.665
$^{208}Pb/^{204}Pb$	38.963	39.086	39.190	38.947	39.062	38.918	39.049
$^{176}Hf/^{177}Hf$	–	–	0.282504	0.282435	0.282404	0.282398	0.282554

Magmatic centre	Elba			Montecristo	Giglio		Larderello	Vercelli Seamount
Rock type	Monzo-granite	Syeno-granite	Andesitic dike	Monzo-granite	Monzogranite	Leucogranite	Granite Monteverdi Drilling	Syenogranite
Data source	9,16,1	19	21	6, 8	6, 7	6, 7	11	3
SiO_2 wt%	67.49	70.79	62.77	70.55	67.63	73.54	70.06	72.60
TiO_2	0.52	0.32	0.66	0.36	0.66	0.21	0.37	0.23
Al_2O_3	15.69	15.74	15.25	14.94	15.56	14.07	15.53	14.23
FeO	2.68	2.18	4.51	2.10	3.73	1.47	2.39	1.59
MnO	0.05	0.04	0.07	0.04	0.08	0.05	<0.02	0.04
MgO_{total}	1.57	0.74	3.79	0.66	1.24	0.38	0.76	0.50
CaO	2.7	1.65	2.42	1.76	2.13	0.82	1.90	1.05
Na_2O	3.53	3.16	3.03	3.71	2.81	2.38	2.94	3.91
K_2O	4.13	4.45	3.08	4.77	4.79	6.04	4.66	4.51
P_2O_5	0.2	0.17	0.17	0.22	0.17	0.14	0.16	0.26
LOI	1.14	1.21	3.32	0.64	1.11	0.89	0.80	0.90
Sc ppm	8.3	5.0	14	6.1	10.7	4	–	–
V	36	24.0	91	–	–	–	34.3	17
Cr	30.6	15.0	190	28	44	4	25.7	12
Co	7.2	4.0	14	5	8.5	4	5.3	–
Ni	13.4	10.0	20	–	–	–	13	–
Rb	275	294	65	357	337	372	238	440
Sr	229	126	316	122	131	65	305	302
Y	20	12.1	18	22	30	21	13.2	19
Zr	–	–	143	116	223	92	148	87
Nb	13.1	11.8	10.5	6	12	11	8.8	–
Cs	41.5	34	2.6	–	–	–	13.7	–
Ba	402	257	425	293	318	74	601	112
La	33.7	28.9	29.2	30	44	16	28	28
Ce	68.9	60.0	56	61	96	39	58.8	36
Nd	29.7	24.8	25.7	27	41	18.6	24	–
Sm	5.96	5.0	5.19	5.8	8.8	4.6	4.7	–
Eu	0.92	0.68	1.18	0.7	1.1	0.48	0.87	–
Tb	0.63	0.48	0.63	0.6	1.0	0.55	0.44	–
Yb	1.35	0.73	1.75	1.7	2.7	2.8	1.17	–

(continued)

Table 2.2 (continued)

Magmatic centre	Elba			Montecristo	Giglio		Larderello	Vercelli Seamount
Rock type	Monzo-granite	Syeno-granite	Andesitic dike	Monzo-granite	Monzogranite	Leucogranite	Granite Monteverdi Drilling	Syenogranite
Lu	0.18	0.09	0.26	0.30	0.45	0.46	0.17	–
Hf	(3.3)	0.33	3.8	–	5.5	2.8	4.35	–
Ta	1.79	2.30	1.03	–	2.5	1.1	1.07	–
Pb	43.7	33	18	–	–	–	44.7	–
Th	19.5	11.6	12.6	30	21	10	21.2	11
U	14.4	7.4	6.17	–	–	–	10.3	7
$^{87}Sr/^{86}Sr$	0.71461	0.722800	0.708129	(0.71466)	(0.7173)	(0.7195)	0.719075	0.71140
$^{143}Nd/^{144}Nd$	0.512182	0.512117	0.512209	–	(0.5122)	(0.5121)	0.512086	–
$^{206}Pb/^{204}Pb$	(18.68)	–	–	–	–	–	–	–
$^{207}Pb/^{204}Pb$	(15.67)	–	–	–	–	–	–	–
$^{208}Pb/^{204}Pb$	(38.87)	–	–	–	–	–	–	–

Source of data: (1) Vollmer (1977); (2) Hawkesworth and Vollmer (1979); (3) Barbieri et al. (1986); (4) Giraud et al. (1986); (5) Pinarelli (1991); (6) Poli (1992); (7) Westerman et al. (1993); (8) Innocenti et al. (1997); (9) Dini et al. (2002); (10) Gasperini et al. (2002); (11) Dini et al. (2005); (12) Conticelli et al. (2007); (13) Cadoux and Pinti (2009); (14) Conticelli et al. (2009a, b); (15) Conticelli et al. (2010); (16) Farina et al. (2010); (17) Prelevic et al. (2010); (18) Conticelli et al. (2011); (19) Farina et al. (2012); (20) Conticelli et al. (2013); (21) Pandeli et al. (2014)

rifting, oceanic convergence, continental collision and post-orogenic extension. Early rifting occurred during Triassic-Jurassic times; continental to littoral clastic sediments, evaporitic deposits and carbonate rocks, as well as ophiolitic sequences of the Liguride units were formed during this phase (e.g. Principi and Treves 1984; Balestrieri et al. 2011). Compression started during Cretaceous in the Liguride units and continued during the Lower Oligocene when the Tuscan units were tectonically emplaced over the Umbria sequences on the east. Successively, the compressive front shifted eastward involving progressively more external sectors of the Adriatic foreland (Brunet et al. 2000; Barchi 2010 with references). Contemporaneously, extension and magmatism affected the backarc area, migrating from west to east behind the shifting compressive front of Northern Apennines (e.g. Civetta et al. 1978; Peccerillo et al. 1987; Serri et al. 1993; Pauselli et al. 2006; Pandeli et al. 2010). According to Boccaletti et al. (1997), extensional tectonic regime west of the compression front was interrupted by Messinian, Late Pliocene and Middle Pleistocene compressive tectonic phases, which would be responsible for interruption of magmatic activity in these periods.

Syn-tectonic terrigenous turbidite and pelagic sedimentation, thrusting and HP–LT to Barrovian metamorphism took place during the compressional tectonic phases. Synorogenic metamorphic rocks are exposed at various localities of the mainland Tuscany, Tuscan archipelago and northeastern Corsica (Alpine Corsica). Peak pressures of 1.3–1.6 GPa at temperatures of about 300–420 °C were found for the *Schistes Lustrés* in the Corsica and Gorgona islands, whereas lower pressures of 0.6–0.85 GPa have been determined for the same units in the Giglio area (Franceschelli et al. 2004 with references), and for Cretaceous metabasites at Elba (P ~ 1 GPa, T = 300–350 °C; Bianco et al. 2015). A younger metamorphic phase took place in the Miocene during extension and rapid exhumation (Balestrieri et al. 2011).

Moho is about 20 km deep and asthenosphere occurs at 40–50 km beneath the Tyrrhenian border of southern Tuscany; crustal thickness increases eastward reaching a maximum of about 40 km beneath the axial zone of the Apennines (e.g. Gianelli et al. 1997; Piromallo and Morelli

2003; Mele and Sandvol 2003; Mele et al. 2006; Di Stefano et al. 2011; Carannante et al. 2013; Buttinelli et al. 2014). A vertical zone with high seismic-wave velocity has been detected beneath the Northern Apennines (e.g. Panza and Mueller 1979; Benoit et al. 2011). This has been interpreted as a remnant lithospheric slab from the Adriatic plate (east of the Apennine chain), which is passively sinking into the upper mantle. According to recent studies, the slab extends to about 300 km depth, with its southern edge at about 43°N, with no deep continuity with any slab segment to the south (Benoit et al. 2011). An additional important geophysical feature is denoted by the occurrence of a high-Vp anomaly at about 40–50 km depth beneath the westernmost sector of the Tuscany Magmatic Province. This may represent the remnants of the European lithosphere subducted during Alpine orogenic phase (Finetti et al. 2001; Pauselli et al. 2006; Di Stefano et al. 2011; Giacomuzzi et al. 2012; Carannante et al. 2013).

Heat flow is high in southern Tuscany (around 100–200 mW/m^2 in some areas), with peaks of 600 mW/m^2 recorded in the Larderello area, i.e. at the site of the well-known active and long exploited geothermal fields (Della Vedova et al. 1991, 2001, 2006; Minissale 1991; Brogi et al. 2005 with references). Geophysical and field data suggest that much of the heat is provided by partially molten dome-shaped intrusions sited at a minimum depth of about 5000 m (Boccaletti et al. 1997; Gianelli et al. 1997).

CO_2 degassing in southern Tuscany is around 4.8×10^6 mol y^{-1} km^{-2}, about five times the baseline of terrestrial emission (Frondini et al. 2008). Geochemical and isotopic data indicate significant contributions from the upper mantle, as testified by He-isotope signatures of gas manifestations, which show values of $R/R_A = 2.4$ at Orciatico, Montecatini Val di Cecina and Larderello (Minissale et al. 2000).

Tuscany hosts an important metallogenic province that is related to magmatism (e.g. Tanelli and Lattanzi 1986; Dini 2003). Main mineralized districts occur at Elba, in the hilly area of southwestern Tuscany known as "Colline Metallifere", and at Monte Amiata. Important quantities of iron, lead, copper, zinc, silver, antimony, mercury, and gold, have been extracted from Etruscan times until a few decades ago when mines were closed.

2.3 Composition and Classification of Tuscany Magmatism

The igneous activity of the Tuscany Province emplaced a wide variety of intrusive and extrusive bodies. Volcanic and dike rocks mostly fall in the fields of shoshonite, latite dacite-trachydacite, and rhyolite (Fig. 2.2a). Plutonic rocks partially overlap volcanic compositions, but show moderate enrichment in potassium and mostly concentrate in the acidic field. They span the alkali feldspar granite to diorite field, according to Q'-ANOR classification scheme of Streckeisen and Le Maitre (1979); granodiorites and monzogranites are largely dominating rock types, whereas mafic samples exclusively occur as microgranular enclaves within plutonic rocks (Fig. 2.2b). Several acid intrusive and extrusive rocks show a peraluminous character (Alumina Saturation Index, ASI[1] > 1)

Mafic rocks are mostly represented by dikes, lavas, and enclaves. They are generally saturated to oversaturated in silica, and show variable enrichments in potassium and K_2O/Na_2O ratios, from calcalkaline to ultrapotassic (Fig. 2.2c). An overview of Tuscany rock compositions is shown by variation diagrams reported in Fig. 2.3. These highlight a wide range of major and trace element abundances, with mafic rocks showing very high concentrations of Cr, Ni and Co, typical of unmodified mantle-derived magmas.

Sr–Nd–Pb–Hf isotope ratios display partially overlapping range of values for the silicic and mafic rocks, and all are closer to crustal than to mantle compositions (Fig. 2.4; Vollmer 1977; Hawkesworth and Vollmer 1979; Conticelli et al. 2002, 2010; Gasperini et al. 2002; Farina et al. 2010). This is obvious for the silicic rocks but is rather surprising for the mafic magmas.

[1] Molar ratio of $Al_2O_3/(CaO + K_2O + Na_2O)$.

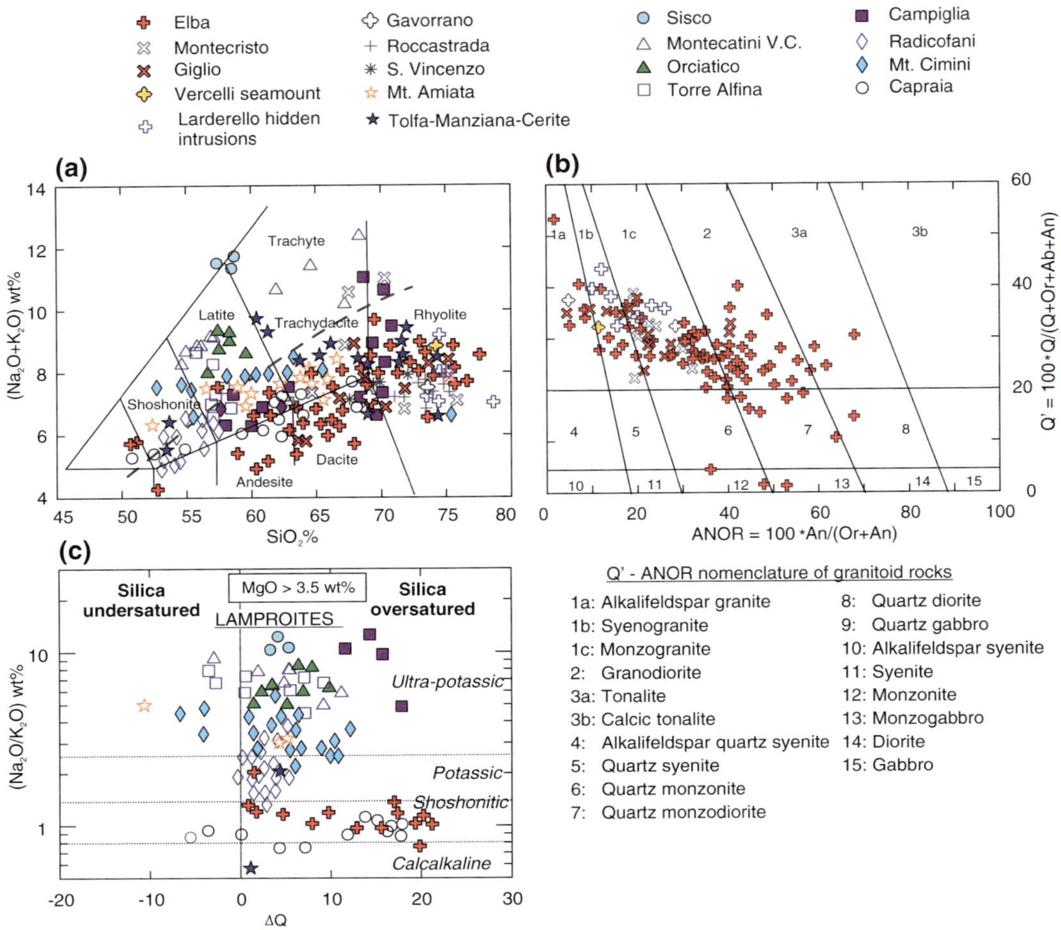

Fig. 2.2 **a** Total-Alkali versus Silica (TAS) classification diagram of Tuscany igneous rocks (Le Maitre 2002). Data recalculated to 100 % on a LOI-free basis. Note that TAS nomenclature applies to volcanic and hypabyssal rocks only. The *dashed line* divides the subalkaline and alkaline fields (Irvine and Baragar 1971). A limited number of representative samples are reported in this and the following plots, to avoid excessive symbol crowding; **b** Q′ versus ANOR classification diagram of Tuscany plutonic rocks; Q, Or, Ab, and An are CIPW normative contents of quartz, orthoclase, albite and anorthite (Steckeisen and Le Maitre 1979); **c** K_2O/Na_2O wt% versus ΔQ diagram for mafic rocks (MgO > 3.5 wt%). ΔQ is the algebraic sum of normative quartz, minus undersaturated minerals (foids and Mg-olivine). Silica undersaturated magmas have ΔQ < 0, whereas oversaturated magmas have ΔQ > 0

Therefore, the Tuscany Magmatic Province is a petrologically anomalous area, where mafic magmas (and their mantle sources) look like the upper continental crust, for radiogenic isotope ratios. Such a statement also applies to incompatible trace elements, as it will be demonstrated later in this chapter.

2.4 Silicic Magmatism

The silicic rocks of the Tuscany province occur as intrusive, hypabyssal and extrusive bodies. Intrusive rocks crop out essentially in the islands of the Tuscan Archipelago (Elba, Montecristo, Giglio), whereas they occur mainly as hidden

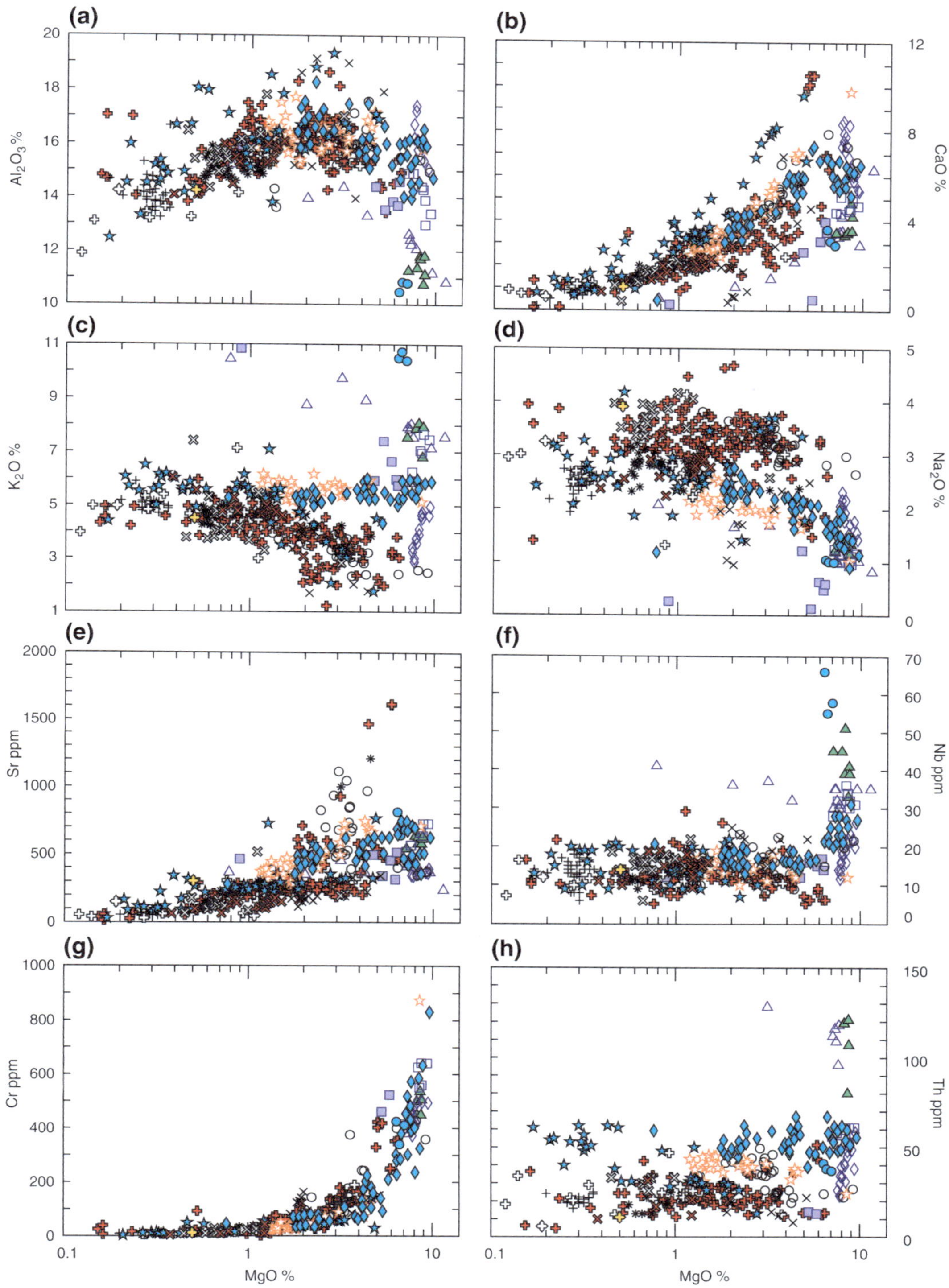

Fig. 2.3 Variation diagrams of selected major and trace elements versus MgO for magmatic rocks of the Tuscany Province. Only selected representative samples from different centres are plotted. Symbols as in Fig. 2.2

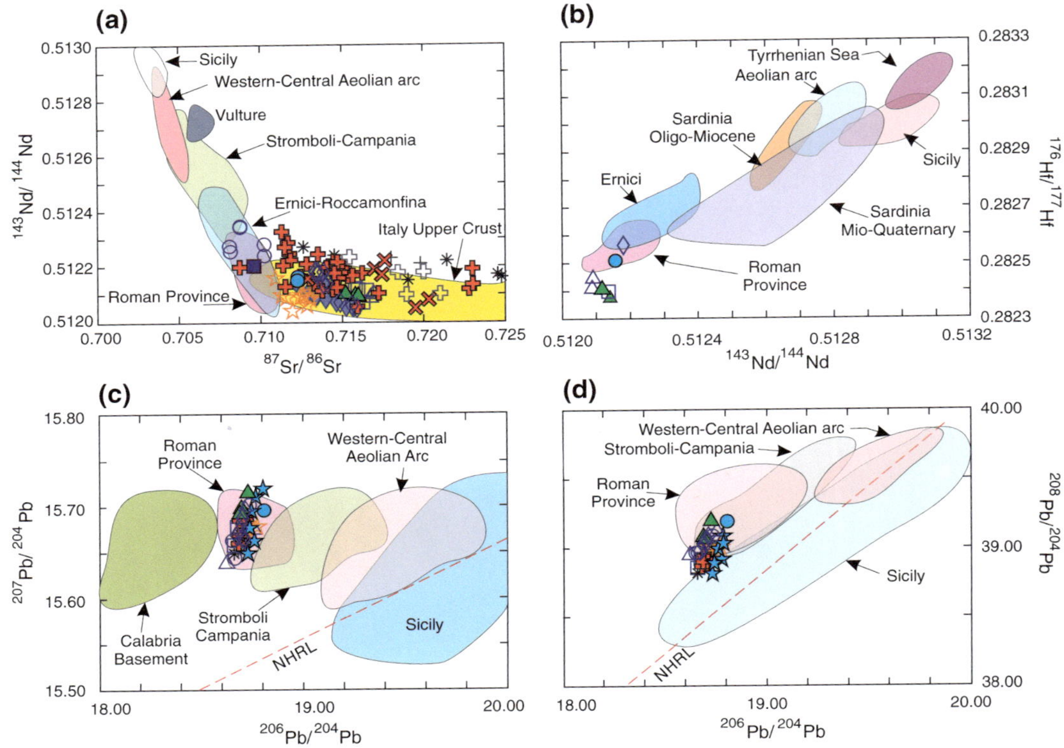

Fig. 2.4 Plots of Sr, Nd, Pb and Hf isotopes of Tuscany magmatic rocks. Compositions of other magmatic provinces (restricted to mafic compositions) from central-southern Italy are shown for comparison. Symbols as in Fig. 2.2

bodies beneath the surface in the mainland of southern Tuscany (e.g. Colline Metallifere, Larderello and Amiata areas). Some seamounts in the northern Tyrrhenian Sea are made of granitoid rocks (e.g. Vercelli; Barbieri et al. 1986; Savelli 2000; Cocchi et al. 2015). Extrusive rocks consist essentially of lava domes and flows, whereas pyroclastic rocks occur only in a few places. Main volcanic silicic centres include San Vincenzo, Roccastrada, Tolfa-Manziana-Cerveteri area and Monte Amiata. Silicic lavas and ignimbrites erupted during the early phases of activity at Cimini volcano where, however, mafic rocks are the most interesting magma types. A few dacites crop out at Capraia, and trachyte veinlets are found within the Montecatini Val di Cecina minette.

2.4.1 Effusive Rocks

San Vincenzo. Rhyolitic lava flows and dome (about 4.5 Ma; Feldstein et al. 1994) crop out at San Vincenzo, on the Tyrrhenian coast of southern Tuscany. Rocks form a small plateau that covers an area of about 10 km^2.

Petrography and mineralogy. Rock texture is porphyritic with phenocrysts of K-feldspar, quartz, plagioclase and biotite, and minor cordierite. Zircon, apatite, epidote, monazite are present as accessories; the groundmass is glassy to microcrystalline. Metasedimentary xenoliths are present in the rhyolites; latite enclaves, megacrysts of diopside ($En_{48}Fs_6Wo_{46}$) and glomeroporphyric aggregates of clinopyroxene, orthopyroxene ($En_{78}Fs_{18}Wo_4$), plagioclase

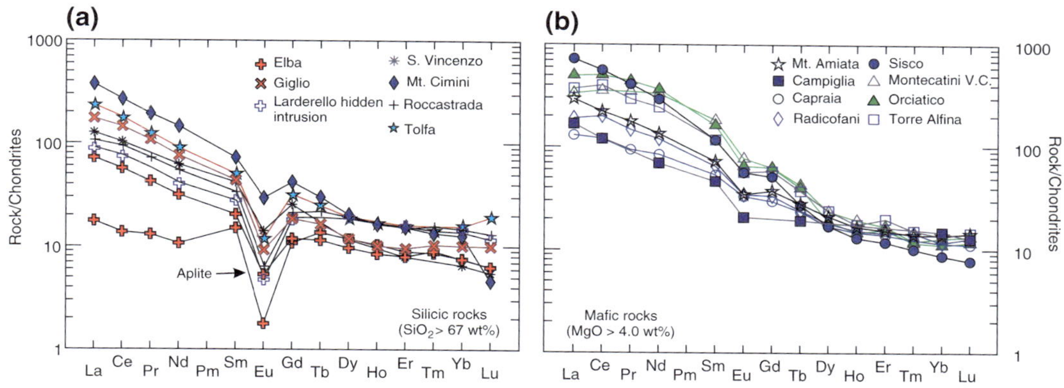

Fig. 2.5 REE patterns of selected silicic **a** and mafic **b** rocks from Tuscany. Normalising values are from Sun and McDonough (1989)

(~$An_{85\text{-}45}$) and biotite are found in some lavas. Plagioclase phenocrysts are strongly zoned (about An_{80} to An_{30} from core to rim), with the highest anorthite contents being found in the rocks that contain latite enclaves. Alkali feldspar ($Or_{56\text{–}78}$) shows minute inclusions of biotite and plagioclase. Biotite occurs as phenocryst and groundmass phase in the rhyolites and in the latite enclaves. Composition is variable (Mg#[2] ~ 0.38–0.78), with maximum Mg# being found in biotites from enclaves (Poli and Perugini 2003a). Cordierite occurs as both euhedral magmatic (Mg# = 0.47–0.59) and anhedral restitic crystals (Pinarelli et al. 1989; Ridolfi et al. 2014).

Petrology and geochemistry. The San Vincenzo rhyolites are peraluminous (ASI = 1.1–1.3), and exhibit important variations for many geochemical parameters. Based on Sr absolute abundances and Sr-isotope ratios, two groups of rocks have been distinguished (Ferrara et al. 1989; Pinarelli et al. 1989). One group of rhyolites, mostly coming from the western sector of the plateau, has relatively lower Sr abundance (Sr ~ 100–150 ppm), and higher initial Sr isotope ratios ($^{87}Sr/^{86}Sr$ ~ 0.7190–0.7248) than the rhyolites from the eastern plateau (Sr ~ 200–300 ppm; $^{87}Sr/^{86}Sr$ ~ 0.7126–0.7154). Nd isotope ratio ($^{143}Nd/^{144}Nd$ ~ 0.51215–0.51227) increases from low-Sr to high-Sr group. Pb-isotope ratios are poorly variable ($^{206}Pb/^{204}Pb$ ~ 18.66–18.74; $^{207}Pb/^{204}Pb$ ~ 15.65–15.68; $^{208}Pb/^{204}Pb$ ~ 38.82–38.95) (Vollmer 1976, 1977; Ferrara et al. 1989; Cadoux et al. 2007). Latite enclaves and pyroxene clots only occur in the high-Sr rhyolites and show high Sr abundances (Sr ~ 1000–1500 ppm), relatively low initial Sr isotope ratio ($^{87}Sr/^{86}Sr$ ~ 0.7081–0.7088) and high $^{143}Nd/^{144}Nd$ in the range 0.51245–0.51252; Ferrara et al. 1989; Feldstein et al. 1994). $\delta^{18}O_{SMOW}$ of whole rocks and separated quartz and feldspar ranges from +11.7 to +14.6 ‰ (Turi and Taylor 1976; Masuda and O'Neil 1994). Data for separated phases highlight marked Sr- and O-isotope disequilibrium. REE have variable abundances, increasing from the low-Sr to the high-Sr group, and show fractionated patterns with negative Eu anomalies (Fig. 2.5a).

Petrogenesis. Geochemical and isotopic data on whole rocks, enclaves and separated minerals indicate that the San Vincenzo rhyolites are mixtures between crustal anatectic and mafic-intermediate magmas. Crustal anatectic melts are represented by low-Sr rocks, whereas the mafic end-members had a composition close to that of the latitic enclaves (Ferrara et al. 1989; Pinarelli et al. 1989; Feldstein et al. 1994). Such a mixing process is well highlighted by the hyperbolic trend defined by the San Vincenzo rocks on $^{87}Sr/^{86}Sr$ versus Sr diagram (Fig. 2.6).

[2]Mg# is the molar ratio of MgO/(MgO + FeO).

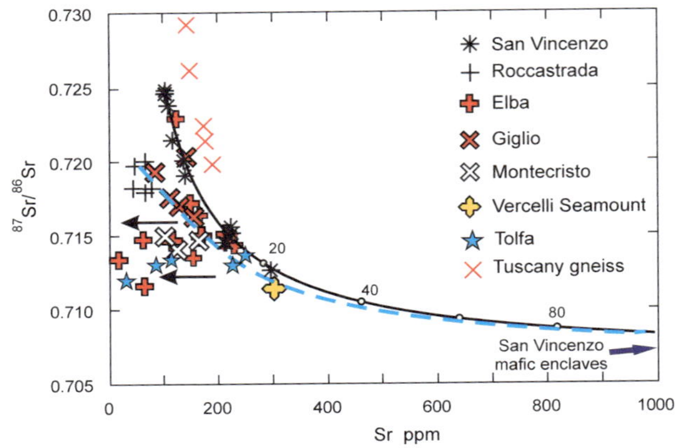

Fig. 2.6 $^{87}Sr/^{86}Sr$ versus Sr for Tuscany silicic rocks. The full line is a mixing trend between the most radiogenic San Vincenzo rhyolite and average mafic enclave from the same centre. Numbers along the line indicate the amounts of mafic melt involved in the mixing. The *blue dashed line* is a mixing trend between Roccastrada rhyolites and San Vincenzo average enclave. *Thin arrows* indicate fractional crystallisation trends. For further explanation, see text

Roccastrada. Rhyolitic lava flows and domes with an age of about 2.5 Ma (Laurenzi et al. 2007) crop out over an area of about 100 km^2 at Roccastrada (Mazzuoli 1967; Pinarelli et al. 1989).

Petrography and mineralogy. The Roccastrada rocks exhibit porphyritic textures with phenocrysts and megacrysts of K-feldspar ($\sim Or_{60-75}$) and quartz, plus minor plagioclase (mostly in the range An_{50-20}), biotite (Mg# ~ 0.50–0.40) and cordierite (Mg# ~ 0.50–0.40). Accessory phases include zircon, apatite, magnetite, and rare garnet. Groundmass texture is generally massive, glassy to perlitic, sometimes microcrystalline. Small xenoliths of metasedimentary origin have been found, whereas there is no evidence for the occurrence of magmatic mafic enclaves.

Petrology and geochemistry. The Roccastrada rhyolites are more strongly peraluminous (ASI ~ 1.2–1.5), richer in silica and Rb and more depleted in Sr and Ba than the San Vincenzo lavas. Composition resembles closely some hidden granites sampled by drilling in the Larderello area (Dini et al. 2005). Radiogenic-isotope signatures are less variable than at San Vincenzo ($^{87}Sr/^{86}Sr \sim 0.7182–0.7198$; $^{143}Nd/^{144}Nd \sim 0.51222$; $^{206}Pb/^{204}Pb \sim 18.70$; $^{207}Pb/^{204}Pb \sim 15.67$; $^{208}Pb/^{204}Pb \sim 38.97$; Vollmer 1976, 1977; Hawkesworth and Vollmer 1979; Cadoux et al. 2007). Oxygen isotopic ratio is high with $\delta^{18}O_{SMOW} \sim +13$ ‰ (Turi and Taylor 1976).

Petrogenesis. The Roccastrada rhyolites are considered as pure crustal anatectic magmas. They formed by melting of metasedimentary rocks of the Tuscany basement (e.g. Paleozoic garnet micaschists) and experienced little or no interaction with mafic magmas (Pinarelli et al. 1989; Poli and Perugini 2003b). Lower Sr isotopic ratios than the low-Sr San Vincenzo rhyolites suggest an isotopically less radiogenic source for Roccastrada.

Tolfa-Manziana-Cerite volcanoes. This volcanic complex is the most southerly exposure of the Tuscany Province. It consists of a series of lava domes containing mafic magmatic enclaves and crustal xenoliths, associated with strongly altered pyroclastic flow deposits (Cimarelli and De Rita 2006a). K/Ar ages of 3.7–1.8 Ma were reported by Clausen and Holm (1990), but more restricted values around 3.5 Ma were found for separated sanidine by Villa et al. (1989).

Petrography and mineralogy. The volcanic rocks from Tolfa-Manziana-Cerite range from trachydacite to rhyolite; magmatic enclaves are shoshonites and latites (Clausen and Holm 1990;

Pinarelli 1991; Bertagnini et al. 1995). Trachydacites have a porphyritic texture with phenocrysts of plagioclase ($\sim An_{60-50}$), sanidine, orthopyroxene (En_{52}), augite and biotite (Mg# = 0.57–0.60) set in a glassy groundmass (De Rita et al. 1994, 1997). Rhyolites are porphyritic with phenocrysts of reversely zoned plagioclase ($\sim An_{35-70}$), sanidine, and some orthopyroxene and quartz set in a microcrystalline to glassy groundmass. Accessory minerals include zircon, apatite and Fe–Ti oxides. Mafic enclaves have rounded shapes and diffused contacts with the host rocks, indicating they were incorporated in the host magmas still in a molten state. They are porphyritic and contain plagioclase ($\sim An_{66-52}$), alkali feldspar ($\sim Or_{54-75}$), diopside to augite clinopyroxene, othopyroxene ($\sim En_{55-45}$) and biotite (Bertagnini et al. 1995).

Petrology and geochemistry. The Tolfa-Manziana-Cerite rocks are moderately to highly silicic (SiO_2 ~ 63–74 wt%). Mafic enclaves display broadly calcalkaline to potassic alkaline compositions. There is a continuous decrease in Al_2O_3, TiO_2, FeO, MgO and CaO, and an increase in some incompatible elements (e.g. Rb, Th) from mafic enclaves to rhyolites. Isotopic compositions of lavas and pyroclastics are moderately variable ($^{87}Sr/^{86}Sr$ = 0.7123–0.7144; $^{206}Pb/^{204}Pb$ = 18.72–18.75; $^{207}Pb/^{204}Pb$ = 15.65–15.70; $^{208}Pb/^{204}Pb$ = 38.79–38.91; Clausen and Holm 1990; Pinarelli 1991). Mafic enclaves show a somewhat larger range of Sr–Pb isotopic ratios ($^{87}Sr/^{86}Sr$ = 0.7079–0.7127; $^{206}Pb/^{204}Pb$ = 18.72–18.79; $^{207}Pb/^{204}Pb$ = 15.67–15.79; $^{208}Pb/^{204}Pb$ = 38.87–39.12). Metasedimentary xenoliths have much more radiogenic Sr isotope ratios ($^{87}Sr/^{86}Sr$ ~ 0.726; Pinarelli 1991)

Petrogenesis. According to Clausen and Holm (1990), the Tolfa-Maziana-Cerite volcanics were generated by fractional crystallisation starting from intermediate parents, which derived directly from melting of subducted upper crustal rocks. Pinarelli (1991) suggests an origin by melting of metasedimentary crustal rocks followed by fractional crystallisation and mixing with various types of mantle-derived mafic magmas, represented by mafic enclaves. These are derived from anomalous metasomatic mantle sources less strongly enriched in incompatible elements than that of the nearby Roman Province (Pinarelli 1991; Bertagnini et al. 1995).

Monte Amiata. This is a 1738 m-high volcano dominated by trachydacitic lava flows and domes with a few late-erupted shoshonites and latites. The volcano is affected by numerous faults, related either to strike-slip regional tectonics (Brogi et al. 2010) or/and to volcanic spreading over its substratum (Borgia et al. 2014). The high elevation is related to both the accumulation of volcanic products around a central crater area and to doming by magma intrusion in an extensional setting (Acocella and Mulugheta 2001; Marroni et al. 2015). K/Ar and $^{40}Ar/^{39}Ar$ ages range from about 0.4 to 0.2 Ma (see Laurenzi et al. 2015) with a most probable age of 0.3 Ma (Cadoux and Pinti 2009). Well-known and long-exploited cinnabar mineralisations occur at Monte Amiata (Mazzuoli and Pratesi 1963; Barberi et al. 1971; Conticelli et al. 2015b with references).

Petrography and mineralogy. The Monte Amiata trachydacites exhibit a porphyritic texture with abundant phenocrysts and megacrysts of sanidine ($\sim Or_{80}$), plagioclase ($\sim An_{80-50}$), orthopyroxene ($\sim En_{55-40}$), high-Ti biotite and diopside to augite clinopyroxene. Groundmass is glassy and contains microlites of clinopyroxene, rare orthopyroxene, and some biotite. Xenocrysts of olivine have been observed. Accessory phases include apatite, zircon, ilmenite, magnetite and perrierite, a hydrous sorosilicate of Ca, Fe, Ti, Th and REE (Cristiani and Mazzuoli 2003; Conticelli et al. 2015b with references). Latites and shoshonites are porphyritic with phenocrysts of reversely zoned plagioclase, diopside, and olivine (up to Fo_{90}), set in a hypocrystaline groundmass containing the same phases plus sanidine and glass. Abundant magmatic mafic enclaves and metamorphic xenoliths occur in the Monte Amiata lavas. Mafic enclaves are particularly abundant in the summit domes (van Bergen et al. 1983; van Bergen and Barton 1984; Conticelli et al. 2015b with references), and show clear textural evidence (rounded shapes, chilled margins, etc.) of being incorporated into the host magma when they were still in a molten state (Conticelli et al. 2010, 2015b). Textures are porphyritic with dominant diopside phenocrysts plus some phlogopite-biotite and olivine ($\sim Fo_{90-82}$) set in a groundmass consisting of sanidine, clinopyroxene, biotite and rare olivine. Xenocrysts of plagioclase, sanidine and orthopyroxene are also present (van Bergen et al. 1983).

Petrology and geochemistry. The Monte Amiata lavas have a potassic petrochemical affinity, with the most mafic samples being ultrapotassic in composition (K_2O/Na_2O wt % > 2.5). All samples, except some enclaves, are oversaturated in silica. There is a decrease in FeO_{total}, MgO, CaO, ferromagnesian trace elements, Sr and Ba with increasing silica. In contrast, incompatible elements (e.g. Th, Rb, Nb) remain virtually constant and are rather scattered in the silicic rocks. $^{87}Sr/^{86}Sr$ ratio ($\sim$0.7116–0.7131 for the lavas and 0.7105–0.7118 for the magmatic enclaves) increases linearly with silica, whereas Nd- and Pb-isotopes are poorly variable ($^{143}Nd/^{144}Nd \sim 0.5121$; $^{206}Pb/^{204}Pb \sim 18.71$–18.77; $^{207}Pb/^{204}Pb \sim 15.67$; $^{208}Pb/^{204}Pb \sim$ 38.98–39.00; Giraud et al. 1986; Cadoux et al. 2007; Cadoux and Pinti 2009; Conticelli et al. 2015b with references).

Petrogenesis. An origin by mixing between crustal anatectic melts and mafic potassic alkaline magmas similar to the nearby Vulsini district has been suggested by some authors (van Bergen 1985; Peccerillo et al. 1987; Cadoux and Pinti 2009). According to Conticelli et al. (2010), the early-erupted silicic magmas derived from a shoshonitic mafic parent by fractional crystallization plus crustal contamination. The intermediate and final magmas are hybrids between early-erupted high silica magmas and ultrapotassic silica-undersaturated Roman-type melts.

Monti Cimini. This volcano contains significant amounts of early-erupted silicic rocks (trachyte and trachydacite), along with late mafic latite, olivine-latite and shoshonite lavas, which represent the petrologically most important products. Both silicic and mafic rocks are descrived in Section 2.5.

2.4.2 Plutonic Rocks

Island of Elba. A large number of intrusive bodies crop out in the Island of Elba, showing various size and composition, from monzogranite and granodiorite to alkali-feldspar granite, aplite, and pegmatite. K/Ar, $^{40}Ar/^{39}Ar$, Rb/Sr and U/Pb ages range from about 8.4 to 6.4 Ma (Poli et al. 1989a; Dini et al. 2002; Rocchi et al. 2002, 2003a; Pandeli et al. 2006; Farina et al. 2010; Poli and Peccerillo 2016). A 5.8 Ma old strongly altered shoshonitic mafic dike has been found at Monte Castello, eastern Elba (Conticelli et al. 2001). Fe–Pb–Sn mineralisation are associated with intrusive magmatism at Elba and have been exploited since Etruscan times until a few decades ago. Except for a few intrusions cropping out in the eastern Elba (Monte Castello dikes and Porto Azzurro granites and associated aplites; Conticelli et al. 2001; Pandeli et al. 2006; Dini et al. 2009; Poli and Peccerillo 2016), the igneous activity is concentrated in the western and central sectors of the island, where the magmatism started with intrusions of a series of laccoliths (Capo Bianco aplite, Portoferraio and San Martino porphyries) that emplaced as separate sheets to form a sort of Christmas-tree structure (Rocchi et al. 2002, 2010; Westerman et al. 2004). This was uplifted and displaced laterally by the emplacement of the Monte Capanne monzogranite stock. Dark coloured dikes (Orano porphyries) represent the final stages of the magmatism in this sector of the island.

Petrography and mineralogy. Textures and modal mineralogy of igneous rocks at Elba differ considerably. The Capo Bianco aplites are white-coloured porphyritic alkali-feldspar granites, locally spotted with dark blue spherical aggregates of tourmaline microcrysts. Portoferraio porphyries are monzogranitic to syenogranitic in composition and show a porphyritic texture with phenocrysts of alkali feldspar, quartz, plagioclase and biotite. San Martino monzogranite porphyries are characterised by alkali feldspar megacrysts plus quartz, plagioclase and biotite phenocrysts. Orano dikes consist of monzodiorites to monzogranites containing phenocrysts of plagioclase, biotite and

Fig. 2.7 Mafic enclaves in the Monte Capanne (Elba) monzogranitic rocks. Photo by G. Poli

rare amphibole set in a fine-grained groundmass. The Monte Capanne monzogranitic stock (about 10 km in diameter) is the largest intrusion and is surrounded by a well-developed thermometamorphic aureole. It exhibits a porphyritic texture with various amounts of centimetre- to decimetre-sized euhedral megacrysts of K-feldspar that are set in a medium- to coarse-grained matrix formed of plagioclase, quartz, K-feldspar and biotite with accessory apatite, zircon, thorite, allanite, monazite, ilmenite, and tourmaline. The intrusion contains abundant centimetre- to metre-sized microgranular calcalkaline-shoshonitic mafic enclaves (Fig. 2.7), and is cut by aplitic and pegmatitic dikes, and by the Orano porphyry system. Large euhedral crystals of quartz, K-feldspar, tourmaline, pollucite and other rare minerals have been recovered from pegmatites (e.g. Tanelli and Poggi 2012). Sedimentary and metasedimentary xenoliths, and microgranular mafic enclaves are found in many intrusions (e.g. Farina et al. 2012).

Petrology and Geochemistry. Major and trace element compositions of igneous rocks at Elba are very variable (Fig. 2.3). Capo Bianco aplites show the highest ASI (~ 1.3–1.6) and the lowest abundances of TiO_2, FeO_{total}, MgO and CaO among the Elba intrusions. Portoferraio porphyries are richer in TiO_2, FeO_{total}, MgO, CaO than Capo Bianco aplite, a trend that continues in the San Martino porphyries and Monte Capanne monzogranites, up to the Orano porphyries and the late mafic intrusions of eastern Elba. Some Orano dikes show strong enrichments in Sr, Ba and LREE (Fig. 2.3e), but not for other incompatible elements, resembling some andesitic lavas from Capraia and the San Vincenzo mafic enclaves (Poli and Peccerillo 2016). REE decrease slightly with silica and show fractionated patterns with negative Eu anomalies, which are stronger in the aplites (Fig. 2.5a). Sr-isotope ratios of main intrusions are in the range $^{87}Sr/^{86}Sr \sim 0.711$–$0.723$, increasing from Orano porphyries to the felsic dikes of Cotoncello, an outcrop sited at the northern border of Monte Capanne. A larger range ($^{87}Sr/^{86}Sr \sim 0.7101$–$0.7324$) has been found for separated phases by Farina et al. (2014). Nd- and Pb-isotope ratios are less variable ($^{143}Nd/^{144}Nd \sim 0.51219$–$0.51238$; $^{206}Pb/^{204}Pb \sim 18.68$–$18.75$; $^{207}Pb/^{204}Pb \sim 15.66$–$15.69$; $^{208}Pb/^{204}Pb \sim 38.87$–$38.96$). Mafic microgranular enclaves generally have radiogenic isotope compositions lying within the field of host rocks, probably indicating some isotopic re-equilibration with the enclosing magmas (Poli 1992; Dini et al. 2002); however, lower Sr-isotope ratios ($^{87}Sr/^{86}Sr \sim 0.7093$) have been found for some enclaves (e.g. Gagnevin et al. 2004). The Monte Castello mafic dike has $^{87}Sr/^{86}Sr \sim 0.708$–$0.709$ (Conticelli et al. 2001).

Petrogenesis. According to most authors (Poli et al. 1989a; Dini et al. 2002; Gagnevin et al. 2004, 2005, 2010), the bulk of Elba magmas was formed by mixing between mafic end-members of ultimate mantle origin and peraluminous silicic melts formed by crustal anatexis. Mafic magmas had a calcalkaline-shoshonitic composition similar to the eastern Elba dikes, whereas the crustal end-member had a composition as the Cotoncello intrusion (e.g. Poli 1992; Gagnevin et al. 2004; Poli and Peccerillo 2016). Other authors (e.g. Farina et al. 2012, 2014) suggest that at least part of the petrological variability of Elba intrusions is primary, i.e. reflects compositions of magmas ascended directly from crustal sources with contrasting metapelitic to metavolcanic compositions. Whatever the case, the Elba magmas, especially Monte Capanne, were affected by significant fractional crystallisation and filter-pressing processes with formation of silicic differentiates (leucogranites, aplites and pegmatites) that show radiogenic Sr-isotope compositions similar to their parent monzogranites and granodiorites (Poli et al. 1989a; Rocchi et al. 2002; Poli and Peccerillo 2016).

Island of Montecristo. This island is situated about 60 km south of Elba, along a N-S trending extensional fault, parallel to the Corsica coast. It consists of a monzogranite stock containing microgranular mafic enclaves, cut by aplite and porphyritic dikes. Age (Rb/Sr) is 7.1 Ma (Innocenti et al. 1997).

Petrography and mineralogy. The Montecristo monzogranites have a porphyritic texture with zoned phenocrysts and/or megacrysts of K-feldspar ($\sim Or_{75-85}$), quartz and plagioclase ($\sim An_{25-20}$) surrounded by a medium-grained matrix formed by the same phases, plus variable amounts of biotite ($X_{Fe} \sim 0.62$–0.65) and accessory apatite, zircon, tourmaline, ilmenite, sphene, allanite, monazite, and rutile. Cordierite has been also observed. Aplite dikes (10–30 cm thick) are rich in tourmaline and contain some muscovite. Porphyritic dikes contain ubiquitous biotite ($X_{Fe} \sim 0.41$–0.59) and plagioclase ($\sim An_{55-40}$) phenocrysts, plus resorbed quartz, alkali-feldspars or magnetite in some outcrops.

Petrology and geochemistry. The Montecristo monzogranites display moderate compositional variation (e.g. $SiO_2 \sim 69$–75 wt%; $K_2O \sim 3.7$–4.8 wt%). Porphyritic dikes have slightly lower silica but remarkably higher K_2O contents than monzogranites. Microgranular mafic enclaves ($SiO_2 \sim 58.5$–70.6 wt%), show a broadly calcalkaline composition (Poli 1992; Innocenti et al. 1997). Initial Sr-isotope ratios of monzogranites are around 0.714–0.715, whereas a porphyritic dike has $^{87}Sr/^{86}Sr = 0.70962$ (Innocenti et al. 1997).

Petrogenesis. The origin of the Montecristo monzogranite is attributed to melting of pelitic crustal sources, followed by mixing with mafic magmas and moderate fractional crystallisation. The porphyritic dikes represent late emplaced batches of magmas characterised by higher potassium enrichment than the main intrusion (Innocenti et al. 1997; Rocchi et al. 2003b).

Island of Giglio. The island consists of a main monzogranitic stock, a small leucocratic monzogranitic body forming the Le Scole islets, and a few sedimentary and metamorphic rocks. Rb/Sr age of intrusive units is about 5 Ma (Westerman et al. 1993). Two main texturally distinct rock types are observed in the monzogranite, respectively representing the core and the margin of the intrusion: the Arenella facies forming the eastern half of the island, and the Pietrabona facies occurring on the western island. Dikes and veins of granites, tourmaline-rich aplites and pegmatites occur in the main intrusion and, to a minor extent, in the Le Scole rocks.

Petrography and mineralogy. The Arenella facies is characterized by isotropic texture with abundant megacrysts of K-feldspar set in a medium-grained homogeneous matrix; the Pietrabona facies is strongly foliated with a preferred orientation of minerals and xenoliths. Modal mineralogy of both rock types consists of plagioclase, K-feldspar, quartz and variable amounts of biotite that decreases from Pietrabona to Arenella facies, plus accessory muscovite, tourmaline, Fe–Ti oxides, apatite, monazite, and zircon. Xenocrystic cordierite, garnet, andalusite, and sillimanite have been also observed. Metamorphic xenoliths and microgranular mafic enclaves are common. The leucocratic monzogranite of Le Scole is weakly porphyritic with K-feldspar megacrysts set in a medium- to fine-grained matrix composed of quartz, plagioclase, alkali feldspar, biotite, tourmaline, and cordierite. No microgranular mafic enclaves are found in the Le Scole rocks.

Petrology and geochemistry. The Giglio rocks display moderately variable major and trace element contents (Fig. 2.3), with the Arenella facies displaying lower FeO, MgO, and higher CaO than Pietrabona. Sr-isotope ratios increase from monzogranites ($^{87}Sr/^{86}Sr = 0.7163$–0.7176) to Le Scole leucocratic rocks ($^{87}Sr/^{86}Sr = 0.7195$–0.7203). $^{143}Nd/^{144}Nd$ ranges from 0.51222 to 0.51205 and shows an opposite tendency. The mafic enclaves have comparable and sometimes lower Sr–isotope compositions than the host rocks ($^{87}Sr/^{86}Sr =$ 0.7128–0.7172; Westerman et al. 1993, 2003).

Petrogenesis. The magmatism of the Giglio island is related to distinct petrogenetic and intrusive events. The main monzogranite intrusion is older and results from crustal melting plus mixing with moderate amounts of mafic magmas. Le Scole rocks probably represent pure crustal melts that underwent little or no interaction with mafic magmas.

Campiglia Marittima. The granitoid intrusions at Campiglia Marittima only crop out over a small area at Botro ai Marmi. Exposed rocks show monzogranite to alkalifeldspar granite composition, strongly depleted in Fe by secondary processes and, therefore, extensively mined as row ceramic material (Lattanzi et al. 2001). K/Ar age is around 5 Ma (Rocchi et al. 2003c). Independent of granitoid intrusions, felsic and mafic porphyritic dikes (4.3 Ma old) occur in the area (Barberi et al. 1967). The Botro ai Marmi rocks contain quartz, orthoclase and some plagioclase with scarce altered biotite. Strong alteration by late magmatic fluids affected the intrusion and the surrounding thermo-metamorphic and sedimentary rocks. These processes generated dramatic modification of rock compositions and were responsible for the formation of Cu–Pb–Zn–Ag mineralisation. Least-altered intrusive rocks are moderately peraluminous (ASI $\sim$ 1.2) and silicic ($SiO_2 \sim 70$ wt%).

Gavorrano. Intrusive rocks at Gavorrano consist of about 4.3 Ma old biotite-bearing monzogranite with K-feldspar megacrysts, and tourmaline-rich alkali feldspar granite. The exposed intrusion is about 3 km long and NNW-SSE oriented (Mazzarini et al. 2004; Musumeci et al. 2005). Major and trace element compositions are rather variable (e.g. $SiO_2 \sim$ 66–75 wt%; Rb = 250–700). Except for Rb, incompatible trace element contents and REE fractionation decrease with increasing silica, suggesting separation of accessory phases such as monazite (Poli et al. 1989a; Rocchi et al. 2003c; Dini et al. 2005). Sr-isotope ratios are in the lower range of Tuscany silicic rocks ($^{87}Sr/^{86}Sr = 0.7139$–0.7149; Ferrara and Tonarini 1985), possibly an effect of interaction between crustal anatectic and relatively unradiogenic mantle-derived magmas.

Hidden plutons. There are several hidden intrusions in southern Tuscany, whose presence has been revealed by geophysical investigation and drillings (e.g. Zitellini et al. 1986; Gianelli and Laurenzi 2001). Northeast of Gavorrano (**Castel di Pietra**), borehole drilling encountered 4.3 Ma plutonic rocks at a depth of about 870–880 m beneath a thick thermometamorphic roof

(Franceschini et al. 2000). Cored samples include porphyritic monzogranites, fine-grained leucocratic syenogranites, and orthopyroxenene-bearing dark granodiorites and monzogranites. Silica (SiO_2 ~ 65–72 wt%) and Sr isotope ratios ($^{87}Sr/^{86}Sr$ ~ 0.7145–0.7222) increase from the dark to leucocratic facies (Dini et al. 2003). At **Monte Spinosa**, a few kilometres south of Botro ai Marmi, syenogranites and monzogranites with SiO_2 = 66–71 wt% have been recovered. In the **Larderello** area, a series of borehole drillings (Monteverdi, Radicondoli, Travale, Carboli) encountered peraluminous muscovite-biotite bearing syenogranites to monzogranites with ages of 3.8–1.3 Ma (Gianelli and Laurenzi 2001; Villa et al. 2001; Dini et al. 2005). Mafic enclaves commonly contained by other Tuscany intrusive rocks have not been observed in these granitoids. According to Dini et al. (2005), the hidden intrusions of the Larderello area represent pure crustal anatectic magmas formed by melting of biotite and muscovite-rich metasedimentary sources. However, Sr-isotope ratios are variable and somewhat lower ($^{87}Sr/^{86}Sr$ = 0.7148–0.7209) than the local basement ($^{87}Sr/^{86}Sr$ ~ 0.7196–0.7318), calling for a less radiogenic component of deep crustal or subcrustal origin.

Seamounts: Several seamounts, partly with granitoid composition, occur between Corsica and Tuscany. These are aligned along a north-south trending ridge, south of Elba and Montecristo islands (Savelli 2000), in a zone characterised by a tectonic style that recalls the Basin and Range province (Carmignani and Kligfield 1990; Marani and Gamberi 2004). A sample dredged from the Vercelli seamount, southeast of this alignment near to the 41° Parallel line, revealed a K/Ar age on feldspar of 7.2 Ma, a tourmaline-biotite syenogranite composition, and initial $^{87}Sr/^{86}Sr$ = 0.71140 (Barbieri et al. 1986; Table 2.2). The Vercelli Seamount is a 8 km long and 3.4 km wide, SW-NE oriented edifice, whose summit rises about 1200 m above its surrounds, reaching 58 m bsl. A highly magnetic body, possibly a mafic lava flow, has been detected near to Vercelli (Cocchi et al. 2015).

2.5 Mafic Magmatism

Mafic magmas (MgO higher than 3.5–4.0 wt%) make up small hypabyssal and effusive bodies, the large volcanoes of Capraia and the younger effusive activity of Monti Cimini. They are also present as enclaves and dikes in several silicic intrusive and extrusive rocks, as recalled earlier in this chapter. Compositions range from calcalkaline and shoshonitic to ultrapotassic lamproitic (Fig. 2.2c). Lamproites are slightly undersaturated to oversaturated in silica and show high MgO and SiO_2 abundances and relatively low concentrations of CaO, Al_2O_3, Na_2O, and FeO_{total} (Peccerillo et al. 1988; Conticelli and Peccerillo 1992; Conticelli et al. 2015a). The most primitive calcalkaline and shoshonitic compositions exhibit higher CaO, Na_2O, FeO_{total} and Al_2O_3 than lamproites, as observed for typical calcalkaline basalts, such as those from the Aeolian Islands (Fig. 2.8).

Sisco. A small sill intruded into Alpine high-pressure metamorphic terrains occurs in northeastern Corsica, south and west of Marine de Sisco along the D80 route (Peccerillo et al. 1988; Marinosci 1993). K/Ar ages yielded values from 15.5 to 13.5 Ma (Civetta et al. 1978; Bellon 1981).

Petrography and mineralogy. The rock is a lamprophyre with a slightly porphyritic to microgranular texture, consisting of altered olivine, phlogopite, sanidine (Or_{95}), Al-poor diopside, and K-richterite. Accessory minerals include titanite, chromite, ilmenite, priderite (K–Ba–Fe^{3+}–Ti oxide), rutile and rare roedderite, a cyclosilicate of Fe, Mg, Na, K (Wagner and Velde 1986).

Petrology and geochemistry. The Sisco rock is a peralkaline (Peralkaline Index, PI[3] > 1; normative acmite around 4–5 wt%) high-silica lamproite exhibiting relatively high concentrations of MgO (~6.5–7.0 wt%), Ni (~250 ppm), Co and Cr, and moderate abundances of Sc (~10 ppm) and V (~90 ppm). Al_2O_3 (~10 wt%) and Na_2O

[3]PI (Peralkaline Index) is the molar ratio of (Na_2O + K_2O)/Al_2O_3.

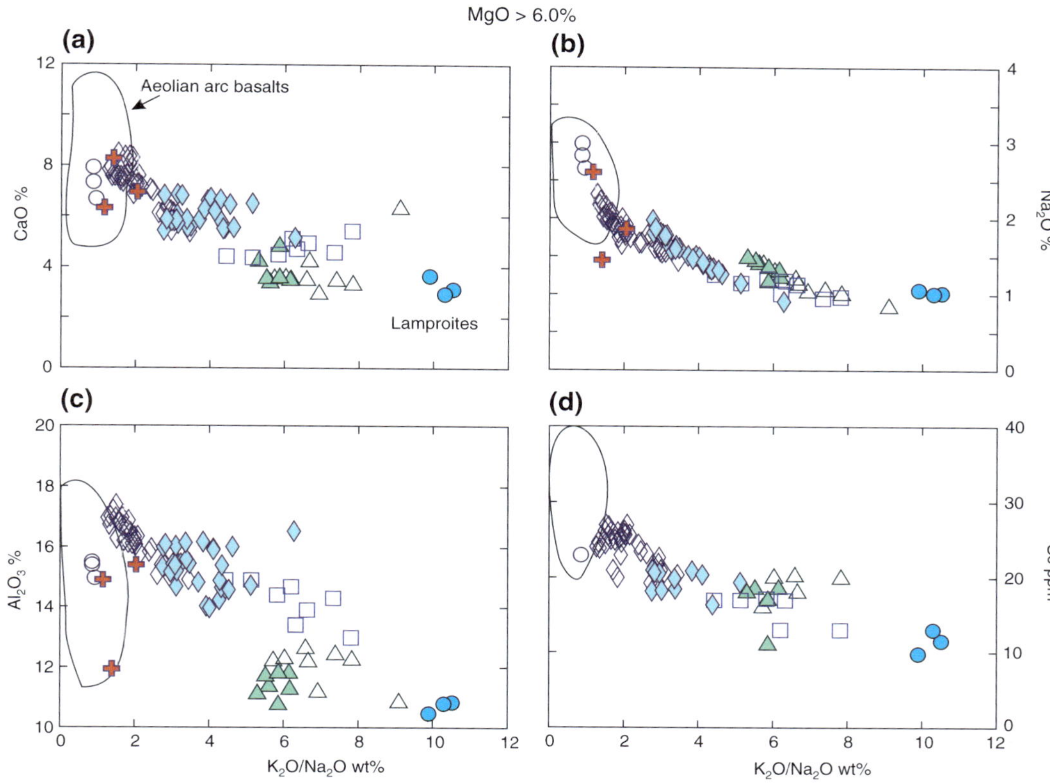

Fig. 2.8 CaO, Na_2O, Al_2O_3, and Sc versus K_2O/Na_2O relationships for the most primitive rocks (MgO > 6.0 wt %) from Tuscany. The field of calcalkaline basalts from the Aeolian arc is reported for comparison (Peccerillo et al. 2013). Symbols as in Fig. 2.2

(~1.0 wt%) are low, a feature that is typical of lamproites. REE are fractionated with a small negative Eu anomaly (Fig. 2.5b). The mantle-normalised incompatible element pattern differs somewhat from those of Tuscany lamproites because of lower enrichments in some LILE (Cs, Rb, Th, U) and the lack of a positive spike of Pb (Fig. 2.9a). The initial Sr isotope ratio is relatively high ($^{87}Sr/^{86}Sr$ ~ 0.7123). Nd, Pb and Hf isotope ratios are poorly to moderately radiogenic ($^{143}Nd/^{144}Nd$ ~ 0.51216; $^{206}Pb/^{204}Pb$ ~ 18.81 $^{206}Pb/^{204}Pb$ ~ 15.69 $^{206}Pb/^{204}Pb$ ~ 39.19; $^{176}Hf/^{177}Hf$ = 0.282504). Initial Os isotope ratio is $^{187}Os/^{188}Os$ = 0.1889 (Conticelli et al. 2007, 2009a, 2010).

Petrogenesis. High MgO, Ni, and Cr abundances clearly indicate a mantle origin for the Sisco lamproite. The low abundances in Al_2O_3 and Na_2O suggest a clinopyroxene-poor harzburgitic source. On the other hand, radiogenic isotope signatures and LILE abundances call for an enriched mantle source, which was affected by metasomatic processes.

Orciatico. The Orciatico outcrop consists of a laccolith several meters thick, and a small vertical feeder dike intruded into Pliocene marly sediments. Glass-rich chilled margins occur at the contact with intruded sediments. Age is 4.1 Ma (Conticelli et al. 1992)

Petrography and mineralogy. Textures are poorly porphyritic with microphenocrysts of olivine (~Fo_{90-75}), phlogopite (Mg# ~ 75–79), and Al-poor diopside (Al_2O_3 ~ 0.5 wt%; Cellai et al. 1994) set in a groundmass consisting of the same phases plus sanidine, glass, and accessory K-richterite, rutile, ilmenite, and Mg-chromite. Some of the high-MgO olivine crystals show evidence of corrosion and kinking, and are

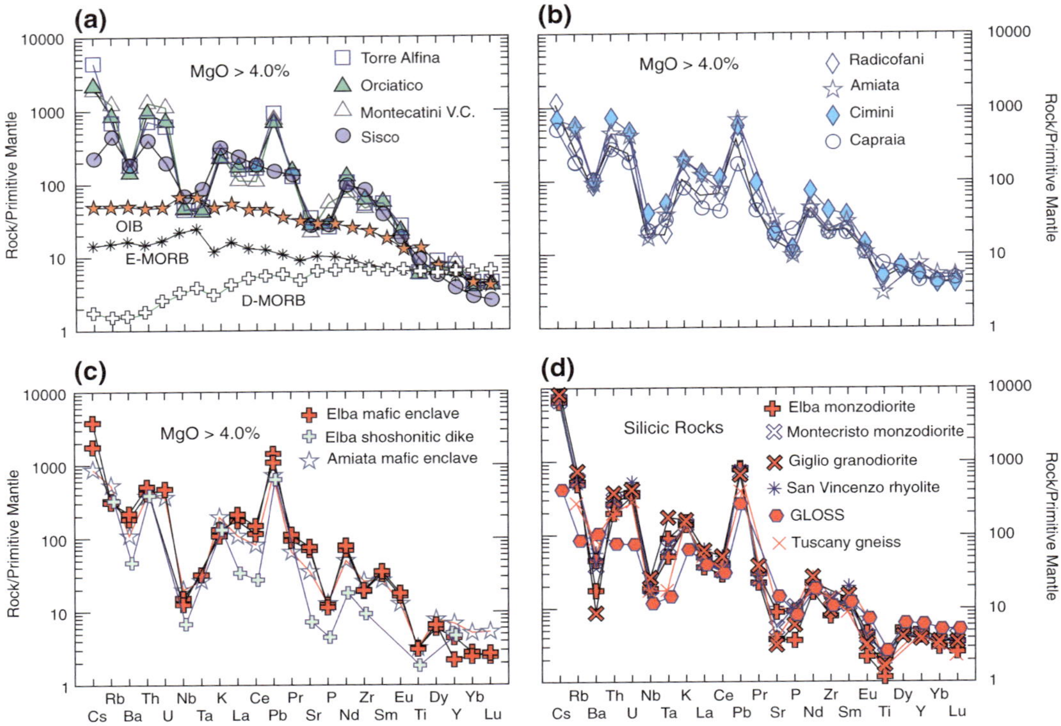

Fig. 2.9 Incompatible element patterns normalised to primordial mantle composition (Sun and McDonough 1989) of representative Tuscany mafic rocks **a**, **b**, of mafic enclaves and shoshonitic dike **c**, and of selected silicic crustal rocks **d**. GLOSS is the average Global Subducted Sediments of Plank and Langmuir (1998); D-MORB, E-MORB, and OIB are averages of depleted and enriched Mid-Ocean Ridge Basalts (Gale et al. 2013) and of Ocean Island Basalts (Sun and McDonough 1989)

probably xenocrysts resulting from disaggregation of high-pressure ultramafic xenoliths (Poli 1985; Wagner and Velde 1986; Peccerillo et al. 1987, 1988; Conticelli and Peccerillo 1992; Conticelli et al. 1992).

Petrology and geochemistry. The Orciatico rocks are high silica lamproites that differ from Sisco, because of a lower peralkaline index (PI ~ 0.9–1.0), K_2O, TiO_2 Nb, and Ta, slightly higher Al_2O_3 and Na_2O, and much higher Cs, Rb, Th, U, and Pb. MgO, Ni, Co, and Cr are in the range of mantle equilibrated melts. The REE pattern is fractionated, with an upward convexity for light REE and a small negative Eu anomaly (Fig. 2.5b). Incompatible element patterns are highly fractionated and contain pronounced positive spikes of Cs, Rb, Th and Pb, and stronger negative anomalies of HFSE than Sisco (Fig. 2.9a). Overall, they closely resemble patterns of some upper crustal rocks such as Tuscany gneiss or crustal anatectic rhyolites (Fig. 2.9d; Peccerillo and Martinotti 2006). Sr–Nd–Pb–Hf–Os isotope ratios are closer to upper crustal than to mantle values, or somewhat intermediate between the two ($^{87}Sr/^{86}Sr$ = 0.7152–0.7159; $^{143}Nd/^{144}Nd$ = 0.5121; $^{206}Pb/^{204}Pb$ = 18.69–18.73; $^{176}Hf/^{177}Hf$ = 0.28240; $^{187}Os/^{188}Os$ = 0.3405; Fig. 2.4; Table 2.2). Oxygen-isotope ratios show large differences between phenocryst and groundmass minerals with $\delta^{18}O_{SMOW}$ ~ +7.2 ‰ in olivine phenocrysts and $\delta^{18}O_{SMOW}$ ~ +11.1 ‰ in groundmass sanidine (Peccerillo et al. 1988; Barnekow 2000; Conticelli et al. 1992, 2007, 2010).

Petrogenesis. The major, trace element and radiogenic isotope signatures of Orciatico lamproite suggest an origin in a mantle source depleted in some major elements (i.e. CaO, Na_2O), but strongly enriched in geochemical components akin to the upper continental crust, such as metapelites.

Montecatini Val di Cecina. The Montecatini subvolcanic body is a 4.1 Ma old plug (Borsi et al. 1967) intruded into Miocene marine and lacustrine sediments and Liguride units. The rock is a phlogopite-rich lamprophyre[4] (minette) permeated by a dense network of leucocratic veinlets.

Petrography and mineralogy. The Montecatini minette has a medium- to fine-grained sometimes porphyritic texture with abundant phlogopite, Al-poor augite (Cellai et al. 1994), K-feldspar, minor altered olivine and accessory apatite, amphibole, Fe–Ti oxides, zircon, thorite, apatite and perrierite (Peccerillo et al. 1988; Conticelli et al. 1992). The leucocratic veinlets consist of dominant sanidine with minor quartz and brown mica, plus accessory apatite. These veinlets likely represent residual melts separated from the host magma during crystallisation.

Petrology and geochemistry. The Montecatini minette has similar major element composition as the Orciatico rock, although the peralkaline index is lower (PI ~ 0.8–0.9) and P_2O_5 is higher. Incompatible element and REE patterns, and Sr–Nd–Pb–Hf–Os isotope ratios are not far from values of Orciatico ($^{87}Sr/^{86}Sr = 0.7169$; $^{143}Nd/^{144}Nd = 0.5121$; $^{206}Pb/^{204}Pb = 18.76$; $^{176}Hf/^{177}Hf = 0.28245$; $^{187}Os/^{188}Os = 0.5502$; Table 2.2). The leucocratic veins have a trachytic composition and similar radiogenic isotopic composition to the host minette (e.g. Peccerillo et al. 1988; Conticelli et al. 1992, 2007, 2010).

Petrogenesis. Similar geochemical compositions and ages of the Montecatini and Orciatico rocks suggest a common origin. Some differences between the two occurrences could result from the separation of felsic veinlets in the Montecatini magma, which somewhat modified the original composition of the lamproitic magma.

Torre Alfina. This is a small volcano (about 0.8 Ma old) formed by a few lava units, one of whom is several hundred meters long, and two necks. The Torre Alfina rocks contain a large number of xenoliths of both crustal and mantle origin, which document a rapid magma ascent on the order of few hours (Conticelli and Peccerillo 1990, 1992; Barnekov 2000; Casagli 2009). The xenoliths, best observed on the front wall of the castle dominating the village of Torre Alfina (Fig. 2.10), include large amounts of crustal rocks (granulites, gneiss, schists, sandstones and marls) and a few cm-sized ultramafic nodules (phlogopite-bearing dunites, spinel harzburgites and lherzolites, and rare phlogopite-rich peridotites). Geothermobarometric investigations on ultramafic xenoliths gave equilibration T-P of about 1050 °C and 1.5 GPa, corresponding to a depth of 50–60 km (Pera et al. 2003). Most of the crustal xenoliths and xenocrysts show evidence of partial melting and reaction with the host magma, exhibiting one of the most compelling cases of wall rock assimilation by a rapidly ascending hot mafic magma (Conticelli 1998).

Petrography and mineralogy. Rocks have aphyric to poorly porphyritic texture and the only phenocryst phase is rare euhedral to skeletal Mg-olivine (up to Fo_{92}). Groundmass phases include olivine, low-Al diopside, phlogopite (Mg# ~ 95), K-feldspar (Or_{82-85}), glass and accessory ilmenite, Ti-magnetite and Mg-chromite. The latter is also found as inclusions within olivine phenocrysts (Conticelli 1998).

Petrology and geochemistry. The Torre Alfina lavas have similar composition to the Montecatini and Orciatico lamproites, except for lower Th and Rb contents and higher $^{187}Os/^{188}Os$ (=0.2756; Conticelli et al. 2007). There are small but significant compositional variations between lava flows and necks.

Petrogenesis. A mantle origin for the Torre Alfina magmas is clearly indicated by both rock geochemistry and occurrence of high-P ultramafic xenoliths. The phlogopite-rich peridotite xenoliths found in the lavas could represent the source of the Torre Alfina magma (Conticelli and Peccerillo 1990). Geochemical differences within lava flows and necks have been suggested to derive from

[4]The term *lamprophyre* is used in this book to indicate dike rocks characterised by mafic mineral phenocrysts (amphibole, biotite, phlogopite etc.) with feldspar and/or feldspatoids confined in the groundmass. They take different names depending on mineralogy (e.g. Gill 2010).

Fig. 2.10 Front wall of the Torre Alfina castle showing large amounts of a wide variety of crustal and mantle xenoliths in the lamproite lavas

assimilation of variable amounts of crustal rocks by ascending magma. Such a process generated minor modification of trace element ratios and isotopic signatures, and a general dilution for several compatible and incompatible elements (e.g. $^{87}Sr/^{86}Sr$ from about 0.7158 to 0.7165, Ni from 350 to 250 ppm; La from 100 to 85 ppm; Conticelli 1998).

Campiglia. Mafic dikes with an age of about 4.3 Ma occur in the Campiglia area, as mentioned earlier in this chapter. These dikes are strongly altered and have porphyritic textures with phenocrysts of clinopyroxene, plagioclase, biotite, alkali feldspar and some corroded quartz set in a groundmass of plagioclase, sanidine, and pyroxene. Major element compositions of least altered samples are characterised by high K_2O and K_2O/Na_2O ratio, resembling high-silica lamproites from Montecatini and Orciatico. However, concentrations of several incompatible elements that are relatively immobile during secondary alteration (e.g. Ti, Nb, LREE, Zr, Hf), as well as Sr isotope ratio ($^{87}Sr/^{86}Sr = 0.7096$) are lower and Nd isotope ratio ($^{143}Nd/^{144}Nd = 0.51220$) is higher at Campiglia, and rather resemble Capraia calcalkaline and shoshonitic rocks (Peccerillo et al. 1987; Conticelli and Peccerillo 1992; Conticelli et al. 2002). This suggests that high enrichments in potassium could be a secondary feature and that the

Campiglia mafic magmas had a calcalkaline to shoshonitic original composition.

Monti Cimini. This is a 1.35–0.94 Ma volcano formed of early trachyte (<20 % normative quartz) and trachydacite (>20 % normative quartz) lava domes, welded ignimbrites and lava flows, followed by latite, olivine-latite and shoshonite lavas (Puxeddu 1971; Lardini and Nappi 1987; Cimarelli and De Rita 2006b; LaBerge et al. 2006; Aulinas et al. 2011; Conticelli et al. 2013 and references therein).

Petrography and mineralogy. Trachytes and trachydacites display porphyritic textures with phenocrysts and megacrysts of sanidine plus plagioclase, biotite, othopyroxene and some clinopyroxene set in a hypocrystalline groundmass containing the same phases plus accessory ilmenite, apatite, zircon, monazite, and perrierite. Latites and shoshonites have aphyric to porphyritic textures with phenocrysts of olivine and minor clinopyroxene set in a groundmass composed of the same phases plus sanidine and accessory chromite, Ti-magnetite and ilmenite. Mineral chemistry is very variable. Olivine ranges from about Fo_{93} to Fo_{40} from phenocryst cores to rims and groundmass and sometimes contain Mg-chromite inclusions; orthopyroxene ranges from enstatite to ferrosilite ($\sim En_{65-45}$), and mica from biotite to phlogopite. Clinopyroxene ranges from augite and diopside to hedenbergite ($Di_{54}Fs_{18}Wo_{28}$, to $Di_{20}Fs_{32}Wo_{46}$ with Mg# = 90–39); their Al_2O_3 content is variable (0.3–7.9 wt%) and minimum values are found in phenocrysts of the most potassic samples, in which Al is not sufficient to fill the tetrahedral site. Ti/Al (>0.2, expressed as atoms per formula unit: afu) of clinopyroxene and Cr/(Cr + Al) of chromite are high (Fig. 2.11) and close to values found in the same phases from Tuscany lamproites (Aulinas et al. 2011; Conticelli et al. 2013).

Petrology and geochemistry. The Monti Cimini rocks have variable K_2O/Na_2O ratio ($\sim$2–5), which decreases with silica. Mg#, Ni, Co and Cr of mafic rocks are high and close to values of primary mantle melts and decrease linearly with increasing silica. Notably, several incompatible elements (e.g. Rb, Nb, Zr, Hf) also show a similar trend. The incompatible element patterns normalised to primitive mantle composition resemble those of Tuscany lamproites. Initial Sr- and Nd-isotope ratios are variable ($^{87}Sr/^{86}Sr \sim 0.7122–0.7157$; $^{143}Nd/^{144}Nd \sim 0.51205–0.51214$), especially in the mafic rocks. Pb isotopic ratios ($^{206}Pb/^{204}Pb = 18.69–18.73$; $^{207}Pb/^{204}Pb = 15.66–15.69$; $^{208}Pb/^{204}Pb = 38.92–39.02$) are not much different from other Tuscany rocks (Aulinas et al. 2011; Conticelli et al. 2013 and references therein).

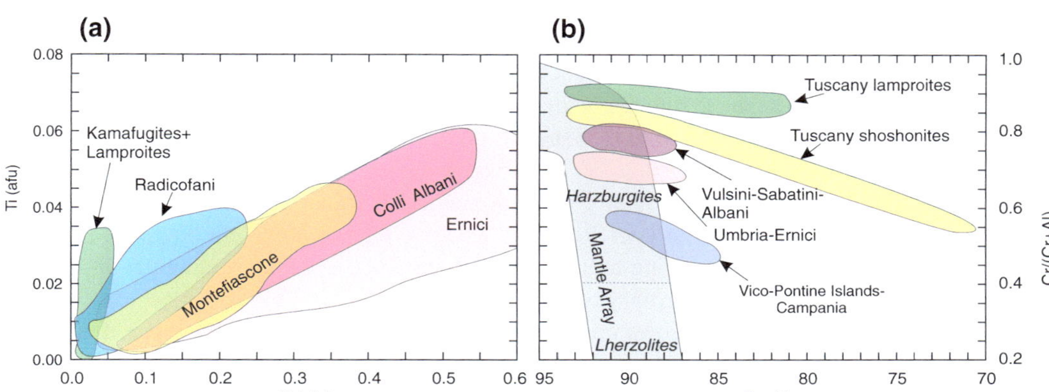

Fig. 2.11 **a** Ti versus Al (atoms per formula units, afu) in clinopyroxenes from some Italy volcanic rocks; **b** Cr/(Cr + Al), atomic ratios of spinel inclusions versus forsterite contents of host olivine for some Italy volcanics. The Mantle Array is from Arai (1994). Redrawn after Nikogossian and van Bergen (2010) and Conticelli et al. (2011, 2013)

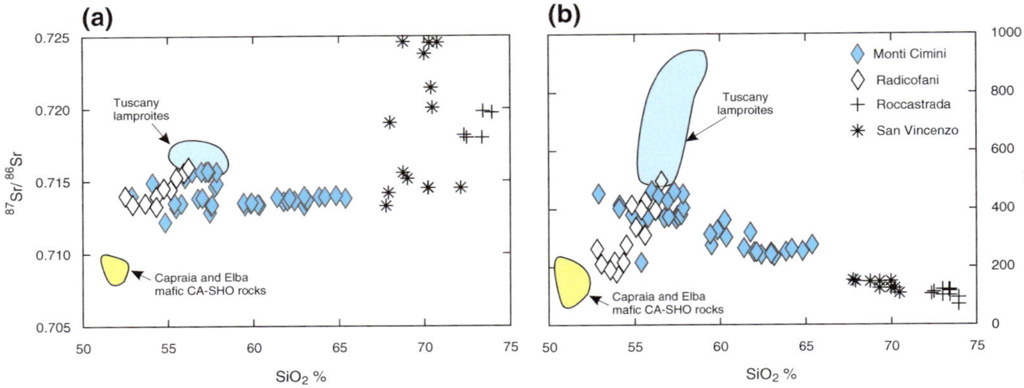

Fig. 2.12 $^{87}Sr/^{86}Sr$ and Zr versus SiO_2 for Monti Cimini and Radicofani rocks

Petrogenesis. The origin of Monti Cimini magmas can be well explained by binary plots of SiO_2 (or MgO) versus $^{87}Sr/^{86}Sr$ and incompatible elements (Fig. 2.12). The most mafic rocks (shoshonites and olivine-bearing latites) plot between the Capraia-Elba calcalkaline-shoshonitic mafic rocks (or also the Roman potassic trachybasalt) and the Tuscany lamproites. In contrast, the more evolved rocks (latites, trachytes and trachydacites) define a distinct trend, which starts from mafic samples and points to silicic compositions such as those of Roccastrada and San Vincenzo rhyolites. These variations suggest that the Cimini mafic magmas were generated either by variable degrees of melting of a heterogeneous mantle source or by mixing between calcalkaline-shoshonitic and lamproitic magmas. The evolved rocks can be explained by mixing between a mafic end-member and crustal anatectic liquids with compositions similar to Roccastrada and San Vincenzo.

Radicofani. The volcano at Radicofani consists of a neck and a few remnants of lava flows with K/Ar and $^{40}Ar/^{39}Ar$ ages clustering around 1.3 Ma (D'Orazio et al. 1991, 1994). In spite of its small size, the Radicofani volcano shows important variations for both geochemistry and mineralogy. Rocks range from basaltic andesite to shoshonite and latite, according to the TAS classification of Le Maitre (2002). Basaltic andesites make up the base of the neck and become more potassium- and silica-rich at the top. Lavas similar to the upper neck crop out at Poggio Sasseta, north of the Radicofani village. Shoshonite lavas occur at the nearby localities of Poggio Casano and Ceppete (D'Orazio et al. 1994).

Petrography and mineralogy. Basaltic andesites at the base of the neck are poorly porphyritic with phenocrysts of plagioclase ($\sim An_{90-80}$), zoned olivine ($\sim Fo_{75-60}$) and augite-diopside clinopyroxene set in a groundmass made of the same phases plus orthopyroxene, ilmenite, sanidine, and accessory biotite, apatite, magnetite and glass (Poli et al. 1984; D'Orazio et al. 1994; Barnekow 2000). A modification in the mineralogy is observed at the top of the neck, where forsterite content of olivine and the modal amount of alkali feldspar in the groundmass increase, whereas plagioclase becomes rare or disappears as a phenocryst phase. Shoshonites are microcrystalline to poorly porphyritic with phenocrysts of olivine ($\sim Fo_{85-70}$) and minor clinopyroxene. Plagioclase is scarce and restricted to the groundmass. This consists of dominant K-feldspar plus Al-poor clinopyroxene, biotite, plagioclase and accessory amounts of olivine, brown amphibole, Fe–Ti oxides, apatite, and glass. Chromite inclusions are common in olivine phenocrysts. The Radicofani volcanics contain a few xenocrystic minerals coming from the wall rocks (corroded quartz and some cordierite) and rare small ultramafic nodules.

Petrology and geochemistry. Major element composition shows poorly variable silica

(SiO_2 ~ 53–56 wt%) and MgO (~8–9 wt%) contents. In contrast, incompatible elements (Rb, Th, Nb, Ta, REE, etc.) and some ferromagnesian elements (Ni, Cr) vary considerably, and are positively correlated with K_2O and SiO_2; in contrast, Na_2O, V, and Sc show a negative trend (D'Orazio et al. 1994; Conticelli et al. 2011). The incompatible element patterns normalised to primordial mantle composition resemble those of Tuscany lamproites, although absolute element enrichment is lower (Fig. 2.9b). $^{87}Sr/^{86}Sr$ ratios (0.7133–0.7159) increase with K_2O and SiO_2, i.e. from basaltic andesites to shoshonites, whereas $^{143}Nd/^{144}Nd$ ratios (0.51205–0.51218) exhibit an opposite trend (D'Orazio et al. 1994; Conticelli et al. 2002). Pb- and Hf-isotope ratios ($^{206}Pb/^{204}Pb$ ~ 18.67–18.72; $^{207}Pb/^{204}Pb$ ~ 15.66–15.69; $^{208}Pb/^{204}Pb$ ~ 38.98–39.08; $^{176}Hf/^{177}Hf$ = 0.28255) fall in the field of the Tuscany ultrapotassic rocks (De Astis et al. 2000a, b; Conticelli et al. 2002; Gasperini et al. 2002). O-isotopes on clinopyroxene and olivine vary from $\delta^{18}O_{SMOW}$ = +6.9 to + 7.8 ‰ (Barnekow 2000).

Petrogenesis. Overall, the compositions of Radicofani magmas are intermediate between moderately potassic rocks such as those from Capraia or some Roman-type trachybasalts, and lamproites (Fig. 2.12). This suggests an origin by mixing between magmas with contrasting calcalkaline-shoshonitic and lamproitic compositions. Such a hypothesis is supported by the variable Ti/Al of clinopyroxenes that encompass lamproitic and Roman-type compositions (Fig. 2.11a). Conticelli et al. (2011) reject the magma mixing hypothesis and suggest that compositional variation is related to melting of a veined lithospheric mantle, with variable contribution from veins and host peridotite.

Island of Capraia. This is the only volcano in the Tuscany Province dominated by calcalkaline rocks. The island was constructed during two distinct phases of activity at about 7.6–7.2 Ma and 4.6 Ma (Borelli et al. 2003; Gasparon et al. 2009). The older phase is dominated by lavas, which make up the bulk of the island. They show intermediate composition and are classified as high-K calcalkaline andesites and dacites, according to the classification scheme of Peccerillo and Taylor (1976). Younger activity is represented by the shoshonitic basalts of Punta dello Zenobito, a monogenetic strombolian centre at the southernmost end of the island (Prosperini 1993; Borelli et al. 2003; Poli and Perugini 2003c).

Petrography and mineralogy. The shoshonitic basalts from Zenobito are vesicular and almost aphyric with a few microphenocrysts of olivine in a groundmass of plagioclase, augite, spinel and rare biotite. High-K andesites and dacites exhibit phenocrysts of strongly zoned plagioclase, augitic clinopyroxene and minor enstatite, olivine and biotite, K-feldspar and amphibole set in a groundmass containing the same phases and glass (Gagnevin et al. 2007; Gasparon et al. 2009).

Petrology and geochemistry. Shoshonitic basalts have variable MgO, Ni, and Cr. CaO (around 7–8 wt%) and Na_2O (around 3 wt%) show some of the highest concentrations among the Tuscany mafic rocks (Fig. 2.8). There is a decrease in ferromagnesian elements and CaO, and an increase in K_2O from basalts to dacites. A group of andesites has an adakite-like composition characterised by selective enrichment in Sr-Ba-LREE and resembles the Orano dikes (Elba) and the San Vincenzo mafic enclaves. Incompatible elements patterns of Capraia mafic rocks are fractionated with strong negative anomalies of HFSE and a positive spike of Pb (Fig. 2.9b), as observed for other Tuscany mafic rocks. Sr-isotope ratios are among the lowest observed in the Tuscany Province, and increase from basalts ($^{87}Sr/^{86}Sr$ ~ 0.7081) to andesites and dacites ($^{87}Sr/^{86}Sr$ ~ 0.7087–0.7102). Sr-isotope variation has been observed among coexisting phenocrysts and within single crystals (Gagnevin et al. 2007), suggesting mixing among isotopically distinct magmas during mineral crystallisation.

Petrogenesis. Geochemical data suggest that the early high-K calcalkaline andesites and dacites evolved from mafic parents by combined fractional crystallisation and mixing (e.g. Gagnevin et al. 2007). However, the occurrence of Ba-Sr-LREE-rich andesites point to the occurrence of a distinct type of intermediate magma

(Poli and Peccerillo 2016). Interaction with lamproitic magma has been suggested to explain trace element enrichments in these rocks (Chelazzi et al. 2006). However, this contrasts with the high Sr and Ba, which show moderate to low concentrations in the lamproites, and suggest a magma origin/evolution at high pressure, outside the stability field of feldspars. The late shoshonites represent a distinct batch of magma with respect to calcalkaline activity. Their relatively high Na_2O and CaO suggest an origin in a clinopyroxene-bearing (i.e. lherzolitic) mantle. Incompatible element patterns and radiogenic isotopes require metasomatic modification of the mantle source by moderate amounts of upper crustal components.

2.6 Petrogenesis of the Tuscany Magmatic Province

According to Peccerillo (2005a, b) and Poli and Peccerillo (2016), the rock compositions of the Tuscany Magmatic Province can be considered as comprised in the space delimited by three petrologically distinct end-members (Fig. 2.13a). One end-member is silicic and displays $K_2O/Na_2O \sim 1$–2. The other two end-members are mafic, show very different K_2O/Na_2O ratios, and are classified as calcalkaline-shoshonitic basalts (CA-SHO), and high-silica lamproites (LMP).

$^{87}Sr/^{86}Sr$ versus. SiO_2 plot indicates distinct isotopic signatures for mafic CA-SHO and LMP magmas, but also reveals that the silicic end-member actually consists of three isotopically distinct groups of rock (Fig. 2.13b). Therefore, it has been stated that the petrogenetic problem of the Tuscany magmatism is that of explaining the origin of the mafic and silicic end-member magmas and of elucidating how the intermediate compositions were generated (Poli and Peccerillo 2016).

2.6.1 Origin of Silicic Magmas

A crustal anatectic origin was suggested by early authors for the Tuscany silicic magmatism (e.g. Marinelli 1975). However, several studies demonstrate that only a few silicic rocks do actually represent pure crustal anatectic melts (e.g. Poli 1992). These are labelled as Group-1 and Group-2 in Fig. 2.13b, and include some of the San Vincenzo lavas (the low-Sr, high-$^{87}Sr/^{86}Sr$ rocks), the Roccastrada rhyolites, and some leucocratic granitoid bodies occurring

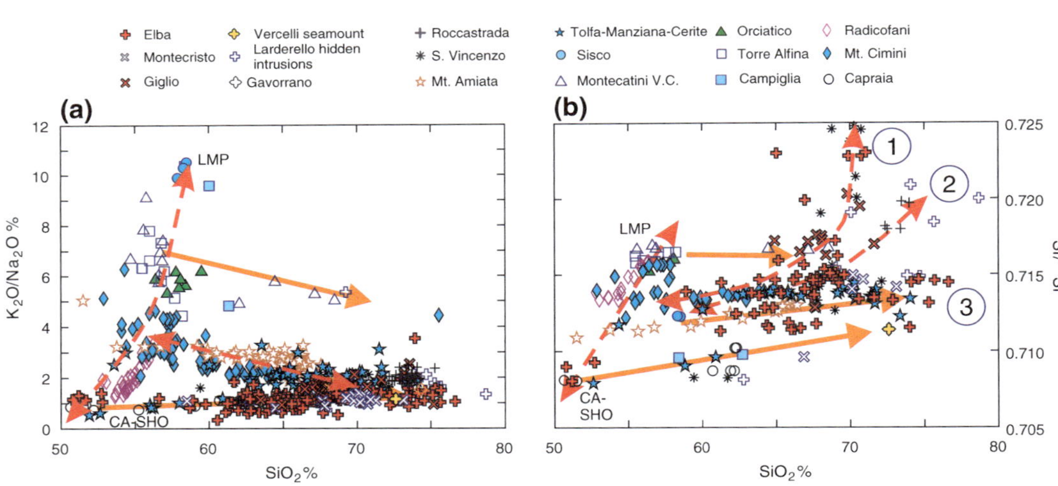

Fig. 2.13 K_2O/Na_2O and $^{87}Sr/^{86}Sr$ versus SiO_2 diagrams for the Tuscany Magmatic Province. *Dashed double-sided arrows* indicate mixing processes between various end-members; *full arrows* indicate fractional crystallisation and AFC. CA-SHO and LMP are calcalkaline-shoshonitic basalts and lamproites. Numbers in panel **b** indicate different groups of silicic rocks. For further explanation, see text

at Elba, Giglio and Larderello (e.g. Poli 1992; Westerman et al. 1993; Dini et al. 2005; Poli and Peccerillo 2016 and references therein). Note that most of these rocks contain little or no mafic enclaves, indicating scarce or no interaction with mafic melts (e.g. Pinarelli et al. 1989; Dini et al. 2005). Several studies have shown that the compositions of unmodified crustal anatectic magmas in Tuscany can be modelled by large degrees (some 40–50 %) of partial melting of garnet micaschists and gneiss, such as those found by deep drilling in Tuscany (e.g. Pinarelli et al. 1989; Poli 1992). However, Poli et al. (2002) and Poli (2004) pointed out that CaO content of many silicic granitoid rocks is exceedingly high for pelite-derived melts and advocated plagioclase-rich sources (e.g. metagreywackes). Petrological data and geochemical modelling suggest that melting occurred in fluid-absent conditions at pressure of at least 0.4–0.6 GPa. The distinct Sr isotope ratios of Group-1 and Group-2 magmas simply indicate isotopic heterogeneities for the crustal sources, an assumption justified by the variable Sr isotope ratios of Tuscany basement rocks (see Fig. 2.6).

The majority of silicic rocks have lower Sr isotope ratios than pure crustal anatectic magmas, and exhibit clear textural and geochemical evidence (e.g. occurrence of microgranular mafic enclaves with chilled margins and crenulated edges, mafic xenocrysts, isotopic disequilibrium among coexisting phases, etc.) of interaction (i.e. mixing or mingling) between crustal anatectic and mafic magmas. The hyperbolic trends between mafic enclaves and rhyolites at San Vincenzo and Roccastrada (Fig. 2.6) have been considered as the most compelling geochemical evidence in favour of this process. Note that many granitoid and volcanic rocks fall on these trends, indicating that mixing was a widespread process operating at a regional scale. However, some rocks plot on the left-hand side of the mixing curves indicating a derivation from hybrid melts by fractional crystallisation.

The nature of mafic end-members involved in the mixing changes from one centre to the other. Poli et al. (2002) and Gagnevin et al. (2011) suggested that mafic calcalkaline melts such as those from Capraia participated in the mixing processes at Elba and Giglio. In other cases (e.g. Monte Amiata, Tolfa-Manziana-Cerite rocks), potassic to ultrapotassic mafic magmas were involved (Poli 2004).

Finally, some silicic intrusions and lavas (e.g. many aplites and leucogranites from Elba, Tolfa lavas) show relatively low $^{87}Sr/^{86}Sr \sim 0.711$–0.715, with respect to pure crustal anatectic melts. These may represent evolved liquids derived from intermediate hybrid parents by crystal fractionation, AFC or filter-pressing (Poli and Peccerillo 2016). These rocks plot on the left-hand side of the hyperbolic mixing trends reported in Fig. 2.6 and define the Group-3 samples in Fig. 2.13b.

In conclusion, the model summarised above supports the hypothesis that silicic rocks in Tuscany are polygenetic, and were originated by crustal melting, acid-mafic magma mixing-mingling, and fractional crystallisation starting from mafic-intermediate parents. The last process determined the formation of leucocratic rocks such as the aplitic and pegmatitic dikes and veins occurring in several intrusions.

Another theory, however, is that compositional variation of silicic rocks in Tuscany is primary, i.e. reflects variable source compositions and melting processes (e.g. Farina et al. 2012, 2014). Factors that favour formation of different magma types during crustal anatexis include composition of phases that participate in the melting reactions and the degree of entrainment within the magma of the peritectic assemblages produced during incongruent melting of the source (e.g. garnet and cordierite produced by fluid-absent melting of biotite). The isotopic variation observed in the minerals and host rocks (e.g. at Monte Capanne; Farina et al. 2014) would reflect the transitions from melting of more strongly radiogenic minerals such as muscovite and biotite, typical of metapelites, to the involvement of less radiogenic phases such as amphibole, typical of metavolcanics. Melting of different rock types could be an effect of temperature variations within the crust. According to this view, microgranular mafic enclaves and late mafic dikes associated with silicic rocks may well represent mantle-derived magmas, but they would have a minor role in determining the observed compositional variations of silicic magmas (Farina et al. 2012, 2014).

2.6.2 Origin of Mafic Magmas

The compositions of mafic rocks in Tuscany are characterised by the coexistence of both mantle and crustal signatures. The high MgO, Ni and Cr concentrations are typical of primitive mantle-derived melts. On the other hand, incompatible element abundances and ratios, high Sr- and low Nd-, Hf and Pb-isotope ratios are close or within the field of upper crustal rocks. Therefore, petrological and geochemical data provide compelling evidence for the participation of both crustal and mantle components to the origin of mafic magmas in Tuscany. The highly primitive signatures and the occurrence of ultramafic xenoliths in some rocks exclude that mantle-crust interaction occurred during magma emplacement to the surface (*magma contamination*), and clearly point to the introduction of upper crustal rocks into the mantle (*mantle contamination* or *metasomatism*). This was likely provided by subduction, but there is debate on the timing of such a process, and on the nature, origin and mechanisms of contamination. Other controversial issues include the composition of pre-metasomatic mantle rocks, the mineralogical-geochemical modifications induced by metasomatism, and the mechanisms of mantle melting (e.g. Peccerillo et al. 1987; Conticelli and Peccerillo 1992; Serri et al. 1993; Conticelli et al. 2010, 2011; Poli and Peccerillo 2016).

The high MgO and K_2O and low CaO, Na_2O, and Al_2O_3 of Tuscany lamproites suggest an origin from an ultramafic rock containing a K-rich phase, most probably phlogopite or K-richterite (e.g. Harlow and Davies 2004), and little Ca- and Al-rich minerals such as clinopyroxene and Al-spinel or garnet. Therefore, a phlogopite-harzburgite is the best candidate as source rock of lamproites, a conclusion that applies to lamproitic occurrences worldwide (e.g. Foley et al. 1987; Mitchell and Bergman 1991; Prelevic et al. 2005).

Such a hypothesis is strongly supported by experimental evidence. Investigation on simplified and natural mantle systems demonstrates that melts derived from a phlogopite-bearing peridotite at 1.0–1.5 GPa are saturated to oversaturated in silica, and become undersaturated in silica with increasing pressure (e.g. Wendlandt and Eggler 1980a,b; Foley 1992; Sato 1997; Melzer and Foley 2000; Conceição and Green 2004; Condamine and Médard 2014). Melting phase relationships demonstrate that compositions similar to Tuscany lamproites (high silica, low CaO and Na_2O), but somewhat less strongly enriched in potassium, can be generated at a pressure of 1.0 GPa and 1150–1200 °C, by about 7–13 % fluid-absent partial melting of phlogopite-harzburgite (Condamine and Médard 2014).

The occurrence of phlogopite or other K-rich phases in the mantle sources of potassic magmas requires contamination or metasomatism by K-rich material. Peccerillo et al. (1988) and Peccerillo and Martinotti (2006) noticed that incompatible element patterns of lamproites (Fig. 2.9) resemble very closely upper crustal siliceous rocks such as gneiss, crustal anatectic rhyolites, and the Global Subducted Sediments (GLOSS; Plank and Langmuir 1998). Therefore, it was suggested that introduction of siliceous upper crustal rocks into the mantle wedge was the cause of metasomatic modification of the lamproitic magma sources in Tuscany. Turbiditic sediments have been proposed as contaminants on the basis of Hf–Nd isotopic data by Prelevic et al. (2010), but other types of silicic rocks, e.g. or pelagic sediments (Gasperini et al. 2002) or material provided by erosion along the subduction channel (Vannucchi et al. 2010, 2012; Remitti et al. 2013), cannot be excluded. The close similarity of incompatible element patterns between the upper crust and Tuscany lamproites suggests that there was a mass transfer of chemical components from sediments to the mantle and hence to magmas, with little element fractionation. Notable exceptions are represented by HREE that are very fractionated in the Tuscany magma, requiring residual garnet during magma genesis. Therefore, it has been concluded that physical mixtures of sediments and harzburgites were the source of lamproitic magmas (Peccerillo and Martinotti 2006). However, the mechanisms of sediment transport and mixing are enigmatic and controversial.

Combined numerical modelling and petrological, geophysical and geochemical studies (e.g. Gerya and Yuen 2003; Gerya et al. 2006; Castro and Gerya 2008; Castro et al. 2010; Marschall and Schumacher 2012) offer an interesting scenario for interaction between mantle and bulk crust above subduction slabs. It has been suggested that a chaotic mixture of sediments, slices of oceanic crust and mantle rocks (subduction mélange) can develop at the interface between the subducting plate and the mantle wedge. Subduction mélange bodies have lower density than the surrounding mantle rocks, if sufficient amounts of sediments are involved (Behn et al. 2011). Therefore, they may rise buoyantly as diapirs or cold plumes from the surface of the slab to the hotter overlying mantle wedge. Here, diapirs undergo release of aqueous fluids and melts that migrate upward inducing metasomatism in the surrounding peridotite. Magmas can originate both in the mélange and in the metasomatic mantle around the diapirs. Tumiati et al. (2013) have shown that phlogopite is a ubiquitous phase in the mantle rocks around subduction mélange bodies and that this mineral melts preferentially to give potassic magmas.

Based on these models, it is has been suggested that mélange diapirs formed by a mixture of mantle harzburgites and subducted silicic sediments represent the source of lamproitic rocks in Tuscany (Peccerillo and Frezzotti 2015; Poli and Peccerillo 2016). Simple mass balance calculation indicates the participation of about 15 % sediment to mixing, in order to explain isotopic compositions of Tuscany lamproites (Peccerillo et al. 1988).

However, the lack of Pb positive spikes and high La/U at Sisco require a somewhat distinct type of contaminant or/and mechanism of element transfer and/or pre-metasomatic sources for this rock (Conticelli et al. 2009a,b). Speculatively, the deficiency of Pb and other LILE relative to Tuscany lamproites may reflect a minor role of transfer by fluid phases during metasomatism.

Calcalkaline and shoshonitic mafic rocks have similar incompatible element patterns but lower absolute element enrichments and less radiogenic Sr-isotope compositions than lamproites. This indicates the participation of a lower amount of crustal material. However, contents of CaO, Al_2O_3, and Na_2O increase from lamproites to calcalkaline-shoshonitic magmas, requiring the participation of other components to melting, which were scarce or absent in the source of lamproites. Therefore, it has been suggested that cpx- and spinel- or garnet-bearing rocks (i.e. spinel or garnet lherzolites rather than harzburgites) are the source of calcalkaline and shoshonitic magmas (e.g. Peccerillo and Frezzotti 2015; Poli and Peccerillo 2016). This is supported by experimental data on phlogopite-rich lherzolite whose melting (around 10–20 %) at about 1 GPa gives liquids very similar to Radicofani shoshonites (Condamine and Médard 2014). The increase from lamproitic to calcalkaline magmas of some ferromagnesian elements that are hosted by clinopyroxene, such as Sc (Fig. 2.8d), agrees with a greater involvement of this phase in the origin of calcalkaline magmas. The relatively low Cr/(Cr + Al) of chromite inclusions in olivine from shoshonitic magmas (Fig. 2.11b) provides further support to a magma origin from lherzolitic sources.

An alternative hypothesis for the formation of calcalkaline to lamproitic magmas has been proposed by Conticelli et al. (2009a, 2011, 2015a with references). According to these authors, the Tuscany mafic magmas were generated in a residual lithospheric mantle cut by veins rich in phlogopite formed by infiltration of supercritical fluids or melts from subducted sediments. Low degrees of partial melting affected prevalently the phlogopitic veins, generating lamproitic magmas. Successively, melting involved the mantle around the veins, which produced depleted melts. These mixed with early formed lamproitic magmas to give diluted calcalkaline and shoshonitic compositions.

This hypothesis, although simple and elegant, fails to explain some first-order petrological and geochemical data. First of all, if the source rocks of lamproites have a phlogopite-harzburgite composition, as generally accepted, progressive melting would dilute concentrations in early liquids not only for potassium, but also for other oxides such as Al_2O_3, CaO, and Na_2O, as clearly demonstrated by the experimental work of Condamine and Médard (2014). However, these oxides increase from lamproitic to calcalkaline magmas in Tuscany (Fig. 2.8). Another objection is that about 80–90 % of a basaltic-like melt is necessary to dilute potassium and incompatible elements contents of lamproitic liquids to the much lower levels of calcalkaline magmas. This requires extensive mantle melting, which is improbable especially for an area where volumes of mafic magmas are very low (Poli and Peccerillo 2016).

Finally, the origin of mafic rocks with intermediate compositions between calcalkaline-shoshonitic basalts and lamproites, sometimes occurring at single volcanoes, can be simply explained by mixing of extreme end-member melts. This hypothesis leads to the conclusion that distinct but closely associated harzburgitic and lherzolitic rocks occurred in the upper mantle beneath some volcanoes. These melted contemporaneously generating calcalkaline-shoshonitic and lamproitic melts that mixed during ascent to give a continuous suite of magmas with variable petrological, geochemical and isotopic characteristics (Poli and Peccerillo 2016). However, the adakite-like Sr-Ba-LREE rich intermediate compositions found at Capraia, Elba-Orano dikes and among San Vincenzo enclaves represent a distinct type of intermediate magma. This could be generated either by basalt melting in a thickened lower crust, along the subducted slab, or within the subduction mélanges, or also by high-pressure fractional crystallisation of mantle melts (e.g. Ribeiro et al. 2016).

2.7 Geodynamic Implications

The mafic rocks of the Tuscany Province show strong geochemical and isotopic differences with respect to other mantle-derived magmas in Central Italy, such as the Roman and Intra-Apennine provinces. This calls for the occurrence of distinct and possibly dyachronous metasomatic processes for these nearby magmatic districts.

It has been long suggested that mantle contamination beneath Central Italy is related to addition of different types of subducted sediments, consisting of siliceous rocks in Tuscany and carbonated pelites (marls) in the Roman and Intra-Apennine provinces (Peccerillo et al. 1988). Such a hypothesis received much support by recent geochemical ad experimental investigation (e.g. Prelevic et al. 2008, 2010; Avanzinelli et al. 2009; Conticelli et al. 2009a, b, 2010, 2011; Grassi et al. 2012). The distinct types of contamination explain the petrological and geochemical diversities of Tuscany mafic magmas with respect to other volcanoes in Central Italy.

According to Serri et al. (1993) siliceous and carbonate-rich sediments were both introduced into the sub-Apennine upper mantle by the west-dipping delaminated Adriatic continental plate from Early Miocene to Present. Such a hypothesis, however, leaves some questions unanswered. In particular, it is not clear why a single subduction process of the same slab provided two contrasting types of contaminants to the mantle sources of contiguous and partially superimposed areas. It also remains unclear why the mantle rocks contaminated by siliceous sediments beneath Tuscany melted much earlier (14.5–0.3 Ma) than the marl-contaminated mantle sources of the Roman and Intra-Apennine provinces (<0.8 Ma old).

According to Peccerillo (1999, 2002) and Peccerillo and Martinotti (2006), the variable ages of magmatism and the distinct nature of metasomatism in Tuscany and in the Roman and Intra-Apennine provinces can be better explained by assuming two mantle contamination events, occurred at different times during the evolution of the Alps-Apennine system.

A most popular hypothesis on the evolution of the Alpine-Apennine orogens submits that Cretaceous to Present convergence between Africa and Europe was first accomplished by east-directed subduction of the Alpine Tethys and European crust beneath the northern African margin (Upper Cretaceous to Eocene: Alpine stage), and successively by a new west-directed subduction process of the Neotethys-Adriatic-Ionian lithosphere (Lutetian to Present: Apennine stage) beneath the southern European margin (e.g. Doglioni et al. 1999; Carminati et al. 2010). More discussion on this topic can be found in Chap. 13.

Peccerillo (1999, 2002) and Peccerillo and Martinotti (2006) proposed that the mantle metasomatic contaminations by siliceous sediments and marls in Central Italy are diachronous and respectively occurred during the Alpine and Apennine subduction stages. The contrasting compositions of contaminants are explained by their different origin, i.e. from the subducted European and Adriatic plates, respectively. According to this hypothesis, the subduction-contamination-magma genesis history in Tuscany can be summarised as follows (Fig. 2.14):

1. Introduction of siliceous upper crustal rocks into the mantle wedge took place during the Cretaceous to Eocene south-eastward subduction of the Alpine Tethys and fragments of European continent beneath the northern African margin (Alpine stage subduction). The ultra-high pressure rocks of the Dora Maira massif represent the most compelling evidence for subduction of upper continental crust (e.g. Chopin 1984; Cadoppi 1990; Compagnoni 2003). Contamination affected the lithospheric mantle of the northern African margin probably along the entire belt going from the Betic Cordillera to Central Italy and the Western Alps (Peccerillo and Martinotti 2006). Mechanical mixing of upper crustal rocks, slices of oceanic crust

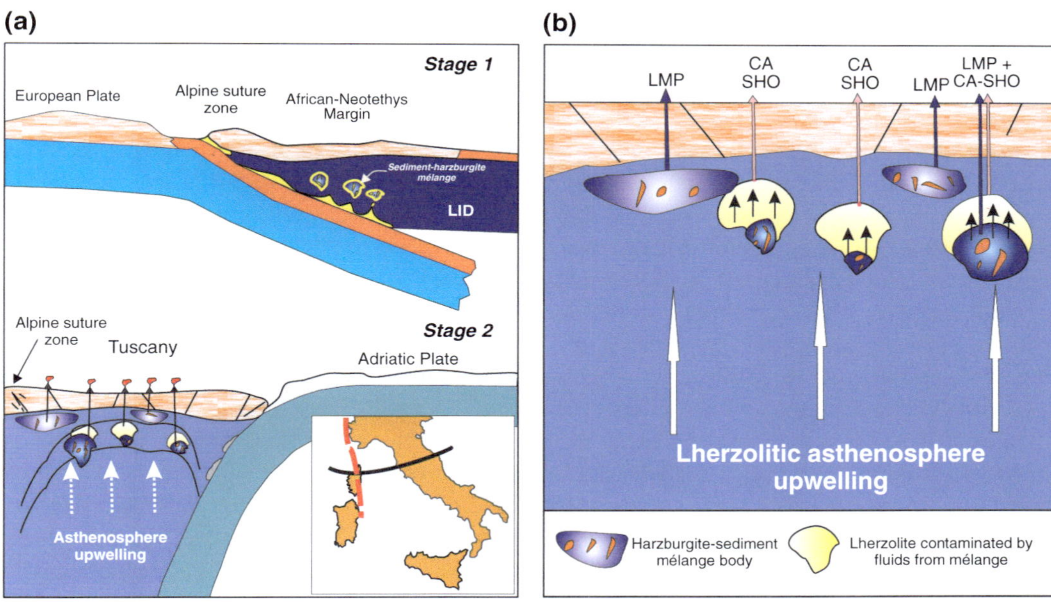

Fig. 2.14 Cartoon showing a two-stage model of mantle contamination and melting for the Tuscany Magmatic Province (modified after Poli and Peccerillo 2016). **a** *Stage 1*: Upper Cretaceous to Eocene "Alpine" subduction-collision between European and African plates brought siliceous upper crustal material into the lithospheric mantle beneath the African margin. Mélange bodies made of metasediments, slices of oceanic crust and harzburgitic mantle rocks were formed above the slab. *Stage 2*: Subduction inversion from Eocene to present generated backarc spreading, lithospheric breakup, and ascent of hot lherzolitic asthenospheric mantle. This induced dehydration and melting of Alpine mélange bodies, and metasomatism in the surrounding lherzolitic rocks. Melting of the subduction mélange bodies (*blue blobs*) gave lamproitic magmas whereas melting of metasomatised lherzolite around mélange bodies (*yellow blobs*) gave calcalkaline to shoshonitic magmas. Location of section (*black line*) and traces of the Alpine suture zone (*red dashed line*) are reported in the *Inset*. **b** Blow-up of the mantle wedge beneath Tuscany showing the origin of lamproitic (LMP) and of calcalkaline-shoshonitic (CA, SHO) magmas

and lithospheric harzburgitic mantle formed subduction mélange bodies above the slab surface. These, however, were unable to ascent much through the rigid lithospheric mantle and escaped melting, likely because of the low temperature regime along the subduction zone.

2. Starting from the Lutetian, a new west-directed subduction process developed with immersion of the Neotethys-Ionian-Adriatic lithosphere beneath the Alpine-Betic retrobelt (Apennine-Maghrebian subduction stage). Backarc extension behind the westward immerging slab prompted stretching and dismembering of the lithosphere contaminated during the Alpine subduction stage (Gueguen et al. 1997, 1998). Spreading also favoured upwelling of hot asthenosphere into the mantle wedge. Therefore, fragments of harzburgite-metapelite mélange bodies that formed during the Alpine stage were embodied into newly emplaced hot asthenospheric rocks that likely had lherzolitic composition. Mélange bodies underwent strong dehydration and melting, with fluid migration both within the mélange and from this to the surrounding newly emplaced lherzolitic mantle. Melting in the mélange body (sediments plus harzburgites) generated lamproitic magmas, whereas melting in metasomatized lherzolites around mélange bodies formed calcalkaline to shoshonitic magmas (Peccerillo and Frezzotti 2015; Poli and Peccerillo 2016). Fluid transfer phenomena may have played a minor role as a contaminant for the Sisco lamproite, a hypothesis suggested by depletion in Pb and other fluid-mobile trace elements at Sisco relative to Tuscany lamproites.
3. Miocene to present subduction of the Adriatic continental crust brought new types of sediments (i.e. marls) down into the mantle wedge, which was formed by newly emplaced lherzolitic rocks ascended from the asthenosphere during backarc spreading. Such a newly contaminated lherzolitic mantle made up the source of the Roman Province. The effects of this new contamination process were particularly important in Latium, but are also recognisable in some late centres of the Tuscany Province (e.g. Radicofani and Cimini; Peccerillo et al. 1987; Peccerillo 1999; Conticelli et al. 2013; Peccerillo and Frezzotti 2015).

A different geodynamic scenario is envisaged by Bell et al. (2013). According to these authors, a pre-Alpine contamination event affected a wide mantle sector in Central Italy and Western Alps. This successively interacted with FOZO-type mantle, producing heterogeneous sources that gave the compositionally variable magmas along the Italian peninsula. Such a hypothesis will be further discussed in Chap. 13 and Appendix 1. Here, it is only recalled that compositions as the Tuscany lamproites are not restricted to Italy but occur at several places along the Alpine-Himalayan belt and that ages and geodynamic evidence strongly supports an Alpine stage contamination for their mantle sources. Moreover, there is significant evidence that isotopic variations along the Italian do not result from simple two-end-member mixing as envisaged by Bell et al. (2004, 2013) and many others. Rather there were different mantle and crustal end-members participating in the mixing, as it will be discussed in Chap. 13 and Appendix 1.

2.8 Summary and Conclusions

The Tuscany Magmatic Province consists of an association of calcalkaline to lamproitic mafic to intermediate magmas and silicic intrusive and effusive rocks. Silicic magmas are polygenetic and have been formed by crustal melting, mixing between crustal anatectic and minor amounts of mafic melts, and fractional crystallisation or AFC starting from intermediate-mafic parents. Mafic melts originated in the mantle but resemble closely some upper crustal rocks in terms of incompatible trace element and radiogenic isotope signatures. The particular composition of these magmas reveals anomalous sources, consisting of upper mantle rocks that underwent contamination by subducted upper crustal siliceous material, such as metapelites.

Mantle contamination beneath Tuscany probably took place during the Late Cretaceous to Eocene subduction of the European plate beneath the African margin. Magmatism is much younger and occurred from Miocene to present during the opening of the northern Tyrrhenian Sea behind the west dipping subducted Adriatic plate. Tuscany magmatism becomes younger from west to the east, following the eastward backarc extension migrating in the same direction (Boccaletti et al. 1990).

The ascent of mafic magmas into the crust induced anatexis and formation of peraluminous, highly silicic magmas. These emplaced either as unmodified melts or mixed with different types of mantle-derived magmas giving hybrid products. Fractional crystallisation of hybrid magmas gave high-silica aplites and pegmatites, which are commonly found in many granitoid bodies in Tuscany.

References

Abbate E, Bortolotti V, Passerini P, Sagri M (1970) Introduction to the geology of Northern Apennines. Sedim Geol 4:521–558

Acocella V, Mulugheta G (2001) Surface deformation induced by pluton emplacement: the case of Amiata (Italy). Phys Chem Earth 26:355–362

Arai S (1994) Compositional variation of olivine–chromian spinel in Mg-rich magmas as a guide to their residual spinel peridotites. J Volcanol Geoth Res 59:279–293

Aulinas M, Gasperini D, Gimeno D, Macera P, Fernandez-Turiel JL, Cimarelli C (2011) Coexistence of calc-alkaline and ultrapotassic alkaline magmas at Mounts Cimini: evidence for transition from the Tuscan to the Roman magmatic provinces (Central Italy). Geol Acta 9:103–125

Avanzinelli R, Lustrino M, Mattei M, Melluso L, Conticelli S (2009) Potassic and ultrapotassic magmatism in the circum-Tyrrhenian region: the role of carbonated pelitic vs. pelitic sediment recycling at destructive plate margin. Lithos 113:213–227

Balestrieri ML, Pandeli E, Bigazzi G, Carosi R, Montomoli C (2011) Age and temperature constraints on metamorphism and exhumation of the syn-orogenic metamorphic complexes of Northern Apennines, Italy. Tectonophysics 509:254–271

Barberi F, Innocenti F, Mazzuoli R (1967) Contributo alla conoscenza chimico-petrografica e magmatologica delle rocce intrusive, vulcaniche e filoniane del Campigliese (Toscana). Mem Soc Geol It 6:643–681

Barberi F, Innocenti F, Ricci CA (1971) La Toscana Meridionale. Il magmatismo. Rend Soc It Min Pet 27:169–210

Barbieri M, Gasparotto G, Lucchini F, Savelli C, Vigliotti L (1986) Contributo allo studio del magmatismo del Mar Tirreno: l'intrusione granitica tardo-Miocenica del monte submarino Vercelli. Mem Soc Geol It 36:41–54

Barchi M (2010) The neogene-quaternary evolution of the Northern Apennines: crustal structure, style of deformation and seismicity. In: Beltrando M, Peccerillo A, Mattei M, Conticelli S, Doglioni C (eds) Journal virtual explorer 36, paper 11 doi:10.3809/jvirtex.2010.00220

Barchi M, Landuzzi A, Minelli G, Pialli G (2001) Outhern Northern Apennines. In: Vai GB, Martini PI (eds) Anatomy of an Orogen. The Apennines and the adjacent Mediterranean basins. Kluwer, Dordrecht, pp 215–254

Barnekow P (2000) Volcanic rocks from central Italy: an oxygen isotopic microanalytical and geochemical study. PhD thesis, University of Gottingen, 99 pp

Behn MD, Kelemen PB, Hirth G, Hacker BR, Massonne H-J (2011) Diapirs as the source of the sediment signature in arc lavas. Nat Geosci 4:641–646

Bell K, Castorina F, Lavecchia G, Rosatelli G, Stoppa F (2004) Is there a mantle plume below Italy? EOS 85:541–547

Bell K, Lavecchia G, Rosatelli G (2013) Cenozoic Italian magmatism—isotope constraints for possible plume-related activity. J South Am Earth Sci 41:22–40

Bellon, H (1981) Chronologie radiométrique (K–Ar) des manifestations magmatiques autour de la Méditerranée occidentale entre 33 Ma. et 1 Ma. In: Wezel FC (ed) Sedimentary basins of mediterranean margins, Tecnoprint Bologna, pp 341–360

Benoit MH, Torpey M, Liszewski K, Levin V, Park J (2011) P and S wave upper mantle seismic velocity structure beneath the northern Apennines: new evidence for the end of subduction. Geochem Geophys Geosyst 12:Q06004. doi:10.1029/2010GC003428

Bertagnini A, De Rita D, Landi P (1995) Mafic inclusions in the silica-rich rocks of the Tolfa-Ceriti-Manziana volcanic district (Tuscan Province, Central Italy): chemistry and mineralogy. Min Pet 54:261–276

Bianco C, Brogi A, Caggianelli A, Giorgetti G, Liotta D, Meccheri M (2015) HP-LT metamorphism in Elba Island: implications for the geodynamic evolution of the inner Northern Apennines (Italy). J Geodyn 91:13–25

Boccaletti M, Calamita F, Deiana G, Gelati R, Massari F, Moratti G, Ricci Lucchi F (1990) Migrating foredeep-thrust belt systems in the northern Apennines and southern Alps. Palaeogeogr Palaeocl 77:3–14

Boccaletti M, Gianelli G, Sani F (1997) Tectonic regime, granite emplacement and crustal structure in the inner zone of the Northern Apennines (Tuscany, Italy): a new hypothesis. Tectonophysics 270:127–143

Borelli E, Groppelli G, Aldighieri B, Battaglia A, Gamba A, Gasparon M, Malara F, Pasquaré G, Serri G, Testa B, Wijbrans J (2003) Evoluzione geologica dell'Isola di Capraia (Arcipelago Toscano) nel quadro della geodinamica del Tirreno settentrionale. In: Capozzi R (ed) Geology of the Tyrrhenian sea and Apennines, Geoacta 2:19–22

Borgia A, Mazzoldi A, Brunori CA, Allocca C, Delcroix C, Micheli L, Vercellino A, Grieco G (2014) Volcanic spreading forcing and feedback in geothermal reservoir development, Amiata Volcano, Italia. J Volcanol Geoth Res 284:16–31

Borsi S, Ferrara G, Tongiorgi E (1967) Determinazione con il metodo K/Ar delle età delle rocce magmatiche della Toscana. Boll Soc Geol It 86:403–410

Brogi A, Lazzarotto A, Liotta D, Gii Ranalli et al (2005) Crustal structures in the geothermal areas of southern Tuscany (Italy): insights from the CROP 18 deep seismic ref lection lines. J Volcanol Geoth Res 148:60–80

Brogi A, Liotta D, Meccheri M, Fabbrini L (2010) Transtensional shear zones controlling volcanic eruptions: the middle Pleistocene Mt. Amiata volcano (inner Northern Apennines, Italy). Terra Nova 22:137–146

Brunet C, Monié P, Jolivet L, Cadet JP (2000) Migration of compression and extension in the Tyrrhenian Sea, insights from $^{40}Ar/^{39}Ar$ ages on micas along a transect from Corsica to Tuscany. Tectonophysics 321: 127–155

Buttinelli M, Chiarabba C, Anselmi M, Bianchi I, De Rita D, Quattrocchi F (2014) Crustal structure of Northern Latium (central Italy) from receiver functions analysis: new evidences of a post-collisional back-arc margin evolution. Tectonophysics 621: 148–158

Cadoppi P (1990) Geologia del basamento cristallino nel settore settentrionale del massiccio Dora Maira, Alpi occidentali. PhD thesis, University of Tourin, Italy, 208 pp

Cadoux A, Pinti DL (2009) Hybrid character and pre-eruptive events of Mt. Amiata volcano (Italy) inferred from geochronological, petr-geochemical and isotopic data. J Volcanol Geoth Res 179:169–190

Cadoux A, Blichert-Toft J, Pinti DL, Albarede F (2007) A unique lower mantle source for Southern Italy volcanics. Earth Planet Sci Lett 259:227–238

Carannante S, Monachesi G, Cattaneo M, Amato A, Chiarabba C (2013) Deep structure and tectonics of the northern-central Apennines as seen by regional-scale tomography and 3-D located earthquakes. J Geophys Res 118:1–13 doi:10.1002/jgrb.50371

Carmignani L, Kligfield R (1990) Crustal extension in the northern Apennines: the transition from compression to extension in the Alpi Apuane complex. Tectonics 9:1275–1305

Carmignani L, Decandia FA, Disperati L, Fantozzi PL, Kligfield R, Lazzarotto A, Liotta D, Meccheri M (2001) Inner Northen Apennines. In: Vai GB, Martini PI (eds) Anatomy of an orogen. The Apennines and the adjacent Mediterranean basins. Kluwer, Dordrecht, pp 197–213

Carminati E, Lustrino M, Cuffaro M, Doglioni C (2010) Tectonics, magmatism and geodynamics of Italy: what we know and what we imagine. In: Beltrando M, Peccerillo A, Mattei M, Conticelli S, Doglioni C (eds) Journal virtual explorer 36, paper 9 doi:10.3809/jvirtex.2010.00226

Casagli A (2009) Studio petrografico e petrologico degli xenoliti peridotitici nelle lave lamproitiche di Torre Alfina (Lazio). MSc thesis, University of Siena, 126 pp

Castro A, Gerya TV (2008) Magmatic implications of mantle wedge plumes: experimental study. Lithos 103:103–148

Castro A, Gerya T, Garcia-Casco A, Fernandez C, Diaz Alvarado J, Moreno-Ventas I, Loew I (2010) Melting relations of MORB–sediment mélanges in underplated mantle wedge plumes. Implications for the origin of cordilleran-type batholiths. J Pet 51:1267–1295

Cellai D, Conticelli S, Menchetti S (1994) Crystal-chemistry of clinopyroxenes from potassic and ultrapotassic rocks in central Italy: implications for their genesis. Contrib Min Pet 116:301–315

Chelazzi L, Bindi L, Olmi F, Menchetti S, Peccerillo A, Conticelli S (2006) A lamproitic component in the high-K calc-alkaline and shoshonitic volcanic rocks of the Capraia Island, Tuscan Magmatic province: evidence from clinopyroxene crystal chemical data. Per Min 75:75–94

Chopin C (1984) Very-high-pressure metamorphism in western Alps: implications for subduction of continental crust. Phil Trans R Soc Lond A321:183–197

Cimarelli C, De Rita D (2006a) Relatively rapid emplacement of dome-forming magma inferred from strain analyses: the case of the acid Latian dome complexes (Central Italy). J Volcanol Geoth Res 158:106–116

Cimarelli C, De Rita D (2006b) Structural evolution of the Pleistocene Cimini trachytic volcanic complex (Central Italy). Bull Volcanol 68: 538–548 doi:10.1007/s00445-005-0028-3

Civetta L, Orsi G, Scandone P, Pece R (1978) Eastwards migration of the Tuscan anatectic magmatism due to anticlockwise rotation of the Apennines. Nature 276:604–606

Clausen C, Holm PM (1990) Origin of acid volcanics of the Tolfa district, Tuscan Province, central Italy: an elemental and Sr-isotopic study. Contrib Min Pet 105:403–411

Cocchi L, Masetti G, Muccini F, Carmisciano C (2015) Geophysical mapping of Vercelli Seamount: implications for Mioceneevolution of the Tyrrhenian back arc basin. Geosci Front 2015:1–15

Compagnoni R (2003) HP metamorphic belt of the western Alps. Episodes 26:200–204

Conceição RV, Green DH (2004) Derivation of potassic (shoshonitic) magmas by decompression melting of phlogopite + pargasite lherzolite. Lithos 72:209–229

Condamine P, Médard E (2014) Experimental melting of phlogopite-bearin mantle at 1GPa: implications for potassic magmatism. Earth Planet Sci Lett 397:80–92

Conticelli S (1998) The effect of crustal contamination on ultrapotassic magmas with lamproitic affinity: mineralogical, geochemical and isotope data from the Torre Alfina lavas and xenoliths, Central Italy. Chem Geol 149:51–81

Conticelli S, Peccerillo A (1990) Petrological significance of high-pressure ultramafic xenoliths from ultrapotassic rocks of Central Italy. Lithos 24:305–322

Conticelli S, Peccerillo A (1992) Petrology and geochemistry of potassic and ultrapotassic volcanism in Central Italy: petrogenesis and inferences on the evolution of the mantle sources. Lithos 28:221–240

Conticelli S, Manetti P, Menichetti S (1992) Mineralogy, geochemistry and Sr isotopes in orendites from South Tuscany: constraints on their genesis and evolution. Eur J Min 4:1359–1375

Conticelli S, Bortolotti V, Principi G, Laurenzi M, Vaggelli G, D'Antonio M (2001) Petrology, mineralogy and geochemistry of a mafic dyke from Monte Castello, Elba Island, Italy. Ofioliti 26:249–262

Conticelli S, D'Antonio M, Pinarelli L, Civetta L (2002) Source contamination and mantle heterogeneity in the genesis of Italian potassic and ultrapotassic volcanic rocks: Sr–Nd–Pb isotope data from Roman Province and Southern Tuscany. Min Pet 74:189–222

Conticelli S, Carlson RW, Widom E, Serri G (2007) Chemical and isotopic composition (Os, Pb, Nd and Sr) of Neogene to quaternary calc-alkalic, shoshonitic, and ultrapotassic mafic rocks from the Italian peninsula: Inferences on the nature of their sources. In: Beccaluva L, Bianchini G, Wilson M (eds) Cenozoic volcanismin the Mediterranean area. Geol Soc Am Spec Paper 418:171–202 doi:10.1130/2007.2418(09)

Conticelli S, Guarnieri L, Farinelli A, Mattei M, Avanzinelli R, Bianchini G, Boari E, Tommasini S, Tiepolo M, Prelevic D, Venturelli G (2009a) Trace elements and Sr–Nd–Pb isotopes of K-rich, shoshonitic, and calc-alkaline magmatism of the western mediterranean region: genesis of ultrapotassic to calc-alkaline magmatic associations in a post-collisional geodynamic setting. Lithos 107:68–92

Conticelli S, Marchionni S, Rosa D, Giordano G, Boari E, Avanzinelli R (2009b) Shoshonite and sub-alkaline magmas from an ultrapotassic volcano: Sr-Nd-Pb isotope data on the Roccamonfina volcanic rocks, Roman Magmatic Province, Southern Italy. Contrib Min Pet 157:41–63

Conticelli S, Laurenzi MA, Giordano G, Mattei M, Avanzinelli R, Melluso L, Tommasini S, Boari E, Cifelli F, Perini G (2010) Leucite-bearing (kamafugitic/leucititic) and -free (lamproitic) ultrapotassic rocks and associated shoshonites from Italy: constraints on petrogenesis and geodynamics. In: Beltrando M, Peccerillo A, Mattei M, Conticelli S, Doglioni C (eds) Journal virtual explorer 36, paper 20 doi:10.3809/jvirtex.2010.00251

Conticelli S, Avanzinelli R, Marchionni S, Tommasini S, Melluso L (2011) Sr-Nd-Pb isotopes from the Radicofani Volcano, Central Italy: constraints on heterogeneities in a veined mantle responsible for the shift from ultrapotassic shoshonite to basaltic andesite magmas in a post-collisional setting. Min Pet 103:123–148

Conticelli S, Avanzinelli R, Poli G, Braschi E, Giordano G (2013) Shift from lamproite-like to leucititic rocks: Sr-Nd-Pb isotope data from the Monte Cimino volcanic complex vs. the Vico stratovolcano, Central Italy. Chem Geol 353:246–266

Conticelli S, Avanzinelli R, Ammannati E, Casalini M (2015a) The role of carbon from recycled sediments in the origin of ultrapotassic igneous rocks in the Central Mediterranean. Lithos 232:174–196

Conticelli S, Boari E, Burlamacchi L et al (2015b) Geochemistry and Sr-Nd-Pb isotopes of Monte Amiata Volcano, Central Italy: evidence for magma mixing between high-K calc-alkaline and leucititic mantle-derived magmas. It J Geosci 134:266–290

Cristiani C, Mazzuoli R (2003) Monte Amiata volcanic products and their inclusions. Per Min 72:169–181

D'Orazio M, Laurenzi MA, Villa IM (1991) $^{40}Ar/^{39}Ar$ dating of a shoshonitic lava flow of the Radicofani volcanic center (Southern Tuscany). Acta Vulcanol 1:63–67

D'Orazio M, Innocenti F, Petrini R, Serri G (1994) Il vulcano di Radicofani nel quadro del magmatismo neogenico-quaternario dell'Appennino Settentrionale. Studi Geol Camerti 1:79

De Astis G, Peccerillo A, Kempton PD, La Volpe L, Wu TW (2000a) Transition from calc-alkaline to potassium-rich magmatism in subduction environments: geochemical and Sr, Nd, Pb isotopic constraints from the island of Vulcano (Aeolian arc). Contrib Min Pet 139:684–703

De Astis G, Peccerillo A, Kempton PD, La Volpe L, Wu TW (2000b) Transition from calc-alkaline to potassium-rich magmatism in subduction environments: geochemical and Sr, Nd, Pb isotopic constraints from the island of Vulcano (Aeolian arc). Contrib Min Pet 139:684–703

De Rita D, Bertagnini A, Carboni G, Ciccacci S, Di Filippo M, Faccenna C, Fredi P, Funiciello R, Landi P, Sciacca P, Vannucci N, Zarlenga F (1994) Geological-petrografical evolution of the Ceriti Mountains area (Latium, Central Italy). Mem Descr Carta Geol It 49:291–322

De Rita D, Bertagnini A, Faccenna C, Landi P, Rosa C, Zarlenga F, Di Filippo M, Carboni MG (1997) Evoluzione geopetrografica-strutturale dell'area tolfetana. Boll Soc Geol It 116:143–175

Della Vedova B, Marson I, Panza G, Suhadolc P (1991) Upper mantle properties of the Tuscan-Tyrrhenian area: a key for understanding the recent tectonic evolution of the Italian region. Tectonophysics 195:311–318

Della Vedova B, Bellani S, Pelli G, Squarci P (2001) Deep temperatures and surface heath distribution. In: Vai GB, Martini PI (eds) Anatomy of an Orogen. The Apennines and the adjacent Mediterranean basins. Kluwer, Dordrecht, pp 65–76

Della Vedova B, Vecellio C, Bellani S, Tinivella U (2006) Thermal modelling of Larderello geothermal field (Tuscany, Italy). Int J Earth Sci 97:317–332

Di Stefano R, Bianchi I, Ciaccio MG, Carrara C, Kissling E (2011) Three-dimensional Moho topography in Italy: new constraints from receiver functions and controlled source seismology. Geochem Geophys Geosyst 12:Q09006. doi:10.1029/2011GC003649

Dini A (2003) Ore deposits, industrial minerals and geothermal resources. In: Poli G, Perugini D, Rocchi S, Dini A (eds) Miocene to recent plutonism and volcanism in the Tuscan Magmatic Region (central Italy), Per Min 72:41–52

Dini A, Innocenti F, Rocchi S, Tonarini S, Westerman S (2002) The magmatic evolution of the late Miocene laccolith-pluton-dyke granitic complex of Elba Island, Italy. Geol Mag 139:257–273

Dini A, Rocchi S, Poli G (2003) Hidden granitoid from boreholes and seamounts. In: Poli G, Perugini D, Rocchi S, Dini A (eds) Miocene to recent plutonism and volcanism in the Tuscan Magmatic Region (central Italy), Per Min 72:133–138

Dini A, Gianelli G, Puxeddu M, Ruggieri G (2005) Origin and evolution of Pliocene-Pleistocene granites from the Larderello geothermal field (Tuscan Magmatic Province, Italy). Lithos 81:1–31

Dini A, Rocchi S, Westerman DS, Farina F (2009) The late Miocene intrusive complex of Elba Island: two centuries of studies from Savi to Innocenti. Acta Vulcanol 20–21:11–32

Doglioni C, Harabaglia P, Merlini S, Mongelli F, Peccerillo A, Piromallo C (1999) Orogens and slabs vs. their direction of subduction. Earth Sci Rev 45:167–208

Farina F, Dini A, Innocenti F, Rocchi S, Westerman DS (2010) Rapid incremental assembly of the Monte Capanne pluton (Elba Island, Tuscany) by downward stacking of magma sheets. Geol Soc Am Bull 122:1463–1479

Farina F, Stevens G, Dini A, Rocchi S (2012) Peritectic phase entrainment and magma mixing in the late Miocene Elba Island laccolith-pluton-dyke complex (Italy). Lithos 153:243–260

Farina F, Dini A, Rocchi S, Stevens G (2014) Extreme mineral-scale Sr isotope heterogeneity in granite by disequilibrium melting of the crust. Earth Planet Sci Lett 399:103–115

Feldstein SN, Halliday AN, Davies GR, Hall CM (1994) Isotope and chemical microsampling constraints on the history of a S-type rhyolite, San Vincenzo, Tuscany (Italy). Geochim Cosmochim Acta 58:943–958

Ferrara G, Tonarini S (1985) Radiometric geochronology in Tuscany: results and problems. Rend Soc It Min Pet 40:111–124

Ferrara G, Petrini R, Serri G, Tonarini S (1989) Petrology and isotope-geochemistry of San Vincenzo rhyolites (Tuscany, Italy). Bull Volcanol 51:379–388

Finetti IR, Boccaletti M, Bonini M, Del Ben A, Geletti R, Pipan M, Sani F (2001) Crustal section based on CROP seismic data across the North Tyrrhenian-Northen Apennines-Adriatic Sea. Tectonophysics 43:135–163

Foley SF (1992) Vein-plus-wall-rock melting mechanisms in the lithosphere and the origin of potassic alkaline magmas. Lithos 28:435–453

Foley SF, Venturelli G, Green DH, Toscani L (1987) The ultrapotassic rocks: characteristics classification, and constraints for petrogenetic models. Earth Sci Rev 24:81–134

Franceschelli M, Gianelli G, Pandeli E, Puxeddu M (2004) Variscan and Alpine metamorphic events in the Northern Apennines (Italy): a review. Per Mineral 73:43–56

Franceschini F, Innocenti F, Marsi A, Tamponi M, Serri G (2000) Petrography and chemistry of the buried Pliocene Castel di Pietra pluton (Southern Tuscany, Italy). Neues Jahrb Geol P M 215:17–46

Frondini F, Caliro S, Cardellini C, Chiodini G, Morgantini N, Parello F (2008) Carbon dioxide degassing from Tuscany and Northern Latium. Global Planet Change 61:89–102

Gagnevin D, Daly JS, Poli G (2004) Petrographic, geochemical and isotopic constraints on magma dynamics and mixing in the Miocene Monte Capanne monzogranite (Elba Island, Italy). Lithos 78:157–195

Gagnevin D, Daly JS, Poli G, Morgan D (2005) Microchemical and Sr isotopic investigation of zoned K-feldspar megacrysts: insights into the petrogenesis of a granitic system and disequilibrium crystal growth. J Petrol 46:1689–1724

Gagnevin D, Waight TE, Daly JS, Poli G, Conticelli S (2007) Insights into magmatic evolution and recharge history in Capraia Volcano (Italy) from chemical and isotopic zoning in plagioclase phenocrysts. J Volcanol Geoth Res 168:28–54

Gagnevin D, Daly JS, Kronz A (2010) Zircon texture and chemical composition as a guide to magmatic processes and mixing in a granitic environment and coeval volcanic system. Contrib Min Pet 159:579–596

Gagnevin D, Daly JS, Horstwood MSA, Whitehouse MJ (2011) In-situ zircon U-Pb, oxygen and hafnium isotopic evidence for magma mixing and mantle metasomatism in the Tuscan Magmatic Province, Italy. Earth Planet Sci Lett 305:45–56

Gale A, Dalton CA, Langmuir CH, Su Y, Schilling J-G (2013) The mean composition of ocean ridge basalts. Gechem Geophys Geosyst 14 doi:10.1029/2012GC004334

Gasparon M, Rosenbaum G, Wijbrans J, Manetti P (2009) The transition from subduction arc to slab tearing: evidence from Capraia Island, northern Tyrrhenian Sea. J Geodyn 47:30–38

Gasperini D, Blichert Toft J, Bosch D, Del Moro A, Macera P, Albarède F (2002) Upwelling of deep mantle material through a plate window: evidence from the geochemistry of Italian basaltic volcanics. J Geophys Res 107(B12):2367. doi:10.1029/2001JB000418

Gerya T, Yuen DA (2003) Rayleigh-Taylor instabilities from hydration and melting propel 'cold plumes' at subduction zones. Earth Planet Sci Lett 212:47–62

Gerya TV, Connolly JAD, Yuen DA, Gorczyk W, Capel AM (2006) Seismic implications of mantle wedge plumes. Phys Earth Planet In 156:59–74

Giacomuzzi G, Civalleri M, De Gori P, Chiarabba C (2012) A 3D Vs model of the upper mantle beneath Italy: Insight on the geodynamics of central Mediterranean. Earth Planet Sci Lett 335-336:105–120

Gianelli G, Laurenzi M (2001) Age and cooling rate of the geothermal system of Larderello. Trans Geoth Resour Counc 25:731–735

Gianelli G, Manzella A, Puxeddu M (1997) Crustal models of the geothermal areas of southern Tuscany. Tectonophysics 281:221–239

Gill R (2010) Igneous rocks and processes. A practical guide. Wiley-Blackwell, Chichester, p 428

Giraud A, Dupuy C, Dostal J (1986) Behaviour of trace elements during magmatic processes in the crust: application to silicic volcanic rocks of Tuscany. Chem Geol 57:269–288

Grassi D, Schmidt MW, Gunther D (2012) Element partitioning during carbonated pelite melting at 8, 13 and 22 GPa and the sediment signature in the EM mantle components. Earth Planet Sci Lett 327–328:84–96

Gueguen E, Doglioni C, Fernandez M (1997) Lithospheric boudinage in the Western Mediterranean back-arc basin. Terra Nova 9:184–187

Gueguen E, Doglioni C, Fernandez M (1998) On the post-25 Ma geodynamic evolution of the western Mediterranean. Tectonophysics 298:259–269

Harlow GE, Davies R (2004) Status report on stability of K-rich phases in the upper mantle. Lithos 77:647–653

Hawkesworth CJ, Vollmer R (1979) Crustal contamination vs. enriched mantle: $^{143}Nd/^{144}Nd$ and $^{87}Sr/^{86}Sr$ evidence from the Italian volcanics. Contrib Min Pet 69:151–165

Innocenti F, Westerman DS, Rocchi S, Tonarini S (1997) The Montecristo monzogranite (Northern Tyrrhenian Sea, Italy): a collisional pluton in an extensional setting. Geol J 32:131–151

Irvine TN, Baragar WRA (1971) A guide to chemical classification of common volcanic rocks. Can J Earth Sci 8:523–548

LaBerge RD, Giordano G, Cas RAF, Ailleres L (2006) Syn-depositional substrate deformation produced by the shear force of a pyroclastic density current: an example from the Pleistocene ignimbrite at Monte Cimino, northern Lazio, Italy. J Volcanol Geoth Res 148:307–320

Lardini D, Nappi G (1987) I cicli eruttivi del complesso vulcanico Cimino. Rend Soc It Min Pet 42:141–153

Lattanzi P, Benvenuti M, Costagliola P, Maineri C, Mascaro I, Tanelli G, Dini A, Ruggieri G (2001) Magmatic versus hydrotermal processes in the formation of raw ceramic material deposits in southern Tuscany. In: Proceeding 10th international symposium water-rock interaction, Villasimius (Italy), 10–15 June 2001, pp 725–728

Laurenzi MA, Balestrieri ML, Bigazzi G, Neto JCH, Iunes PJ, Norelli P, Oddone M, Osorio Araya AM, Viramonte JG (2007) New constraints on ages of glasses proposed as reference materials for fission-track dating. Geostand Geoanal Res 31:105–124

Laurenzi MA, Braschi E, Casalini M, Conticelli S (2015) New $^{40}Ar/^{39}Ar$ dating and revision of the geochronology of the Monte Amiata Volcano, Central Italy. It J Geosci 134:255–265

Le Maitre RW (ed) (2002) A classification of igneous rocks and glossary of terms. Cambridge University Press, Cambridge 252 pp

Marani MP, Gamberi F (2004) Distribution and nature of submarine volcanic landforms in the Tyrrhenian Sea: the arc versus the back-arc. Mem Descr Carta Geol It 44:109–126

Marinelli G (1975) Magma evolution in Italy. In: Squyres CH (ed) Geology of Italy. Petrol Expl Soc Libya, Tripoli, pp 165–219

Marinosci I (1993) Studio petrografico comparativo tra la minette del filone strato di Sisco (Corsica Settentrionale) e le selagiti del volterrano toscano. Geol Romana 29:237–247

Marroni M, Moratti G, Costantini A et al (2015) Geology of the Monte Amiata region, Southern Tuscany, Central Italy. It J Geosci 134:200–218

Marschall HR, Schumacher JC (2012) Arc magmas sourced from mélange diapirs in subduction zones. Nat Geosci 5:862–867

Masuda H, O'Neil R (1994) Oxygen-isotope heterogeneity of phenocrysts in rhyolite from San Vincenzo, Italy, by laser microprobe analysis. Geochem J 28:377–385

Mazzarini F, Corti G, Musumeci G, Innocenti F (2004) Tectonic control on laccolith emplacement in the northern Apennines fold-thrust belt: the Gavorrano intrusion (southern Tuscany, Italy). In: Breitkreuz C, Petford N (eds) Physical geology of high-level magmatic systems, Geol Soc London Spec Publ 234:151–161

Mazzuoli R (1967) Le vulcaniti di Roccastrada (Grosseto). Atti Soc Tosc Sci Nat A84:315–373

Mazzuoli R, Pratesi M (1963) Rilevamento e studio chimico-petrografico delle rocce vulcaniche del Monte Amiata. Atti Soc Tosc Sci Nat A70:355–429

Mele G, Sandvol E (2003) Deep crustal roots beneath the northern Apennines inferred from teleseismic receiver functions. Earth Planet Sci Lett 211:69–78

Mele G, Sandvol E, Cavinato GP (2006) Evidence of crustal thickening beneath the central Apennines (Italy) from teleseismic receiver functions. Earth Planet Sci Lett 249:425–435

Melzer S, Foley SF (2000) Phase relations and fractionation sequences in potassic magma series modelled in the system $CaMgSi_2O_6$-$KAlSiO_4$-Mg_2SiO_4-F at 1 bar to 18 kbar. Contrib Min Pet 138:186–197

Minissale A (1991) The Larderello geothermal field: a review. Earth Sci Rev 31:133–151

Minissale A, Magro G, Martinelli G, Vaselli O, Tassi GF (2000) Fluid geochemical transect in the Northern Apennines (central-northern Italy): fluid genesis and migration and tectonic implications. Tectonophysics 319:199–222

Mitchell RH, Bergman SC (1991) Petrology of lamproites. Plenum, New York 447 pp
Musumeci G, Mazzarini F, Corti G, Barsella M, Montanari D (2005) Magma emplacement in a thrust ramp anticline: the Gavorrano granite (northern Apennines, Italy). Tectonics 24, TC6009 doi:10.1029/2005TC001801
Nikogossian IK, van Bergen MJ (2010) Heterogeneous mantle sources of potassium-rich magmas in central-southern Italy: melt inclusion evidence from Roccamonfina and Ernici (Mid Latina Valley). J Volcanol Geoth Res 197:279–302
Pandeli E, Santo AP, Morelli M, Orti L (2006) Petrological and geological data of porphyritic dikes from the Capo Arco area (Eastern Elba Island, northern Tyrrhenian Sea). Per Min 75:241–154
Pandeli E, Bartolini C, Dini A, Antolini E (2010) New data on the paleogeography of Southern Tuscany (Italy) since Late Miocene time. Int J Earth Sci 99:1357–1381
Pandeli E, Santo AP, Candido MR, Petrone CM, Giusti R (2014) The calc-alkaline Monte Capo Stella dykes in the ophiolitic unit of the Elba Island (Italy): Geological setting and compositional characterization. Ofioliti 39:79–93
Panza GF, Mueller S (1979) The plate boundary between Eurasia and Africa in the Alpine area. Mem Soc Geol It 33:43–50
Pauselli C, Barchi MR, Federico C, Magnani MB, Minelli G (2006) The crustal structure of the northern Apennines (central Italy): an insight by the CROP03 seismic line. Am J Sci 306:428–450
Peccerillo A (1999) Multiple mantle metasomatism in central-southern Italy: geochemical effects, timing and geodynamic implications. Geology 27:315–318
Peccerillo A (2002) Plio-quaternary magmatism in central-southern Italy: a new classification scheme for volcanic provinces and its geodynamic implications. In: Barchi RM, Cirilli S, Minelli G (eds) Geological and geodynamic evolution of the Apennines. Boll Soc Geol It (Spec Vol 1):113–127
Peccerillo A (2005a) Plio-Quaternary volcanism in Italy. Petrology, geohemistry, geodinamics. Springer, Berlin, p 365
Peccerillo A (2005b) On the nature of carbonate-rich volcanic rocks in Central Italy. A reply to comments by Woolley et al. Per Min 74:195–204
Peccerillo A, Frezzotti ML (2015) Magmatism, mantle evolution and geodynamics at the converging plate margins of Italy. J Geol Soc Lond 172:407–427
Peccerillo A, Martinotti G (2006) The Western Mediterranean lamproitic magmatism: origin and geodynamic significance. Terra Nova 18:109–117
Peccerillo A, Taylor SR (1976) Geochemistry of Eocene calc-alkaline volcanic rocks of the Kastamonu area, northern Turkey. Contrib Min Pet 58:63–81
Peccerillo A, Conticelli S, Manetti P (1987) Petrological characteristics and the genesis of recent magmatism of Southern Tuscany and Northern Latium. Per Min 56:157–172
Peccerillo A, Poli G, Serri G (1988) Petrogenesis of orenditic and kamafugitic rocks from Central Italy. Can Min 26:45–65
Peccerillo A, De Astis G, Faraone D, Forni F, Frezzotti ML (2013) Compositional variations of magmas in the Aeolian arc: implications for petrogenesis and geodynamics. In: Lucchi F, Peccerillo A, Keller J, Tranne CA, Rossi PL (eds) The Aeolian islands volcanoes. Geol Soc London Memoirs 37:491–510
Pera E, Mainprice D, Burlini L (2003) Anisotropic seismic properties of the upper mantle beneath the Torre Alfina area (Northern Apennines, Central Italy). Tectonophysics 370:11–30
Pinarelli L (1991) Geochemical and isotopic (Sr, Pb) evidence of crust-mantle interaction in silicicic melts. The Tolfa-Cerveteri-Manziana volcanic complex (Central Italy): a case history. Chem Geol 92:177–195
Pinarelli L, Poli G, Santo A (1989) Geochemical characterization of recent volcanism from the Tuscan Magmatic Province (Central Italy): the Roccastrada and San Vincenzo centers. Per Min 58:67–96
Piromallo C, Morelli A (2003) P wave tomography of the mantle under the Alpine-Mediterranean area. J Geophys Res 108(B2):2065. doi:10.1029/2002JB001757
Plank T, Langmuir CH (1998) The chemical composition of subducting sediment and its consequences for the crust and mantle. Chem Geol 145:325–394
Poli G (1985) Phase composition in volcanic rocks from Soth Tuscany magmatic area (Central Italy): inferences on conditions of magma crystallization. Neues Jb Miner Abh 151:141–161
Poli G (1992) Geochemistry of Tuscan Archipelago granitoids, Central Italy: the role of hybridization processes in their genesis. J Geol 100:41–56
Poli G (2004) Genesis and evolution of Miocene-quaternary intermediate-acid rocks from the Tuscan Magmatic Province. Per Min 73:187–214
Poli G, Peccerillo A (2016) The upper miocene magmatism of the Elba Island: compositional characteristics, petrogenesis and implications for the origin of the Tuscany Magmatic Province. Min Pet 110:421–445
Poli G, Perugini D (2003a) San Vincenzo volcanites. In: Poli G, Perugini D, Rocchi S, Dini A (eds) Miocene to recent plutonism and volcanism in the Tuscan Magmatic Region (central Italy), Per Min 72:141–155
Poli G, Perugini D (2003b) Roccastrada volcanites. In: Poli G, Perugini D, Rocchi S, Dini A (eds) Miocene to recent plutonism and volcanism in the Tuscan magmatic region (central Italy), Per Min 72:157–168
Poli G, Perugini D (2003c) The Island of Capraia. In: Poli G, Perugini D, Rocchi S, Dini A (eds) Miocene to recent plutonism and volcanism in the Tuscan magmatic region (central Italy), Per Min 72:195–201
Poli G, Frey FA, Ferrara G (1984) Geochemical characteristics of the south Tuscany (Italy) volcanic

province, constraints on lava petrogenesis. Chem Geol 43:203–221

Poli G, Manetti P, Tommasini S (1989) A petrological review on Miocene-Pliocene intrusive rocks from Southern Tuscany and Tyrrhenian Sea (Italy). Per Min 58:109–126

Poli G, Peccerillo A, Donati C (2002) The Plio-Quaternary acid magmatism of Southern Tuscany. In: Barchi RM, Cirilli S, Minelli G (eds) Geological and geodynamic evolution of the Apennines. Boll Soc Geol It Spec 1:143–151

Poli G, Perugini D, Rocchi S, Dini A (eds) (2003) Miocene to Recent plutonism and volcanism in the Tuscan Magmatic Province. Per Mineral 72:244 pp

Prelevic D, Foley SF, Cvetkovic V, Romer RL, Downes H (2005) Tertiary ultrapotassic volcanism in Serbia: constraints on petrogenesis and mantle source characteristics. J Pet 46:1443–1487

Prelevic D, Foley SF, Romer R, Conticelli S (2008) Mediterranean tertiary lamproites derived from multiple source components in postcollisional geodynamics. Geochim Cosmochim Acta 72:2125–2156

Prelevic D, Stracke A, Foley SF, Romer RL, Conticelli S (2010) Hf isotope compositions of Mediterranean lamproites: mixing of melts from asthenosphere and crustally contaminated mantle lithosphere. Lithos 119:297–312

Principi G, Treves B (1984) Il sistema corso-appenninico come prisma d'accrezione: Riflessi sul problema generale del limite Alpi-Appennini. Mem Soc Geol It 28:549–576

Prosperini N (1993) Petrologia e geochimica delle rocce dell'isola di Capraia (Arcipelago Toscano, Italia): un vulcano calcalcalino di origine complessa. BSc thesis, University of Perugia, 149 pp

Puxeddu M (1971) Studio chimico-petrografico delle vulcaniti del M. Cimino (Viterbo). Atti Soc Tosc Sci Nat A78:329–394

Remitti F, Balestrieri ML, Vannucchi P, Bettelli G (2013) Early exhumation of underthrust units near the toe of an ancient erosive subduction zone: A case study from the Northern Apennines of Italy. Geol Soc Am Bull 125:1820–1832

Ribeiro JM, Maury RC, Grégpire M (2016) Are adakites slab melts or high-pressure fractionatedmantle melts? J Petrol 57, doi: 10.1093/petrology/egw023 (in press)

Ridolfi F, Renzulli A, Acosta-Vigil A (2014) On the stability of magmatic cordierite and new thermobarometric equations for cordierite-saturated liquids. Contrib Min Pet 167:996. doi:10.1007/s00410-014-0996-4

Rocchi S, Westerman DS, Dini A, Innocenti F, Tonarini S (2002) Two-stage growth of laccoliths at Elba Island, Italy. Geology 30:983–986

Rocchi S, Dini A, Innocenti F, Tonarini S Westerman DS, (2003a) The Elba Island: intrusive magmatism. In: Poli G, Perugini D, Rocchi S, Dini A (eds) Miocene to recent plutonism and volcanism in the tuscan magmatic region (central Italy), Per Mineral 72:73–104

Rocchi S, Westerman DS, Innocenti F (2003b) Montecristo island: intrusive magmatism. In: Poli G, Perugini D, Rocchi S, Dini A (eds) Miocene to recent plutonism and volcanism in the Tuscan magmatic region (central Italy), Per Min 72:105–118

Rocchi S, Dini A, Mazzarini F, Poli G (2003c) Campiglia Marittima and Gavorrano intrusive magmatism. In: Poli G, Perugini D, Rocchi S, Dini A (eds) Miocene to Recent plutonism and volcanism in the Tuscan Magmatic region (central Italy), Per Min 72:127–132

Rocchi S, Westerman DS, Dini A, Farina F (2010) Intrusive sheets and sheeted intrusions at Elba Island, Italy. Geosphere 6:225–236

Sato K (1997) Melting experiments on a synthetic olivine lamproite compositionup to 8 GPa: implication to its petrogenesis. J Gephys Res 102:14751–14764

Savelli C (2000) Subduction-related episodes of K-alkaline magmatism (15–0.1 Ma) and geodynamic implications in the north Tyrrhenian, central Italy region: a review. J Geodyn 30:575–591

Serri G, Innocenti F, Manetti P (1993) Geochemical and petrological evidence of the subduction of delaminated Adriatic continental lithosphere in the genesis of the Neogene-quaternary magmatism of central Italy. Tectonophysics 223:117–147

Streckeisen AL, Le Maitre RW (1979) Chemical approximation to modal QAPF classification of the igneous rocks. Neues Neues Jb Min 136:169–206

Sun SS, McDonough WF (1989) Chemical and isotopic systematics of oceanic basalts: implications for mantle composition and processes. In: Saunders AD, Norry MJ (eds) Magmatism in ocean basins. Geol Soc London Spec Publ 42:313–345

Tanelli G, Lattanzi P (1986) Metallogeny and mineral exploration in Tuscany: state of the art. Mem Soc Geol It 31:299–304

Tanelli G, Poggi L (2012) La Collezione Elbana. In: Pratesi G (ed) The museum of natural history of the University of Florence, the mineralogical and lithological collections. Florence Univ Press 4:67–81

Tarquini S, Vinci S, Favalli M, Doumaz F, Fornaciai A, Nannipieri L (2012) Release of a 10-m-resolution DEM for the Italian territory: comparison with global-coverage DEMs and anaglyph-mode exploration via the web. Comput Geosci 38:168–170

Tumiati S, Fumagalli P, Tiraboschi C, Poli S (2013) An experimental study on COH-bearing Peridotite up to 3.2 GPa and implications for crust-mantle recycling. J Pet 54:453–479

Turi B, Taylor HP (1976) Oxygen isotope studies of potassic volcanic rocks of the Roman Province, central Italy. Contrib Min Pet 55:1–31

van Bergen MJ (1985) Common trace element characteristics of crustal- and mantle-derived K-rich magmas from Monte Amiata (central Italy). Chem Geol 48:125–135

van Bergen MJ, Barton M (1984) Complex interaction of aluminous metasedimentary xenoliths and siliceous magma: an example from Mt. Amiata (Central Italy). Contrib Min Pet 86:374–385

van Bergen MJ, Ghezzo C, Ricci CA (1983) Minette inclusions in the rhyodacitic lavas of Mt. Amiata (central Italy): mineralogical and chemical evidence of

mixing between Tuscan- and Roman-type magmas. J Volcanol Geoth Res 19:1–35
Vannucchi P, Remitti F, Bettelli G, Boschi C, Dallai L (2010) Fluid history related to the early Eocene-middle Miocene convergent system of the Northern Apennines (Italy): Constraints from structural and isotopic studies. J Geophys Res 115, B05405 doi:10.1029/2009JB006590
Vannucchi P, Sage F, Morgan JP, Remitti F, Collot JY (2012) Toward a dynamic concept of the subduction channel at erosive convergent margins with implications for interplate material transfer. Geochem Geophys Geosyst 13:Q02003. doi:10.1029/2011GC003846
Villa IM, Giuliani O, De Grandis G, Cioni R (1989) Datazioni K/Ar dei vulcani di Tolfa e Manziana. Boll Gruppo Naz Vulcanol 2:1025–1026
Villa IM, Ruggieri G, Puxeddu M (2001) Geochronology of magmatic and hydrothermal micas from Larderello geothermal field. In: Cidu R (ed) Proceedings 10th international symposium water-rock interaction, Balkema 1, pp 1589–1592
Vollmer R (1976) Rb-Sr and U-Th-Pb systematics of alkaline rocks: the alkaline rocks from Italy. Geochim Cosmochim Acta 40:283–295
Vollmer R (1977) Isotopic evidence for genetic relations between acid and alkaline rocks in Italy. Contrib Min Pet 60:109–118
Wagner C, Velde D (1986) The mineralogy of K-richterite-bearing lamproites. Am Min 71:17–37
Wendlandt RF, Eggler DH (1980a) The origins of potassic magmas: 1. Melting relations in the systems $KAlSiO_4$ -Mg_2SiO_2 and $KAlSiO_4$ -MgO -SiO_2 -CO_2 to 30 kilobars. Am J Sci 280:385–420
Wendlandt RF, Eggler DH (1980b) The origins of potassic magmas: 2. Stability of phlogopite in natural spinel lherzolite an in the $KAlSiO_4$-MgO-SiO_2-H_2O-CO_2 at high pressures and high temperatures. Am J Sci 280:421–458
Westerman DS, Innocenti F, Tonarini S, Ferrara G (1993) The Pliocene intrusion of the Island of Giglio. Mem Soc Geol It 49:345–363
Westerman DS, Innocenti F, Rocchi S (2003) Giglio island: intrusive magmatism. In: Poli G, Perugini D, Rocchi S, Dini A (eds) Miocene to recent plutonism and volcanism in the Tuscan Magmatic Region (central Italy), Per Min 72:119–126
Westerman DS, Dini A, Innocenti F, Rocchi S (2004) Rise and fall of a nested Christmas-tree laccolith complex, Elba Island, Italy. In: Breitkruez C, Petford N (eds) Physical geology of high-level magmatic systems. Geol Soc Lond Spec Publ 234:195–213
Zitellini N, Trincardi F, Marani M, Fabbri A (1986) Neogene tectonics of the northern Tyrrhenian Sea. Giorn Geol 48:25–40

The Intra-Apennine Province 3

Abstract

The Intra-Apennine Province (IAP) consists of various small outcrops of pyroclastic rocks plus a few lavas at San Venanzo and Cupaello (Umbria). Pyroclastic rocks often contain silicate and carbonate material; silicate fraction is melilititic to phonolitic in composition. Lavas have olivine melilitite and kalsilite melilitite compositions. Overall, lavas and silicate pyroclasts exhibit an ultrapotassic kamafugitic affinity, high enrichments in LILE (e.g. Cs, Rb, LREE) and low abundance of Ta and Nb. The rocks of the Intra-Apennine Province have very similar Sr–Nd–Pb isotope ratios and incompatible trace element signatures to the mafic rocks of the nearby Roman Province, suggesting a close genetic relationship and geodynamic significance. These compositions are best explained by assuming a magma origin in an anomalous upper mantle that was enriched in incompatible elements by subducted marly sediments. Interaction between kamafugitic melts and sedimentary carbonates probably occurred during magma ascent. This process generated an increase of $\delta^{18}O$ ‰ and CaO content of magmas, but had little effects on incompatible trace element ratios and radiogenic isotope signatures. The carbonate-rich rocks occurring in the IAP are mixtures of carbonate and silicate components. The very nature (i.e. magmatic vs. secondary, and mantle-derived vs. crust-derived) of these carbonates has been much debated and, perhaps, different types of calcite occur in the single rocks.

Keywords

Intra-Apennine volcanism · Kamafugite · Melilitite · Carbonatite · Mantle contamination · Adriatic plate · Northern Apennines

A. Peccerillo, *Cenozoic Volcanism in the Tyrrhenian Sea Region*,
Advances in Volcanology, DOI 10.1007/978-3-319-42491-0_3

3.1 Introduction

The Intra-Apennine Magmatic Province (IAP) consists of several small pyroclastic centres and few lavas scattered along the axial zones of the Northern Apennines, from Umbria to the nearby regions of Abruzzi and Latium (Fig. 3.1). Volumes of igneous rocks are very small, but compositions are strikingly different from other volcanic areas in Italy. Pyroclastic rocks crop out at various localities, including San Venanzo, Cupaello, Perugia, Pietrafitta, Polino, Acquasparta, and Oricola. Other pyroclastic outcrops represent fallout deposits from the nearby volcanoes of the Roman Province (e.g. Zanon 2005), and will not be considered. Lavas crop out only at San Venanzo and Cupaello. A particular outcrop of melilititic rocks occurs at Colle Fabbri, but its origin is controversial. The IAP has been also indicated as the Umbria Province by Washington (1906) and as the Intramontane Ultra-alkaline Province (IUP) or the Umbria-Latium Ultra-alkaline District (ULUD) by Lavecchia and Stoppa (1996) and Stoppa et al. (2005). Information on age, petrology and volcanology of the main centres of the IAP is given in Table 3.1.

3.2 Regional Geology

Most of the IAP centres are situated in an area west of the Northern Apennine thrust front, an arcuate structure around Tuscany, Umbria and northern Latium (Fig. 3.1). Regional geology is characterised by thick piles of sedimentary rocks, sitting on the metamorphic basement of the Adriatic plate (e.g. Keller et al. 1994; Ciarapica and Passeri 1998; Decandia et al. 1998; Barchi et al. 2001; Barchi 2010). Permian-Triassic phyllites and quartzites (Verrucano Group) encountered by borehole drillings represent the oldest rocks in the region. These are overlain by sedimentary sequences that include Late Triassic evaporites, Late Triassic to Oligocene marls, limestones and dolostones, and Miocene mainly terrigenous turbiditic rocks. Messinian evaporites occur in the external (eastern) sector of the area. Plio-Pleistocene mainly continental sands, clay and conglomerates infill extensional basins.

The Intra-Apennine area was affected by late Triassic to Jurassic rifting, with sedimentation of both complete and reduced successions inside progressively deepening extensional basins. Starting in Cretaceous time, sedimentation occurred in syn-orogenic basins. Compressional tectonics related to the formation of the Apennine chain began during the Miocene. The compression front progressively shifted eastward, contemporaneously with terrigenous foredeep sedimentation and with backarc extension (see Chap. 2). Compression was much less intensive than in Tuscany, and generated moderate thrusting and folding. Plio-Pleistocene rifting behind the eastward migrating compression front formed small subsiding extensional basins, which were sites of lacustrine to fluviatile sedimentation, and of the episodic IAP volcanic activity at their border faults (Pauselli et al. 2006; Barchi 2010).

Crustal thickness in the IAP is slightly greater than in Tuscany, increasing across the Apennine chain up to about 40–50 km, and decreasing towards the Adriatic foreland (Scarascia et al. 1994; Piromallo and Morelli 2003; Pauselli et al. 2006; Di Stefano et al. 2009, 2011). By contrast, there is a continuous increase of lithospheric thickness from about 40 km in Tuscany to about 80 km in the Adriatic Sea coast. The entire area is characterised by negative gravimetric anomalies and low heat flow (less than 70 mWm^{-2}; Della Vedova et al. 2001). The Intra-Apennine volcanic area is also the site of intense deep CO_2 degassing reaching fluxes of about 0.8 t/d km^2 (Chiodini et al. 2008).

3.3 Compositional Characteristics of Intra-Apennine Magmatism

The IAP is made up of dominant pyroclastic rocks and a few lava flows and dikes. Compositions are variable, partially due to the effect of deuteric processes. The best preserved rocks are

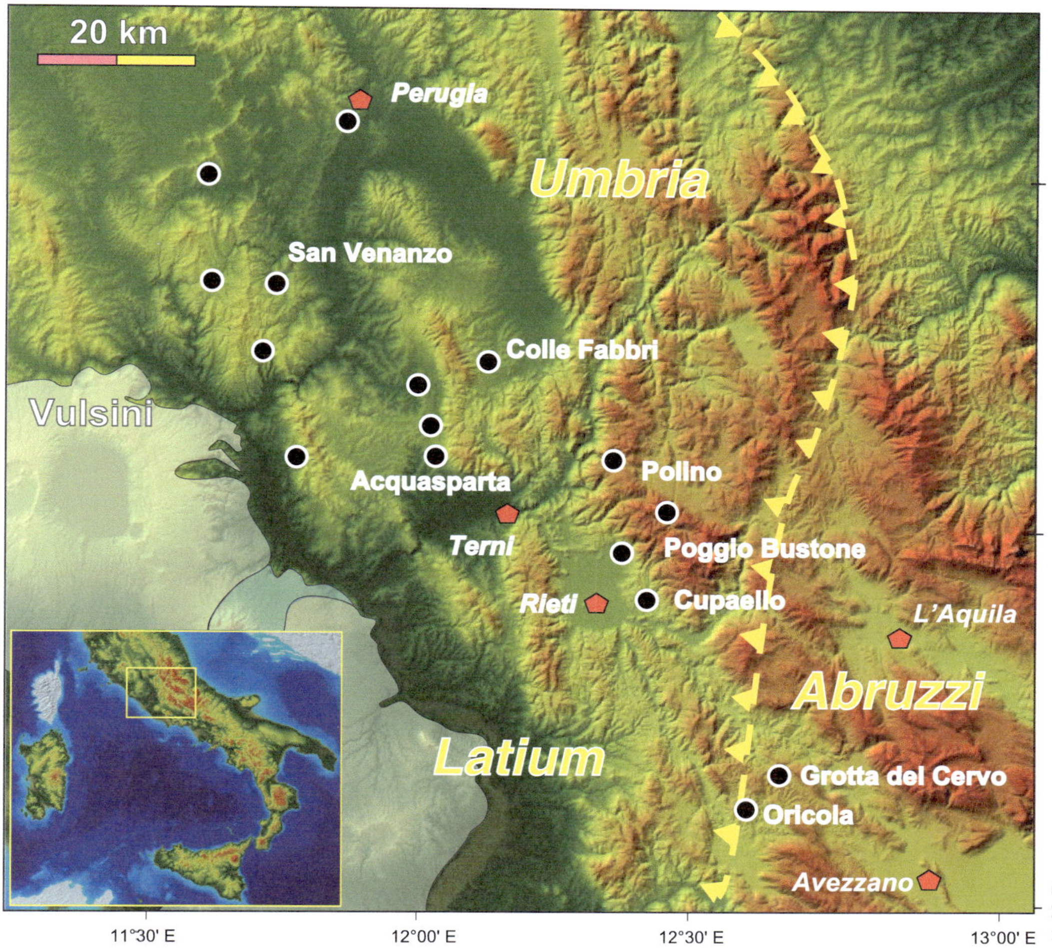

Fig. 3.1 Location of the main outcrops of the Intra-Apennine Magmatic Province. The *dashed line* is the Olevano-Antrodoco outer front of the Northern Apennines thrusts. Relief map from Tarquini et al. (2012)

strongly silica undersaturated, sometime peralkaline and have modal melilite and kalsilite. Lavas consist of melilitites, with major element composition typical of ultrapotassic rocks with a kamafugitic affinity (Sahama 1974; Gallo et al. 1984; Peccerillo et al. 1988; Conticelli et al. 2015). The pyroclastic rocks range from melilitite to trachyphonolite (Stoppa and Lavecchia 1992; Stoppa and Wooley 1997; Stoppa et al. 2005). Several pyroclastic occurrences contain abundant calcite and have been classified as carbonatites (Stoppa and Woolley 1997; Stoppa et al. 2005; Martin et al. 2012).

3.3.1 San Venanzo

This volcano consists of three main centres (San Venanzo, Pian di Celle and Celli) whose explosive and effusive products cover an area of about 0.5 km^2 (Stoppa 1996; Zanon 2005). San Venanzo centre is a maar depression occurring south of the San Venanzo village. Pian di Celle is a composite tuff ring formed of pyroclastic products, two lava flows, and a feeder dike. Lava flows are 3–8 m thick and have a total volume of about 2.2×10^6 m^3. Celli is an isolated small tuff cone with a bulk volume of about

Table 3.1 Petrological characteristics and ages of the main centres of the Intra-Apennine magmatic province

Magmatic centres	Age (in Ma)	Volcanology	Petrology-geochemistry
San Venanzo	0.26	– Three eruptive centres formed by two lava flows and dike, and pyroclastic rocks. Lava contains pegmatoid veins	– Olivine melilitite lavas with ultrapotassic kamafugitic affinity; slightly different compositions for pyroclastic rocks. High enrichment in LILE. $^{87}Sr/^{86}Sr \sim 0.7104$; $^{143}Nd/^{144}Nd \sim 0.5121$
Cupaello	0.64	– Lava flow overlying pyroclastic rocks, some of which contain abundant carbonates	– Kalsilite melilitite lava with ultrapotassic kamafugitic affinity and carbonate-rich pyroclastic rocks. Higher incompatible element abundances and radiogenic Sr than at San Venanzo ($^{87}Sr/^{86}Sr \sim 0.7113$; $^{143}Nd/^{144}Nd \sim 0.5121$). Pyroclasts have similar signatures as lavas, but are depleted in incompatible elements
Polino	0.25	– Two small diatremes filled with a chaotic breccia made of phlogopite- and calcite-rich carbonatitic blocks, lapilli and a fine matrix	– Carbonatitic blocks showing incompatible element patterns alike to the Cupaello carbonate-rich pyroclasts. Sr-Nd isotope ratios similar to San Venanzo
Oricola-Carsoli	0.53	– Tuff ring made of breccia, surge and fall deposits	– Carbonatite to phonolitic foidite juvenile clasts showing high contents of incompatible trace elements, with calcite being more enriched than the silicate juvenile fraction
Grotta del Cervo	0.5 (?)	– Massive pyroclastic rocks containing ashes, lapilli, blocks and ultramafic micronodules	– Similar composition of juvenile pyroclasts to the San Venanzo lavas
Acquasparta	0.39	– Various small deposits from different monogenitic centres	– Juvenile pyroclasts with foidite to tephriphonolite composition
Colle Fabbri	0.8	– Breccia intruded by sills and dikes and by a final massive body	– Granular to vesicular igneous-like rocks. Massive facies have melilitolite compositions. Magmatic origin has been questioned

0.14×10^6 m^3 (Zanon 2005). Lavas are olivine melilitites locally known as *venanzite*. Volcanic activity took place during distinct phases whose products are separated by a paleosol (Zanon 2005). $^{40}Ar/^{39}Ar$ dating on minerals separated from Pian di Celle volcanics yields an age of 265 ka (Laurenzi et al. 1994).

Petrography and mineralogy. The San Venanzo lava and dike rocks have a poorly porphyritic holocrystalline texture and contain microphenocrysts of olivine ($\sim Fo_{92-80}$) set in a groundmass composed of olivine ($\sim Fo_{80}$), melilite, leucite, phlogopite (Mg $\neq$ = 0.90), Al-poor diopside ($Al_2O_3 \sim 0.5$–1.3 wt%; Cellai et al. 1994), kalsilite, spinel, monticellite and calcite. Spinel, perovskite and apatite are present as accessory phases (Cundari and Ferguson 1991). The lava flow contains coarse-grained pegmatoid veins formed of leucite, melilite, spinel, olivine ($\sim Fo_{65}$), phlogopite (Mg $\neq$ = 0.81), kalsilite, calcite, clinopyroxene and apatite. The juvenile pyroclastic material has a less mafic composition than the lavas (Stoppa 1996). Fragments of ultramafic rocks and mafic minerals have been found in the pyroclastic deposits and have been interpreted as mantle debris (Stoppa et al. 2005). Abundant isolated crystals include kinked olivine (Fo > 92), diopside (Cr_2O_3 = 0.8–1.2 wt%) and phlogopite (Mg# = 0.90). Calcite is an important component of the San Venanzo rocks, particularly of the pyroclastic deposits. This carbonate material is considered to be of magmatic origin by Stoppa (1996) and Stoppa and Woolley (1997), who suggest a carbonatitic affinity for pyroclastic rocks. A large number of uncommon mineral phases, such as götzenite $(Na,Ca)_{3.5}Ti_{0.5}(Si_2O_7)(F,O)_2$, khibinskite $K_4Zr_2(Si_4O_{14})$ and Zr-cuspidine $Na_{0.5}Ca_3Zr_{0.5}(Si_2O_7)F_{1.5}O_{0.5}$ have been found in the San Venanzo volcanics (e.g. Sharygin et al. 1996).

Petrology and geochemistry. Compositions of San Venanzo lavas are characterised by a strong undersaturation in silica, yielding high levels of normative leucite and kalsilite (Gallo et al. 1984; Peccerillo et al. 1988). Major elements exhibit many features, such as high contents of MgO, CaO and K_2O, and moderate to low Na_2O and Al_2O_3 (Fig. 3.2), which are typical of kamafugites. However, TiO_2 is low (Fig. 3.2a; Table 3.2). Abundances of Cr, Ni and Co are in the range of mantle-equilibrated melts, whereas V and Sc are lower than typical values of mafic rocks (Fig. 3.3; Table 3.2). Boron contents are high (B ~ 100–200 ppm; Vaselli and Conticelli 1990). REE patterns are fractionated and exhibit a small negative Eu anomaly (Fig. 3.4a), which cannot be due to separation of feldspars as these phases are absent in the San Venanzo lavas. Mantle-normalised incompatible element patterns are fractionated, and contain positive spikes of Cs, U, Th and Pb, along with negative anomalies of Ba and HFSE (Fig. 3.4b). These patterns closely resemble those of ultrapotassic mafic rocks from the nearby Roman Province (see Chap. 4), and are strikingly different from patterns of kamafugites from the Virunga region in East Africa, and of OIB (Fig. 3.4b; e.g. Sun and McDonough 1989; Rogers et al. 1992; Tappe et al. 2003; Chakrabarti et al. 2009). The pegmatoid veins in the Pian di Celle lava have lower silica, MgO, Cr and Rb contents, and much higher abundances of TiO_2 and incompatible trace elements than the host lavas. Their origin probably relates to fluid enrichment processes and/or magma unmixing. Pyroclastic scoriae have similar compositions to lavas for most elements, but are somewhat enriched in silica and depleted in alkalies, CaO and Sr. Radiogenic isotopic ratios of San Venanzo volcanics (Fig. 3.5) are close to values observed in the Roman Province ($^{87}Sr/^{86}Sr$ ~ 0.7104; $^{143}Nd/^{144}Nd$ ~ 0.5121; $^{206}Pb/^{204}Pb$ ~ 18.73; $^{207}Pb/^{204}Pb$ ~ 15.66; $^{208}Pb/^{204}Pb$ ~ 39.00; $^{176}Hf/^{177}Hf$ ~ 0.28249; $^{187}Os/^{188}Os$ = 0.1888; Conticelli et al. 2002, 2007; Author's unpublished data). However, very high O-isotope ratios have been measured for lavas ($\delta^{18}O_{SMOW}$ ~ +11 to +12; Holm and Munksgaard 1982; Taylor et al. 1984) and for separated olivine ($\delta^{18}O_{SMOW}$ = +13; Author's unpublished data), a feature that is not observed in the mafic Roman rocks. Very high values have been measured for carbonates from lavas and pyroclastic rocks ($\delta^{18}O_{SMOW}$ = +23 to +26 ‰; Turi 1969; Stoppa and Wooley 1997).

3.3.2 Cupaello

The Cupaello volcanic centre consists of a single 750 m long lava flow overlying a sequence of pyroclastic material composed of altered ashes, breccias, lapilli and scoriae (Stoppa and Cundari 1995). $^{40}Ar/^{39}Ar$ dating on separated kalsilite and phlogopite yields an age of 639 ka (Laurenzi et al. 1994).

Petrography and mineralogy. The lava has a poorly porphyritic hypocrystalline texture and contains phenocrysts of Al-poor diopside (Al_2O_3 ~ 0.3–0.4 wt%) and phlogopite (Mg = 0.93–0.80) set in a groundmass of kalsilite, melilite, monticellite and glass (Cundari and Ferguson 1991; Stoppa and Cundari 1995; Stoppa and Woolley 1997). The rock has been petrographically classified as a pyroxene-kalsilite melilitite, and is also known by the local name of *coppaellite*. The Cupaello lava has been affected to various extents by syn- to post-eruptive modification, which led to the formation of secondary phases such as zeolites and chlorite. The pyroclastic rocks consist of ashes and lapilli, the latter being mainly represented by fragmented melilitite and crystal debris. Calcite is an

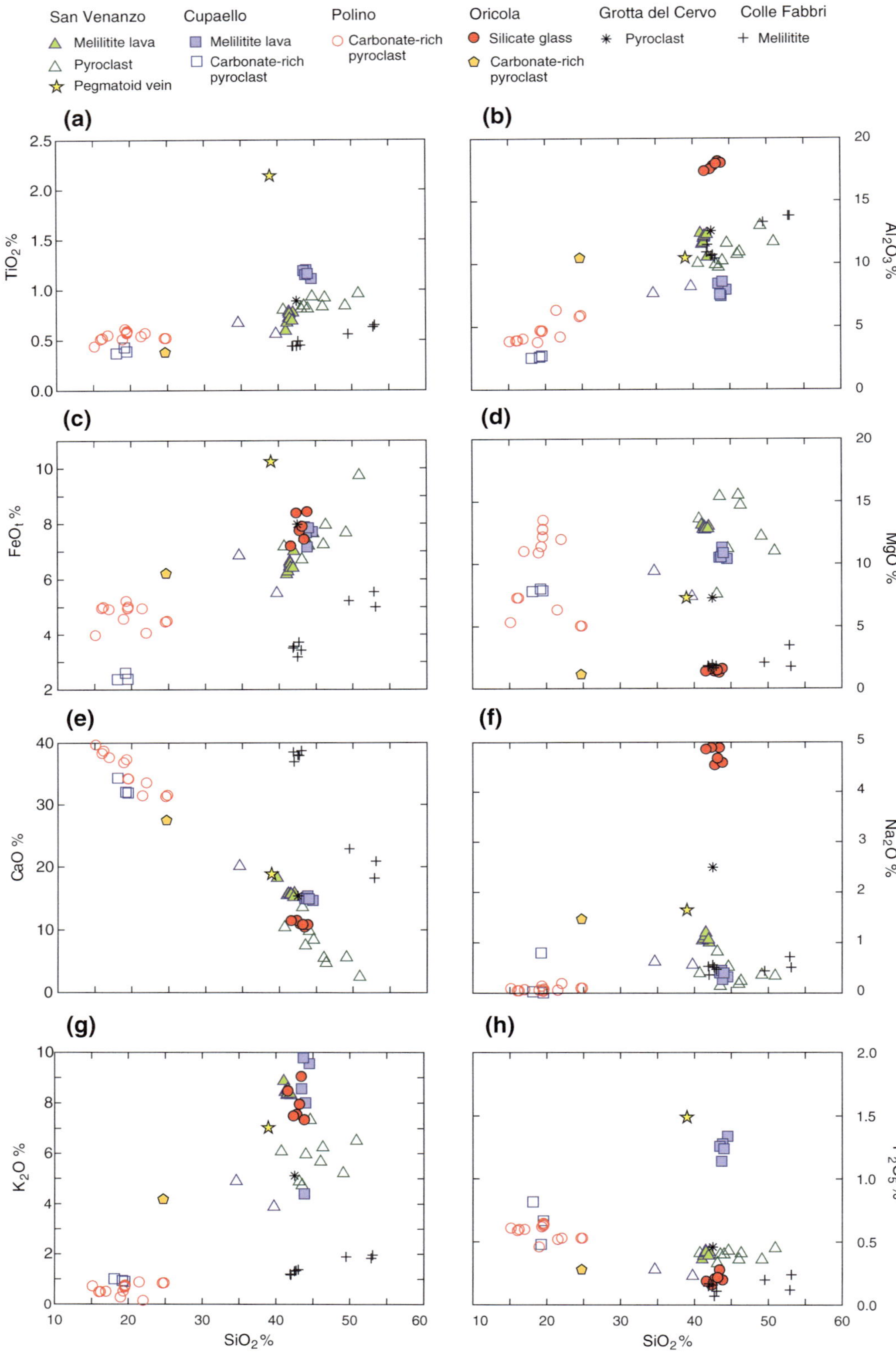

Fig. 3.2 Major element variations of the Intra-Apennine volcanic rocks

Table 3.2 Major, trace element and initial radiogenic isotopic compositions for representative samples from the Intra-Apennine magmatic province

Volcano	San Venanzo	San Venanzo	San Venanzo	Cupaello	Cupaello	Polino	Grotta del Cervo	Colle Fabbri	Oricola	Oricola
Rock type	Melilitite lava	Pegmatoid vein	Calcite-bearing pyroclast	Lava	Calcite-rich pyroclast	Calcite-rich pyroclast	Pyroclastic bomb	Average melilitolite	Carbonatite	Silicate glass
Data source	2,5,9	1	3	2,5,10	3,4	3,4	6	9	7	8
SiO_2 wt%	41.88	38.96	39.70	44.49	19.20	16.00	42.50	42.42	24.7	43.15
TiO_2	0.69	2.14	0.56	1.11	0.43	0.51	0.89	0.45	0.38	–
Al_2O_3	12.28	10.44	8.16	7.92	2.60	3.90	12.60	10.82	10.45	18
FeO_{total}	2.21	9.89	5.12	7.13	2.33	4.60	7.36	1.84	5.72	7.91
MnO	0.11	0.18	0.10	0.11	0.06	0.07	0.13	1.64	1.19	0.3
MgO	12.78	7.30	7.39	10.38	8.03	7.31	7.28	0.07	1.12	1.46
CaO	15.21	18.85	18.20	14.68	32.02	38.30	15.40	1.78	27.5	10.89
Na_2O	1.06	1.65	0.56	0.32	0.80	0.05	2.49	38.00	1.47	4.68
K_2O	8.36	7.03	3.89	9.55	0.94	0.49	5.11	0.49	4.18	7.96
P_2O_5	0.39	1.49	0.23	1.34	0.48	0.59	0.46	1.28	0.28	0.22
CO_2	–	–	11.30	–	27.20	24.00	–	0.13	–	–
LOI	0.83	1.72	3.95	2.48	5.32	2.93	3.29	0.94	19.6	–
Sc ppm	21	58	–	17	5.3	10	27	11	–	–
V	110	430	107	72	53	47	260	95	–	–
Cr	880	75	528	65	33	490	61	94	16	–
Co	42	38	49	35	22	34	53	10	–	–
Ni	141	40	137	87	30	342	52	47	27	–
Rb	432	277	320	509	139	51	453	61	186	50.7
Sr	1706	2330	939	3758	1160	1897	1200	588	2395	1151
Y	27	95	22	44	15	22	25	30	20	7.5
Zr	319	952	237	848	286	337	267	85	104	84
Nb	15	39	12	47	20	16	16	11	25	15.3

(continued)

Table 3.2 (continued)

Volcano	San Venanzo	San Venanzo	San Venanzo	Cupaello	Cupaello	Polino	Grotta del Cervo	Colle Fabbri	Oricola	Oricola
Rock type	Melilitite lava	Pegmatoid vein	Calcite-bearing pyroclast	Lava	Calcite-rich pyroclast	Calcite-rich pyroclast	Pyroclastic bomb	Average melilitolite	Carbonatite	Silicate glass
Cs	33	–	–	74	39	5	39	4	–	–
Ba	501	2044	534	3980	895	2216	649	5201	2945	1147
La	77	236	63	257	95	111	–	33	170	65.2
Ce	176	482	135	526	184	232	–	53	316	110.3
Nd	94	214	70	248	82	113	–	27	88	32
Sm	16.6	52	14.2	39	15.1	22.2	–	5.5	10.4	4.1
Eu	3.0	8.1	2.75	6.8	2.9	4.72	–	1.09	2.7	0.8
Tb	1.4	3.9	1.30	2.5	1.2	1.70	–	0.93	1.1	0.3
Yb	2.4	6.7	2.00	2.9	0.9	1.50	–	2.23	1.9	0.5
Lu	(0.33)	1.0	0.29	(0.36)	0.13	0.22	–	0.37	0.3	0.1
Hf	8.8	25	–	22.9	7.0	8.0	13.5	3	6	1.4
Ta	0.92	2.3	–	3.3	1.0	0.9	–	–	0.9	0.7
Pb	(25)	–	21	(133)	50	35	43	7	129	50.5
Th	30	112	26	124	42	50	25	16.7	73	29
U	9.5	–	6.3	27.9	4.1	8.8	7.0	2.55	76	33.7
$^{87}Sr/^{86}Sr$	0.71041	–	–	0.71128	(0.71007)	(0.71040)	–	0.710625	0.7109	–
$^{143}Nd/^{144}Nd$	0.51208	–	–	0.51212	(0.51188)	(0.51200)	–	0.512193	0.51207	–
$^{206}Pb/^{204}Pb$	18.735	–	–	18.740	–	–	–	–	–	–
$^{207}Pb/^{204}Pb$	15.658	–	–	15.663	–	–	–	–	–	–
$^{208}Pb/^{204}Pb$	38.931	–	–	38.953	–	–	–	–	–	–
$^{176}Hf/^{177}Hf$	0.28249	–	–	–	–	–	–	–	–	–
$\delta^{18}O_{SMOW}$	(+13ol)	–	–	(+11cpx)	–	(+23calc)	–	–	+30	–

Numbers in parentheses refer to data obtained on different, though similar rock samples from the same locality as those analysed for the other elements. Source of data: (1) Peccerillo et al. (1988); (2) Conticelli and Peccerillo (1992); (3) Stoppa and Woolley (1997); (4) Castorina et al. (2000); (5) Conticelli et al. (2002); (6) Stoppa et al. (2002); (7) Stoppa (2007); (8) Stoppa et al. (2005); (9) Stoppa and Sharygin (2009); (10) Author's un published data

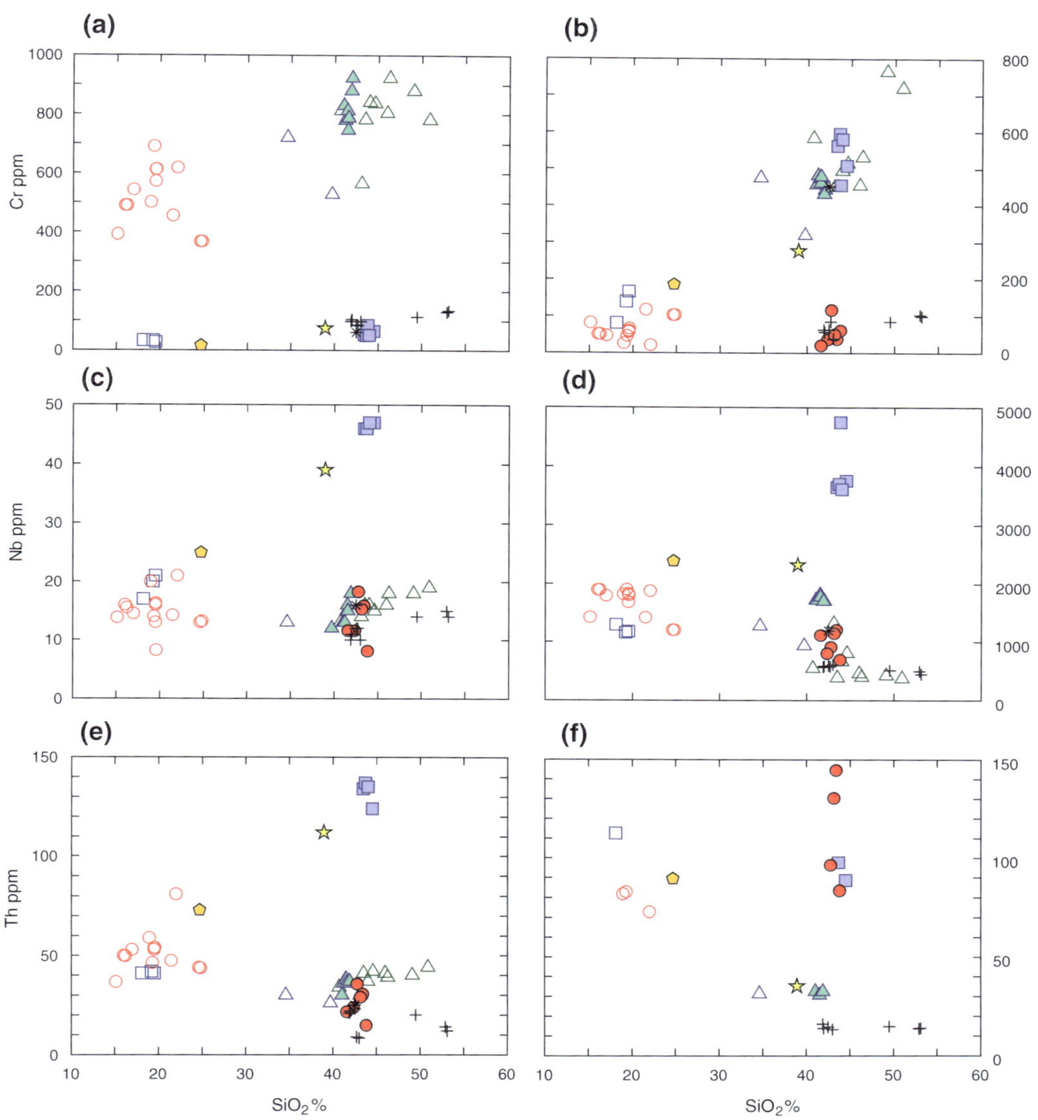

Fig. 3.3 Trace element variations for the Intra-Apennine volcanic rocks. Symbols as in Fig. 3.2

important phase (up to more than 50 % of total rock volume) of the Cupaello pyroclastic rocks. Accessory phases in the Cupaello rocks include magnetite, monticellite, apophyllite, spinel, khibinskite, goetzenite, perovskite and apatite.

Petrology and geochemistry. The Cupaello lava is peralkaline and strongly undersaturated in silica. As for typical kamafugitic rocks, Al_2O_3 and Na_2O are low and CaO is very high (Fig. 3.2). K_2O is variable ($\sim$3–10 wt%), but the freshest samples have high concentration of K_2O, indicating secondary loss of potassium in the altered rocks (Peccerillo et al. 1988). Ferromagnesian trace elements (Cr, Ni, Co, V and Sc) have much lower abundances than in the San Venanzo lavas, whereas LILE and HFSE are higher (Fig. 3.3). Incompatible element and REE patterns are similar to those of San Venanzo, but

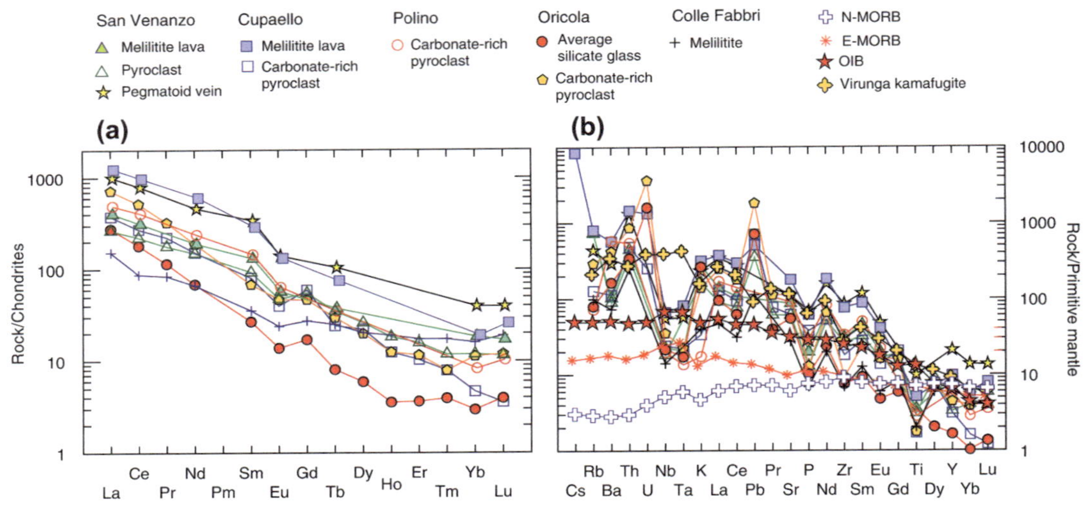

Fig. 3.4 Chondrite normalised REE and mantle normalised incompatible element patterns of Intra-Apennine volcanic rocks. Patterns of average mafic rocks from Virunga (East Africa; Chakrabarti et al. 2009), Normal and Enriched Mid Ocean Ridge Basalts (N-MORB, E-MORB; Gale et al. 2013), and Oceanic Island Basalts (OIB; Sun and McDonough 1989) are also reported

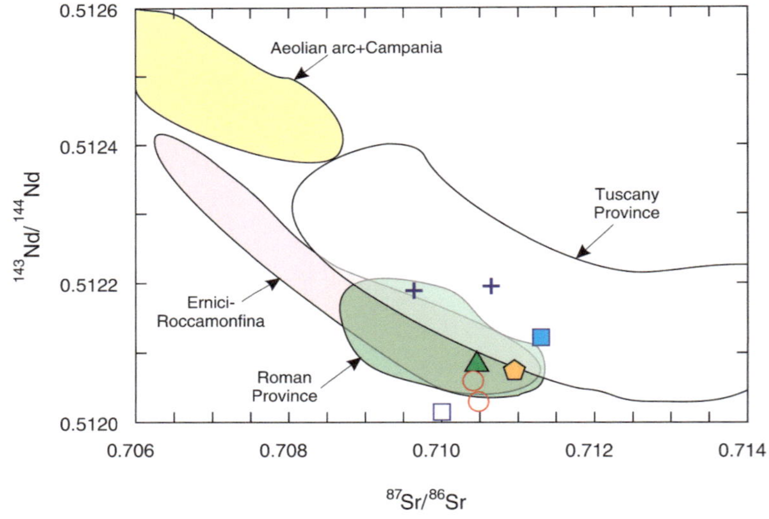

Fig. 3.5 Sr–Nd isotope compositions of Intra-Apennine rocks. Symbols as in Fig. 3.4. Fields of nearby volcanic provinces (restricted to mafic rocks) are reported for comparison

fractionation and element enrichment are stronger at Cupaello. The carbonate-rich pyroclastic rocks have REE and incompatible element patterns that are similar to the associated kamafugitic lavas, but with much lower element concentrations (Fig. 3.4). Radiogenic isotopic compositions of Cupaello lavas are slightly but significantly different from those of San Venanzo ($^{87}Sr/^{86}Sr \sim 0.7113$; $^{143}Nd/^{144}Nd \sim 0.51212$; $^{206}Pb/^{204}Pb \sim 18.74$; $^{207}Pb/^{204}Pb \sim 15.66$; $^{208}Pb/^{204}Pb \sim 38.95$; $^{187}Os/^{188}Os = 0.1926$). A sample of carbonate-rich pyroclast shows much lower Nd isotope ratio ($^{143}Nd/^{144}Nd \sim 0.51188$; Stoppa 2007). O-isotope ratios are very high, and values of $\delta^{18}O_{SMOW} = +11.9$ to +14.4 ‰ have been determined for whole rock and separated diopside phenocrysts (Holm and Munksgaard 1982, 1986; Taylor et al. 1984). The carbonate-rich pyroclastic rocks have lower Sr and Nd isotopic ratios than the associated lava (Fig. 3.5). Calcite from these rocks has $\delta^{18}O_{SMOW}$ ~ +21 to +23 ‰. Some crystals are enriched in Sr, Ba and REE (e.g. Stoppa and Woolley 1997).

3.3.3 Polino

The Polino centre consists of layered lapilli tuffs, and a 40 m across diatreme cutting Liassic limestones. The diatreme is filled with tuffisite formed of massive phlogopite- and carbonate-rich igneous bocks, concentric carbonate-rich lapilli and bombs, country rock fragments and abundant fine-grained matrix. Some lapilli consisting of olivine and phlogopite may represent mantle material (Stoppa and Lupini 1993; Rosatelli et al. 2010). A $^{40}Ar/^{39}Ar$ age of 246 ka has been measured for phlogopites separated from the massive blocks (Laurenzi et al. 1994).

Petrography and mineralogy. Massive carbonate-rich blocks are classified as carbonatites, based on modal mineralogy. Calcite is the main phase, followed by phlogopite, olivine, perovskite, Zr-schorlomite, and apatite set in a micro-cryptocrystalline groundmass (Stoppa and Lupini 1993). Olivine (Fo_{94-93}) occurs as anhedral to subhedral crystals, sometimes rounded, kinked and surrounded by a reaction rim of monticellite. Crystals contain minute inclusions of Cr-spinel. The phlogopite laths are generally kinked, have high Mg# (92–94), low Ti and Fe, and high Ni and Cr. Calcite occurs as groundmass mineral of lapilli and bombs, filling of veins and vugs, and as matrix mineral of tuffisite. The different types of calcite show variable enrichments in Sr, Ba and REE, and stable isotope signatures (Rosatelli et al. 2010). Fragments of dunite and glimmerite nodules occurring in the tuffisite are generally rounded, fine- to medium-grained, and exhibit granoblastic texture; phlogopite and olivine have similar texture and compositions to that of the isolated xenocrysts.

Petrology and geochemistry. The Polino carbonatitic blocks are depleted in silica (SiO_2 ~ 15–25 wt%) and rich in CaO (~30–40 wt%) and, to a lower extent, in MgO (~7–13 wt%). By contrast, K_2O (~0.2–0.7 wt%) and Na_2O (<0.2 wt%) are low. Overall, the Polino rocks very closely resemble the carbonate-rich pyroclastics from Cupaello. However, concentrations of the ferromagnesian trace elements Ni and Cr are much higher at Polino. Radiogenic isotope ratios are similar to those of San Venanzo, with $^{87}Sr/^{86}Sr$ ~ 0.70104 and $^{143}Nd/^{144}Nd$ ~ 0.51203 (Castorina et al. 2000; Conticelli et al. 2002). Oxygen isotope ratios of calcite shows $\delta^{18}O_{SMOW}$ ~ +22 to +24 ‰ and $\delta^{13}C_{PDB}$ ~ −8 to −13 ‰ (Rosatelli et al. 2010). Stable isotope signatures, incompatible element abundances, and textural characteristics of calcite suggest the occurrence of both primary magmatic and secondary carbonates in the Polino rocks (Rosatelli et al. 2010).

3.3.4 Oricola

Monogenetic pyroclastic centres (K/Ar age = 530 ka) composed of altered breccia, surge and fall deposits occur at Oricola-Carsoli, about 40 km southwest of L'Aquila (Bosi et al. 1991; Barbieri et al. 2002; Stoppa et al. 2005; Stoppa 2007). The pyroclastic rocks at Oricola make up a tuff ring composed of interlayered breccia and surge beds covered by pyroclastic flow, fall and surge products.

Petrography and mineralogy. The juvenile lapilli from Oricola range from carbonatite to phonolitic foidite with a kamafugitic affinity. Carbonatitic flattened and welded lapilli prevail at the base, whereas silicate clasts dominate the top of the sequence (Stoppa et al. 2005). Mineral phases include Ca-rich olivine (Fo_{87}), diopside, leucite, phlogopite, alkali-feldspar, and accessory garnet, apatite and spinel.

Petrology and geochemistry. Analyses of Oricola phases (Stoppa et al. 2005) highlight strong LREE-enrichment in calcites, with total

LREE up to 1150 ppm, and average Ba and Sr abundances of about 3800 and 4400 ppm, respectively. Silicate glasses from the top of the sequence show strongly fractionated REE patterns with negative Eu anomaly (Fig. 3.4a); mantle normalised incompatible element patterns are fractionated and contain negative anomalies of HFSE and positive spikes of U and Pb. Similar REE and incompatible element patterns are observed in the bulk carbonatite clasts, although these have higher absolute element abundances (Fig, 3.4). Sr–Nd isotopic ratios reported by Stoppa (2007) are similar to Umbria kamafugites ($^{87}Sr/^{86}Sr = 0.7109$; $^{143}Nd/^{144}Nd = 0.51207$).

3.3.5 Intra-Apennine Pyroclastic Deposits

There is a large number of pyroclastic deposits scattered through the internal zones of the Apennines. Composition, grain size and depositional characteristics are variable and some of the fine-grained clasts may have a remote origin. The best-known locally originated pyroclastic deposits occur in Umbria at Perugia, Acquasparta, Norcia and Pietrafitta, and in the Latium and Abruzzi regions at Campo Imperatore, Grotta del Cervo, Oricola-Carsoli, Poggio Bustone and several other localities (Bosi et al. 1991; Bosi and Locardi 1992; Lavecchia et al. 2002). Stratigraphic relationships and radiometric dating indicate an age of a few hundreds thousand years for all these outcrops.

A sequence of pumiceous lapilli makes up a small outcrop completely covered by a thick soil at Pian di Massiano, Perugia. At Acquasparta, phreatomagmatic deposits of ashes and lapilli erupted over a distance of about 10 km from N-S aligned monogenetic centres. $^{40}Ar/^{39}Ar$ dating on separated phases has yielded an age of 390 ka for these deposits. Mineral phases include diopside, olivine (up to Fo_{93}), sanidine, Ti-rich phlogopite, leucite, apatite, sphene and Fe–Ti oxides. Overall, this mineral assemblage is believed to reveal a kamafugitic affinity (Brozzetti and Stoppa 1995). However, analyses of some scoriae and glass inclusions in clinopyroxenes indicate compositions ranging from foidite to tephriphonolite, with similar major element contents as the magmas from the Roman Province (Author's unpublished data).

In the Abruzzi region, the pyroclastic rocks display variable depositional and structural characteristics. Some occurrences contain coarse-grained material (up to 10 cm in diameter), which indicates nearby vents. Pyroclastic rocks containing coarse-grained juvenile material are present at Grotta del Cervo, a locality not far from L'Aquila. Mineral phases from this outcrop are calcite, olivine (Fo_{88}), diopside, phlogopite, garnet, haüyne, leucite, kalsilite, apatite and Ti-magnetite. Major and trace element composition of juvenile lapilli is similar to San Venanzo kamafugitic lavas (Stoppa et al. 2002).

3.3.6 Colle Fabbri

A small outcrop of melilite-bearing rocks with igneous-type textures is present at Colle Fabbri. The outcrop is located at the south-western edge of the Umbria Valley graben, and rests over Lower Pleistocene conglomerates and clay beds (Stoppa 1988). The centre is formed of a breccia apron intruded by dikes and by a final massive plug. Evidence of thermal metamorphism has been observed in the country rocks (Stoppa and Rosatelli 2009). A whole rock K/Ar age determination yielded a value of 790 $\pm$ 90 ka (Laurenzi et al. 1994).

Petrography and mineralogy. The dike rocks have medium grained, equigranular texture, and consist of euhedral melilite crystals with wollastonite inclusions. Leucite and kalsilite occur as intergranular minerals, along with Ti-garnet, Ti-clinopyroxene, magnetite, perovskite, rankinite, apatite, Fe–Ni sulphides, carbonates, zeolites and scattered anorthite crystals (Stoppa and Rosatelli 2009). The plug has textures ranging from microgranular to porphyritic and vesicular, passing from the centre to the border. Modal mineralogy consists of various proportions of melilite, wollastonite, clinopyroxene, leucite, plus accessory perovskite, garnet, magnetite, rankinite ($Ca_3Si_2O_7$), Si-bearing apatite, Fe–Ni sulfides, calcite and zeolites (Stoppa and Sharygin 2009). Ocelli filled with calcite are observed in the border rocks. The Colle Fabbri rocks are classified as melilitolites (Stoppa and Rosatelli 2009).

Petrology and geochemistry. Major element compositions (Fig. 3.2) are rather variable for a small outcrop (e.g. $SiO_2 \sim 43$–60 wt%; CaO ~ 8–40 wt%; $Na_2O \sim 0.1$–1.1 wt%; $K_2O \sim 0.2$–2.5 wt%; Melluso et al. 2003). However, new analyses by Stoppa and Sharygin (2009) greatly restrict these ranges and only data from these authors are reported in the variation diagrams. Glasses included in some minerals or occurring in the border facies are acidic, potassic (K_2O = 5.5–8.5 wt%) and depleted in Na_2O (~0.5–1.0 wt%). Chondrite-normalised REE profiles are different from other IAP rocks (Fig. 3.4a). $^{87}Sr/^{86}Sr$ ratios range from 0.7077 to 0.7119 and $^{143}Nd/^{144}Nd$ ratios vary from 0.51218 to 0.51223 (Melluso et al. 2003; Stoppa and Sharygin 2009).

Petrogenesis. Colle Fabbri is thought to be a member of the carbonatite-melilitite mantle-derived association of the Intra-Apennine Magmatic Province (Stoppa and Woolley 1997; Stoppa and Sharygin 2009). However, REE and incompatible element contents and fractionation are much lower than other melilititic rocks from IAP (Figs. 3.3 and 3.4). Melluso et al. (2003) pointed out that both the mineral paragenesis and the major and trace element geochemistry of Colle Fabbri rocks are unusual and thermodynamically incompatible with any magmatic system. It was observed that abundances for several major and trace elements closely resemble those of marly sediments. Therefore, it was concluded that the Colle Fabbri rocks are not magmatic, but were generated by decarbonation and melting of marls (*paralava*). Burning of lignite, abundantly present in the recent sediments in the Colle Fabbri area, was suggested to be responsible for marl melting. This same conclusion was also applied to another igneous-like rock cropping out at Ricetto, a few hundred km south of Colle Fabbri (Melluso et al. 2003). Such a hypothesis was challenged by Capitanio (2005) and Stoppa et al. (2005) who suggested an igneous origin for this outcrop. Recent geochemical and mineral chemistry data by Stoppa and Sharygin (2009) were interpreted to support the hypothesis that the Colle Fabbri rocks represent a melilitite magma similar to other IAP occurrences, which digested pelite country-rock at temperatures exceeding 1230 °C. This, however, does not explain some compositional features of Colle Fabbri rocks, such as the flat HREE patterns.

3.4 Petrogenesis of the Intra-Apennine Province

The main questions concerning the petrogenesis of the IAP relate to the origin and evolution of the kamafugitic magmas, best represented by the lavas from San Venanzo and Cupaello, and to the nature and significance of carbonate-rich pyroclastic rocks that occur at several centres.

3.4.1 Kamafugitic Rocks

The high MgO, Ni and Cr contents of IAP kamafugites attest to a mantle origin (e.g. Peccerillo et al. 1988; Cundari and Ferguson 1991). However, several lines of evidence also attest wall rock assimilation. The San Venanzo and Cupaello volcanoes are monogenetic centres developed on a thick sedimentary sequence composed mostly of limestones and marls. The scarcely porphyritic to aphyric texture of kamafugitic lavas indicates a near liquidus temperature during eruption (i.e. about 1250 °C; Cundari and Ferguson 1991), which makes it likely the incorporation and dissolution of carbonate wall rocks. The high CaO abundances observed in these rocks could partially depend on carbonate assimilation (see Carter and Dasgupta 2015). An additional piece of evidence for this process comes from high O-isotope ratios of whole rocks, separated silicate minerals, and of calcite occurring in IAP rocks (Turi et al. 1986; Stoppa and Woolley 1997; Stoppa and Sharygin 2009;

Rosatelli et al. 2010). However, Holm and Munskgaard (1982) suggested that high O-isotope values denote an anomalous mantle source strongly modified by the addition of subducted upper crustal material.

Whatever the case, an important point to emphasize is that limestone assimilation does not produce dramatic modification of incompatible element ratios and radiogenic isotope signatures in the kamafugitic magmas. Such behaviour is a consequence of the much lower abundances of incompatible element in the limestone than in ultrapotassic magmas, a compositional contrast that effectively buffers incompatible element and radiogenic isotope variations during assimilation. It has been concluded, therefore, that, whereas high O-isotope composition (and possibly CaO concentration) of kamafugites might be effects of magma assimilation processes, radiogenic isotope and trace element ratios reflect compositions of mantle sources (Peccerillo 1998).

Experimental investigations of silica undersaturated kamafugitic rocks both from Umbria and other worldwide occurrences suggest a magma origin by melting of phlogopite-rich wherlite or clinopyroxenite at pressures around 2 GPa (Wendland and Eggler 1980a, b; Edgar 1987; Cundari and Ferguson 1991; Gupta and Fyfe 2003; Gupta 2015). These compositions are anomalous for upper mantle and require metasomatic modification to generate abundant clinopyroxene and phlogopite. Trace element and radiogenic isotope compositions of IAP kamafugites are halfway between typical mantle and upper crustal values. This brought to conclusion that the mantle modification was accomplished by addition of upper crustal material. However, radiogenic isotope ratios are less close to upper crustal values than observed in the mafic rocks of the Tuscany Province, where a pelitic nature for mantle contaminant has been proposed (see Chap. 2). Therefore, a mantle contamination by subducted marly sediments was suggested by Peccerillo et al. (1988), a hypothesis supported by recent geochemical investigation (e.g. Avanzinelli et al. 2009). Grassi and Schmidt (2011a,b) and Grassi et al. (2012) demonstrated that melting of subducted carbonated pelites at high pressure (8–22 GPa) generates K-rich carbonate melts with high K/Na ratios and exhibiting many trace element signatures akin to Central Italy ultrapotassic rocks. The interaction between these carbonate K-rich melts and peridotite is able to produce ultramafic rocks rich in phlogopite and clinopyroxene, whose melting gives ultrapotassic magmas. Such a process might also explain the high LILE/HFSE ratios of the IAP kamafugites (e.g. Th/Ta $\sim$ 35–50), which are much higher than any common igneous or sedimentary rock (e.g. Taylor and McLennan 1985; Plank and Langmuir 1998). Ti-perovskite has been found as a residual phase during melting of carbonated pelites at 22 GPa. This phase has very high solid/carbonate-liquid partition coefficients ($K_{s/l}$) for Nb and other HFSE (e.g. $K_{Nb/carbonate\ melt} = 28$), whereas partition coefficients for Th and LREE are low (e.g. $K_{Th/carbonate\ melt} = 0.2$). Therefore, the occurrence of residual perovskite during sediment melting yields extremely LILE/HFSE fractionated liquids, a feature that is transferred to mantle wedge and hence to the magmas.

The pre-metasomatic mantle rocks of IAP kamafugites probably had a residual harzburgitic composition as suggested by the low Al_2O_3 and Na_2O of lavas. These rocks were modified by carbonate-rich supercritical fluids or melts from marls, which increased modal clinopyroxene of harzburgite forming lherzolites or clinopyroxenites, thereby creating the conditions for the generation of CaO-rich lavas (Chin et al. 2014).

An alternative hypothesis for the origin of kamafugitic magmas emphasises the strongly undersaturated and ultrapotassic composition of these magmas and their connection with carbonate-rich pyroclastics, which are believed to represent mantle-derived carbonatites (Stoppa and Woolley 1997; Bell et al. 2013 with references). It is stated that a similar rock association is found in some intraplate settings such as East Africa, concluding that the IAP magmatism is related to emplacement of deep plumes in an intraplate setting (e.g. Cundari and Ferguson 1991; Castorina et al. 2000; Bell et al. 2004,

2013). According to this hypothesis, the crustal-like Sr-Nd isotope signatures of the kamafugitic magmas would be related to an ancient metasomatic modification of mantle rocks, which remained isolated for a long period of time (1-2 Ga) in order to acquire high Sr and low Nd isotopic ratios. Such a hypothesis does not consider the utterly different trace element signatures of Africa and Central Italy rocks (Fig. 3.4b), which call for a distinct origin and geodynamic significance. Further discussion of this issue is reported in Chap. 13.

3.4.2 Carbonate-rich Rocks

The volcanic rocks from IAP commonly contain calcite, which occurs in the lavas but is particularly abundant (sometimes exceeding 50 % by volume) in the pyroclastic rocks. The latter consist of mixtures of carbonate and silicate components, which show textural evidence of being solidified from coexisting silicate and carbonate melts (Stoppa and Woolley 1997; Stoppa et al. 2005). There is a growing consensus that carbonate component in the IAP rocks is magmatic in origin, and derived from unmixing of mantle-originated carbonate-rich kamafugitic magmas (Stoppa and Cundari 1995; Stoppa and Woolley 1997; Bailey and Collier 2000; Stoppa et al. 2005; Martin et al. 2012).

The nature of the IAP carbonate-rich rocks has been questioned, at least for the San Venanzo and Cupaello occurrences, which are the only centres where there is an association of carbonate-bearing pyroclastics and silicate kamafugite lavas. The main question has been the one of understanding whether the abundant calcite occurring in pyroclastic rocks is entirely magmatic in origin and crystallised from unmixed mantle-derived melts, or, instead, it comes from wall rocks by some type of syn- to post-magmatic process. Such a discussion has been the subject of several papers and will not be repeated here (e.g. Peccerillo 1998, 2005b, 2006; Bailey 2005; Woolley et al. 2005; Bell and Kjarsgaard 2006). In short, evidence in favour of an origin of calcite from sedimentary wall rocks include the very thick (several thousand metres) carbonate substratum, the high oxygen isotope ratios of carbonates for all the IAP volcanics, and the barren nature of the carbonate component at San Venanzo and Cupaello that contrasts with the high enrichments typical of carbonatitic melts (e.g. Peccerillo 1998). On the other hand, textural relationships between carbonate and silicate fractions, high incompatible element contents in some calcite crystals, and occurrence of flattened lapilli and bombs with carbonate compositions at some volcanoes support that carbonates crystallised from melts (e.g. Stoppa and Woolley 1997; Stoppa et al. 2005; Rosatelli et al. 2010).

In conclusion, evidence for the origin of carbonates in the IAP rocks is somewhat contradictory, and perhaps different types of calcite are present in the rocks (e.g. Stoppa et al. 2005; Rosatelli et al. 2010). Carbonates melts could be formed within the crust, as suggested by several classical and recent studies (e.g. Wyllie and Tuttle 1960; Peccerillo et al. 2010; Gozzi et al. 2014), and in an upper mantle strongly contaminated by marly sediments (e.g. Grassi et al. 2012; Tumiati et al. 2013).

Whatever the case, the carbonate-rich rocks from the IAP exhibit the same incompatible element pattern and radiogenic isotope signatures of the associated kamafugitic lavas, and both resemble closely the ultrapotassic mafic rocks from the nearby Roman Province. Therefore, geochemical evidence strongly suggests an origin in the same type of enriched mantle sources and the same geodynamic significance for all these rocks.

3.5 Geodynamic Implications

Two conflicting hypotheses have been proposed for the genesis of IAP magmas, as discussed in the previous section. One suggests recent source contamination by marly sediments, whereas an alternative theory invokes old metasomatic modification of deep mantle material. According to the former hypothesis, mantle contamination occurred as a consequence of subduction of the Adriatic plate beneath Central Italy (e.g. Serri et al. 1993; Peccerillo and Frezzotti 2015). Various mechanisms can be envisaged for slab-to-mantle wedge element transfer. One possibility is that marly sediments were brought to high depths along the subduction zone, where they released K-rich carbonate melts (e.g. Peccerillo et al. 1988; Grassi et al. 2012). These melts contaminated the mantle wedge peridotite, which successively melted to give kamafugitic magmas. Another scenario is that the subduction of great quantities of sediments induced gravitational instability at the slab-mantle interface, with the formation of subduction mèlange diapirs consisting of mantle rocks and sediments (e.g. Peccerillo and Frezzotti 2015). Mélange bodies ascended to the shallower and hotter corner of the mantle wedge, releasing supercritical fluids and melts that contaminated the surrounding peridotite. Formation of kamafugitic magmas occurred by melting of mélange diapirs or/and surrounding rocks, during post-collisional stages of Northern Apennine evolution, because of temperature increase subsequent to slab breakoff. Experimental evidence demonstrates that carbonatite magmas can be generated by melting of sediment-mantle mélange rocks (Tumiati et al. 2013; Bulatov et al. 2014), thereby supporting the possibility of a direct mantle provenance for carbonate-rich magmas in the IAP.

Low Na_2O and Al_2O_3 contents of kamafugitic magmas suggest a harzburgitic pre-metasomatic source that had been depleted in its basaltic components prior to metasomatism. As suggested for the Tuscany region, fragments of harzburgite could derive from stretching and boudinage of old Alpine lithospheric mantle, during the opening of the northern Tyrrhenian Sea basin (e.g. Doglioni et al. 1998).

Alternative hypotheses suggest that the IAP rocks would represent a magmatism emplaced in an intra-continental rifting zone by melting of a deep rooted mantle plume head (e.g. Lavecchia and Stoppa 1990, 1996; Stoppa et al. 2002; Bell et al. 2013). Such a hypothesis, however, strongly contrasts with the clear arc geochemical signatures of the IAP and other magmas in Central Italy, as recalled earlier.

3.6 Summary and Conclusions

The volcanic centres of the IAP are made up of pyroclastic rocks and a few lava flows and dikes. Lavas have olivine melilitite and kalsilite melilitite composition and show an ultrapotassic kamafugitic affinity. Pyroclastic rocks range from melilitite and phonolite to carbonate-rich compositions.

Kamafugitic lavas have radiogenic isotope and incompatible trace element signatures similar to the mafic rocks of the nearby Roman Province, suggesting a close genetic relationship and geodynamic significance. These compositions are best explained by an origin of magmas in anomalous upper mantle sources that were enriched in incompatible elements by subducted marly sediments. A harzburgite composition is assumed as pre-metasomatic mantle rocks to explain some petrological characteristics of IAP kamafugites, such as low contents of Na_2O and Al_2O_3. Harzburgites were modified and transformed to clinopyroxenites by Ca-rich components coming from subducted marls. Interaction between kamafugitic magmas and crustal rocks (mostly sedimentary carbonates) probably occurred during magma ascent. This process generated an increase of $\delta^{18}O$ (up to + 14 ‰) and, possibly, CaO content of magmas, but had little effects on incompatible trace element ratios and radiogenic isotope signatures.

The carbonate-rich rocks occurring in the IAP are mixtures of carbonate and silicate components. The very nature (i.e. magmatic vs. secondary, and mantle-derived vs. crust-derived) of these carbonates has been much debated and, perhaps, different types of calcite occur in the rock. In any case, carbonate-rich rocks and kamafugitic lavas have similar trace element and radiogenic isotope signatures, suggesting an origin from a common source in the same geodynamic setting.

References

Avanzinelli R, Lustrino M, Mattei M, Melluso L, Conticelli S (2009) Potassic and ultrapotassic magmatism in the circum-Tyrrhenian region: the role of carbonated pelitic vs. pelitic sediment recycling at destructive plate margin. Lithos 113:213–227

Bailey DK (2005) Carbonate volcanics in Italy: numerical tests for the hypothesis of lava-sedimentary limestone mixing. Per Min 74:205–208

Bailey DK, Collier JD (2000) Carbonatite-melilitite association in the Italian collision zone and the Uganda rifted craton: significant common factors. Min Mag 64:675–682

Barbieri M, Barbieri M, D'Orefice M, Graciotti R, Stoppa F (2002) Il vulcanismo monogenico medio-pleistocenico della Conca di Carsoli (L'Aquila). Geol Rom 36:13–31

Barchi M (2010) The neogene-quaternary evolution of the Northern Apennines: crustal structure, style of deformation and seismicity. In: Beltrando M, Peccerillo A, Mattei M, Conticelli S, Doglioni C (eds) J virtual explorer 36, paper 11 doi:10.3809/jvirtex.2010.00220

Barchi M, Landuzzi A, Minelli G, Pialli G (2001) Outhern Northern Apennines. In: Vai GB, Martini PI (eds) Anatomy of an orogen. The Apennines and the adjacent Mediterranean basins. Kluwer, Dordrecht, pp 215–254

Bell K, Kjarsgaard B (2006) Discussion of Peccerillo (2004) "Carbonate-rich pyroclastic rocks from central Apennines: carbonatites or carbonated rocks?". Per Min 75:85–92

Bell K, Castorina F, Lavecchia G, Rosatelli G, Stoppa F (2004) Is there a mantle plume below Italy? EOS 85:541–547

Bell K, Lavecchia G, Rosatelli G (2013) Cenozoic Italian magmatism—isotope constraints for possible plume-related activity. J South Am Earth Sci 41:22–40

Bosi C, Locardi E (1992) Vulcanismo meso-pleistocenico nell'Appennino laziale-abruzzese. Studi Geol Camerti Spec Issue CROP 11:319–325

Bosi C, Cittadini A, De Casa G, Casa G, Messina P, Palieri L (1991) Dati preliminari su alcune successioni tufitiche pleistoceniche dell'Appennino abruzzese. Studi Geol Camerti, Spec Issue CROP 11:313–318

Brozzetti F, Stoppa F (1995) Le piroclastiti medio-pleistoceniche di Massa Martana-Acquasparta (Umbria): caratteri strutturali e vulcanologici. Il Quaternario 8:95–110

Bulatov VK, Brey GP, Girnis AV, Gerdes A, Höfer HE (2014) Carbonated sediment–peridotite interaction and melting at 7.5–12 GPa. Lithos 200–201:368–385

Capitanio F (2005) The Ricetto and Colle Fabbri wollastonite and melilitite-bearing rocks of the Central Apennine, Italy. Comment on the paper of Melluso et al. (2003). Am Min 90:1934–1939

Carter LB, Dasgupta R (2015) Hydrous basalt–limestone interaction at crustal conditions: Implications for generation of ultracalcic melts and outflux of CO_2 at volcanic arcs. Earth Planet Sci Lett 427:202–214

Castorina F, Stoppa F, Cundari A, Barbieri M (2000) An enriched mantle source for Italy's melilitite-carbonatite association as inferred by its Nd–Sr isotope signature. Min Mag 64:625–639

Cellai D, Conticelli S, Menchetti S (1994) Crystal-chemistry of clinopyroxenes from potassic and ultrapotassic rocks in central Italy: implications for their genesis. Contrib Min Pet 116:301–315

Chakrabarti R, Basu AR, Santo AP, Tedesco D, Vaselli O (2009) Isotopic and geochemical evidence for a heterogeneous mantle plume origin of the Virunga volcanics, Western rift, East African Rift system. Chem Geol 259:273–289

Chin EJ, Lee C-TA, Barnes JD (2014) Thickening, refertilization, and the deep lithosphere filter in continental arcs: constraints from major and trace elements and oxygen isotopes. Earth Planet Sci Lett 397:184–200

Chiodini G, Valenza M, Cardellini C, Frigeri A (2008) A new web-based catalog of Earth degassing sites in Italy. Eos 89:341–342

Ciarapica G, Passeri L (1998) Evoluzione paleogeografica degli Appennini. Atti Tic Sci Terra 40:233–290

Conticelli S, Peccerillo A (1992) Petrology and geochemistry of potassic and ultrapotassic volcanism in Central Italy: petrogenesis and inferences on the evolution of the mantle sources. Lithos 28:221–240

Conticelli S, D'Antonio M, Pinarelli L, Civetta L (2002) Source contamination and mantle heterogeneity in the genesis of Italian potassic and ultrapotassic volcanic rocks: Sr–Nd–Pb isotope data from Roman Province and Southern Tuscany. Min Pet 74:189–222

Conticelli S, Carlson RW, Widom E, Serri G (2007) Chemical and isotopic composition (Os, Pb, Nd and Sr) of Neogene to Quaternary calc-alkalic, shoshonitic, and ultrapotassic mafic rocks from the Italian peninsula: inferences on the nature of their sources. In: Beccaluva L, Bianchini G, Wilson M (eds) Cenozoic volcanismin the mediterranean area. Geol Soc Am Spec Paper 418:171–202 doi:10.1130/2007.2418(09)

Conticelli S, Avanzinelli R, Ammannati E, Casalini M (2015) The role of carbon from recycled sediments in the origin of ultrapotassic igneous rocks in the Central Mediterranean. Lithos 232:174–196

Cundari A, Ferguson AK (1991) Petrogenetic relationships between melilitite and lamproite in the Roman Comagmatic region: the lavas of San Venanzo and Cupaello. Contrib Min Pet 107:343–357

Decandia FA, Lazzarotto A, Liotta D, Cernobori L, Nicolich R (1998) The CROP 03 traverse: insights on post-collisional evolution of Northern Apennines. Mem Soc Geol It 52:427–439

Della Vedova B, Bellani S, Pelli G, Squarci P (2001) Deep temperatures and surface heath distribution. In: Vai GB, Martini PI (eds) Anatomy of an orogen. The Apennines and the adjacent mediterranean basins. Kluwer, Dordrecht, pp 65–76

Di Stefano R, Kissling E, Chiarabba C, Amato A, Giardini D (2009) Shallow subduction beneath Italy: three-dimensional images of the adriatic-European-Tyrrhenian lithosphere system based on high-quality P wave arrival times. J Geophys Res 114:B05305. doi:10.1029/2008JB005641

Di Stefano R, Bianchi I, Ciaccio MG, Carrara C, Kissling E (2011) Three-dimensional moho topography in Italy: new constraints from receiver functions and controlled source seismology. Geochem Geophys Geosyst 12:Q09006. doi:10.1029/2011GC003649

Doglioni C, Mongelli F, Pialli G (1998) Boudinage of the Alpine belt in the Apenninic back-arc. Mem Soc Geol It 52:457–468

Edgar AD (1987) The genesis of alkaline magmas with emphases on their source region: inferences from experimental studies. In: Fitton JG, Upton BGJ (eds) Alkaline igneous rocks. Geol Soc Spec Publ 30:205–220

Gale A, Dalton CA, Langmuir CH, Su Y, Schilling J-G (2013) The mean composition of ocean ridge basalts. Gechem Geophys Geosyst 14 doi:10.1029/2012GC004334

Gallo F, Giammetti F, Venturelli G, Vernia L (1984) The kamafugitic rocks from San Venanzo and Cupaello, central Italy. Neues Jb Miner Monat 1984:198–210

Gozzi F, Gaeta M, Freda C, Mollo S, Di Rocco T, Marra F, Dallai L, Pack A (2014) Primary magmatic calcite reveals origin from crustal carbonate. Lithos 190–191:191–203

Grassi D, Schmidt MW (2011a) Melting of carbonated pelites at 8–13 GPa: generating K-rich carbonatites for mantle metasomatism. Contrib Min Pet 162:169–191

Grassi D, Schmidt MW (2011b) The melting of carbonated pelites from 70 to 700 km depth. J Pet 52:765–789

Grassi D, Schmidt MW, Gunther D (2012) Element partitioning during carbonated pelite melting at 8, 13 and 22 GPa and the sediment signature in the EM mantle components. Earth Planet Sci Lett 327–328:84–96

Gupta AK (2015) Origin of potassium-rich silica-deficient igneous rocks. Springer, New Delhi 536 pp

Gupta AK, Fyfe WS (2003) The young potassic rocks. ANE Books, New Delhi 370 pp

Holm PM, Munksgaard NC (1982) Evidence for mantle metasomatism: an oxygen and strontium isotope study of the Vulsinian district, central Italy. Earth Planet Sci Lett 60:376–388

Holm PM, Munksgaard NC (1986) Reply to: a criticism of the Holm-Munksgaard oxygen and strontium isotope study of the Vulsinian District, Central Italy. Earth Planet Sci Lett 78:454–459

Keller JVA, Minelli G, Pialli G (1994) Anatomy of late orogenic extension: the northern Apennines case. Tectonophysics 238:275–294

Laurenzi MA, Villa I, Stoppa F (1994) Eventi ignei monogenici e depositi piroclastici nel Distretto Ultra-alcalino Umbro-laziale (ULUD): revisione, aggiornamento e comparazione dei dati cronologici. Plinius 12:61–65

Lavecchia G, Stoppa F (1990) The Tyrrhenian zone: a case of lithosphere extension control of intra-continental magmatism. Earth Planet Sci Lett 99:336–350

Lavecchia G, Stoppa F (1996) The tectonic significance of Italian magmatism: an alternative view to the popular interpretation. Terra Nova 8:435–446

Lavecchia G, Creati N, Boncio P (2002) The Intramontane Ultra-alkaline Province (IUP) of Italy: a brief review with considerations on the thickness of the underlying lithosphere. In: Barchi RM, Cirilli S, Minelli G (eds) Geological and geodynamic evolution of the Apennines. Boll Soc Geol It (Spec Vol 1):87–98

Martin LHJ, Schmidt MW, Mattsson HB, Ulmer P, Hametner K, Günther D (2012) Element partitioning between immiscible carbonatite–kamafugite melts with application to the Italian ultrapotassic suite. Chem Geol 320–321:96–112

Melluso L, Conticelli S, D'Antonio M, Mirco NP, Saccani E (2003) Petrology and mineralogy of wollastonite- and melilitite-bearing paralavas from the central Apennines, Italy. Am Min 88:1287–1299

Pauselli C, Barchi MR, Federico C, Magnani MB, Minelli G (2006) The crustal structure of the northern Apennines (central Italy): an insight by the CROP03 seismic line. Am J Sci 306:428–450

Peccerillo A (1998) Relationships between ultrapotassic and carbonate-rich volcanic rocks in central Italy: petrogenetic implications and geodynamic significance. Lithos 43:267–279

Peccerillo A (2005) On the nature of carbonate-rich volcanic rocks in Central Italy. A reply to comments by Woolley et alii. Per Min 74:195–204

Peccerillo A (2006) Carbonatites vs. carbonated rocks in central Italy. A reply to comments by Bell and Kjarsgaard. Per Min 75:93–100

Peccerillo A, Frezzotti ML (2015) Magmatism, mantle evolution and geodynamics at the converging plate margins of Italy. J Geol Soc Lond 172:407–427

Peccerillo A, Poli G, Serri G (1988) Petrogenesis of orenditic and kamafugitic rocks from Central Italy. Can Min 26:45–65

Peccerillo A, Federico M, Barbieri M, Brilli M, Wu TW (2010) Interaction between ultrapotassic magmas and carbonate rocks: evidence from geochemical and isotopic (Sr, Nd, O) compositions of granular lithic clasts from the Alban Hills Volcano, Central Italy. Geochim Cosmochim Acta 74:2999–3022

Piromallo C, Morelli A (2003) P wave tomography of the mantle under the Alpine-mediterranean area. J Geophys Res 108(B2):2065. doi:10.1029/2002JB001757

Plank T, Langmuir CH (1998) The chemical composition of subducting sediment and its consequences for the crust and mantle. Chem Geol 145:325–394

Rogers NW, De Mulder M, Hawkesworth CJ (1992) An enriched mantle source for potassic basanites: evidence from Karisimbi volcano, Virunga volcanic province, Rwanda. Contrib Min Pet 111:543–556

Rosatelli G, Wall F, Stoppa F, Brilli M (2010) Geochemical distinctions between igneous carbonate, calcite cements, and limestone xenoliths (Polino carbonatite,

Italy): spatially resolved LAICPMS analyses. Contrib Min Pet 160:645–661

Sahama TG (1974) Potassium-rich alkaline rocks. In: Sorensen H (ed) The alkaline rocks. Wiley, London, pp 94–109

Scarascia S, Lozej A, Cassinis R (1994) Crustal structure of Ligurian, Tyrrhenian and Ionian seas and adjacent onshore areas interpreted from wide-angle seismic profiles. Boll Geofis Teor Appl 36:1–19

Serri G, Innocenti F, Manetti P (1993) Geochemical and petrological evidence of the subduction of delaminated Adriatic continental lithosphere in the genesis of the Neogene-Quaternary magmatism of central Italy. Tectonophysics 223:117–147

Sharygin VV, Stoppa F, Kolesov BA (1996) Zr–Ti disilicates from the Pian di Celle volcano, Umbria, Italy. Eur J Min 8:1199–1212

Stoppa F (1988) L'euremite di Colle Fabbri (Spoleto): un litotipo ad affinità carbonatitica in Italia. Boll Soc Geol It 107:239–248

Stoppa F (1996) The San Venanzo maar and tuff ring: eruptive behaviour of a carbonatite-melilitite volcano. Bull Volcanol 57:563–577

Stoppa F (2007) CO_2 magmatism in Italy: from deep carbon to carbonatite volcanism. In: Vladykin VN (ed) Alkaline magmatism, its sources and plumes, pp 109–126

Stoppa F, Cundari A (1995) A new Italian carbonatite occurrence at Cupaello (Rieti) and its genetic significance. Contrib Min Pet 122:275–288

Stoppa F, Lavecchia G (1992) Late Pleistocene ultra-alkaline magmatic activity in the Umbria-Latium region: an overview. J Volcanol Geoth Res 52:277–293

Stoppa F, Lupini L (1993) Mineralogy and petrology of the Polino monticellite calciocarbonatite (Central Italy). Min Pet 49:213–231

Stoppa F, Rosatelli G (2009) Ultramafic intrusion triggers hydrothermal explosions at Colle Fabbri (Spoleto, Umbria), Italy. J Volcanol Geoth Res 187:85–92

Stoppa F, Sharygin VV (2009) Melilitolite intrusion and pelite digestion by high temperature kamafugitic magma at Colle Fabbri, Spoleto, Italy. Lithos 112:306–320

Stoppa F, Woolley AR (1997) The Italian carbonatites: field occurrence, petrology and regional significance. Min Pet 59:43–67

Stoppa F, Wooley AR, Cundari A (2002) Extension of the melilitite-carbonatite province in the Apennines of Italy: the kamafugite of Grotta del Cervo. Min Mag 66:555–574

Stoppa F, Rosatelli G, Wall F, Jeffries T (2005) Geochemistry of carbonatite–silicate pairs in nature: a case history from Central Italy. Lithos 85:26–47

Sun SS, McDonough WF (1989) Chemical and isotopic systematics of oceanic basalts: implications for mantle composition and processes. In: Saunders AD, Norry MJ (eds) Magmatism in ocean basins. Geol Soc London Spec Publ 42:313–345

Tappe S, Foley SF, Pearson DG (2003) The kamafugites of Uganda: a mineralogical and geochemical comparison with their Italian and Brasilian counterpart. Per Min, Spec Issue 72:51–77

Tarquini S, Vinci S, Favalli M, Doumaz F, Fornaciai A, Nannipieri L (2012) Release of a 10-m-resolution DEM for the Italian territory: comparison with global-coverage DEMs and anaglyph-mode exploration via the web. Comput Geosci 38:168–170

Taylor SR, McLennan SM (1985) The continental crust: its composition and evolution. An examination of the geochemical record preserved in sedimentary rocks. Blackwell, Oxford 312 pp

Taylor HP, Turi B, Cundari A (1984) $^{18}O/^{16}O$ and chemical relationships in K-rich volcanic rocks from Australia, East Africa, Antarctica and San Venanzo-Cupaello, Italy. Earth Planet Sci Lett 69: 263–276

Tumiati S, Fumagalli P, Tiraboschi C, Poli S (2013) An experimental study on COH-bearing Peridotite up to 3.2 GPa and implications for crust-mantle recycling. J Petrol 54:453–479

Turi B (1969) La composizione isotopica dell'ossigeno e del carbonio dei carbonati presenti nelle vulcaniti di San Venanzo (Umbria). Per Min 38:589–603

Turi B, Taylor HP, Ferrara G (1986) A criticism of the Holm-Munksgaard oxygen and strontium isotope study of the Vulsinian district, central Italy. Earth Planet Sci Lett 78:447–453

Vaselli O, Conticelli S (1990) Boron, cesium and lithium distribution in some alkaline potassic volcanics from central Italy. Min Petrogr Acta 33:189–204

Washington HS (1906) The Roman comagmatic region. Carnegie Inst Wash Publ 57: 199 pp

Wendlandt RF, Eggler DH (1980a) The origins of potassic magmas: 1. Melting relations in the systems $KAlSiO_4$ -Mg_2SiO_2 and $KAlSiO_4$ -MgO -SiO_2 -CO_2 to 30 kilobars. Am J Sci 280:385–420

Wendlandt RF, Eggler DH (1980b) The origins of potassic magmas: 2. Stability of phlogopite in natural spinel lherzolite an in the $KAlSiO_4$-MgO-SiO_2-H_2O-CO_2 at high pressures and high temperatures. Am J Sci 280:421–458

Woolley AR, Bailey DK, Castorina F, Rosatelli G, Stoppa F, Wall F (2005) Reply to: "Carbonate-rich pyroclastic rocks from central Apennines: carbonatites or carbonated rocks? A commentary" by A. Peccerillo. Per Min 74:183–194

Wyllie PJ, Tuttle OF (1960) The system CaO–CO_2–H_2O and the origin of carbonatites. J Pet 1:1–46

Zanon V (2005) Geology and volcanology of San Venanzo volcanic field (Umbria, Central Italy). Geol Mag 142:683–698

The Roman Province

4

Abstract

The Roman Magmatic Province (about 0.8–0.02 Ma) consists of about 900 km^3 of dominant pyroclastic deposits and minor lavas building up the large volcanoes of Vulsini, Vico, Sabatini and Colli Albani (Alban Hills). Rocks range from potassic (K-trachybasalt to trachyte) to Roman-type ultrapotassic (leucite tephrite and leucitite to leucite phonolite) in composition. Felsic magmas were generated by fractional crystallisation of mafic melts. Evolution processes took place in large shallow-level magma chambers whose post-eruption collapse formed huge calderas and volcano-tectonic depressions. Potassic and ultrapotassic mafic magmas in the Roman Province display variable degrees of enrichment in incompatible elements. In contrast, radiogenic isotopic ratios are less variable ($^{87}Sr/^{86}Sr \sim 0.7090$ to 7110, $^{143}Nd/^{144}Nd \sim 0.5121$, $^{206}Pb/^{204}Pb \sim 18.80$, $^{176}Hf/^{177}Hf \sim 0.28258$). Trace element and radiogenic isotope evidence suggests an origin of primary magmas in an anomalous but rather homogenous mantle source that has undergone variable degrees of partial melting at a pressure increasing from potassic to ultrapotassic magmas. Metasomatised phlogopite-bearing lherzolite or wehrlitic-clinopyroxenite mantle rocks are likely sources of Roman magmas. Mantle contamination was provided by carbonated pelites (marls), a process that took place during the Miocene to present subduction of the Adriatic continental plate beneath Northern Apennines.

Keywords

Roman province · Potassic rocks · Ultrapotassic rocks · Vulsini · Vico · Sabatini · Colli Albani (Alban Hills) · Mantle contamination · Adriatic plate · Northern Apennines · Geodynamics

A. Peccerillo, *Cenozoic Volcanism in the Tyrrhenian Sea Region*, Advances in Volcanology, DOI 10.1007/978-3-319-42491-0_4

4.1 Introduction

Early in the 20th century, the Roman Province was defined by Washington (1906) as the large region of potassium-rich volcanism, extending from southern Tuscany to the Campania area. The Roman Province described here is more restricted in extent, and only includes the belt of potassium-rich volcanoes running from southern Tuscany to the southern outskirts of Rome, parallel to the border of the Tyrrhenian Sea. This is also called as the Latium Province. The re-defined Roman Province consists of the two large volcanic complexes of Monti Vulsini and Monti Sabatini and of the Vico and Colli Albani (Alban Hills) stratovolcanoes (Fig. 4.1). These volcanoes together erupted about 900 km^3 of pyroclastics and lavas over a time span from about 800 ka to

Fig. 4.1 1 Location map of Roman volcanoes. Relief map from Tarquini et al. (2012)

less than 20 ka. In the north, the Roman Province is superimposed over the Tuscany magmatic rocks and there is evidence of hybridism between these two magma types in some places (van Bergen et al. 1983; Peccerillo et al. 1987; Conticelli et al. 2013). The southern border of the Roman Province is marked by an important tectonic structure, which has been named the Ancona-Anzio line (Castellarin et al. 1982; Locardi 1988) or the Olevano-Antrodoco line (e.g. Turtù et al. 2013). This is an approximately N-S trending tectonic line that crosses the Italian peninsula and divides the Northern Apennines from Abruzzi-Latium sequences of Central Apennines (Mattei et al. 1995). It represents the eastern limb of the Northern Apennines thrust front, but develops along a pre-orogenic structure of the Adriatic plate (e.g. Tavarnelli et al. 2004; Satolli and Calamita 2008; Turtù et al. 2013).

The prevalent volcanism has been explosive, with numerous plinian eruptions and associated caldera and volcano-tectonic collapses. The rocks consist of large-volume pyroclastic deposits and minor lava flows. Ignimbrites are particularly abundant and fascinating, from both a geological and historical viewpoint. These consist of chaotic deposits of scoriae, pumices, lithics and crystals, embedded in fine-grained ashes, showing variable thickness (*high-aspect ratio ignimbrites*; Walker 1983) and degrees of cementation. Many outcrops are well lithified (*sillar facies*), and tend to form small plateau surrounded by steep cliffs, upon which several historical towns (e.g. Orvieto, Civita di Bagnoregio, Pitigliano, etc.) have been constructed. Lithified pyroclastic rocks are common in central-southern Italy (e.g. Orvieto-Bagnoregio ignimbrite at Vulsini, Tufo Rosso a Scorie Nere ignimbrite at Vico, Campanian Ignimbrite and Neapolitan Yellow Tuff at Campi Flegrei, etc.) and have been exploited as building material, locally known as "peperino", since Etruscan and ancient Roman times. Many of the best known Etruscan necropolises have been carved into these rocks and various edifices from villages and towns, including the cities of Rome and Naples, are constructed by blocks of lithified tuffs. The lithification of pyroclastic deposits is the result of intensive post-depositional zeolitization of glassy matrix, under the effect of circulating water. Zeolite formation reduces the porosity and increases strength and cohesiveness of the rock. Water may derive from geothermal systems around the conduits (e.g. hydrovolcanic explosions) or from the external environment (meteoric or groundwater; e.g. Hall 1998; De Gennaro et al. 1999; Bear et al. 2009).

Petrologically, the Roman mafic volcanic rocks are mostly ultrapotassic and undersaturated in silica, but nearly saturated potassic rocks occur in the Vulsini district and at Vico (Fig. 4.2).

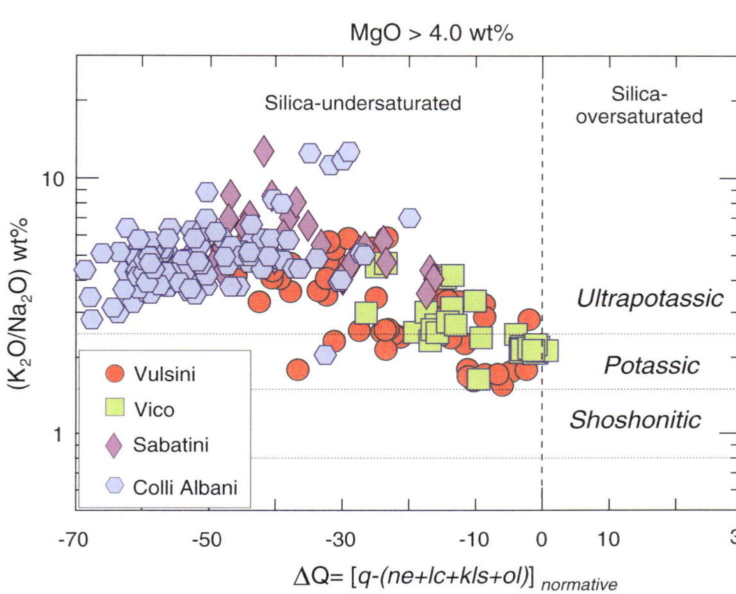

Fig. 4.2 K_2O/Na_2O versus ΔQ plot for mafic volcanics (MgO > 4.0 wt %) from the Roman Province. ΔQ is the algebraic sum of normative quartz minus normative undersaturated minerals (foids, Mg-olivine). See Chap. 1 for further explanation. Rocks with LOI > 3.0 wt% have not been plotted to avoid altered samples, which may have suffered loss or gain of alkalis

Evolved tephriphonolitic, phonolitic and trachytic rocks predominate over mafic types, except at Colli Albani volcano, which is completely built up by tephrites, leucitites and phonotephrites. Many rocks, especially scoriae and pumices, exhibit low alkali contents, high LOI (Loss on Ignition) and Na_2O/K_2O ratios, revealing modification by secondary processes. These rocks will not be considered in the following discussion. Th, U, Large Ion Lithophile Elements (LILE) and LREE show invariably high concentrations, which become extreme in the most evolved rocks.

The peculiar major and trace element compositions of Roman magmas and the secondary processes affecting many lavas and pyroclastics are the cause for the generation of a large number of exotic mineral phases (e.g. Federico 1995). Some of these are unique to Roman rocks and contributed to expand considerably the number of newly discovered mineral phases in the last decades (e.g. Della Ventura et al. 1999, 2002; Federico and Peccerillo 2002; Callegari et al. 2012).

The main petrological, geochronological and volcanological features of Roman volcanoes are summarised in Table 4.1. Representative whole rock analyses are given in Table 4.2.

4.2 Regional Geology

The volcanoes of the Roman Province developed in the northern-central Latium, an area that is in structural continuity with southern Tuscany, having suffered similar geodynamic evolution, at

Table 4.1 Petrological-volcanological characteristics and ages of volcanoes in the Roman Province

Volcano	Age (in Ma)	Volcanology	Petrology
Vulsini	>0.6 to 0.13	– Partially superimposed multicentre volcanic complexes with calderas (Montefiascone, Latera), developed around the volcano-tectonic depression of the Bolsena Lake. Dominant pyroclastic fall deposits and ignimbrites, minor pyroclastic surges and lavas	– Potassic (KS: trachybasalt to trachyte) and ultrapotassic (HKS: leucitite and leucite tephrite to phonolite) compositions. Some melilitite. High enrichments in incompatible elements and radiogenic Sr ($^{87}Sr/^{86}Sr \sim 0.709$–0.711)
Vico	0.42 to 0.095	– Stratovolcano with a central polygenetic caldera, an intra-caldera cone and a few small circum-caldera centres, made up of dominant pyroclastic fall deposits and ignimbrites, and minor lavas	– Early latite-trachyte to rhyolite magmas, followed by HKS cone-forming rocks (phonotephrite to phonolite) and final post-caldera KS (trachybasalt, latite) and HKS (tephriphonolites). High enrichments in incompatible elements and radiogenic Sr ($^{87}Sr/^{86}Sr \sim 0.708$–0.711) increasing with silica and from KS to HKS
Sabatini	0.8 to 0.09	– Two main multicentre complexes (Sacrofano and Bracciano) with several calderas, aligned in an E-W direction. Dominant pyroclastic fall deposits and ignimbrites, and minor lavas	– Dominant HKS (leucite phonotephrite to phonolite) and very minor KS (latites and trachytes) exhibiting high enrichments in incompatible elements and $^{87}Sr/^{86}Sr$ around 0.7095–0.711
Colli Albani (Alban Hills)	0.6 to <0.02 Historical activity (?)	– Stratovolcano constructed during three periods of activity. High-magnitude explosive eruptions and minor lava effusion built up the main cone and a large central caldera. The following activity mainly occurred inside the caldera and at pericaldera centres. Final hydrovolcanic eruptions on the SW flank of the main cone	– HKS (leucitite, leucite tephrite and phonotephrite) pyroclastic fall, flow and surges, hydromagmatic products, and minor lavas. Higher degrees of silica undersaturation than other Roman volcanoes, with occurrence of a few melilite-bearing rocks. $^{87}Sr/^{86}Sr \sim 0.7095$–0.711, decreasing from older to younger products

Table 4.2 Major, trace element and initial radiogenic isotopic compositions for representative samples from the Roman Province. Numbers in parentheses refer to data obtained on distinct though similar rock samples from the same locality as those analysed for the other elements

Volcano	**Vulsini**					**Vico**		
Rock type	Leucite Tephrite	Trachybasalt	Shoshonite	Phonolite	Melilitite (Kamafugite)	Trachybasalt	Latite	Phonolite
Data source	5	5	5	7	3	5	2	2
SiO_2 wt%	46.14	51.60	53.95	60.56	41.70	49.29	57.07	57.09
TiO_2	0.83	0.91	0.71	0.48	0.91	0.75	0.62	0.61
Al_2O_3	17.82	18.11	16.99	14.27	14.90	16.17	16.66	18.50
FeO_{total}	8.09	9.10	5.53	2.97	8.03	7.5	4.70	4.19
MnO	0.15	0.14	0.11	0.14	0.16	0.15	0.10	0.15
MgO	5.69	4.14	7.44	0.84	4.90	7.2	5.67	0.84
CaO	11.54	8.48	7.40	3.02	15.20	11.32	6.05	3.72
Na_2O	1.89	1.92	2.58	5.4	1.67	1.78	2.71	3.51
K_2O	6.24	5.53	4.16	9.8	8.21	4.4	5.74	9.32
P_2O_5	0.50	0.42	0.22	0.06	0.77	0.29	0.16	0.08
LOI	0.73	0.65	0.84	2.47	2.30	0.79	0.30	1.72
Sc ppm	(32)	(23)	(23)	8	–	–	20	2
V	302	257	168	61	–	250	138	115
Cr	20	29	356	8	7	249	247	8
Ni	40	28	192	1	39	34	108	15
Co	37	21	99	34	32	66.9	21	5
Cs	26	18.2	15	725	–	344	–	–
Rb	402	376	312	216	635	1047	423	509
Sr	1395	1268	552	26	2547	22	601	1595
Y	34	28.4	27	451	43	205	53	65
Zr	262	268	263	50	429	11.2	349	604

(continued)

Table 4.2 (continued)

Volcano	**Vulsini**					**Vico**		
Rock type	Leucite Tephrite	Trachybasalt	Shoshonite	Phonolite	Melilitite (Kamafugite)	Trachybasalt	Latite	Phonolite
Nb	12	13.9	16	90	19	20.1	26	35
Ba	870	847	426	37	1628	526	425	896
La	71	68	66	102	161	48	105	167
Ce	155	142	123	214	336	102	186	309
Pr	17.7	17.6	13.3	18	40	12.5	–	–
Nd	70	67	49	68	152	50.6	58	96
Sm	12.7	12.4	8.2	10.6	27	8.91	14.2	21
Eu	2.67	2.56	1.59	1.43	21.6	1.92	1.65	3.15
Gd	9.57	8.2	5.86	6.64	17.5	6.49	–	–
Tb	1.24	1.07	0.85	0.8	–	0.88	1.2	1.6
Yb	2.42	2.1	2.29	2.53	3.19	1.76	2.6	3.6
Lu	0.34	0.32	0.36	0.38	0.37	0.26	0.43	0.57
Hf	5.98	6.64	6.23	9.17	9.8	5.3	8	12.2
Ta	0.59	0.84	0.96	2.25	1.0	0.7	1.8	2.6
Pb	38	23	42	185	89	21.8	101	142
Th	30	29.6	51	87.4	76	26.5	103	107
U	6.8	5.94	6.9	22.9	16.4	5.19	25	18
$^{87}Sr/^{86}Sr$	0.71020	0.710318	0.70994	0.710990	0.71042	0.708335	0.710133	0.71052
$^{143}Nd/^{144}Nd$	0.51213	0.512146	0.51220	0.512110	0.51213	0.512218	0.512177	0.512113
$^{206}Pb/^{204}Pb$	18.779	18.717	18.740	–	–	18.815	18.728	18.740
$^{207}Pb/^{204}Pb$	15.653	15.656	15.663	–	–	15.646	15.665	15.679
$^{208}Pb/^{204}Pb$	39.015	38.994	39.039	–	–	39.047	39.029	39.010
$^{176}Hf/^{177}Hf$	0.282573	0.282579	0.282604	–	–	0.282601	–	–

Table 4.2 (continue)

Volcano	**Vico**		**Sabatini**			**Colli Albani**		
Rock type	Phonotephrite	Rhyolite	Phonotephrite	Tephriphonolite	Phonolite	Tephriphonolite	Leucite tephrite	Leucitite
Data source	2	2	5	1	1	6	6	6
SiO_2 wt%	50.80	69.10	49.52	53.48	56.99	49.90	45.70	43.70
TiO_2	0.99	0.20	0.67	0.58	0.41	0.72	0.90	0.71
Al_2O_3	18.15	14.30	16.55	19.55	20.6	14.80	14.70	16.20
FeO_{total}	8.55	1.63	6.97	5.84	3.14	6.75	7.25	8.64
MnO	0.14	0.09	0.13	0.12	0.12	0.12	0.14	0.16
MgO	3.40	0.29	6.73	1.61	0.53	5.88	8.91	3.85
CaO	7.40	1.58	10.03	5.73	3.29	9.17	12.30	10.80
Na_2O	1.85	2.6	1.43	2.13	3.35	1.91	1.15	2.50
K_2O	7.54	7.43	6.71	9.4	10.87	8.59	7.21	8.84
P_2O_5	0.56	0.02	0.62	0.36	0.1	0.69	0.54	0.77
LOI	0.58	2.7	0.64	1.2	0.6	0.98	0.89	3.36
Sc ppm	16.6	2.3	–	13.7	1.4	29	–	10
V	209	23	272	213	–	265	243	312
Cr	13	11	258	16	12	68	418	11
Ni	26	19	76	11	7	61	125	27
Co	27	2.2	33	20	7.2	31	35.2	31
Cs	–	–	31	–	–	47.4	–	19.6
Rb	497	730	524	658	393	554	429	292
Sr	1719	209	1728	2114	1989	1548	1658	1618
Y	56	69	30	43	54	45	27	50
Zr	368	363	299	415	436	291	213	444
Nb	17	37	13	17	29	15	13	34
Ba	2152	109	1013	1577	1879	1388	1329	2599
La	121	176	90	134	156	116	107	172

(continued)

Table 4.2 (continued)

Volcano	Vico		Sabatini			Colli Albani		
Rock type	Phonotephrite	Rhyolite	Phonotephrite	Tephriphonolite	Phonolite	Tephriphonolite	Leucite tephrite	Leucitite
Ce	230	297	182	274	295	240	197	327
Pr		–	23	–	–	29.6	–	38.5
Nd	86	80	88	103	92	108	90	144
Sm	19.6	15.4	17	19.5	17.8	19.6	19	22.4
Eu	3.62	1.18	3.27	3.3	3	3.88	3	4.42
Gd	–	–	11.2	–	–	15.3	–	17.8
Tb	1.6	1.1	1.28	1.6	1.8	1.61	1.2	1.85
Yb	1.8	3.9	2.14	2.7	3.1	2.5	2.1	2.9
Lu	0.29	0.63	0.28	0.58	0.49	0.37	0.24	0.36
Hf	7.9	11.3	7.1	7.4	7.7	9	7.4	9.01
Ta	0.99	3	0.66	1	1.3	0.7	0.85	1.59
Pb	222	261	45	–	–	66	90	192
Th	47	165	45	75	90	44	45	87.1
U	6	49	10	–	–	10.2	7	9.3
$^{87}Sr/^{86}Sr$	0.711684	0.711427	0.710086	0.71036	0.71031	0.710312	0.710393	0.710587
$^{143}Nd/^{144}Nd$	0.512147	0.512113	0.512106	–	–	0.512104	0.512104	0.51211
$^{206}Pb/^{204}Pb$	18.70	–	18.797	–	–	18.761	18.765	18.758
$^{207}Pb/^{204}Pb$	15.66	–	15.674	–	–	15.689	15.681	15.685
$^{208}Pb/^{204}Pb$	38.91	–	39.090	–	–	39.03o	39.006	39.023
$^{176}Hf/^{177}Hf$	–	–	0.282562	–	–	–	–	–

Source of data: (1) Conticelli et al. (1997); (2) Perini et al. (2000, 2004); (3) Di Battistini et al. (2001); (4) Conticelli et al. (2002); (5) Gasperini et al. (2002); (6) Boari et al. (2009); (7) Author's unpublished data least from Oligo-Miocene. Late Miocene-Quaternary extensional tectonics related to the eastward migration of Apennine arc and to the opening of the Tyrrhenian Sea, generated NW-SE trending basins, bounded by normal faults and intersected by NE-SW strike-slip faults (Bartolini et al. 1982; Acocella and Funiciello 2006). Both fault systems represent zones of crustal weakness along which Roman potassic magmas were intruded.

The pre-volcanic substratum consists of Tuscany, Liguride and Umbria-Marche allochthonous sequences (see Chaps. 2 and 3), and of neo-autochthonous Miocene to Quaternary shallow marine to continental clays, sands, and conglomerates filling post-orogenic extensional basins (Buonasorte et al. 1987; Barberi et al. 1994). The Umbria-Marche sequences, mainly consisting of carbonate rocks, define eastward convex regional structures in the Northern Apennine until the Tyrrhenian Sea border (e.g. Barchi et al. 2001).

The Roman Province is characterised by about 22–25 km thick crust (Piromallo and Morelli 2003; Butttinelli et al. 2014), with Moho deepening toward the east. Heath flow is high (100 m Wm^{-2} at a regional scale), locally reaching 600 m Wm^{-2} (Mongelli and Zito 1991; Mongelli et al. 1991). Seismicity is shallow and mostly concentrated east of the volcanic area (internal zones of Apennines) or along the southern segment of the Olevano-Antrodoco line. Shallow seismicity has been detected in the Latera area (Vulsini district) and at Colli Albani (Chiarabba et al. 1995, 2010a, b).

The entire volcanic region is also characterised by a belt of positive gravity anomaly, which decreases going eastward up to a belt of negative anomaly along the external zones of Apennines (e.g. Scarascia et al. 1994). The Roman volcanic Province is an area of high CO_2 degassing, mostly originating from the mantle (e.g. Gambardella et al. 2004; Chiodini et al. 1999, 2008). Gaseous flow decreases toward the Apennines where its entrapment at specific sites may have triggered important seismic events (e.g. Miller et al. 2004). The Roman Province is also characterised by high natural radiation background and radon emission, which are related to the high K, Th and U contents of volcanic rocks (e.g. Cinelli et al. 2014).

4.3 Vulsini District

Monti Vulsini district consists of various multi-center volcanic complexes, which developed in the northern sector of the Roman Province over an area of about 2200 km^2. Rocks consist of prevailing pyroclastic products and minor lavas with potassic to ultrapotassic affinity. Eruptions occurred at over 100 different centres, including two calderas and the large polygenic volcano-tectonic depression of the Bolsena Lake. The volcanism developed along a Tortonian to Pleistocene horst and graben system (Siena-Radicofani and Paglia-Tevere grabens, Cetona and Razzano horsts). The pre-volcanic rocks include Liguride, Tuscan and Umbria sequences and the underlying metamorphic basement. Most of these rocks are found as xenoliths in several pyroclastic units, especially at Montefiascone and Latera.

Ages determined on exposed rocks range from about 0.6–0.13 Ma. However, fragments of ultrapotassic rocks and constituent minerals (leucite, clinopyroxene) have been found in Early Pleistocene continental deposits near to Orvieto, shifting the onset of volcanism to about 1.7 Ma (Bizzarri et al. 2003, 2010).

4.3.1 Volcanology and Stratigraphy

The Vulsini volcanic district is characterized by an overall shield-like landform, on which the 19 km wide Bolsena Lake volcano-tectonic depression, the Latera-Vepe caldera and the Montefiascone cone and caldera stand out (Fig. 4.3a). There are many minor eruptive centres (strombolian cones, tuff cones, tuff rings), mostly aligned along faults around the Bolsena Lake and the Latera caldera (e.g. Vezzoli et al. 1987; Freda et al. 1990).

The Vulsini district (Fig. 4.3b) consists of five major volcanic complexes: Paleo-Vulsini,

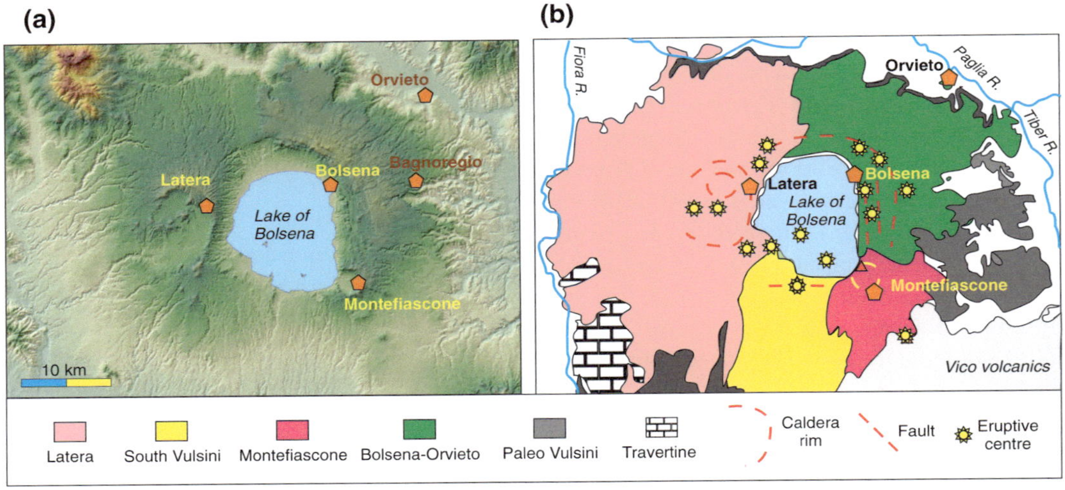

Fig. 4.3 **a** *Color shaded* relief map of the Vulsini volcanoes (Tarquini et al. 2007, 2012); **b** Schematic geological map, simplified after Vezzoli et al. (1987)

Bolsena-Orvieto, Montefiascone, Southern Vulsini and Latera (Nappi et al. 1987, 1995, 1998; Vezzoli et al. 1987; Palladino et al. 1994). The Southern Vulsini complex has been included in the Vulsini Fields lithostratigraphic unit by Palladino et al. (2010) and Acocella et al. (2012).

The Paleo-Vulsini activity (about 0.6–0.49 Ma) in the north-eastern sector of the Bolsena Lake emplaced fall deposits, ignimbrites ("Basal Ignimbrites" or "Nenfri") and lava flows, which crop out at the margins of the Vulsini district, directly overlying the sedimentary bedrocks (Nappi et al. 1987, 1995; Vezzoli et al. 1987). Ignimbrite eruptions were responsible for early stages of caldera collapse in the Bolsena area. Magma compositions range from trachyte to phonotephrite and phonolite. Some of the massive lavas containing megacrysts of leucite have been quarried from Etruscan and Roman times to Middle Age for the production of millstones, which were exported in several places around the Mediterranean Sea (e.g. Renzulli et al. 2002; Santi et al. 2004; Antonelli and Lazzarini 2010).

The activity of the Bolsena-Orvieto complex (about 0.49–0.3 Ma) was concentrated along the north-eastern rim of the protocaldera, with emplacement of lavas, pyroclastic fall and surge deposits, and of the Orvieto-Bagnoregio pyroclastic flow (296–294 ka, Turbeville 1992a; 318 ka, Barberi et al. 1994; 333 ka, Nappi et al. 1995), whose eruption caused a further collapse of the Bolsena caldera. The Orvieto-Bagnoregio deposit is a high-aspect ratio ignimbrite showing variable mechanical properties from loose pyroclastic material to fairly cemented *sillar facies* rock (e.g. Varekamp 1979; Tommasi et al. 2015). Some cemented outcrops show very rough columnar jointing and sometimes form small plateaux (*mesa* and *butte*), bounded by steep cliffs, on which the towns of Orvieto, Bagnoregio and Civita di Bagnoregio are located (Fig. 4.4). The rocks of the Bolsena-Orvieto complex display dominant leucite tephrite to trachyte and phonolite compositions, with minor latite and shoshonite. The Orvieto-Bolsena stage is coeval with the early activity of the nearby Vico volcano, which started at about 0.42 Ma (Cioni et al. 1987; Laurenzi and Villa 1987).

The Montefiascone complex (about 0.3 to 0.2 Ma; Nappi et al. 1987, 1995; Brocchini et al. 2000) is a large cone whose summit rises to 400 m above its surroundings, at the south-eastern corner of the Bolsena Lake. It consists of several coalescing eruptive centres and a 2.5 km wide caldera. Rocks include lava flows, ignimbrites, hydrovolcanic products and strombolian scoriae,

Fig. 4.4 Civita di Bagnoregio "The city that dies" sitting over a small butte of the Orvieto-Bagnoregio ignimbrite. The underlying rocks consist of a series of stratified pyroclastic, mostly fall, deposits and a few paleosols overlying Pliocene argillaceous sediments

mostly with a leucite-tephrite and basanite composition. Melilite-bearing lavas and kalsilite-melilitolite ejecta have been also found (Di Battistini et al. 2001).

The Southern Vulsini complex consists of prevailing phonotephritic and phonolitic products with ages of 0.4–0.13 Ma (Palladino et al. 1994; Nappi et al. 1995). The latest activity constructed the Bisentina and Martana islands, two tuff cones inside the Bolsena Lake (Fig. 4.5).

The Latera volcano (about 0.38–0.15 Ma; Turbeville 1992a) is composed of several pyroclastic units and some lava flows, cropping out inside and around an 8 km across polyphasic caldera (Latera-Vepe; Nappi et al. 1991; Palladino and Simei 2005a, b). Rock compositions range from trachybasalt and phonotephrite to trachyte and phonolite (Nappi et al. 1987; Conticelli et al. 1991; Landi 1987; Turbeville 1993; Renzulli et al. 1995).

4.3.2 Petrography and Mineral Chemistry

The volcanic rocks from Vulsini range from mafic to felsic and are saturated to undersaturated in silica (Figs. 4.2 and 4.6). Although there is a continuum in potassium enrichments, two main series of rocks have been distinguished. One consists of moderately potassic and nearly saturated trachybasalts to trachytes, and is known as the Roman-type Potassic Series (KS) or Saturated Series. Another rock series is ultrapotassic in composition and consists of leucitites, leucite tephrites, leucite basanites to leucite phonolites, and is known as the Roman-type High Potassium Series (HKS). KS and HKS erupted through the entire evolution history of Vulsini volcanoes; however, KS rocks prevail during the latest stages of activity at Latera (e.g. the Lamone trachybasalt lava flow). A third minor group

Fig. 4.5 Panoramic view of the Lake of Bolsena showing the Montefiascone caldera in the foreground, the tuff cones of Bisentina and Martana islands, and the eastern margin of the Latera caldera in the far distance

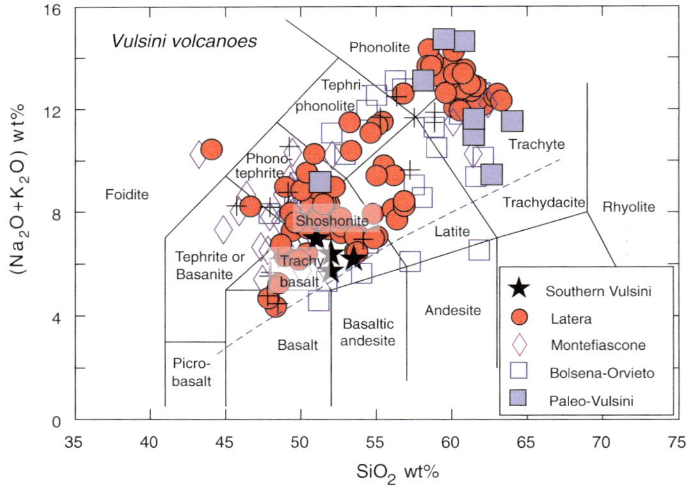

Fig. 4.6 Total alkali versus silica diagram (TAS; Le Maitre 2002) for Vulsini volcanic rocks. *Crosses* indicate samples of unknown stratigraphic position. Data normalised to 100 % on a LOI-free basis. The *dashed line* is the divide between the alkaline and subalkaline fields (Irvine and Baragar 1971)

consists of melilite-bearing rocks from Montefiascone; these have major element compositions resembling ultrapotassic kamafugites and fall in the tephrite or in the foidite field on the TAS diagram (Di Battistini et al. 2001).

Lava textures range from almost aphyric to strongly porphyritic, sometimes with megacrysts of leucite. Scoria and pumices are glassy with moderate to low phenocryst contents. Mafic KS rocks contain phenocrysts of clinopyroxene, olivine, and plagioclase set in a groundmass containing the same phases plus Fe–Ti oxides, some alkali-feldspar, and brown mica. Evolved KS rocks contain phenocrysts of sanidine, plagioclase, clinopyroxene, biotite and rare amphibole set in a groundmass consisting of the same phases and Fe–Ti oxides. Leucite occurs as phenocrysts, xenocrysts and in the groundmass of several KS rocks, e.g. at Latera (Landi 1987). Apatite is the main accessory mineral. Garnet has been sometimes observed (e.g. Nappi et al. 1987).

The most mafic HKS rocks contain olivine and clinopyroxene phenocrysts and microphenocrysts; leucite and plagioclase appear in the more evolved samples. The groundmass consists of clinopyroxene, plagioclase, leucite, some alkali-feldspar, and brown mica. Felsic rocks contain phenocrysts of clinopyroxene, leucite, plagioclase, sanidine, biotite and rare amphibole; groundmass consists of the same phases plus Fe–Ti oxides, sporadic haüyne, nepheline, amphibole, garnet, apatite, and titanite (e.g. Holm 1982; Holm et al. 1982; Rogers et al. 1985 and references therein). Melilititic lavas from Montefiascone have a poorly porphyritic texture with phenocrysts of leucite, clinopyroxene, and melilite, surrounded by a groundmass of clinopyroxene, olivine, Fe–Ti oxides, brown mica, some amphibole and carbonates (Di Battistini et al. 1998). Olivine (Fo_{92-60}), sometimes with inclusions of Cr-spinel, becomes progressively Mg-poor from mafic to intermediate rocks. Mg-rich olivine is found in some trachytes, where it coexists with Fe-rich (about Fo_{33}) crystals (Holm 1982); CaO is variable (0.3–1.5), the highest values being found in olivine phenocrysts from Montefiascone melilite-bearing rocks (Di Battistini et al. 1998). Clinopyroxene is mainly diopside with variable Fe content (e.g. Nappi et al. 1987; Gentili et al. 2014). Compositional zoning is often present in single crystals, as revealed by strong colour modification in thin section, from colourless diopside to green Fe-rich compositions. These variations in single pyroxene crystals are common in the ultrapotassic rocks from Central Italy and have been explained as resulting from P_{H_2O} variations during crystallisation or from mixing between potassic magmas with different degrees of evolution (Thompson 1977; Brooks and Printzlau 1978; Dolfi and Trigila 1978; Barton et al. 1982; Varekamp and Kalamarides 1989). Al/Ti ratios (expressed as atoms per formula unit, afu) of clinopyroxenes are high ($\sim$8–10; see Chap. 2, Fig. 2.11) relative to clinopyroxenes of mafic rocks from the Tuscany and IAP magmatic provinces (e.g. Kamenetsky et al. 1995; Di Battistini et al. 1998; Conticelli et al. 2013; Gentili et al. 2014). Brown mica ranges from phlogopite to biotite and is sometimes enriched in Ba and TiO_2. Leucite is often transformed to analcime, which implies a modification of K/Na ratios for the bulk rocks.

Numerous xenoliths of magmatic, metamorphic and sedimentary origin occur in the Vulsini rocks, especially in the pyroclastic deposits. Most ejecta represent bedrocks; others are skarns, intrusive equivalents of erupted lavas, or cumulate lithologies (e.g. Turbeville 1992b; Di Battistini et al. 1998).

4.3.3 Petrology and Geochemistry

Major and trace element geochemistry of Vulsini rocks shows an overall decrease in CaO, and ferromagnesian elements (Ni, Cr, Co, V, Sc), and an increase in K_2O and incompatible elements (e.g. Zr, Th, LREE, etc.) with decreasing MgO. Na_2O also increases, but with some scattering. TiO_2, Al_2O_3 and FeO_{total} roughly show a bell-shaped trend (Fig. 4.7).

There is an overall increase in incompatible element concentration from KS to HKS rocks, which is particularly evident when only mafic rocks (MgO > 4 wt%) are considered (Fig. 4.8a).

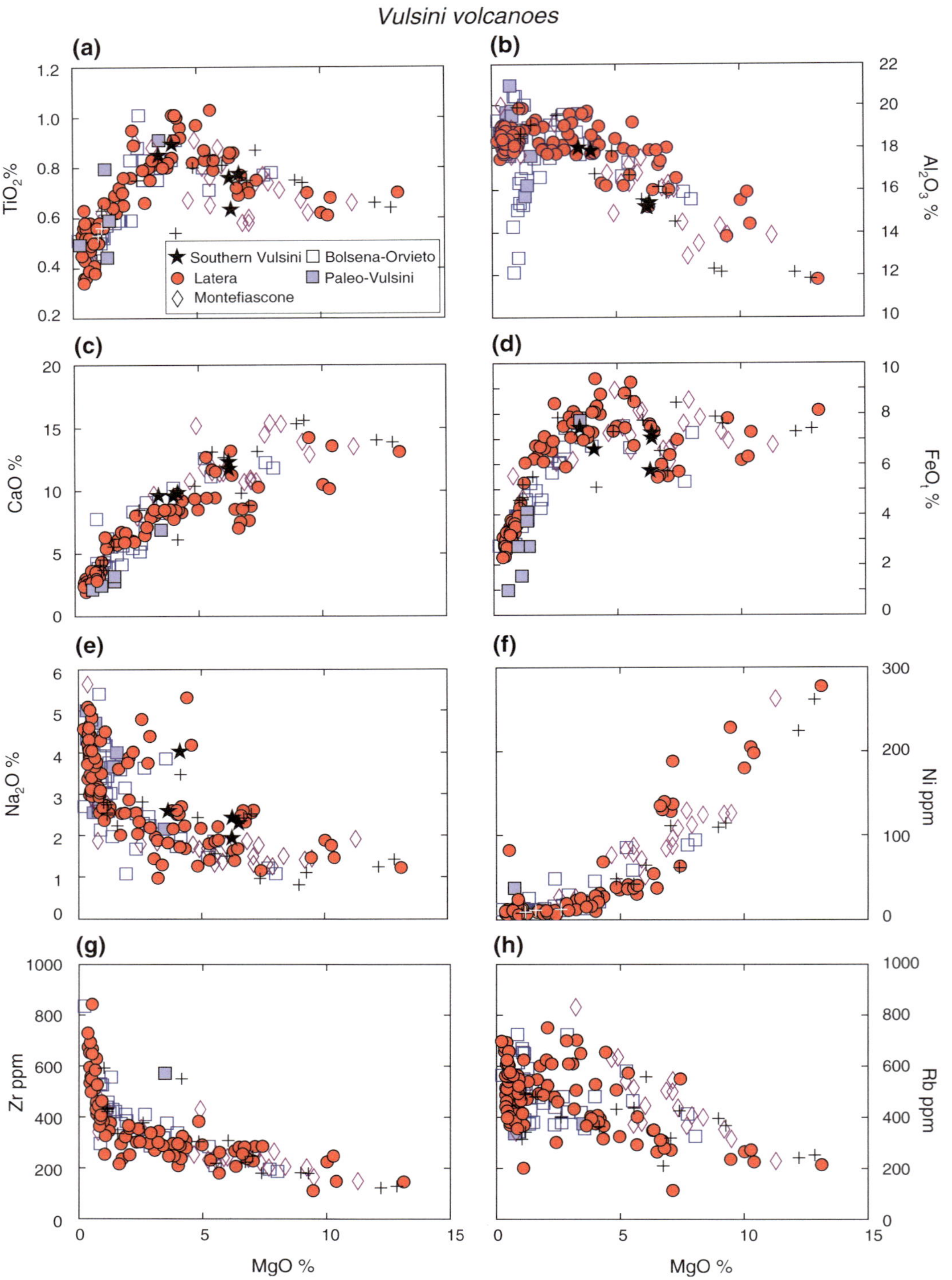

Fig. 4.7 Variation diagrams of selected major and trace elements versus MgO for the Vulsini rocks. *Crosses* indicate samples of unknown stratigraphic position

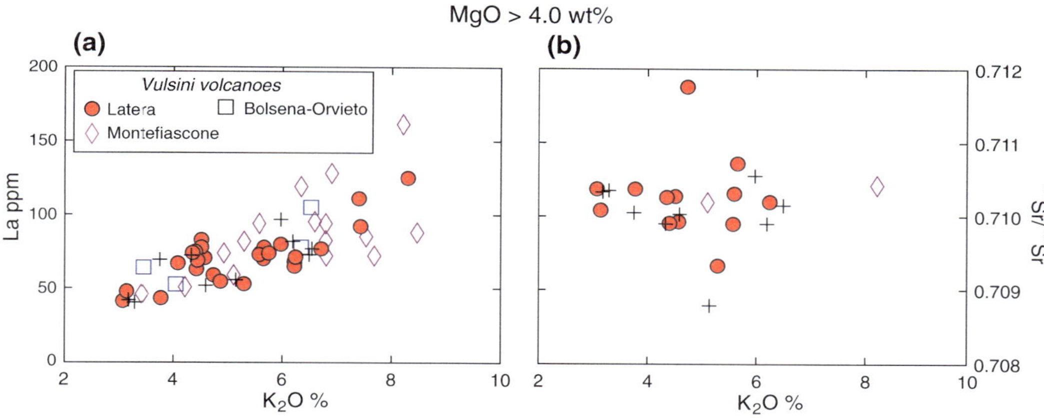

Fig. 4.8 La and $^{87}Sr/^{86}Sr$ versus K_2O diagrams for the Vulsini mafic rocks (MgO > 4.0 wt%). *Crosses* indicate samples of unknown stratigraphic position

REE patterns of both KS and HKS mafic rocks are fractionated and display a small negative Eu anomaly (Fig. 4.9a); note that the most primitive rocks have no plagioclase on liquidus, suggesting that Eu anomalies are not due to plagioclase fractionation but are features of primary melts.

Mantle normalised incompatible element patterns display strong fractionation with negative anomalies of Ba and HFSE (Fig. 4.9b), and positive spikes of Rb, Th, light REE and Pb. These resemble closely the IAP kamafugites, whereas are somewhat different from Tuscany mafic K-rich rocks, in which Nb and Ta negative anomalies are smaller and LREE/Sr ratios are higher.

Sr isotope ratios cluster around 0.7100 to 0.7110, with a few values falling outside this interval (e.g. Holm and Munksgaard 1982; Rogers et al. 1985; Ferrara et al. 1986; Varekamp and Kalamarides 1989). These ratios are comparable to IAP kamafugites but are lower than those of the Tuscany K-rich rocks (Fig. 4.10a). Unlike other volcanic regions in Italy (e.g. Ernici, Roccamonfina, eastern Aeolian arc), there is no systematic isotopic variation among mafic

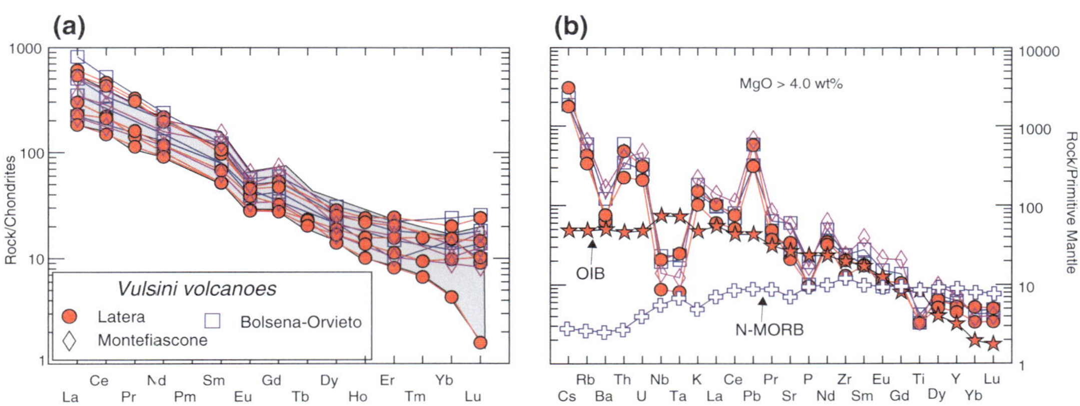

Fig. 4.9 **a** REE patterns of the Vulsini rocks. The *shaded area* encloses patterns of mafic rocks; **b** Incompatible element patterns of mafic rocks from Vulsini. Compositions of average Normal Mid-Ocean Ridge Basalts (N-MORB; Gale et al. 2013) and Ocean Island Basalts (OIB; Sun and McDonough 1989) are reported for comparison

rocks with different enrichments in potassium (Fig. 4.8b). $^{143}Nd/^{144}Nd$ ranges from about 0.5121 to 0.5122. Pb isotope ratios show values close to those of Tuscany mafic rocks (Fig. 4.10b; $^{206}Pb/^{204}Pb \sim 18.73$ to 18.77; $^{207}Pb/^{204}Pb \sim 15.65$ to 15.71; $^{208}Pb/^{204}Pb \sim 38.93$ to 39.15). A much wider range of Pb isotopic compositions has been found on melt inclusions from Latera olivine phenocrysts ($^{206}Pb/^{204}Pb \sim 18.30$ to 18.93; $^{207}Pb/^{204}Pb \sim 14.59$ to 15.79; $^{208}Pb/^{204}Pb \sim 37.04$ to 40.73; Nikogosian et al. 2016) $^{176}Hf/^{177}Hf$ analyses on a few samples yielded values of about 0.28257 (Gasperini et al. 2002).

Oxygen isotope data for whole rocks range from $\delta^{18}O_{SMOW} \sim +8$ ‰ to +12 ‰, and increase with silica (Holm and Munksgaards 1982; Varekamp and Kalamarides 1989). Lower values of $\delta^{18}O \sim +6.5$ ‰ to +7.9 ‰ have been observed in clinopyroxene and MgO-rich olivine from Latera and Montefiascone by Varekamp and Kalamarides (1989) and Barnekow (2000). He-isotope data reported by Martelli et al. (2004) for olivine and clinopyroxene are low, in the range $R/R_A = 0.62$–1.06.

4.3.4 Evolution of Vulsini Magmas

Although scattered, major and trace element variations for the Vulsini rocks define overall trends that support an evolution dominated by fractional crystallisation starting from different types of parental melts. Holm et al. (1982) suggested that the evolution of HKS magmas was dominated by separation of olivine and clinopyroxene in the mafic range, and of variable proportions of clinopyroxene, leucite, plagioclase, sanidine, and some apatite in the evolved compositions. In contrast, KS magmas were suggested to have evolved from nearly saturated trachybasalts by separation of olivine, clinopyroxene, and feldspar. Variations of $\delta^{18}O$‰ and $^{87}Sr/^{86}Sr$ were interpreted as evidence for crustal assimilation. There is ample mineralogical and geochemical evidence that mixing between differently evolved melts was an additional first-order evolutionary process at Vulsini (e.g. Varekamp and Kalamarides 1989).

The high potassium and incompatible element contents, radiogenic isotope signatures and $^3He/^4He$ values of mafic magma at Vulsini suggest an origin in an anomalous upper mantle. Variable potassium contents of mafic rocks likely reflect the occurrence of different types of parental magmas. The positive correlation of potassium versus incompatible elements and the lack of systematic differences of isotopic compositions between potassic and ultrapotassic mafic rocks (Fig. 4.8) are suggestive of a magma genesis by different degrees of partial melting of a broadly homogeneous mantle source. This is

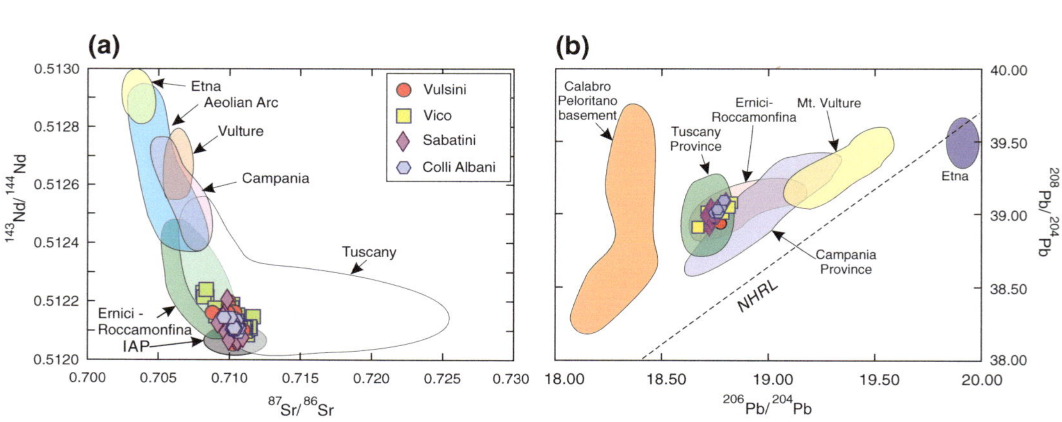

Fig. 4.10 Sr–Nd–Pb isotope diagram for the volcanic rocks of the Roman Province. Compositions of other provinces from central-southern Italy (restricted to mafic rocks) and of the Calabro-Peloritano crystalline basement are reported for comparison

supported by the similar shape of REE and incompatible element patterns for all the mafic rocks (Fig. 4.9). Also the melilite-bearing rocks from Montefiascone share these characteristics, indicating a similar source. However, mafic rocks from Latera show some major and trace element abundances and ratios (e.g. relatively lower CaO and higher Ce/Sr than other rocks with similar MgO contents) that recall Tuscany shoshonites. This suggests a somewhat different source for these volcanics with respect to other Vulsini magmas. Pb-isotope variability observed for melt inclusions in olivine phenocrysts highlight a very heterogeneous and much more contaminated mantle source for Latera than for other magmas in the area (Nikogosian et al. 2016).

4.4 Vico Volcano

The Vico volcano is located south of Monti Vulsini along the southern extension of the Siena-Radicofani, Paglia-Tevere and Monte Razzano graben-horst system (Locardi et al. 1977). It is partially superimposed on the Monti Cimini rocks (about 1.3–0.9 Ma old), which belong to the Tuscany Magmatic Province (see Chap. 2). The volcano erupted silica-oversaturated to undersaturated potassic and ultrapotassic magmas over a time span of about 0.42–0.1 Ma (e.g. Laurenzi and Villa 1987; Bertagnini and Sbrana 1986; Barberi et al. 1994). The pre-volcanic rocks of the Vico region consist of Tuscan allochthonous sequences (flysch, carbonates and low-grade metamorphic rocks) and of neo-autochthonous sediments, which crop out at the margins of the volcano and have been found by borehole drilling (Sollevanti 1983).

4.4.1 Volcanology and Stratigraphy

Vico consists of a large volcanic cone, a central caldera, and a small number of post-caldera centres, the largest being the intra-caldera cone of Monte Venere (Fig. 4.11). Volcanological, petrological and ^{40}Ar/^{39}Ar–K/Ar data (Sollevanti 1983; Laurenzi and Villa 1987; Barberi et al. 1994) allowed to identify three main periods of activity (Perini et al. 2000, 2004; Bear et al. 2009). Period I (about 420 to 400 ka) is composed mainly of latite-trachyte, and rhyolite pyroclastic fall deposits and minor lavas, which crop out mainly in the northern and eastern sectors of the volcano. Period II (305 to 138 ka) was dominated by the effusion of intermediate to felsic leucite-bearing lavas, and by emplacement of voluminous leucite-bearing pyroclastic flow and fall deposits, which built up the main cone

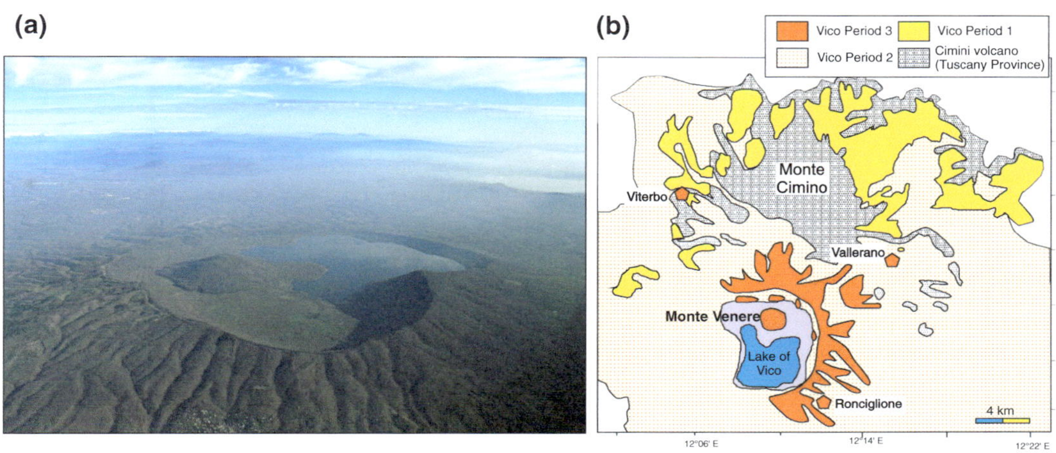

Fig. 4.11 **a** Aerial view of Vico volcano, showing the wide polygenetic caldera and the Monte Venere intra-caldera cone. Photo courtesy: James T. Kirk; **b** Schematic geological map, simplified after Perini et al. (2004)

(305–258 ka), and generated a progressive collapse of the central caldera (Locardi 1965). The post-caldera activity (Period III, about 138 to 95 ka) produced small volumes of both leucite-free and leucite-bearing mafic lavas, scoriae, and phreatomagmatic products, mostly scattered along the caldera rim. The volcanism ended with the formation of the intra-caldera leucite-bearing lava and scoria cone of Monte Venere (Bertagnini and Sbrana 1986).

4.4.2 Petrography and Mineral Chemistry

According to the TAS classification diagram, the Vico rocks span the K-trachybasalt, phonotephrite, tephriphonolite, phonolite, latite and trachyte fields, reaching rhyolitic composition for some rocks (Fig. 4.12; Cundari 1975; Barbieri et al. 1988; Perini et al. 2003, 2004). The trachybasalt to phonolite group is mildly to strongly undersaturated in silica and commonly contains leucite. The other rocks are leucite-free, saturated to oversaturated in silica.

Vico volcanics are porphyritic with phenocryst contents ranging from about 10 to 60 vol %. Groundmass is generally holocrystalline in the lavas, and mostly glassy in the juvenile pyroclasts. Zoned clinopyroxene (diopside to rare augite) is a ubiquitous phenocryst and groundmass phase (Perini et al. 2000). Al/Ti (afu) ratios are high, as for other clinopyroxenes from the Roman Province (Conticelli et al. 2013). Plagioclase ($\sim An_{95-30}$) occurs in variable amounts as a phenocryst phase and in the groundmass of almost all the Vico volcanics, except some trachybasalts. Olivine, sometimes with Mg-chromite inclusions, is a common phenocryst of several mafic to intermediate rocks. It shows very variable forsterite contents (Fo_{91-24}), with the highest values being observed in some latites. Leucite occurs as a phenocryst in the evolved undersaturated products and in the groundmass of potassic trachybasalts. Haüyne has been observed in some samples. Sanidine (Or_{98-53}) is a phenocryst and megacryst phase of trachytes and a groundmass mineral of phonolites and latites. Phlogopite (Mg# = 0.95–0.62) is commonly present as a phenocryst in several leucite-free rocks. Magnesiohastingsite amphibole occurs in some trachytes. Accessory minerals include apatite, Fe–Ti oxides and sporadic zircon and titanite (Perini et al. 2004).

Like other Roman volcanics, the Vico pyroclastic rocks contain xenoliths of various origins, including bedrock fragments, intrusive equivalents of lavas, and cumulate rocks. These xenoliths often contain exotic minerals, such as Zr–Ti–Th–U-REE rich phases (e.g. britholite,

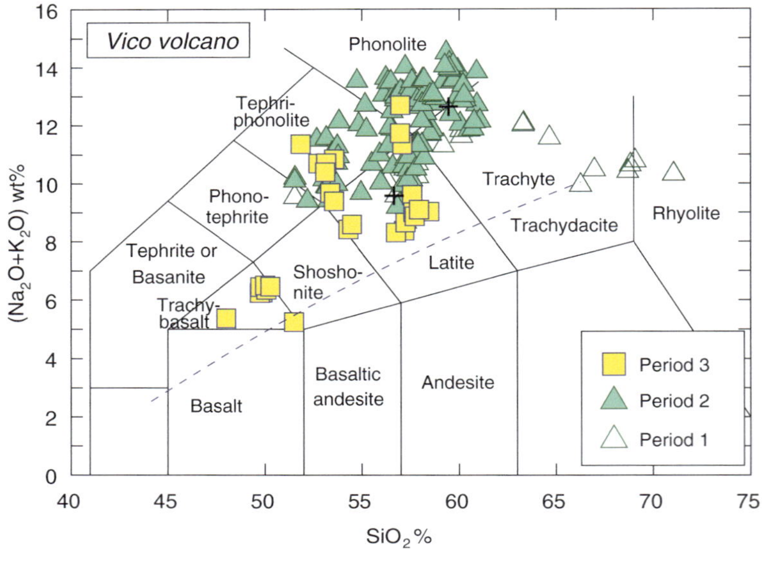

Fig. 4.12 Total alkali versus silica (TAS) diagram for Vico volcanic rocks. *Crosses* are samples of unknown stratigraphic position. Data recalculated to 100 % on a LOI-free basis. The *dashed line* is the divide between the alkaline and subalkaline fields (Irvine and Baragar 1971)

baddeleyite, pyrochlore), which have been deposited by late-magmatic fluids rich in incompatible elements (e.g. Della Ventura et al. 1999).

4.4.3 Petrology and Geochemistry

Major elements of Vico rocks display scattered distributions (Fig. 4.13). However, rocks from different periods of activity define more coherent groups, which suggest the occurrence of different magma types in the volcanic systems at various stages of activity.

Trace elements also are scattered, with different levels of enrichment for rocks of various periods (Fig. 4.13). Incompatible element versus incompatible element plots show rather smooth positive trends (Fig. 4.13g). There is a positive correlation between incompatible elements and $K_2O\%$, a feature that is better observed in the mafic rocks (Fig. 4.14a). REE patterns are fractionated with small negative Eu anomalies, which increase slightly in the felsic rocks (Fig. 4.15a). Incompatible element patterns of mafic rocks show enrichments in LILE and depletion in HFSE (Fig. 4.15b). Period III latites have strong positive spikes of U, Th, LREE and Pb, which recall compositions of some Tuscany mafic rocks.

The Vico volcanics show a larger range of Sr-Nd isotope ratios than other Roman volcanoes ($^{87}Sr/^{86}Sr \sim 0.7081$ to 0.7117; $^{143}Nd/^{144}Nd \sim 0.51209$ to 0.51223; Fig. 4.10a). There is an overall increase of Sr isotope ratios versus potassium (Fig. 4.14b), and a decrease with time (Barbieri et al. 1988; Perini et al. 2004), with the youngest Vico rocks having the lowest $^{87}Sr/^{86}Sr$ values yet documented in the Roman Province (Figs. 4.10a and 4.13h). Pb-isotope compositions show moderate variability (Fig. 4.10b; $^{206}Pb/^{204}Pb \sim 18.67$ to 18.83; $^{207}Pb/^{204}Pb \sim 15.62$ to 15.68; $^{208}Pb/^{204}Pb \sim 38.91$ to 39.08) and resemble closely those of other potassic rocks in Central Italy (Conticelli et al. 2002; Perini et al. 2004). $\delta^{18}O_{SMOW}$ determined on whole rocks and separated leucite and sanidine by Turi and Taylor (1976) have high values, from about +9.0 ‰ to +10.2 ‰.

4.4.4 Evolution of Vico Magmas

Early petrological studies (Cundari and Mattias 1974; Cundari 1975) interpreted the Vico volcanic suite as having formed from one parent magma by fractional crystallisation, with separation of plagioclase, clinopyroxene, leucite, phlogopite and Fe–Ti oxides. Villemant and Palacin (1987) accepted fractional crystallisation, but argued that the very high enrichment in incompatible elements (e.g. Th and Pb) in some phonolites require an additional evolutionary process such as element transfer by gaseous phases (Villement and Fléhoc 1989). Barbieri et al. (1988) noticed a wide Sr isotope variation in the Vico rocks and an overall decrease in radiogenic Sr and silica with time, suggesting that various batches of magmas were emplaced into the shallow level reservoir at different stages of volcanic activity; these magmas become more mafic and isotopically more primitive with time, until the late trachybasalts and latites were erupted. According to Barbieri et al. (1988), fractional crystallisation, crustal assimilation and mixing operated during evolution of various magma batches, with mixing being particularly evident at the transition between pre- and post-caldera activity. Perini et al. (2004) stated the transition from trachytes to rhyolites during Period I occurred primarily by fractional crystallisation, with separation of plagioclase, sanidine, clinopyroxene, Fe–Ti oxides, apatite, sphene, and zircon. Period II was characterised by fractional crystallisation of clinopyroxene, plagioclase and alkali-feldspar, with generation of abundant phonolites with extreme enrichments in incompatible element. During Period III, two

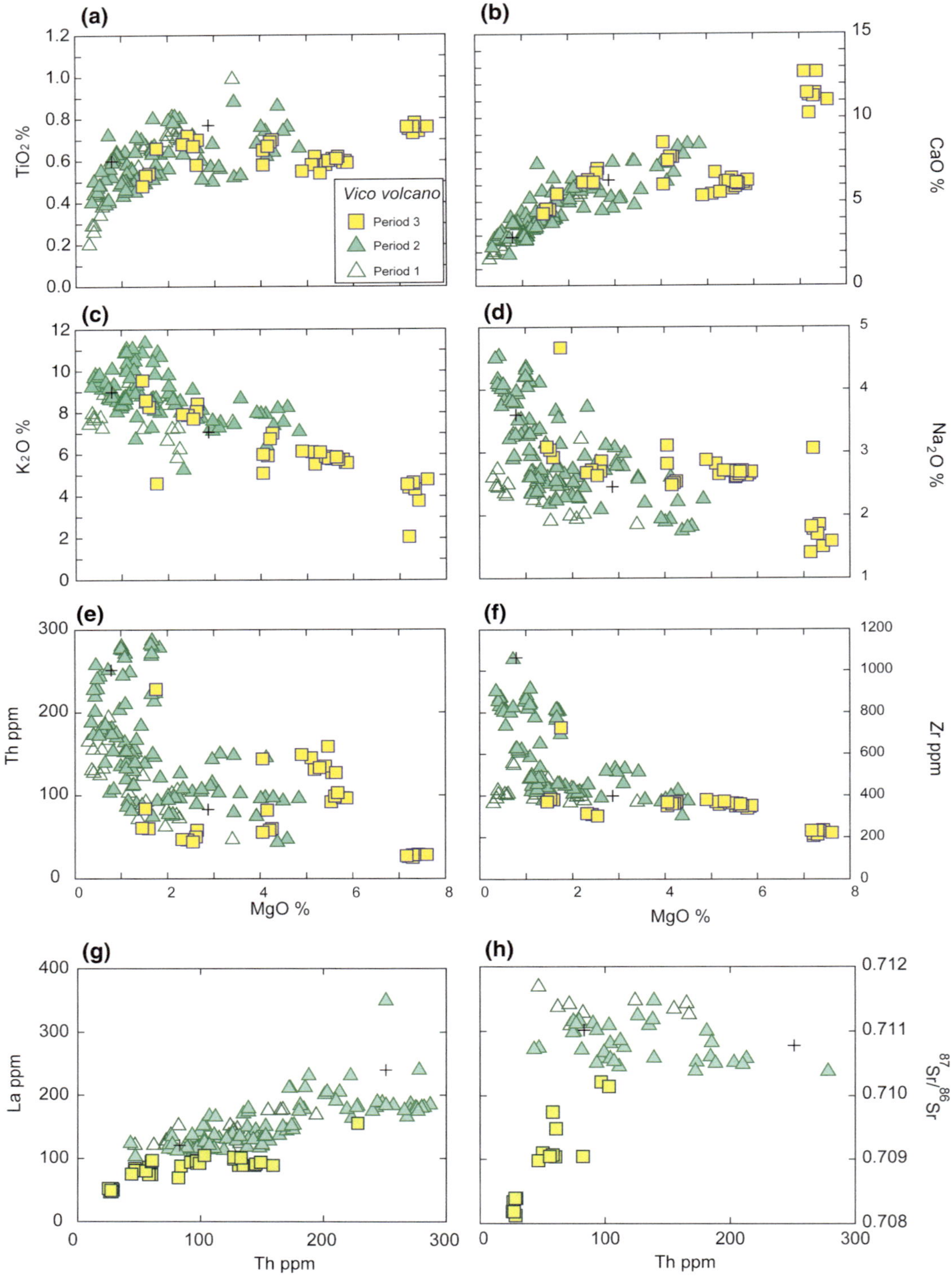

Fig. 4.13 Variation diagrams for the Vico volcano. *Crosses* are samples of unknown stratigraphic position

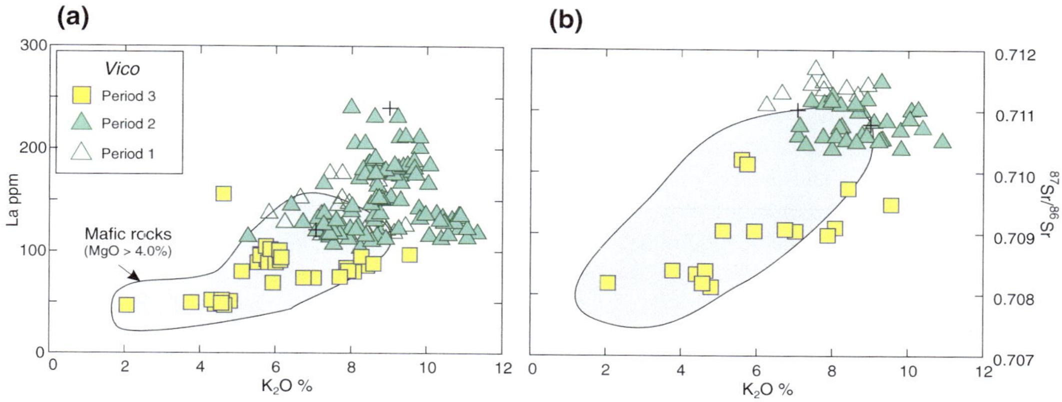

Fig. 4.14 Variation of La and $^{87}Sr/^{86}Sr$ versus K_2O wt% for the Vico volcanics. The *shaded area* is the field of mafic rocks. *Crosses* are samples of unknown stratigraphic position

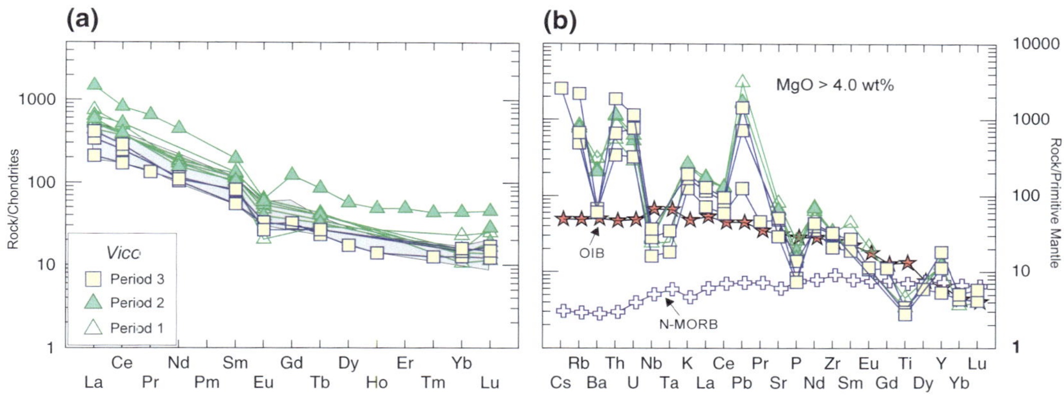

Fig. 4.15 **a** REE patterns of the Vico rocks. The *shaded area* encloses patterns of mafic rocks; **b** Incompatible element patterns of mafic rocks from Vico. OIB and N-MORB composition (Sun and Mcdonough 1989; Gale et al. 2013) are reported for comparison

distinct types of magmas (latite and trachybasalts) were emplaced. These probably mixed shortly before eruption, as already suggested by Barbieri et al. (1988) and as supported by mineral-host rock isotopic disequilibria (Perini et al. 2004). However, most authors agree that the variable compositions of mafic rocks cannot be related to shallow level evolution and mostly reflect source heterogeneity (Barbieri et al. 1988; Perini et al. 2004).

4.5 Sabatini District

The Sabatini Volcanic District developed between about 0.8–0.6 Ma and 90 ka over an area of about 2000 km^2, including a large part of the city of Rome (e.g. Karner et al. 2001; Sottili et al. 2010). Volcanism was predominantly explosive and generated widespread pyroclastic deposits and some lava flows, which were emitted from a large number of centres including some calderas. The volcanism occurred along a zone of crosscutting NW-SE and NE-SW faults. The Sabatini rocks rest over the same type of bedrocks as Vico, and also overlie the acid volcanics of Tolfa-Manziana-Cerite complex of the Tuscany Province (e.g. Funiciello et al. 1976; Di Filippo 1993; De Rita et al. 1983, 1993).

4.5.1 Volcanology and Stratigraphy

The Sabatini district consists of two main multicentre complexes, the Sacrofano volcano in the east and the Bracciano volcano in the west, both developed around a large E-W trending fracture zone characterised by several caldera collapses (Fig. 4.16; De Rita et al. 1983, 1996; Cioni et al. 1993). Various phases of eruptive activity have been recognised on the basis of field and geochronological data (De Rita et al. 1983, 1996; Conticelli et al. 1997; Karner et al. 2001; Sottili et al. 2004, 2010, 2012; Masotta et al. 2010; Del Bello et al. 2014; Marra et al. 2014). The early eruptions took place in the eastern sector (Morlupo) and at an inferred Southern Sabatini centre, and produced voluminous pyroclastic deposits (e.g. Tufo Giallo della Via Tiberina; Masotta et al. 2010). Ages range from about 0.59–0.38 Ma, but older activity is testified by tephra layers in the fluvial deposits of the Tiber river, which yielded ages of 0.8–0.61 Ma (Marra et al. 2014 with references). Successively, intense explosive eruptions took place at Sacrofano and Bracciano (mostly 0.32–0.2 Ma; Sottili et al. 2010) where caldera collapses occurred. Final activity (about 0.25–0.09 Ma) was concentrated between the Bracciano and Baccano calderas with the emission of strombolian scoriae and lavas, and hydrovolcanic products from monogenetic cones and maars (Fornaseri 1985; De Rita et al. 1993; Villa 1993; Karner et al. 2001; Sottili et al. 2012).

4.5.2 Petrography and Mineral Chemistry

Major element studies (Cundari 1979; Conticelli et al. 1997; Del Bello et al. 2014) have shown that the Sabatini volcanics consist almost entirely of undersaturated ultrapotassic rocks, ranging from tephrite to phonolite (Fig. 4.17). A few trachytes and latites have been erupted locally (e.g. Morlupo and Vigna di Valle lava flows).

Rock textures are porphyritic with variable amounts of phenocrysts (about 5–40 vol%) set in holocrystalline to hypocrystalline groundmasses. Zoned diopside to hedenbergite clinopyroxene is an abundant and ubiquitous phenocryst phase. Leucite is a common groundmass mineral and occurs as a phenocryst in some phonotephrites and tephriphonolites. Olivine (Fo_{90-30}) is present as microphenocrysts, sometimes with euhedral chromite inclusions, and in the groundmass of leucite tephrites. Plagioclase ($\sim An_{93-53}$) and Fe–Ti oxides are phenocryst and microphenocryst phases in tephriphonolites and phonotephrites; sanidine and haüyne occur in the phonolites. Phlogopite phenocrysts are rare. Groundmass generally contains clinopyroxene, leucite and opaque minerals plus K-feldspar, plagioclase, some glass, and sporadic phlogopite and brown amphibole. Apatite is a common accessory mineral. Sphene is found in some leucite-haüyne phonolites. The rare latites have strongly porphyritic textures, with abundant phenocrysts of plagioclase and phlogopite, and minor clinopyroxene and K-feldspar set in a groundmass consisting of clinopyroxene, sanidine, hyalophane (Ba-rich K-feldspar), plagioclase, opaque minerals, some nepheline, and apatite. Large leucite phenocrysts or xenocrysts are also present. Trachytes are characterised by a trachytic texture, with phenocrysts and microphenocrysts of sanidine and zoned dark green clinopyroxene surrounded by a groundmass of sanidine, clinopyroxene and opaque minerals (Cundari 1979; Conticelli et al. 1997).

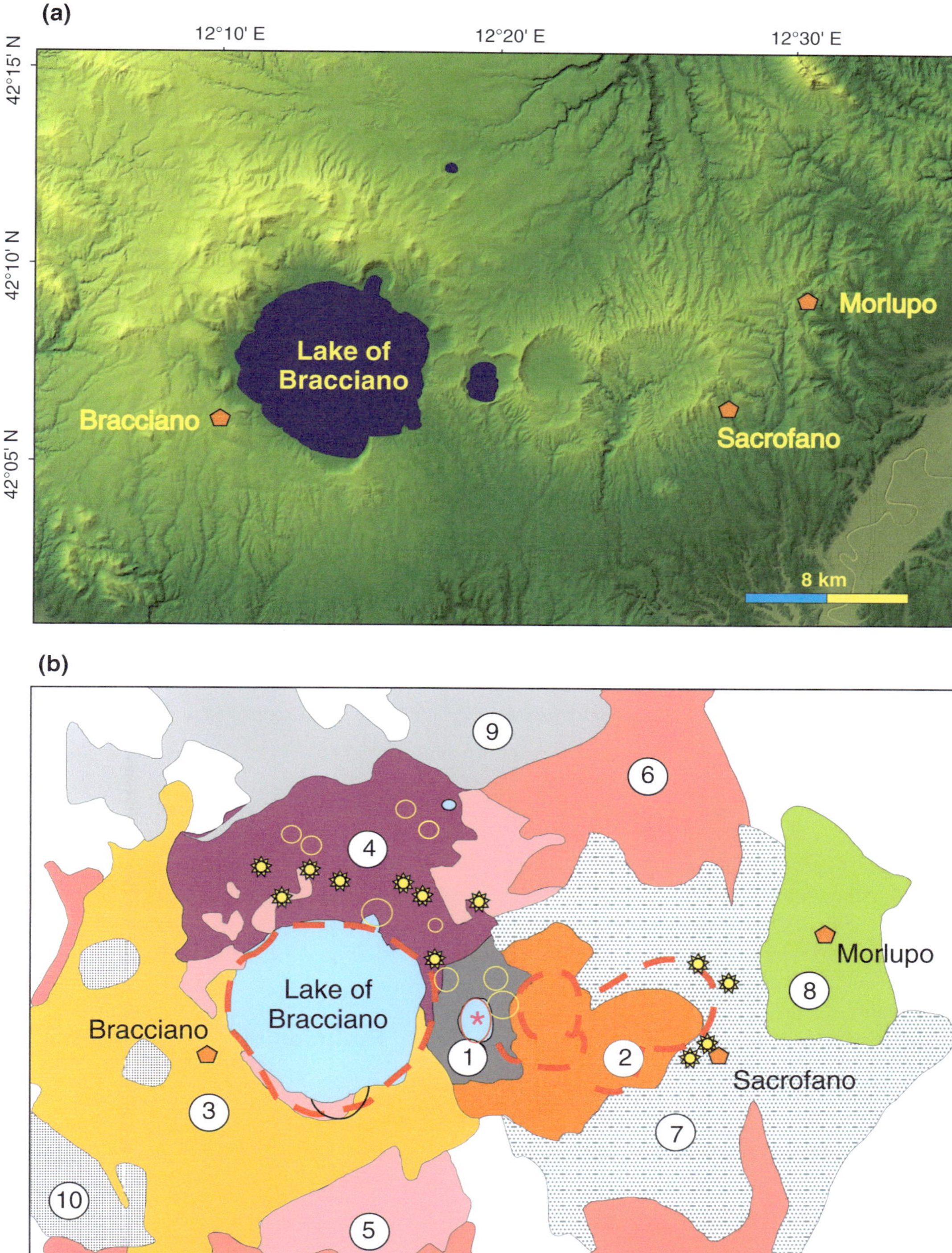

Fig. 4.16 **a** *Colour shaded* relief map of Sabatini volcanoes (Tarquini et al. 2012); **b** Schematic geological map of Sabatini district, simplified after De Rita et al. (1993). *1* Final stage volcanics; *2* Baccano pyroclastics; *3* Western lava sequence; *4* Northern volcanics; *5* Main lava flows; *6* Tufo Rosso a Scorie Nere (Red Tuff with Black Scoriae); *7* Sacrofano volcanics; *8* Morlupo volcanics; *9* Vico volcanics; *10* Manziana-Ceriti dome complex (Tuscany Province). *Thick dashed line* caldera rim; *circle* crater; *star* scoria cone

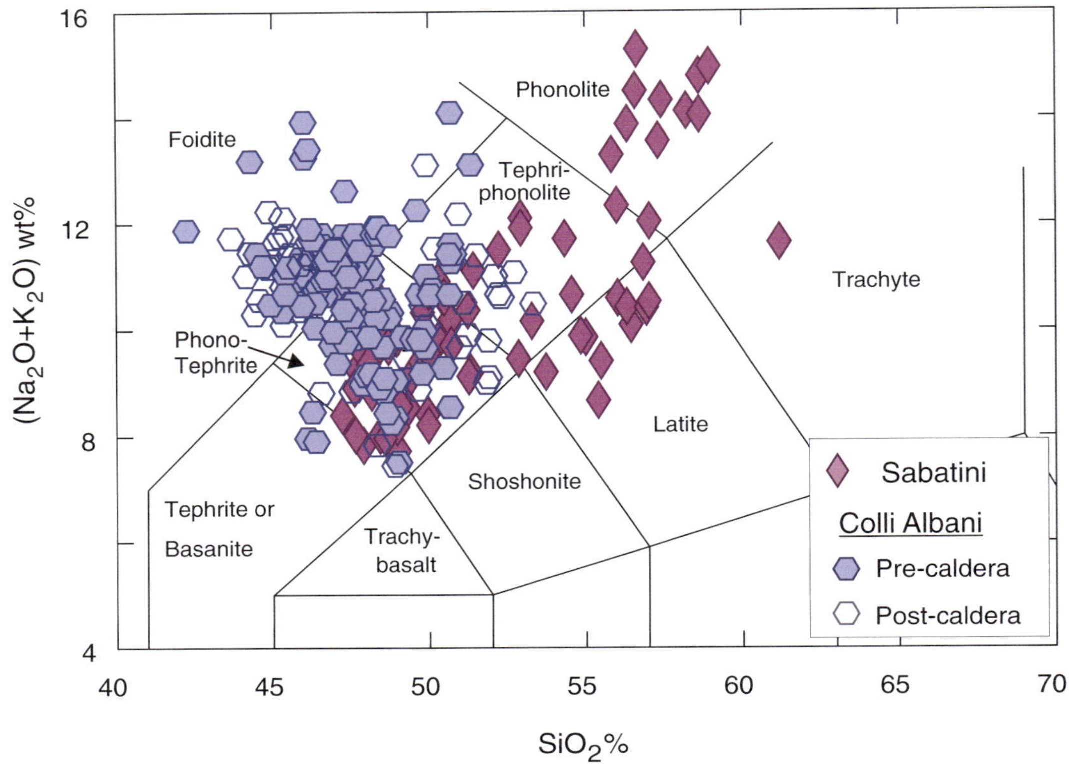

Fig. 4.17 Total alkali versus silica diagram for Sabatini and Colli Albani. Data normalised to 100 % on a LOI-free basis

4.5.3 Petrology and Geochemistry

Major and trace element versus MgO diagrams display rather scattered distribution for several elements (Fig. 4.18). Conticelli et al. (1997) recognised two groups of rocks, characterised by distinct enrichments in Ba, LREE, and other incompatible elements. The highly enriched rocks commonly have leucite phenocrysts, whereas the less enriched group contains plagioclase as phenocryst and leucite in the groundmass. The depleted group mostly erupted at small post-calderas monogenetic centres, whereas the rocks enriched in incompatible elements were emplaced during pre- to syn-calderas activity. All the rocks have fractionated REE with small negative Eu anomalies (Fig. 4.19a). Incompatible element patterns of mafic rocks (Fig. 4.19b) resemble closely other Roman mafic volcanics.

Sr-isotope ratios range from $^{87}Sr/^{86}Sr \sim$ 0.7095–0.711, and exhibit a tendency to increase with increasing abundances of incompatible elements in the mafic rocks. Nd–Pb–Hf isotope ratios fall within the field of other Roman volcanoes (Fig. 4.10; Table 4.2).

4.5.4 Evolution of Sabatini Magmas

Trends of major element variation have been interpreted to indicate a fractional crystallisation evolutionary mechanism for the bulk of the Sabatini rocks (Cundari 1979). Olivine and clinopyroxene were dominant fractionating phases in the mafic magmas, whereas feldspar separation had a major role in evolved melts (Del Bello et al. 2014). Conticelli et al. (1997) suggested that all the rocks derived from a single type of parent magma. According to this hypothesis, the Ba-REE-rich series was subjected to continuous fractional crystallisation and input of mafic magma. Such a process was able to generate strong enrichments in some incompatible elements, leaving CaO and MgO at relatively high levels. In contrast, the less enriched group was subjected to a process in which continuous mixing of mafic magma played a minor role, but there was significant crustal assimilation. This generated an increase of Sr isotope ratios and moderate incompatible element enrichment in the evolved magmas. The latites and trachytes were probably derived from a parent magma less enriched in potassium than the other Sabatini volcanics. However, no evidence for this mafic melt has been found at Sabatini. Polybaric fractional crystallisation and modest (less than 5 %) assimilation of carbonate wall rocks were suggested by Del Bello et al. (2014) for some late erupted magmas.

4.6 Colli Albani (Alban Hills)

The Colli Albani (Alban Hills), also known as Vulcano Laziale, is a large stratovolcano with central calderas, located a few kilometres south-east of Rome (Fig. 4.20). It was constructed at the southern end of the Olevano-Antrodoco line (OAL), i.e. at the boundary between structurally and stratigraphically different zones of the Apennine chain (e.g. Parotto and Praturlon 1975; Funiciello and Parotto 1978; De Rita et al. 1995; Meloni et al. 1997). Here, pre- and syn-orogenic terrains of the Sabina Units occurring west of the OAL (Upper Triassic-Lower Jurassic carbonate platform and Lower Jurassic-Middle Miocene carbonate pelagic sediments) are tectonically superimposed over Mesozoic-Cenozoic Latium-Abruzzi carbonate platforms of Central Apennines, which crop out east of the OAL (Mattei et al. 1995). The post-orogenic rocks consist of Upper Miocene-Pleistocene marine and continental clastic sediments disconformably covering the older units.

The Colli Albani area is affected by regional NW-SE faulting, which cuts the N-S trending OAL. The volcano basically developed at the intersection between the two systems.

The Moho beneath the Colli Albani volcano is about 22 km deep. The upper mantle is characterised by relatively low Vp, which has been attributed to upwelling of hydrated asthenosphere (Di Stefano et al. 2009). The crust does not contain important discontinuities, except for a high-Vp dome-shaped body detected at 3–5 km depth, likely representing solidified magmatic material (Feuillet et al. 2004; Bianchi et al. 2008).

Colli Albani is also the only volcano in Central Italy where there are seismic activity and ground uplift, possibly testifying to the persistence of an active magmatic body at depth (Feuillet et al. 2004; Riguzzi et al. 2009). Seismicity is NW-SE oriented and is concentrated at about 3–5 km depth beneath the western rim of the caldera, at the limit between the high-Vp dome-shaped body and the sedimentary units (Feuillet et al. 2004; Chiarabba et al. 2010b).

4.6.1 Volcanology and Stratigraphy

The Colli Albani volcano consists of dominant pyroclastic rocks and minor lavas, which were erupted from about 0.6 Ma to a few thousand years before present, possibly until early Roman times (Soligo and Tuccimei 2010). The volcanic rocks crop out over an area of about 1000 km^2, including the eastern outskirts of Rome.

Because of its proximity to a highly populated area, the Colli Albani volcano has long attracted the attention of scientists, who reported on important field data and clarified several first-order volcanological, petrological and morphological features of the volcano (e.g. Sabatini 1900). However, the extensive study by Fornaseri et al. (1963) laid the foundation for modern petrological and volcanological investigation and most of their findings are still valid today.

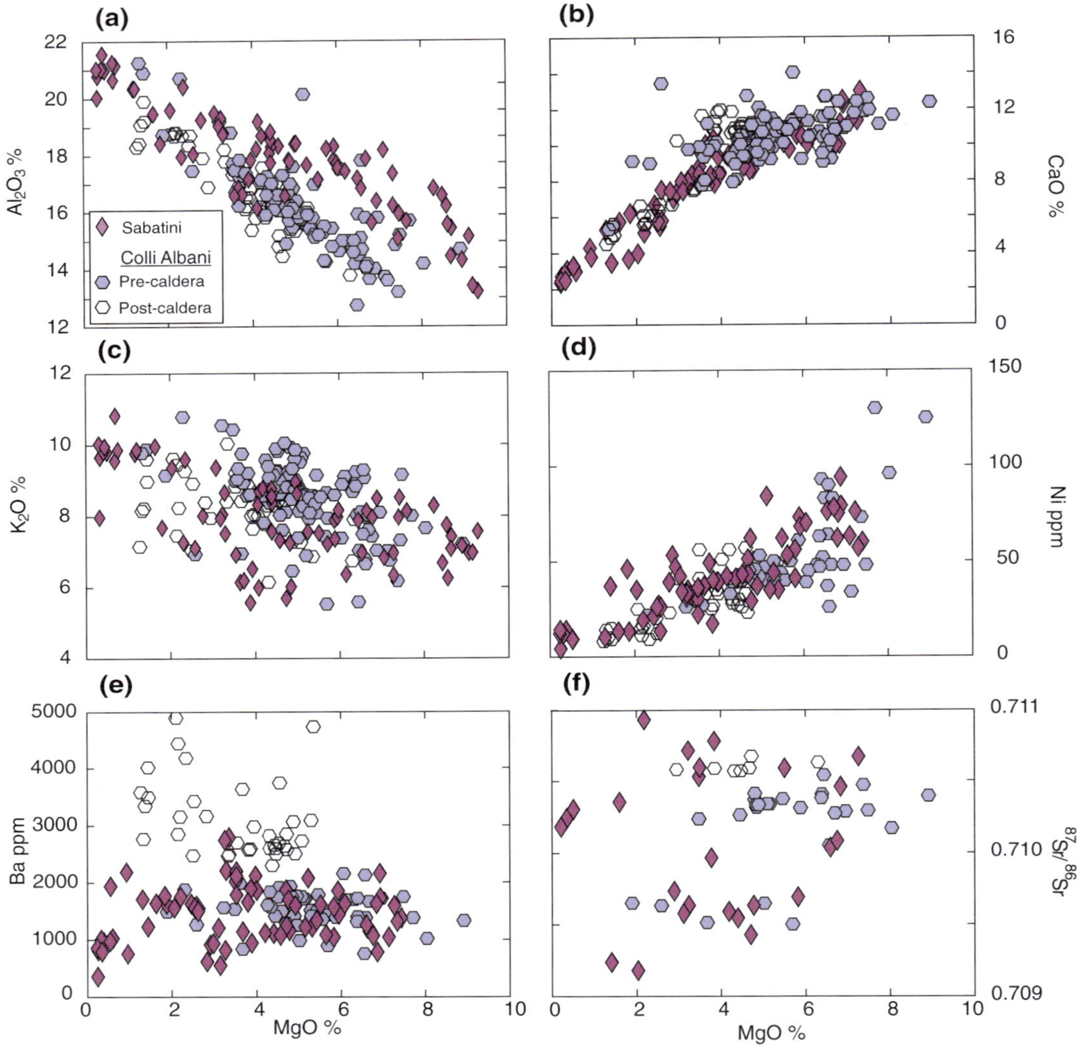

Fig. 4.18 Variation diagrams for Sabatini and Colli Albani volcanics

Volcanological and geochronological studies (Fornaseri et al. 1963; De Rita et al. 1995; Karner et al. 2001; Marra et al. 2003; Funiciello and Giordano 2010) recognized extreme variations for eruptive style, from plinian to strombolian and hawaiian, which took place during various phases of activity (Fornaseri et al. 1963; De Rita et al. 1988; Giordano 2010). Here the stratigraphic scheme of Giordano (2010) is summarized (Fig. 4.20b):

1. The Vulcano Laziale Period (about 0.6–0.355 Ma), which built up the main cone and a large central caldera;
2. The Tuscolano-Artemisio-Faete Period (0.355–0.18 Ma), during which various peri-caldera and extra-caldera centres were constructed, along with the intra-caldera Faete cone;
3. The Via dei Laghi Period (0.2 Ma to 5.8 ka), characterised by phreatomagmatic explosions along the western rim of the caldera.

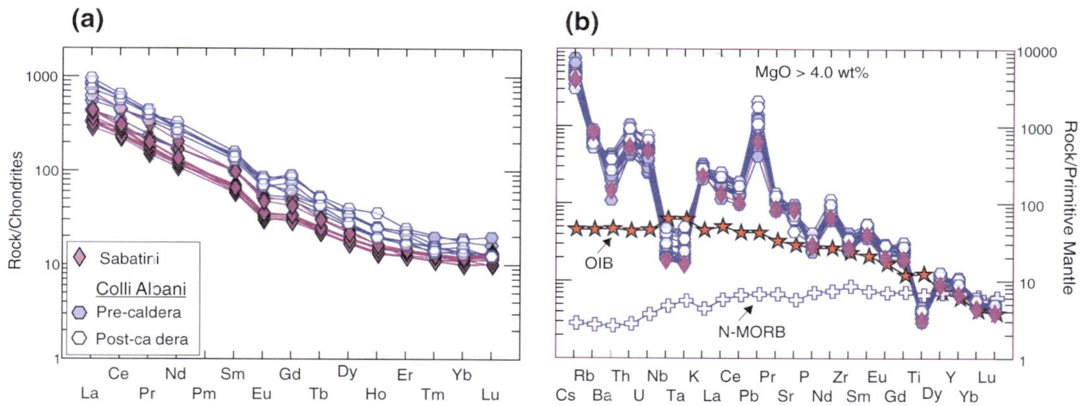

Fig. 4.19 **a** REE patterns normalised to chondrites for the Sabatini and Colli Albani volcanics; **b** Incompatible element patterns of mafic rocks. The average OIB and N-MORB (Sun and McDonough 1989; Gale et al. 2013) are reported for comparison

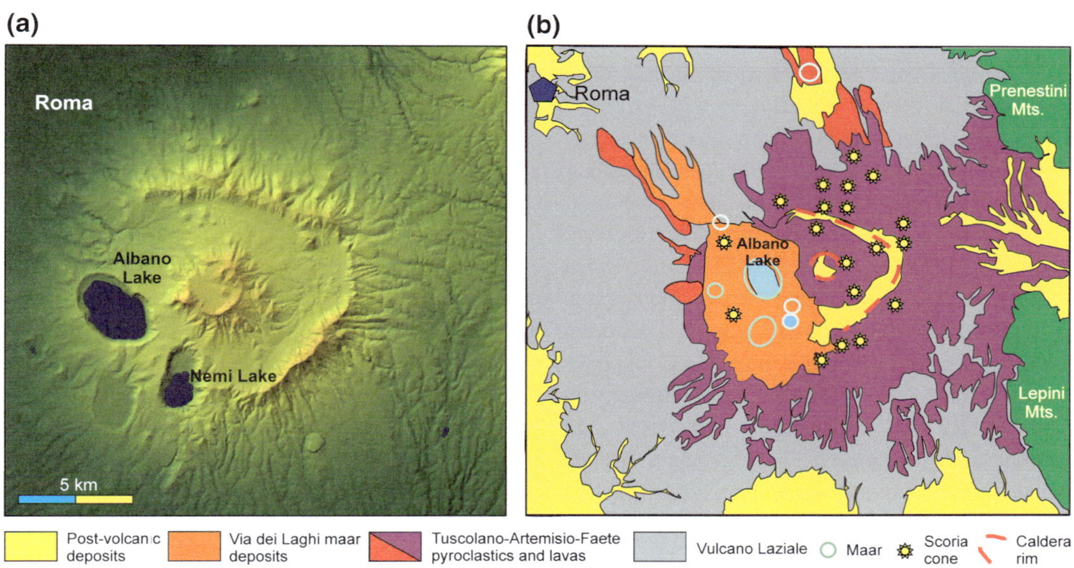

Fig. 4.20 **a** *Colour shaded* relief map of the Colli Albani volcano (Tarquini et al. 2012); **b** Schematic geological map of Colli Albani volcano, simplified after Giordano (2010)

The Vulcano Laziale Period erupted about 280 km^3 of volcanic material, during at least seven large-volume ignimbrite eruptions (VEI[1] = 5–7), which spread over an area of about 1600 km^2. The latest ignimbrite eruption of this period occurred at about 355 ka, and emplaced the Villa Senni Tuff, a 25 km^3 pyroclastic flow deposit (Freda et al. 1997; Vinkler et al. 2012). Each major ignimbrite eruption was followed by strombolian activity and lava effusions mostly from peri-caldera fractures (e.g. Vallerano-Acquacetosa lavas, De Rita et al. 1995; Marra et al. 2003).

[1]VEI = Volcanic Explosivity Index (Newhall and Self 1982) is a measure of explosive magnitude of volcanic eruptions. It ranges from zero (effusive Hawaiian eruptions, with a maximum of 10^4 m^3 of erupted tephra) to 8 (cataclysmic eruptions with up to 10^{13} m^3 of tephra).

The Tuscolano-Artemisio-Faete activity outpoured about 6 km^3 of volcanic products, which built up the Monte Faete lava and scoria cone, and several peri-caldera and extra caldera pyroclastic and lava centres (e.g. Osa lava at 297 ka; Capo di Bove lava at 277 ka; Karner et al. 2001; Marra et al. 2003). A small nested caldera was produced by the Faete cone eruptions.

The final Via dei Laghi hydromagmatic activity erupted about 1 km^3 of pyroclastic surge, flow and lahar deposits, and very few lavas from several explosion craters (e.g. Nemi and Albano lakes) mainly located along the southwestern rim of the Vulcano Laziale caldera. According to volcanological and $^{40}Ar/^{39}Ar$ geochronological study of Freda et al. (2006) the activity took place during three main phases, at about 69 ± 1 ka, 39 ± 1 ka and 36 ± 1 ka. Villa et al. (1999) measured a $^{40}Ar/^{39}Ar$ age of 26 ± 1 ka for late pyroclastic flow deposits from the Albano crater. Finally, a ^{14}C age of 5.8 ± 0.1 ka has been found in a paleosoil underlying one of the uppermost laar deposits of the Albano Lake, shifting the latest activity of Colli Albani to prehistoric times (De Benedetti et al. 2008). However, integrated $^{40}Ar/^{39}Ar$ and ^{14}C geochronological studies gave inconsistent results for the two methods, with ^{14}C ages being scattered and much younger than $^{40}Ar/^{39}Ar$ ages. It was therefore suggested that scattered and contradictory ^{14}C ages might result from contamination of the paleosols by younger organic carbon from the overlying soil horizons (Giaccio et al. 2009).

There is considerable debate on the present state of the Colli Albani volcano. As mentioned earlier, shallow seismicity and ground deformation have been registered. Moreover, historical documents by Roman authors (e.g. Titus Livius, Pliny the Elder, etc.) report on phenomena, such as rain of stones, coloured vapour emission, sudden explosions, which can be ascribed to volcanic activity. Moreover, a catastrophic flood from the Albano area has been described in the classical Roman literature. According to some authors, these events were related to water uprise and spillover at the crater lake, and prompted the excavation (IV century BC) of a drain tunnel through the flank of the Albano crater, with the objective of keeping the water level 70 m below the rim (e.g. Funiciello et al. 2003). Therefore, based on archaeological and historical records and on seismicity, the Colli Albani volcano is now considered in a quiescent state (Montone et al. 1995; Voltaggio and Barbieri 1995; De Benedetti et al. 2008; Giordano 2010 with references).

4.6.2 Petrography and Mineral Chemistry

The Colli Albani rocks range from ultrapotassic tephrite to foidite (leucitite) and tephriphonolite, exhibiting a stronger degree of silica undersaturation than other Roman volcanics (Figs. 4.2 and 4.17). Lava flows are more mafic (i.e. contain higher MgO) than pyroclastic rocks from the same eruptive phase, but ranges in silica and alkalies are similar (Trigila et al. 1995).

Most lava flows exhibit aphyric to poorly porphyritic textures. Main phenocryst phases include clinopyroxene and leucite; olivine and Fe–Ti oxides occur in minor amounts. Groundmass contains the same phases plus nepheline, melilite, phlogopite and some calcite. Notably, plagioclase is virtually absent in the Colli Albani rocks, except for rare microcrysts in the groundmass of some lavas.

Pyroclastic rocks include scoriae and pumices, which show variably porphyritic textures with dominant leucite phenocrysts and minor clinopyroxene, some mica, and rare olivine. Groundmass is microcrystalline to glassy and often contains secondary zeolites and clay minerals (Fornaseri et al. 1963; Trigila et al. 1995). Accessory phases include Fe–Ti oxides, apatite and rare garnet.

Clinopyroxene is Ca-rich and ranges from diopside to hedenbergite. Normal and reverse zoning is commonly observed (e.g. Aurisicchio et al. 1988; Trigila et al. 1995; Boari et al. 2009). Al/Ti (a.f.u.) ratios are high (about 7 to 20), as typically observed in the Roman Province (see Fig. 2.11a). Olivine exhibits a strong compositional variation (Fo_{92-33}), becoming enriched in FeO from phenocryst cores to rims, and to groundmass. Some olivine phenocrysts contain chromite at their cores. Cr# [molar Cr/(Cr + Al)] is around 0.78, similar to chromite from other Roman rocks and lower than values observed in the Tuscany lamproites and Intra-Apennine kamafugites (see Fig. 2.11b; Perini and Conticelli 2002; Boari and Conticelli 2007; Conticelli et al. 2010a, b). Mica is generally phlogopite with variable Ti and Ba contents (Trigila et al. 1995). Leucite has a poorly variable nearly stoichiometric composition, nepheline is characterised by K/Na ratio around 2.7, and melilite has a fairly constant akermanite-rich composition (e.g. Boari et al. 2009).

The Colli Albani pyroclastic deposits contain abundant xenoliths. These include fragments of lava flows and microgranular rocks that represent intrusive equivalents of erupted lavas, cumulate lithologies, and skarns. Some xenoliths are rich in kalsilite and have a composition resembling kamafugitic lavas of the Umbria volcanoes (Federico et al. 1994 and references therein). Xenoliths often contain minerals rich in incompatible elements such as uranpyrochlore, Th-rich britholite, cuspidine, baddeleyite, and perovskite (Washington 1906; Fornaseri et al. 1963; Federico et al. 1994; Federico 1995; Federico and Peccerillo 2002; Gaeta et al. 2009; De Benedetti et al. 2010; Peccerillo et al. 2010).

4.6.3 Petrology and Geochemistry

Variation diagrams of major and trace elements versus MgO at Colli Albani (Fig. 4.18) show scattering, with overall positive correlations for CaO, FeO_{total} and ferromagnesian trace elements (Cr, Ni, Co, etc.), negative correlations for Na_2O, K_2O, Al_2O_3 and incompatible elements (Th, La, Ta, etc.). Inter-element variation diagrams (not shown) show smooth trends for some elements (e.g. Th vs. Zr and Ta) but scattering for others (e.g. Th vs. Rb). Many correlations improve if rocks from single volcanic edifices are considered (Boari et al. 2009). The pre-caldera lavas are enriched in incompatible elements relative to post-caldera samples, at the same MgO level (Fig. 4.18; Table 4.2). REE and incompatible element patterns resemble closely those of other Roman ultrapotassic rocks (Fig. 4.19). Sr, Nd and Pb isotopic ratios for whole rocks are moderately variable (Fig. 4.10; $^{87}Sr/^{86}Sr \sim 0.7095$ to 0.7107; $^{143}Nd/^{144}Nd \sim 0.51209$ to 0.51215; $^{206}Pb/^{204}Pb \sim 18.76$; $^{207}Pb/^{204}Pb \sim 15.70$; $^{208}Pb/^{204}Pb \sim 39.03$; e.g. Conticelli et al. 2002, 2010b; Boari et al. 2009; Gaeta et al. 2011). Sr-isotopes for separated clinopyroxenes show somewhat wider range of values with $^{87}Sr/^{86}Sr$ ratio mostly falling between 0.7085 and 0.7112 (Freda et al. 2006; Gaeta et al. 2006). O-isotope compositions for whole rocks and separated phenocrysts range from $\delta^{18}O_{SMOW} \sim +5.4$ to +8.3 ‰. There is an overall negative trend of $\delta^{18}O$ versus MgO contents for clinopyroxenes from juvenile pyroclasts (Dallai et al. 2004). He-isotope data reported by Martelli et al. (2004) for olivine and clinopyroxene are in the range $R/R_A = 0.44$–1.73.

4.6.4 Evolution of Colli Albani Magmas

Trigila et al. (1995) suggested that the entire rock suite from Colli Albani could be explained by about 60 % fractional crystallisation of a mafic tephrite, dominated by the separation of clinopyroxene, plus minor leucite and olivine. Clinopyroxene-dominated fractionation was favoured by assimilation of carbonate wall rocks (e.g. Iacono Marziano et al. 2007) and prevented magma evolution to phonolitic compositions. Peccerillo et al. (1984) argued that distinct trace element trends for pre- and post-caldera rocks indicate the occurrence of two magma suites, possibly derived from slightly different parental magmas. In contrast, experimental evidence favours the hypothesis that distinct trends may be related to variable chemical-physical conditions during fractionation (Trigila et al. 1995). Boari et al. (2009) suggested that different elemental trends were particularly evident in the post-caldera magmas, indicating the occurrence of different batches of magmas ponding and differentiating in various distinct reservoirs. Magma-wall rock interaction is also a main evolutionary process at Colli Albani. In particular, assimilation of carbonate rocks has been amply documented by geochemical and experimental studies, and by stable isotope geochemistry on lavas, mineral phases and xenoliths (e.g. Federico et al. 1994; Dallai et al. 2004; Iacono Marziano et al. 2007; Freda et al. 2008; Conte et al. 2009; Mollo et al. 2010; Peccerillo et al. 2010; Mollo and Vona 2014). This process stabilised clinopyroxene and probably also generated carbonate melts in the crust (Conte et al. 2009; Peccerillo et al. 2010; Gozzi et al. 2014). Carbonate wall-rock assimilation is also an excellent and simple explanation for the high degree of silica undersaturation of the Colli Albani magmas, which is stronger than in other Roman volcanoes. It might also be a main cause of extensive CO_2 degassing in the area (Iacono Marziano et al. 2007).

4.7 Petrogenesis of the Roman Province

The petrogenesis of potassium-rich magmatism of the Roman Province has been one of the most debated subjects in igneous petrology. As for other Italian potassic rocks, controversial issue is the occurrence of lavas bearing both mantle and crustal signatures, such as high MgO, Ni, Cr contents and undersaturation in silica, along with elevated K_2O, large-ion lithophile elements and radiogenic Sr. Some of these contrasting features were noticed by early scholars (e.g. Sabatini 1900; Washington 1906) and have prompted numerous studies over the last century.

According to Rittman (1933), the peculiar composition of potassium-rich rocks from Central Italy (and of the entire Mediterranean Series, as the potassic rock association was named in the past) was related to assimilation (or syntexis) of carbonate rocks by a trachytic magma. This hypothesis, first suggested by Daly (1918) for the undersaturated magmas in general, explained the high potassium contents of Roman rocks as inherited from the trachyte magma, whereas the silica undersaturation was related to carbonate assimilation. Carbonate syntexis was rejected by Savelli (1967), who demonstrated that ultrapotassic rocks from Vesuvio and other Central Italy volcanoes have higher incompatible element abundances than both limestones and any common magma, including trachytes. This conclusion brought full discredit on carbonate assimilation as an important petrogenetic process in the Roman volcanoes, a conclusion that recent studies recognised as unduly categorical (e.g. Iacono Marziano et al. 2007; Freda et al. 2008).

Ideas suggesting potassium enrichment by gaseous transfer operating on normal basaltic magmas inside the volcano plumbing systems became quite popular for some time. However, it was not explained how this randomly operating process could generate potassic magmas with basically homogeneous compositional features at a regional scale.

A genesis by interaction between mantle-derived magmas and upper crustal rocks, such as schists and gneiss, was also suggested by some authors (e.g. Turi and Taylor 1976). Such a hypothesis, however, conflicts with many petrological and geochemical characteristics of Roman rocks, including strong undersaturation in silica and high Ni and Cr contents of mafic rocks. Therefore, it has been concluded that fractional crystallisation and wall rock assimilation have been working in the Roman volcanoes, and often had important roles during shallow-level magma evolution. However, they are unable to explain the main compositional characteristics of the mafic (i.e. parental) magmas, whose first-order petrological and geochemical characteristics must be related to mantle rather than to intra-crustal processes (e.g. Conticelli and Peccerillo 1992; Peccerillo 2002 and references therein).

4.7.1 Magma Evolution Processes

The magmas erupted at Roman volcanoes are dominated by phonolitic, tephriphonolitic and trachytic pyroclastics and lavas. It is amply acknowledged that these rocks represent daughter products generated by evolution of mafic parents. However, mafic rocks are scarce in the Roman volcanoes, which makes magma evolution an intriguing petrological and volcanological process in the Roman Province.

4.7.1.1 Fractional Crystallization and Magma Mixing

Fractional crystallisation mostly took place in huge chemically zoned shallow-level magma chambers, which were tapped by high-magnitude explosive eruptions and subsequently collapsed to form calderas and volcano-tectonic depressions (e.g. Palladino et al. 2014). Mass balance calculations demonstrate that some 50-60 % fractionation is necessary to have felsic compositions (i.e. trachytes and phonolites) from mafic parents (i.e. K-trachybasalts and K-tephrites). Since the Roman Province mostly consists of evolved rocks, it is obvious that the largest fraction of the parental magmas did not erupt at the surface but remained as melts and crystallised rocks at depth. This implies that the amount of potassic magmatism produced in the Roman Province is much larger than the huge volumes of rocks exposed at the surface.

Magma mixing has also been extensively working in the Roman volcanoes. Such a process is demonstrated by several petrographic and geochemical features, such as the common occurrence of clinopyroxene phenocrysts showing normal (i.e. diopside to hedenbergite from core to rim) and reverse chemical zoning and isotopic disequilibrium with the coexisting matrix (e.g. Thompson 1977; Varekamp and Kalamarides 1989). Elemental scattering and linear inter-element trends observed for some rock suites may depend on mixing among compositionally different melts.

4.7.1.2 Assimilation of Wall Rocks

Interaction between magma and wall rocks has been a common process in the Roman volcanoes. Therefore, a core problem is the question of how much this interaction modified the pristine compositional characteristics of magmas.

Geochemical studies have shown that assimilation of crustal rocks by ultrapotassic magmas has a dilution effect on concentrations of many trace elements (e.g. Conticelli 1998; Peccerillo 1998, 2004). This is obvious since the Roman magmas have higher incompatible element contents the most common silicate and carbonate crustal rocks. Compositional contrast also effectively buffers variations of incompatible element ratios and radiogenic isotope signatures, and it has been amply demonstrated that small to moderate (some 10–20 %) assimilation of siliceous upper crustal rocks (schists, gneiss, granitoids) does not affect dramatically these parameters for potassic magmas (e.g. Conticelli and Peccerillo 1992; Conticelli 1998).

However, assimilation of carbonate rocks is able to modify significantly some petrological characteristics of potassic magmas. It promotes crystallization of Ca-rich pyroxene and favours desilication of magma, increasing the degree of silica undersaturation (e.g. Iacono Marziano et al. 2007; Freda et al. 2008; Conte et al. 2009; Mollo et al. 2010; Mollo and Vona 2014; Pichavant et al. 2014). According to Conte et al. (2009), variable degrees of carbonate assimilation associated with fractional crystallisation may generate different suites of derivative liquids that mimic those formed by fractional crystallisation of different mafic parents. Incompatible element contents are strongly diluted by carbonate syntexis, whereas incompatible element ratios are poorly affected. Radiogenic isotopes have somewhat different behaviour during carbonate assimilation. Nd–Pb isotope ratios remain basically unmodified, because of the low amounts of elemental Nd and Pb in the carbonate rocks. In contrast, Sr-isotope ratios may change significantly, due to rather high Sr contents in the assimilant (e.g. Melluso et al. 2003). However, an important consequence of carbonate syntexis is an increase of oxygen isotope ratios in the magmas. Keeping these effects in mind, it can be concluded that some high degrees of silica undersaturation, such as at Colli Albani, may well depend on carbonate wall-rock assimilation (Iacono Marziano et al. 2007). This also explains the high $\delta^{18}O$‰ of clinopyroxene and olivine at this volcano (e.g. Dallai et al. 2004: Freda et al. 2006), as well as the moderately variable $^{87}Sr/^{86}Sr$ at nearly constant Nd-isotope ratios noticed by several studies (e.g. Boari et al. 2009).

4.7.2 Genesis of Primary Magmas

4.7.2.1 Evidence from Major Elements and Normative Compositions

As in the case of the Tuscany and Intra-Apennine magmatic provinces, the potassium-rich composition of Roman mafic rocks requires derivation from mantle sources containing phases rich in potassium, such as phlogopite. However, unlike the Tuscany ultrapotassic rocks, concentrations of both CaO and Na_2O are high in the Roman magmas, suggesting participation to the melt of mineral phases containing these elements (e.g. clinopyroxene). Therefore, it has been concluded that the mafic magmas of the Roman Province derived from phlogopite-bearing lherzolites, wehrlite or clinopyroxenite (e.g. Conticelli and Peccerillo 1992). The variable enrichments in potassium and incompatible elements for the KS and HKS suggest different relative roles of potassic and non-potassic phases during melting.

Experimental studies on simplified phlogopite-bearing mantle systems at high pressure (Wendlandt and Eggler 1980) have shown that low degrees of partial melting at 1 to about 2.5 GPa under H_2O-poor conditions generate silica saturated to undersaturated potassic liquids with increasing pressure. The presence of CO_2 does not alter this pattern, but similar liquids can be obtained at lower temperatures. Further studies demonstrate that melting of phlogopite-bearing peridotite under F-rich conditions gives silica saturated to undersaturated melts with increasing pressure (>1.2 GPa for harzburgite and >2.0 GPa for lherzolite; Melzer and Foley 2000 and references therein). Multiple-saturation investigations on liquidus phase relationships of natural volcanic rocks at high pressures with the presence of $H_2O + CO_2$ suggest that potassic undersaturated magmas can be formed by partial melting of phlogopite-bearing lherzolite or wehrlite at pressures around 2.5–3.5 GPa (e.g. Edgar et al. 1976; Thibault et al. 1992). A role for $H_2O + CO_2$ fluid phase at high pressure has been recognised by several other studies (e.g. Conceição and Green 2000, 2004; Thomsen and Schmidt 2008a, b). Overall, experimental evidence favours the hypothesis that the Roman KS and HKS primitive magmas are generated by melting of phlogopite-bearing clinopyroxenen-rich upper mantle rocks, at variable depth and H_2O/CO_2 pressure.

A somewhat different hypothesis for the origin of KS and HKS primary magmas is that the variable potassium and incompatible element enrichments reflect different amounts of phlogopite entering the melt, as a result of variable degrees of partial melting (Conticelli et al. 2004, 2010a). According to this hypothesis, metasomatic melts or fluids percolating through the mantle (most likely the rigid lithosphere) crystallise as phlogopite-rich veins across normal mantle rocks. Melting of these veins, e.g. triggered by an increase in temperature or decompression, produce melts whose potassium and incompatible element concentrations is simply determined by the relative amounts of veins and wall rocks participating in the melting. Early formed liquids consist of almost pure vein material and are more enriched in potassium than late melts, whose composition is heavily affected by melting of relatively depleted mantle around the veins. The eruption of less potassic magmas during late stages of activity at Vulsini-Latera and Vico would be evidence in favour of this hypothesis.

In conclusion, the studies summarised above have produced a consensus that the Roman magmas were generated in a phlogopite-bearing upper mantle and that variable enrichment in potassium likely reflects different degrees of (polybaric) partial melting. The occurrence of phlogopite within the upper mantle requires metasomatic modifications. Obviously, the high volumes of potassic magmas in this province demand metasomatism over very large masses of mantle rocks. Thus, the debate has shifted in recent years to the problem of the very nature of the processes responsible for phlogopite crystallisation in the mantle, i.e. the origin, nature, age and physical mechanism of mantle metasomatism in the Roman Province (e.g. Peccerillo and Frezzotti 2015).

4.7.2.2 Evidence from Trace Element and Isotope Geochemistry

The incompatible element patterns of Roman mafic rocks resemble those of pelites or other upper crustal silicic rocks (see Chap. 2, Fig. 2.9 d), although potassic magmas show higher LILE/HFSE ratios and more fractionated HREE. Negative anomalies of Eu in the chondritic patterns of REE also recall the upper crust (Taylor and McLennan 1985). Most likely, these features have been inherited from mantle sources, since evolution processes do not change significantly the ratios among incompatible elements and there is no plagioclase on liquidus for many mafic rocks, which could justify negative Eu anomalies. Therefore, most authors agree that parental magmas originated in upper mantle sources that underwent contamination by upper crustal material (e.g. Cox et al. 1976; Hawkesworth and Vollmer 1979; Peccerillo 1985).

Trace element and radiogenic isotope data furnish important constraints on the nature of the crustal contaminant. Most of the Roman mafic magmas have lower LREE/Sr and much higher LILE/HFSE ratios than pelites, metapelites and granites. Therefore, incompatible element ratios rule out bulk mixing between mantle and pelitic rocks or granites, as suggested for the Tuscany Province, and require some additional components or/and processes. Radiogenic isotope ratios also point to the same conclusion. For instance, $^{87}Sr/^{86}Sr$ ratios of Roman mafic rocks, although anomalous for mantle-derived magmas, are lower than most pelites and their metamorphic equivalents (e.g. schists and gneisses from Tuscany metamorphic basement) and are somewhat midway between these and carbonate rocks from Northern Apennines (e.g. Melluso et al. 2003).

The poorly variable $^{87}Sr/^{86}Sr$ and relatively low LREE/Sr ratios of Roman mafic rocks reveal a mantle contamination by a crustal rock, which had higher Sr abundances than metapelites and rather constant $^{87}Sr/^{86}Sr$ buffered at around 0.709–0.710. Such a composition is typical of carbonated pelites or marls. These rocks have similar incompatible element and REE patterns to pelites, but have lower LREE/Sr and Sr isotope ratios as an effect of the carbonate component (e.g. Melluso et al. 2003). The hypothesis of mantle wedge contamination by marly sediments, early suggested on the sole basis of petrological and geochemical evidence (Peccerillo et al. 1988; Conticelli and Peccerillo 1992) has found a wide acceptance and received much support by further geochemical and experimental investigation (e.g. Avanzinelli et al. 2009; Grassi and Schmidt 2011a, b). In particular, Thomsen and Schmidt (2008a, b) demonstrated that carbonate melts derived from marls at mantle depths have high K/Na ratios and LILE abundances and ratios closely resembling Roman magmas.

The high fractionation of HREE of Roman rocks is simply explained by assuming residual garnet either in the sources of metasomatic fluids or/and within the mantle during formation of potassic magmas. With regard to the very high values of LILE/HFSE, it has been noticed that these cannot derive from any common crustal rock and again require element fractionation at some stage of the potassic magma generation or during mantle evolution (e.g. Conticelli and Peccerillo 1992; see Chap. 3). This fractionation could result from the presence of a residual HFSE-rich phase either inside the mantle during the formation of the potassic magmas (e.g. Foley and Wheller 1990), or within the upper crustal rocks during separation of metasomatic fluids (Grassi et al. 2012). Many authors interpret the low HFSE concentrations of Roman magmas, close to MORB values (Figs. 4.9, 4.15, 4.19), as an evidence for a depleted nature of pre-metasomatic mantle rocks (e.g. Serri 1990). This is supported by high Cr/(Cr + Al) values of chromite inclusions inside olivine phenocrysts from many Roman volcanoes (Conticelli et al. 2011, 2013).

Finally, the Roman mafic rocks have similar REE and incompatible element patterns, as well as radiogenic isotopic signatures as the IAP kamafugites. This clearly supports the same type of source metasomatism for the two magmatic provinces. However, the IAP kamafugites have stronger degree of silica undersaturation than Roman magmas. Moreover, they show depletion in Al_2O_3 and Na_2O, which is not observed in the Roman mafic rocks. Whereas higher undersaturation of IAP magmas may depend on the amount of intra-crustal carbonate assimilation, low Al_2O_3 and Na_2O contents likely reflects pre-metasomatic mantle mineralogy. Accordingly, harzburgitic and lherzolitic pre-metasomatic sources have been respectively suggested for the Roman Province and the IAP, as also discussed in Chap. 3.

4.8 Age of Mantle Metasomatism and Geodynamic Implications

The age of mantle contamination in Central Italy is a key issue for geodynamic interpretation. Some authors (e.g. Castorina et al. 2000) suggest that contamination by material with high Rb/Sr and low Sm/Nd took place some 1.5–2.0 Ga ago, and the anomalous mantle remained isolated to develop the present day isotopic signatures. Conticelli et al. (2004) objected that this is unlikely in a tectonically unstable area such as the Mediterranean. However, storage of contaminated material in the deep mantle and its emplacement as a plume overcomes this objection (Bell et al. 2004, 2013). Therefore, old contamination implies a plume mechanism.

The suggestion of very old ages for mantle contamination stems from the steep trend of Rb/Sr versus $^{87}Sr/^{86}Sr$ defined by the volcanic rocks from central-southern Italy. If interpreted as an isochron, this trend would imply an aging of about 1.5 Ga within the mantle for material with variable Rb/Sr (Castorina et al. 2000). However, it has been noticed that, if considered in insulation, the Roman Province defines a rather flat trend of Rb/Sr versus $^{87}Sr/^{86}Sr$ (Peccerillo 2002). Therefore, if this trend is considered as an isochron, it clearly suggests a recent mantle contamination for the Roman Province. A young metasomatism has been also substantiated by Th-U disequilibrium studies (Villemant and Fléhoc 1989; Avanzinelli et al. 2008).

A young mantle metasomatic event by upper crustal rocks requires a recent subduction process. Such a hypothesis was first suggested by Thompson (1977) and has been successively constrained by a wealth of trace element and isotopic studies (e.g. Peccerillo et al. 1984; Peccerillo 1985; Rogers et al. 1985; Conticelli et al. 2002; Avanzinelli et al. 2008). Metasomatism in the Roman Province has been related to addition of marly sediments to the upper mantle by the Adriatic plate (e.g. Peccerillo et al. 1988; Serri 1990; Doglioni et al. 1999). The subduction of the Adriatic plate has been going on from Miocene to Present and involved about 170 km of lithosphere (e.g. Carminati et al. 2005, 2012), carrying huge amounts of upper crustal material into the mantle wedge. Such process generated widespread modification of mantle rocks thus explaining the very voluminous potassic magmatism of the Roman Province. Introduction of carbonate sediments may also explain the highly explosive nature of Roman volcanoes, whose fluid load was provided by CO_2 liberated from marls.

In the view of this hypothesis, the metasomatic contamination of the Roman Province is younger than the one that affected the mantle source of Tuscany magmas. As discussed in Chap. 2, the Tuscany metasomatism took place during the Alpine phases of convergence between Africa and Europe, and was accomplished by introduction of siliceous upper crustal material into the mantle wedge above the east-directed subduction zone beneath the African margin (e.g. Doglioni et al. 1998). Formation of Tuscany magmas took place at a later time, during the Miocene to present back-arc opening behind the eastward retreating Apennine compression front. The Roman magmatism also postdates the contamination of its mantle source, being formed during Quaternary times, mostly from about 0.8 Ma to the present. Such a late onset of magmatism has been attributed to modification of thermal regime within the mantle wedge. Thermal modification could be a consequence of dying out of active subduction and/or possible slab vertical breakoff and its migration toward the north, following the northward drift of the Adriatic slab (Malusà et al. 2015, 2016). This generated an increase in temperature inside the mantle wedge triggering melting (Peccerillo 1990; Peccerillo and Frezzotti 2015). This model is schematically shown in Fig. 4.21.

4.9 Physical Models of Mantle Metasomatism

Various scenarios, not necessarily alternative, can be envisaged for the physical processes of element transfer from subducted sediments to the mantle wedge. One possibility is element transfer by supercritical fluids and/or H_2O-rich melts, which migrate through the mantle wedge and

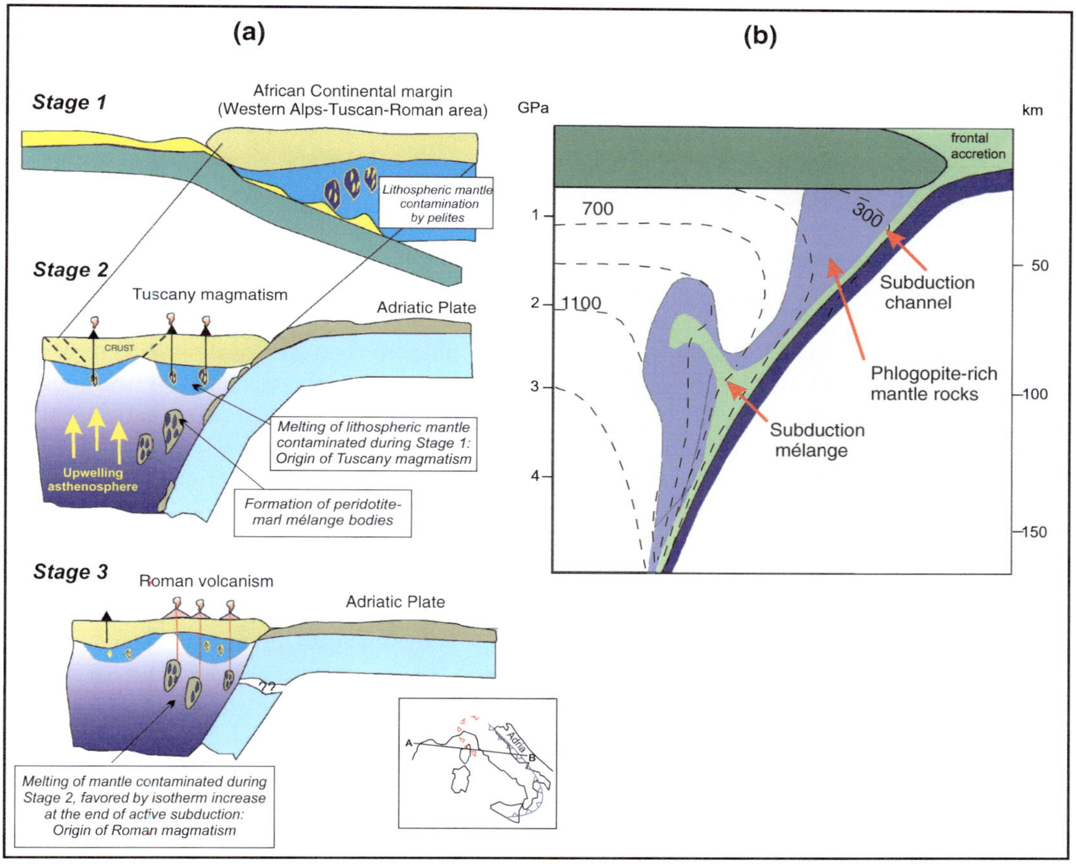

Fig. 4.21 Schematic model of mantle contamination beneath Central Italy, along the section A–B indicated in the *inset*. **a** Stage 1 Contamination of lithospheric mantle by siliceous sediments above the east-dipping Alpine subduction zone (see Chap. 2); Stage 2; Miocene to Quaternary backarc opening behind west-dipping Adriatic slab, asthenospheric mantle upwelling, formation of Tuscany magmas, and mantle contamination by marls from the Adriatic plate, with formation of marl-mantle mélange bodies; Stage 3. Increase of the upper mantle temperatures at the end of active subduction process and/or slab break-off and migration toward the north, melting of marl-mantle mélange bodies and surrounding rocks, and generation of Roman magmas. *Inset* map: position of Alpine (pre-Oligocene, *red dashed line*) and Plio-Quaternary (*black full line*) collision zones. **b** Model of the upper mantle structure above the Adriatic slab, simplified after Peccerillo and Frezzotti (2015). For further explanation, see text

react with peridotite to give phlogopite as pervasive mineral phase or as veins (e.g. Grassi et al. 2012; Martin et al. 2012). According to experimental studies, phengite is the main carrier of potassium and other LILE along subduction zones. This mineral is abundant in subducted sediments and survives breakdown or melting until a depth of more than 300 km (Schmidt et al. 2004). Other experiments on carbonate-saturated pelites found that decarbonation and melting require very high temperatures (Grassi and Schmidt 2011a, b), which are hardly found in most subduction zones (Van Keken et al. 2002; Abers et al. 2006; Arcay et al. 2007). As a consequence, subducted marls would carry their CO_2 and K_2O and LILE load deep into the mantle wedge. The implication of these experiments for the Roman Province is that the mantle contamination is a deep process. These experimental data would also explain the absence of calcalkaline rocks in the Roman Province, whose origin requires a release of fluids at relatively shallow depths along subduction zones (Grove et al. 2012). Note, however, that experimental studies on hydrous carbonated gabbros indicate that melting can occur at P $\sim$ 4 GPa and T $\sim$ 800–900 °C, to give hydrous carbonatitic melts at much lower pressure than indicated by previous studies (Poli 2015). These P-T conditions are typical of hot subduction zones but may also be found in subduction system where an increase in temperature is provided by slab breakoff.

Another possibility is a process of physical mixing between slab rock and peridotite, prior to magma formation (e.g. Gerya et al. 2006; Castro and Gerya 2008; Marschall and Schumacher 2012). According to this scenario, subducted sediments mix with peridotite to form mélange bodies at the slab-mantle interface. These have a lower density than the surrounding peridotite and ascent as diapirs or "cold plumes" toward the hotter corner of the mantle wedge. Here, plumes release aqueous fluids and melts, which flux the mantle peridotite, producing geochemical and mineralogical anomalies. Tumiati et al. (2013) demonstrated that phlogopite is a ubiquitous phase around mélange bodies and this phase disappears early during melting, to give potassic magmas. Formation of subduction mélange requires high amounts of slab sediments to mix with mantle rocks (e.g. Hacker 2008; Behn et al. 2011), a condition that was likely for the Roman Province.

4.10 Summary and Conclusions

The Roman Magmatic Province is made of about 900 km^3 of potassium-rich rocks, mostly phonolitic and trachytic pyroclastic deposits, with minor mafic lavas and scoriae. Petrological and geochemical studies suggest that felsic magmas were derived from mafic parents by evolution processes dominated by extensive (some 50–60 %) fractional crystallisation and mixing, with some crustal contamination and possibly gaseous transfer.

Mafic magmas in the Roman Province range from saturated (potassic trachybasalts) to undersaturated in silica (potassic tephrites and foidites) and display variable degrees of enrichment in potassium and incompatible elements. In contrast, radiogenic isotopic ratios are poorly variable, with most rocks clustering around $^{87}Sr/^{86}Sr \sim 0.7090$ to 7110, $^{143}Nd/^{144}Nd \sim 0.5121$, $^{206}Pb/^{204}Pb \sim 18.80$, and $^{176}Hf/^{177}Hf \sim 0.28258$. This supports an origin of primary magmas in a rather homogenous mantle source by variable degrees of partial melting at pressure increasing from saturated to undersaturated compositions.

The hypothesis that best explains the compositional features of Roman primary magmas suggests a genesis by polybaric melting of lherzolitic or wehrlitic-clinopyroxenite mantle rocks enriched in phlogopite by metasomatic processes. Mantle contamination was provided by carbonated pelites (marls), a process that took place during the Miocene to present subduction of the Adriatic continental plate beneath Northern Apennines.

The Roman mafic rocks resemble closely the Intra-Apennine kamafugites for trace elements and radiogenic isotopes, suggesting the same type of metasomatic modifications for their mantle sources. In contrast, there are significant petrological and geochemical differences with the Tuscany ultrapotassic rocks, indicating distinct nature and age of metasomatic events.

References

Abers AA, van Keken PE, Kneller EA, Ferris A, Stachnik JC (2006) The thermal structure of subduction zones constrained by seismic imaging: implications for slab dehydration and wedge flow. Earth Planet Sci Lett 241:387–397

Acocella V, Funiciello R (2006) Transverse systems along the extensional Tyrrhenian margin of Central Italy and their influence on volcanism. Tectonics 25: TC2003. doi:10.1029/2005TC001845

Acocella V, Palladino DM, Cioni R, Russo P, Simei S (2012) Caldera structure, amount of collapse, and erupted volumes: the case of Bolsena caldera, Italy. Geol Soc Am Bull 124:1562–1576

Antonelli F, Lazzarini L (2010) Mediterranean trade of the most widespread Roman volcanic millstones from Italy and petrochemical markers of their raw materials. J Archaeol Sci 37:2081–2092

Arcay D, Tric E, Doin MP (2007) Slab surface temperature in subduction zones: influence of the interplate decoupling depth and upper plate thinning processes. Earth Planet Sci Lett 255:324–338

Aurisicchio C, Federico M, Gianfagna A (1988) Clinopyroxene chemistry of the High-Potassium suite from the Alban Hills, Italy. Mineral Petrol 39:1–19

Avanzinelli R, Elliott T, Tommasini S, Conticelli S (2008) Constraints on the genesis of the potassium-rich Italian volcanics from U/Th disequilibrium. J Petrol 49:195–223

Avanzinelli R, Lustrino M, Mattei M, Melluso L, Conticelli S (2009) Potassic and ultrapotassic magmatism in the circum-Tyrrhenian region: the role of carbonated pelitic vs. pelitic sediment recycling at destructive plate margin. Lithos 113:213–227

Barberi F, Buonasorte G, Cioni R, Fiordelisi A, Foresi L, Iaccarino S, Laurenzi MA, Sbrana A, Vernia L, Villa IM (1994) Plio-Pleistocene geological evolution of the geothermal area of Tuscany and Latium. Mem Descr Carta Geol It 49:77–134

Barbieri M, Peccerillo A, Poli G, Tolomeo L (1988) Major, trace element and Sr isotopic composition of lavas from Vico volcano (Central Italy) and their evolution in an open system. Contrib Mineral Petrol 99:485–497

Barchi M, Landuzzi A, Minelli G, Pialli G (2001) Outhern Northern Apennines. In: Vai GB, Martini PI (eds) Anatomy of an Orogen. The Apennines and the adjacent Mediterranean basins. Kluwer, Dordrecht, pp 215–254

Barnekow P (2000) Volcanic rocks from central Italy: an Oxygen isotopic microanalytical and geochemical study. PhD Thesis, University of Gottingen, 99 pp

Bartolini C, Bernini M et al (1982) Carta neotettonica dell'Appennino settentrionale. Note illustrative. Boll Soc Geol It 101:523–549

Barton M, Varekamp JC, Van Bergen MJ (1982) Complex zoning of clinopyroxenes in the lavas of Vulsini, Latium, Italy: evidence for magma mixing. J Volcanol Geoth Res 14:361–388

Bear AN, Giordano G, Giampaolo C, Cas RAF (2009) Volcanological constraints on the post-emplacement zeolitisation of ignimbrites and geoarchaeological implications for Etruscan tomb construction (6th–3rd century BC) in the Tufo Rosso a Scorie Nere, Vico Caldera, Central Italy. J Volcanol Geoth Res 183:183–200

Behn MD, Kelemen PB, Hirth G, Hacker BR, Massonne H-J (2011) Diapirs as the source of the sediment signature in arc lavas. Nature Geosci 4:641–646

Bell K, Castorina F, Lavecchia G, Rosatelli G, Stoppa F (2004) Is there a mantle plume below Italy? EOS 85:541–547

Bell K, Lavecchia G, Rosatelli G (2013) Cenozoic Italian magmatism—Isotope constraints for possible plume-related activity. J South Am Earth Sci 41:22–40

Bertagnini A, Sbrana A (1986) Il vulcano di Vico: stratigrafia del complesso vulcanico e sequenze eruttive delle formazioni piroclastiche. Mem Soc Geol It 35:699–713

Bianchi I, Piana Agostinetti N, De Gori P, Chiarabba C (2008) Deep structure of the Colli Albani volcanic district (Central Italy) from receiver functions analysis. J Geophys Res 113:B09313. doi:10.1029/2007JB005548

Bizzarri R, Ambrosetti P, Argenti P, Gatta GD, Baldanza A (2003) L'affioramento del Caio (Lago di Corbara, Orvieto, Italia Centrale) nell'ambito dell'evoluzione paleogeografica Plio-Pelistocenica della Valle del Tevere: evidenze sedimentologiche e stratigrafiche. Il Quaternario 16:241–255

Bizzarri R, Baldanza A, Petrelli M, Famiani F, Peccerillo A (2010) Early Pleistocene distal pyroclastic-fallout material in continental and marine deposits of Western Umbria (Italy): chemical composition, provenance and correlation potential. Il Quaternario 23:245–250

Boari E, Conticelli S (2007) Mineralogy and petrology of Mg-rich calc-alkalic, potassic, and ultrapotassic associated rocks: the Middle Latin Valley monogenetic volcanoes, Roman Magmatic Province, Southern Italy. Can Mineral 45:1443–1469. doi:10.2113/gscanmin.45.6.1000

Boari E, Avanzinelli R, Melluso L, Giordano G, Mattei M, Morra V, Conticelli S (2009) Isotope geochemistry (Sr-Nd-Pb) and petrogenesis of leucite-bearing volcanic rocks from "Colli Albani" volcano, Roman Magmatic Province, Central Italy: inferences on volcano evolution and magma genesis. Bull Volcanol 71:977–1005

Brocchini D, Di Battistini G, Laurenzi MA, Vernia L, Bargossi GM (2000) New $^{40}Ar/^{39}Ar$ datings on the southeastern sector of the Vulsinian volcanic district (Central Italy). Boll Soc Geol It 119:113–120
Brooks CK, Printzlau I (1978) Magma mixing in mafic alkaline volcanic rocks: the evidence from relict phenocryst phases and other inclusions. J Volcanol Geoth Res 4:315–331
Buonasorte G, Fiordelisi A, Pandeli E, Rossi U, Sollevanti F (1987) Stratigraphic correlations and structural setting of the pre-neoautochthonous sedimentary sequences of Northen Latium. Per Mineral 56:111–122
Buttinelli M, Chiarabba C, Anselmi M, Bianchi I, De Rita D, Quattrocchi F (2014) Crustal structure of Northern Latium (central Italy) from receiver functions analysis: new evidences of a post-collisional back-arc margin evolution. Tectonophysics 621:148–158
Callegari AM, Boiocchi M, Bellatreccia F, Caprilli E, Medenbach O, Cavallo A (2012) Capranicaite, (K)(Ca, Na)$Al_4B_4Si_2O_{18}$: a new inosilicate from Capranica, Italy, with a peculiar topology of the periodic single chain [Si_2O_6]. Mineral Mag 75:33–43
Carminati E, Negredo AM, Valera JL, Doglioni C (2005) Subduction-related intermediate-depth and deep seismicity in Italy: insights from thermal and rheological modelling. Phys Earth Planet In 149:65–79
Carminati E, Lustrino M, Doglioni C (2012) Geodynamic evolution of the central and western Mediterranean: tectonic vs. igneous constraints. Tectonophysics 579:173–192
Castellarin A, Colacicchi R, Praturlon A, Cantelli C (1982) The Jurassic-Lower Pliocene history of the Ancona-Anzio line (Central Italy). Mem Soc Geol It 24:243–260
Castorina F, Stoppa F, Cundari A, Barbieri M (2000) An enriched mantle source for Italy's melilitite-carbonatite association as inferred by its Nd-Sr isotope signature. Mineral Mag 64:625–639
Castro A, Gerya TV (2008) Magmatic implications of mantle wedge plumes: experimental study. Lithos 103:103–148
Chiarabba C, Amato A, Fordelisi A (1995) Upper crustal tomographic images of the Amiata-Vulsini geothermal region, central Italy. J Geophys Res 100:4053–4066
Chiarabba C, Bagh S, Bianchi I, De Gori P, Barchi M (2010a) Deep structural heterogeneities and the tectonic evolution of the Abruzzi region (Central Apennines, Italy) revealed by microseismicity, seismic tomography, and teleseismic receiver functions. Earth Planet Sci Lett 295:462–476
Chiarabba C, Giordano G, Mattei M, Funiciello R (2010b) The three-dimensional structure of the Colli Albani volcano. In: Funiciello R, Giordano G (eds) The Colli Albani Volcano (Geol Soc London, Spec IAVCEI Publ 3):29–41
Chiodini G, Frondini F, Kerrick DM, Rogie J, Parello F, Peruzzi L, Zanzari AR (1999) Quantification of deep CO_2 fluxes from Central Italy. Examples of carbon balance for regional aquifers and soil diffuse degassing. Chem Geol 159:205–222
Chiodini G, Valenza M, Cardellini C, Frigeri A (2008) A new web-based catalog of Earth degassing sites in Italy. Eos 89:341–342
Cinelli G, Tositti L, Capaccioni B, Brattich E, Mostacci D (2014) Soil gas radon assessment and development of a radon risk map in Bolsena, Central Italy. Environ Geochem Health 37:305–319
Cioni R, Sbrana A, Bertagnini A, Buonasorte G, Landi P, Rossi U, Salvati L (1987) Tephrostratigraphic correlations in the Vulsini, Vico and Sabatini volcanic successions. Per Mineral 56:137–155
Cioni R, Laurenzi MA, Sbrana A, Villa IM (1993) $^{40}Ar/^{39}Ar$ chronostratigraphy of the initial activity of the Sabatini volcanic complex (Italy). Boll Soc Geol It 112:251–263
Conceição RV, Green DH (2000) Behaviour of the cotectic curve En-Ol in the leucite-olivine-quartz under dry conditions to 2 GPa. Geochem Geophys Geosyst, paper number 2000GC000071
Conceição RV, Green DH (2004) Derivation of potassic (shoshonitic) magmas by decompression melting of phlogopite+pargasite lherzolite. Lithos 72:209–229
Conte AM, Dolfi D, Gaeta M, Misiti V, Mollo S, Perinelli C (2009) Experimental constraints on evolution of leucite-basanite magma at 1 and 10^{-4} GPa: implications for parental compositions of Roman high-potassium magmas. Eur J Mineral 21:763–782
Conticelli S (1998) The effect of crustal contamination on ultrapotassic magmas with lamproitic affinity: mineralogical, geochemical and isotope data from the Torre Alfina lavas and xenoliths, Central Italy. Chem Geol 149:51–81
Conticelli S, Peccerillo A (1992) Petrology and geochemistry of potassic and ultrapotassic volcanism in Central Italy: petrogenesis and inferences on the evolution of the mantle sources. Lithos 28:221–240
Conticelli S, Francalanci L, Santo AP (1991) Petrology of the final stage Latera lavas: mineralogical, geochemical and Sr- isotopic data and their bearing on the genesis of some potassic magmas in Central Italy. J Volcanol Geoth Res 46:187–212
Conticelli S, Francalanci L, Manetti P, Cioni R, Sbrana A (1997) Petrology and geochemistry of the ultrapotassic rocks from the Sabatini volcanic district, Central Italy: the role of evolutionary processes in the genesis of variably enriched alkaline magmas. J Volcanol Geoth Res 75:107–136
Conticelli S, D'Antonio M, Pinarelli L, Civetta L (2002) Source contamination and mantle heterogeneity in the genesis of Italian potassic and ultrapotassic volcanic rocks: Sr-Nd-Pb isotope data from Roman Province and Southern Tuscany. Mineral Petrol 74:189–222
Conticelli S, Melluso L, Perini G, Avanzinelli R, Boari E (2004) Petrologic, geochemical and isotopic characteristics of potassic and ultrapotassic magmatism in central-southern Italy: inferences on its genesis and on the nature of mantle sources. Per Mineral 73:135–164

Conticelli S, Laurenzi MA, Giordano G, Mattei M, Avanzinelli R, Melluso L, Tommasini S, Boari E, Cifelli F, Perini G (2010a) Leucite-bearing (kamafugitic/leucititic) and -free (lamproitic) ultrapotassic rocks and associated shoshonites from Italy: constraints on petrogenesis and geodynamics. In: Beltrando M, Peccerillo A, Mattei M, Conticelli S, Doglioni C (eds) J Virtual Explorer 36, paper 20. doi:10.3809/jvirtex.2010.00251

Conticelli S, Boari E, Avanzinelli R, De Benedetti AA, Giordano G, Mattei M, Melluso L, Morra V (2010b) Geochemistry, isotopes and minerl chemistry of the Colli Albani volcanic rock: constraints on magma genesis and evolution. In: Funiciello R, Giordano G (eds) The Colli Albani Volcano. Geological Society of London, Spec IAVCEI Publ 3:107–140

Conticelli S, Avanzinelli R, Marchionni S, Tommasini S, Melluso L (2011) Sr-Nd-Pb isotopes from the Radicofani Volcano, Central Italy: constraints on heterogeneities in a veined mantle responsible for the shift from ultrapotassic shoshonite to basaltic andesite magmas in a post-collisional setting. Mineral Petrol 103:123–148

Conticelli S, Avanzinelli R, Poli G, Braschi E, Giordano G (2013) Shift from lamproite-like to leucititic rocks: Sr-Nd-Pb isotope data from the Monte Cimino volcanic complex vs. the Vico stratovolcano, Central Italy. Chem Geol 353:246–266

Cox HG, Hawkesworth CJ, O'Nions RK, Appleton JD (1976) Isotopic evidence for the derivation of some Roman Region volcanics from anomalously enriched mantle. Contrib Mineral Petrol 56:173–180

Cundari A (1975) Mineral chemistry and petrogenetic aspects of the Vico lavas, Roman Volcanic Region, Italy. Contrib Mineral Petrol 53:129–144

Cundari A (1979) Petrogenesis of leucite-bearing lavas in the Roman volcanic region, Italy. The Sabatini lavas. Contrib Mineral Petrol 70:9–21

Cundari A, Mattias PP (1974) Evolution of the of the Vico lavas, Roman Volcanic Region, Italy. Bull Volcanol 38:98–114

Dallai L, Freda C, Gaeta M (2004) Oxygen isotope geochemistry of pyroclastic clinopyroxene monitors carbonate contributions to Roman-type ultrapotassic magmas. Contrib Mineral Petrol 148:247–263

Daly RA (1918) Genesis of alkaline rocks. J Geol 26:97–134

De Benedetti AA, Funiciello R, Giordano G, Diano G, Caprilli E, Paterne A (2008) Volcanology, history and myths of the Lake Albano maar (Colli Albani volcano, Italy). J Volcanol Geoth Res 176:387–406

De Benedetti AA, Caprilli A, Rossetti F, Giordano G (2010) Metamorphic, metasomatic and intrusive xenoliths of the Colli Albani volcano and their significance for the reconstruction of the volcano plumbing system. In: Funiciello R, Giordano G (eds) The Colli Albani Volcano (Geol Soc London, Spec IAVCEI Publ 3):153–176

De Gennaro M, Incoronato A, Mastrolorenzo G, Adabbo M, Spina G (1999) Depositional mechanisms and alteration processes in different types of pyroclastic deposits from Campi Flegri volcanic field (Southern Italy). J Volcanol Geoth Res 91:303–320

De Rita D, Funiciello R, Rossi U, Sposato A (1983) Structure and evolution of the Sacrofano-Baccano caldera, Sabatini volcanic complex, Rome. J Volcanol Geoth Res 17:219–236

De Rita D, Funiciello R, Parotto M (1988) Carta geologica del Complesso vulcanico dei Colli Albani. - Progetto Finalizzato Geodinamica, subproject Sorveglianza dei vulcani attivi e rischi vulcanico. CNR, Roma

De Rita D, Funiciello R, Corda L, Sposato A, Rossi U (1993) Volcanic units. In: Di Filippo M (ed) Sabatini volcanic complex (Quad Ric Sci CNR Rome, 114, 11):33–79

De Rita D, Faccenna C, Funiciello R, Rosa C (1995) Stratigraphy and volcano-tectonics. In: Trigila R (ed) The volcano of the Alban Hills. University La Sapienza, Rome, pp 33–71

De Rita D, Di Filippo M, Rosa C (1996) Structural evolution of the Bracciano volcano-tectonic depression, Sabatini volcanic district, Italy. In: Mc Giure WJ, Jones AP, Neuberg J (eds) Volcano instability on the Earth and other planets (Geol Soc Spec Publ 110): 225–236

Del Bello E, Mollo S, Scarlato P, von Quadt A, Forni F, Bachmann O (2014) New petrological constraints on the last eruptive phase of the Sabatini Volcanic District (Central Italy): clues from mineralogy, geochemistry, and Sr-Nd isotopes. Lithos 205:28–38. doi:10.1016/j.lithos.2014.06.015

Della Ventura G, William CT, Cabella R, Oberti R, Caprilli E, Bellatreccia F (1999) Britholite-hellandite intergrowths and associated REE-minerals from the alkali syenitic ejecta of the Vico volcanic complex (Latium, Italy): petrological implications bearing on REE mobility in volcanic systems. Eur J Mineral 11:843–854

Della Ventura G, Bonazzi P, Oberti R, Ottolini L (2002) Ciprianiite and mottanaite-(Ce), two new minerals of the hellandite group from Latium (Italy). Am Mineral 87:739–744

Di Battistini G, Montanini A, Vemia L, Bargossi GM, Castorina F (1998) Petrology and geochemistry of ultrapotassic rocks from the Montefiascone volcanic complex (Central Italy): magmatic evolution and petrogenesis. Lithos 43:169–195

Di Battistini G, Montanini A, Vernia L, Venturelli G, Tonarini S (2001) Petrology of melilite-bearing rocks from the Montefiascone Volcanic complex (Roman Magmatic Province): new insights into the ultrapotassic volcanism of Central Italy. Lithos 59:1–24

Di Filippo M (ed) (1993) Sabatini volcanic complex. Quad Ric Sci CNR, Rome, 114, 11:109 pp

Di Stefano R, Kissling E, Chiarabba C, Amato A, Giardini D (2009) Shallow subduction beneath Italy: three-dimensional images of the Adriatic-European-Tyrrhenian lithosphere system based on high-quality P wave arrival times. J Geophys Res 114:B05305. doi:10.1029/2008JB005641

Doglioni C, Mongelli F, Pialli G (1998) Boudinage of the Alpine belt in the Apenninic back-arc. Mem Soc Geol It 52:457–468

Doglioni C, Harabaglia P, Merlini S, Mongelli F, Peccerillo A, Piromallo C (1999) Orogens and slabs vs. their direction of subduction. Earth Sci Rev 45:167–208

Dolfi D, Trigila R (1978) Clinopyroxenes from potassic lavas of Central Italy Quaternary volcanism. Progr Exp Petrol, NERC, UK, Rep 4:18–22

Edgar AD, Green DH, Hibberson WO (1976) Experimental petrology of highly potassic magmas. J Petrol 17:339–356

Federico M (1995) Mineralogy. In: Trigila R (ed) The volcano of the Alban Hills. University La Sapienza, Rome, pp 73–93

Federico M, Peccerillo A (2002) Mineral chemistry and petrogenesis of granular ejecta from the Alban Hills volcano (Central Italy). Mineral Petrol 74:223–252

Federico M, Peccerillo A, Barbieri M, Wu TW (1994) Mineralogical and geochemical study on granular xenoliths from the Alban Hills volcano (Central Italy): bearing on evolutionary processes in potassic magma chambers. Contrib Mineral Petrol 115:384–401

Ferrara G, Preite-Martinez M, Taylor HP, Tonarini S, Turi B (1986) Evidence for crustal assimilation, mixing of magmas, and a ^{87}Sr-rich upper mantle. Contrib Mineral Petrol 92:269–280

Feuillet N, Nostro C, Chiarabba C, Cocco M (2004) Coupling between earthquake swarms and volcanic unrest at the Alban Hills Volcano (Central Italy) modeled through elastic stress transfer. J Geophys Res 109:B02308. doi:10.1029/2003JB002419

Foley SF, Wheller GE (1990) Parallels in the origin of the geochemical signature of island-arc volcanics and continental potassic igneous rocks: the role of residual titanates. Chem Geol 85:1–18

Fornaseri M (1985) Geochronology of volcanic rocks from Latium (Italy). Rend Soc It Mineral Petrol 40:73–106

Fornaseri M, Scherillo A, Ventriglia U (1963) La regione vulcanica dei Colli Albani. CNR, Rome (561 pp)

Freda C, Palladino DM, Pignatti S, Trigila R, Onorati G, Pascolieri M (1990) Vocano-tectonic scenario of Vulsini volcanos (Central Italy) from LANDSAT-SS images and digital elevations data. J Photogram Rem Sensing 45:316–328

Freda C, Gaeta M, Palladino DM, Trigila R (1997) The Villa Senni eruption (Alban Hills, Central Italy): the role of H_2O and CO_2 on the magma chamber evolution and on the eruptive scenario. J Volcanol Geoth Res 78:103–120

Freda C, Gaeta M, Karner DB, Marra F, Renne PR, Taddeucci J, Scarlato P, Christensen JN, Dallai L (2006) Eruptive history and petrologic evolution of the Albano multiple maar (Alban Hills, Central Italy). Bull Volcanol 68:567–591

Freda C, Gaeta M, Misiti V, Mollo S, Dolfi D, Scarlato P (2008) Magma–carbonate interaction: an experimental study on ultrapotassic rocks from Alban Hills (Central Italy). Lithos 101:397–415

Funiciello R, Giordano G (eds) (2010) The Colli Albani volcano. Geological Society of London, Spec IAVCEI Publ 3, 393 pp

Funiciello R, Parotto M (1978) Il substrato sedimentario nell'area dei Colli Albani: considerazioni geodinamiche e paleogeografiche sul margine tirrenico dell'Appennino centrale. Geol Rom 17:233–287

Funiciello R, Locardi E, Parotto M (1976) Lineamenti geologici dell'area Sabatina Orientale. Mem Soc Geol It 95:831–849

Funiciello R, Giordano G, De Rita D (2003) The Albano maar lake (Colli Albani volcano, Italy): recent volcanic activity and evidences of pre-Roman age catastrophic lahar events. J Volcanol Geoth Res 123:43–61

Gaeta M, Freda C, Christensen JN, Dallai L, Marra F, Karner DB, Scarlato P (2006) Time dependent geochemistry of clinopyroxene from the Alban Hills (Central Italy): clues to the source and evolution of ultrapotassic magmas. Lithos 86:330–346

Gaeta M, Di Rocco T, Freda C (2009) Carbonate assimilation in open magmatic systems: the role of melt-bearing skarns and cumulate-forming processes. J Petrol 50:361–385

Gaeta M, Freda C, Marra F, Di Rocco T, Gozzi F, Arienzo I, Giaccio B, Scarlato P (2011) Petrology of the most recent ultrapotassic magmas from the Roman Province (Central Italy). Lithos 127:298–308

Gale A, Dalton CA, Langmuir CH, Su Y, Schilling J-G (2013) The mean composition of ocean ridge basalts. Gechem Geophys Geosyst 14. doi:10.1029/2012GC004334

Gambardella B, Cardellini C, Chiodini G, Frondini F, Marini L, Ottonello G, Zuccolini MV (2004) Fluxes of deep CO_2 in the volcanic areas of central-southern Italy. J Volcanol Geoth Res 136:31–52

Gasperini D, Blichert Toft J, Bosch D, Del Moro A, Macera P, Albaréde F (2002) Upwelling of deep mantle material through a plate window: evidence from the geochemistry of Italian basaltic volcanics. J Geophys Res 107(B12):2367. doi:10.1029/2001JB000418

Gentili S, Comodi P, Nazzareni S, Zucchini A (2014) The Orvieto-Bagnoregio Ignimbrite: pyroxene crystal-chemistry and bulk phase composition of pyroclastic deposits, a tool to identify syn- and post-depositional processes. Eur J Mineral 26:743–756

Gerya TV, Connolly JAD, Yuen DA, Gorczyk W, Capel AM (2006) Seismic implications of mantle wedge plumes. Phys Earth Planet In 156:59–74

Giaccio B, Marra F, Hajdas I, Karner DB, RennePR Sposato A (2009) ^{40}Ar/^{39}Ar and ^{14}C geochronology of the Albano maar deposits: Implications for defining the age and eruptive style of the most recent explosive activity at Colli Albani Volcanic District, Central Italy. J Volcanol Geoth Res 185:203–213

Giordano G (2010) Stratigraphy, volcano tectonics and evolution of the Colli Albani volcanic field. In: Funiciello R, Giordano G (eds) The Colli Albani Volcano. Geological Society of London, Spec IAVCEI Publ 3:43–98

Gozzi F, Gaeta M, Freda C, Mollo S, Di Rocco T, Marra F, Dallai L, Pack A (2014) Primary magmatic calcite reveals origin from crustal carbonate. Lithos 190–191:191–203

Grassi D, Schmidt MW (2011a) Melting of carbonated pelites at 8–13 GPa: generating K-rich carbonatites for mantle metasomatism. Contrib Mineral Petrol 162:169–191

Grassi D, Schmidt MW (2011b) The melting of carbonated pelites from 70 to 700 km depth. J Petrol 52:765–789

Grassi D, Schmidt MW, Gunther D (2012) Element partitioning during carbonated pelite melting at 8, 13 and 22 GPa and the sediment signature in the EM mantle components. Earth Planet Sci Lett 327–328:84–96

Grove TL, Till CB, Krawczynski MJ (2012) The role of H_2O in subduction zone magmatism. Annu Rev Earth Planet Sci 40:413–439

Hacker BR (2008) H_2O subduction beyond arc. Geochem Geophys Geosyst 9:Q03001. doi:10.1029/2007GC001707

Hall A (1998) Zeolitisation of volcaniclastic sediments: the role of temperature and pH. J Sediment Res 68:739–745

Hawkesworth CJ, Vollmer R (1979) Crustal contamination vs. enriched mantle: $^{143}Nd/^{144}Nd$ and $^{87}Sr/^{86}Sr$ evidence from the Italian volcanics. Contrib Mineral Petrol 69:151–165

Holm PM (1982) Mineral chemistry of perpotassic lavas of the Vulsininan district, the Roman Province, Italy. Mineral Mag 46:379–386

Holm PM, Munksgaard NC (1982) Evidence for mantle metasomatism: an oxygen and strontium isotope study of the Vulsinian district, Central Italy. Earth Planet Sci Lett 60:376–388

Holm PM, Lou S, Nielsen A (1982) The geochemistry and petrogenesis of the Vulsinian district, Roman Province, Central Italy. Contrib Mineral Petrol 80:367–378

Iacono-Marziano G, Gaillard F, Pichavant M (2007) Limestone assimilation and the origin of CO_2 emission at the Alban Hills (Central Italy): constraints from experimental petrology. J Volcanol Geoth Res 166:91–105

Irvine TN, Baragar WRA (1971) A guide to chemical classification of common volcanic rocks. Can J Earth Sci 8:523–548

Kamenetsky V, Métrich N, Cioni R (1995) Potassic primary melts of Vulsini (Roman Province)—evidence from mineralogy and melt inclusions. Contrib Mineral Petrol 120:186–196

Karner DB, Marra F, Renne PR (2001) The history of the Monti Sabatini and Alban Hills volcanoes: groundwork for assessing volcanic-tectonic hazards for Rome. J Volcanol Geoth Res 107:185–219

Landi P (1987) Stratigraphy and petrochemical evolution of Latera Volcano (Central Italy). Per Mineral 56:210–224

Laurenzi MA, Villa IM (1987) $^{40}Ar/^{39}Ar$ chronostratigraphy of Vico ignimbrites. Per Mineral 56:285–293

Le Maitre RW (ed) (2002) A classification of igneous rocks and glossary of terms. Cambridge University Press, Cambridge (252 pp)

Locardi E (1965) Tipi ignimbritici di magmi mediterranei: le ignimbriti del vulcano di Vico. Atti Soc Tosc Sci Nat A72:55–174

Locardi E (1988) The origin of the Apenninic arc. Tectonophysics 146:105–123

Locardi E, Funiciello R, Lombardi G, Parotto M (1977) The main volcanic groups of Latium (Italy): relations between structural evolution and petrogenesis. Geol Rom 15:279–300

Malusà MG, Faccenna C, Baldwin SL, Fitzgerald PG, Rossetti F, Balestrieri ML, Danisik M, Ellero A, Ottria G, Piromallo C (2015) Contrasting styles of (U) HP rock exhumation along the Cenozoic Adria–Europe plate boundary (Western Alps, Calabria, Corsica). Geochem Geophys Geosyst 16. doi:10.1002/2015GC005767

Malusà MG, Danisik M, Kuhlemann J (2016) Tracking the Adriatic-slab travel beneath the Tethyan margin of Corsica-Sardinia by low-temperature thermochronometry. Gondwana Res 31:135–149

Marra F, Freda C, Scarlato P, Taddeucci J, Karner DB, Renne PR, Gaeta M, Palladino DM, Trigila R, Cavarretta G (2003) Post-caldera activity in the Alban Hills volcanic district (Italy): $^{40}Ar/^{39}Ar$ geochronology and insight into magma evolution. Bull Volcanol 65:227–247

Marra F, Sottili G, Gaeta M, Giaccio B, Jicha B, Masotta M, Palladino DM, Deocampo DM (2014) Major explosive activity in the Monti Sabatini Volcanic District (Central Italy) over the 800–390 ka interval: geochronological-geochemical overview and tephrostratigraphic implications. Quarter Sci Rev 94:74–101

Marschall HR, Schumacher JC (2012) Arc magmas sourced from mélange diapirs in subduction zones. Nature Geosci 5:862–867

Martelli M, Nuccio PM, Stuart FM, Burgess R, Ellam RM, Italiano F (2004) Helium-strontium isotope constraints on mantle evolution beneath the Roman Comagmatic Province, Italy. Earth Planet Sci Lett 224:295–308

Martin LHJ, Schmidt MW, Mattsson HB, Ulmer P, Hametner K, Günther D (2012) Element partitioning between immiscible carbonatite–kamafugite melts with application to the Italian ultrapotassic suite. Chem Geol 320–321:96–112

Masotta M, Gaeta M, Gozzi F, Marra F, Palladino DM, Sottili G (2010) H_2O- and temperature-zoning in magma chambers: The example of the Tufo Giallo della Via Tiberina eruptions (Sabatini Volcanic District, central Italy). Lithos 118:119–130. doi:10.1016/j.lithos.2010.04.004

Mattei M, Funiciello R, Kissel C (1995) Paleomagnetic and structural evidence of Neogene block rotation in the Central Apennines, Italy. J Geophys Res 100:17863–17883
Melluso L, Conticelli S, D'Antonio M, Mirco NP, Saccani E (2003) Petrology and mineralogy of wollastonite- and melilite-bearing paralavas from the central Apennines, Italy. Am Mineral 88:1287–1299
Meloni A, Alfonsi L, Florindo F, Sagnotti L, Speranza F, Winkler A (1997) Neogene and Quaternari geodynamic evolution of the Italian peninsula: the contribution of paleomagnetic data. Ann Geophys 40:705–727
Melzer S, Foley SF (2000) Phase relations and fractionation sequences in potassic magma series modelled in the system $CaMgSi_2O_6$-$KAlSiO_4$-Mg_2SiO_4-F at 1 bar to 18 kbar. Contrib Mineral Petrol 138:186–197
Miller SA, Collettini C, Chiaraluce L, Cocco M, Barchi MR, Kraus JB (2004) Aftershocks driven by a high-pressure CO_2 source at depth. Nature 427:724–727
Mollo S, Vona A (2014) The geochemical evolution of clinopyroxene in the Roman Province: a window on decarbonation from wall-rocks to magma. Lithos 192–195:1–7
Mollo S, Gaeta M, Freda C, Di Rocco T, Misiti V, Scarlato P (2010) Carbonate assimilation in magmas: a reappraisal based on experimental petrology. Lithos 114:503–514
Mongelli F, Zito G (1991) Flusso di calore nella regione Toscana. Studi Geol Camerti, Spec Issue CROP 03:91–98
Mongelli F, Zito G, Della Vedova B, Pellis G, Squarci P, Taffi L (1991) Geothermal regime of Italy and surrounding seas. In: Cermàk V, Rybach L (eds) Exploration of the deep continental crust. Springer, Berlin, pp 380–394
Montone P, Amato A, Chiarabba C, Buonasorte G, Fiordelisi A (1995) Evidence of active extension in Quaternary volcanoes of central Italy from breackouts analysis and seismiciy. Geophys Res Lett 22:1909–1912
Nappi G, Renzulli A, Santi P (1987) An evolutionary model for the Paleo-Bolsena and Bolsena volcanic complexes: a structural and petrographic study. Per Mineral 56:241–267
Nappi G, Renzulli A, Santi P (1991) Evidence of incremental growth in the Vulsinian calderas (Central Italy). J Volcanol Geoth Res 47:13–31
Nappi G, Renzulli A, Santi P, Gillot P-Y (1995) Geological evolution and geochronology of the Vulsini volcanic district (Central Italy). Boll Soc Geol It 114:599–613
Nappi G, Antonelli F, Coltorti M, Milani A, Renzulli A, Siena F (1998) Volcanological and petrological evolution of the Eastern Vulsini District, Central Italy. J Volcanol Geoth Res 87:211–232
Newhall CG, Self S (1982) The volcanic explosivity index (VEI): an estimate of explosive magnitude for historical volcanism. J Geophys Res 87:1231–1238
Nikogosian I, Ersoy O, Whitehouse M, Mason PRD, deHoog JCM, Wortel R, van Bergen MJ (2016) Multiple subduction imprints in the mantle below Italy detected in a single lava flow. Earth Planet Sci Lett 449:12–19
Palladino DM, Simei S (2005a) Eruptive dynamics and caldera collapse during the Onano eruption, Vulsini, Italy. Bull Volcanol 67:423–440
Palladino DM, Simei S (2005b) The Latera volcanic complex (Vulsini, Central Italy): eruptive activity and caldera evolution. Acta Vulcanol 17:75–80
Palladino DM, Agosta E, Freda C, Spaziani S, Trigila R (1994) Geopetrographic and volcanologic study of southern Vulsini: the Valentano-Marta-La Rocca sector. Mem Descr Carta Geol It 49:255–276
Palladino DM, Simei S, Sottili G, Trigila R (2010) Integrated approach for the reconstruction of stratigraphy and geology of Quaternary volcanic terrains: an application to the Vulsini Volcanoes (Central Italy). In: Groppelli G, Viereck-Goette L (eds) Stratigraphy and geology of volcanic areas (Geol Soc Am Spec Paper 464):63–84
Palladino DM, Gaeta M, Giaccio B, Sottili G (2014) On the anatomy of magma chamber and caldera collapse: the example of trachy-phonolitic explosive eruptions of the Roman Province. J Volcanol Geoth Res 281:12–26
Parotto M, Praturlon A (1975) Geological summary of Central Apennines. In: "Structural Model of Italy" (Quad Ric Sci CNR, Rome, 90):257–311
Peccerillo A (1985) Roman Comagmatic Province (Central Italy): evidence for subduction-related magma genesis. Geology 13:103–106
Peccerillo A (1990) On the origin of italian potassic magmas: comments. Chem Geol 85:183–196
Peccerillo A (1998) Relationships between ultrapotassic and carbonate-rich volcanic rocks in central Italy: petrogenetic implications and geodynamic significance. Lithos 43:267–279
Peccerillo A (2002) Plio-Quaternary magmatism in central-southern Italy: a new classification scheme for volcanic provinces and its geodynamic implications. In: Barchi RM, Cirilli S, Minelli G (eds) Geological and geodynamic evolution of the Apennines. Boll Soc Geol It (Spec Vol 1):113–127
Peccerillo A (2004) Carbonate-rich pyroclastic rocks from central Apennines: carbonatites or carbonated rocks? A commentary. Per Mineral 73:165–175
Peccerillo A, Frezzotti ML (2015) Magmatism, mantle evolution and geodynamics at the converging plate margins of Italy. J Geol Soc London 172:407–427
Peccerillo A, Poli G, Tolomeo L (1984) Genesis, evolution and tectonic significance of K-rich volcanics from the Alban Hills (Roman Comagmatic Region) as inferred from trace element geochemistry. Contrib Mineral Petrol 86:230–240
Peccerillo A, Conticelli S, Manetti P (1987) Petrological characteristics and the genesis of recent magmatism of Southern Tuscany and Northern Latium. Per Mineral 56:157–172

Peccerillo A, Poli G, Serri G (1988) Petrogenesis of orenditic and kamafugitic rocks from Central Italy. Can Mineral 26:45–65

Peccerillo A, Federico M, Barbieri M, Brilli M, Wu TW (2010) Interaction between ultrapotassic magmas and carbonate rocks: Evidence from geochemical and isotopic (Sr, Nd, O) compositions of granular lithic clasts from the Alban Hills Volcano, Central Italy. Geochim Cosmochim Acta 74:2999–3022

Perini G, Conticelli S (2002) Crystallization conditions of leucite-bearing magmas and their implications on the magmatological evolution of ultrapotassic magmas: the Vico volcano, Central Italy. Mineral Petrol 74:253–276

Perini G, Conticelli S, Francalanci L, Davidson JP (2000) The relationship between potassic and calc-alkaline post-orogenic magmatism at Vico volcano, Central Italy. J Volcanol Geoth Res 95:247–272

Perini G, Tepley FG, Davidson JP, Conticelli S (2003) The origin of K-feldspar megacrysts hosted in alkaline potassic rocks from central Italy: a track for low-pressure processes in mafic magmas. Lithos 66:223–240

Perini G, Francalanci L, Davidson JP, Conticelli S (2004) The petrogenesis of Vico Volcano, Central Italy: an example of low scale mantle heterogeneity. J Petrol 45:139–182

Pichavant M, Scaillet B, Pommier A, Iacono-Marziano G, Cioni R (2014) Nature and evolution of primitive Vesuvius magmas: an experimental study. J Petrol 55:2281–2309

Piromallo C, Morelli A (2003) P wave tomography of the mantle under the Alpine-Mediterranean area. J Geophys Res 108(B2):2065. doi:10.1029/2002JB001757

Poli S (2015) Carbon mobilized at shallow depths in subduction zones by carbonatitic liquids. Nature Geosci 8:633–636

Renzulli A, Upton BGJ, Nappi G (1995) Magma chamber processes preceding the Pitigliano formation eruption (Latera volcanic complex, Central Italy): evidence from cognate plutonic clasts. Acta Vulcanol 7:55–74

Renzulli A, Santi P, Nappi G, Luini M, Vitali D (2002) Provenance and trade of volcanic rock millstones from Etruscan-Celtic and Roman archaeological sites in Central Italy. Eur J Mineral 14:175–183

Riguzzi F, Pietrantonio G, Devoti R, Atzori S, Anzidei M (2009) Volcanic unrest of the Colli Albani (Central Italy) detected by GPS monitoring test. Phye Earth Planet In 177:79–87

Rittmann A (1933) Die geologische bedingte evolution und differentiation des Somma-Vesuvius magmas. Zeitschr Vulkanol 15:8–94

Rogers NW, Hawkesworth CJ, Parker RJ, Marsh JS (1985) Geochemistry of potassic lavas from Vulsini, Central Italy, and implications for mantle enrichment processes beneath the Roman region. Contrib Mineral Petrol 90:244–257

Sabatini V (1900) I Vulcani dell'Italia Centrale e i loro prodotti. I: Vulcano Laziale. Tipografia Nazionale G Bertero, Roma 392 pp

Santi P, Antonelli F, Renzulli A, Pensabene P (2004) Leucite phonolite millstones from the Orvieto production centre: new data and insights into the Roman trade. Per Mineral 73:57–69

Satolli S, Calamita F (2008) Differences and similarities between the central and the southern Apennines (Italy): examining the Gran Sasso versus the Matese-Frosolone salients using paleomagnetic, geological, and structural data. J Geophys Res 113: B10101. doi:10.1029/2008JB005699

Savelli C (1967) The problem of rock assimilation by Somma-Vesuvius magma. Composition of Somma-Vesuvius lavas. Contrib Mineral Petrol 16:328–353

Scarascia S, Lozej A, Cassinis R (1994) Crustal structure of Ligurian, Tyrrhenian and Ionian seas and adjacent onshore areas interpreted from wide-angle seismic profiles. Boll Geofis Teor Appl 36:1–19

Schmidt MW, Vielzeuf D, Auzanneau E (2004) Melting and dissolution of subducting crust at high pressures: the key role of white mica. Earth Planet Sci Lett 228:65–84

Serri G (1990) Neogene-Quaternary magmatism of the Tyrrhenian region: characterization of the magma sources and geodynamic implications. Mem Geol Soc It 41:219–242

Soligo M, Tuccimei P (2010) Geochronology of Colli Albano volcano. In: Funiciello R, Giordano G (eds) The Colli Albani Volcano. Geological Society of London, Spec IAVCEI Publ 3:99–106

Sollevanti F (1983) Geologic, volcanologic and tectonic setting of the Vico-Cimini area, Italy. J Volcanol Getherm Res 17:203–217

Sottili G, Palladino DM, Zanon V (2004) Plinian activity during the early eruptive history of the Sabatini Volcanic District, Central Italy. J Volcanol Geoth Res 135:361–379

Sottili G, Palladino DM, Marra F, Jicha B, Karner DB, Renne PF (2010) Geochronology of the most recent activity in the Sabatini Volcanic District, Roman Province, Central Italy. J Volcanol Geoth Res 196:20–30

Sottili G, Palladino DM, Gaeta M, Masotta M (2012) Origins and energetics of maar volcanoes: examples from the ultrapotassic Sabatini Volcanic District (Roman Province, Central Italy). Bull Volcanol 74:163–186

Sun SS, McDonough WF (1989) Chemical and isotopic systematics of oceanic basalts: implications for mantle composition and processes. In: Saunders AD, Norry MJ (eds) Magmatism in ocean basins (Geol Soc London Spec Publ 42):313–345

Tarquini S, Isola I, Favalli M et al (2007) TINITALY/01: a new Triangular Irregular Network of Italy. Ann Geophys 50:407–425

Tarquini S, Vinci S, Favalli M, Doumaz F, Fornaciai A, Nannipieri L (2012) Release of a 10-m-resolution DEM for the Italian territory: comparison with global-coverage DEMs and anaglyph-mode exploration via the web. Comput Geosci 38:168–170

Tavarnelli E, Butler RWH, Decandia FA, Calamita F, Grasso M, Alvarez W, Renda P (2004) Implications of fault reactivation and structural inheritance in the Cenozoic tectonic evolution of Italy. In: Special volume Italian geological society 32th IGC Florence, pp 209–222

Taylor SR, McLennan SM (1985) The continental crust: its composition and evolution. An examination of the geochemical record preserved in sedimentary rocks. Blackwell, Oxford (312 pp)

Thibault Y, Edgar AD, Lloyd FE (1992) Experimental investigation of melts from a carbonated phlogopite lherzolite: implications for metasomatism in the continental lithospheric mantle. Am Mineral 77:784–794

Thompson RN (1977) Primary basalts and magma genesis. III. Alban Hills, Roman Comagmatic Province, Central Italy. Contrib Mineral Petrol 50:91–108

Thomsen TB, Schmidt MW (2008a) The biotite to phengite reaction and mica-dominated melting in fluid+carbonate-saturated pelites at high pressures. J Petrol 49:1889–1914

Thomsen TB, Schmidt MW (2008b) Melting of carbonated pelites at 2.5–5.0 GPa, silicate-carbonatite liquid immiscibility, and potassium-carbon metasomatism of the mantle. Earth Planet Sci Lett 267:17–31

Tommasi P, Verrucci L, Rotona T (2015) Mechanical properties of a weak pyroclastic rock and their relationship with microstructure. Can Geotech J 52:211–223

Trigila R, Agosta E, Currado C, De Benedetti AA, Freda C, Gaeta M, Palladino DM, Rosa C (1995) Petrology. In: Trigila R (ed) The volcano of the Alban Hills. University La Sapienza, Rome, pp 95–165

Tumiati S, Fumagalli P, Tiraboschi C, Poli S (2013) An experimental study on COH-bearing Peridotite up to 3.2 GPa and implications for crust-mantle recycling. J Petrol 54:453–479

Turbeville BN (1992a) $^{40}Ar/^{39}Ar$ ages and stratigraphy of the Latera caldera, Italy. Bull Volcanol 55:110–118

Turbeville BN (1992b) Relationships between chamber margin accumulates and pore liquids: evidence from arrested in situ processes in ejecta, Latera caldera, Italy. Contrib Mineral Petrol 110:429–441

Turbeville BN (1993) Petrology and petrogenesis of the Latera caldera, Central Italy. J Petrol 34:77–123

Turi B, Taylor HP (1976) Oxygen isotope studies of potassic volcanic rocks of the Roman Province, Central Italy. Contrib Mineral Petrol 55:1–31

Turtù A, Satolli S, Maniscalco R, Calamita F, Speranza F (2013) Understanding progressive-arc- and strike-slip-related rotations in curve-shaped orogenic belts: the case of the Olevano-Antrodoco-Sibillini thrust (Northern Apennines, Italy). J Geophys Res, Solid Earth 118:1–15

van Bergen MJ, Ghezzo C, Ricci CA (1983) Minette inclusions in the rhyodacitic lavas of Mt. Amiata (central Italy): mineralogical and chemical evidence of mixing between Tuscan- and Roman-type magmas. J Volcanol Geoth Res 19:1–35

van Keken PE, Kiefer B, Peacock SM (2002) High-resolution models of subduction zones: implications for mineral dehydration reactions and the transport of water into the deep mantle. Geochem Geophys Geosyst 3: 1056 doi:10.1029/2001GC000256

Varekamp JC (1979) Geology and petrology of the Vulsinian volcanic area (Latium, Italy). Geologica Ultrajectina 22:384

Varekamp JC, Kalamarides R (1989) Hybridization processes in leucite tephrites from Vulsini, Italy, and the evolution of the Italian potassic suite. J Geophys Res 94:4603–4618

Vezzoli L, Conticelli S, Innocenti F, Landi P, Manetti P, Palladino DM, Trigila L (1987) Stratigraphy of the Latera Volcanic complex: proposal for a new nomenclature. Per Mineral 56:89–110

Villa IM (1993) Geochronology. In: Di Filippo M (ed) Sabatini volcanic complex (Quad Ric Sci CNR, Rome, 114, 11):33–79

Villa IM, Calanchi M, Dielli E, Lucchini F (1999) Age and evolution of the Albano crater lake (Roman Volcanic Province). Acta Vulcanol 11:305–310

Villemant B, Fléhoc C (1989) U-Th fractionation by fluids in K-rich magma genesis: the Vico volcano, Central Italy. Earth Planet Sci Lett 91:312–326

Villemant B, Palacin P (1987) Differentiation magmatique et mechanismes de concentration del l'uranium: exemple du vulcanisme du Latium (Italie centrale). Bull Minéral 110:319–333

Vinkler AP, Cashman K, Giordano G, Groppelli G (2012) Evolution of the mafic Villa Senni caldera-forming eruption at Colli Albani volcano, Italy, indicated by textural analysis of juvenile fragments. J Volcanol Geoth Res 235–236:37–54

Voltaggio M, Barbieri M (1995) Geochronology. In: Trigila R (ed) The volcano of the Alban Hills. University La Sapienza, Rome, pp 167–192

Walker GPL (1983) Ignimbrite types and ignimbrite problems. J Volcanol Geoth Res 17:65–88

Washington HS (1906) The Roman Comagmatic Region. Carnegie Inst Washington Publ 57 (199 pp)

Wendlandt RF, Eggler DH (1980) The origins of potassic magmas: 2. Stability of phlogopite in natural spinel lherzolite an in the $KAlSiO_4$-MgO-SiO_2-H_2O-CO_2 at high pressures and high temperatures. Am J Sci 280:421–458

The Ernici-Roccamonfina Province 5

Abstract

The Ernici-Roccamonfina Province is sited on the western side of Central Apennines, between the Ancona-Anzio (or Olevano-Antrodoco) and the Ortona-Roccamonfina tectonic lines. Volcanic rocks exhibit a very wide range of compositions, from calcalkaline (CA) to Roman-type ultrapotassic (HKS) and kamafugitic (KAM). Both at Ernici and Roccamonfina, calcalkaline to potassic rocks were emplaced later than ultrapotassic volcanics. Abundances of incompatible elements and Sr-isotope ratios increase strongly from calcalkaline to ultrapotassic rocks (e.g. Th $\sim$10 to 90 ppm, $^{87}Sr/^{86}Sr$ $\sim$ 0.7067 to 0.7112), whereas Nd-Pb isotopes show an opposite trend. The Ernici volcanoes (about 0.6 to 0.3 Ma) consist of several monogenetic centres that erupted mafic CA to HKS and KAM pyroclastics and lavas. Roccamonfina (about 0.5 to 0.15 Ma) is a HKS stratovolcano with a large central caldera and post-caldera calcalkaline to potassic lava domes and flows. The ultrapotassic rocks from Ernici-Roccamonfina have compositions similar to volcanoes of the Roman Province, whereas calcalkaline to potassic rocks are geochemically akin to Campania. These variations reflect a heterogeneous mantle source, which was modified by two compositionally (and temporally?) distinct metasomatic events. An episode of metasomatism took place by addition of marly sediments from the Adriatic subducted lithosphere and was responsible for contamination of the mantle source of ultrapotassic magmas. This event was followed/accompanied by the arrival of new subduction-related material with distinct compositional features, which probably originated from the northern limb of the Southern Apennine subduction slab.

Keywords

Ernici • Middle Latina Valley • Roccamonfina • Calcalkaline rocks • Potassic rocks • Ultrapotassic rocks • Kamafugites • Mantle metasomatism • Ancona-Anzio line • Olevano-Antrodoco line • Ortona-Roccamonfina line • Adriatic plate • Ionian plate • Central Apennines

A. Peccerillo, *Cenozoic Volcanism in the Tyrrhenian Sea Region*,
Advances in Volcanology, DOI 10.1007/978-3-319-42491-0_5

5.1 Introduction

Several small monogenetic centres and a medium-size stratovolcano crop out in the Middle Latina Valley and at Roccamonfina, between the Olevano-Antrodoco (or Ancona-Anzio) and the Ortona-Roccamonfina tectonic lines, two main fault systems across the Apennines (Fig. 5.1). The Middle Latina Valley centres are also known as Ernici volcanoes, a name early used by Washington (1906) and somewhat incorrect, inasmuch as Monti Ernici are sited some 15–20 km NE of the volcanic area, and consist entirely of Mesozoic carbonates. However, for respect to tradition, and in support of simplicity, the shorter name of Ernici is used in this book.

The Ernici volcanoes consist of about twenty pyroclastic and lava centres scattered through the central sector of NW-SE trending Latina Valley graben (Angelucci et al. 1974). K/Ar ages are about 0.7 to 0.1 Ma (Basilone and Civetta 1975) and $^{40}Ar/^{39}Ar$ ages range from 630 to 260 ka (Boari et al. 2009). The Roccamonfina volcano is a stratocone with a large central caldera, made up of lava flows, domes and pyroclastic deposits, emplaced between about 550 and 150 ka (Rouchon et al. 2008). Overall, volumes of erupted rocks at Ernici-Roccamonfina are modest, largely lower than those of the nearby Roman and Campania provinces.

Early studies (e.g. Appleton 1972; Civetta et al. 1981) recognised two distinct rock series in the Ernici and Roccamonfina volcanoes: a potassic series (KS) and a high-potassium series (HKS), showing contrasting enrichments in potassium, incompatible elements, and radiogenic Sr. However, recent studies found a larger spectrum of petrological characteristics, with rocks ranging from calcalkaline (CA) and shoshonitic (SHO) to potassic, Roman-type (HKS) and kamafugitic (KAM) ultrapotassic (Frezzotti et al. 2007; Rouchon et al. 2008; Boari et al. 2009). A similar or perhaps larger range of compositions has been found for melt inclusions in early-crystallized mineral phases (Nikogossian and van Bergen 2010). The rock associations occurring at

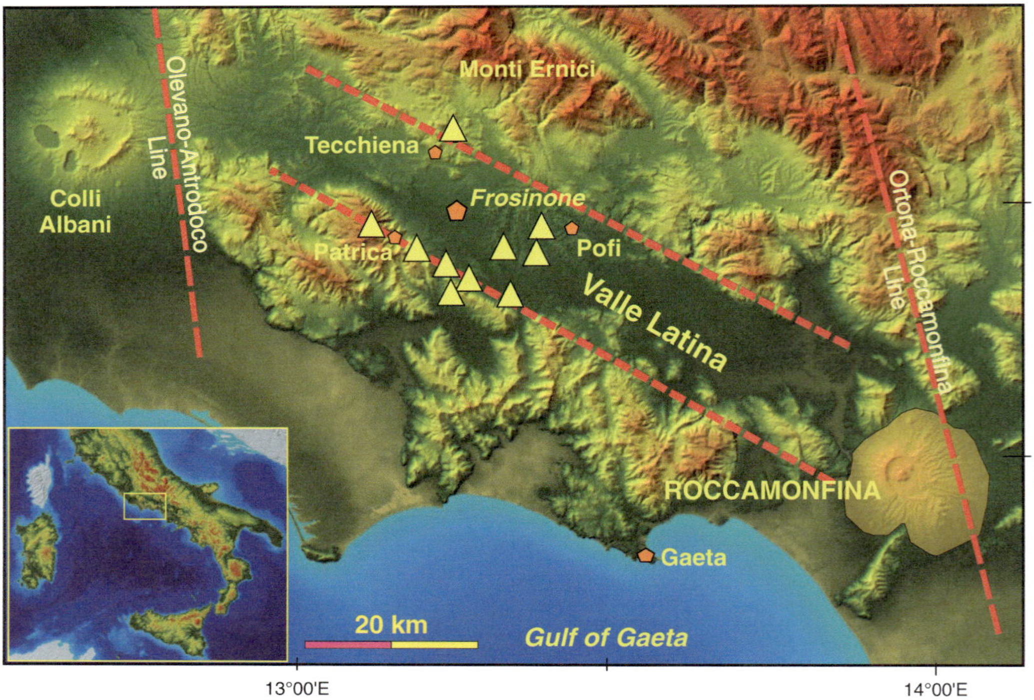

Fig. 5.1 Location map of Ernici (*full triangles*) and Roccamonfina volcanoes

Table 5.1 Age, volcanology and petrology of Ernici and Roccamonfina volcanoes

Volcano	Age (in Ma)	Volcanology	Petrology-Geochemistry
Ernici (Middle Latina Valley)	0.63–0.27	– About twenty monogenetic cones of strombolian scoriae, hydrovolcanic surge and flow deposits and some lava flows	– Calcalkaline basalt, K-trachybasalt, tephrite, phonotephrite, leucitite, and melilitite rocks, with subalkaline to alkaline potassic and ultrapotassic affinity (HKS and kamafugites). Large variation of incompatible element abundances and Sr isotope ratios (~0.706-0.711), both increasing from subalkaline to ultrapotassic rocks
Roccamonfina	0.55–0.15	– Stravolcano with a main central caldera and eccentric cones, consisting of alternating lava flows. Post-caldera lava flows and domes	– Subalkaline (calcalkaline basalt) to alkaline potassic (KS: trachybasalt to trachyte) and ultrapotassic (HKS: tephrite to phonolite) rocks. Incompatible element abundances and radiogenic isotope signatures similar to Ernici

Ernici-Roccamonfina and their wide geochemical and isotopic variation are not encountered elsewhere in central-southern Italy. This brought Peccerillo and Panza (1999) and Peccerillo (2002) to consider Ernici-Roccamonfina as a distinct magmatic province.

Information on volcanology and petrology is summarised in Table 5.1. Major, trace element and isotopic data for representative samples are reported in Table 5.2.

5.2 Regional Geology

The volcanoes of Ernici and Roccamonfina occur on the Tyrrhenian side of the Central Apennines, in a sector delimited by the Olevano-Antrodoco line (OAL) or Ancona-Anzio line in the northwest and by the Ortona-Roccamonfina line (ORL) in the southeast (Locardi 1988; Milano et al. 2008; Satolli and Calamita 2008). The pre-volcanic sequences consist of Early Mesozoic to Quaternary paleogeographic-structural units made up of several thousand meters-thick Upper Triassic to Miocene carbonates and Eocene-Miocene silico-clastic formations. These are overlain by east-verging allochthonous Miocene limestones, marly limestones and flysch sequences, intercalated with Messinian and Pliocene deposits. Foredeep sediments consist of Pliocene to Lower Pleistocene terrigenous deposits (e.g. Accordi and Carbone 1988). The OAL is the transpressional tectonic boundary between the Sabina units and the underlying Central Apennine units, whereas the ORL is the transpressional boundary between the Central Apennine units and the underlying Apulian carbonate platforms (Patacca et al. 2008; Satolli and Calamita 2008).

As elsewhere along the chain, various tectonic phases affected the Central Apennines. Upper Cretaceous to Lower Pliocene compressional regime generated fragmentation of the carbonate platform, and intensive thrusting and folding. Successively, Middle Pliocene to Quaternary extension generated intensive NW-SE faulting and development of graben-horst systems, initially along the Tyrrhenian Sea border and successively shifting toward the east (e.g. Patacca et al. 1990; D'Agostino et al. 2001).

Satolli and Calamita (2008) consider the sector between the OAL and the ORL as a discrete junction area between the Central and the Southern Apennines. This hypothesis is supported by paleomagnetic data, which demonstrate an independent Miocene to Quaternary block rotation of this region, with respect to both the Northern and Southern Apennines (Mattei et al. 1995; Meloni et al. 1997). Locardi (1988) suggested that the

Table 5.2 Composition of representative samples from Ernici and Roccamonfina volcanoes

Ernici				
Rock type	CA basalt	Trachybasalt	Tephrite	Kamafugite
Data source	3,1	3	1	3
SiO_2 wt%	48.62	49.57	47.34	46.03
TiO_2	0.76	0.81	0.81	1.06
Al_2O_3	16.25	17.84	17.41	16.46
Fe_2O_3	3.1	3.5	3.37	4.59
FeO	4.72	4.28	4.2	3.18
MnO	0.15	0.15	0.15	0.14
MgO	9.33	6.55	6.50	4.61
CaO	12.14	11.02	10.62	10.55
Na_2O	2.59	2.48	2.15	2.17
K_2O	0.73	3.39	5.79	9.77
P_2O_5	0.21	0.22	0.53	0.97
LOI	1.4	0.2	1.14	0.47
Sc ppm	33	26	–	20
V	220	228	263	242
Cr	563	167	284	13
Co	35.8	32	31.7	28.5
Ni	97	41	66.4	34
Rb	157	180	392	434
Sr	1003	1076	1692	2553
Y	20	21	31	43
Zr	114	132	263	516
Nb	6	9	12.8	36
Cs	6.99	10.6	22.6	53.9
Ba	466	563	1039	3938
La	33	42	85	242
Ce	69	85	168	478
Nd	30	31	78	181
Sm	6.82	8.2	14.4	33.7
Eu	1.72	2.02	2.97	7.03
Gd	6.45	7.8	9.97	28.8
Tb	0.88	1.03	1.37	3.02
Yb	1.91	2.27	2.13	3.01
Lu	0.29	0.35	0.29	0.42
Hf	3.27	3.79	6.07	14.9
Ta	0.46	0.62	0.56	1.07
Pb	22	26	42	141
Th	8.5	13.7	30.2	93.5

(continued)

Table 5.2 (continued)

Ernici

Rock type	CA basalt	Trachybasalt	Tephrite	Kamafugite
U	2.45	4.3	6.4	7.6
$^{87}Sr/^{86}Sr_i$	0.706532	0.706971	0.709764	0.711171
$^{143}Nd/^{144}Nd$	0.512373	0.512341	0.512144	0.512120
$^{206}Pb/^{204}Pb$	18.940	18.8987	18.886	18.726
$^{207}Pb/^{204}Pb$	15.675	15.6811	15.673	15.670
$^{208}Pb/^{204}Pb$	39.030	39.025	39.075	38.969
$^{176}Hf/^{177}Hf$	(0.282764)	–	0.282590	–

Roccamonfina

Rock type	Trachy-basalt	Sho shonite	Latite	Tephrite	Tephriphonolite	Phonolite
Data source	4	4	4	4	4	2
SiO_2 wt%	47.28	55.19	58.00	46.33	52.57	56.93
TiO_2	0.91	0.73	0.62	0.97	0.58	0.40
Al_2O_3	15.80	18.96	18.69	15.26	20.08	20.02
Fe_2O_3	2.48	4.14	2.86	4.73	1.32	3.60
FeO	5.96	2.46	2.60	4.10	3.98	–
MnO	0.15	0.16	0.13	0.15	0.12	0.10
MgO	9.35	3.04	2.68	6.61	1.75	0.84
CaO	12.05	6.24	5.57	12.59	5.16	3.63
Na_2O	1.32	2.89	2.97	1.60	2.41	3.23
K_2O	1.29	4.45	4.86	6.63	10.41	9.35
P_2O_5	0.30	0.24	0.20	0.62	0.36	0.16
LOI	3.22	1.36	0.83	0.58	1.01	1.16
Sc ppm	44	–	–	–	–	–
V	–	189	159	312	141	96.6
Cr	350	9	20	164	3	–
Co	43	19.9	15.2	34.5	15.4	5.17
Ni	74	11	12	72	11	–
Rb	189	162	188	322	606	437
Sr	819	898	773	1784	1670	1761
Y	30	31	31	39	38	29
Zr	146	199	211	226	256	310
Nb	16	17	19	10	18	20
Cs	18.0	4.98	7.06	31.0	46.3	43.5
Ba	841	656	561	1460	1280	1451
La	41	53	65	105	98	103
Ce	77	109	121	211	185	189
Nd	36	43	45	90	66	68

(continued)

Table 5.2 (continued)

Roccamonfina						
Rock type	Trachy-basalt	Sho shonite	Latite	Tephrite	Tephriphonolite	Phonolite
Sm	7.66	8.65	8.57	18.4	11.5	11.07
Eu	1.88	1.97	1.91	3.70	2.31	2.58
Gd	–	6.85	6.67	12.9	7.44	8.41
Tb	0.70	1.07	1.06	1.81	1.08	1.03
Yb	1.73	2.80	2.91	2.16	2.02	2.45
Lu	–	0.424	0.435	0.31	0.29	0.38
Hf	2.8	5.04	5.14	6.38	4.3	6.28
Ta	0.90	0.96	1.06	0.54	0.87	1.25
Pb	8.7	25	24	51	52	78
Th	7.8	19.1	21.2	36.7	40.4	48.32
U	1.7	5.6	7.1	8.7	9.8	9.98
$^{87}Sr/^{86}Sr$	0.706662	0.707030	0.706975	0.709529	0.709255	–
$^{143}Nd/^{144}Nd$	0.512377	0.512369	0.512382	0.512152	0.512134	–
$^{206}Pb/^{204}Pb$	19.153	19.008	–	18.788	18.805	–
$^{207}Pb/^{204}Pb$	15.680	15.693	–	15.689	15.685	–
$^{208}Pb/^{204}Pb$	39.199	39.168	–	39.048	39.048	–

Source of data: (1) Gasperini et al. (2002); (2) Rouchon et al. (2008); (3) Boari et al. (2009); (4) Conticelli et al. (2009)

Central Apennines represent a segment of an orogenic arc that developed as an independent sector in the Middle Miocene between the OAL and the ORL. According to this author, changes in the degree of rotation along the Apennines disrupted the orogenic belt and resulted in the formation of various arc segments bordered by dextral transcurrent faults.

The Ernici-Roccamonfina zone has a crustal thickness of about 25 km. According to Panza et al. (2004), the uppermost mantle is characterised by a thin layer of material with relatively low S-wave velocity (Vs ~ 3.95 km/s), which passes to a thick lid with S-wave velocities decreasing with depth from Vs ~4.65 to ~4.40 km/s. Such an upper mantle structure seems unique in the circum-Tyrrhenian area (Panza et al. 2004). A new model by Panza et al. (2007) shows a more homogenous structure with an S-wave velocity of about 4.20 to 4.40 km/s, decreasing from Moho to about 250–300 km depth. Vertical sections of dVs and d(Vp/Vs) parallel to the axis of the Apennine chain by Giacomuzzi et al. (2012) highlight a mechanical discontinuity beneath the Central Apennines. This consists of relatively low-velocity mantle rocks extending from 100 to 300 km depth, which separate two high-velocity bodies imaged beneath the Northern and the Southern Apennines. A mantle structural discontinuity has been also suggested by other authors (Benoit et al. 2011; Rosenbaum and Piana Agostinetti 2015), thus reinforcing the idea of Ernici-Roccamonfina as a geological distinct sector in the Italian peninsula, clearly indicated by geochemical data of volcanic rocks.

The Ernici area and other zones of Central Apennines are characterised by high heath flow, with peaks exceeding 300 mW/m^2 (Chiodini et al. 2013). These values are somewhat surprising, since volumes of erupted magmas are small and the area has been traditionally considered as relatively cold.

5.3 Ernici

5.3.1 Volcanology and Stratigraphy

The volcanic activity has been characterised by moderately explosive and effusive eruptions that generated strombolian scoriae, pyroclastic flow and surge deposits, and a few lavas. The most important eruptive centres include Patrica, Tecchiena, Colle Castellone, Pofi, Villa Santo Stefano, and Giuliano di Roma. Most of these volcanoes consist of mafic ultrapotassic rocks (HKS); potassic (KS), shoshonitic (SHO) and calcalkaline (CA) basalts occur in a few places south of Ceccano (Fig. 5.2). Ultrapotassic lavas with kamafugitic (KAM) affinity are found at Colle Castellone (Boari et al. 2009).

According to K/Ar dating by Basilone and Civetta (1975) the volcanism developed between 0.7 and 0.1 Ma. The most K-rich products erupted first (about 0.7 to 0.2 Ma), followed by the less potassic volcanics (about 0.2 to 0.1 Ma; Civetta et al. 1981). $^{40}Ar/^{39}Ar$ dating by Boari et al. (2009) basically confirms the decrease in potassium with time, but restricts the time span to about 633 to 269 ka.

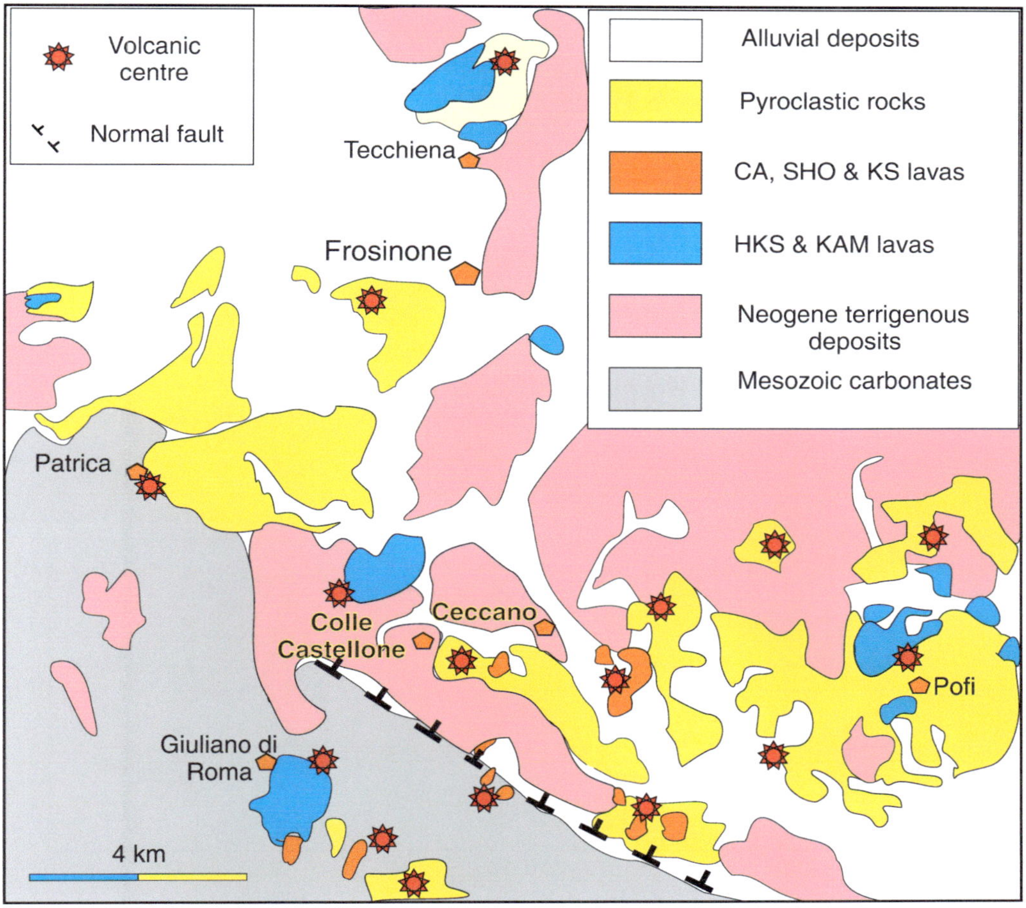

Fig. 5.2 Schematic geological map of Ernici volcanoes. Simplified after Boari et al. (2009)

5.3.2 Petrography and Mineral Chemistry

The Ernici volcanics define a vertical trend on a TAS diagram, spanning the basalt, K-trachybasalt, shoshonite, leucite tephrite and foidite compositional fields, and straddling the boundary between subalkaline and alkaline rocks (Fig. 5.3a). Subalkaline mafic rocks are saturated to slightly undersaturated calcalkaline basalts (Fig. 5.3b). The alkaline rocks are mildly to strongly undersaturated in silica and range from shoshonitic to ultrapotassic, the latter spanning Roman-type and kamafugitic compositions (Boari and Conticelli 2007; Frezzotti et al. 2007; Boari et al. 2009).

The calcalkaline basalts have sub-aphyric to porphyritic texture with phenocrysts of diopside, olivine ($\sim Fo_{92-55}$) and plagioclase set in a groundmass made of the same phases, plus a few phlogopite, Fe-Ti oxides, and glass. The K-trachybasalts and shoshonites have aphyric to sparsely porphyritic textures with phenocrysts of zoned diopside, Mg-rich olivine (up to Fo_{91-83}) and rare microphenocrysts of plagioclase set in a groundmass consisting of strongly zoned plagioclase ($\sim An_{80-35}$), diopside, olivine ($\sim Fo_{75-50}$), magnetite, rare leucite and alkali feldspar, and sporadic nepheline (Civetta et al. 1981; Boari and Conticelli 2007; Frezzotti et al. 2007; Boari et al. 2009; Nikogossian and van Bergen 2010). Olivine phenocrysts often contain spinel inclusions. The HKS leucite-tephrites and leucitites generally exhibit poorly porphyritic textures with about 10–20 vol.% phenocrysts of clinopyroxene, minor leucite and olivine (up to Fo_{90}) commonly rimmed by clinopyroxene, and sporadic microphenocrysts of andesine plagioclase. Clinopyroxene may also be present as 3–5 cm-long megacrysts with compositional zoning from diopside to hedenbergite, which appears colourless to green in thin section. The groundmass consists of leucite, clinopyroxene and opaque minerals. Nepheline, K-feldspar, brown mica and hastingsitic amphibole are also observed. Rocks with kamafugitic affinity are moderately porphyritic with phenocrysts of diopside, leucite, and minor olivine ($\sim Fo_{91-81}$). Melilite occurs in the groundmass and as phenocryst in a few rocks. Other phases include monticellite, apatite, nepheline, phlogopite and Ba-rich sanidine (Boari et al. 2009; Nikogossian and van Bergen 2010).

Clinopyroxene from Ernici rocks is generally diopside with rare hedenbergite. High wollastonite contents are found in clinopyroxenes from kamafugites. Ti/Al ratios are generally low (around 0.1, atoms per formula units), but clinopyroxene from kamafugites show somewhat

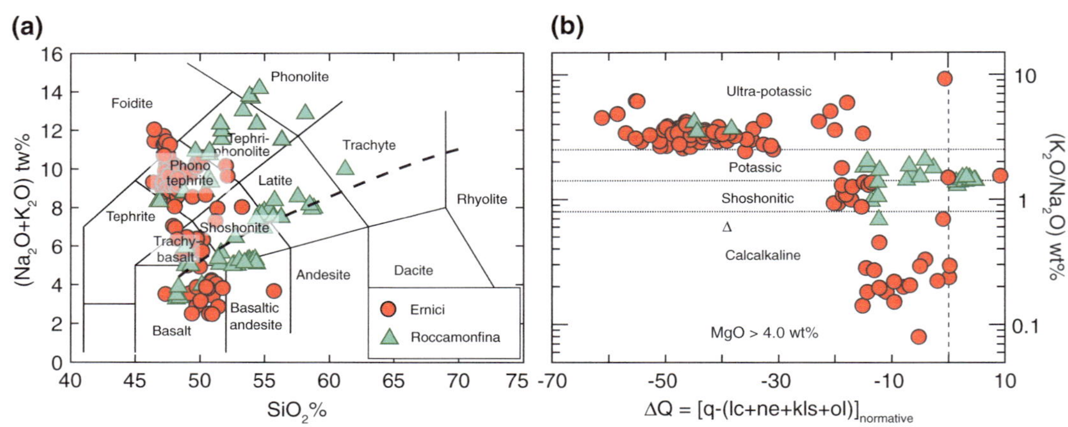

Fig. 5.3 **a** TAS diagram for the Ernici and Roccamonfina volcanic rocks. The *dashed line* is the divide between alkaline and subalkaline fields of Irvine and Baragar (1971); **b** K_2O/Na_2O versus ΔQ diagram for the mafic rocks (MgO > 4 wt%). ΔQ is the algebraic sum of normative quartz minus normative undersaturated minerals (see Chap. 1)

higher values, plotting close to Tuscany-Umbria lamproite-kamafugite clinopyroxenes. Cr/(Cr + Al) ratios of spinel-group minerals included in olivine phenocrysts range from about 0.8 to 0.5, and decrease from calcalkaline to ultrapotassic rocks (Nikogossian ad van Bergen 2010). Overall, their compositions fall in the field of spinels from Umbria kamafugites (Boari and Conticelli 2007).

5.3.3 Petrology and Geochemistry

The Ernici rocks exhibit low SiO_2 contents and moderate to high MgO (about 3 to 9 wt%). Overall, the CA and SHO-KS rocks have higher MgO and ferromagnesian element contents than HKS volcanics (Fig. 5.4). Such a relationship is also observed for melt inclusions contained in early crystallised mafic phases (Nikogossian and van Bergen 2010). LIL element contents increase from CA to HKS and KAM rocks. There is an erratic behaviour of some mobile trace elements (e.g. Cs and Rb; Frezzotti et al. 2007; Table 5.2) in the KS and CA basalts, which is not seen in the melt inclusions and likely derives from secondary modification. REE are variably fractionated with LREE/HREE ratios increasing from CA and KS to HKS and KAM. Most rocks exhibit a negative Eu anomaly, which is stronger in the ultrapotassic samples (Fig. 5.5a). Incompatible element patterns are strongly fractionated and show relative depletions of HFSE and Ba, and positive spikes for LILE, LREE, and Pb (Fig. 5.5b).

Sr-isotope ratios display a large range of values ($^{87}Sr/^{86}Sr$ = 0.7062–0.7112). The lowest ratios ($^{87}Sr/^{86}Sr$ ~0.7062 to 0.7076) are found in the CA, SHO and KS rocks, which partially overlap compositions of the Campania Province. The HKS and KAM rocks show much more radiogenic compositions ($^{87}Sr/^{86}Sr$ ~0.7095–0.7112), falling in the same field of Roman rocks. Nd-isotope ratios increase from HKS-KAM to CA rocks (Fig. 5.6a), the latter showing slightly lower values than Campania mafic volcanics. Pb isotopic ratios (Fig. 5.6b) exhibit moderate variation, with $^{206}Pb/^{204}Pb$ ~18.70 to 18.90, $^{207}Pb/^{204}Pb$ ~15.67 to 15.73 and $^{208}Pb/^{204}Pb$ ~39.00 to 39.18, bridging the Roman and Campania compositional fields (Civetta et al. 1981; Conticelli et al. 2002; Frezzotti et al. 2007; Boari et al. 2009). HKS-KAM rocks have slightly less radiogenic Pb isotope compositions than CA-SHO volcanics. $^{176}Hf/^{177}Hf$ ratio is higher for the KS than HKS (Table 5.2; Gasperini et al. 2002). Oxygen isotopic ratios of whole rocks are variable ($\delta^{18}O_{SMOW}$ ~+5.8 to +8.5), with most values clustering around $\delta^{18}O$ ~+8.0 (Turi et al. 1991). Data on olivine and clinopyroxene phenocrysts separated from rocks with different potassium contents show $\delta^{18}O_{ol}$ ~+5.50‰ to +5.84‰ and $\delta^{18}O_{cpx}$ ~+5.55‰ to +6.50‰, relative to SMOW (Frezzotti et al. 2007). He isotope analyses of clinopyroxene and olivine from two shoshonitic samples by Martelli et al. (2004) yielded values of R/R_A = 2.12–2.87.

5.3.4 Evolution of Ernici Magmas

Decrease in MgO, Ni, and Cr abundances from CA-SHO-KS to ultrapotassic volcanics may suggest a derivation of ultrapotassic magmas from less potassic parents by evolution processes such as fractional crystallisation or AFC. However, petrological, geochemical and isotopic evidence argues against this hypothesis (Civetta et al. 1981; Frezzotti et al. 2007; Boari et al. 2009). In fact, assimilation of large volumes of crustal material (about 50 to 60%) would be necessary to drive isotopic compositions of CA and SHO-KS rocks to those of HKS and KAM volcanics. Such a heavy assimilation would yield oversaturation in silica and a much more dramatic decrease in MgO and compatible element contents than observed in the Ernici rocks. The monogenetic nature of the volcanic centres also supports rapid ascent of magmas from the mantle with little storage and evolution within the crust (e.g. Boari and Conticelli 2007). Therefore, it has been concluded that the compositional variability shown by the Ernici rocks reflects the occurrence of different primary melts. Such a deduction leaves the absence of high-MgO ultrapotassic rocks an unsolved thought-provoking problem, probably related to mantle metasomatic effects.

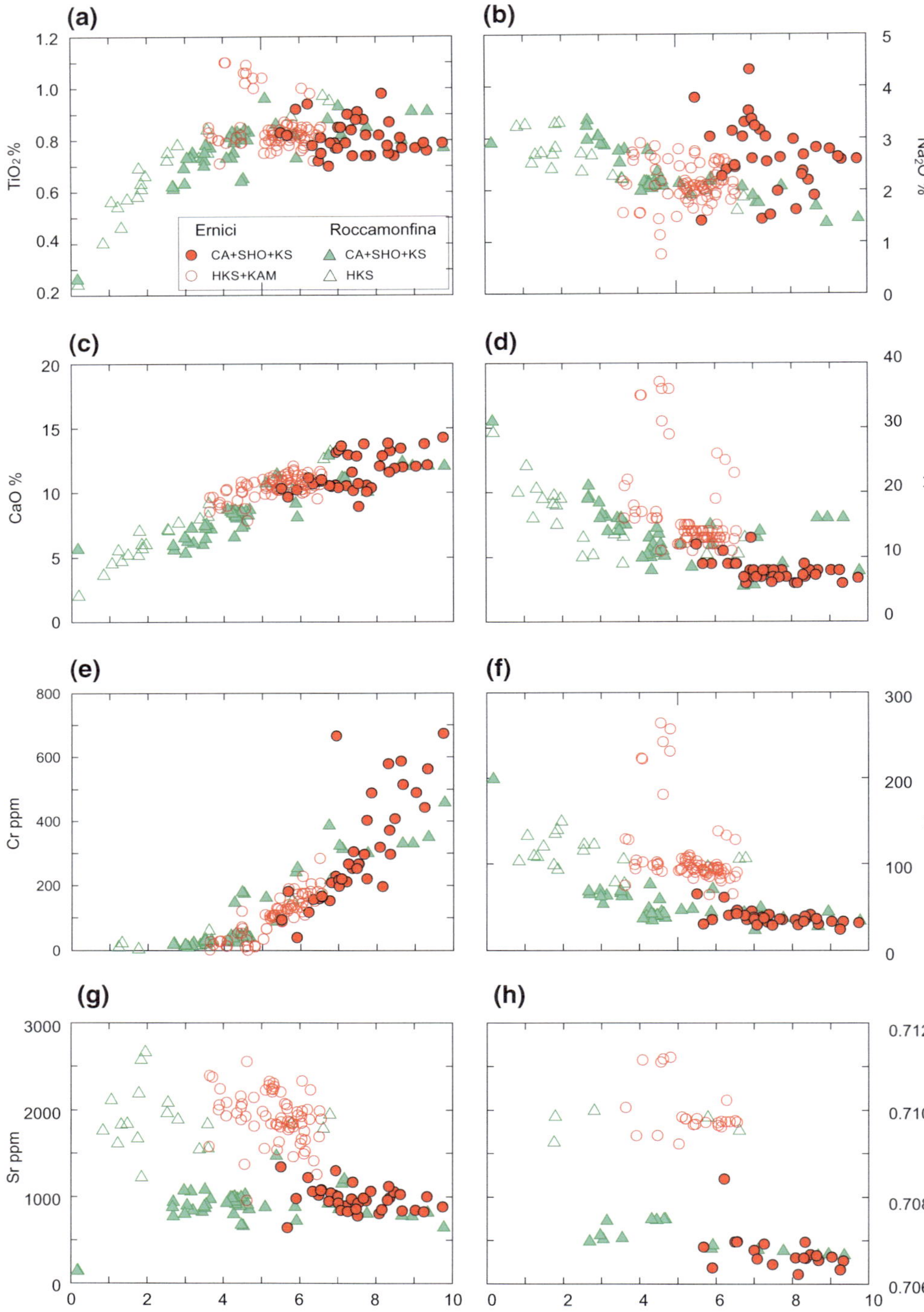

Fig. 5.4 Variation diagrams of selected major and trace elements and $^{87}Sr/^{86}Sr$ versus MgO for Ernici and Roccamonfina volcanics

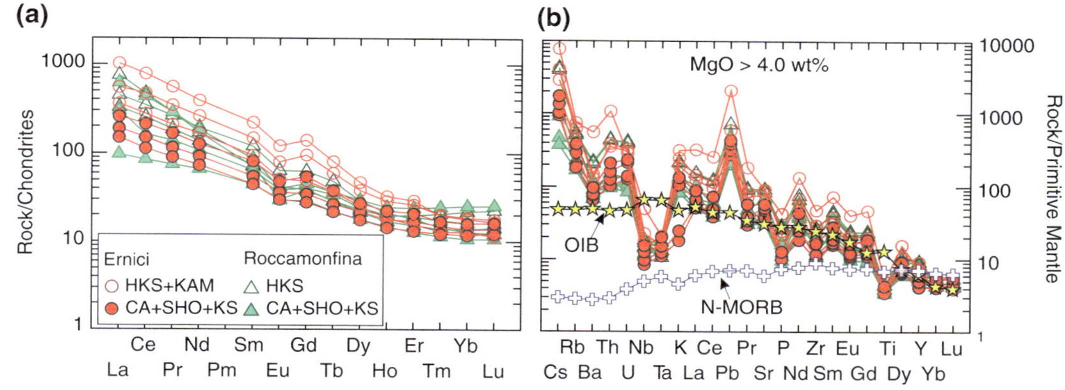

Fig. 5.5 **a** REE patterns of rocks from Ernici and Roccamonfina volcanoes; **b** Incompatible element patterns of mafic rocks. Average OIB and N-MORB (Sun and Mcdonough 1989; Gale et al. 2013) is reported for comparison

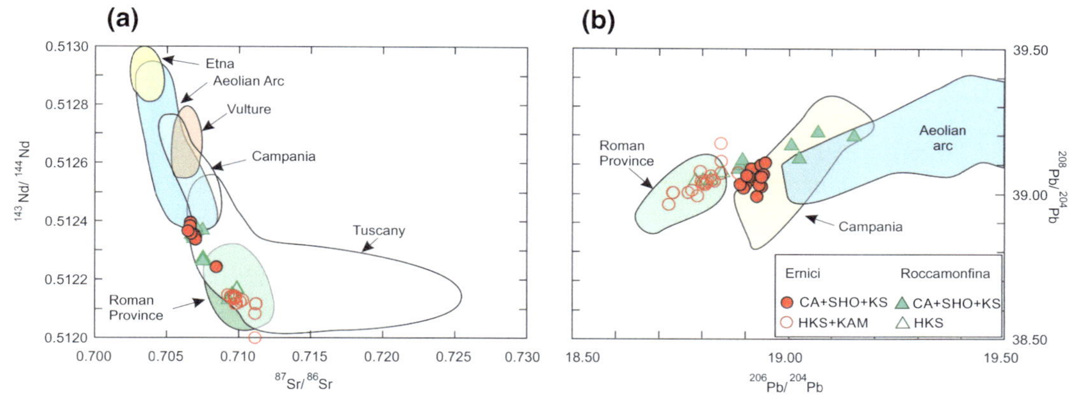

Fig. 5.6 Sr, Nd and Pb isotope variations for the Ernici and Roccamonfina volcanics. The fields of mafic rocks of other magmatic provinces are also shown

Major and trace element variations within single series are likely the consequence of shallow level processes, including moderate degrees of fractional crystallisation and mixing between magmas with variable enrichments in potassium (Civetta et al. 1981; Frezzotti et al. 2007; Boari et al. 2009). Variation of oxygen isotope compositions also suggests episodic assimilation of carbonate wall rocks (Frezzotti et al. 2007). Positive correlations between MgO and compatible trace elements, such as Cr, Ni, and Sc, suggest separation of olivine and clinopyroxene. In contrast, negative correlations between MgO and Al_2O_3 and Sr rule out significant plagioclase fractionation, in agreement with the few plagioclase phenocrysts occurring in the rocks, especially in the HKS. The limited role for plagioclase separation means that the negative Eu anomaly observed in most samples is not related to fractional crystallisation but is a feature of primary melts.

5.4 Roccamonfina

5.4.1 Volcanology and Stratigraphy

Roccamonfina is an asymmetric truncated composite cone, with a base diameter of about 15 km and a 5 km wide NW-SE-elongated summit caldera that is breached on its eastern side (Fig. 5.7 a). The caldera floor, sited at about 600 m above

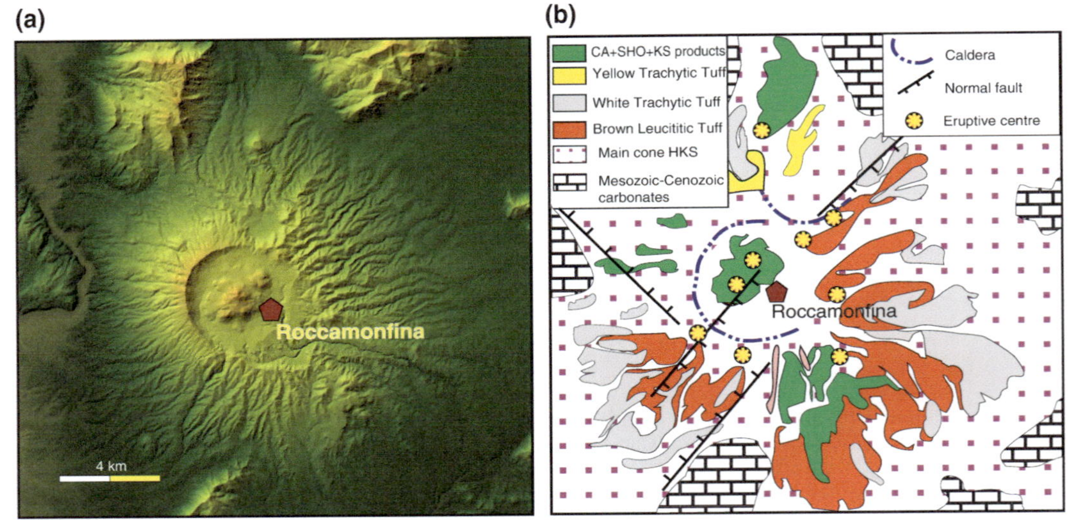

Fig. 5.7 **a** Relief shaded map of the Roccamonfina volcano; **b** Schematic geological map of the Roccamonfina volcano. Simplified after Conticelli et al. (2009)

sea level, hosts several lava flows and domes, which reach a maximum altitude of about 1000 m. The volcano is composed of lava flows, domes and pyroclastic deposits, emitted both from central and parasitic vents. Ages range from about 0.6 to 0.05 Ma (Luhr and Giannetti 1987; De Rita and Giordano 1996; De Rita et al. 1997; Giannetti and De Casa 2000), but a shorter time interval of about 550 to 150 ka has been suggested by Rouchon et al. (2008) on the basis of new K/Ar dating. The volcano developed within the Garigliano graben, a NW-SE trending extensional area cut by NE-SW faults (Chiesa et al. 1995; De Rita and Giordano 1996). Volcanic rocks rest over Mesozoic-Cenozoic carbonate and Miocene flysch sequences. V_S model published by Nunziata and Gerecitano (2012) indicates a 5 km thick sedimentary succession, which overlies a high-velocity body (Vs = 3.8 km/s) probably representing a solidified magmatic chamber, and a low-velocity layer that extends to the nearby Campania volcanic area.

Volcanology and stratigraphy of the Roccamonfina volcano have been the subjects of several papers (e.g. Giannetti and Luhr 1983; Luhr and Giannetti 1987; Cole et al. 1992; Chiesa et al. 1995; Giannetti and De Casa 2000; Rouchon et al. 2008; Conticelli et al. 2009 and references therein). According to De Rita and Giordano (1996) the volcano developed during three main epochs. During the first epoch (about 600 to 400 ka) explosive and effusive eruptions of phonotephritic and tephriphonolitic lava flows and pyroclastics formed the main cone and the central caldera. Successively, large phonotephritic-phonolitic and trachytic pyroclastic sequences were deposited (about 385–230 ka), the most prominent of which (Brown Leucitic Tuff and the White Trachytic Tuff) caused further collapse of the central caldera (Giannetti and Luhr 1983; Luhr and Giannetti 1987). The latest activity (250–59 ka) emplaced KS, SHO and CA products mainly inside the central caldera, and the White Trachytic Tuff on the northern sector of the main cone, where the Conca caldera was formed (Fig. 5.7b). The Roccamonfina volcanics are covered discontinuously by the tephra of Campanian Ignimbrite (39 ka), a large pyroclastic flow from Campi Flegrei (see Chap. 7).

Rouchon et al. (2008) recognised five main stages of activity. The initial stages constructed two superimposed HKS stratovolcanoes and ended with the eruption of the Brown Leucitic Tuff and a sector collapse at about 350 ka. The following stage was responsible for the emplacement of the Lower White Trachytic Tuff at about 331 ka, and of trachybasaltic to trachytic lavas. Successively, basaltic–shoshonitic parasitic cones and lava flows were emplaced, and the Upper White Trachytic Tuff was deposited by a sub-plinian eruption, between 275 and 230 ka. The final activity emplaced the Yellow Trachytic Tuff at 230 ka and constructed the SHO and CA lava domes/flows and pyroclastics at about 150 ka.

5.4.2 Petrography and Mineral Chemistry

The Roccamonfina magmas range in composition from subalkaline to transitional and alkaline ultrapotassic (Fig. 5.3). The subalkaline-transitional rocks mainly consist of SHO-KS trachybasalts, shoshonites, latites and trachytes, with a few CA basalts. The HKS rocks range from leucite-tephrite to phonolite.

The HKS mafic rocks are generally porphyritic with variable amounts of olivine, clinopyroxene and leucite phenocrysts set in a groundmass containing the same phases plus plagioclase, some nepheline and sanidine. Tephriphonolites and phonotephrites contain clinopyroxene, plagioclase, leucite and biotite phenocrysts, whereas phonolites are generally dominated by sanidine, some clinopyroxene, leucite and glass (Chiesa et al. 1995: Conticelli et al. 2009). Accessory phases include Fe-Ti oxides, apatite and sphene.

The CA to KS basalts are variably porphyritic with ubiquitous clinopyroxene phenocrysts plus olivine (up to Fo_{93-79}), and some plagioclase (about An_{90-80}) set in a groundmass composed of the same phases plus Fe-Ti oxides. Biotite is present in small amounts. The evolved SHO-KS rocks have porphyritic textures with phenocrysts of clinopyroxene, plagioclase, biotite and sanidine. Plagioclase is abundant in the latites and decreases in the trachytes, where sanidine dominates phenocryst mineralogy. Accessory phases include Fe-Ti oxides, apatite and in some cases amphibole, garnet and sphene (e.g. Ghiara and Lirer 1977; Giannetti and Luhr 1983; Luhr and Giannetti 1987; Giannetti and Ellam 1994; Conticelli et al. 2009; Nikogossian and van Bergen 2010). Xenoliths of metamorphic and sedimentary wall rocks and cumulate lithologies are found at Roccamonfina (Appleton 1972; Giannetti and Luhr 1990).

In general, there is a strong compositional variability for single mineral phases in the Roccamonfina rocks. Plagioclase is generally zoned and ranges from bytownite to oligoclase. Colourless to green diopside to hedenbergite clinopyroxenes occur in many rocks. Olivine is generally rich in MgO, even in the evolved rocks, where crystals with Fo_{90-85} have been found; NiO content is high, reaching 0.5 wt% in some cases. Cr/(Cr+Al) ratios of spinel inclusions in olivine phenocrysts are similar to Ernici, and decrease from calcalkaline to ultrapotassic rocks (Nikogossian and van Bergen 2010).

5.4.3 Petrology and Geochemistry

Variation diagrams of major and trace elements versus MgO (Fig. 5.4) show positive correlations for CaO, and ferromagnesian elements, and opposite trends for incompatible elements. As for Ernici, the CA, SHO and KS rocks show the highest MgO contents among the exposed volcanics. An overall positive correlation is observed for LILE and LREE versus K_2O in rocks with similar MgO contents, whereas HFSE (e.g. Nb) remain nearly constant. However, Giannetti and Ellam (1994) recognised some exceptions to this general rule.

The REE patterns of mafic rocks are fractionated, with increasing LREE contents and La/Yb ratios from CA to HKS. Negative Eu anomalies are observed in several mafic rocks, especially for the HKS (Fig. 5.5a). REE patterns of some trachytes (not shown) have an upward concave pattern with low concentrations of intermediate REE. Incompatible element patterns of mafic rocks are fractionated, with relative depletion in the HFSE and Ba, and positive spikes for LILE, LREE and Pb (Fig. 5.5b).

Sr-isotope ratios range from about 0.7067 to 0.7099 (Fig. 5.4h). Similar compositions have been found for melt inclusions in olivine phenocrysts by Koornneef et al. (2015). Nd-isotope ratios range from about 0.5121 to 0.5124 (Fig. 5.6a). As in the case of Ernici, Sr-isotope ratios decrease and Nd-isotope increase from ultrapotassic to CA-SHO-KS rocks, spanning the compositions of Roman and Campania provinces. Pb-isotope ratios display more radiogenic compositions in the CA-SHO-KS than in the HKS rocks (Fig. 5.6b; Hawkesworth and Vollmer 1979; Vollmer and Hawkesworth 1980; Conticelli et al. 2009). Whole rock oxygen isotope ratios are variable ($\delta^{18}O‰$ ~+6.6 to +10.3) with most values around $\delta^{18}O‰$ ~+8.0 relative to SMOW (Taylor et al. 1979).

5.4.4 Evolution of Roccamonfina Magmas

The Roccamonfina magmas underwent a more complex evolutionary history than Ernici. Geochemical data suggest that the evolution of KS and HKS magmas was dominated by fractional crystallisation, starting from K-trachybasalt and tephrite parents. According to Conticelli et al. (2009), mafic HKS magmas experienced dominant fractionation of clinopyroxene, plus some leucite, and accessory magnetite and apatite, whereas plagioclase-clinopyroxene fractionation was responsible for the evolution of felsic magmas. The KS magma evolution was dominated by fractionation of clinopyroxene-plagioclase-olivine-magnetite in the mafic range, followed/joined by biotite and K-feldspar in the evolved compositions (Appleton 1972).

There is also ample geochemical and petrographic evidence for mixing and wall-rock assimilation during the evolution of Roccamonfina magmas (e.g. Giannetti and Luhr 1983; Luhr and Giannetti 1987). Mixing between compositionally contrasting magma types is suggested by geochemical and isotopic disequilibria within single samples, and by the coexistence of mafic and felsic juvenile clasts in many pyroclastic deposits. Assimilation of crustal rocks is indicated by the increase in Sr and O isotopic ratios from mafic to felsic CA-SHO-KS rocks (Taylor et al. 1979; Conticelli et al. 2009).

As in the case of Ernici, the wide compositional variation of mafic magmas cannot be explained by evolution processes, but point to the occurrence of distinct parental melts. Isotopic variations also exclude a derivation of HKS magma from a potassic parent by high-pressure eclogite fractionation (Appleton 1972).

5.5 Petrogenesis of the Ernici-Roccamonfina Province

The Ernici and Roccamonfina volcanoes are characterised by an association of calcalkaline, shoshonitic, potassic and ultrapotassic rocks. The latter have HKS and kamafugitic affinity, resembling closely the analogous volcanic rocks from the Roman and Intra-Apennine provinces (Figs. 5.6 and 5.8). Therefore, as with these two provinces, an origin by melting of phlogopite-rich lherzolitc to clinopyroxenite mantle contaminated by supercritical fluids or melts from subducted marls, can be envisaged for Ernici-Roccamonfina ultrapotassic magmas (see discussion in Chap. 3 and 4).

The origin of mafic calcalkaline, shoshonitic and potassic magmas is more controversial. Basically, two main hypotheses have been proposed. One suggests higher degrees of partial melting of the same source of ultrapotassic magmas (e.g. Boari et al. 2009), whereas another hypothesis envisages polybaric melting of compositionally heterogeneous upper mantle sources (e.g. Peccerillo 2002; Frezzotti et al. 2007; Conticelli et al. 2009; Nikogossian and van Bergen 2010).

Boari et al. (2009) believe that the Ernici primary magmas were generated by melting of upper mantle rocks that had undergone a two-stage metasomatism, with an early pervasive infiltration of melts from an oceanic slab, followed by a second stage influx of melts from subducted marly sediments. The marl-derived melts concentrated along particular pathways

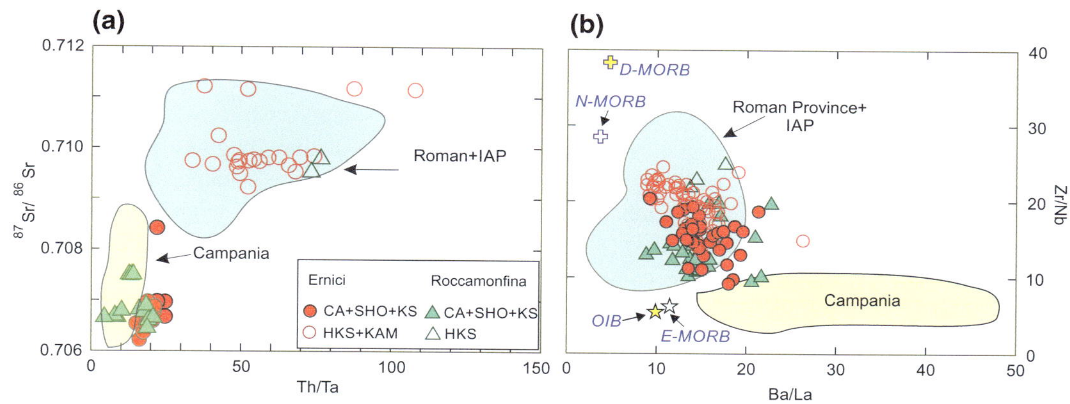

Fig. 5.8 Plots of incompatible element and Sr isotope ratios of the mafic rocks (MgO > 4%) from Ernici and Roccamonfina volcanoes, compared with the mafic volcanics from the Campania, Roman and Intra-Apennine (IAP) provinces. *Stars* indicate average Enriched MORB and OIB; *crosses* are average Normal MORB and Depleted MORB (Sun and Mcdonough 1989; Gale et al. 2013)

inside the mantle wedge, generating a network of phlogopite-rich clinopyroxenite vein throughout the peridotite. Melting processes, possibly triggered by slab breakoff or asthenospheric upwelling, first affected the vein network, generating early KAM and HKS magmas. Successively, melting involved the mantle surrounding the veins, giving late KS, SHO and CA magmas by dilution of early formed melts. The vein plus wall rock melting process requires high amounts of dilution by about 90% of a liquid with a composition as the Aeolian arc calcalkaline basalt, in order to drive Sr-Nd isotope composition of HKS and KAM magmas to those of the calcalkaline and shoshonitic basalts.

An alternative hypothesis is that the Ernici-Roccamonfina parent magmas originated in a variably metasomatised upper mantle (e.g. Peccerillo 1985, 2001; Giannetti and Ellam 1994; Frezzotti et al. 2007; Nikogossian and van Bergen 2010). The KS-SHO-CA primary melts were generated at relatively shallow depth where a moderately contaminated phlogopite-peridotite produced saturated or slightly undersaturated liquids showing relatively poorly enriched compositions. In contrast, the HKS primary magmas were generated at higher pressure by melting of more strongly contaminated mantle rocks, with undersaturated compositions being favoured by incongruent melting of phlogopite at high pressure (Wendlandt and Eggler 1980). According to Giannetti and Ellam (1994), the HKS parental melts were generated in a phlogopite-rich peridotite, whereas the KS and SHO-CA magmas come from shallower amphibole-bearing peridotite. Based on major and trace element study of melt inclusions and host mafic minerals, Nikogossian and van Bergen (2010) come to the conclusion that distinct primary magmas were related to progressive melting of various types of veins. Veining was formed by infiltration of two distinct types of melts, with silicate and carbonate composition, respectively. Variable proportions of clinopyroxene, phlogopite and amphibole crystallised along the veins, whose melting gave different types of magmas. The mantle zoning beneath Ernici-Roccamonfina could reflect percolation of metasomatic agents released by subducted upper crustal material, a process that is able to generate strong vertical compositional gradients above subduction zones (e.g. Chin et al. 2014). An alternative possibility is that there were multiple metasomatic events in the area, as suggested by several geochemical studies of whole rocks and melt inclusions (Peccerillo 1999, 2002; Nikogossian and van Bergen 2010).

Conticelli et al. (2009) accept a compositionally zoned mantle source for Roccamonfina volcano and propose that the transition from ultrapotassic to calcalkaline composition reflects an inflow of OIB-type mantle from the foreland into the source region of primary magmas. However, the similar HFSE concentrations (e.g. Nb, Ta; Figs. 5.4d and 5.5b) in all the mafic rocks from Ernici-Roccamonfina argue against a role of OIB in the origin of CA to KS magmas, since OIB-type components are enriched in HFSE.

As for the pre-metasomatic mantle composition, a depleted nature is suggested by the low abundances of Ta and Nb and by HFSE/HFSE ratios (e.g. Zr/Nb) in the mafic rocks, which are close to average N-type MORB (Figs. 5.5b and 5.8). Such a conclusion is supported by compositions of chromite inclusion and host olivine reported by Boari and Conticelli (2007), although Nikogossian and van Bergen (2010) interpret Cr# variability of spinels as an effect of metasomatism.

5.6 Mantle Metasomatism and Geodynamic Implications

Several lines of evidence indicate that the mantle metasomatic processes in the Roman and Intra-Apennine provinces are compositionally distinct from the Campania Province (e.g. Peccerillo 2002; De Astis et al. 2006). Therefore, the large range of mafic rock compositions at Ernici-Roccamonfina, which span the Roman and Campania volcanics, led to the conclusion that two distinct episodes of mantle metasomatism, with Roman and Campanian affinities, took place in this volcanic district (Peccerillo 1999, 2002; Peccerillo and Panza 1999). As discussed in Chap. 4, mantle metasomatism in the Roman Province was accomplished by marly sediments of the Adriatic continental plate. In contrast, the mantle metasomatism in the Campania Province is related to infiltration of fluids originated from an oceanic slab, with some contributions by sediments (Peccerillo 2002; De Astis et al. 2006; see Chap. 7). It has been suggested that metasomatic fluids were provided by the Ionian subduction system that is still presently active beneath the southern Tyrrhenian Sea (Peccerillo and Frezzotti 2015 with references). The occurrence of two distinct types of metasomatic processes at Ernici-Roccamonfina generated a heterogeneous mantle, which was the source of compositionally distinct magmas. Since the CA, SHO and KS rocks are younger than the HKS volcanics, it has been suggested that the Roman-type "Adriatic" metasomatism might have preceded the Campanian-type "Ionian" event (Peccerillo 1999; Peccerillo and Panza 1999).

A role for both the Ionian and Adriatic subduction systems in the metasomatism of the Ernici-Roccamonfina magma sources is compatible with geophysical evidence. As recalled earlier, the upper mantle beneath this area show significant differences with respect to the nearby Roman and Campania provinces. Tomographic models by Giacomuzzi et al. (2012) show relatively low seismic-wave velocities, which contrast with the occurrence of contiguous high-velocity vertical bodies beneath the Northern and Southern Apennine arcs. In this view, there is a gap in the Apennine subduction zone with a window beneath Ernici-Roccamonfina. The absence of a dehydrating slab could be an explanation for the scarcity of volcanism in this region. Compositional different melts or supercritical fluids that were released from the Northern Apennine and Southern Apennine subduction systems were responsible for mantle metasomatism.

A detailed geodynamic scenario that brought to the particular setting for the Ernici-Roccamonfina zone is still lacking. It can be recalled, however, that the Miocene to present evolution of the Tyrrhenian Sea-Apennine system has been characterised by an early stage of E-W extension, which changed during Pliocene to NW-SE in the south and NE-SW in the north (e.g. Sartori 2003). Such a modification has been related to the different resistance opposed to slab retreat by the Adriatic continental plate and by the Ionian oceanic plate. Basically, the strong laterally variable nature of the foreland prompted splitting of the subduction zone, with the formation of two independent slab segments exhibiting different dipping direction and degrees of rollback (e.g. Carminati et al. 1998, 2012). Consequently, an early process of west-directed subduction of a continuous Adriatic-Ionian plate was followed by a double system of west-directed subduction of the northward migrating Adriatic plate, and northwest-directed subduction of a clockwise rotated and southeast retreating Ionian plate (e.g. Malusà et al. 2015). In this context, the formation of a window in the slab was inescapable and the Ernici-Roccamonfina volcanism would be located above the slab tear zone in which both the subduction systems might have a role in mantle metasomatism.

5.7 Summary and Conclusions

Volcanism in the Ernici-Roccamonfina area is characterised by extremely variable petrological, incompatible element and radiogenic isotope signatures. At Ernici several monogenetic volcanoes were formed by the eruption of mafic magmas, with CA to HKS and KAM affinities. At Roccamonfina, effusive and explosive activity built up a stratocone with a large central caldera and various intra-caldera and flank cones and domes. Roccamonfina rocks range from mafic to felsic, with CA to HKS compositions.

The ultrapotassic rocks of Ernici and Roccamonfina volcanoes have compositions similar to volcanoes of the IAP and Roman provinces, whereas CA-SHO-KS rocks are geochemically similar to Campania volcanoes. These variations probably reflect a heterogeneous mantle source, which was modified by two compositionally (and temporally?) distinct metasomatic events. An episode of metasomatism took place by infiltration of fluids released from the Adriatic subducted lithosphere and was responsible for contamination of the mantle source of ultrapotassic magmas. This event was followed/accompanied by the arrival of new subduction-related material with distinct compositional features. This may have been provided by the oceanic Ionian slab and associated sediments.

References

Accordi G, Carbone F (eds.) (1988) Lithofacies map of Latium-Abruzzi and neighbouring areas, scale 1:500,000. Quad Ric Sci CNR, Rome, 114, 5, 223 pp

Angelucci A, Brotzu P, Civitelli G, Morbidelli L, Traversa G (1974) Il vulcanismo Pleistocenico della Media Valle Latina (Lazio). Caratteristiche petrografiche e geologiche dei principali affioramenti lavici. Geol Rom 13:83–123

Appleton JD (1972) Petrogenesis of potassium-rich lavas from the Roccamonfina Volcano, Roman Region, Italy. J Petrol 13:425–456

Basilone P, Civetta L (1975) Datazione K/Ar dell'attivita vulcanica dei Monti Ernici (Latina). Rend Soc It Mineral Petrol 31:175–179

Benoit MH, Torpey M, Liszewski K, Levin V, Park J (2011) P and S wave upper mantle seismic velocity structure beneath the northern Apennines: New evidence for the end of subduction. Geochem Geophys Geosyst 12:Q06004. doi:10.1029/2010GC003428

Boari E, Conticelli S (2007) Mineralogy and petrology of Mg-rich calc-alkalic, potassic, and ultrapotassic associated rocks: the Middle Latin Valley monogenetic volcanoes, Roman Magmatic Province, Southern Italy. Can Mineral 45:1443–1469. doi:10.2113/gscanmin.45.6.1000

Boari E, Tommasini S, Laurenzi MA, Conticelli S (2009) Transition from ultrapotassic kamafugitic to sub-alkaline magmas: Sr, Nd, and Pb isotope, trace element and $^{40}Ar/^{39}Ar$ age data from the Middle Latin Valley volcanic field, Roman Magmatic Province, Central Italy. J Petrol 50:1327–1357

Carminati E, Wortel MJR, Spakman W, Sabadini R (1998) The role of slab detachment processes in the opening of the western-central Mediterranean basins: some geological and geophysical evidence. Earth Planet Sci Lett 160:651–665

Carminati E, Lustrino M, Doglioni C (2012) Geodynamic evolution of the central andwestern Mediterranean: tectonic vs. igneous constraints. Tectonophysics 579:173–192

Chiesa S, Floris B, Gillot P-Y, Prosperi L, Vezzoli L (1995) Il Vulcano di Roccamonfina. Lazio Meridionale. ENEA, Rome, pp 128–150

Chin EJ, Lee C-TA, Barnes JD (2014) Thickening, refertilization, and the deep lithosphere filter in continental arcs: Constraints from major and trace elements and oxygen isotopes. Earth Planet Sci Lett 397:184–200

Chiodini G, Cardellini C, Caliro S, Chiarabba C, Frondini F (2013) Advective heat transport associated with regional Earth degassing in central Apennine (Italy). Earth Planet Sci Lett 373:65–74

Civetta L, Innocenti F, Manetti P, Peccerillo A, Poli G (1981) Geochemical characteristics of potassic volcanics from Mt. Ernici (Southern, Latium, Italy). Contrib Mineral Petrol 78:37–47

Cole PD, Guest JE, Duncan AM, Chester DK, Bianchi R (1992) Post-collapse volcanic history of calderas on a composite volcano: an example from Roccamonfina volcano, southern Italy. Bull Volcanol 54:253–266

Conticelli S, D'Antonio M, Pinarelli L, Civetta L (2002) Source contamination and mantle heterogeneity in the genesis of Italian potassic and ultrapotassic volcanic rocks: Sr-Nd-Pb isotope data from Roman Province and Southern Tuscany. Mineral Petrol 74:189–222

Conticelli S, Marchionni S, Rosa D, Giordano G, Boari E, Avanzinelli R (2009) Shoshonite and sub-alkaline magmas from an ultrapotassic volcano: Sr-Nd-Pb isotope data on the Roccamonfina volcanic rocks, Roman Magmatic Province, Southern Italy. Contrib Mineral Petrol 157:41–63

D'Agostino N, Giuliani R, Mattone M, Bonci L (2001) Active crustal extension in the central Apennines (Italy) inferred from GPS measurements in the interval 1994–1999. Geophys Res Lett 28:2121–2124

De Astis G, Kempton PD, Peccerillo A, Wu TW (2006) Trace element and isotopic variations from Mt. Vulture to Campanian volcanoes: constraints for slab detachment and mantle inflow beneath southern Italy. Contrib Mineral Petrol 151:331–351

De Rita D, Giordano G (1996) Volcanological and structural evolution of Roccamonfina volcano (Italy): origin of the summit caldera. In: Mc Guire WJ, Jones AP, Neuberg J (eds) Volcano Instability on the Earth and Other Planets, Geol Soc (Spec Publ 110): 209–224

De Rita D, Giordano G, Milli S (1997) Forestepping-backstepping stacking pattern of volcanic sequences: Roccamonfina volcano, Italy. J Volcanol Geoth Research 80:155–178

Frezzotti ML, De Astis G, Dallai L, Ghezzo C (2007) Coexisting calc-alkaline and ultrapotassic magmatism at Monti Ernici, Mid Latina Valley (Latium, central Italy). Eur J Mineral 19:479–497

Gale A, Dalton CA, Langmuir CH, Su Y, Schilling J-G (2013) The mean composition of ocean ridge basalts. Gechem Geophys Geosyst 14 doi:10.1029/2012GC004334

Gasperini D, Blichert Toft J, Bosch D, Del Moro A, Macera P, Albarède F (2002) Upwelling of deep mantle material through a plate window: evidence from the geochemistry of Italian basaltic volcanics. J Geophys Res 107(B12):2367. doi:10.1029/2001JB000418

Ghiara MR, Lirer L (1977) Mineralogy and geochemistry of the "low potassium" series of the Roccamonfina volcanic suite (Campania, South Italy). Bull Volcanol 40:39–56

Giacomuzzi G, Civalleri M, De Gori P, Chiarabba C (2012) A 3D Vs model of the upper mantle beneath Italy: Insight on the geodynamics of central Mediterranean. Earth Planet Sci Lett 335–336:105–120

Giannetti B, De Casa G (2000) Stratigraphy, chronology, and sedimentology of ignimbrites from the white trachytic tuff, Roccamonfina volcano, Italy. J Volcanol Geoth Res 96:243–295

Giannetti B, Ellam R (1994) The primitive lavas of Rocccamonfina volcano, Roman region, Italy: new coinstraints on melting processes and source mineralogy. Contrib Mineral Petrol 116:21–31

Giannetti B, Luhr JF (1983) The White Trachytic Tuff of Roccamonfina volcano (Roman Region, Italy). Contrib Mineral Petrol 84:235–252

Giannetti B, Luhr JL (1990) Phlogopite-clinopyroxenite nodules from the high-K magmas, Roccamonfina Volcano, Italy: evidence for a low-pressure metasomatic origin. Earth Planet Sci Lett 101:404–424

Hawkesworth CJ, Vollmer R (1979) Crustal contamination vs. enriched mantle: $^{143}Nd/^{144}Nd$ and $^{87}Sr/^{86}Sr$ evidence from the Italian volcanics. Contrib Mineral Petrol 69:151–165

Irvine TN, Baragar WRA (1971) A guide to chemical classification of common volcanic rocks. Can J Earth Sci 8:523–548

Koornneef JM, Nikogosian I, van Bergen MJ, Smeets R, Bouman C, Davies GR (2015) TIMS analysis of Sr and Nd isotopes in melt inclusions from Italian potassium-rich lavas using prototype 10^{13} Ω amplifiers. Chem Geol 397:14–23

Locardi E (1988) The origin of the Apenninic arc. Tectonophysics 146:105–123

Luhr JF, Giannetti B (1987) The Brown Leucitic Tuff of Roccamonfina Volcano (Roman Region, Italy). Contrib Mineral Petrol 95:420–436

Malusà MG, Faccenna C, Baldwin SL, Fitzgerald PG, Rossetti F, Balestrieri ML, Danisik M, Ellero A, Ottria G, Piromallo C (2015) Contrasting styles of (U) HP rock exhumation along the Cenozoic Adria-Europe plate boundary (Western Alps, Calabria, Corsica). Geochem Geophys Geosyst 16 doi:10.1002/2015GC005767

Martelli M, Nuccio PM, Stuart FM, Burgess R, Ellam RM, Italiano F (2004) Helium-strontium isotope constraints on mantle evolution beneath the Roman Comagmatic Province, Italy. Earth Planet Sci Lett 224:295–308

Mattei M, Funiciello R, Kissel C (1995) Paleomagnetic and structural evidence of Neogene block rotation in the Central Apennines, Italy. J Geophys Res 100:17863–17883

Meloni A, Alfonsi L, Florindo F, Sagnotti L, Speranza F, Winkler A (1997) Neogene and Quaternari geodynamic evolution of the Italian peninsula: the contribution of paleomagnetic data. Ann Geophys 40:705–727

Milano G, Di Giovanbattista R, Ventura G (2008) Seismic activity in the transition zone between Southern and Central Apennines (Italy): Evidences of longitudinal extension inside the Ortona-Roccamonfina tectonic line. Tectonophysics 457:102–110

Nikogossian IK, van Bergen MJ (2010) Heterogeneous mantle sources of potassium-rich magmas in central-southern Italy: melt inclusion evidence from Roccamonfina and Ernici (Mid Latina Valley). J Volcanol Geoth Res 197:279–302

Nunziata C, Gerecitano F (2012) V_S crustal models of the Roccamonfina volcano and relationships with Neapolitan volcanoes (southern Italy). Int J Earth Sci 101:1371–1383

Panza GF, Pontevivo A, Sarao' A, Aoudia A, Peccerillo A (2004) Structure of the lithosphere-asthenosphere and volcanism in the Tyrrhenian Sea and surroundings. Mem Descr Carta Geol It 64:29–56

Panza GF, Peccerillo A, Aoudia A, Farina B (2007) Geophysical and petrological modeling of the structure and composition of the crust and upper mantle in complex geodynamic settings: The Tyrrhenian Sea and surroundings. Earth Sci Rev 80:1–46

Patacca E, Sartori R, Scandone P (1990) Tyrrhenian basin and Apenninic arcs: kinematic relations since Late Tortonian times. Mem Soc Geol It 45:425–451

Patacca E, Scandone P, Di Luzio E, Cavinato GP, Parotto M (2008) Structural architecture of the central Apennines: interpretation of the CROP 11 seismic profile from the Adriatic coast to the orographic divide. Tectonics 27, TC3006 doi:10.1029/2005TC001917

Peccerillo A (1985) Roman Comagmatic Province (Central Italy): evidence for subduction-related magma genesis. Geology 13:103–106

Peccerillo A (1999) Multiple mantle metasomatism in central-southern Italy: geochemical effects, timing and geodynamic implications. Geology 27:315–318

Peccerillo A (2001) Geochemistry of Quaternary magmatism in central-southern Italy: genesis of primary melts and interaction with crustal rocks. Geochem Int 39:521–535

Peccerillo A (2002) Plio-Quaternary magmatism in central-southern Italy: a new classification scheme for volcanic provinces and its geodynamic implications. In: Barchi RM, Cirilli S, Minelli G (eds) Geological and geodynamic evolution of the Apennines. Boll Soc Geol It (Spec Vol 1):113–127

Peccerillo A, Frezzotti ML (2015) Magmatism, mantle evolution and geodynamics at the converging plate margins of Italy. J Geol Soc London 172:407–427

Peccerillo A, Panza GF (1999) Upper mantle domains beneath central-southern Italy: petrological, geochemical and geophysical constraints. Pure Appl Geophys 156:421–443

Rosenbaum G, Piana Agostinetti N (2015) Crustal and upper mantle responses to lithospheric segmentation in the northern Apennines. Tectonics 34 doi:10.1002/2013TC003498

Rouchon V, Gillot P-Y, Quidelleur X, Chiesa S, Floris B (2008) Temporal evolution of the Roccamonfina volcanic complex (Pleistocene), Central Italy. J Volcanol Geoth Res 177:500–514

Sartori R (2003) The Thyrrhenian backarc basin and subduction of the Ionian lithosphere. Episodes 26:217–221

Satolli S, Calamita F (2008) Differences and similarities between the central and the southern Apennines (Italy): Examining the Gran Sasso versus the Matese-Frosolone salients using paleomagnetic, geological, and structural data. J Geophys Res 113: B10101. doi:10.1029/2008JB005699

Sun SS, McDonough WF (1989) Chemical and isotopic systematics of oceanic basalts: implications for mantle composition and processes. In: Saunders AD, Norry MJ (eds) Magmatism in ocean basins. Geol Soc London (Spec Publ 42):313–345

Taylor HP, Giannetti B, Turi B (1979) Oxygen isotope geochemistry of the potassic igneous rocks from the Roccamonfina volcano, Roman Comagmatic Region, Italy. Earth Planet Sci Lett 46:81–106

Turi B, Taylor HP, Ferrara G (1991) Comparisons of $^{18}O/^{16}O$ and $^{87}Sr/^{86}Sr$ in volcanic rocks from the Pontine Islands, M. Ernici and Campania with other areas in Italy In: Taylor HP, O'Neil JR, Kaplan IR (eds) Stable Isotope Geochemistry: A Tribute to Samuel Epstein. Geochem Soc Spec Publ 3:307–324

Vollmer R, Hawkesworth CJ (1980) Lead isotopic composition of the potassic rocks from Roccamonfina (south Italy). Earth Planet Sci Lett 47:91–101

Washington HS (1906) The Roman Comagmatic Region. Carnegie Inst. Wash. Publ. 57:199 pp

Wendlandt RF, Eggler DH (1980) The origins of potassic magmas: 2. Stability of phlogopite in natural spinel lherzolite an in the $KAlSiO_4$-MgO-SiO_2-H_2O-CO_2 at high pressures and high temperatures. Am J Sci 280:421–458

6 The Pontine Islands

Abstract

The Pontine archipelago consists of several volcanic islands and islets sited at the eastern end of the 41° Parallel line. Ages of exposed products range from about 4.1 Ma to less than 130 ka, and decrease eastward. Calcalkaline rhyolites, peralkaline rhyolites, and trachytes occur in the western islands of Ponza, La Botte, Palmarola and Zannone, where mafic-intermediate calcalkaline to potassic 4.0-3.9 Ma old volcanism has been recently discovered below sea level. A suite of basalt-trachybasalt to phonolite rocks makes up the eastern islands of Ventotene and Santo Stefano. Trachytes and peralkaline rhyolites derived from moderately potassic mafic parents by fractional crystallisation plus crustal assimilation. Calcalkaline rhyolites were formed either by crustal melting or by AFC, starting from mafic-intermediate calcalkaline parents. Basalts and trachybasalts from Ventotene have variable isotopic signatures, which reflect open-system intra-crustal evolution processes and/or source heterogeneity. The mafic rocks from Ventotene have geochemical and isotopic signatures that resemble the moderately potassic rocks from Ernici-Roccamonfina, suggesting similar type of mantle metasomatism and geodynamic significance. The Pontine volcanoes represent the northern termination of a Plio-Pleistocene volcanic alignment that extends from Anchise and the western Aeolian seamounts in the Southern Tyrrhenian Sea to offshore Central Italy. These represent remnant arcs that were constructed across the Tyrrhenian Sea during the eastward migration of the Ionian subduction system.

Keywords

Pontine volcanoes · Hyaloclastites · Calcalkaline rhyolites · Peralkaline rocks · Potassic rocks · Remnant arc · 41° Parallel line · Southern tyrrhenian sea · Geodynamics

A. Peccerillo, *Cenozoic Volcanism in the Tyrrhenian Sea Region*,
Advances in Volcanology, DOI 10.1007/978-3-319-42491-0_6

6.1 Introduction

The Pontine archipelago consists of five islands and a number of islets, roughly aligned in a WNW direction, offshore the Campania-Latium coast (Fig. 6.1). The islands are sited on the edge of the continental shelf, about 60–70 km south of the Gulf of Gaeta coast, and are mainly formed of volcanic rocks, except for Zannone and Palmarola, where Paleozoic to Mesozoic metamorphic and sedimentary units, and Upper Pliocene argillaceous marine sediments crop out. Cenozoic limestones have been found by drilling at Ponza (De Rita et al. 1986, 2004; Bellucci et al. 1997, 1999).

The Pontine Islands have been previously considered as belonging to the Campania Province (Peccerillo 2005), but they are much older and have distinct petrological-geochemical compositions than the Campania rocks, thus identifying a separate magmatic province. The western islands (Ponza, Palmarola, Zannone and La Botte islet) rise from the top of a large structural high on the edge of the continental shelf. Volcanism is 4.2–1.0 Ma old, and emplaced an association of subaerial metaluminous trachyte-rhyolite and peralkaline rhyolites, plus submarine mafic-intermediate calcalkaline to potassic rocks (Barberi et al. 1967; Vezzoli 1988; Conte and Dolfi 2002; Cadoux et al. 2005; Conte et al. 2016). The eastern group (Ventotene and Santo Stefano) has an age of 0.8 to less than 0.13 Ma, and consists of basalt-trachybasalt to phonolite rocks (Métrich 1985; Métrich et al. 1988). A NW-SE elongated volcanic ridge has been recently imaged by high-resolution multibeam bathymetry and geophysical data on the edge of the continental shelf about 30 km north of the Ventotene island (Cuffaro et al. 2016).

The geological setting of the Gulf of Gaeta is characterised by structural low and highs bordered by ESE-WNW to E-W and NE-SW faults, related to the intersection of the Ortona-Roccamonfina

Fig. 6.1 Location map of the Pontine Islands

Table 6.1 Ages, volcanology and petrology of the Pontine Islands

Volcano	Age (in Ma)	Volcanology	Petrology-Geochemistry
Ponza	4.2–1.0	– Mostly submarine lava domes, dykes, breccias, hyaloclastites and minor pyroclastic rocks	– Pliocene calcalkaline rhyolites followed by Pleistocene comendite and trachytes. Variable Sr isotopic ratios ($^{87}Sr/^{86}Sr$ = 0.7085–0.7139), increasing from trachytes to rhyolites and comendite. Submarine mafic-intermediate calcalkaline to potassic rocks
La Botte	1.2	– Trachytic-phonolitic neck rising from a larger platform sited about 150 m below sea level	– Major and trace element composition similar to Ponza trachytes
Palmarola	1.64–1.54	– Lava domes, breccia, and hyaloclastites emplaced in a submarine environment	– Peralkaline rhyolites with highly radiogenic Sr-isotope compositions
Zannone	?	– Cryptodome	– Calcalkaline rhyolites compositionally similar and probably coeval to Ponza
Ventotene and Santo Stefano	0.8 to<0.13	– Remnants of a largely collapsed stratovolcano with caldera (Ventotene), and a flank lava dome (Santo Stefano)	– Basalt, K-trachybasalt to phonolite with $^{87}Sr/^{86}Sr$ ~ 0.7070 to 0.708, similar to Roccamonfina

with the 41° Parallel lines. The latter is a transfer fault marked by E–W striking magnetic and free-air gravity anomaly, which separates the northern and the southern Tyrrhenian domains (Boccaletti et al. 1990a, b; Serri 1990; Bruno et al. 2000). Crustal thickness is about 20 km (Panza et al. 2007).

A summary of the main petrological-volcanological characteristics and ages of the Pontine volcanoes are reported in Table 6.1. Representative geochemical data are given in Table 6.2.

6.2 Western Pontine Islands: Ponza, La Botte, Palmarola and Zannone

Ponza: Ponza is largely constructed of rhyolitic obsidianaceous lava flows and hyaloclastites, with minor dykes and pyroclastics overlying older domes. Younger trachytic lavas are only exposed as a dome and minor pyroclastics at Monte La Guardia promontory, in the southernmost end of the island (Scutter et al. 1998). Extensive K/Ar dating indicates that volcanism took place between 4.2 and 1.0 Ma, during two phases of rhyolitic activity followed by the late emplacement of La Guardia trachytes (Cadoux et al. 2005). Modification of magma compositions was accompanied by an uplift of the island, with shifting of volcanism from submarine to subaerial (Bellucci et al. 1997, 1999; De Rita et al. 2001, 2004).

Rhyolites from Ponza are calcalkaline, although a peralkaline comendite occurrence is reported by Conte and Dolfi (2002) among Pleistocene products. Textures are generally vitrophyric, with scarce phenocrysts (<10 vol.%) of plagioclase ($\sim An_{60-30}$), sanidine ($\sim Or_{70-75}$), biotite ($\sim Mg\#_{45}$), orthopyroxene ($\sim Mg\#_{41}$), diopside to hedenbergite clinopyroxene, and sporadic hornblende set in a glassy to microgranular groundmass. Comendite is also scarcely porphyritic with microphenocrysts of alkali-feldspar (Or_{35-30}) set in a groundmass containing alkali-feldspar, alkali-amphibole, mica, quartz, bastnäsite (Ce fluoride carbonate), and zircon (Conte and Dolfi 2002).

Trachytes are dark to light coloured rocks with scarcely to moderately porphyritic textures and phenocrysts of plagioclase ($\sim An_{65-30}$), alkali-feldspar ($\sim Or_{33-63}$), clinopyroxene ($Wo_{47-46}En_{41-39}Fe_{13-14}$) and biotite ($\sim Mg\#_{40-70}$). Groundmass is mostly formed of alkali-feldspar, some clinopyroxene and Fe–Ti oxides (Bellucci et al. 1999; Conte and Dolfi 2002; Paone 2013).

Table 6.2 Major, trace element and initial radiogenic isotope ratios for representative samples from Pontine Islands. Data in parentheses refer to compositions determined on distinct, though compositionally similar samples from the same locality as those analysed for the other elements

Volcano	**Palmarola**	**Ponza**	**Ponza**	**Ponza**	**Ponza**	**Zannone**	**Ventotene**	**Ventotene**	**Ventotene**
Rock type	Rhyolite	Rhyolite	Rhyolite	Comendite	Trachyte	Rhyolite	Basalt	Trachybasalt	Phonolite
Age	Pliocene	Pliocene	Pliocene	Pleistocene	Pleistocene	–	–	–	–
–	1,2	3	1,2,3	3	1,2,3	3	4,5	4,5	6
SiO_2 %	74.49	73.71	74.56	70.48	57.98	69.28	48.83	49.25	58.21
TiO_2	0.07	0.06	0.05	0.15	0.76	0.44	1.14	1.15	0.33
Al_2O_3	13.59	12.67	13.27	12.58	18.92	15.64	17.07	17.99	21.46
FeO_{total}	1.02	1.13	1.03	4.66	5.16	2.24	8.43	8.60	2.28
MnO	–	–	–	0.08	0.11	0.05	0.15	0.15	0.18
MgO	–	0.04	0.11	0.28	1.36	1.56	6.42	4.93	0.29
CaO	0.33	0.76	0.93	0.27	3.53	2.4	11.59	10.57	1.14
Na_2O	4.64	3.42	3.35	5.81	4.74	3.37	2.80	2.94	7.39
K_2O	4.92	3.87	5.22	4.44	6.53	4.95	1.67	2.45	7.31
P_2O_5	–	0.01	–	0.02	0.28	0.07	0.23	0.3	0.33
LOI	0.69	4.15	1.26	0.72	0.55	1.65	1.04	1.04	0.51
Sc ppm	–	–	–	–	–	–	31	24	0.99
V	–	–	1.56	–	62.71	–	203	218	–
Cr	–	–	–	–	–	–	190	35	10
Co	–	–	0.51	–	6.46	–	35	32	1.06
Ni	–	–	–	–	–	–	57	34	6
Rb	447	254	249	1750	277	232	70	111	525
Sr	8	90	97	10	591	270	612	816	–
Y	44	20	11	241	23	–	25	24	–
Zr	261	107	119	2600	255	192	101	120	817
Nb	56	12	12	511	24	11	8	10	–
Cs	4.93	51	9.53	150	18.7	–	–	–	57

(continued)

Table 6.2 (continued)

Volcano	Palmarola	Ponza	Ponza	Ponza	Ponza	Zannone	Ventotene	Ventotene	Ventotene
Rock type	Rhyolite	Rhyolite	Rhyolite	Comendite	Trachyte	Rhyolite	Basalt	Trachybasalt	Phonolite
Ba	11	413	444	15	706	540	295	472	24
La	83	33	48	315	50	39	21	28	160
Ce	153	56	85	490	98	67	44	56	310
Nd	54	18	26	141	41	–	27	31	–
Sm	9.85	3.3	4.25	27.0	7.16	–	6.4	6.7	15.2
Eu	0.05	0.39	0.42	0.73	1.96	–	1.72	1.83	0.91
Gd	7.44	2.55	2.94	23.80	4.96	–	–	–	–
Tb	1.14	0.41	0.47	4.12	0.73	–	–	–	1.08
Er	3.97	–	1.57	–	2.15	–	–	–	–
Yb	4.77	2.13	1.64	26.60	2.32	–	1.95	1.93	5.0
Lu	0.71	0.29	0.28	3.52	0.31	–	0.31	0.28	–
Hf	9.23	2.9	4.06	64.8	5.59	–	2.2	2.7	17.7
Ta	5.22	1.7	1.52	35.6	1.66	–	0.52	0.8	3.02
Pb	186	–	55	–	32	1.3	13	10	–
Th	77.69	26	29.89	359	17.94	25	4.9	7.7	133
U	19.45	5.4	7.88	57.3	5.04	–	1.1	–	29
$^{87}Sr/^{86}Sr$	–	0.710456	(0.710456)	0.713865	(0.708495)	–	0.70709	0.70758	–
$^{143}Nd/^{144}Nd$	–	0.512252	(0.512252)	0.512390	(0.512401)	–	0.51242	0.512296	–
$^{206}Pb/^{204}Pb$	18.810	–	18.781	–	18.758	–	18.496	18.759	–
$^{207}Pb/^{204}Pb$	15.671	–	15.673	–	15.670	–	15.638	15.659	–
$^{208}Pb/^{204}Pb$	39.008	–	38.963	–	39.015	–	38.451	38.903	–

Source of data: (1) Cadoux et al. (2005); (2) Cadoux et al. (2007); (3) Conte and Dolfi (2002, Unpublished data); (4) D'Antonio et al. (1996); (5) D'Antonio et al. (1999); (6) Métrich (1985)

Data on recently found submarine mafic-intermediate rocks can be found in Conte et al. (2016)

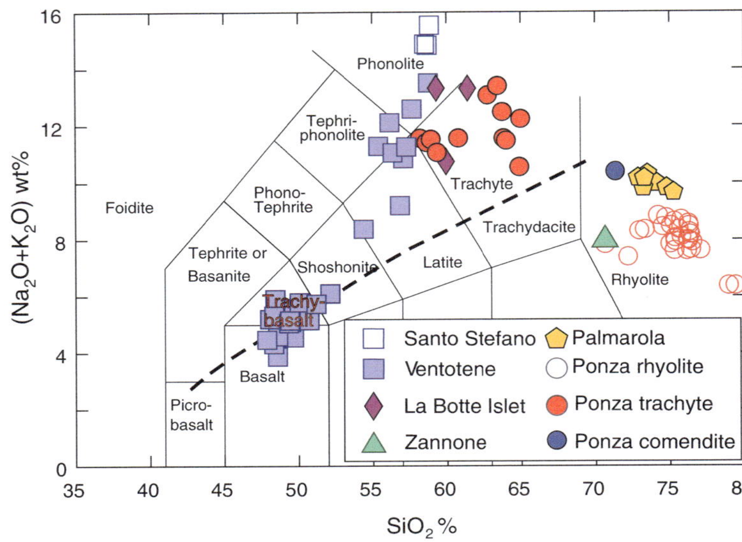

Fig. 6.2 TAS classification diagram (Le Maitre 2002) for the Pontine Islands volcanic rocks. The *dashed line* is the divide between subalkaline and alkaline fields of Irvine and Baragar (1971). Data have been recalculated to 100 % on a LOI-free basis

Many samples from Ponza show high values of LOI, which might result from secondary processes. However, geochemical data on mobile elements such as alkalies seem to exclude this hypothesis (Cadoux et al. 2005). Trachytes and rhyolites define compositionally distinct groups of rocks (Fig. 6.3). The comendite is strongly enriched in La, Rb, Th and other incompatible elements, but is depleted in Ba and Sr with respect to both trachytes and rhyolites. Two trends can be distinguished within trachytes for some major and trace elements (e.g. La, Rb vs. SiO_2), one pointing to the highly enriched Ponza comendite and the other to Palmarola peralkaline rhyolites (Fig. 6.3). REE patterns are variably fractionated, with negative Eu anomalies (Fig. 6.4a). Volatile elements concentrations (F, Cl) range from several hundred to several thousand ppm (Conte and Dolfi 2002).

$^{87}Sr/^{86}Sr$ ratios range from about 0.7085 to 0.714 and increase from trachytes to rhyolites and comendite (Figs. 6.3h and 6.5a). Nd isotopic ratio is higher in trachytes and comendite ($^{143}Nd/^{144}Nd$ ~ 0.51240) than in the rhyolites ($^{143}Nd/^{144}Nd$ ~ 0.51225; Conte and Dolfi 2002). Pb isotope ratios ($^{206}Pb/^{204}Pb$ ~ 18.80; $^{207}Pb/^{204}Pb$ ~ 16.80; $^{208}Pb/^{204}Pb$ ~ 39.00; Cadoux et al. 2007) fall in the compositional field of the Roman and Tuscany provinces, and have lower values than rocks from Campania (Fig. 6.5b, c). Oxygen isotope composition of whole rocks and separated feldspars ranges from $\delta^{18}O_{SMOW}$ ~ +7.3‰ to +11.1‰, increasing from trachytes to rhyolites (Turi et al. 1991).

The steep decrease in Sr-Ba and the increase in incompatible element contents with a small increase in silica within the trachytic rocks, clearly suggest an evolution by alkali feldspar-dominated fractional crystallisation, with the comendite probably representing the final residual liquid. A role of element enrichment by fluid-transfer is envisaged by Belkin et al. (1996) and Conte and Dolfi (2002) to explain some extreme element enrichments, especially in the comendite. Establishing the nature and origin of parental melt of trachytes has been hampered by the absence of exposed intermediate-mafic products in the western Pontine Islands (Conte and Dolfi 2002; Cadoux et al. 2005). However, basalt to trachyandesite glass inclusions in plagioclase from Monte La Guardia trachytic centre and some submarine mafic-intermediate rocks recently studied by Conte et al. (2016) may represent the parent liquids of trachyte-comendite suite (Fedele et al. 2003). Rhyolites represent a compositional distinct group of rocks, whose origin might be related to either crustal melting (e.g. Paone 2013) or fractional

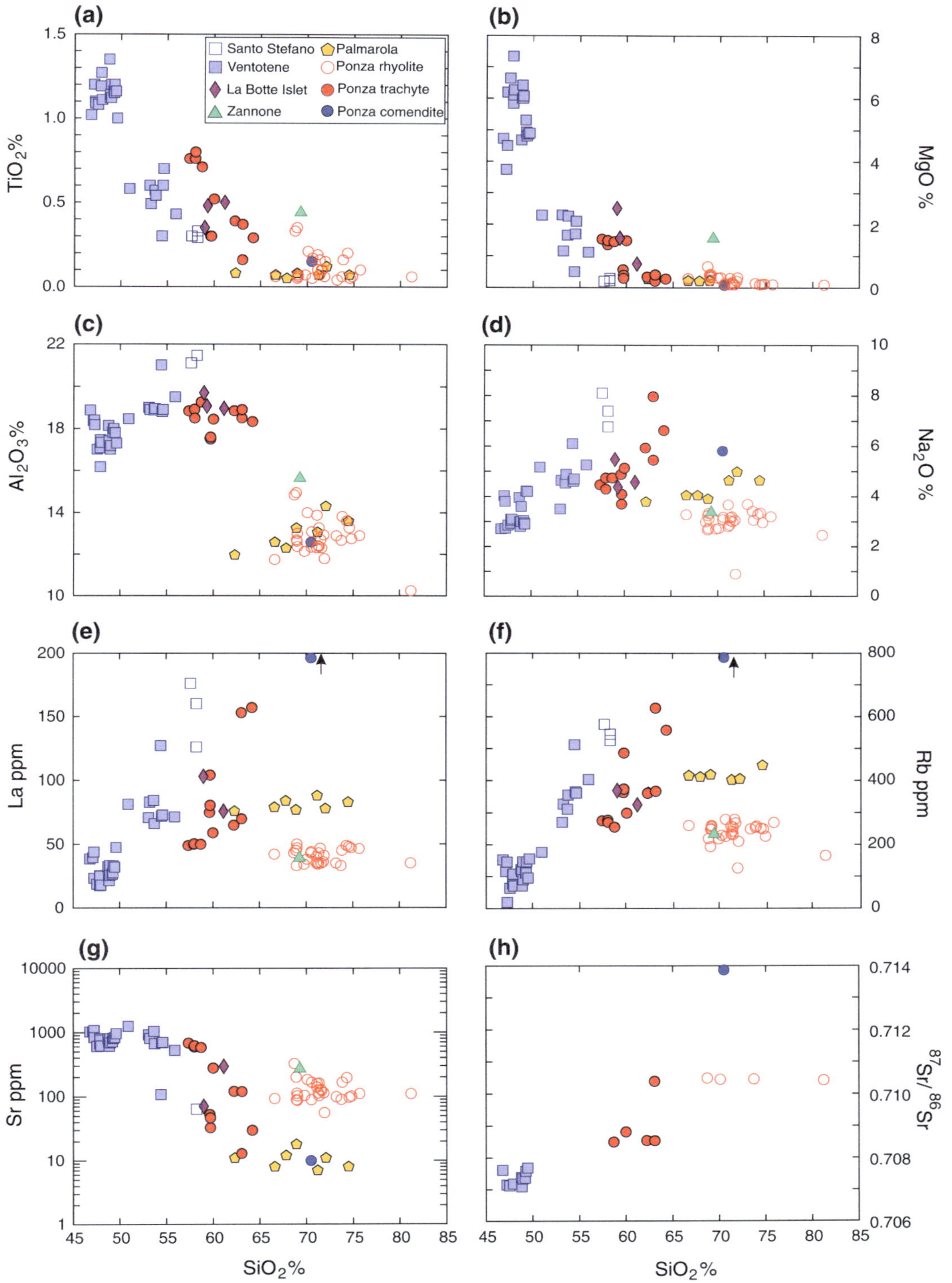

Fig. 6.3 Harker diagrams of the Pontine volcanics

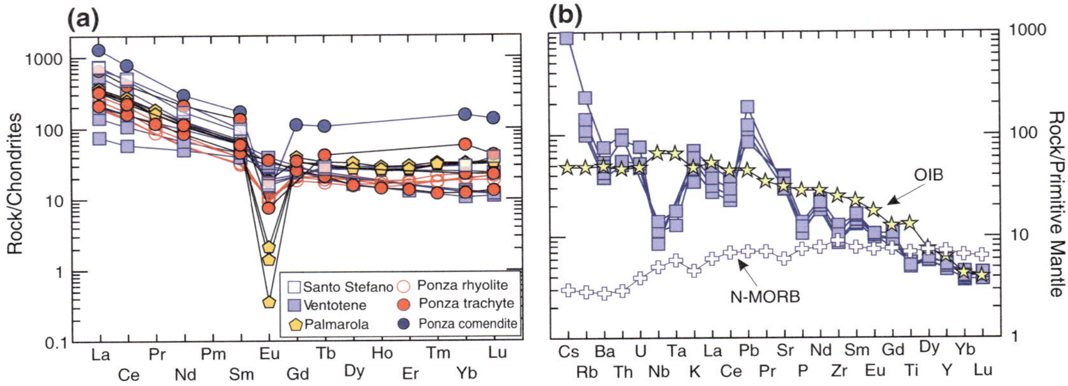

Fig. 6.4 **a** REE patterns of rocks from the Pontine Islands; **b** Incompatible element patterns of selected mafic rocks from Ventotene. Average OIB and N-MORB compositions (Sun and Mcdonough 1989; Gale et al. 2013) are reported for comparison

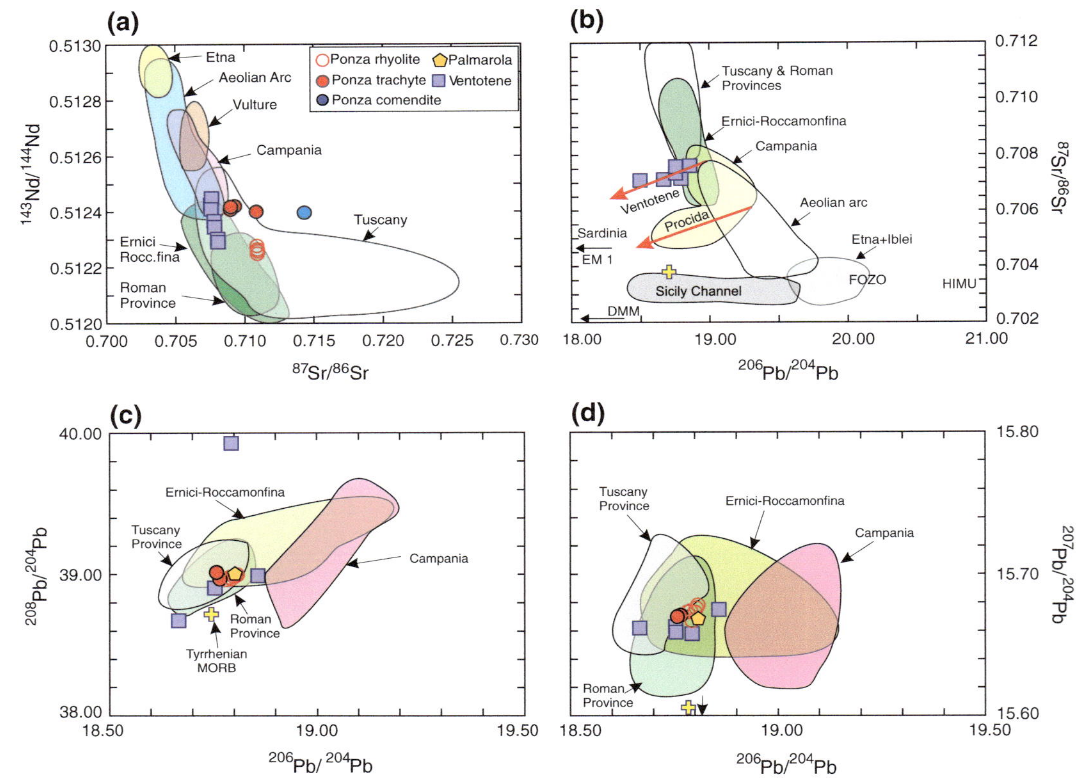

Fig. 6.5 Sr, Nd and Pb isotopic variations of rocks from the Pontine Islands. Fields of other volcanic provinces (restricted to mafic rocks) and some mantle end-members (EM1, HIMU, ZOZO, DMM; Stracke et al. 2005) are reported for comparison. Composition of Tyrrhenian MORB is indicated with a cross

crystallisation plus crustal assimilation, starting from a calcalkaline intermediate-mafic parent. Such magma does not occur either in the outcrops or among melt inclusions in the western Pontine Islands, but has been recently found among submarine samples by Conte et al. (2016).

Zannone: Volcanism in this island is represented by a rhyolitic cryptodome intruded into deformed and tilted Permian-Triassic sandstones and phyllite, Triassic dolomites, and Cretaceous-Miocene sediments (De Rita et al. 2004). The Zannone rhyolites are generally altered by high-temperature fluids, which prevented radiometric dating. They show poorly porphyritic textures with phenocrysts of alkali feldspar and biotite set in a devitrified groundmass. Analysis of the least altered samples and data on immobile elements indicate a composition similar to the Ponza rhyolites.

La Botte: La Botte islet is a Pleistocene trachytic-phonolitic neck, a few tens of metres across, situated about 12 km SE of Ponza. A single K/Ar age determination yielded a value of 1.2 Ma (Savelli 1988). Rocks from La Botte straddle the boundary between trachytes and phonolites (Fig. 6.2). Textures are poorly porphyritic with phenocrysts of alkali feldspar plus some clinopyroxene and biotite set in a trachytic to pilotaxitic groundmass mostly made of alkali feldspar (Conte AM, personal communication). Major and trace element compositions are similar to Ponza trachytes (Fig. 6.3).

Palmarola: Volcanic rocks at Palmarola occur as domes and associated breccias and hyaloclastites, emplaced during two stages of activity separated by deposition of marine sediments. K/Ar dating by Cadoux et al. (2005) indicates ages of about 1.6–1.5 Ma. The Palmarola rocks are peralkaline rhyolites with Peralkaline Index = 1.04–1.11. They show higher Na_2O and incompatible element contents, but lower Sr, Ba, and LILE/HFSE ratios than Ponza rhyolites (Cadoux et al. 2005). In contrast, variation diagrams highlight an affinity with some Ponza trachytes. REE patterns are fractionated with strong negative Eu anomalies (Fig. 6.4a). Overall, geochemical data suggest an origin of Palmarola rhyolites by feldspar-dominated fractional crystallisation, starting from a parental magma similar to some Ponza trachytes.

6.3 Ventotene and Santo Stefano

The islands of Ventotene and Santo Stefano are the emergent remnants of a large collapsed stratovolcano, rising more than 700 above sea floor. Ventotene represents the south-eastern sector of the caldera rim and consists of a sequence of thin lava flows and pyroclastic deposits. Santo Stefano is a flank lava dome, partially covered by pyroclastic products. Pyroclastic rocks at Ventotene and Santo Stefano include fall, flow and surge deposits, and contain lava lithics, cumulate xenoliths and skarns (Perrotta et al. 1996). K/Ar ages range from about 0.8 to less than 0.13 Ma. The Santo Stefano lava dome records the oldest exposed activity. The suite of basal lava flows at Ventotene was erupted mainly between 0.8 and 0.5 Ma (Métrich et al. 1988). The caldera collapse occurred around 0.5 Ma (Perrotta et al. 1996). Rock compositions range from basalt and trachybasalt to phonolite (Fig. 6.2). Mafic rocks have a porphyritic texture with phenocrysts of clinopyroxene, plagioclase ($\sim An_{90-50}$), and olivine (about Fo_{85-65}), set in holocrystalline to hypocrystalline groundmass consisting of the same phases plus Fe–Ti oxides, alkali-feldspars and rare biotite. Latites have similar phenocryst assemblage as trachybasalts but also contain biotite and zoned Cr-diopside to hedenbergite clinopyroxene. Phonolites have phenocrysts of sanidine, plagioclase ($\sim An_{85-35}$), potassic ferropargasitic amphibole, and Fe–Ti oxides sometimes replacing biotite. Haüyne and nepheline are found in some latites and phonolites (Barberi et al. 1967; Métrich 1985; Métrich et al. 1988; D'Antonio and Di Girolamo 1994; Bellucci et al. 1999; D'Antonio et al. 1999; Paone 2013). Mafic rocks are undersaturated in silica (10–20 % normative nepheline), exhibit low to moderate enrichment in potassium, and plot in the fields of basalt and trachybasalt on the TAS diagram (Fig. 6.2). Major elements show an increase in Al_2O_3 and alkalies, and a decrease in MgO, CaO, FeO_{total} and TiO_2 from basalt to phonolite

(Fig. 6.3). Incompatible trace elements (Th, Rb, LREE, Nb, etc.) increase from basalts to phonolites, whereas Sr and Ba show an opposite trend (Métrich 1985). REE are fractionated with significant negative Eu anomalies in the phonolites (Fig. 6.4a). Incompatible element patterns of mafic rocks have negative anomaly of Ba and large negative anomalies of HFSE, resembling closely the Ernici-Roccamonfina trachybasalts (Fig. 6.4 b). Initial Sr-Nd isotope compositions in the mafic rocks vary significantly ($^{87}Sr/^{86}Sr \sim 0.7070$ to 0.7077; $^{143}Nd/^{144}Nd \sim 0.51228$ to 0.51244) and partially overlap the compositional fields of Campania an Ernici-Roccamonfina (Fig. 6.5a). Slightly higher values of $^{87}Sr/^{86}Sr = 0.7070$–0.7089 have been found for two phonolite samples by Barberi et al. (1967). There is a well-defined negative trend of MgO versus $^{87}Sr/^{86}Sr$ in the mafic rocks, which is mirrored by a positive trend for Nd isotope ratios (not shown). Pb-isotope ratios ($^{206}Pb/^{204}Pb \sim 18.50$ to 18.86; $^{207}Pb/^{204}Pb \sim$ 15.64 to 15.70; $^{208}Pb/^{204}Pb \sim 38.45$ to 39.09; De Vivo et al. 1995; D'Antonio et al. 1996, 1999) define a trend towards the Plio-Quaternary EM1-type rocks from Sardinia on $^{87}Sr/^{86}Sr$ versus $^{206}Pb/^{204}Pb$ diagram (Fig. 6.5b). Whole rock oxygen isotope ratios ($\delta^{18}O_{SMOW} \sim +5.9‰$ to +7.6‰) increase from basalts to felsic rocks (Turi et al. 1991).

Overall, major and trace element data suggest an evolution in an open system for the Ventotene-Santo Stefano magmas, with dominant fractional crystallisation of mafic minerals and plagioclase, accompanied by magma mixing (D'Antonio and Di Girolamo 1994; D'Antonio et al. 1999). High Sr-isotope ratios observed in the phonolites as well as the negative correlation of MgO and ferromagnesian elements against Sr-isotope ratios in the mafic rocks, strongly suggest significant interaction with crustal rocks during magma evolution.

6.4 Petrogenesis and Geodynamic Significance of Pontine Islands

The origin and geodynamic significance of magmatism in the Pontine Islands are still poorly understood. Mafic rocks from Ventotene show trace element and radiogenic isotope signatures that are different from those of the Campania volcanoes (D'Antonio et al. 1999), and resemble more closely the potassic rocks from Ernici-Roccamonfina for LILE/LILE, LILE/HFSE, HFSE/HFSE and radiogenic isotope ratios (Fig. 6.6). Therefore, as with the latter province, an origin in a depleted mantle source contaminated by supercritical fluid or melts from oceanic slab plus sediments can be envisaged for Ventotene primary magmas.

The rocks from the western Pontine Islands have a diverse origin. Geochemical and melt inclusion data suggest that the Ponza trachytes evolved from moderately potassic mafic parents

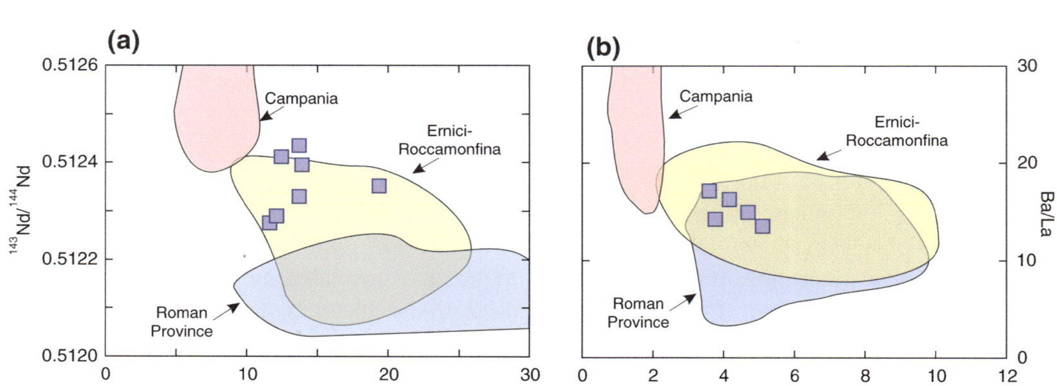

Fig. 6.6 Variation of incompatible element and Nd-isotope ratios in the mafic rocks (MgO > 4 wt%) from Ventotene. Compositions of nearby magmatic provinces are shown for comparison

(Fedele et al. 2003; Paone 2013); comendites could represent the final step of such a process. However, gaseous transfer phenomena might be responsible for some extreme element enrichments in the comendite (Conte and Dolfi 2002).

The peralkaline rocks from Palmarola show trace element evidence suggesting a derivation by fractional crystallisation from parent magmas similar to Ponza trachytes (Fig. 6.3). However, compositional differences for some key trace element ratios, suggest somewhat distinct parental magmas (Cadoux et al. 2005).

Finally, the Ponza rhyolites define a compositionally distinct group of rocks, which appears to have little affinity with other magmas from the same region. According to Paone (2013), such a compositional contrast is related to a different origin for rhyolites, which would be formed by crustal melting. However, the Ponza rhyolites have similar composition as some acid calcalkaline rocks from the Aeolian arc, although Sr-isotope ratios are much higher at Ponza. This may suggest a derivation of Ponza rhyolites by fractional crystallisation plus possible crustal assimilation, starting from a mafic-intermediate calcalkaline parent.

New data recently published by Conte et al. (2016) on submarine mafic-intermediate rocks from the western arc provide support to the hypothesis of an origin of silicic rocks by evolutionary processes dominated by fractional crystallization. These submarine samples have high-K calcalkaline to potassic petrological affinity and are likely parental to various silicic rocks cropping out at the surface. Submarine rocks show variable but overall high Sr- and low Nd-isotope ratios ($^{87}Sr/^{86}Sr \sim 0.7059$ to 0.712; $^{143}Nd/^{144}Nd \sim 0.51253$ to 0.51215), which suggest anomalous mantle sources and significant shallow level interaction between magma and crust for the western Pontine Islands (Conte et al. 2016).

According to several authors (Locardi 1991, 1993; Argnani and Savelli 1999; Savelli 2001) the Pontine Islands, as well as the buried volcanoes of the Campania Plain (see Chap. 7), represent the northern end of a remnant volcanic arcs developed during Plio-Pleistocene across the southern Tyrrhenian Sea, and left behind by southeastward migration of the Ionian subduction system (Fig. 6.7). The arc extended from Anchise volcano, west of Ustica, in the south, to the Pontine Islands in the north; the Garibaldi (Glauco) seamount and the ODP Site 650 are located in the central sector of the arc, although geochronological and geochemical data are insufficient to support this hypothesis.

The Pontine Island magmas have clear island-arc geochemical signatures (i.e. high LILE/HFSE ratios), which support a subduction-related origin at the northern end of cross-Tyrrhenian remnant arcs. However, Cadoux et al. (2005) noticed a decrease of LILE/HFSE ratios with time in the rhyolites and trachytes from Palmarola and Ponza. This led to the suggestion that an OIB-type component was progressively involved in the origin of magmatism. Such a conclusion is supported by Sr-Pb isotope data of the Ventotene mafic rocks, which define a trend towards EM1-type Plio-Quaternary OIB rocks occurring in Sardinia (Fig. 6.5b). A similar trend is also shown by Procida-Vivara basalts (Campania Province), which, however, have lower Sr isotope ratios than Ventotene.

6.5 Summary and Conclusions

The Pontine Islands are located offshore the Gulf of Gaeta (southern Latium), along the eastern segment of the 41° Parallel tectonic line. Volcanic activity developed between about 4.1 Ma and less than 130 ka and is younger in the eastern (Ventotene and Santo Stefano) than in the western islands (Ponza, Palmarola, Zannone, La Botte). Calcalkaline rhyolites, peralkaline rhyolites, and trachytes occur at Ponza, La Botte, Palmarola and Zannone, where mafic-intermediate calcalkaline to potassic volcanism (4.0-3.9 Ma old) has been recently discovered below sea level. A suite of basalt-trachybasalt to phonolite rocks makes up Ventotene and Santo Stefano.

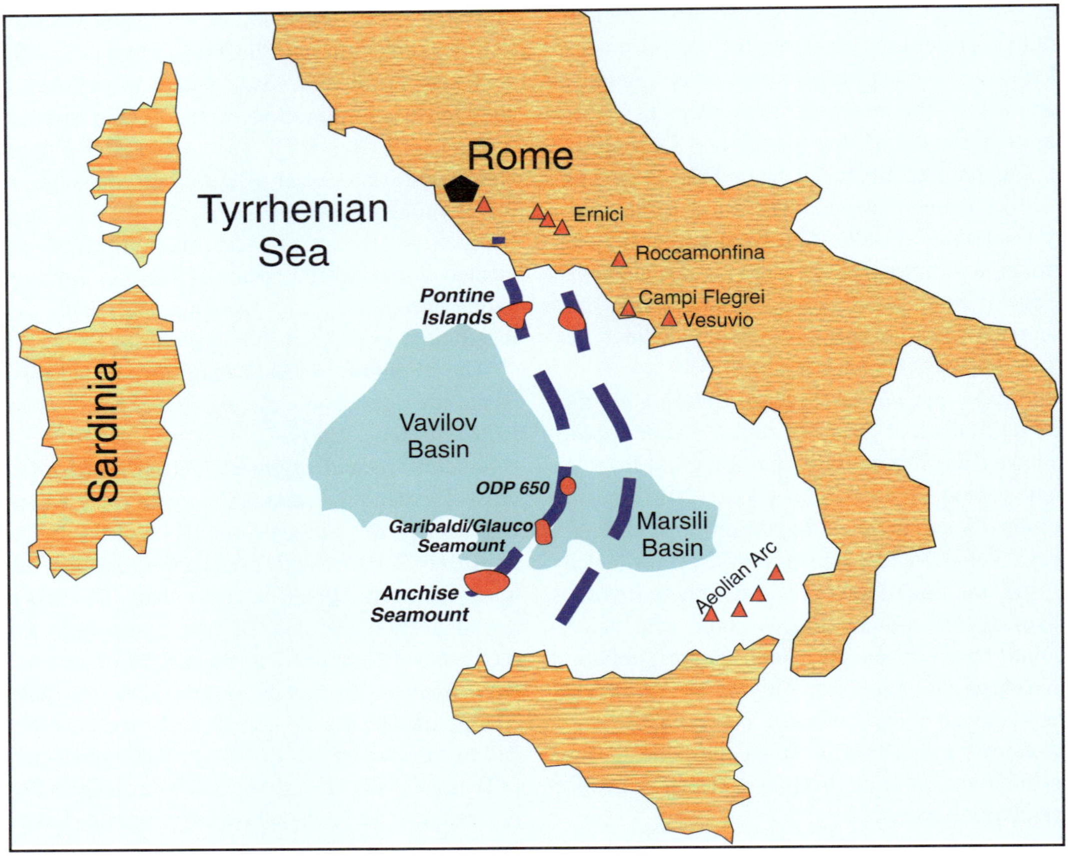

Fig. 6.7 Approximate position of Plio-Quaternary remnant volcanic arcs across the Southern Tyrrhenian Sea, according to Locardi (1991) and Argnani and Savelli (1999)

The Pontine magmas have a complex origin and geodynamic significance, which deserves further investigation to be better understood. Trachytes and peralkaline rhyolites likely derived from moderately potassic mafic parents by fractional crystallisation plus crustal assimilation. Calcalkaline rhyolites were formed either by crustal melting or by fractional crystallization or AFC, starting from mafic-intermediate calcalkaline parents. Basalts and trachybasalts from Ventotene have variable isotopic signatures, which reflect open-system intra-crustal evolution processes and/or source heterogeneity.

All the magmas from the Pontine Islands show island-arc geochemical signatures, i.e. high LILE/HFSE ratios. However, there is a significant decrease of these ratios with time, which has been interpreted as indicating a role of OIB-type mantle sources in the origin of younger magmas. The basalts and trachybasalts from Ventotene have geochemical and isotopic signatures similar to mafic potassic rocks from the Ernici-Roccamonfina Province. Therefore, a similar type of mantle metasomatism is suggested by geochemical data for all these volcanoes.

The Pontine Islands, along with the buried volcanoes of the Campanian Plain and, possibly, the calcalkaline-shoshonitic Ernici and Roccamonfina rocks may represent the northern termination of Plio-Pleistocene volcanic remnant-arcs extending from the Anchise and western Aeolian seamount in the southern Tyrrhenian Sea to Ponza, Ventotene, and the Campanian Plain. These were constructed across the Tyrrhenian Sea during the eastward migration of the Ionian subduction system.

References

Argnani A, Savelli C (1999) Cenozoic volcanism and tectonics in the southern Tyrrhenian Sea: space-time distribution and geodynamic significance. Geodynamics 27:409–432

Barberi F, Borsi S, Ferrara G, Innocenti F (1967) Contributo alla conoscenza vulcanologica e magmatologica delle Isole dell'Archipelago Pontino. Mem Soc Geol It 6:581–606

Belkin HE, De Vivo B, Lima A, Török K (1996) Magmatic (silicates/saline/sulfur-richlC02) immiscibility and zirconium and rare-earth element enrichment from alkaline magma chamber margins: evidence from Ponza Island, Pontine Archipelago, Italy. Eur J Mineral 8:1401–1420

Bellucci F, Grimaldi M, Lirer L, Rapolla A (1997) Stucture and geological evolution of the island of Ponza, Italy; inferences from geological and gravimetric data. J Volcanol Geoth Res 79:87–96

Bellucci F, Lirer M, Munno R (1999) Geology of Ponza, Ventotene and Santo Stefano islands (with 1:15.000 scale geological map). Acta Vulcanol 11:197–222

Boccaletti M, Calamita F, Deiana G, Gelati R, Massari F, Moratti G, Ricci Lucchi F (1990a) Migrating foredeep-thrust belt systems in the northern Apennines and southern Alps. Palaeogeogr Palaeocl 77:3–14

Boccaletti M, Nicolich R, Tortorici L (1990b) New data and hypothesis on the development of the Tyrrhenian Basin. Palaeogeogr Palaeocl 77:15–40

Bruno PP, Di Fiore V, Ventura G (2000) Seismic study of the "41st Parallel" fault system offshore the Campanian-Latial continental margin, Italy. Tectonophysics 324:37–55

Cadoux A, Pinti DL, Aznar C, Chiesa S, Gillot P-Y (2005) New chronological and geochemical constraints on the genesis and evolution of Ponza and Palmarola volcanic islands (Tyrrhenian Sea, Italy). Lithos 81:121–151

Cadoux A, Blichert-Toft J, Pinti DL, Albarede F (2007) A unique lower mantle source for Southern Italy volcanics. Earth Planet Sci Lett 259:227–238

Conte AM, Dolfi D (2002) Petrological and geochemical characteristics of Plio-Pleistocene volcanics from Ponza Island (Tyrrhenian Sea, Italy). J Volcanol Geoth Res, in press

Conte AM, Perinelli C, Bianchini G, Natali C, Martorelli E, Chiocci FL (2016) New insights on the petrology of submarine volcanics from the Western Pontine Archipelago (Tyrrhenian Sea, Italy). J Volcanol Geoth Res, in press

Cuffaro M, Martorelli E, Bosman A et al (2016) The Ventotene volcanic ridge: a newly explored complex in the central Tyrrhenian Sea (Italy). Bull Volcanol 78:86

D'Antonio M, Di Girolamo P (1994) Petrological and geochemical study of mafic shoshonitic volcanics from Procida-Vivara and Ventotene islands (Campanian Region, South Italy). Acta Vulcanol 5:69–80

D'Antonio M, Tilton GR, Civetta L (1996) Petrogenesis of Italian alkaline lavas deduced from Pb-Sr-Nd isotope relationships. In: Basu A, Hart SR (eds) Isotopic studies of crust-mantle evolution. Am Geophys Un Mon 95:253–267

D'Antonio M, Civetta L, Di Girolamo P (1999) Mantle source heterogeneity in the Campanian region (south Italy) as inferred from geochemical and isotopic features of mafic volcanic rocks with shoshonitic affinity. Mineral Petrol 67:163–192

De Rita D, Funiciello R, Pantosti D, Salvini F, Sposato A, Velona M (1986) Geological and structural characteristics of the Pontine islands (Italy) and implications with the evolution of the Tyrrhenian margin. Mem Soc Geol It 36:55–65

De Rita D, Giordano G, Cecili A (2001) A model for submarine rhyolite dome growth: Ponza Island (central Italy). J Volcanol Geoth Res 107:221–239

De Rita D, Fabbri M, Cimarelli C (2004) Evoluzione pleistocenica del margine tirrenico dell'Italia centrale tra eustatismo, vulcanismo e tettonica. Il Quaternario 17:523–536

De Vivo B, Török K, Ayuso RA, Lima A, Lirer R (1995) Fluid inclusion evidence for magmatic silicate/saline/CO_2 immiscibility and geochemistry of alkaline xenoliths from Ventotene Island, Italy. Geochim Cosmochim Acta 59:2941–2953

Fedele L, Bodnar RJ, De Vivo B, Tracy R (2003) Melt inclusion geochemistry and computer modeling of trachyte petrogenesis at Ponza, Italy. Chem Geol 194:81–104

Gale A, Dalton CA, Langmuir CH, Su Y, Schilling J-G (2013) The mean composition of ocean ridge basalts. Gechem Geophys Geosyst 14 doi:10.1029/2012GC004334

Irvine TN, Baragar WRA (1971) A guide to chemical classification of common volcanic rocks. Can J Earth Sci 8:523–548

Le Maitre RW (ed) (2002) A classification of igneous rocks and glossary of terms. Cambridge University Press, Cambridge 252 pp

Locardi E (1991) Geodinamica delle strutture profonde dell'Appennino centro-meridionale. Mem Soc Geol It 47:325–332

Locardi E (1993) Dynamics of deep structures in the Tyrrhenian-Apennines area and its relation to neotectonics. Il Quaternario 6:59–66

Métrich N (1985) Mecanismes d'evolution a l'origine des magmas potassiques d'Italie centrale et meridionale.

Exemples du Mt. Somme-Vesuve, des Champs Phlegreens et de l'Ile de Ventotene. PhD Thesis, University Paris-Sud, Orsay, 336 pp

Métrich N, Santacroce R, Savelli C (1988) Ventotene, a potassic quaternary volcano in central Tyrrhenian Sea. Rend Soc It Mineral Petrol 43:1195–1213

Panza GF, Peccerillo A, Aoudia A, Farina B (2007) Geophysical and petrological modeling of the structure and composition of the crust and upper mantle in complex geodynamic settings: The Tyrrhenian Sea and surroundings. Earth Sci Rev 80:1–46

Paone A (2013) Petrogenesis of trachyte and rhyolitemagmas on Ponza Island (Italy) and its relationship to the Campanian magmatism. J Volcanol Geoth Res 267:15–29

Peccerillo a (2005) Plio-Quaternary volcanism in Italy. Petrology, geochemistry, geodynamics. Springer, Berlin, 365 pp

Perrotta A, Scarpati C, Giacomelli L, Capozzi AR (1996) Proximal depositional facies from a caldera-forming eruption: the Parata Grande Tuff at Ventotene Island (Italy). J Volcanol Geoth Res 71:207–228

Savelli C (1988) Late Oligocene to recent episodes of magmatism in and around the Tyrrhenian Sea: implications for the processes of opening in a young inter-arc basin of intra-orogenic (Mediterranean) type. Tectonophysics 146:163–181

Savelli C (2001) Two-stage progression of volcanism (8–0 Ma) in the central Mediterranean (southern Italy). J Geodyn 31:393–410

Scutter CR, Cas RAF, Moore CL, De Rita D (1998) Facies architecture and origin of a submarine rhyolitic lava flow-dome complex, Ponza, Italy. J Geophys Res 103:27551–27566

Serri G (1990) Neogene-Quaternary magmatism of the Tyrrhenian region: characterization of the magma sources and geodynamic implications. Mem Geol Soc It 41:219–242

Stracke A, Hofmann AW, Hart SR (2005) FOZO, HIMU, and the rest of the mantle zoo. Geochem Geophys Geosyst 6:Q05007

Sun SS, McDonough WF (1989) Chemical and isotopic systematics of oceanic basalts: implications for mantle composition and processes. In: Saunders AD, Norry MJ (eds) Magmatism in ocean basins. Geol Soc London Spec Publ 42:313–345

Turi B, Taylor HP, Ferrara G (1991) Comparisons of $^{18}O/^{16}O$ and $^{87}Sr/^{86}Sr$ in volcanic rocks from the Pontine Islands, M. Ernici and Campania with other areas in Italy In: Taylor HP, O'Neil JR, Kaplan IR (eds) Stable Isotope Geochemistry: A Tribute to Samuel Epstein. Geochem Soc (Spec Publ 3):307–324

Vezzoli L (1988) Attività esplosiva alcalino-potassica pleistocenica dell'isola di Ponza. Boll Gruppo Naz Vulcanol 4:584–599

7 The Campania Province

Abstract

The Campania Province (about 0.2 Ma to 1944 AD) consists of stratovolcanoes and multi-centre volcanic complexes (Vesuvio, Campi Flegrei, Procida-Vivara, Ischia) that are built up by mafic to felsic alkaline potassic magmas. Leucite-bearing ultrapotassic rocks are restricted to Somma-Vesuvio. Older lavas (about 2 Ma) with calcalkaline affinity are buried beneath the Campanian Plain. The rocks of the Campania Province define various suites that were generated by polybaric fractional crystallisation, mixing and assimilation of different types of wall rocks, starting from trachybasalt parents. The leucite-bearing rocks of Vesuvio also derive from trachybasalt, but extensive carbonate assimilation induced strong silica undersaturation and crystallisation of leucite. Therefore, primary ultrapotassic magmas, such as those occurring in the Roman Province, seem to be lacking in Campania. In contrast, there is a strong geochemical similarity between Campania and Stromboli, Aeolian arc. Petrological and geochemical data, therefore, argue against the commonly held view that the Campania volcanoes represent the southern extension of the Roman Province and rather indicate they represent the northernmost end of the eastern Aeolian arc. The Campania and Stromboli rocks have intermediate OIB-arc geochemical signatures, which suggest melt generation in a hybrid source, containing both OIB and subduction components. The arc-type components have been provided by fluids released by the subducted Ionian oceanic slab and associated sediments. The OIB-type component is attributed to the inflow of asthenospheric mantle from the foreland onto the subducting and southeastward retreating Ionian slab. Migration of asthenospheric mantle was favoured by opening of a slab window along the Apulian-Ionian plate.

A. Peccerillo, *Cenozoic Volcanism in the Tyrrhenian Sea Region*, Advances in Volcanology, DOI 10.1007/978-3-319-42491-0_7

Keywords
Campania province · Somma-Vesuvio · Campi Flegrei · Phlegraean Fields · Procida · Vivara · Ischia · Potassic rocks · Carbonate assimilation · Southern Apennines · Ionian plate · Southern Tyrrhenian Sea · Slab window

7.1 Introduction

The Campania Province is the southernmost sector of the Plio-Quaternary volcanic belt along the Italian peninsula. It is formed by the dormant volcanoes of Somma-Vesuvio, Ischia and Campi Flegrei (Phlegraean Fields), and by the islands of Procida and Vivara (Fig. 7.1). A number of actively degassing submarine centres in the Bay of Naples (e.g. Gaia Bank, Pentapalummo Bank, Secca di Ischia, etc.) and the Plio-Pleistocene volcanic rocks buried beneath the Campania Plain are also included in the Campania Province (Di Girolamo 1978; Passaro et al. 2014).

Volcanic rocks range from mafic to felsic and the mafic rocks have silica undersaturated potassic to ultrapotassic compositions (Fig. 7.2). Older lavas (about 2 Ma) buried beneath the Campanian Plain have oversaturated calcalkaline basalt to andesite compositions. Lava xenoliths from the pyroclastic deposits from Procida also have calcalkaline petrological affinity.

The Campania Province hosts two of the most famous active (though presently dormant) volcanoes: Vesuvio and Campi Flegrei. They are sited in an intensively populated area and, because of their violent explosive activity, are considered among the highest risk volcanoes on Earth. This has prompted numerous studies, and some thousand papers have been published on these volcanoes

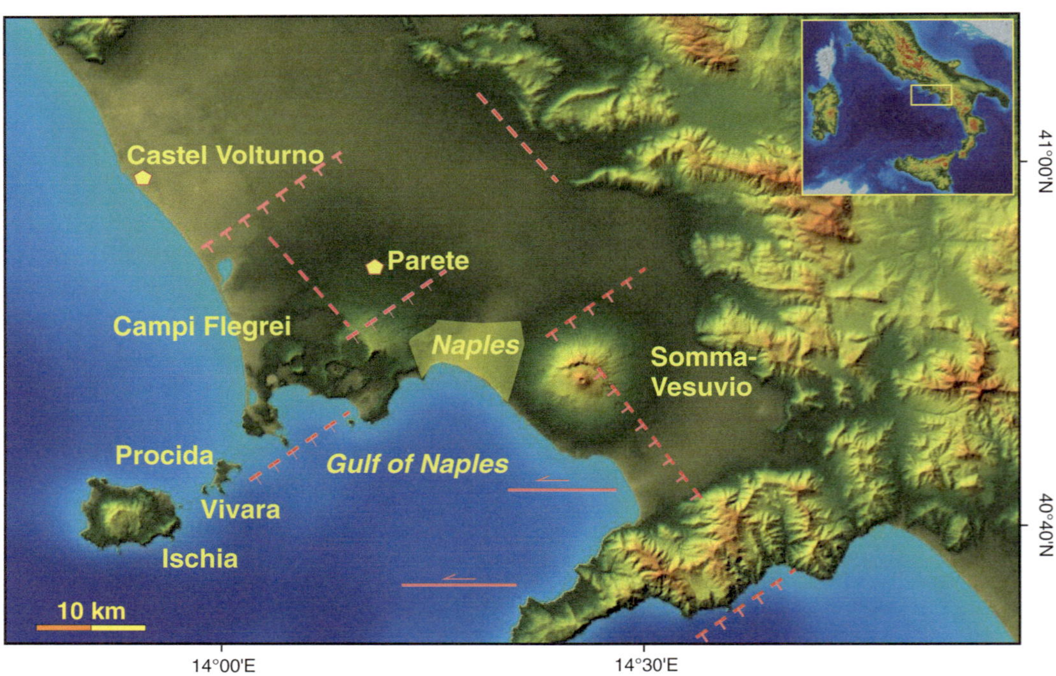

Fig. 7.1 Location of the Campania volcanoes. Extensional and transfer faults are indicated with *dashed* and *full lines*

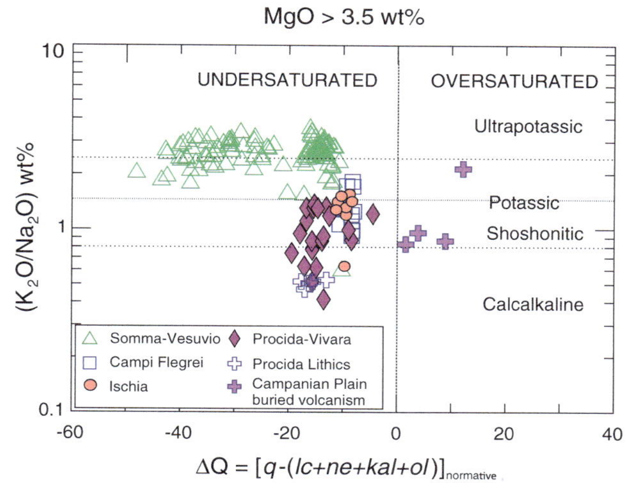

Fig. 7.2 K_2O/Na_2O versus ΔQ diagram for mafic volcanic rocks from the Campania Province. ΔQ is the algebraic sum of normative quartz minus undersaturated minerals (see Chap. 1). A limit for mafic composition at $MgO = 3.5$ wt% has been set to include samples from Ischia and Campi Flegrei

during the last two-three decades. The wealth of data increased considerably our understanding of volcanism in the area, but also expanded enormously the number of hypotheses on magma origin-evolution and modalities of eruption, ideas that can be only partially considered in this review.

A summary of ages and compositional characteristics of volcanism in Campania is given in Table 7.1. Data for representative rocks are reported in Table 7.2.

7.2 Regional Geology

Volcanism of the Campania Province developed in a rapidly subsiding area, between the Southern Apennines fold-thrust belt and the Tyrrhenian backarc basin. The area is affected by NE-SW and NW-SE main faulting, and volcanic centres are located at the intersection between the two systems (e.g. Bianco et al. 1998; Bruno et al. 2003; Acocella 2010; Milia and Torrente 2003, 2015).

The Southern Apennine chain consists of a number of NE-verging thrusts, locally covered by Plio-Quaternary autochthonous shallow marine and continental sediments. Tectonic units are formed by a wide variety of rock types (limestones, dolostones, flysch sequences, sandstones, marls, ophiolitic rocks, etc.) that range in age from Upper Triassic to Miocene (e.g. Ippolito et al. 1975; Grasso 2001). They mostly represent basinal sequences formed on the border of the African plate, which were delaminated and superimposed over the Apulia foreland. The latter is located east of the Apennine chain and contain Mesozoic to Tertiary platform carbonates (e.g. Ogniben et al. 1975; Grasso 2001; Patacca and Scandone 2001; Vai and Martini 2001).

Thrusting of tectonic units in the Southern Apennines occurred during Upper Oligocene to Lower Pleistocene, and had a northeast vergence. Starting from Pleistocene, thrusting rotated toward the southeast, in contrast with Northern Apennine where northeast displacement persisted until recently (Patacca et al. 1990). Latest compression tectonic phases in the Southern Apennines were related to subduction of the Ionian seafloor beneath the Calabrian-Tyrrhenian microplate (Gvirtzman and Nur 1999). Such a process generated backarc extension and block rotation in the Campania Province, with formation of Quaternary NE-trending normal faults, NW-trending transtensional faults, and E-W trending left-lateral faults (e.g. Cello and Mazzoli 1999; Milia and Torrente 2003).

The Campania Province is characterised by high heath flow (> 100 mW/m^2) with peaks up to 600 mW/m^2 at Ischia and Campi Flegrei (Della Vedova et al. 2001; Carlino et al. 2012). The thickness of the lithosphere is about 50 km and increases eastward reaching more than 110 km in

Table 7.1 Age, volcanology and petrology-geochemistry of Campania volcanoes

Volcano	Age	Volcanology	Petrology-Geochemistry
Somma-Vesuvio	~30 ka to 1944 AD	– Stratovolcano (Mount Somma) with multiple caldera and an intracaldera cone (Vesuvio). Early effusion of lava flows forming the Somma cone, followed by plinian eruption and by a period of moderately explosive to effusive activity ended in 1944 AD	– Poorly undersaturated trachybasalt to trachyte, and moderately to strongly undersaturated leucite-bearing foidites, phonotephrites to phonolites. Different trends for major and trace elements but similar radiogenic isotope ratios ($^{87}Sr/^{86}Sr$ ~ 0.7062 to 0.7080) shown by the various rock suites
Campi Flegrei	About 0.2 Ma to 1538 AD (Monte Nuovo eruption)	– Multicentre volcanic complex with two nested calderas, several monogenetic pyroclastic cones and maars made up of largely prevailing pyroclastic rocks. Two high magnitude caldera-forming plinian eruptions emplaced the widespread Campanian Ignimbrite (39 ka) and Neapolitan Yellow Tuff (15 ka) pyroclastic deposits	– Dominant trachyte phonolite rocks, with minor younger latites and shoshonites, showing similar radiogenic isotope signatures as Somma-Vesuvio, but increasing Sr-isotope ratios from the older silicic products to younger latites and shoshonites
Island of Ischia	150 ka to 1302 AD	– Intensely faulted and eroded stratovolcano consisting of prevailing pyroclastic rocks and minor lavas. Major explosive eruption occurred at 55 ka, with emplacement of Monte Epomeo Green Tuff. Morphology dominated by the 787 m high central resurgent and tilted intra caldera block of Mt. Epomeo, and by minor highs along the coastline	– Dominant trachyte to phonolite compositions, with minor late erupted latites and shoshonites. $^{87}Sr/^{86}Sr$ ratios (~ 0.7060 to 0.7076) slightly lower than Campi Flegrei and Somma-Vesuvio
Procida and Vivara islands	About 70 to 14 ka	– Coalescing hydrovolcanic tuff cones and rings formed of ashes, scoriae, pumices, and lithics	– Juvenile trachybasalt to trachyte pyroclastic rocks. Calcalkaline basalt lithic clasts. $^{87}Sr/^{86}Sr$ (0.7052 to 0.7073) increase from mafic to felsic rocks
Plio-Pleistocene buried volcanism	~2 Ma	– About 1300 m thick sequence of lavas encountered by deep drillings at Parete and Castel Volturno, in the Campanian Plain	– Altered calcalkaline basalt to andesite lavas with $^{87}Sr/^{86}Sr$ ~ 0.7051–0.7080

the Apulia foreland (Calcagnile and Panza 1981). The depth of the Moho increases from about 20–25 km at Ischia and along the Tyrrhenian Sea coast to 35 km beneath the internal zones of the Apennines, to decrease to about 25–30 km beneath the Apulia foreland and the southern Adriatic Sea (e.g. Locardi and Nicolich 1988; Piromallo and Morelli 2003; Mele et al. 2006; Panza et al. 2007; Nunziata 2010; Nunziata and Costanzo 2010; Carminati and Doglioni 2012). In contrast with other zones of the Apennine chain, the sector running from the Campania Province to Apulia is characterised by a moderate elevation and by a positive Bouguer anomaly, which crosses the Apennines connecting the larger positive anomalies of the Tyrrhenian Sea and Apulia (Turco and Zuppetta 1998; Tiberti et al. 2005).

Table 7.2 Selected compositions for rocks from the Campania Province. Numbers in parentheses refer to compositions determined on distinct, though compositionally similar samples from the same locality as those analysed for the other elements

	Somma-Vesuvio						**Campi Flegrei**			
Rock	Trachybasalt	Trachyte	Shoshonite	Tephriphonolite	Phonotephrite	Phonolite	Shoshonite	Latite	Trachyte	Phonolite
Reference	11	8,9	16	5	5	12	4,7,10	4,7,10	15	15
SiO_2 wt%	50.17	57.70	52.40	52.30	48.20	57.02	52.08	55.52	60.10	56.40
TiO_2	1.02	0.39	0.78	0.56	0.87	0.08	0.81	0.60	0.43	0.40
Al_2O_3	16.52	17.80	18.20	18.40	17.00	20.10	17.11	18.59	17.20	18.70
Fe_2O_3	8.39	3.77	7.90	4.86	7.74	1.88	7.73	5.65	2.93	4.07
MnO	0.14	0.15	0.15	0.14	0.14	0.17	0.13	0.12	0.18	0.26
MgO	6.14	0.55	2.46	2.76	4.55	0.04	3.94	1.51	0.33	0.33
CaO	10.15	3.30	7.73	6.52	10.10	1.54	8.50	4.21	1.57	2.03
Na_2O	2.29	3.46	3.09	4.10	2.24	8.13	3.65	4.06	5.23	6.61
K_2O	4.13	7.78	4.81	7.14	6.40	6.60	3.59	6.60	6.80	7.17
P_2O_5	0.58	0.13	0.51	0.36	0.63	0.01	0.57	0.32	0.05	0.06
LOI	0.63	3.97	1.11	1.69	1.11	3.76	1.50	1.94	4.26	2.70
Sc ppm	–	2.5	12	13	–	–	15	6.0	–	4
V	252	–	–	–	–	16	195	115	31	11
Cr	148	–	21	65	81	16	16	–	–	–
Co	30	3	17	13	26	0	19	7	–	–
Ni	57	3	9	22	15	1	14	–	6	4
Rb	189	315	240	325	293	548	190	175	430	420
Sr	735	715	950	600	1060	29	919	874	56	20
Y	24	40	–	30	29	–	26	28	69	71
Zr	177	355	265	315	221	883	188	275	672	733
Nb	22	46	30	53	32	94	22	37	96	131
Cs	8.7	18.9	13	25.1	16	–	9.0	13.8	39.4	36.5
Ba	1482	1050	1400	1100	2130	40	1591	1509	47	23
La	35	68	56	93	51	70	50	64	116	114

(continued)

Table 7.2 (continued)

Rock	Somma-Vesuvio Trachybasalt	Trachyte	Shoshonite	Tephriphonolite	Phonotephrite	Phonolite	Campi Flegrei Shoshonite	Latite	Trachyte	Phonolite
Ce	72	124	107	157	103	155	100	123	208	225
Pr	8.9	–	–	–	–	12	12.0	13.3	21.8	22.6
Nd	37	46	46	51	47	33	45	47	75	77
Sm	7.6	9.6	10.0	8.4	9.9	4.0	9.0	8.0	12.6	13.7
Eu	2.30	1.97	2.00	1.72	2.52	0.50	2.30	2.20	1.39	1.30
Gd	6.40	–	6.40	–	7.49	4.40	7.00	6.00	12.90	13.90
Tb	0.81	1.03	1.00	0.75	1.10	0.40	1.00	0.90	1.60	1.85
Dy	4.80	–	–	–	–	1.80	5.00	5.00	8.95	10.20
Ho	0.84	–	–	–	–	0.30	0.90	0.90	1.72	1.98
Er	2.30	–	–	–	–	1.10	2.20	2.50	5.09	5.78
Tm	0.32	–	0.30	–	0.35	0.20	0.30	0.40	0.79	0.86
Yb	2.00	3.02	3.00	2.06	2.17	1.40	2.00	2.30	5.33	5.81
Lu	0.27	0.42	0.00	0.25	0.27	0.20	0.30	0.40	0.84	0.90
Hf	4.40	7.23	6.00	5.80	–	17.00	4.40	6.00	16.2	17.0
Ta	1.50	2.89	2.00	2.26	1.50	0.50	1.40	2.30	5.32	5.93
Pb	21	44	26	57	33	123	32	43	62	65
Th	12	29	20	38	19.9	92	15.3	23.7	65	57
U	3.9	8.1	5.0	11.5	6.5	–	4.3	8.0	19.1	18.9
$^{87}Sr/^{86}Sr$	0.706661	0.707653	0.70679	0.707449	0.707308	0.707182	0.708070	0.707350	0.707533	0.707284
$^{143}Nd/^{144}Nd$	0.512549	0.51243	0.51243	0.512405	0.512406	0.51247	0.512400	0.512540	0.512463	0.512518
$^{206}Pb/^{204}Pb$	19.088	18.956	18.883	18.978	19.005	–	18.969	19.040	–	–
$^{207}Pb/^{204}Pb$	15.694	15.636	15.570	15.673	15.663	–	15.677	15.673	–	–
$^{208}Pb/^{204}Pb$	39.227	38.981	38.857	39.104	39.070	–	39.074	39.126	–	–

(continued)

Table 7.2 (continued)

	Ischia				Procida					Parete 2 well
Rock	Shoshonite	Latite	Trachyte	Phonolite	Basalt lithic	Trachybasalt	Shoshonite	Trachyte	Trachybasalt	Andesite
Reference	2,3	18	18	14,19	20	6	13,17	13	6	1
SiO_2 wt%	52.60	57.54	61.32	58.25	48.35	48.87	52.10	58.90	48.87	52.93
TiO_2	1.36	1.06	0.70	0.56	1.24	1.12	1.27	0.45	1.12	0.78
Al_2O_3	18.69	17.15	18.07	17.81	15.48	18.37	15.80	19.00	18.37	17.8
Fe_2O_3	8.66	6.21	3.27	3.19	7.88	8.64	10.20	4.16	8.64	7.38
MnO	0.14	0.13	0.15	0.29	0.14	0.14	0.15	0.13	0.14	0.09
MgO	3.25	2.82	0.80	0.54	9.54	5.35	4.71	1.73	5.35	4.49
CaO	7.52	4.81	1.66	1.01	12.01	11.34	9.08	3.14	11.34	9.79
Na_2O	3.61	4.07	5.96	7.67	2.85	3.50	2.82	2.61	3.50	2.31
K_2O	3.89	5.41	6.15	6.32	1.47	2.16	3.42	9.52	2.16	2.21
P_2O_5	0.62	0.32	0.11	0.10	0.28	0.50	0.44	0.37	0.50	–
LOI	0.43	0.48	1.82	4.11	0.77	1.00	3.65	1.44	1.00	1.00
Sc ppm	20	23	16	–	37	–	20	5.6	–	24
V	168	146	46	25	207	239	237	74	239	–
Cr	11	37	7	–	421	20	5	4	20	50
Co	22	13	1	1	42	34	27	5	34	27
Ni	–	25	18	–	150	31	23	5	31	18
Rb	106	184	298	469	46	68	131	204	68	74
Sr	711	525	130	10	492	943	814	694	943	664
Y	39	29	55	80	21	25	24	16	25	23
Zr	165	221	540	945	102	131	133	140	131	54
Nb	32	34	78	117	12	17	20	27	17	7
Cs	5	7.84	14.08	28.7	–	8.8	5.5	6.7	8.8	5.2
Ba	1350	1017	272	8.4	521.2	1156	1310	998	1156	465
La	42	54	88	185	14	33	41	41	33	23.7
Ce	97	101	167	331	32	70	79	74	70	45.2
Pr	–	10.86	16.94	34.1	4.19	–	9.88	8.47	–	–

(continued)

Table 7.2 (continued)

	Ischia				Procida					Parete 2 well
Rock	Shoshonite	Latite	Trachyte	Phonolite	Basalt lithic	Trachybasalt	Shoshonite	Trachyte	Trachybasalt	Andesite
Nd	45	50	79	110	18	38	38	30	38	–
Sm	8.0	8.7	13.3	18.6	4.4	8.2	7.6	5.4	8.2	3.8
Eu	2.80	2.32	1.72	1.10	1.47	2.32	2.29	2.03	2.32	1.2
Gd	–	7.11	12.78	–	4.58	–	7.09	5.03	–	–
Tb	0.90	0.85	1.53	2.30	0.72	0.93	0.86	0.60	0.93	0.6
Dy	–	6.10	10.09	13.30	4.11	5.00	4.54	3.21	5.00	–
Ho	–	1.14	1.97	2.60	0.77	–	0.81	0.58	–	–
Er	–	2.97	5.29	7.80	2.13	–	2.31	1.68	–	–
Tm	–	0.38	0.82	1.20	0.28	–	0.28	0.23	–	–
Yb	2.50	3.32	5.90	8.60	1.83	1.86	1.90	1.60	1.86	3.2
Lu	0.51	0.36	0.83	1.37	0.27	0.25	0.28	0.25	0.25	0.5
Hf	3.90	4.97	11.43	19.70	2.70	3.22	3.00	3.00	3.22	3.2
Ta	1.90	1.68	4.26	6.80	0.86	1.05	1.00	1.60	1.05	–
Pb	–	32	32	68	13	15	15	28	15	–
Th	9.5	16	43	67	2.88	8.1	9.9	12	8.1	9
U	2.5	4.0	9.0	18.3	1.0	2.6	2.5	3.6	2.6	–
$^{87}Sr/^{86}Sr$	0.70608	0.706314	0.706083	0.70646	0.70514	0.706485	0.706556	0.707329	0.706485	0.7060
$^{143}Nd/^{144}Nd$	–	0.51254	0.512556	0.512529	0.51272	0.512568	0.512536	0.512471	0.512568	–
$^{206}Pb/^{204}Pb$	–	–	–	–	18.675	19.300	19.305	–	19.300	–
$^{207}Pb/^{204}Pb$	–	–	–	–	15.637	15.699	15.798	–	15.699	–
$^{208}Pb/^{204}Pb$	–	–	–	–	38.682	39.392	39.445	–	39.392	–

Source of data: (1) Albini et al. (1980); (2) Poli et al. (1987); (3) Civetta et al. (1991a); (4) D'Antonio et al. (1999b); (5) Somma et al. (2001); (6) De Astis et al. (2004); (7) Tonarini et al. (2004); (8) Paone (2006); (9) Piochi et al. (2006b); (10) D'Antonio et al. (2007); (11) Di Renzo et al. (2007); (12) Aulinas et al. (2008); (13) Avanzinelli et al. (2008); (14) Brown et al. (2008); (15) Pabst et al. (2008); (16) Paone (2008); (17) Conticelli et al. (2010a, b); (18) D'Antonio et al. (2013); (19) Brown et al. (2014); (20) Mazzeo et al. (2014)

7.3 Somma-Vesuvio

Somma-Vesuvio is a 1281 m high composite volcano (Somma) with a large polygenetic asymmetrical caldera, in which the active cone of Vesuvio has grown during the last two thousand years. The caldera has a horseshoe shape with a maximum diameter of about 3 km. It is delimited by a steep northeastern wall, and by an open southwest rim, where the volcano slopes gently toward the Tyrrhenian coast (Fig. 7.3).

The Somma-Vesuvio volcano sits over a 2 km thick sequence of sedimentary and volcanoclastic material, which overlies some 6–8 km thick sequences of Mesozoic carbonate rocks, and the Hercynian crystalline basement (e.g. Berrino et al. 1998; Tondi and De Franco 2003). The Moho is sited at a depth of about 30 km (e.g. De Natale et al. 2006).

7.3.1 Volcanology and Stratigraphy

Somma-Vesuvio is a relatively young system, developed during the last 30 ka, although much older volcanic activity (some 0.4 Ma) is highlighted by deep borehole drillings in the area (Brocchini et al. 2001). A summary of the

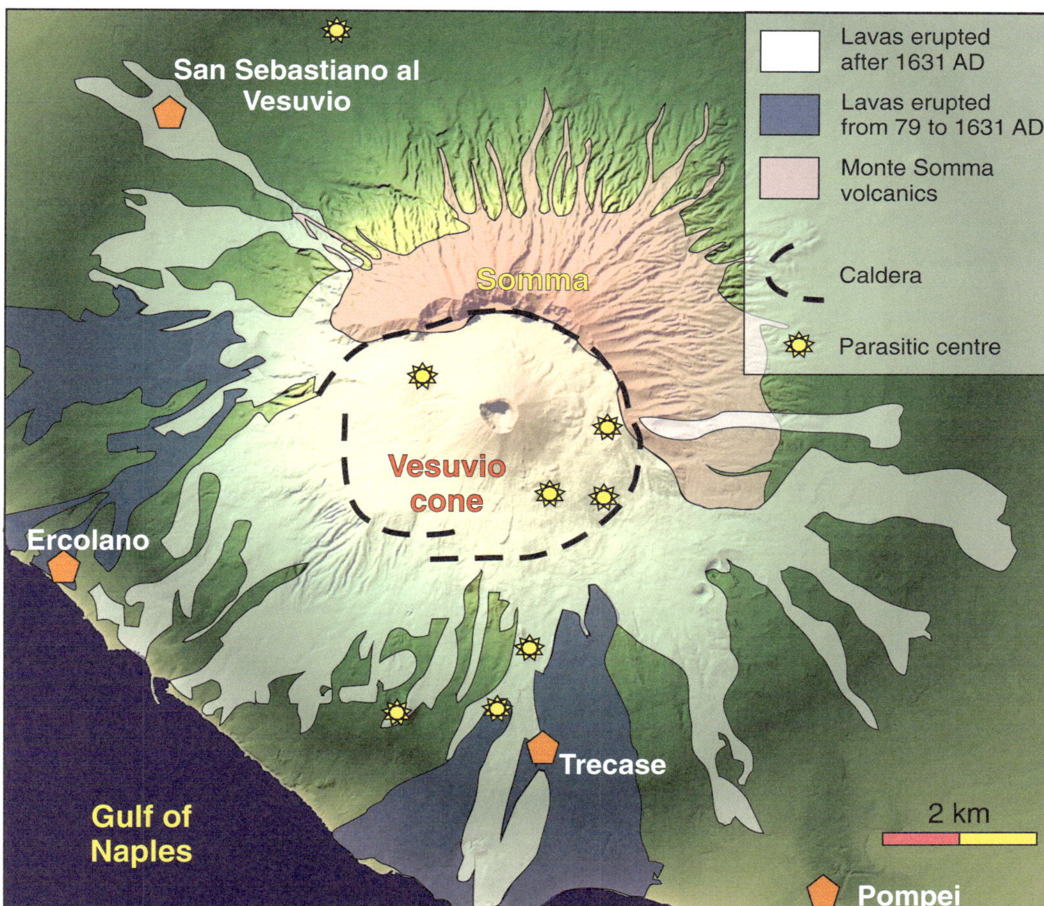

Fig. 7.3 Sketch geological map of Somma-Vesuvio volcano. Undifferentiated pyroclastics cover of the lower flank of the cone and surrounding areas is not reported. Simplified after the Geological map of Somma-Vesuvio in Santacroce (1987)

volcanic activity at Somma-Vesuvio is schematically shown in Fig. 7.4a (Bertagnini et al. 1998; Principe et al. 2004; Piochi et al. 2006a; Santacroce et al. 2008; Cioni et al. 2008; Sulpizio et al. 2010).

The earliest eruptions of Somma volcano were prevailingly effusive and emplaced a suite of mafic potassic lava flows and scoriae, which crop out along the walls of the caldera and at a few parasitic cones (Santacroce 1987; MacDonald et al. 2015). Some pyroclastic deposits (e.g. 33 ka-old Codola eruption; Giaccio et al. 2008) are interposed between the Somma lavas and the 39 ka-old Campanian Ignimbrite (Campi Flegrei), but some authors doubt that these pyroclastics originated from Somma (Santacroce et al. 2008).

Starting from about 22 ka until 79 AD, the style of volcanic activity became more explosive,

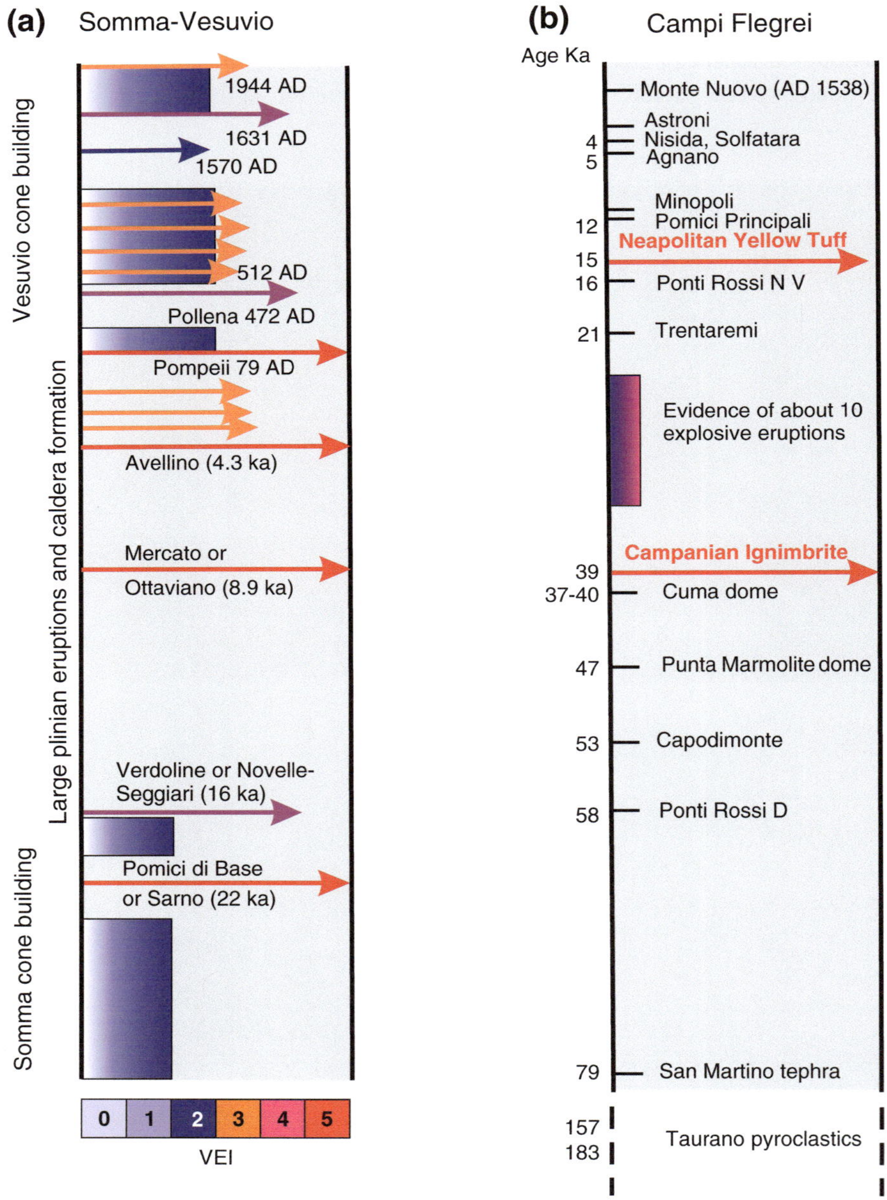

Fig. 7.4 Simplified chronograms of the Somma-Vesuvio and Campi Flegrei activity. VEI is the Volcanic Explosivity Index of Newhall and Self (1982)

and plinian eruptions took place, affecting latite, trachyte, phonolite and tephriphonolite magmas. Plinian events were preceded by long periods of quiescence and were probably associated with caldera collapses (Sheridan et al. 1981; Santacroce et al. 2003). However, some authors believe that the caldera might result from several mechanisms, including rock removal by explosion and gravimetric flank failure, an hypothesis supported by the occurrence of extensive debris flow deposits in the southern sector of the volcano (e.g. Ventura et al. 1999; Rolandi et al. 2004). Main plinian eruptions with VEI = 4-5 occurred at about 22 ka (Pomici di Base or Sarno), 16 ka (Pomici Verdoline or Novelle-Seggiari-Bosco), 8.9 ka (Mercato-Ottaviano), 4.3 ka (Avellino) and 79 AD (Pompeii). Moderately explosive activity (VEI = 2-3) took place between Avellino and Pompeii eruptions (Santacroce et al. 2008 with references; Andronico and Cioni 2002).

The 79 AD eruption is one of the most famous historical volcanic events and was responsible for destruction of several ancient Roman cities and villas, notably Pompeii, Herculaneum and Oplontis (e.g. Sigurdsson et al. 1985; Lirer et al. 1997; Giacomelli et al. 2003). The account of this eruption given by Pliny the Younger in two letters to Cornelius Tacitus is the first accurate description of an eruption and is considered the first scientific document in the history of volcanology.

Vesuvio started to grow after the 472 AD "Pollena" eruption inside the Somma caldera (Santacroce 1987; Rolandi et al. 1998; Santacroce et al. 2003), and was built up by intense mildly explosive and effusive activity, and by a few high-magnitude plinian-subplinian events. The 1631 AD plinian eruption marked the onset of a period of nearly continuous open-conduit magma emission, which lasted until 1944 AD. Most eruptions of this period began with lava outpouring, followed by explosive phases with formation of lava fountains, steady columns and phreatovolcanism (e.g. Arrighi et al. 2001; Marianelli et al. 2005). Vesuvio is presently in a quiescent state, with CO_2 degassing mostly from the crater area (Frondini et al. 2004), and shallow-focus earthquakes with maximum magnitude M = 3-4 (Bianco et al. 1998; De Lorenzo et al. 2006).

The internal structure of the Somma-Vesuvio volcano has been the subject of many studies, because of its importance for understanding magma dynamics and forecasting future behaviour (e.g. Marianelli et al. 1999; De Natale et al. 2006; Pappalardo and Mastrolorenzo 2010). Based on petrological-volcanological data, some authors suggest that a shallow magma chamber has been operating in conditions of alternated periods of open or obstructed conduit (e.g. Joron et al. 1987; Santacroce et al. 1993, 2003; Cioni et al. 1995, 1997; Marianelli et al. 2005). This reservoir was periodically replenished by mafic magmas from depth. When the conduit was obstructed, magma injection generated expansion of magma chamber and extensive fractional crystallisation, with the generation of felsic melts. These accumulated at the top of the reservoir, which became chemically zoned (Cioni et al. 1995). Stronger explosions occurred after quiescent periods, reflecting the recharge of the system. In these cases the magnitude of explosions correlates very significantly, though not exclusively, with the duration of repose time (Santacroce et al. 2003). During periods of open conduit, the magma chamber was smaller and the arrival of fresh magma resulted in either quiet lava effusions or moderately explosive eruptions. The 1944 eruption marks the transition from open to obstructed conduit conditions.

However, it has been argued that geophysical studies are unable to reveal a magma reservoir of significant size at shallow depth (e.g. Zollo et al. 1998; Auger et al. 2001; Scarpa et al. 2002). The high rigidity body detected by seismic studies beneath the crater may represent crystallised magmas along the conduit rather than a truly crystallised magma chamber. Instead, a low seismic wave velocity (magma) body has been detected at depths of about 8 km and 10-12 km (Zollo et al. 1996, 1998; Auger et al. 2001). A deeper low-Vs layer has been imaged at the regional scale and has been interpreted as the source of all Campania volcanoes (De Natale et al. 2006; Nunziata 2010).

Based on gephysical evidence as well as on fluid inclusion geothermobarometry, rock geochemistry, and mineral chemistry data, several authors envisage a multiple-chamber structure for Somma-Vesuvio. Belkin et al. (1985) and Belkin and De Vivo (1993) found a distribution of CO_2 fluid inclusion densities suggesting an entrapment at about 4–5 km, at 8–10 km and at 12 km. These were interpreted as sites of magma ponding and crystallisation beneath the volcano. Pappalardo et al. (2004) and Pappalardo and Mastrolorenzo (2010) suggest that two magma chambers have fed the Vesuvio activity. The shallower reservoir, located at the discontinuity between the carbonate platform and the Hercynian basement (some 7–8 km) fed the plinian and sub-plinian eruptions, whereas a 14–16 km deep reservoir(s) was responsible for the effusive to moderately explosive activity such as that occurred between 1631–1944 AD. Two long-lived magma storage zones, at about 8 and 12 km, and a shallower reservoir at about 5 km were suggested by Piochi et al. (2006a, b) on the basis of geochemical, geophysical and fluid inclusion data. The deepest chambers were the source of 1944 AD eruption and inter-plinian and sub-plinian events, whereas the shallower reservoir was believed to originate the plinian activity, such as Pompeii. In contrast, Paone (2006) envisage a deep storage for magmas of plinian eruptions.

In conclusion, the ideas on the plumbing structure of Somma-Vesuvio volcano are many and somewhat contradictory. Most models accept a multiple magma chamber system, which underwent significant modification with time. According to Borgia et al. (2005), such an evolution is related to a slow but progressive spreading of the volcano over its plastic substratum. This expansion determined variation of the plumbing system, which had important effects on the nature of magma evolution and hence on the eruption style.

The present-day shallow system beneath the crater consists of crystallised magmas hosting low electrical resistivity material, likely brines (e.g. Chiodini et al. 2001; Pommier et al. 2010).

7.3.2 Petrography and Mineral Chemistry

Joron et al. (1987) distinguished three main rock series at Somma-Vesuvio. These are characterised by different ages, degrees of silica undersaturation and enrichment in alkalies (Fig. 7.5). The rocks older than about 17 ka mostly consist of poorly silica undersaturated to oversaturated rocks ranging from potassic trachybasalt to trachyte. These include the lavas and scoriae of the Somma stratovolcano, as well as the pyroclastic products of Pomici di Base (Basal Pumices) and

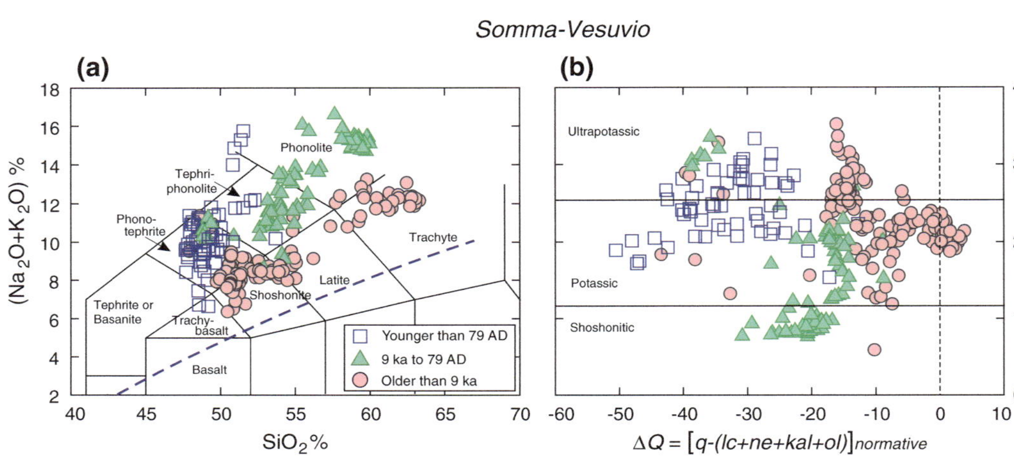

Fig. 7.5 a TAS classification diagram of Somma-Vesuvio rocks; the *dashed line* is the divide between the alkaline and subalkaline rocks (Irvine and Baragar 1971). Data normalised to 100 % on a LOI-free basis; **b** ΔQ versus K_2O/Na_2O diagram for the Somma-Vesuvio mafic to felsic rocks. Samples with LOI > 4.0 wt% have not been plotted because of possible alteration

Pomici Verdoline (Greenish Pumices) eruptions. Three other eruptions are included in this group (Codola, Schiava and Taurano), although the provenance of these pyroclastics has been questioned (Santacroce et al. 2008).

A second series (about 17 ka to 79 AD) essentially consists of phonotephrites, tephriphonolites and phonolites with intermediate degrees of silica undersaturation. These rocks were mainly emplaced by three main plinian events of Mercato (8.9 ka), Avellino (4.4 ka), and Pompeii (79 AD), and by several minor eruptions between Avellino and Pompeii (Andronico and Cioni 2002).

Finally, a third series, younger than 79 AD, shows strongly undersaturated leucite-bearing phonotephrite, tephriphonolite to phonolite composition. However, trachybasalts are also found, e.g. among the products of the 1906 eruption (Santacroce et al. 1993). Ayuso et al. (1998) also recognised three series on the basis of trace element concentrations, although the time limit between the two older series was set at about 9 ka, a limit that is also accepted in this book.

The rocks belonging to the poorly undersaturated older series display variably porphyritic textures, with phenocryst contents decreasing from mafic to felsic volcanics. Plagioclase and clinopyroxene phenocrysts are ubiquitous; olivine and leucite occur in the mafic rocks, whereas biotite, K-feldspar and some amphibole are observed in the latites and trachytes. Groundmasses range from holocrystalline to hypocrystalline and contain the same phases as the phenocrysts plus Fe–Ti oxides, accessory garnet, apatite and variable amounts of glass. Trachytic pumices are strongly vesicular and range from aphyric to poorly porphyritic with a few phenocrysts of sanidine, plagioclase, green clinopyroxene and Fe–Ti oxides (Joron et al. 1987).

Phonotephrites of the mildly undersaturated series are strongly porphyritic with phenocrysts of plagioclase, clinopyroxene and leucite set in a groundmass formed by the same phases and glass. Tephriphonolites include lavas and pumices exhibiting variably porphyritic textures, with phenocrysts of clinopyroxene (green to colourless in thin section), plagioclase, leucite, alkali feldspar, biotite and Fe–Ti oxides, set in a glassy groundmass. Melanite garnet and amphibole have been observed. Phonolites consist of pumices with a poorly porphyritic to aphyric vesicular texture; phenocrysts are represented by clinopyroxene, sanidine, some biotite and leucite set in a glassy groundmass; minor phases include garnet, olivine and calcic plagioclase; nepheline is observed in some rocks.

The highly undersaturated younger rocks show more strongly porphyritic textures than older lavas and pyroclastics. Phenocrysts of mafic rocks consist of dominant clinopyroxene, and minor leucite and olivine; groundmass is holocrystalline to hypocrystalline and contains the same phases plus plagioclase, opaques and some brown mica. Tephriphonolites are strongly porphyritic with phenocrysts of leucite, clinopyroxene and plagioclase, and sporadic olivine and biotite set in holocrystalline to hypocrystalline groundmass formed by the same phases plus Fe–Ti oxides and apatite. Phonolites consist of vesicular porphyritic pumices made of leucite and clinopyroxene phenocrysts and of a glassy matrix containing plagioclase, sanidine, garnet, amphibole, Fe–Ti oxides and sporadic haüyne.

Clinopyroxene mainly shows Cr- and Fe-diopside to hedenbergitic compositions. Zoning, both normal and reverse, is common (e.g. Joron et al. 1987; Cioni et al. 1998; Morgan et al. 2004; Aulinas et al. 2008; MacDonald et al. 2015). Ti/Al ratios are low (e.g. Lima et al. 1999), as typically observed for clinopyroxenes from the Roman Province (e.g. Boari et al. 2009a, b). Olivine is scarce and exhibits a large compositional variation both in the phenocrysts (up to about Fo_{90}) and in the groundmass (about Fo_{70-45}). Some highly forsteritic crystals (up to Fo_{94}) probably represent xenocrysts from disaggregated skarns. Plagioclase is strongly zoned and exhibits a wide range of compositions ($\sim An_{90-50}$). Sanidine is more sodic in the slightly undersaturated latites and trachytes than in the highly undersaturated tephriphonolites and phonolites. Leucite shows an almost stoichiometric composition. Mica is rich in fluorine and ranges from phlogopite to biotite, the latter being typical of felsic products. Amphibole is potassic ferropargasite (e.g. Marianelli et al. 1995; Cioni et al. 1997; Lima et al. 2007).

Several ejected blocks are found in the Somma-Vesuvio pyroclastic deposits. These include lavas, sedimentary carbonates, skarns, and mafic and ultramafic xenoliths (e.g. Joron et al. 1987; Del Moro et al. 2001; Klébesz et al. 2012). Dunitic ejecta have been interpreted as either cumulates, sometimes affected by post crystallisation deformation, or possibly mantle material that survived magma chamber processes (Cigolini 1999). Xenoliths contain several mineral species that, along with those found in the lavas and fumaroles, have been the subject of study since the early history of mineralogy (see Zambonini 1935). More than two hundred different mineral species have been found at Somma-Vesuvio and some of them are typical or exclusively found at this volcano (e.g. Russo and Punzo 2004).

7.3.3 Petrology and Geochemistry

Variation diagrams of major elements for Somma-Vesuvio rocks show a decrease in TiO_2, P_2O_5, MgO, CaO, and FeO_{total}, and an increase in K_2O and Na_2O with increasing silica (Figs. 7.6). LILE (Rb, Cs, LREE, etc.) and HFSE (Nb, Zr, etc.) increase with silica, whereas ferromagnesian trace elements (Sc, Cr, etc.), Sr and Ba have an opposite behaviour (Fig. 7.7). In general terms, the different rock series define distinct trends for most major and trace elements, as pointed out by many studies (e.g. Joron et al. 1987; Ayuso et al. 1998; Santacroce et al. 2008). Note, however, that all the trends originate from a single mafic composition. The mildly undersaturated rocks, especially the Avellino and Mercato pumices, define very sharp enrichments for Na_2O and incompatible elements, and strong depletion in Ba and Sr. Volatile elements (F, Cl, B) have high concentrations, reaching several thousand ppm in most rocks (e.g. Somma et al. 2001).

REE patterns are variably fractionated, with small negative Eu anomalies, which increase in trachytes and phonolites (Fig. 7.8a). HREE depletion is observed in the phonolites of the mildly undersaturated series. Patterns of incompatible elements normalised to primordial mantle composition for mafic rocks are fractionated and contain positive spikes of Pb and negative anomalies of HFSE (Fig. 7.8b). However, in contrast with other potassic rocks in Central Italy, Ba negative anomalies are lacking, HFSE depletion is less marked, and Ta-Nb abundances are much higher than average N-MORB and resemble OIB.

Sr, Nd and Pb isotopic compositions (Fig. 7.9) show similar range of values for the three rock series ($^{87}Sr/^{86}Sr \sim 0.7062$ to 0.7080; $^{143}Nd/^{144}Nd \sim 0.5122$ to 0.5126; $^{206}Pb/^{204}Pb \sim 18.88$ to 19.12; $^{207}Pb/^{204}Pb \sim 15.57$ to 15.75; $^{208}Pb/^{204}Pb \sim 38.86$ to 39.36; Cioni et al. 1995; Ayuso et al. 1998; Somma et al. 2001; Paone 2006). Significant isotopic variations have been observed within single cycles of volcanic activity or eruptions, indicating repeated episodes of wall rock assimilation plus mixing with fresh magmas (Cortini and Hermes 1981; Civetta and Santacroce 1992; Somma et al. 2001; Cortini et al. 2004; Piochi et al. 2006a, b). A trend towards Tyrrhenian MORB and Plio-Quaternary Sardinia compositions is observed for Pb isotopes (Fig. 7.9b, c, d). $^{176}Hf/^{177}Hf$ isotopic ratios determined on two samples yielded values of about 0.28278 (Gasperini et al. 2002).

Oxygen isotope ratios of Vesuvio whole rocks show wide variations ($\delta^{18}O_{SMOW} \sim +7.0$ ‰ to +10.0 ‰) and are negatively correlated with MgO (Ayuso et al. 1998; Paone 2008). Values for separated olivine and clinopyroxene are much lower, but still variable with $\delta^{18}O_{SMOW} \sim +5.5$ ‰ to +7.6 ‰ (Dallai et al. 2011). Helium isotope analyses on clinopyroxene and olivine from historical lavas gave values of $R/R_A \sim 2.2$ to 2.7, close to ratios of fumaroles (Tedesco et al. 1990; Graham et al. 1993; Martelli et al. 2004, 2008). $\delta^{11}B$ clusters around −6 to −8‰ relative

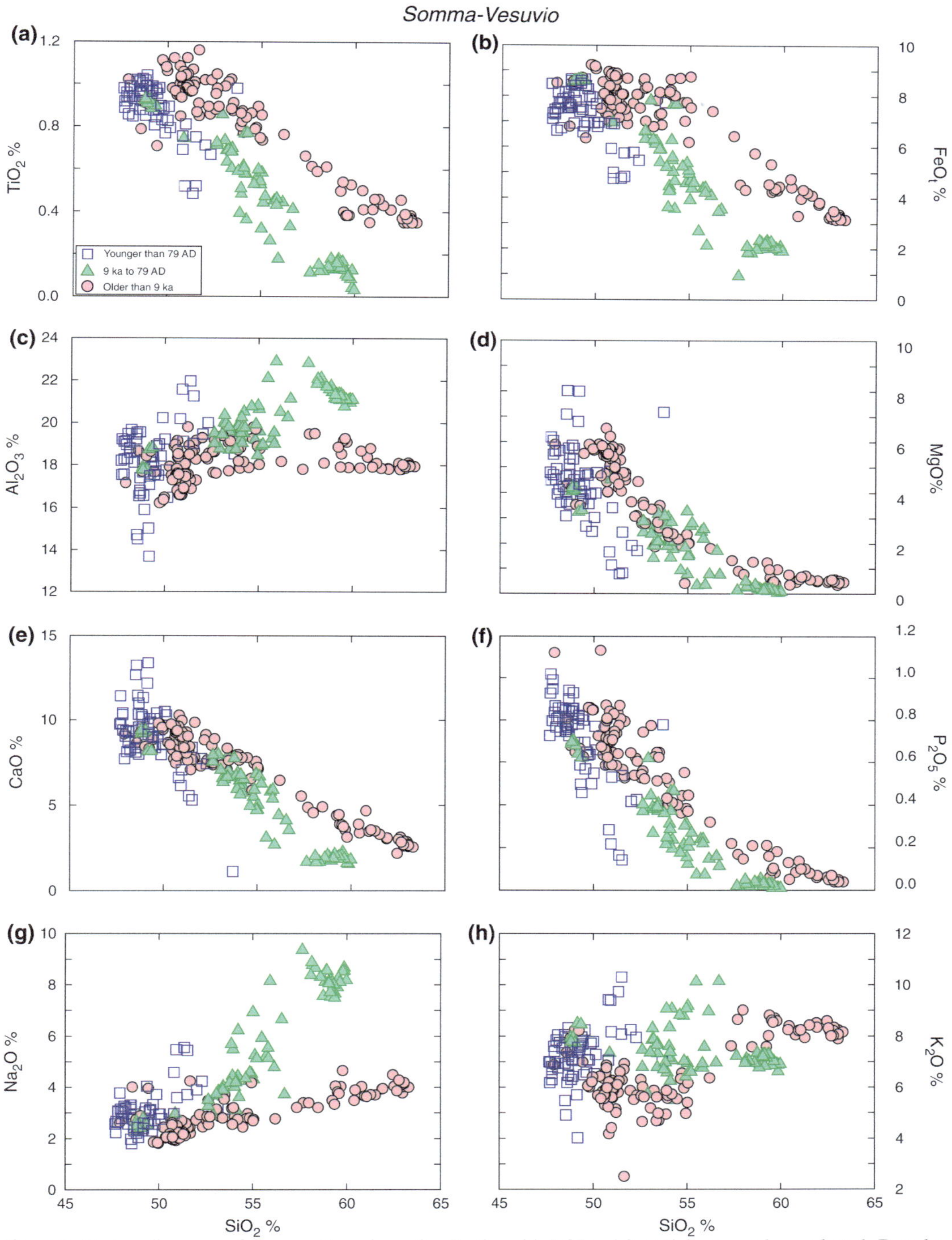

Fig. 7.6 Harker diagrams of Somma-Vesuvio rocks. Rocks with LOI > 4.0 wt% have not been plotted. Data have been normalised to 100 % on a LOI-free basis

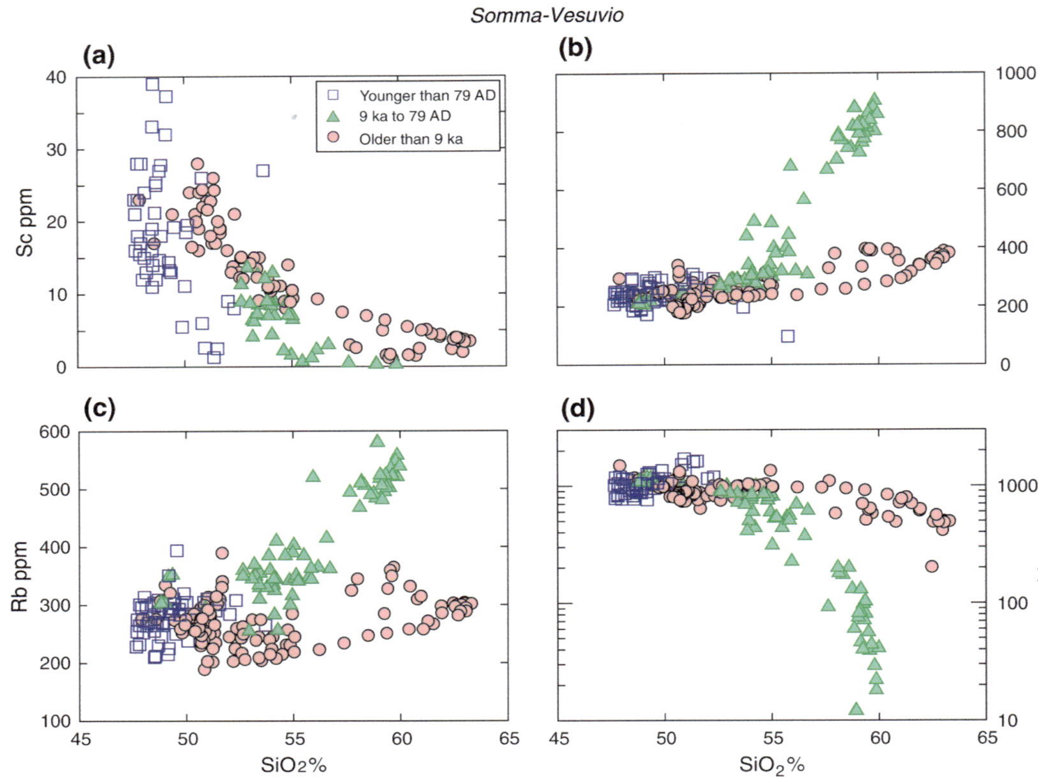

Fig. 7.7 Variation diagrams of selected trace elements for the Somma-Vesuvio rocks

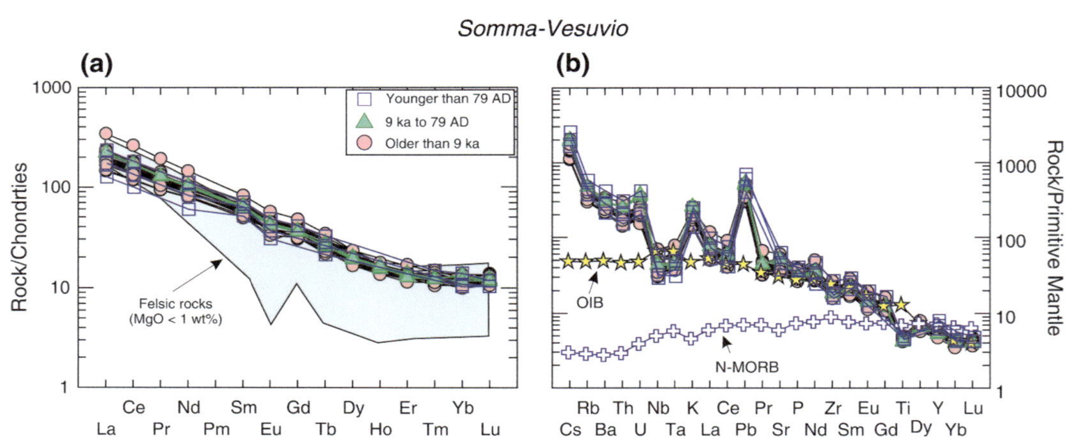

Fig. 7.8 **a** REE patterns of Somma-Vesuvio rocks. The *grey area* encloses felsic compositions; **b** Incompatible element patterns of mafic rocks normalised to primitive mantle composition. Patterns of average N-MORB and OIB (Sun and McDonough 1989; Gale et al. 2013) are also reported

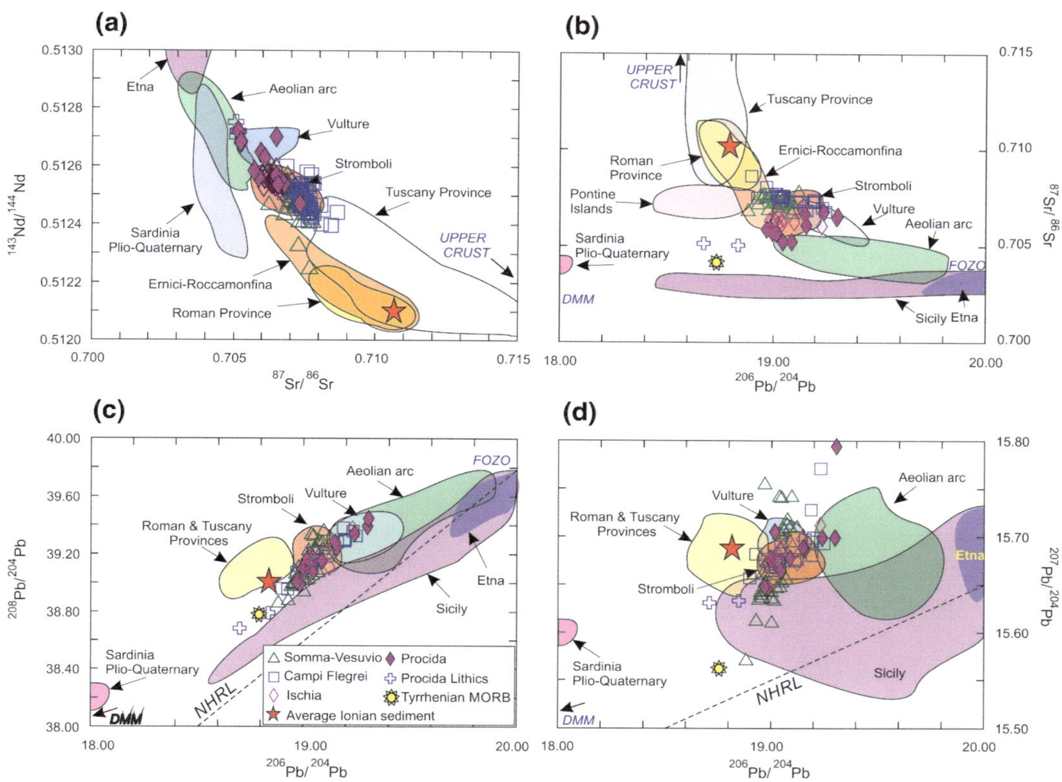

Fig. 7.9 Sr-Nd-Pb isotopic variation of Campania volcanoes. The composition of other Italian magmatic provinces and the Tyrrhenian MORB (ODP Site 655) are also shown. NHRL is the Northern Hemisphere Reference Line of Hart (1984). The *red star* is the average sediment from the Ionian Sea (Site 374; Klaver et al. 2015)

to the NBS SRM 951 standard (Di Renzo et al. 2007; Paone 2008).

7.3.4 Evolution of Somma-Vesuvio Magmas

Major and trace element variations suggest that fractional crystallisation was the main mechanism of magma evolution at Somma-Vesuvio. The various trends of major and trace elements have been interpreted to reveal either the presence of distinct types of parental magmas or an evolution from a single parent by separation of variable proportions of clinopyroxene and feldspar, possibly as a consequence of different conditions of magma ponding and crystallisation (Joron et al. 1987; Trigila and De Benedetti 1993). The occurrence of mafic melt inclusions in olivine and clinopyroxene exhibiting K-trachybasalt to K-tephrite compositions supports the former hypothesis (Marianelli et al. 1995; Lima et al. 2007). However, Pichavant et al. (2014) furnish compelling experimental evidence in favour of a single K-trachybasalt parent for all the Somma-Vesuvio magmas. According to these authors, assimilation of carbonate wall rocks favoured crystallisation of clinopyroxene, generating the highly undersaturated younger series (Iacono Marziano et al. 2008). Note that significant carbonate assimilation was demonstrated by oxygen isotope investigation by Dallai et al. (2011) for magmas of the last 4 ka. In contrast, processes of fractional crystallisation of clinopyroxene and feldspar, plus mixing and assimilation of Hercynian silicate rocks generated the poorly and mildly undersaturated older series. Di Renzo et al. (2007) also found that the older series evolved by AFC processes with assimilation of deep Hercynian basement rocks, whereas the younger Vesuvio series mostly underwent assimilation of carbonate rocks at shallower depths. The obvious implication of a single trachybasalt parental magma at Somma-Vesuvio is that ultrapotassic undersaturated mafic magmas of mantle origin are absent in Campania. This would represent additional evidence in favour of the hypothesis that the Campania volcanoes are genetically (and geodynamically) distinct from the Roman Province (Peccerillo 2002).

Trace element data support multiple magma evolution processes and allow understanding the role of single mineral phases during fractionation (Fig. 7.10). Positive trends of Sr and Ba versus Th and other incompatible elements for the strongly undersaturated younger series, favour pyroxene-dominated fractional crystallisation processes, since all these elements have low cpx/liquid partition coefficents. In contrast, the trends of older rocks highlight an important role of feldspar fractionation, which induced a sharp decrease in Sr and Ba with increasing Th and other incompatible elements (Fig. 7.10a, b). Variation diagrams of incompatible elements and La/Yb versus K_2O show that some rocks of the mildly undersaturated series (essentially Mercato and Avellino) define nearly vertical trends. These rocks are also strongly enriched in Na_2O. Most of these trends could be explained by fractional crystallisation dominated by separation of K-feldspar, which buffered potassium at a

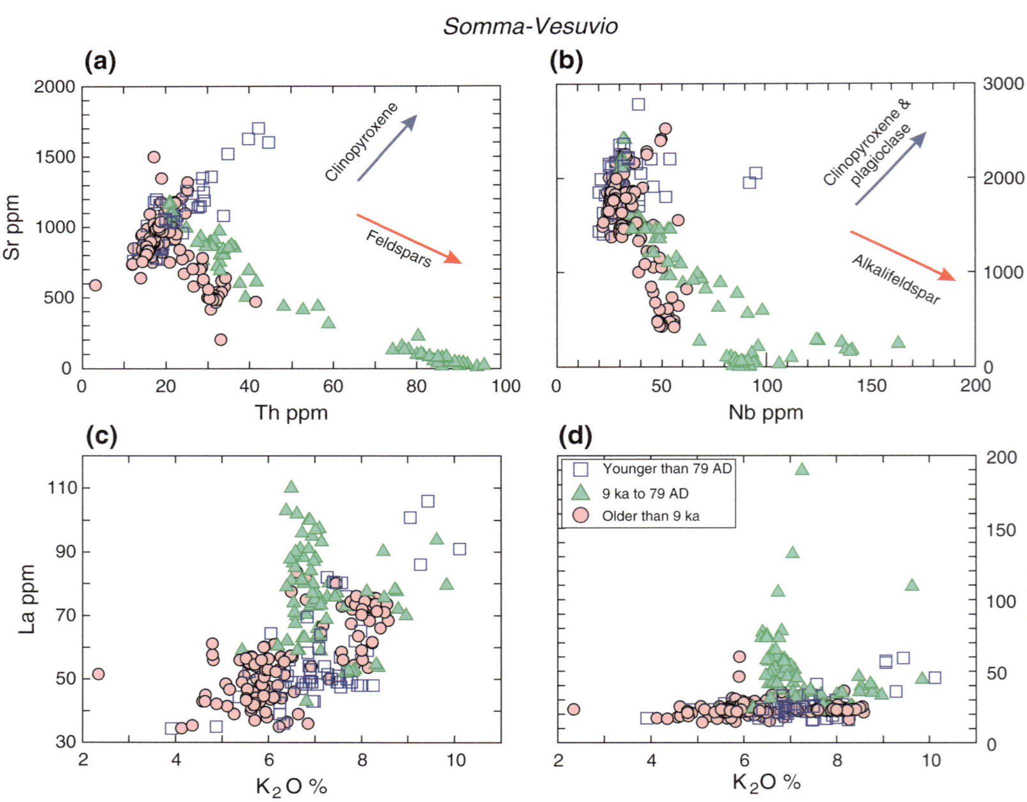

Fig. 7.10 Inter-element variation diagram for Somma-Vesuvio rocks. *Arrows* indicate qualitative element variation determined by fractionation of some major mineral phases

constant concentration, determining an increase in sodium and incompatible elements. However, the increase of La/Yb cannot depend on these factors and require some additional process such as fractionation of accessory garnet (Aulinas et al. 2008) or mobile element enrichments by gaseous transfer. The scattered Tb/Yb ratios and the positive correlation of F versus La/Yb and Na_2O seem the favour the second hypothesis.

Finally, a large number of data (e.g. compositionally zoned phenocrysts, contrasting types of clinopyroxenes coexisting in the same rocks, geochemical and isotopic variations within juvenile clasts from single eruption, mineral/host rock isotopic disequilibrium, etc.) indicates that mixing was a main evolutionary process at Somma-Vesuvio (e.g. Civetta et al. 1991a; Marianelli et al. 1999; Piochi et al. 2006a, b). The transition from pumices to scoriae from the base to the top of several pyroclastic fall deposits including the 79 AD eruption (e.g. Cioni et al. 1995, 2008 with references), indicates that magmas with variable compositions coexisted inside zoned reservoirs. Felsic and mafic magmas were tapped sequentially during various explosive events and underwent mixing before eruption.

7.4 Campi Flegrei (Phlegraean Fields)

The Campi Flegrei (Phlegraean Fields) is a multi-centre volcanic complex with two nested calderas sited northwest of Somma-Vesuvio, along the northern coast of the Gulf of Naples (Fig. 7.11). The volcanic area is characterised by strong tectonic subsidence at the intersection of Plio-Quaternary NE-SW and NW-SE trending fault systems. Rocks cropping out at the surface or cut by deep drillings consist of a complex

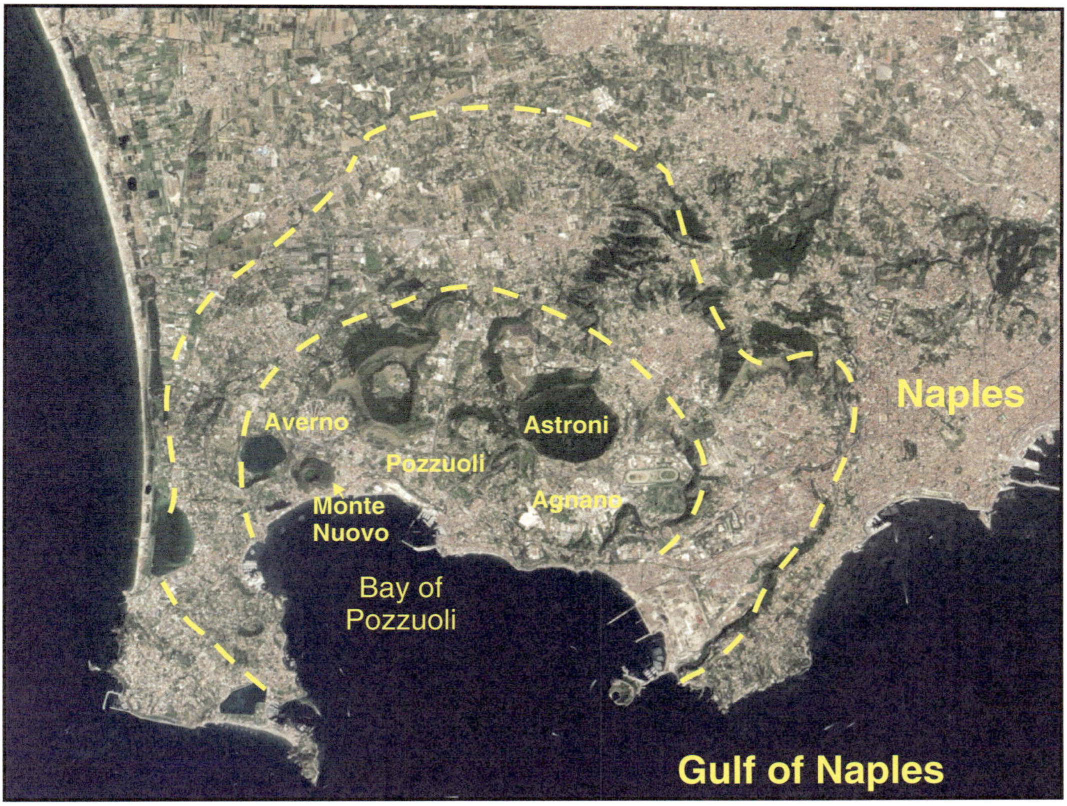

Fig. 7.11 Aerial view of Campi Flegrei, with the city of Naples. *Dashed lines* indicate approximate positions of caldera rims. Photo courtesy of NASA Earth Observatory

series of pyroclastic deposits and fossiliferous marine sediments overlying silicoclastic rocks and lavas that sit over thermometamorphic rocks occurring at a depth of about 3000 m. Several dikes and sills cut the sedimentary deposits (Piochi et al. 2014). Notably, no thick sequences of carbonate rocks seem to be present beneath Campi Flegrei. A 3 km deep drilling is currently in progress in the center of the calderas (Campi Flegrei Deep Drilling Project, CFDDP), aimed at collecting key information on the shallow plumbing structure of the volcano.

Volcanic rocks are dominantly pyrolastic, with largely subordinate lava flows and domes (Rittmann 1950; Rosi and Sbrana 1987). Campi Flegrei is well known for its spectacular fumarolic activity and for bradyseismic inflation and deflation of the ground that has been going on during the last 2000 years (Woo and Kilburn 2010). The best known evidence of soil uplift is observed on a building known as "Serapeo", where boreholes left by marine molluscs cover about one-half of three columns, testifying submersion below sea level for some time (Fig. 7.12).

7.4.1 Volcanology and Stratigraphy

The Campi Flegrei volcanic complex is formed by two nested resurgent calderas and by several monogenetic, mostly phreatomagmatic, centres (Fig. 7.11). The calderas are related to gigantic explosive eruptions that deposited the Campanian Ignimbrite and the Neapolitan Yellow Tuff (Orsi et al. 1995; Civetta et al. 1997). The Campanian Ignimbrite (CI) eruption (39–40 ka; Gebauer et al. 2014 and references therein) is a phreatoplinian event that outpoured large volumes of phonolitic-trachytic pyroclastics (from 54 to some 700 km^3 according to various estimates; see Arienzo et al. 2011; Scarpati et al. 2014 and references therein). It is one of the biggest eruptions in the Tyrrhenian Sea region, whose products have been found over wide areas

Fig. 7.12 Columns of the Temple of Serapis (later re-interpred as a market building) with lithodome boreholes testifying to a higher sea level at some stage in the past (Photo by F. Lucchi)

including the eastern Mediterranean Sea and Russia (Scarpati et al. 2014 with references). The Breccia Museo, a deposit of juvenile material (pumices, scoriae and obsidians) and a wide variety of intrusive, volcanic and sedimentary lithics, is believed to represent the proximal facies of Campanian Ignimbrite (e.g. Fedele et al. 2008; Gebauer et al. 2014). The Neapolitan Yellow Tuff eruption (NYT), discharged some 40 km^3 of latite to trachyte-phonolite pyroclastics about 15 ka before present, and was responsible for formation of an internal about 10 km across resurgent caldera partly occupied by the Bay of Pozzuoli (Orsi et al. 1995; Deino et al. 2004; Sacchi et al. 2014). The deposits of these eruptions rest over older pyroclastic products and a few lavas, for which $^{40}Ar/^{39}Ar$ datings yielded ages between 205 and 157 ka (Taurano outcrop, east of Naples; De Vivo et al. 2001) and between 78 and 40 ka (outcrops both outside and inside the city of Naples; D'Antonio et al. 2007; Scarpati et al. 2013).

The volcanism younger than 15 ka is concentrated inside the calderas and is characterised by several phreatomagmatic explosive events that generated a large number of craters and cones (Agnano, Solfatara, Averno etc.), which pockmark Campi Flegrei and represent a main morphological feature of the area (e.g. Orsi et al. 1996; Fedele et al. 2011; Santacroce et al. 2003; Tomlinson et al. 2012). The latest eruption formed the Monte Nuovo trachytic-phonolitic cone (Fig. 7.11). It took place in 1538 AD between September 29th and October 7th, after two years of local ground uplift, and increased seismicity and fumarolic gas emission (e.g. Piochi et al. 2005). A simplified chronostratigraphic scheme of the Campi Flegrei activity is reported in Fig. 7.4b, based on data by Cassignol and Gillot (1982), Civetta et al. (1991b), Di Vito et al. (1999), De Vivo et al. (2001), Pappalardo et al. (1999, 2002), Orsi et al. (2004), Isaia et al. (2009), Smith et al. (2011), Fedele et al. (2011), Scarpati et al. (2013).

After the Monte Nuovo eruption, the Campi Flegrei volcano has undergone a number of crises characterised by ground inflation, intense shallow seismicity, modification of gas temperature and chemistry, but no eruptions have occurred (e.g. Del Gaudio et al. 2010). The main unrest events in the past 50 years, took place in 1969–1972 and 1982–1984 and were characterised by intense shallow seismicity and ground uplift of 170–180 cm (e.g. Orsi et al. 1996). Successively, there was a period of subsidence, followed by new inflation between 2005 and 2014 with a maximum uplift of 23 cm. These unrest episodes have been related to pulses of fluids coming from deep magma bodies and to long-term heating of rocks, rather than to intrusion of magmas (e.g. Caliro et al. 2014: Chiodini et al. 2015). De Vivo et al. (2010) suggest that deep fluids entrapped beneath an impermeable horizon generate ground uplift, whereas deflation occurs when fluids find their way to the surface through fractures in the impermeable layer.

The magmatic system of the Campi Flegrei volcano is suggested to consist of a shallow 3–4 km deep trachytic reservoir, which has been periodically refilled by a mafic magma chamber sited at a depth of about 8 km, where a low-Vp and high-Vp/Vs layer has been detected (Zollo et al. 2008; De Siena et al. 2010; D'Antonio 2011). This reservoir is hosted in the Hercynian crystalline basement (e.g. Pappalardo et al. 2002) and is currently expanding, causing extension in the overlying brittle zone and, therefore, creating favourable conditions to the ascent of fluids, and possibly of magmas, at shallow levels (Carlino et al. 2015).

7.4.2 Petrography and Mineral Chemistry

The Campi Flegrei pyroclastics and lavas range in composition from shoshonite and latite to dominant trachyte and phonolite, with the absence of basaltic rocks (Fig. 7.13). A few phonolites are peralkaline (Armienti et al. 1983; Rosi and Sbrana 1987; Melluso et al. 2012). The rocks older than 39 ka and the Campanian Ignimbrite have compositions straddling the limit between trachyte and phonolite; the Neapolitan Yellow Tuff ranges from tephriphonolite-latite to trachyte and phonolite (Fig. 7.13; Orsi et al. 1995; Wohletz et al. 1995; Civetta et al. 1997; Pappalardo et al. 1999, 2002). Rocks younger

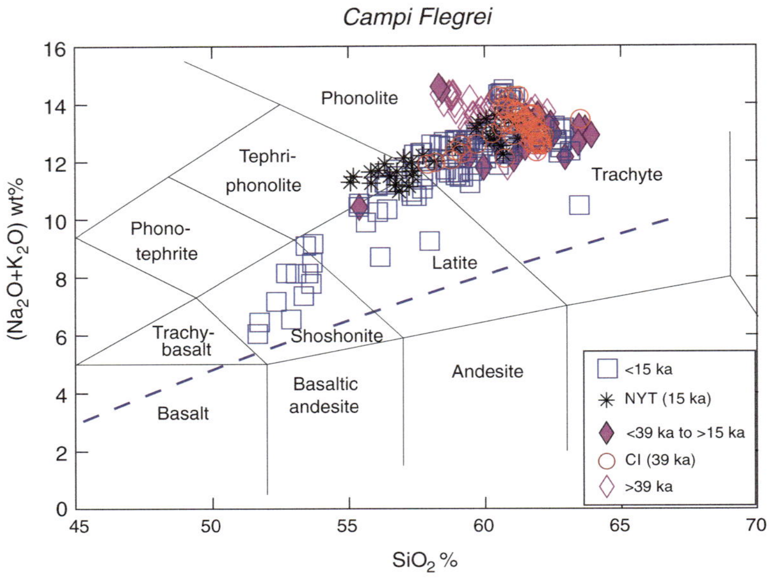

Fig. 7.13 TAS classification diagram for Campi Flegrei volcanic rocks. The *dashed line* is the divide between subalkaline and alkaline rocks. *CI* Campanian Ignimbrite; *NYT* Neapolitan Yellow Tuff. Compositions recalculated to 100 % on a LOI-free basis

than 15 ka cover a wider range of compositions, from shoshonite to trachyte and phonolite (D'Antonio et al. 1999a, b; D'Antonio 2005; Smith et al. 2011).

Rock textures range from porphyritic to subaphyric, with phenocryst contents decreasing from shoshonites to trachytes and phonolites (Armienti et al. 1983). Lavas generally display higher phenocryst contents than scoriae and pumices with similar major element composition (e.g. Piochi et al. 2008; Melluso et al. 2012). Latites contain clinopyroxene, plagioclase, some olivine and biotite phenocrysts, set in a hypocrystalline groundmass made of abundant sanidine, Fe–Ti oxides and glass. Trachytes and phonolites range from holocrystalline to glassy, with phenocrysts of alkali feldspar, minor clinopyroxene, biotite, plagioclase, Fe–Ti oxides and rare amphibole, nepheline and sodalite-group minerals. Accessory phases include apatite, britholite, zircon, spinel and titanite (Armienti et al. 1983; Rosi and Sbrana 1987; Pappalardo et al. 2002; Melluso et al. 2012). Leucite has been only observed in a few rocks (Astroni and Agnano Montespina eruptions; e.g. Tonarini et al. 2009).

Clinopyroxene ranges from colourless Mg-rich to green Fe-diopside and hedenbergite, with colour/chemical zoning often observed in single crystals (e.g. Pappalardo et al. 2002; Fowler et al. 2007). A complete variation from diopside to hedenbergite and aegirine is found in some lavas; these also contain phlogopite, aenigmatite and fayalite (Melluso et al. 2012). Olivine (up to Fo_{90}) is commonly altered to iddingsite. Plagioclase (about An_{90-20}) is generally zoned and is mantled by sanidine in the most evolved rocks (Armienti et al. 1983). Alkali feldspar has variable K_2O content, with Or_{30-90} (Fowler et al. 2007; Melluso et al. 2012).

7.4.3 Petrology and Geochemistry

The Campi Flegrei rocks show scattered distribution of many major and trace elements, especially among felsic rocks (Fig. 7.14). The mafic post-NYT rocks show more smooth variation trends.

REE patterns are fractionated, with negative Eu anomalies in the felsic rocks (Fig. 7.15a). Incompatible element patterns of mafic rocks normalised to primordial mantle composition (Fig. 7.15b) show Pb spikes and moderate negative anomalies of Ta and Nb.

Sr–Nd–Pb isotopic ratios (Fig. 7.9) are variable ($^{87}Sr/^{86}Sr \sim 0.7067$ to 0.7086; $^{143}Nd/^{144}Nd \sim 0.5124$ to 0.5126; $^{206}Pb/^{204}Pb \sim 18.90$ to 19.25;

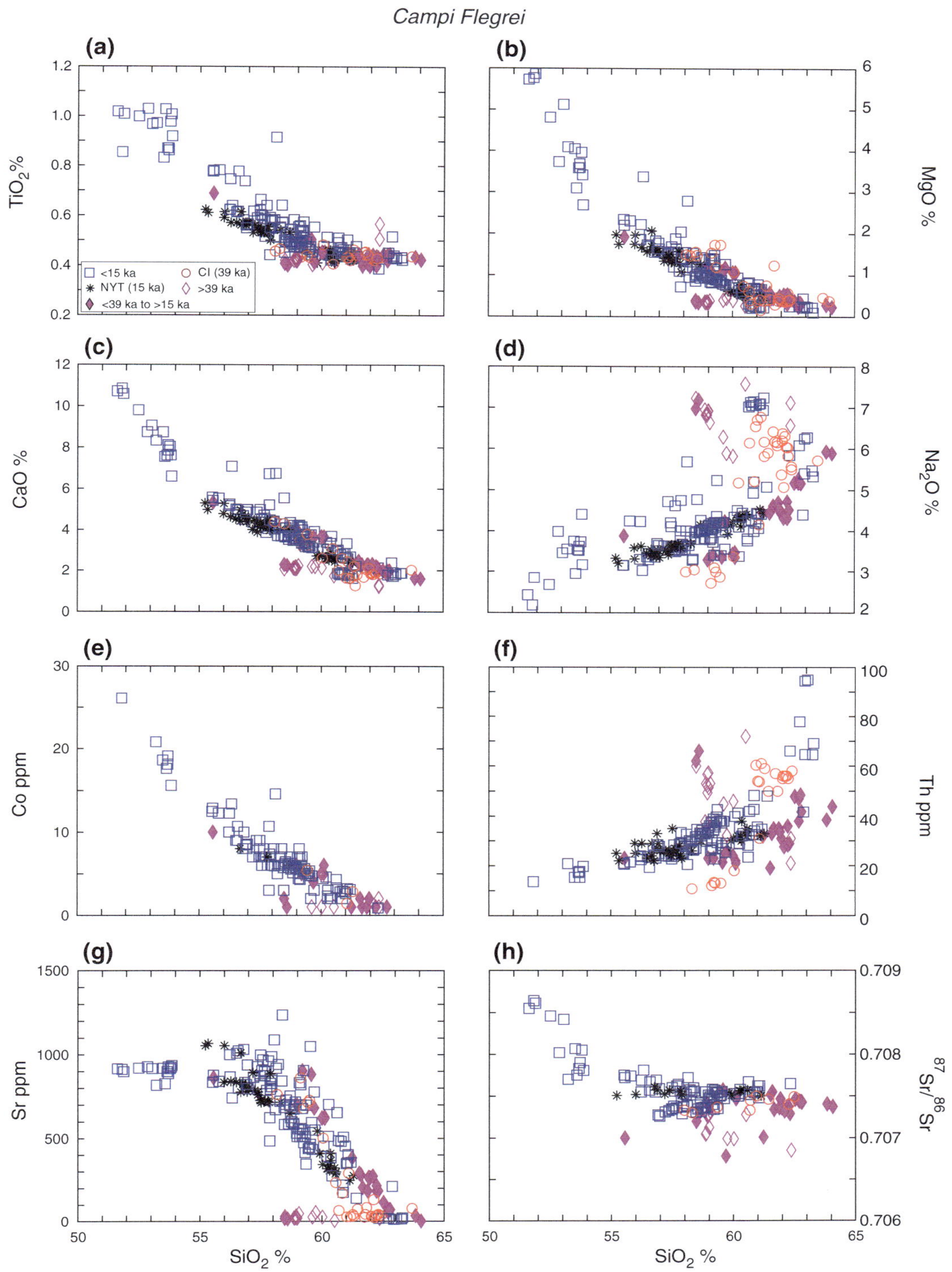

Fig. 7.14 Geochemical variation diagrams of Campi Flegrei rocks. Major elements recalculated to 100 % on a LOI-free basis

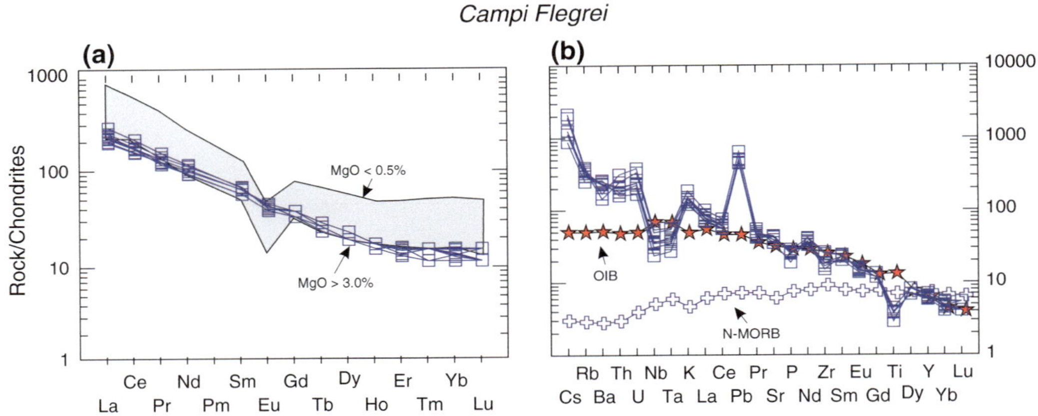

Fig. 7.15 **a** REE patterns of selected Campi Flegrei rocks. The *grey area* includes felsic compositions; **b** Incompatible element patterns for Campi Flegrei mafic rocks. Patterns of average N-MORB and OIB (Sun and McDonough 1989; Gale et al. 2013) are also shown

$^{207}Pb/^{204}Pb$ ~ 15.66 to 15.77; $^{208}Pb/^{204}Pb$ ~ 38.95 to 39.38; e.g. Vollmer 1976; Cortini and Hermes 1981; D'Antonio et al. 1996, 2007; De Vita et al. 1999; Pappalardo et al. 1999, 2002). Significantly higher $^{87}Sr/^{86}Sr$ (and lower Nd–Pb isotope ratios) is observed in the younger shoshonite-latites than in the phonolite-trachytes (Fig. 7.14h). Oxygen isotope ratios of clinopyroxenes range from $\delta^{18}O_{SMOW} = 6.6$ to 7.0 ‰, whereas values of alkali feldspars range from 8.1 to 8.5 ‰ (Pappalardo and Mastrolorenzo 2012). Boron isotopic compositions display $\delta^{11}B \sim -6.5$ to −10.6 ‰ (Tonarini et al. 2004, 2009; Di Renzo et al. 2011). He-isotope data reported by Martelli et al. (2004) for olivine and clinopyroxene show $R/R_A = 2.53$–3.02, slightly higher than at Vesuvio.

7.4.4 Evolution of Campi Flegrei Magmas

Scattered major and trace element distribution at Campi Flegrei reveals that magmas experienced a complex interplay of various evolution processes, including fractional crystallisation, mixing, and assimilation of wall rocks. These processes affected distinct batches of magmas, which were emplaced at different times beneath the volcano (e.g. Armienti et al. 1983). The poorly variable Sr-isotope ratios for some pyroclastic deposits such as CI and NYT (Fig. 7.14h) suggest scarce magma-wall rocks interaction for these eruptions. According to Gebauer et al. (2014), rocks crystallised along the walls of magma chamber(s) acted as a carapace that isolated the CI and NYT melts from host rocks, preventing assimilation. However, isotopic variations observed for the bulk of Campi Flegrei rocks clearly indicate an important role of crustal assimilation for late-emplaced latites and shoshonites (Pappalardo et al. 2002; Di Renzo et al. 2011; Gebauer et al. 2014). Pappalardo et al. (2002) suggested that crustal contamination was a result of the longer time spent by younger magmas in the deep reservoirs. However, if the post-NYT rocks are considered separately, no significant increase of Sr-isotope ratios with time is observed (e.g. Di Renzo et al. 2011). Alternatively, variable degrees of crustal contamination could depend on the distinct compositions of wall rocks that hosted the phonolite-trachyte and the shoshonite-latite magma chambers (e.g. D'Antonio et al. 2007; Arienzo et al. 2009; Tonarini et al. 2009; D'Antonio 2011; Melluso et al. 2012). Finally, the very significant increase of Sr- and a decrease of Nd–Pb isotopic ratios with increasing MgO (not shown) might suggests that variable degrees of contamination could be related to the degree of magma evolution, with the hotter mafic magmas being able to dissolve higher amounts of wall rocks than the cooler felsic ones. A similar process has been demonstrated to occur at several volcanoes, such as the Alicudi and Filicudi islands, in the Aeolian arc (see Chap. 9).

Finally, percolation of fluids through magma chambers also played a role in determining the composition of some magmas. Anomalous enrichments of some volatile elements (K, Sb, Cl and F) in felsic rocks could be related to gaseous transfer processes (Villemant 1988).

7.5 Ischia Island

The Island of Ischia (Fig. 7.16) is the remnant of a large volcano, located offshore the Campi Flegrei, at the intersection of NE-SW and NW-SE regional faults. The morphology of the island is dominated by the central volcano-tectonic horst of Monte Epomeo, by the Ischia graben in the northeast, and by numerous gravitationally collapsed areas (e.g. Rittmann 1930; Vezzoli 1988). Monte Epomeo is a tilted resurgent block, whose uplift of about 900 m in the last 33 ka has been related to intrusion of a trachytic magma body whose solidified roof is revealed by a high Vs layer at about 700 m below sea level (Carlino et al. 2012; Strollo et al. 2015). A caldera rim has been inferred from seismic and gravity data along the periphery of the island (Sbrana and Toccaceli 2011). Ischia is subjected to intensive diffuse CO_2 soil degassing (Inguaggiato et al. 2000) with emission of about 15 kg s^{-1} for the entire island (46 km^2; Pecoraino et al. 2005).

7.5.1 Volcanology and Stratigraphy

The Ischia Island was constructed during several phases of prevailingly explosive activity, volcano-tectonic collapses and erosion (e.g. Gillot et al. 1982; Poli et al. 1987; Civetta et al. 1991c; Brown et al. 2008, 2014). The lowest exposed rocks (>150 ka) consist of pyroclastic deposits and intercalated lava flows and paleosols. Younger rocks (about 150 to 75 ka) consist of pyroclastic deposits and lava domes cropping out along a semicircular structure, probably a

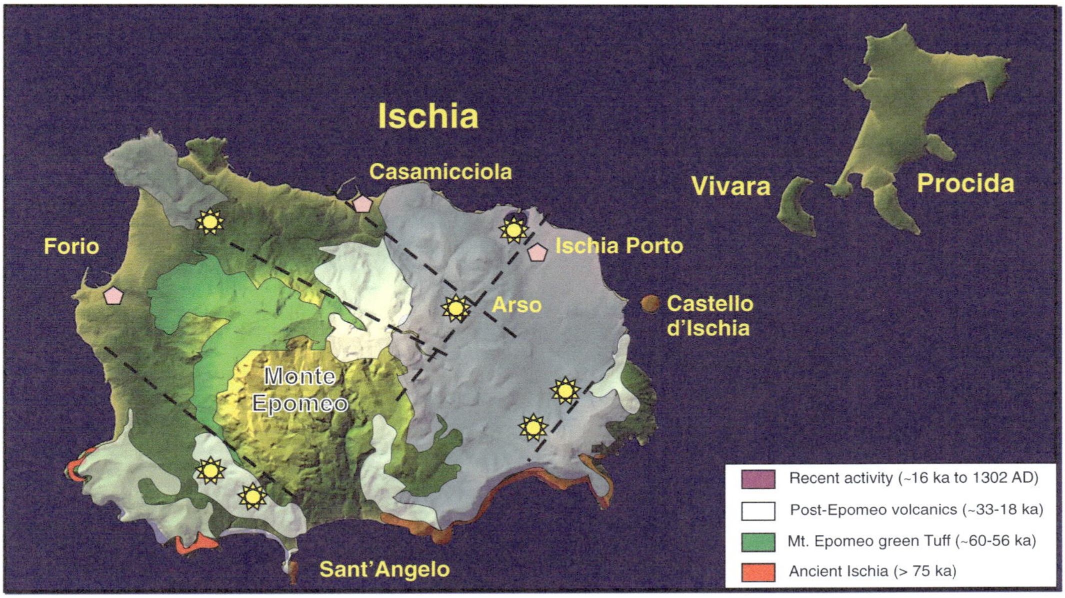

Fig. 7.16 Ischia, Procida and Vivara islands. Schematic rocks distribution of main eruptive periods at Ischia is simplified after the geological map in Sbrana and Toccaceli (2011) and Melluso et al. (2014). *Stars* are some Ischia volcanic centres

caldera rim. Eruption of pumice fall, block and ash flows, and ignimbrites took place between 75 and 70 ka. A main phase of activity occurred at 60–55 ka with the eruption the Monte Epomeo Green Tuff (MEGT) ignimbrite and the associated caldera collapse. This was followed by caldera resurgence and explosive and effusive eruptions at several centres. Volcanism ended with the emplacement of lavas and some pyroclastics from monogenetic centres along extensional faults of the Ischia graben (16 ka to 1302 AD). The latest eruption occurred in the Arso area, eastern Ischia, and emplaced mafic lavas and scoriae.

7.5.2 Petrography and Mineral Chemistry

The Ischia rocks consist of dominant trachytes, minor phonolites and latites, and rare shoshonites (Poli et al. 1987; Civetta et al. 1991c; Brown et al. 2008, 2014). Basaltic-trachybasaltic rocks are absent (Fig. 7.17).

Textures are generally porphyritic with variable relative amounts of alkali feldspar, plagioclase, clinopyroxene, biotite, and Fe–Ti oxides phenocrysts. Groundmass ranges from microcrystalline to holohyaline, with glass contents increasing from intermediate to felsic rocks (Di Girolamo et al. 1995; Piochi et al. 1999 (Rittmann 1930; Chiesa and Poli 1988; Crisci et al. 1989; D'Antonio et al. 2013). Olivine ($\sim Fo_{90-75}$) with inclusions of chromiferous spinel is present as phenocryst and in the groundmass of shoshonites and sometimes shows kinking; zoned Mg–Fe-diopside is a main phenocryst of the intermediate rocks. Plagioclase/alkali feldspar ratio decreases from intermediate to felsic rocks and both phases show large compositional variation (An_{95-25}; Or_{74-10}). Sodalite-group minerals occur in some trachytes and phonolites. Apatite and titanite are common accessories; aenigmatite, amphibole, aegirine, and Zr-Ca–Na-REE-F silicates (rinkite, kochite, hainite, hiortdahlite, etc.) have been found in the groundmass of peralkaline trachyphonolites (Melluso et al. 2014).

7.5.3 Petrology and Geochemistry

The Ischia rocks have slightly higher Na_2O/K_2O ratio than Campi Flegrei, indicating a weak sodic

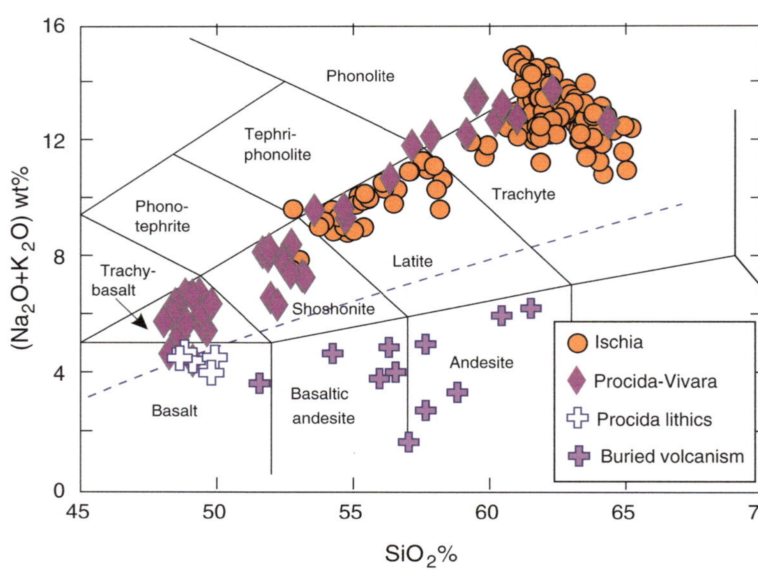

Fig. 7.17 TAS classification diagram for Ischia, Procida-Vivara, and for volcanism buried beneath the Campanian Plain. The *dashed line* is the divide between alkaline and subalkaline rocks (Irvine and Baragar 1971). Data have been recalculated to 100 % on a LOI-free basis

affinity; some trachyphonolites show peralkaline compositions (Melluso et al. 2014). Felsic rocks show scattering for several major and trace elements (Fig. 7.18), but some inter-element diagrams (not shown) exhibit rather smooth trends.

REE patterns are variably fractionated, with negative Eu anomalies in the evolved rocks (Fig. 7.19a). As for other Campania mafic volcanics, incompatible element patterns of least evolved rocks exhibit moderate negative anomalies of HFSE, no Ba anomalies and positive spikes of Pb (Fig. 7.19b). Sr isotopic ratios range from about 0.7060–0.7070 (Fig. 7.18h), and disequilibrium among coexisting minerals has been observed for some rocks (e.g. Civetta et al. 1991c; Brown et al. 2008, 2014). $^{143}Nd/^{144}Nd$ ranges from 0.5125 to 0.5126. Pb isotopic ratios are close to those of other Campania volcanoes ($^{206}Pb/^{204}Pb \sim 18.98$ 19.23; $^{207}Pb/^{204}Pb \sim 15.65$ 15.71; $^{208}Pb/^{204}Pb \sim 39.05$ 39.37; Vezzoli 1988; D'Antonio et al. 2007). Oxygen isotopic data on olivine, clinopyroxenes and alkali feldspars range from $\delta^{18}O_{SMOW} \sim +5.5$ to +7.7 (Turi et al. 1991; D'Antonio et al. 2013). Boron isotope ratios show $\delta^{11}B = -2.8$ to -9.5 ‰ (Tonarini et al. 2004; D'Antonio et al. 2007). He-isotope ratios for olivine and clinopyroxene ($R/R_A = 3.23$–3.45) show high values relative to Flegrei and Somma-Vesuvio (Martelli et al. 2004).

7.5.4 Evolution of Ischia Magmas

The Ischia rocks mostly represent evolved compositions from mafic potassic parents. However, primitive rocks are absent at Ischia, and melt inclusion data indicate maximum MgO contents around 4.0 wt% (Moretti et al. 2013). Magma evolution was dominated by fractional crystallisation, with separation of mafic minerals and plagioclase in the intermediate compositions and alkali feldspars in the felsic magmas. Fractional crystallisation was accompanied by mixing among compositionally different but comagmatic melts (Melluso et al. 2014). Based on melt-inclusion data, Moretti et al. (2013) envisage a major crystallisation zone at about 8–10 km depth, as suggested for other Campania volcanoes. CO_2 had a major role during magma evolution and was probably responsible for fluid mobile element enrichments during differentiation. Plots of rock compositions against time have shown replicating element variation trends for different eruptive phases, indicating repeated episodes of magma ponding and crystallisation, interrupted by magma mixing and eruption (Poli et al. 1987; Civetta et al. 1991c). Mixing between compositionally different types of magmas is also supported by isotopic disequilibria observed among coexisting phases in single rocks. Isotope data also highlight interaction between magmas and wall rocks (D'Antonio et al. 2013).

7.6 Procida and Vivara

The Island of Procida and the nearby islet of Vivara are sited between Campi Flegrei and Ischia (Figs. 7.1 and 7.16). Procida consists of a number of coalescing explosive monogenetic cones (Terra Murata, Pozzo Vecchio, Fiumicello, Solchiaro), whereas Vivara is the remnant of a tuff cone. A few of the Procida-Vivara rocks have been dated, and ages of volcanic activity from about 70 to 14 ka has been inferred essentially from stratigraphic relationships with deposits of Campi Flegrei and Ischia eruptions, and ^{14}C dating of paleosols (e.g. D'Antonio and Di Girolamo 1994; D'Antonio et al. 1999a; De Astis et al. 2004; Fedele et al. 2006).

7.6.1 Volcanology and Stratigraphy

The material erupted at Procida and Vivara consists of scoriae, hyaloclastites, pumices and lithics, which are interfingered with and sometimes hardly distinguishable from pyroclastic deposits from Ischia and Campi Flegrei. Basaltic lava lithics, notably from the Solchiaro centre, represent an older and compositionally distinct cycle of volcanic activity that is not exposed at the surface.

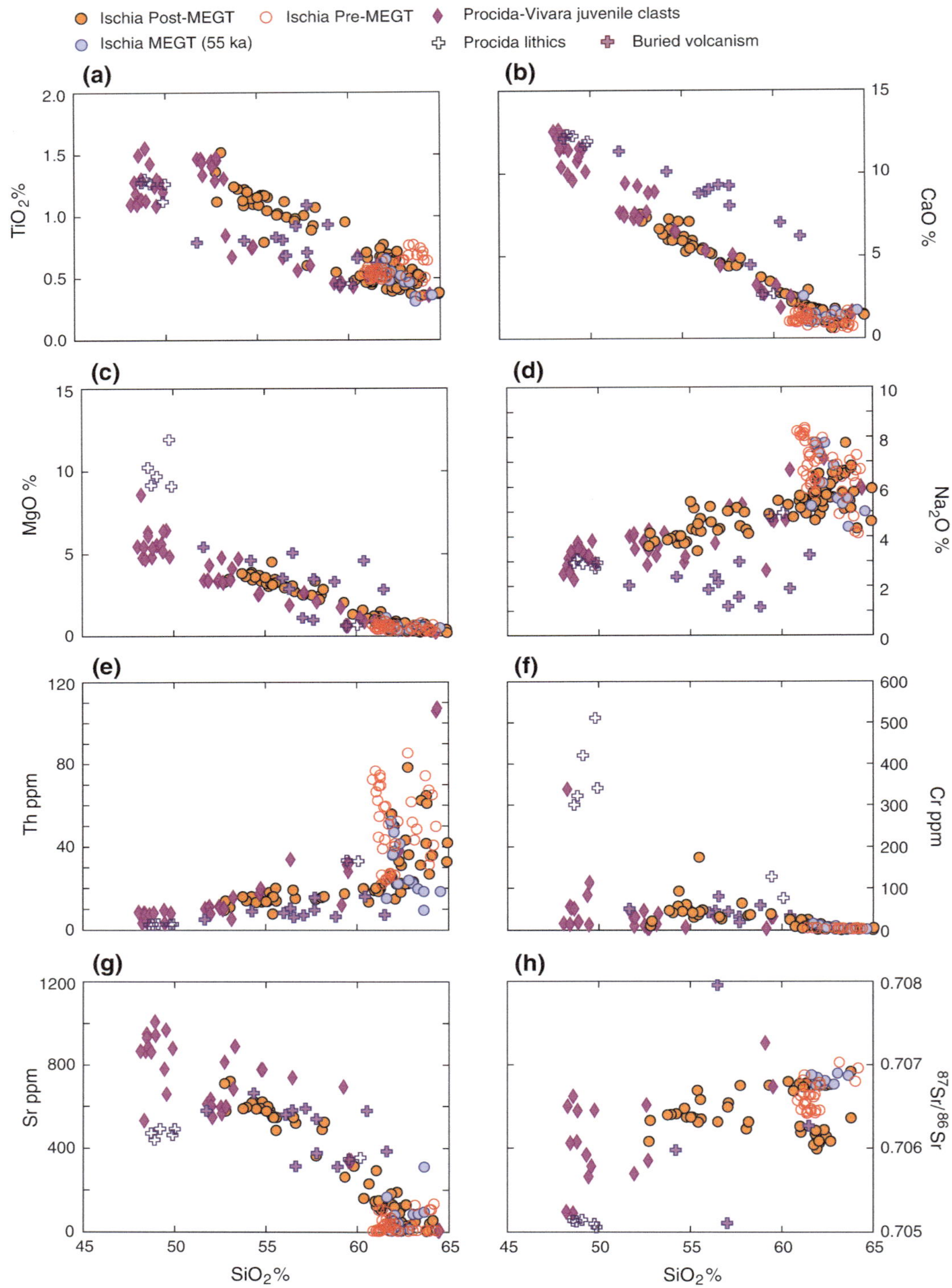

Fig. 7.18 Geochemical variation diagrams for Ischia, Procida-Vivara, and the buried volcanism beneath the Campanian Plain. MEGT is Monte Epomeo Green Tuff. Major elements have been normalised to 100 % on a LOI-free basis

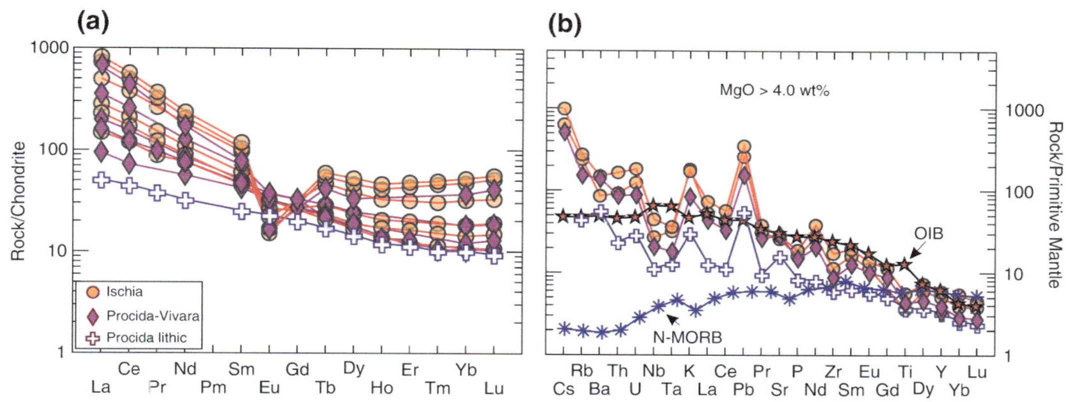

Fig. 7.19 **a** REE patterns of representative Ischia and Procida-Vivara rocks; **b** Incompatible element patterns of mafic rocks. OIB and MORB are reported for comparison (Sun and McDonough 1989; Gale et al. 2013)

7.6.2 Petrography and Mineral Chemistry

The Procida-Vivara juvenile pyroclasts are all potassic alkaline and range from trachybasalt to trachyte (Fig. 7.17). Rock textures are porphyritic and variably vesicular. Mafic scoriae contain phenocrysts of diopside, plagioclase ($\sim An_{90-80}$) and some olivine ($\sim Fo_{90-80}$), set in a matrix made up of the same phases plus Ti-magnetite, rare alkali feldspar and glass. Chromite inclusions are sometimes observed in olivine. Trachytes consist of dominant sanidine phenocrysts, with minor green clinopyroxene and andesine plagioclase (Di Girolamo and Stanzione 1973; D'Antonio and Di Girolamo 1994; De Astis et al. 2004, 2006; Fedele et al. 2006).

Lithic clasts are porphyritic with phenocrysts of Mg-rich olivine (Fo_{87-80}) and diopside, set in a matrix containing plagioclase, Ti-magnetite, rare alkali feldspar and glass. Some olivine crystals contain inclusions of Mg–Cr-spinel.

7.6.3 Petrology and Geochemistry

The Procida-Vivara juvenile pyroclasts show decreasing abundances of TiO_2, MgO, FeO_t, ferromagnesian trace elements and Sr, and increasing K_2O, Na_2O and incompatible elements with silica (Fig. 7.18). REE and incompatible element patterns are similar to Ischia (Fig. 7.19). Sr–Nd isotope ratios show more primitive compositions than other Campania volcanics ($^{87}Sr/^{86}Sr = 0.7052$ to 0.70733; $^{143}Nd/^{144}Nd = 0.5125$ to 0.5127), whereas Pb isotope ratios mostly overlap ($^{206}Pb/^{204}Pb \sim 18.97$ to 19.31; $^{207}Pb/^{204}Pb \sim 15.65$ to 15.80; $^{208}Pb/^{204}Pb \sim 38.99$ to 39.45). Oxygen isotopic ratios of clinopyroxenes and alkali feldspars increase from about $\delta^{18}O_{SMOW} = +5.5$ ‰ to $+8.1$ ‰ passing from basalts to felsic rocks (Turi et al. 1991). $\delta^{11}B$‰ is in the range -5.29 to -8.89 (Tonarini et al. 2004). He-isotope ratios for olivine show the highest values ($R/R_A = 4.76$–5.21) in the Campania Province (Martelli et al. 2004).

Lithic lava clasts have MgO-rich composition and represent the most primitive rocks in the Campania Province. They show a broadly calcalkaline affinity, but are undersaturated in silica (Fig. 7.2) and have been classified as K-basalts (e.g. Mazzeo et al. 2014). Ni and Cr are high, nearly in the range of mantle-equilibrated melts. REE and incompatible element patterns are fractionated and not much different from the associated trachybasaltic scoriae, although incompatible element abundances are lower in the lava lithics (Figs. 7.18 and 7.19). Sr-isotope ratios are the lowest and Nd isotope ratios are the highest observed in the Campania Province ($^{87}Sr/^{86}Sr \sim 0.7051$; $^{143}Nd/^{144}Nd \sim 0.5127$).

Pb isotope ratios show relatively low $^{206}Pb/^{204}Pb$ ratios, which are shifted towards Plio-Quaternary EM1-type OIB from Sardinia. $\delta^{11}B$‰ have values of −3.6 and −4.8 (Tonarini et al. 2004).

7.6.4 Evolution of Procida and Vivara Magmas

Major element trends suggest fractional crystallisation as a main evolutionary process at Procida-Vivara; variable isotopic compositions and relatively high $^{87}Sr/^{86}Sr$ in some evolved samples indicate assimilation of wall rocks during evolution (D'Antonio et al. 1999a). Overall, juvenile scoriae show primitive compositions (high MgO, Ni, Cr, etc.) that testify to rapid ascent of mafic magma from source to the surface. Relatively low Sr- and high Nd-isotope ratios of these rocks also reveal a less enriched source at Procida-Vivara than at other Campania volcanoes.

7.7 Buried Volcanism Beneath the Campanian Plain

The onset of volcanic activity in Campania is much older than indicated by rocks cropping out at the surface. Evidence for ancient volcanism is provided by deep borehole drillings in the Vesuvio area and in the Campanian Plain, as well as by lithics found at Procida.

Deep borehole drilling south of Vesuvio (Trecase well, Fig. 7.3) reached the sedimentary bedrocks after crossing several levels of tuffs and lavas ranging in composition from leucite tephrite to phonolite. The deepest recovered volcanic products are 0.4 Ma old leucite tephritic lavas that testify to ancient ultrapotassic magmatism in the Somma-Vesuvio area.

Drilling in the Campanian Plain, at Parete and Castel Volturno (Fig. 7.1), north of the Campi Flegrei caldera, encountered a thick sequence of calcalkaline rocks underlying the Campi Flegrei volcanic suites (Di Girolamo et al. 1976; Barbieri et al. 1979; Albini et al. 1980). At Parete, a huge complex of calcalkaline basalts to andesites was found at depths of 250 to 1900 m. A single K/Ar age measurement on the deepest drilled rock yielded a value of 2 ± 0.4 (Barbieri et al. 1979). At Castel Volturno similar rocks were drilled at a depth of about 2000 m, intercalated with Pleistocene marine sediments. Buried volcanics show porphyritic textures with phenocrysts of zoned plagioclase, clino- and orthopyroxene and sparse biotite set in a groundmass consisting of the same phases plus Fe–Ti oxides. They are oversaturated in silica and show moderate enrichments in incompatible elements. Sr isotopic ratio varies considerably and partially overlaps values of the Campania Province ($^{87}Sr/^{86}Sr$ = 0.7051–0.7081). REE are fractionated with a flat HREE pattern (Barbieri et al. 1979; Albini et al. 1980).

The importance of buried volcanism beneath the Campanian Plain is high because it testifies to a temporal transition from calcalkaline to potassic-ultrapotassic magmatism in the Campania area. A similar evolution has been observed at Stromboli, where early calcalkaline activity has been followed by shoshonitic and potassic alkaline magmatism (see Chap. 9).

7.8 Petrogenesis of Campania Magmas

The Campania Province is composed of a wide variety of rocks, which range from mafic to felsic and from mildly to strongly undersaturated in silica, the latter being restricted to Somma-Vesuvio. Geochemical and petrological data suggest that complex evolutionary processes, starting from trachybasaltic parents, generated the volcanic suites at Campi Flegrei, Ischia and Procida-Vivara. Geochemical and experimental petrology evidence suggests that a trachybasalt magma was also parental to all the Somma-Vesuvio rocks (e.g. Mormone et al. 2011; Pichavant et al. 2014). Therefore, experimental and geochemical data fit the hypothesis that all the Campania rocks derived from broadly similar mildly alkaline potassic parents, and that compositional variations at various centres are related to different roles of evolutionary processes, such as polybaric fractional crystallisation, absolute amounts and relative proportions of assimilated carbonate and silicate wall rocks, and mixing. Gaseous flux, especially CO_2, may have been an additional factor of magma modification in some centres (e.g. Villemant 1988; Mormone et al. 2011; Moretti et al. 2013 with references). MgO versus $^{87}Sr/^{86}Sr$ relationships best highlight contrasting styles of magma evolution in the Campania volcanoes (Fig. 7.20a). The negative trend of Procida-Vivara is typical of AFC processes, whereas the positive correlation of Campi Flegrei suggests higher degrees of assimilation for mafic than for felsic magmas. In contrast, scattering at Vesuvio and Ischia suggest an irregular interaction between magmas and wall rocks. Variable composition of the substratum that host magma chambers is a main factor of contrasting evolutionary processes. In particular, the occurrence of thick sequences of carbonate rocks beneath Somma-Vesuvio favoured the formation of strongly undersaturated leucite-bearing suites by magma-wall rock interaction (e.g. Iacono Marziano et al. 2008; Pichavant et al. 2014). In contrast, assimilation of Hercynian basement at other centres, especially Campi Flegrei, yielded significant radiogenic isotope variation but did not lead to undersaturation in silica.

Incompatible element patterns show significant similarities among trachybasalts from various centres. However, there are remarkable variations of incompatible element abundance and radiogenic isotope compositions among various volcanoes, with a general decrease of incompatible element contents and Sr-isotope ratios from Somma-Vesuvio and Campi Flegrei to Procida-Vivara and Ischia. Such a trend becomes more cogent if the Procida lithics are also considered. Overall, these data suggest that the primary magmas in the Campania Province were generated in mantle sources that underwent modification by a single type of metasomatic agent, but the western sector was less intensely modified than Campi Flegrei and Somma-Vesuvio.

Morris et al. (1993) suggested that the Be and B contents and Be/B ratios of Somma-Vesuvio are different from those found in arc rocks, arguing against a subduction-related enrichment for mantle sources. Such a hypothesis was shared by other authors (e.g. Ayuso et al. 1998). However, all the rocks from Vesuvio and other Campania volcanoes show island arc geochemical signatures given by negative anomalies of HFSE in their incompatible element patterns. This supports a subduction-related origin for primary magmas (Di Girolamo 1978; Serri 1990; Beccaluva et al. 1991; Peccerillo 1999, 2001; De Astis et al. 2000). In this view, trace element and radiogenic isotope variations of mafic rocks in Campania reflect variable degree of modification by a single type of subduction-related fluid, which imparted homogenous trace element

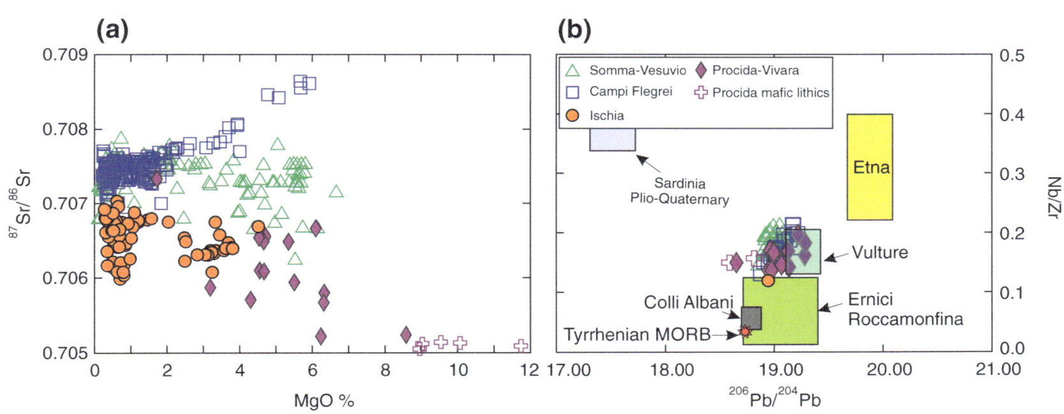

Fig. 7.20 **a** Variation diagram of $^{87}Sr/^{86}Sr$ versus MgO% for Campania volcanics; **b** Nb/Zr versus $^{206}Pb/^{204}Pb$ diagram for the Campania mafic rocks (MgO > 3.5 %), compared with other volcanoes from central-southern Italy

patterns to the magma sources, but also generated heterogeneities in the degree of trace element and radiogenic isotope enrichments.

The Campania rocks exhibit relatively higher contents of HFSE, especially Ta and Nb, and lower LILE/HFSE ratios than other mafic rocks in Central Italy. Since HFSE are relatively immobile during arc metasomatism (e.g. Kessel et al. 2005), the observed abundances likely reflect the composition of pre-metasomatic mantle rocks. High contents in Ta and Nb, relative to other incompatible elements are typical of OIB rocks (e.g. Sun and McDonough 1989), and it has been suggested that OIB-type components have been involved in the origin of the Campania Province (e.g. Serri 1990; Beccaluva et al. 1991; De Astis et al. 2006; D'Antonio et al. 2007; Mazzeo et al. 2014). This conclusion is supported by HFSE/HFSE ratios of Campania volcanoes (e.g. Nb/Zr), which are closer to Etna OIBs, compared with other Central Italy volcanoes (Fig. 7.20b). In contrast, values of Roman and Ernici-Roccamonfina province plot with the Tyrrhenian MORB, indicating distinct and more depleted pre-metasomatic mantle sources (see Chaps. 4 and 5).

Campi Flegrei, Somma-Vesuvio and Ischia have Sr–Nd–Pb isotope compositions that plot along a trend connecting OIB rocks of Etna (Sicily) and the upper continental crust (Fig. 7.9). Such a relationship agrees with a role of OIB-type mantle, but also requires the involvement of the upper crustal components in the genesis of the Campania Province. As discussed for other Central Italy volcanoes, mantle-crust interaction can be accomplished by both introduction of sediments into the mantle wedge (source contamination) and wall rocks assimilation during magma ascent to the surface (magma contamination). However, it has been amply demonstrated that isotopic compositions as those of the Campania rocks cannot derive entirely from magma contamination, and are mostly related to mantle-upper crust mixing. Mantle contamination by upper crustal rocks, particularly subducted sediments, is strongly supported by high B and low $\delta^{11}B$‰ of Campania volcanics (e.g. Tonarini et al. 2004, 2009). Sr–Nd–He–B isotope signatures are much closer to mantle end-member than observed for other volcanic rocks in Central Italy. This indicates a lesser role for crustal contamination in Campania. Geochemical modelling suggests that fluids originated from an oceanic-type slab, plus a contribution by 2–4 % sediments decreasing from Vesuvio-Campi Flegrei to Procida-Ischia, are able to explain the whole range of isotopic compositions of mafic rocks from the Campania Province (D'Antonio et al. 2007, 2013; Mazzeo et al. 2014).

When plotted on Pb-Sr diagram, the Procida lithics depart from the main trend between Etna and upper crust, pointing towards low $^{206}Pb/^{204}Pb$ compositions (Fig. 7.9b). A similar trend, but at higher Sr isotope values, is observed at Ventotene. These trends require a role for a mantle component with low $^{206}Pb/^{204}Pb$, such as that of the Sardinia Plio-Quaternary magmatism (e.g. Gasperini et al. 2000, 2002; Lustrino et al. 2000). Therefore, mantle material coming from the west could be involved in the evolution of mantle sources of the Procida magmas. Such a process could be related to the particular position of Procida-Vivara volcanoes on the edge of the Tyrrhenian Sea basin.

7.9 Geodynamic Significance

The hypothesis of an OIB-type mantle source contaminated by subduction-related fluids raises the problem of the origin of these two components and the way they interacted in the mantle wedge. A particularly important point to consider for discussion is that the Campania volcanoes show close similarities to shoshonitic and potassic rocks from Stromboli, for both incompatible element and radiogenic isotope ratios (Peccerillo 2001). Such an affinity is best shown by comparing incompatible element patterns of mafic potassic rocks (Fig. 7.21). For clarity, only one sample from each volcano is shown in this diagram, but the statistical significance of these

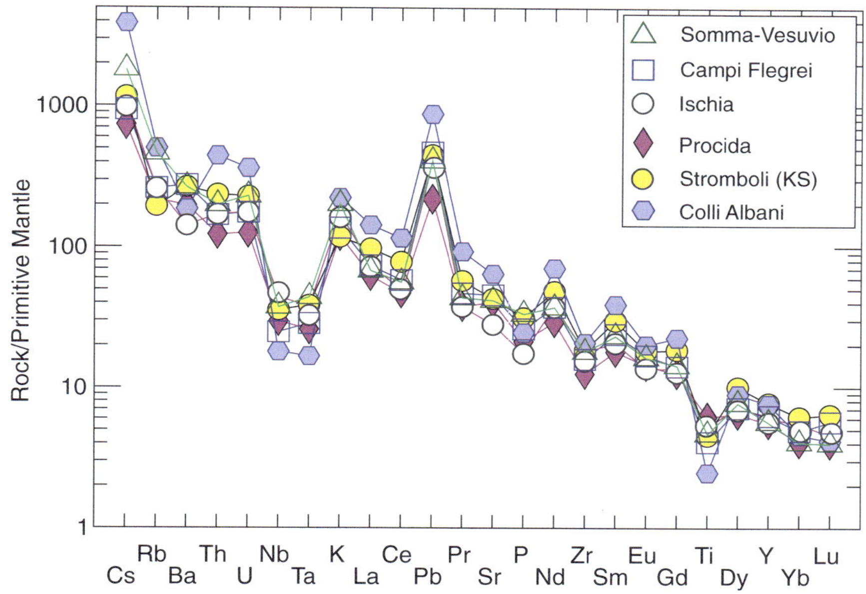

Fig. 7.21 Incompatible element patterns of representative Campania mafic rocks, Stromboli potassic basalts (KS), and Colli Albani tephrites

selected samples can be verified by inspection of more spiderdiagrams in this Chapter and in Chap. 9.

It is evident that the Stromboli potassic alkaline basalt has an almost identical pattern to samples from Somma-Vesuvio, Campi Flegrei and Ischia, and that these are different from Colli Albani or other Roman volcanoes, as amply discussed earlier. These data, along with radiogenic isotopes, clearly suggest that magmas from Stromboli and Campania are genetically related and were formed in the same type of source, which had a similar evolution i.e. an interaction between OIB mantle material and slab-derived fluids.

The Stromboli volcano is sited in the eastern Aeolian arc, a region characterised by a narrow seismic Benioff-Wadati zone related to the subduction of the Ionian oceanic plate beneath the southern Tyrrhenian Sea (e.g. Orecchio et al. 2014 and references therein). The seismic plane extends north-westward reaching the southern limb of the Campania Province where a few earthquake foci at a depth of about 350 have been registered. The close compositional similarity between Campania volcanoes and Stromboli, therefore, fits well a mantle evolution scenario in which a single type of OIB-type peridotitic mantle was contaminated by the same type of subducted material released from the Ionian slab (Peccerillo 2001).

Campania volcanoes are closer to the Apulian than to the Ionian foreland, and an origin of fluids from the Apulia subduction zone has been suggested (Chiarabba et al. 2008; Moretti et al. 2013). However, in such a case, distinct compositions would be expected for the Campania and Stromboli magmas.

Therefore, the geodynamic model that better explains the compositional characteristics of the Campania Province and its similarity with Stromboli is that the source contamination for all these volcanoes is related to fluids coming from the subducted Ionian plate and associated sediments. According to this model (Fig. 7.22a) a single continuous slab was initially subducting beneath the southern Apennines and the eastern Aeolian arc, from Apulia to the Ionian Sea area. Collision of the Apulia block was followed by an along-strike tear in the slab (Fig. 7.22b), which propagated southward, according to a mechanism of slab tear-off as described by Wortel and Spakman (2000) and Spakman and Wortel

(2004). The southern segment of the slab remained attached to the Ionian foreland, retreating south-eastward (e.g. Gvirtzman and Nur 1999; Rosenbaum and Lister 2004a, b; Panza et al. 2007; Peccerillo et al. 2013). Dehydration and melting of the oceanic slab and associated sediments produced contamination and melting of the Stromboli and Campania mantle sources (Fig. 7.22c). The slab tear-off could have occurred at about 0.8 Ma when compression phases in the western Apulia come to an end and distension tectonics favoured the Vulture magmatism (Patacca and Scandone 2001).

Such a model is also able to explain some apparently minor regional geochemical variations. Inspection of spiderdiagrams (Fig. 7.21), highlights higher Rb, Cs and K at Vesuvio than at Stromboli. If not acquired during shallow-level magma evolution, these variations could be related to phengite breakdown in the deepest part of the subduction zone. It is well established that this mineral is a common phase of metasediments, is the main carrier of LILE, and becomes unstable at deep mantle levels where it breaks down to release its load of K, Rb and Cs (Schmidt et al. 2004).

A last point to clarify is the origin of the OIB-type pre-metasomatic mantle components of the Campania and Stromboli magmas. The occurrence of an OIB-type mantle in this region has been explained by a process of mantle inflow from the foreland onto the sinking and retreating Ionian slab, through the window opened by horizontal slab-tear after the Apulia collision. Sinking and south-eastward retreat of the slab promoted suctioning of asthenospheric Ionian mantle around the margin of the slab (De Astis et al. 2000; Peccerillo 2001; Peccerillo and Frezzotti 2015). A slab window has been clearly evidenced by geophysical data (e.g. Chiarabba et al. 2008; Neri et al. 2009; Giacomuzzi et al. 2012). Therefore, the OIB component beneath the Stromboli and Campania volcanoes would be allochthonous in origin.

A different hypothesis on the origin of OIB-type components invokes a deep mantle upwelling beneath the southern Tyrrhenian Sea and contamination by subducting Ionian plate (Gasperini et al. 2002; Cadoux et al. 2007). According to this view, deep mantle material would have been emplaced at shallow mantle levels through zones of slab discontinuity.

The Campania Province was preceded by an older cycle of volcanism that had much less potassic composition. Such an older calcalkaline activity is testified by the occurrence of thick

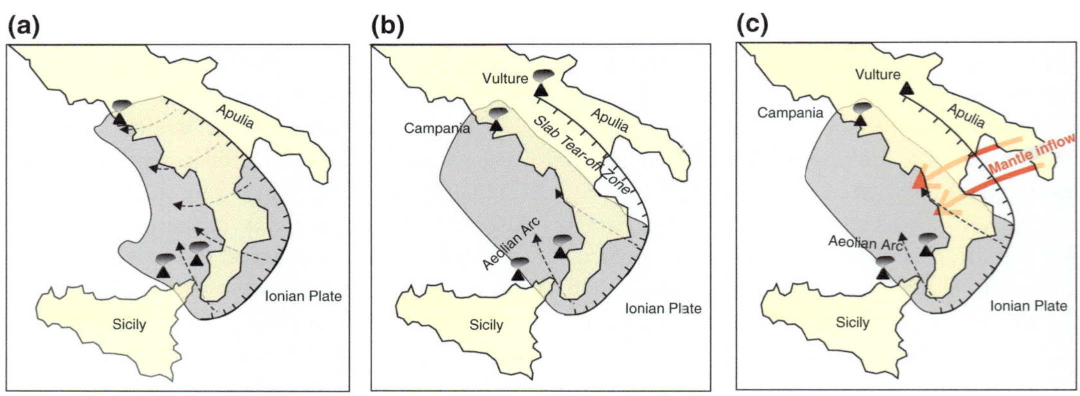

Fig. 7.22 Geodynamic evolution model for the southern Italian Peninsula. **a** A continuous subduction zone of the Apulian-Ionian plate was active until about 0.8 Ma; **b** Slab breakoff in the Apulian sector was associated to extension or transtension at the contact between Apulia and the southern Apennines, with the formation of Vulture volcano; active subduction continued in the Ionian sector; **c** Sinking and rollback of the narrow Ionian slab prompted suctioning of asthenosphere from the Apulia foreland; such an allochthonous mantle material was contaminated by fluids from the subducting slab generating a hybrid OIB-arc mantle wedge, whose melting gave the Campania and Stromboli magmas

2 Ma-old basalt to andesite lava sequences encountered at depths of 300–1800 beneath the Campanian Plain, and by lava xenoliths in the Procida pyroclastic deposits. The geodynamic significance of the increase in potassium with time for the Campania area is poorly known and it could be speculatively related to a modification of the slab geometry and/or composition after the Apulia-Ionian slab breakoff.

7.10 Summary and Conclusions

The Campania volcanoes are formed by alkaline potassic to ultrapotassic magmas. An older volcanic cycle with calcalkaline affinity is buried beneath the Campanian Plain.

The rocks of the Campania Province range from mafic to felsic and define various rock suites with contrasting geochemical and isotopic variations. However, a single type of trachybasalt parental melt is suggested for all the rock suites by experimental petrology and geochemical evidence. Therefore, primary ultrapotassic rocks, such as those occurring in the Roman Province and at Ernici-Roccamonfina, seem to be lacking in Campania. This supports the hypothesis that the Campania volcanism is genetically and geodynamically distinct from the Central Italy potassic alkaline provinces.

There is a large compositional overlap for several incompatible trace element and radiogenic isotopic ratios among mafic rocks from Campania and Stromboli. Geochemical data, therefore, provide an important evidence for a genetic and geodynamic link between Campania and the eastern Aeolian arc volcanoes.

Both Campania and Stromboli rocks have geochemical signatures intermediate between OIB- and arc-type magmas, which suggest melt generation in a hybrid source, consisting of OIB-type mantle rocks and subduction-related components. The eastern Aeolian arc develops over the Ionian Wadati-Benioff zone extending from Calabria to the southern edge of the Campania Province. This has led to suggestion that the arc-related geochemical components in the Campania and Stromboli magmas have been released by the subducted Ionian oceanic slab and associated sediments.

The OIB-type mantle component of the Campania and Stromboli volcanoes is probably provided by the inflow of asthenospheric mantle from the foreland onto the subducting and southeast retreating Ionian slab. Migration of asthenospheric mantle was favoured by the opening of a window generated by slab breakoff along the Apulian-Ionian plate. Deep mantle rocks emplaced into the mantle wedge through slab discontinuities are also possible source of OIB-type components. Pb isotope data on the Island of Procida also suggest a contribution by mantle components coming from the Tyrrhenian Sea or Sardinia.

References

Acocella V (2010) Evaluating fracture patterns within a resurgent caldera: Campi Flegrei, Italy. Bull Volcanol 72:623–638

Albini A, Cristofolini R, Di Girolamo P, Stanzione D (1980) Rare-earth and other trace-element distributions in the calc-alkaline volcanic rocks from deep boreholes in the Phlegrean Fields, Campania (south Italy). Chem Geol 28:123–133

Andronico D, Cioni R (2002) Contrasting styles of Mount Vesuvius activity in the period between the Avellino and Pompeii Plinian eruptions, and some implications for assessment of future hazards. Bull Volcanol 64:372–391

Arienzo I, Civetta L, Heumann A, Worner G, Orsi G (2009) Isotopic evidence for open system processes within the Campanian Ignimbrite (Campi Flegrei-Italy) magma chamber. Bull Volcanol 71:285–300

Arienzo I, Neumann A, Worner G, Civetta L, Orsi G (2011) Processes and timescales of magma evolution prior to the Campanian Ignimbrite eruption (Campi Flegrei, Italy). Earth Planet Sci Lett 306:217–228

Armienti P, Barberi F, Bizouard H, Clocchiatti R, Innocenti F, Métrich N, Rosi M, Sbrana A (1983) The Phlegraean fields: magma evolution within a shallow chamber. J Volcanol Geoth Res 17:289–311

Arrighi S, Principe C, Rosi M (2001) Violent strombolian and subplinian eruptions at Vesuvius during post-1631 activity. Bull Volcanol 63:126–150

Auger M, Gasparini P, Virieux J, Zollo A (2001) Seismic evidence of an extended magmatic sill under Mt. Vesuvius. Science 294:1510–1512

Aulinas M, Civetta L, Di Vito MA, Orsi G, Gimeno D, Fernandez-Turiel JL (2008) The "Pomici di mercato" Plinian eruption of Somma-Vesuvius: magma chamber processes and eruption dynamics. Bull Volcanol 70:825–840

Avanzinelli R, Elliott T, Tommasini S, Conticelli S (2008) Constraints on the genesis of the potassium-rich Italian volcanics from U/Th disequilibrium. J Petrol 49:195–223

Ayuso RA, De Vivo B, Rolandi G, Seal RR II, Paone A (1998) Geochemical and isotopic (Nd-Pb-Sr-O) variations bearing on the genesis of volcanic rocks from Vesuvius, Italy. J Volcanol Geoth Res 82:53–78

Barbieri M, Di Girolamo P, Locardi E, Lombardi G, Stanzione D (1979) Petrology of the calc-alkaline volcanics of the Parete-2 well (Campania, Italy). Per Mineral 48:53–74

Beccaluva L, Di Girolamo P, Serri G (1991) Petrogenesis and tectonic setting of the Roman volcanic province, Italy. Lithos 26:191–221

Belkin HE, De Vivo B (1993) Fluid inclusion studies of ejected nodules from plinian eruptions of Mt. Somma-Vesuvius. J Volcanol Geoth Res 58:98–100

Belkin HE, De Vivo B, Roedder E, Cortini M (1985) Fluid inclusion geobarometry from ejected Mt. Somma-Vesuvius nodules. Am Mineral 70:288–303

Berrino G, Corrado G, Riccardi U (1998) Sea gravity data in the Gulf of Naples: a contribution to delineating the structural pattern of the Vesuvian area. J Volcanol Geoth Res 82:139–150

Bertagnini A, Landi P, Rosi M, Vigliargio A (1998) The Pomici di Base plinian eruption of Somma-Vesuvius. J Volcanol Geoth Res 83:219–239

Bianco F, Castellano M, Milano G, Ventura G, Vilardo G (1998) The Somma-Vesuvius stress field induced by regional tectonics: evidences from seismological and mesostructural data. J Volcanol Geoth Res 82:199–218

Boari E, Avanzinelli R, Melluso L, Giordano G, Mattei M, Morra V, Conticelli S (2009a) Isotope geochemistry (Sr- Nd-Pb) and petrogenesis of leucite-bearing volcanic rocks from "Colli Albani" volcano, Roman Magmatic Province, Central Italy: inferences on volcano evolution and magma genesis. Bull Volcanol 71:977–1005

Boari E, Tommasini S, Laurenzi MA, Conticelli S (2009b) Transition from ultrapotassic kamafugitic to sub-alkaline magmas: Sr, Nd, and Pb isotope, trace element and $^{40}Ar/^{39}Ar$ age data from the Middle Latin Valley volcanic field, Roman Magmatic Province, Central Italy. J Petrol 50:1327–1357

Borgia A, Tizzani P, Solaro G, Manzo M, Casu F, Luongo G, Pepe A, Berardino P, Fornaro G, Sansosti E, Ricciardi GP, Fusi N, Di Donna G, Lanari R (2005) Volcanic spreading of Vesuvius, a new paradigm for interpreting its volcanic activity. Geophys Res Lett 32. doi:10.1029/2004gl022155

Brocchini D, Principe C, Castradori D, Laurenzi MA, Gorla L (2001) Quaternary evolution of the southern sector of the Campanian Plain and early Somma-Vesuvius activity: insights from the Trecase 1 well. Mineral Petrol 73:67–91

Brown RJ, Orsi G, de Vita S (2008) New insights into Late Pleistocene explosive volcanic activity and caldera formation on Ischia (Southern Italy). Bull Volcanol 70:583–603

Brown RJ, Civetta L, Arienzo I, D'Antonio M, Moretti R, Orsi G, Tomlinson EL, Albert PG, Menzies MA (2014) Geochemical and isotopic insights into the assembly, evolution and disruption of a magmatic plumbing system before and after a cataclysmic caldera-collapse eruption at Ischia volcano (Italy). Contrib Mineral Petrol 168:1035

Cadoux A, Blichert-Toft J, Pinti DL, Albarede F (2007) A unique lower mantle source for Southern Italy volcanics. Earth Planet Sci Lett 259:227–238

Calcagnile G, Panza GF (1981) The main characteristics of the lithosphere-asthenosphere system in Italy and surrounding regions. Pure Appl Geophys 119:865–879

Caliro S, Chiodini G, Paonita A (2014) Geochemical evidences of magma dynamics at Campi Flegrei (Italy). Geochim Cosmochim Acta 132:1–15

Carlino S, Somma R, Troise C, De Natale G (2012) The geothermal exploration of Campanian volcanoes: historical review and future development. Renew Sust Energ Rev 16:1004–1030

Carlino S, Kilburn CRJ, Tramelli A, Troise C, Somma R, De Natale G (2015) Tectonic stress and renewed uplift at Campi Flegrei caldera, southern Italy: new insights from caldera drilling. Earth Planet Sci Lett 420:23–29

Carminati E, Doglioni C (2012) Alps vs. Apennines: the paradigm of a tectonically asymmetric Earth. Earth Sci Rev 112:67–96

Cassignol C, Gillot P-Y (1982) Range and effectiveness of unspiked potassium-argon dating: experimental groundwork and application. In: Odin GS (ed) Numerical dating in stratigraph. Wiley, New York, pp 160–179

Cello G, Mazzoli S (1999) Apennine tectonics in southern Italy. A review. J Geodyn 27:191–211

Chiarabba C, De Gori P, Speranza F (2008) The southern Tyrrhenian subduction zone: deep geometry, magmatism and Plio-Pleistocene evolution. Earth Planet Sci Lett 268:408–423

Chiesa S, Poli S (1988) Island of Ischia. Petrology. In: Vezzoli L (ed) Island of Ischia. Quad Ric Sci CNR, Rome, 114, 10:72–93

Chiodini G, Marini L, Russo M (2001) Geochemical evidence for the existence of high-temperature hydrothermal brines at Vesuvio volcano, Italy. Geochim Cosmochim Acta 65:2129–2147

Chiodini G, Vandemeulebrouck J, Caliro S, D'Auria L, De Martino P, Mangiacapra A, Petrillo Z (2015) Evidence of thermal-driven processes triggering the 2005–2014 unrest at Campi Flegrei caldera. Earth Planet Sci Lett 414:58–67

Cigolini C (1999) "High pressure" dunitic ejecta of kimberlitic affinity in recent pyroclastic deposits from Mount Vesuvius: inference on their genesis and evolution. Atti Acc Sc Fis Torino 133:149–158

Cioni R, Civetta L, Marianelli P, Métrich N, Santacroce R, Sbrana A (1995) Compositional layering and syneruptive mixing of a periodically refilled shallow magma chamber: the AD 79 Plinian eruption of Vesuvius. J Petrol 36:739–776

Cioni R, Marianelli P, Santacroce R (1997) Thermal and compositional evolution of the shallow magma chambers of Vesuvius: evidence from pyroxene phenocrysts and melt inclusions. J Geophys Res 103:18277–18294

Cioni R, Marianelli P, Santacroce R (1998) Thermal and compositional evolution of the shallow magma chambers of Vesuvius: evidence from pyroxene phenocrysts and melt inclusions. J Geophys Res 103:18277–18294

Cioni R, Bertagnini A, Santacroce R, Andronico D (2008) Explosive activity and eruption scenarios at Somma-Vesuvius (Italy): towards a new classification scheme. J Volcanol Geoth Res 178:331–346

Civetta L, Santacroce R (1992) Steady state magma supply in the last 3400 years of Vesuvius activity. Acta Vulcanol 2:147–159

Civetta L, Galati R, Santacroce R (1991a) Magma mixing and convective compositional layering within the Vesuvius magma chamber. Bull Volcanol 53:287–300

Civetta L, Carluccio E, Innocenti F, Sbrana A, Taddeucci G (1991b) Magma chamber evolution under the Phlegraean Field during the last 10 ka: trace element and isotope data. Eur J Mineral 3(415):428

Civetta L, Gallo G, Orsi G (1991c) Sr- and Nd-isotope and trace-element constraints on the chemical evolution of the magmatic system of Ischia (Italy) in the last 55 ka. J Volcanol Geoth Res 46:213–230

Civetta L, Orsi G, Pappalardo L, Fisher RV, Heiken G, Ort M (1997) Geochemical zoning, mingling, eruptive dynamics depositional processes—the Campanian Ignimbrite, Flegrei caldera, Italy. J Volcanol Geoth Res 75:183–219

Conticelli S, Laurenzi MA, Giordano G, Mattei M, Avanzinelli R, Melluso L, Tommasini S, Boari E, Cifelli F, Perini G (2010a) Leucite-bearing (kamafugitic/leucititic) and -free (lamproitic) ultrapotassic rocks and associated shoshonites from Italy: constraints on petrogenesis and geodynamics. In: Beltrando M, Peccerillo A, Mattei M, Conticelli S, Doglioni C (eds) J Virtual Explorer 36, paper 20. doi:10.3809/jvirtex.2010.00251

Conticelli S, Boari E, Avanzinelli R, De Benedetti AA, Giordano G, Mattei M, Melluso L, Morra V (2010b) Geochemistry, isotopes and minerl chemistry of the Colli Albani volcanic rock: constraints on magma genesis and evolution. In: Funiciello R, Giordano G (eds) The Colli Albani Volcano. (Geol Soc London, Spec IAVCEI Publ 3):107–140

Cortini M, Hermes OD (1981) Sr isotopic evidence for a multi-source origin of the potassic magmas in the Neapolitan area (South Italy). Contrib Mineral Petrol 77:47–55

Cortini M, Ayuso RA, De Vivo B, Holden P, Somma R (2004) Isotopic composition of Pb and Th in interplinian volcanics from Somma-Vesuvius volcano, Italy. Mineral Petrol 80:83–96

Crisci GM, De Francesco AM, Mazzuoli R, Poli G, Stanzione D (1989) Geochemistry of recent volcanics of Ischia Island, Italy: evidences for fractional crystallisation and magma mixing. Chem Geol 78:15–33

D'Antonio M (2005) Calderas and magmatic feeding systems: examples from Campi Flegrei (Southern Italy). Acta Volcanol 17:53–66

D'Antonio M (2011) Lithology of the basement underlying the Campi Flegrei caldera: volcanological and petrological constraints. J Volcanol Geoth Res 200:91–98

D'Antonio M, Di Girolamo P (1994) Petrological and geochemical study of mafic shoshonitic volcanics from Procida-Vivara and Ventotene islands (Campanian Region, South Italy). Acta Vulcanol 5:69–80

D'Antonio M, Tilton GR, Civetta L (1996) Petrogenesis of Italian alkaline lavas deduced from Pb-Sr-Nd isotope relationships. In: Basu A, Hart SR (eds) Isotopic Studies of Crust-Mantle Evolution. Am Geophys Un Mon 95:253–267

D'Antonio M, Civetta L, Di Girolamo P (1999a) Mantle source heterogeneity in the Campanian region (south Italy) as inferred from geochemical and isotopic features of mafic volcanic rocks with shoshonitic affinity. Mineral Petrol 67:163–192

D'Antonio M, Civetta L, Orsi G, Pappalardo L, Piochi M, Carandente A, De Vita S, Di Vito MA, Isaia R (1999b) The present state of the magmatic system of the Campi Flegrei caldera based on the reconstruction of its behaviour in the past 12 ka. J Volcanol Geoth Res 91:247–268

D'Antonio M, Tonarini S, Arienzo I, Civetta L, Di Renzo V (2007) Components and processes in the magma genesis of the Phlegrean Volcanic District, southern Italy. In: Beccaluva L, Bianchini G, Wilson M (eds.) Cenozoic Volcanism in the Mediterranean Area: Geol Soc Am Spec Paper 418:203–220

D'Antonio M, Tonarini S, Arienzo I, Civetta L, Dallai L, Moretti R, Orsi G, Andria M, Trecalli A (2013) Mantle and crustal processes in the magmatism of the Campania region: inferences from mineralogy, geochemistry, and Sr-Nd-O isotopes of young hybrid volcanics of the Ischia island (South Italy). Contrib Mineral Petrol 165:1173–1194

Dallai L, Cioni R, Boschi C, D'Oriano C (2011) Carbonate-derived CO_2 purging magma at depth: influence on the eruptive activity of Somma-Vesuvius, Italy. Earth Planet Sci Lett 310:84–95

De Astis G, Peccerillo A, Kempton PD, La Volpe L, Wu TW (2000) Transition from calc-alkaline to potassium-rich magmatism in subduction environments: geochemical and Sr, Nd, Pb isotopic constraints from the island of Vulcano (Aeolian arc). Contrib Mineral Petrol 139:684–703

De Astis G, Pappalardo L, Piochi M (2004) Procida volcanic history: new insights into the evolution of the Phlegraean Volcanic District (Campania region, Italy). Bull Volcanol 66:622–641

De Astis G, Kempton PD, Peccerillo A, Wu TW (2006) Trace element and isotopic variations from Mt. Vulture to Campanian volcanoes: constraints for slab

detachment and mantle inflow beneath southern Italy. Contrib Mineral Petrol 151:331–351
De Lorenzo S, Di Renzo V, Civetta L, D'Antonio M, Gasparini P (2006) Thermal model of the Vesuvius magma chamber. Geophys Res Lett 33. doi:10.1029/2006gl026587
De Natale G, Troise C, Pingue F, Mastrolorenzo G, Pappalardo L (2006) The Somma-Vesuvius volcano (Southern Italy): structure, dynamics and hazard evaluation. Earth Sci Rev 74:73–111
De Siena L, Del Pezzo E, Bianco F (2010) Seismic attenuation imaging of Campi Flegrei: evidence of gas reservoirs, hydrothermal basins and feeding systems. J Geophys Res 115:B09312. doi:10.1029/2009JB006938
De Vita S, Orsi G, Civetta L, D'Antonio M et al (1999) The Agnano-Monte Spina eruption (4.1 ka) in the resurgent, nested Campi Flegrei caldera (Italy). J Volcanol Geoth Res 91:269–301
De Vivo B, Rolandi G, Gans PB, Calvert A, Bohrson WA, Spera FJ, Belkin HE (2001) New constraints on the pyroclastic eruptive history of the Campanian volcanic Plain (Italy). Mineral Petrol 73:47–65
De Vivo B, Petrosino P, Lima A, Rolandi G, Belkin HE (2010) Research progress in volcanology in the Neapolitan area, Southern Italy: a review and some alternative views. Mineral Petrol 99:1–28
Deino AL, Orsi G, de Vita S, Piochi M (2004) The age of the Neapolitan Yellow Tuff caldera-forming eruption (Campi Flegrei caldera Italy) assessed by $^{40}Ar/^{39}Ar$ dating method. J Volcanol Geoth Res 133:157–170
Del Gaudio C, Aquino I, Ricciardi GP, Ricco C, Scandone R (2010) Unrest episodes at Campi Flegrei: a reconstruction of vertical ground movements during 1905–2009. J Volcanol Geoth Res 195:48–56
Del Moro A, Fulignati P, Marianelli P, Sbrana A (2001) Magma contamination by direct wall rock interaction: constraints from xenoliths from the walls of a carbonate-hosted magma chamber (Vesuvius 1944 eruption). J Volcanol Geoth Res 112:15–24
Della Vedova B, Bellani S, Pelli G, Squarci P (2001) Deep temperatures and surface heath distribution. In: Vai GB, Martini PI (eds) Anatomy of an Orogen. The Apennines and the adjacent Mediterranean basins. Kluwer, Dordrecht, pp 65–76
Di Girolamo P (1978) Geotectonic settings of Miocene Quaternary volcanism in and around the eastern Tyrrhenian Sea border (Italy) as deduced from major element geochemistry. Bull Volcanol 41:229–250
Di Girolamo P, Stanzione D (1973) Lineamenti geologici e petrologici dell'isola di Procida. Rend Soc It Mineral Petrol 29:81–125
Di Girolamo P, Nardi G, Rolandi G, Stanzione D (1976) Occurrence of calc-alkaline two-pyroxene andesites from deep bore-holes in the Phlegraean Fields. Petrographic and petrochemical data. Rend Accad Sci Fis Mat Napoli 43:1–29
Di Girolamo P, Melluso L, Morra V, Secchi FAG (1995) Evidence of interaction between mafic and differentiated magmas in the youngest phase of activity at Ischia island (Italy). Per Mineral 64:393–411
Di Renzo V, Di Vito MA, Arienzo I, Carandente A, Civetta L, D'Antonio M, Giordano F, Orsi G, Tonarini S (2007) Magmatic history of Somma-Vesuvius on the basis of new geochemical and isotopic data from a deep borehole (Camaldoli della Torre). J Petrol 48:753–784
Di Renzo V, Arienzo I, Civetta L, D'Antonio M, Tonarini S, Di Vito MA, Orsi G (2011) The magmatic feeding system of the Campi Flegrei caldera: architecture and temporal evolution. Chem Geol 281:227–241
Di Vito MA, Isaia R, Orsi G, Southon J, de Vita S, D'Antonio M, Pappalardo L, Piochi M (1999) Volcanism and deformation since 12,000 years at the Campi Flegrei caldera (Italy). J Volcanol Geoth Res 91:221–246
Fedele L, Morra V, Perrrotta A, Scarpati C (2006) Volcanological and geochemical features of the products of the Fiumicello eruption, Procida island, Campi Flegrei (Southern Italy). Per Mineral 75:43–71
Fedele L, Scarpati C, Lanphere M, Melluso L, Morra V, Perrotta A, Ricci G (2008) The Breccia Museo formation, Campi Flegrei, southern Italy: geochronology, chemostratigraphy and relationship with the Campanian Ignimbrite eruption. Bull Volcanol 70:1189–1219
Fedele L, Insinga DD, Calvert AT, Morra V, Perrotta A, Scarpati C (2011) $^{40}Ar/^{39}Ar$ dating of tuff vents in the Campi Flegrei caldera (southern Italy): toward a new chronostratigraphic reconstruction of the Holocene volcanic activity. Bull Volcanol 73:1323–1336
Fowler SJ, Spera F, Bohrson W, Belkin HE, De Vivo B (2007) Phase equilibria constraints on the chemical and physical evolution of the campanian ignimbrite. J Petrol 48:459–493
Frondini F, Chiodini G, Caliro S, Cardellini C, Granieri D, Ventura G (2004) Diffuse CO_2 degassing at Vesuvio, Italy. Bull Volcanol 66:642–651
Gale A, Dalton CA, Langmuir CH, Su Y, Schilling J-G (2013) The mean composition of ocean ridge basalts. Gechem Geophys Geosyst 14. doi:10.1029/2012GC004334
Gasperini D, Blichert-Toft J, Bosch D, Del Moro A, Macera P, Télouk P, Albaréde F (2000) Evidence from Sardinian basalt geochemistry for recycling of plume heads into the Earth's mantle. Nature 408:701–704
Gasperini D, Blichert Toft J, Bosch D, Del Moro A, Macera P, Albaréde F (2002) Upwelling of deep mantle material through a plate window: evidence from the geochemistry of Italian basaltic volcanics. J Geophys Res 107(B12):2367. doi:10.1029/2001JB000418
Gebauer SK, Schmitt AK, Pappalardo L, Stockli DF, Lovera OM (2014) Crystallization and eruption ages of Breccia Museo (Campi Flegrei caldera, Italy) plutonic clasts and their relation to the Campanian ignimbrite. Contrib Mineral Petrol 167:953
Giaccio B, Isaia R, Fedele FG, Di Canzio E, Hoffecker J, Ronchitelli A, Sinitsyn A, Anikovich M, Lisitsyn SN

(2008) The Campanian Ignimbrite and Codola tephra layers: two temporal/stratigraphic markers for the Early Upper Palaeloithic in southern Italy and eastern Europe. J Volcan Geoth Res 177:208–226

Giacomelli L, Perrotta A, Scandone R, Scarpati C (2003) The eruption of Vesuvius of 79 AD and its impact on human environment in Pompei. Episodes 26:234–237

Giacomuzzi G, Civalleri M, De Gori P, Chiarabba C (2012) A 3D Vs model of the upper mantle beneath Italy: insight on the geodynamics of central Mediterranean. Earth Planet Sci Lett 335–336:105–120

Gillot P-Y, Chiesa S, Pasquare G, Vezzoli L (1982) <33000 yr K/Ar dating of the volcano-tectonic horst of the Isle of Ischia, Gulf of Naples. Nature 299:242–245

Graham D, Allard P, Kilburn CRJ, Spera F, Lupton JE (1993) Helium isotopes in some historical lavas from Mount Vesuvius. J Volcanol Geoth Res 58:359–366

Grasso M (2001) The Apenninic-Maghrebian orogen in Southern Italy, Sicily and adjacent areas. In: Vai GB, Martini PI (eds) Anatomy of an Orogen. The Apennines and adjacent Mediterranean basins. Kluwer, Dordrecht, pp 255–286

Gvirtzman Z, Nur A (1999) The formation of Mount Etna as the consequence of slab rollback. Nature 401:782–785

Hart SR (1984) A large-scale isotopic anomaly in the Southern Hemisphere mantle. Nature 309:753–757

Iacono Marziano G, Gaillard F, Pichavant M (2008) Limestone assimilation by basaltic magmas: an experimental re-assessment and application to Italian volcanoes. Contrib Mineral Petrol 155:719–738

Inguaggiato S, Pecoraino G, D'Amore F (2000) Chemical and isotopical characterisation of fluid manifestations of Ischia Island (Italy). J Volcanol Geoth Res 99:151–178

Ippolito F, D'Argenio B, Pescatore T, Scandone P (1975) Structural-stratigraphic units and tectonic framework of Southern Apennines. In: Squyres CH (ed) Geology of Italy. Petrol Expl Soc Libya, Tripoli, pp 317–328

Irvine TN, Baragar WRA (1971) A guide to chemical classification of common volcanic rocks. Can J Earth Sci 8:523–548

Isaia R, Marianelli P, Sbrana A (2009) Caldera unrest prior to intense volcanism in Campi Flegrei (Italy) at 4.0 ka BP: implications for caldera dynamics and future eruptive scenarios. Geophys Res Lett 36. doi:10.1029/2009gl040513

Joron JL, Métrich N, Rosi M, Santacroce R, Sbrana A (1987) Chemistry and petrography. In: Santacroce R (ed) Somma-Vesuvius. Quad Ric Sci CNR, Rome, 114, 8:105–174

Kessel R, Schmidt MW, Ulmer P, Pettke T (2005) Trace element signatures of subduction-zone fluids, melts and supercritical liquids at 120–180 km depth. Nature 437:724–727

Klaver M, Djuly T, de Graaf S, Sakes S, Wijbrans J, Davies G, Vroon P (2015) Temporal and spatial variations in provenance of Eastern Mediterranean Sea sediments: implications for Aegean and Aeolian arc volcanism. Geochim Cosmochim Acta 153:149–168

Klébesz R, Bodnar RJ, De Vivo B, Torok K, Lima A, Petrosino P (2012) Composition and origin of nodules from the approximate to 20 ka Pomici di Base (PB)-Sarno eruption of Mt. Somma–Vesuvius, Italy. Centr Eur J Geosci 4:324–337

Lima A, Belkin HE, Tiiriik K (1999) Understanding Vesuvius magmatic processes: evidence from primitive silicate-melt inclusions in medieval scoria clinopyroxenes (Terzigno Formation). Mineral Petrol 65:185–206

Lima A, De Vivo B, Fedele L, Sintoni F, Milia A (2007) Geochemical variations between the 79 AD and 1944 AD Somma–Vesuvius volcanic products: constraints on the evolution of the hydrothermal system based on fluid and melt inclusions. Chem Geol 237:401–417

Lirer L, Munno R, Postiglione I, Vinci A, Vitelli L (1997) The AD 79 eruption as a future explosive scenario in the Vesuvian area: evaluation of asssociated risk. Bull Volcanol 59:112–124

Locardi E, Nicholich R (1988) Geodinamica del Tirreno e dell'Appennino centro-meridionale: la nuova carta della Moho. Mem Soc Geol It 41:121–140

Lustrino M, Melluso L, Morra V (2000) The role of lower continental crust and lithospheric mantle in the genesis of Plio-Pleistocene volcanic rocks from Sardinia (Italy). Earth Planet Sci Lett 180:259–270

Macdonald R, Bagiński B, Rolandi G, De Vivo B, Kopczyńska A (2015) Petrology of parasitic and eccentric cones on the flanks and base of Somma-Vesuvius. Mineral Petrol 110:65–85

Marianelli R, Métrich N, Santacroce R, Sbrana A (1995) Mafic magma batches at Vesuvius: a glass inclusion approach to the modalities of feeding stratovolcanoes. Contrib Mineral Petrol 120:159–169

Marianelli P, Métrich N, Sbrana A (1999) Shallow and deep reservoirs involved in magma supply of the 1944 eruption of Vesuvius. Bull Volcanol 61:48–63

Marianelli P, Sbrana A, Métrich N, Cecchetti A (2005) The deep feeding system of Vesuvius involved in recent violent strombolian eruptions. Geophys Res Lett 32. doi:10.1029/2004gl021667

Martelli M, Nuccio PM, Stuart FM, Burgess R, Ellam RM, Italiano F (2004) Helium-strontium isotope constraints on mantle evolution beneath the Roman Comagmatic Province, Italy. Earth Planet Sci Lett 224:295–308

Martelli M, Nuccio PM, Stuart FM, Di Liberto V, Ellam RM (2008) Constraints on mantle source and interactions from He-Sr isotope variation in Italian Plio-Quaternary volcanism. Geochem Geophys Geosyst 9:Q02001. doi:10.1029/2007GC001730

Mazzeo FC, D'Antonio M, Arienzo I, Aulinas M, Di Renzo V, Gimeno D (2014) Subduction-related enrichment of the Neapolitan volcanoes (Southern Italy) mantle source: new constraints on the characteristics of the slab-derived components. Chem Geol 386:165–183

Mele G, Sandvol E, Cavinato GP (2006) Evidence of crustal thickening beneath the central Apennines (Italy) from teleseismic receiver functions. Earth Planet Sci Lett 249:425–435

Melluso L, De Gennaro R, Fedele L, Franciosi L, Morra V (2012) Evidence of crystallization in residual, Cl-F-rich, agpaitic, trachyphonolitic magmas and primitive Mg-rich basalt-trachyphonolite interaction in the lava domes of the Phlegrean Fields (Italy). Geol Mag 149:532–550
Melluso L, Morra V, Guarino V, De Gennaro R, Franciosi L, Grifa C (2014) The crystallization of shoshonitic to peralkaline trachyphonolitic magmas in a H_2O-Cl-F-rich environment at Ischia (Italy), with implications for the feeder system of the Campania Plain volcanoes. Lithos 210:242–259
Milia A, Torrente MM (2003) Late-Quaternary volcanism and transtensional tectonics in the Bay of Naples, Campanian continental margin, Italy. Mineral Petrol 79:49–65
Milia A, Torrente MM (2015) Tectono-stratigraphic signature of a rapid multistage subsiding rift basin in the Tyrrhenian-Apennine hinge zone (Italy): a possible interaction of upper plate with subducting slab. J Geodyn 86:42–60
Moretti R, Arienzo I, Orsi G, Civetta L, D'Antonio M (2013) The deep plumbing system of Ischia: a physico-chemical window on the fluid-saturated and CO_2-sustained Neapolitan volcanism (Southern Italy). J Petrol 54:951–984
Morgan DJ, Blake S, Rogers NW, DeVivo B, Rolandi G, Macdonald R, Hawkesworth CJ (2004) Time scales of crystal residence and magma chamber volume from modelling of diffusion profiles in phenocrysts: Vesuvius 1944. Earth Planet Sci Lett 222:933–946
Mormone A, Piochi M, Bellatreccia F, De Astis G, Moretti R, Della Ventura G, Cavallo A, Mangiacapra A (2011) A CO_2-rich magma source beneath the Phlegraean Volcanic District (Southern Italy): evidence from a melt inclusion study. Chem Geol 287:66–80
Morris J, Ryan J, Leeman WP (1993) Be isotope and B-Be investigations of the historic eruptions of Mt. Vesuvius. J Volcanol Geoth Res 58:345–358
Neri G, Orecchio B, Totaro C, Falcone G, Presti D (2009) Subduction beneath Southern Italy close the ending: results from seismic tomography. Seism Res Lett 80:63–70
Newhall CG, Self S (1982) The volcanic explosivity index (VEI): an estimate of explosive magnitude for historical volcanism. J Geophys Res 87:1231–1238
Nunziata C (2010) Low shear-velocity zone in the Neapolitan-area crust between the Campi Flegrei and Vesuvio volcanic areas. Terra Nova 22:208–217
Nunziata C, Costanzo MR (2010) Low V–S crustal zones in the Campanian Plain (Southern Italy). Mineral Petrol 100:215–225
Ogniben L, Parotto M, Praturlon A (eds) (1975) Structural model of Italy. Quad Ric Sci CNR, Rome, 90, 502 pp
Orecchio B, Presti D, Totaro C, Neri G (2014) What earthquakes say concerning residual subduction and STEP dynamics in the Calabrian Arc region, South Italy. Geophys J Int 199:1929–1942
Orsi G, Civetta L, D'Antonio M, Di Girolamo P, Piochi M (1995) Step-filling and development of a three-layers magma chamber: the Neapolitan Yellow Tuff case history. J Volcanol Geoth Res 67:291–312
Orsi G, De Vita S, Di Vito M (1996) The restless, resurgent Campi Flegrei nested caldera (Italy): constraints on its evolution and configuration. J Volcanol Geoth Res 74:179–214
Orsi G, Di Vito MA, Isaia R (2004) Volcanic hazard assessment at the restless Campi Flegrei caldera. Bull Volcanol 66:514–530
Pabst S, Wörner G, Civetta L, Tesoro R (2008) Magma chamber evolution prior to the Campanian Ignimbrite and Neapolitan Yellow Tuff eruptions (Campi Flegrei, Italy). Bull Volcanol 70:961–976
Panza GF, Peccerillo A, Aoudia A, Farina B (2007) Geophysical and petrological modeling of the structure and composition of the crust and upper mantle in complex geodynamic settings: the Tyrrhenian Sea and surroundings. Earth Sci Rev 80:1–46
Paone A (2006) The geochemical evolution of the Mt. Somma-Vesuvius volcano. Mineral Petrol 87:53–80
Paone A (2008) Fractional crystallization models and B-Be–Li systematics at Mt Somma-Vesuvius volcano (Southern Italy). Int J Earth Sci 97:635–650
Pappalardo L, Mastrolorenzo G (2010) Short residence times for alkaline Vesuvius magmas in a multi-depth supply system: evidence from geochemical and textural studies. Earth Planet Sci Lett 296:133–143
Pappalardo L, Mastrolorenzo G (2012) Rapid differentiation in a sill-like magma reservoir: a case study from the Campi Flegrei caldera. Nature Sci Rep 2:712. doi:10.1038/srep00712
Pappalardo L, Civetta L, D'Antonio M, Deino A, Di Vito MA, Orsi G, Carandente A, De Vita S, Isaia R, Piochi M (1999) Chemical and Sr-isotopical evolution of the Phlegraean magmatic system before the Campanian Ignimbrite (37 ka) and the Neapolitan Yellow Tuff (12 ka) eruptions. J Volcanol Geoth Res 91:141–166
Pappalardo L, Piochi M, D'Antonio M, Civetta L, Petrini R (2002) Evidence for multistage magmatic evolution during the past 60 kyr at Campi Flegrei (Italy) deduced from Sr, Nd, and Pb isotope data. J Petrol 43:1415–1434
Pappalardo L, Piochi M, Mastrolorenzo G (2004) The 3800 yr BP–1944 AD magma plumbing system of Somma–Vesuvius: constraints on its behavior and present state through a review of isotope data. Ann Geophys 47:1471–1483
Passaro S, Genovese S, Sacchi M, Barra M et al (2014) First hydroacoustic evidence of marine, active fluid vents in the Naples Bay continental shelf (Southern Italy). J Volcanol Geoth Res 285:29–35
Patacca E, Scandone P (2001) Late thrust propagation and sedimentary response in the thrust-belt-foredeep system of the southern Apennines (Pliocene-Pleistocene). In: Vai GB, Martini PI (eds) Anatomy of an Orogen. The Apennines and adjacent Mediterranean basins. Kluwer, Dordrecht, pp 401–440
Patacca E, Sartori R, Scandone P (1990) Tyrrhenian basin and Apenninic arcs: kinematic relations since Late Tortonian times. Mem Soc Geol It 45:425–451

Peccerillo A (1999) Multiple mantle metasomatism in central-southern Italy: geochemical effects, timing and geodynamic implications. Geology 27:315–318

Peccerillo A (2001) Geochemical similarities between Vesuvius, Phlegraean Fields and Stromboli volcanoes: petrogenetic, geodynamic and volcanological implications. Mineral Petrol 73:93–105

Peccerillo A (2002) Plio-Quaternary magmatism in central-southern Italy: a new classification scheme for volcanic provinces and its geodynamic implications. In: Barchi RM, Cirilli S, Minelli G (eds) Geological and geodynamic evolution of the Apennines (Boll Soc Geol It Spec Vol 1):113–127

Peccerillo A, Frezzotti ML (2015) Magmatism, mantle evolution and geodynamics at the converging plate margins of Italy. J Geol Soc London 172:407–427

Peccerillo A, De Astis G, Faraone D, Forni F, Frezzotti ML (2013) Compositional variations of magmas in the Aeolian arc: implications for petrogenesis and geodynamics. In: Lucchi F, Peccerillo A, Keller J, Tranne CA, Rossi PL (eds) The Aeolian Islands Volcanoes (Geol Soc London Memoirs 37):491–510

Pecoraino G, Brusca L, D'Alessandro W, Giammanco S, Inguaggiato S, Longo M (2005) Total CO_2 output from Ischia Island volcano (Italy). Geochem J 39:451–458

Pichavant M, Scaillet B, Pommier A, Iacono-Marziano G, Cioni R (2014) Nature and evolution of primitive Vesuvius magmas: an experimental study. J Petrol 55:2281–2309

Piochi M, Civetta L, Orsi G (1999) Mingling in the magmatic system of Ischia (Italy) in the past 5 ka. Mineral Petrol 66:227–258

Piochi M, Mastrolorenzo G, Pappalardo L (2005) Magma ascent and eruptive processes from textural and compositional features of Monte Nuovo pyroclastic products, Campi Flegrei, Italy. Bull Volcanol 67:663–678

Piochi M, Ayuso RA, De Vivo B, Somma R (2006a) Crustal contamination and crystal entrapment during polybaric magma evolution at Mt. Somma-Vesuvius volcano, Italy: Geochemical and Sr isotope evidence. Lithos 86:303–329

Piochi M, Mastrolorenzo G, Pappalardo L (2006b) Magma ascent and eruptive processes from textural and compositional features of Monte Nuovo pyroclastic products, Campi Flegrei, Italy (vol 67, pg 663, 2005). Bull Volcanol 68:496

Piochi M, Polacci M, De Astis G, Zanetti A, Mangiacapra A, Vannucci R, Giordano D (2008) Texture and composition of pumices and scoriae from the Campi Flegrei caldera (Italy): implications on the dynamics of explosive eruptions. Geochem Geophys Geosyst 9: Q03013. doi:10.1029/2007gc001746

Piochi M, Kilburn CRJ, Di Vito MA, Mormone A, Tramelli A, Troise C, De Natale G (2014) The volcanic and geothermally active Campi Flegrei caldera: an integrated multidisciplinary image of its buried structure. Int J Earth Sci 103:401–421

Piromallo C, Morelli A (2003) P wave tomography of the mantle under the Alpine-Mediterranean area. J Geophys Res 108(B2):2065. doi:10.1029/2002JB001757

Poli S, Chiesa S, Gillot P-Y, Gregnanin A, Vezzoli L (1987) Chemistry versus time in the volcanic complex of Ischia (Gulf of Naples, Italy): evidence of successive magmatic cycles. Contrib Mineral Petrol 95:322–335

Pommier A, Tarits P, Hautot S, Pichavant M, Scaillet B, Gaillard F (2010) A new petrological and geophysical investigation of the present-day plumbing system of Mount Vesuvius. Geochem Geophys Geosyst 11. doi:10.1029/2010gc003059

Principe C, Tanguy JC, Arrighi S, Paiotti A, Le Goff M, Zoppi U (2004) Chronology of Vesuvius activity from AD 79 to 1631 based on archeomagnetism of lavas and historical sources. Bull Volcanol 66:703–724

Rittmann A (1930) Geologie der Insel Ischia. Zeitschr Vulkanol 6:265 pp

Rittmann A (1950) Sintesi geologica dei Campi Flegrei. Boll Soc Geol It 69:117–177

Rolandi G, Petrosino P, Mc Geehin J (1998) The interplinian activity at Somma-Vesuvius in the last 3500 years. J Volcanol Geoth Res 82:19–52

Rolandi G, Munno R, Postiglione I (2004) The AD 472 eruption of the Somma volcano. J Volcanol Geoth Res 129:291–319

Rosenbaum G, Lister GS (2004a) Neogene and Quaternary rollback evolution of the Tyrrhenian Sea, the Apennines and the Sicilian Maghrebides. Tectonics, 23, TC1013. doi:10.1029/2003TC001518

Rosenbaum G, Lister GS (2004b) Formation of arcuate orogenic belts in the western Mediterranean region. In: Sussman AJ, Weil AB (eds) Orogenic curvature: integrating paleomagnetic and structural analyses (Geol Soc Am Spec Paper 383):41–56

Rosi M, Sbrana A (eds) (1987) The Phlegraean Fields. Quad Ric Sci CNR, Rome, 114, 10:175 pp

Russo M, Punzo I (2004) I minerali del Somma-Vesuvio. Ass Micromineral It, Cremona, Italy 320 pp

Sacchi M, Pepe F, Corradino M, Insinga DD, Molisso F, Lubritto C (2014) The Neapolitan Yellow Tuff caldera offshore the Campi Flegrei: Stratal architecture and kinematic reconstruction during the last 15 ky. Mar Geol 354:15–33

Santacroce R (ed) (1987) Somma-Vesuvius. Quad Ric Sci CNR, Rome, 114, 8, 249 pp

Santacroce R, Bertagnini A, Civetta L, Landi P, Sbrana A (1993) Eruptive dynamics and petrogenetic processes in a very shallow magma reservoir: the 1906 eruption of Vesuvius. J Petrol 34:383–425

Santacroce R, Cristofolini R, La Volpe L, Orsi G, Rosi M (2003) Italian active volcanoes. Episodes 26:227–234

Santacroce R, Cioni R, Marianelli P et al (2008) Age and whole rock–glass compositions of proximal pyroclastics from the major explosive eruptions of Somma-Vesuvius: a review as a tool for distal tephrostratigraphy. J Volcanol Geoth Res 177:1–18

Sbrana A, Toccaceli RM (2011) Note illustrative carta geologica di Ischia. Regione Campania, 215 pp

Scarpa R, Tronca F, Bianco F, Del Pezzo E (2002) High-resolution velocity structure beneath Mount Vesuvius from seismic array data. Geophys Res Lett 29. doi:10.1029/2002gl015576

Scarpati C, Perrotta A, Lepore S, Calvert A (2013) Eruptive history of Neapolitan volcanoes: constraints from $^{40}Ar/^{39}Ar$ dating. Geol Mag 150:412–425

Scarpati C, Sparice D, Perrotta A (2014) A crystal concentration method for calculating ignimbrite volume from distal ash-fall deposits and a reappraisal of the magnitude of the Campanian Ignimbrite. J Volcanol Geoth Res 280:67–75

Schmidt MW, Vielzeuf D, Auzanneau E (2004) Melting and dissolution of subducting crust at high pressures: the key role of white mica. Earth Planet Sci Lett 228:65–84

Serri G (1990) Neogene-Quaternary magmatism of the Tyrrhenian region: characterization of the magma sources and geodynamic implications. Mem Geol Soc It 41:219–242

Sheridan MF, Barberi F, Rosi M, Santacroce R (1981) A model of plinian eruptions of Vesuvius. Nature 289:282–285

Sigurdsson H, Carey S, Cornell W, Pescatore T (1985) The eruption of Vesuvius in AD 79. Natl Geogr Res 1:332–387

Smith VC, Isaia R, Pearce NJG (2011) Tephrostratigraphy and glass compositions of post-15 kyr Campi Flegrei eruptions: implications for eruption history and chronostratigraphic markers. Quat Sci Rev 30:3638–3660

Somma R, Ayuso RA, De Vivo B, Rolandi G (2001) Major, trace element and isotope geochemistry (Sr-Nd-Pb) of interplinian magmas from Mt. Somma-Vesuvius (Southern Italy). Mineral Petrol 73:121–143

Spakman W, Wortel R (2004) A tomographic view on Western Mediterranean Geodynamics. In: Cavazza W, Roure F, Spakman W, Stampfli GM, Ziegler P (eds) The TRANSMED Atlas, The Mediterranean Region from Crust to Mantle, pp 31–52

Strollo R, Nunziata C, Iannotta A, Iannotta D (2015) The uppermost crust structure of Ischia (Southern Italy) from ambient noise Rayleigh waves. J Volcanol Geoth Res 297:39–51

Sulpizio R, Cioni R, Di Vito MA, Mele D, Bonasia R, Dellino P (2010) The Pomici di Avellino eruption of Somma-Vesuvius (3.9 ka bp). Part I: stratigraphy, compositional variability and eruptive dynamics. Bull Volcanol 72:539–558

Sun SS, McDonough WF (1989) Chemical and isotopic systematics of oceanic basalts: implications for mantle composition and processes. In: Saunders AD, Norry MJ (eds) Magmatism in ocean basins (Geol Soc London Spec Publ 42):313–345

Tedesco D, Allard P, Sano Y, Wakita H, Pece R (1990) Helium-3 in subaerial and submarine fumaroles of Campi Flegrei caldera, Italy. Geochim Cosmochim Acta 54:1105–1116

Tiberti MM, Orlando L, Di Bucci D, Bernabini M, Parotto M (2005) Regional gravity anomaly map and crustal model of the Central-Southern Apennines (Italy). J Geodyn 40:73–91

Tomlinson EL, Arienzo I, Civetta L, Wulf S, Smith VC, Hardiman M, Lane CS, Carandente A, Orsi G, Rosi M, Muller W, Menzies MA (2012) Geochemistry of the Phlegraean Fields (Italy) proximal sources for major Mediterranean tephras: implications for the dispersal of Plinian and co-ignimbritic components of explosive eruptions. Geochim Cosmochim Acta 93:102–128

Tonarini S, Leeman WP, Civetta L, D'Antonio M, Ferrara G, Necco A (2004) B/Nb and $\delta^{11}B$ systematics in the Phlegrean Fields District, Italy. J Volcanol Geoth Res 133:123–139

Tonarini S, D'Antonio M, Di Vito MA, Orsi G, Carandente A (2009) Geochemical and B-Sr-Nd isotopic evidence for mingling and mixing processes in the magmatic system that fed the Astroni volcano (4.1–3.8 ka) within the Campi Flegrei caldera (Southern Italy). Lithos 107:135–151

Tondi R, De Franco R (2003) Three-dimensional modeling of Mount Vesuvius with sequential integrated inversion. J Geophys Res 108(B5). doi:10.1029/2001jb001578

Trigila R, De Benedetti AA (1993) Petrogenesis of Vesuvius historical lavas constrained by Pearce element ratios analysis and experimental phase equilibria. J Volcanol Geoth Res 58:315–343

Turco E, Zuppetta A (1998) A kinematic model for the Plio-Quaternary evolution of the Tyrrhenian-Apenninic system: implications for rifting processes and volcanism. J Volcanol Geoth Res 82:1–18

Turi B, Taylor HP, Ferrara G (1991) Comparisons of $^{18}O/^{16}O$ and $^{87}Sr/^{86}Sr$ in volcanic rocks from the Pontine Islands, M. Ernici and Campania with other areas in Italy. In: Taylor HP, O'Neil JR, Kaplan IR (eds) Stable isotope geochemistry: a tribute to Samuel Epstein (Geochem Soc Spec Publ 3):307–324

Vai GB, Martini PI (eds) (2001) Anatomy of an Orogen. The Apennines and adjacent Mediterranean basins. Kluwer, Dordrecht 632 pp

Ventura G, Vilardo G, Bruno PP (1999) The role of flank failure in modifying the shallow plumbing system of volcanoes: an example from Somma-Vesuvius, Italy. Geophys Res Lett 26:3681–3684

Vezzoli L (ed) (1988) Island of Ischia. Ric Sci CNR Rome, 114, 10:133 pp

Villemant B (1988) Trace element evolution in the Phlegrean Fields (Central Italy): fractional crystallization and selective enrichment. Contrib Mineral Petrol 98:169–183

Vollmer R (1976) Rb-Sr and U-Th-Pb systematics of alkaline rocks: the alkaline rocks from Italy. Geochim Cosmochim Acta 40:283–295

Wohletz K, Orsi G, De Vita S (1995) Eruptive mechanisms of the Neapolitan-Yellow-Tuff interpreted from stratigraphic, chemical, and granulometric data. J Volcanol Geoth Res 67:263–290

Woo JYL, Kilburn CRJ (2010) Intrusion and deformation at Campi Flegrei, southern Italy: sills, dikes, and regional extension. J Geophys Res, Solid Earth 115. doi:10.1029/2009jb006913

Wortel MJR, Spakman W (2000) Subduction and slab detachment in the Mediterranean-Carpathian region. Science 290:1910–1917

Zambonini F (1935) Mineralogia Vesuviana. Rend Reg Accad Sci Fis Mat Napoli, Ser 2 20:1–463

Zollo A, Gasparini P, Virieux J, Le Meur H et al (1996) Seismic evidence for a low-velocity zone in the upper crust beneath Mount Vesuvius. Science 274:592–594

Zollo A, Gasparini P, Virieux J, Biella G, Boschi E et al (1998) An image of Mt. Vesuvius obtained by 2D seismic tomography. J Volcanol Geoth Res 82:161–173

Zollo A, Maercklin N, Vassallo M, Dello Iacono D, Virieux J, Gasparini P (2008) Seismic reflections reveal a massive melt layer feeding Campi Flegrei caldera. Geophys Res Lett 35. doi:10.1029/2008gl034242

8 The Apulian Province (Mount Vulture)

Abstract

Mount Vulture (0.75–0.14 Ma) is an isolated stratovolcano with a central caldera sited east of the Apennine chain near to the Apulia foreland, away from other Italian recent volcanoes. Mount Vulture is built up of undersaturated alkaline rocks that show high enrichments in both Na and K. Carbonatitic pyroclastics and a few lavas were erupted late in the history of the volcano. Silicate rocks range from foidite, melilitite and haüynite to haüyne-bearing phonolite. Incompatible element patterns show arc-type negative anomalies of HFSE but also contain relative depletion of K and Rb, a typical feature of OIB magmas. He-isotope ratios ($R/R_A \sim 6.0$) are much higher than other volcanoes in the Italian peninsula and resemble Etna. Carbonatites have similar radiogenic isotope ratios as the silicate rocks supporting an origin by unmixing. Overall, geochemical data suggest a magma origin within an OIB-type upper mantle that was contaminated by subduction-related fluids. OIB components were provided by the asthenospheric mantle from the foreland, whereas the arc-type signatures were furnished by fluids released from the detached Apulia-Ionian slab and associated sediments. Magmas were generated by redox melting in a post-collisional setting above the detached and sinking plate after slab breakoff.

Keywords

Mount Vulture · Melilitite · Haüynite · Carbonatite · Alkaline rocks · Redox melting · Silicate-carbonate liquid unmixing · Ionian plate · Apulia · Geodynamics

A. Peccerillo, *Cenozoic Volcanism in the Tyrrhenian Sea Region*, Advances in Volcanology, DOI 10.1007/978-3-319-42491-0_8

8.1 Introduction

The Apulian Province, also referred to as the Lucanian Province (Conticelli et al. 2010), consists of a single volcano (Mount Vulture) and a few peripheral minor centres. Vulture is the only young Italian volcano sited east of the Apennine chain, near to the western boundary of the Apulia foreland (Fig. 8.1). Activity was characterised by emplacement of silica undersaturated alkaline lavas and pyroclastics, which, in contrast with other alkaline volcanoes in central-southern Italy, are highly enriched in both sodium and potassium. These characteristics led Washington (1906) to establish a separate magmatic province for Mount Vulture, an assessment validated by the finding of carbonatitic rocks among the latest products of this volcano (Stoppa and Woolley 1997; Stoppa and Principe 1997; D'Orazio et al. 2007).

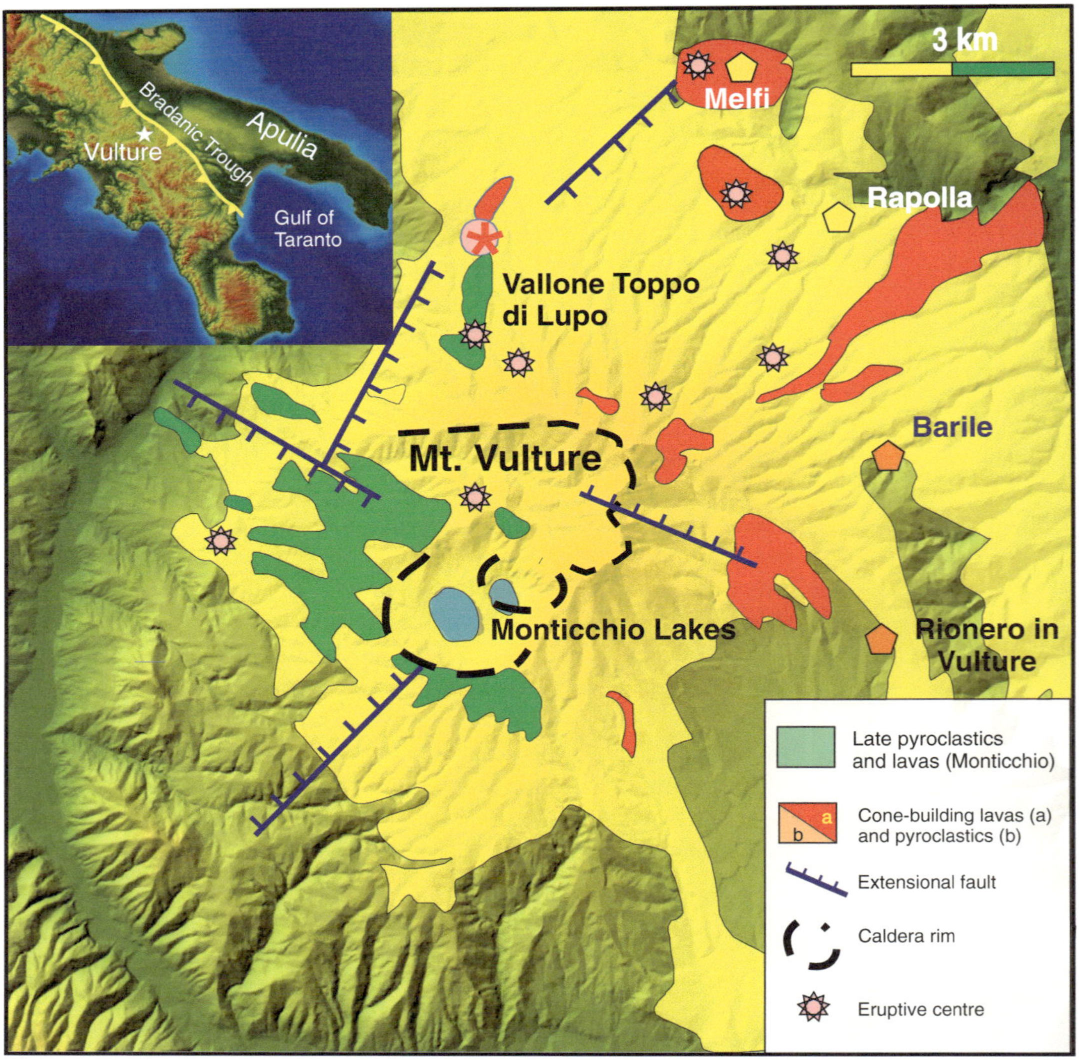

Fig. 8.1 Schematic geological map of Vulture volcano, simplified after Giannandrea et al. (2006). Inset shows position of Vulture volcano and the main structural units in the region

8.2 Regional Geology

Mount Vulture is located at the eastern margin of the Southern Apennine thrust front, near to the Bradanic Trough and Apulia. The latter is a portion of the Adriatic foreland, a promontory of African plate delimited by the Apennine chain on the west and by the Dinarids on the east. Field, geophysical, and exploration well data indicate homogeneous structural-lithological features for Apulia, characterised by some 7000 m of sedimentary rocks overlying a Paleozoic basement. The sedimentary sequence consists of Permian to Early Triassic terrigenous deposits (Verrucano auct.), Triassic dolomites and evaporites, Lower Jurassic to Cretaceous limestones, dolomitic limestones and dolostones, overlain by thin discontinuous Paleocene-Oligocene organogenic deposits and calcarenite, and Neogene-Quaternary carbonate-terrigenous sediments (e.g. Doglioni et al. 1994; Del Ben et al. 2015).

The Bradanic Trough (Fossa Bradanica) is a narrow NW-SE trending sedimentary foredeep basin located between the southern Apennines thrust front and Apulia. The Bradanic basin is filled with several kilometres thick sequences of Pliocene–Lower Pleistocene terrestrial and marine sands and shales, which rest over the westward dipping Apulian foreland. The contact between Apulia carbonate platform and the overlying foredeep sediments occurs at about 5 km depth (Boenzi et al. 1987), where the magma chamber of Vulture volcano probably developed.

The Vulture area experienced an early phase of compressional tectonics related to Adria-Southern Apennine convergence, which lasted until Middle Pleistocene when an inversion of stress regime generated NE-trending extensional and strike-slip faults (e.g. Bonini et al. 2011; Palladino 2011 and references therein). Such a modification started about 0.8 Ma before present, almost contemporaneously with the onset of Vulture volcanism (e.g. Cello et al. 1982) and with deposition of undeformed foredeep sequences on the outer thrust front (e.g. Patacca and Scandone 2001). A later onset of tectonic inversion is recognised in the Gulf of Taranto south of Vulture area, where extension started at about 0.2–0.3 Ma (Pieri et al. 1997). Contemporaneously, there was an uplift of the Apulia foreland, which contrasts with subsidence in the central Adriatic area still continuing at present.

Moho depth beneath Apulia is about 30 km and increases going towards the southern Apennine chain. The thickness of the lithosphere ranges from 120 to 70 km and decreases southward, a feature that has been attributed to erosion by Adriatic asthenosperic mantle flow through the slab window opened by horizontal tear along the Adriatic-Ionian slab (Miller and Piana Agostinetti 2011; see Chap. 7).

8.3 Mount Vulture

8.3.1 Volcanology and Stratigraphy

Mount Vulture is a 1326 m high composite cone with two large calderas (Fig. 8.1), constructed at the intersection between the NE-SW trending Ofanto graben and NW-SE faults (Giannandrea et al. 2006). The volcanic sequence consists of dominant pyroclastic rocks and minor lava flows, which cover an area of about 150 km^2. Activity took place between about 0.75 and 0.14 Ma (La Volpe et al. 1984; Villa and Buettner 2009). The oldest rocks consist of dykes, domes, and ignimbrites, whose emplacement was followed by voluminous central-type activity that built up the main cone between 600–550 ka and generated a central polygenetic caldera. Rock compositions of this phase range from tephrite and basanite to phonotephrite, and include a few melilitite, melafoidites and haüynite or haüynophyre. The latter is a variety of foidite with haüyne as dominant foid, which forms a large 0.56 Ma old peripheral outcrop at Melfi. The latest history of Vulture volcano is characterised by a long period of quiescence, followed by explosive and effusive eruptions of carbonatite-melilitite magmas at about 140 ka, and by the formation of two intra-caldera maars occupied by the Lakes of Monticchio (Stoppa and Principe 1997; D'Orazio et al. 2007).

The Vulture volcano is affected by intense CO_2 degassing with a total budget of 4.85×10^8 mol/yr. This high value supports the existence of actively degassing mantle-derived melts at shallow depth (Nuccio et al. 2014; Caracausi et al. 2015).

The pyroclastic deposits of Mount Vulture, especially around the Monticchio maars, contain abundant ultramafic nodules and megacrysts of clinopyroxene, amphibole, and phlogopite. Some xenoliths represent cumulate rocks whereas others are fragments of lithospheric mantle (Stoppa and Principe 1997; Jones et al. 2000; Downes 2001; Downes et al. 2002).

8.3.2 Petrography and Mineral Chemistry

The Vulture volcano erupted both silicate and carbonate magmas. Silicate volcanics exhibit variable K_2O/Na_2O and degree of silica undersaturation (Fig. 8.2a). They range from foidite, tephrite and basanite to tephriphonolites and phonolite (Fig. 8.2b), with minor melafoidites and melilitites. Some samples plotting in the field of basalts have high LOI contents (>3.5 %) and have likely undergone secondary loss of alkalis. Bindi et al. (1999) distinguished two series of rocks: a feldspar-bearing basanite-tephrite to phonolite suite and a volumetrically subordinate feldspar-free melilitite, melafoidite, and haüynophyre series. Basanites contain phenocrysts of MgO-rich olivine ($\sim Fo_{90-85}$) with Cr-spinel inclusions, colourless to pale green diopside, and haüyne set in a glassy to microcrystalline groundmass containing magnetite, plagioclase, Sr-Ba-rich anorthoclase and rare phlogopite. Tephrites and phonolitic tephrites contain phenocrysts of complexly zoned diopside to hedenbergite clinopyroxene, haüyne, and leucite with some plagioclase, amphibole and biotite; groundmass consists of plagioclase, anorthoclase, clinopyroxene, feldspathoids and accessory apatite; corroded olivine has been also observed. Phonolites show phenocrysts of Ba-Sr-rich alkali feldspar, haüyne, green clinopyroxene, leucite and some melanite garnet, magnetite, and apatite. Melilitite is characterised by phenocrysts of MgO-rich melilite, Ti-rich clinopyroxene, and Fe–Ti oxides, set in a holocrystalline groundmass composed of the same phases plus nepheline, leucite, haüyne, apatite, perovskite, and schorlomite garnet; calcite, clinopyroxene, nepheline, and magnetite are also present. Melafoidite (Santa Caterina locality) shows clinopyroxene phenocrysts surrounded by a

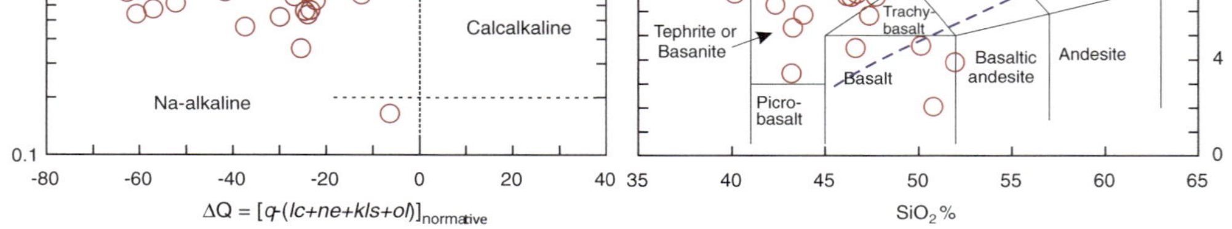

Fig. 8.2 **a** K_2O/Na_2O versus ΔQ diagram for mafic silicate volcanic rocks from the Vulture volcano (MgO > 4 wt%). ΔQ is the algebraic sum of normative quartz minus undersaturated minerals (see Chap. 1); **b** TAS diagram for the Vulture silicate rocks

matrix made of clinopyroxene, nepheline, leucite, haüyne, Fe–Ti oxides, gehlenite-rich melilite, and apatite. The Melfi haüynophyre consists of abundant haüyne, clinopyroxene, leucite, and apatite phenocrysts set in a holocrystalline groundmass of the same phases plus nepheline, Na–Fe-Sr-rich melilite and magnetite (De Fino et al. 1986; Melluso et al. 1996; Beccaluva et al. 2002).

Melilitite-carbonatite tuff sequences and lavas erupted late in the history of Vulture volcano. Pyroclastic rocks consist of a mixture of silicate and carbonate material. Important mineral phases include melilite, diopside, phlogopite, Sr, Ba, REE-rich calcite, apatite, pargasitic amphibole, spinel, perovskite and haüyne (Stoppa and Principe 1997). The carbonatite dikes and lavas cropping out at Vallone Toppo di Lupo have poorly porphyritic textures with phenocrysts of calcite, melilite, apatite and Ti-magnetite set in a matrix made up of the same phases plus monticellite and Ca–Fe-Ti-Nb-LREE oxides. Melilite shows high contents of gehlenite and soda-melilite components and resembles closely melilite from the associated silicate lavas. Calcite is the only carbonate phase occurring in the rock, which is classified as alvikite (D'Orazio et al. 2007).

Mantle xenoliths from the Monticchio pyroclastic deposits consist of spinel lherzolites, harzburgites, wehrlites, dunites and some pyroxenites (Jones et al. 2000; Downes et al. 2002). They are fresh, porphyroclastic in texture and contain olivine (Fo_{90-92}), orthopyroxene ($\sim En_{90}$), Mg–Cr-rich clinopyroxene, glass, some carbonates, chrome-spinel, phlogopite, and accessory apatite and sulphides. Geothermobarometric measurements yielded equilibration P-T conditions of 14–22 kbar and 1050–1150 °C (Jones et al. 2000). Rare relics of Mg–Ca garnet may testify a deeper origin (Stoppa et al. 2009).

8.3.3 Petrology and Geochemistry

Major elements are rather scattered in the Vulture silicate rocks, whereas more linear trends are observed for carbonatites (Fig. 8.3). Ni and Cr show a wide range of values with some silicate lavas exhibiting concentrations close to those of mantle-equilibrated melts. The Vulture volcanics are highly enriched in volatile elements, especially sulfur and chlorine, whose concentrations reaches a per cent level. Cl and S show smooth positive correlations with Na_2O in the silicate rocks (see De Fino et al. 1986), which contrast with the scattering observed for other elements. REE are fractionated (Fig. 8.4a) with small negative Eu anomalies and very high values of La/Yb ratios (up to 300) in some phonolites. Carbonatites show similar patterns to mafic silicate rocks, but enrichment in LREE is higher (D'Orazio et al. 2007; Mongelli et al. 2013). Incompatible element patterns of silicate rocks (Fig. 8.4b) are fractionated and show OIB-type concentrations of Ta and Nb and moderate HFSE negative anomaly, a feature that is also observed in the high-pressure ultramafic xenoliths (Downes et al. 2002). Notably, there are negative spikes of Rb and K, which are not found in other alkaline rocks from the Italian peninsula, and are typically observed in some OIB-type Na-alkaline magmas, such as those from Iblei and Etna (see Chap. 10). Carbonatites have higher contents of several incompatible elements than silicate rocks (e.g. Sr, Ba, Nb, LREE), but are depleted in Rb, K, Zr and Ti. These compositions strikingly contrast with those observed in the San Venanzo and Cupaello carbonate-rich rocks, which are moderately to strongly depleted in all the incompatible elements relative to the associated kamafugitic lavas (see Chap. 3).

$^{87}Sr/^{86}Sr$ (0.7052–0.7070) and $^{143}Nd/^{144}Nd$ (0.5126–0.5127) show similar values for silicate rocks and carbonatites, and resemble closely the Campania Province (Fig. 8.5a). Compositions of ultramafic xenoliths are somewhat more primitive, with $^{87}Sr/^{86}Sr$ and $^{143}Nd/^{144}Nd$ of separated and leached clinopyroxenes ranging between 0.7042–0.7058 and 0.51260–0.5131, respectively (Downes et al. 2002). Lead isotope ratios (Fig. 8.5b) are comparable to slightly more radiogenic than Campania volcanoes ($^{206}Pb/^{204}Pb$ = 19.14–19.55;

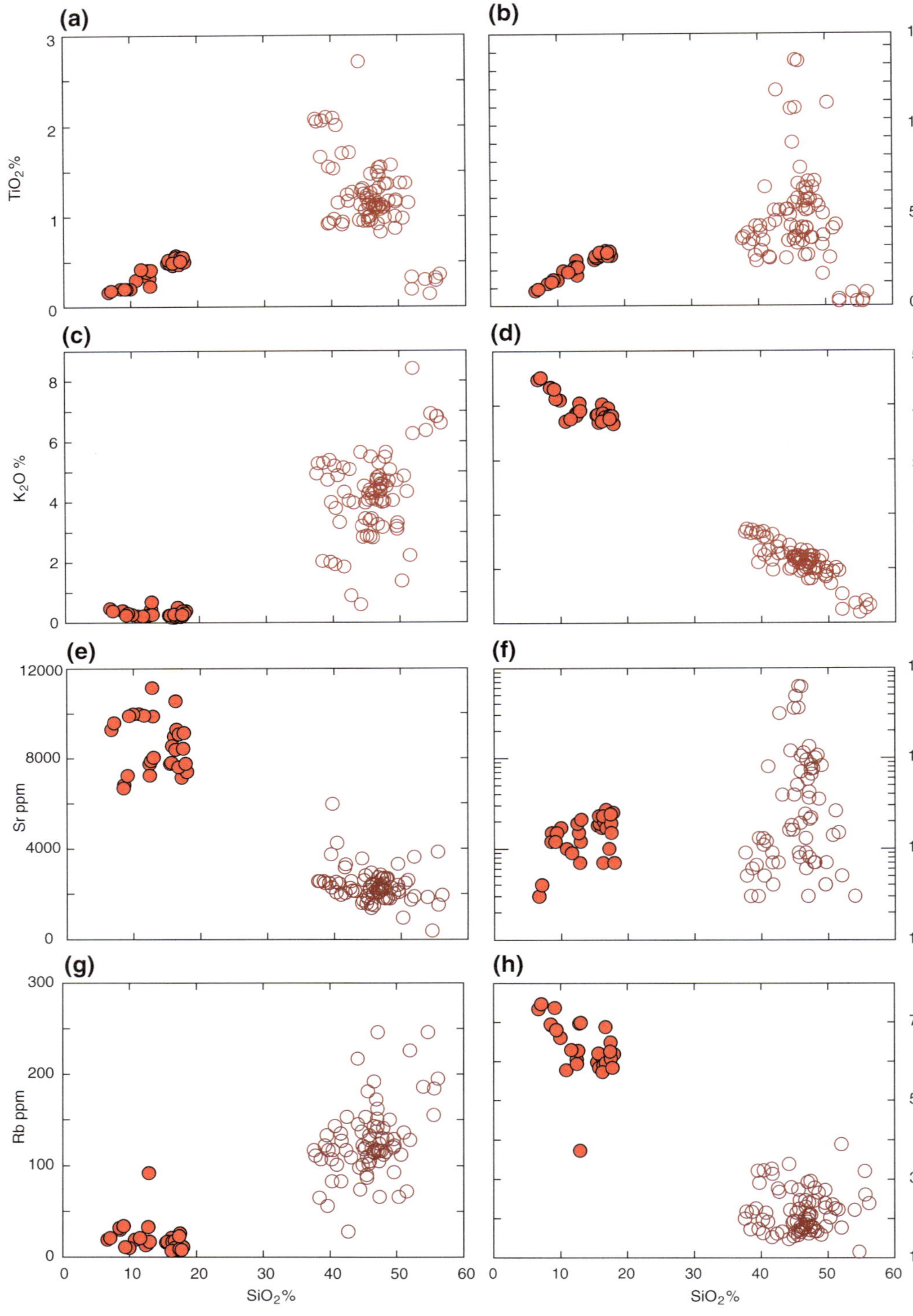

Fig. 8.3 Variation diagram of selected major and trace elements for Mount Vulture rocks. *Open circles* silicate rocks; *full circles* carbonatites

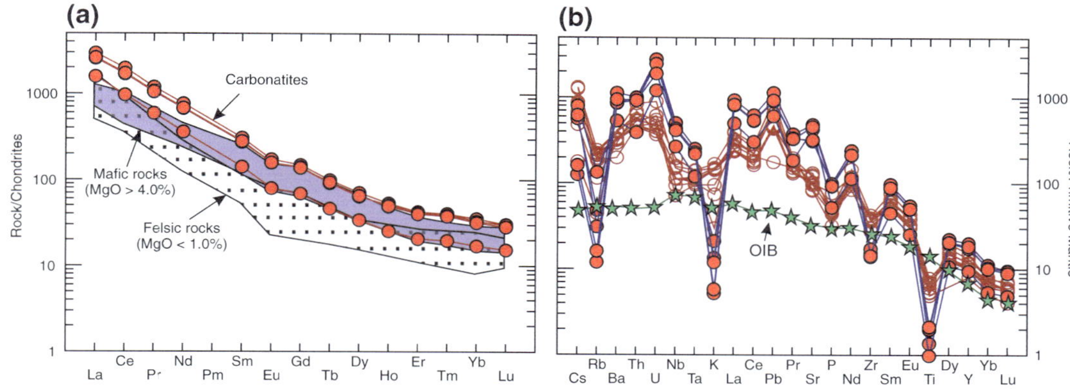

Fig. 8.4 **a** REE patterns of Mount Vulture rocks; **b** Incompatible element patterns of mafic silicate rocks (*open circles*) and carbonatites (*full circles*). OIB composition is from Sun and McDonough (1989)

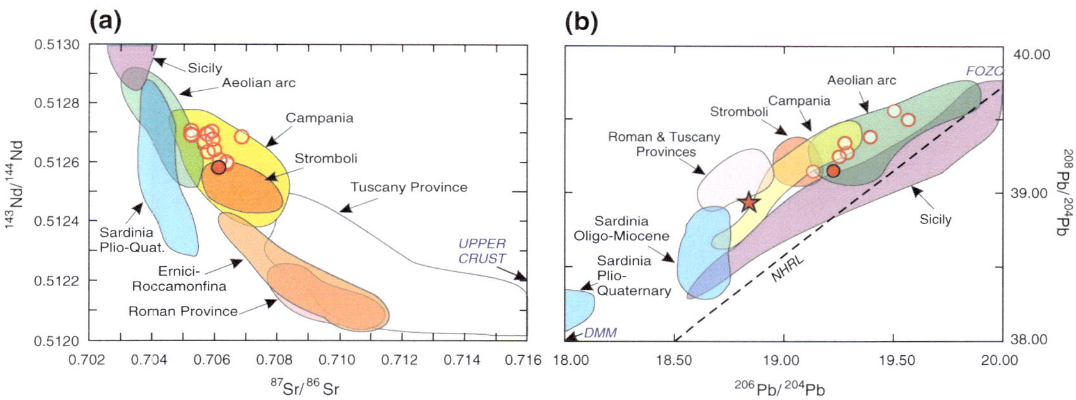

Fig. 8.5 Sr-Nd–Pb isotope compositions of Vulture volcanics. *Open circles* silicate rocks; *full circles* carbonatites. Fields of other Italian magmatic provinces are shown for comparison. The *red star* is the average sediment from the Ionian Sea (DSDP Site 374; Klaver et al. 2015)

$^{207}Pb/^{204}Pb$ = 15.67–15.72; $^{208}Pb/^{204}Pb$ = 39.17–39.56; De Astis et al. 2006; D'Orazio et al. 2007). Stable isotope studies are scanty. Cavarretta and Lombardi (1990) and Marini et al. (1994) found $\delta^{34}S$ = +4.0‰ to +6.6‰ relative to VCDT. Preliminary oxygen isotopic investigation on clinopyroxenes from mafic rocks yielded values of $\delta^{18}O_{SMOW}$ around +5.4‰ to +5.8‰ (Stoppa and Principe 1997; Dallai, personal communication). Calcite from alvikite lavas has $\delta^{18}O_{SMOW}$ = +10.2‰ to +12.7‰, which is much lower than values observed in the Central Apennines carbonate-rich volcanics. Carbon isotope signatures of calcite show $\delta^{13}C_{PDB}$ = −6.7‰ to −9.2‰. Boron isotope values fall in the range $\delta^{11}B$ = 5.6‰ to −9.6‰ for both silicate rocks and carbonatites (Stoppa and Principe 1997; D'Orazio et al. 2007). $^{3}He/^{4}He$ on a few samples show $R/R_A \sim 6.0$, much higher than Campania volcanoes and comparable to Etna (Martelli et al. 2008).

8.3.4 Petrogenesis of Vulture Magmas

De Fino et al. (1986) suggested that the rock series from foidite to phonolite could be related to combined fractional crystallisation and mixing, mainly with separation of clinopyroxene, biotite, Fe–Ti oxides and accessory apatite. They also suggested that the positive correlations between halogens and Na_2O could result from assimilation of evaporitic rocks, a process, however, believed as limited by Marini et al. (1994) on the basis of sulphur isotope data. Melluso et al. (1996) and Beccaluva et al. (2002) accepted shallow level fractional crystallisation as a main evolutionary mechanism, and advocate basanitic melts as parents of the basanite-phonolite suite. Fractionation was dominated by olivine and clinopyroxene in the primitive basanites progressively joined by clinopyroxene, plagioclase and alkali feldspar. A distinct origin by unmixing of a sulphur-rich carbonatite melt has been proposed for haüynophires by Panina and Stoppa (2009).

Beccaluva et al. (2002) recognise different types of primary magmas at Vulture, suggesting that the melilitite-basanite-foidite association derives from variable degrees of partial melting of a heterogeneous mantle source, at less than 90–100 km depth. A similar magma segregation depth (about 2.2 to 3.5 GPa) is also inferred by Jones et al. (2000) and Solovova et al. (2005), on the basis of geobarometric investigations on xenoliths. Pressure conditions of magma segregation correspond to the middle-lower lithosphere in the area and support the currently held hypotheses of a lithospheric origin for worldwide undersaturated Na-alkaline magmatism (e.g. Pilet et al. 2011; Dai et al. 2014). The variable K/Na ratios of Vulture mafic rocks require the occurrence of both Na- and K-rich phases such as phlogopite and amphibole in the source, with different relative amounts of these phases entering into the melt. Amphibole occurs as megacryst in the Vulture volcanics but has not been observed as primary phase in the mantle xenoliths (Jones et al. 2000).

There is a general agreement that the mantle source of Vulture magmas has been metasomatically modified by fluids/melts coming from depth, although the nature and origin of these fluids are controversial. Based on petrology/geochemistry of xenoliths, Downes et al. (2002) suggest that the lithospheric mantle in the area was affected by an early episode of melt extraction, followed by metasomatic enrichments by subduction-related silicate melts. Beccaluva et al. (2002) advocate a two-stage enrichment process, with early infiltration of intraplate-type Na-silicate-carbonatite fluid, followed by a second subduction-related metasomatic enrichment. Other authors propose metasomatism by carbonatitic melts coming from a mantle plume (e.g. Bell et al. 2004, 2013) or directly from the core-mantle boundary (Stoppa et al. 2009).

It has been previously shown that trace element data of lavas and mantle xenoliths clearly indicate the coexistence of both OIB- and arc-type signatures at Vulture. Subduction-related geochemical characteristics include negative anomalies of HFSE, and enrichment in some LILE; OIB-type components are highlighted by depletion on K and Rb relative to other LILE, high absolute enrichments in HFSE (e.g. Nb, Ta), and high He-isotope ratios. These data lead to conclusion of magma origin in OIB-type mantle sources, contaminated by subduction-related fluids or melts.

The occurrence of coexisting arc- and OIB-type characteristics has been also envisaged for the Campania volcanoes (Chap. 7). However, Campania magmas are enriched in Rb and K relative to other LILE, as observed for many arc rocks. Moreover, He-isotope ratios are much higher at Vulture than in the Campania rocks and fumaroles. Therefore, OIB-type signatures are more evident at Vulture, calling for a greater role of OIB-type components.

The origin of Vulture carbonatites has been generally attributed to unmixing from a melilitite silicate melt. Liquid immiscibility is strongly supported by coexisting carbonate and silicate phases separated by a meniscus-shaped boundary in some melt inclusions from Monticchio phenocrysts (Solovova et al. 2005). Pressures of 0.2 and 2.0 GPa have been suggested for this process by D'Orazio et al. (2007) and Rosatelli et al. (2007). In contrast, a direct origin from the upper mantle has been envisioned by Stoppa et al. (2009). Enrichments in Sr, Ba, and LREE of carbonatites relative to associated silicate rocks are well explained by unmixing since these elements preferentially partition into carbonate with respect to silicate melts. However, relative depletion of K and Rb in carbonatites is somewhat problematic, since carbonate/silicate melt partition coefficients for these elements are not far from unity (Martin et al. 2012).

8.4 Geodynamic Significance of Mount Vulture

The Vulture volcano is sited away from other Italian Quaternary volcanoes, in a very particular setting between the Apulia foreland and the eastern side of the Apennine chain. Coexistence of island-arc and OIB-type geochemical signatures reflect such a particular geodynamic setting. However, the association of carbonatites and melilitites is typical of cratonic areas and is rare or absent in subduction-related environments. These unique features make the geodynamic significance of Mount Vulture a particularly challenging issue, but most of the suggested hypotheses are rather speculative and inadequately supported by data.

According to D'Orazio et al. (2007), Mount Vulture developed over an along-dip tear of the Adriatic-Apulian slab, resulted from the differential retreat of the foreland at different sides of the Tremiti-Mattinata-Ofanto transfer zone. According to these authors, the hybrid OIB-orogenic compositional characteristics of Vulture magmas could be explained by a model that envisages (i) flowing of upper mantle material from the west against the retreating Adriatic slab, (ii) gradual ascent along the vertical slab fracture zone, (iii) contamination by slab-derived fluids, (iv) decompression melting at shallow levels, and (v) ascent of magmas along the transfer fault.

The occurrence of vertical tear faults in the subducting lithosphere has been suggested at several places along the Apennine-Maghrebide system, as a result of different rates of slab rollback (Rosenbaum et al. 2008). The volcanic rocks erupted above these slab discontinuities show geochemical evidence of an OIB-type component, which is provided by foreland asthenospheric mantle ascending through the slab tear faults. However, an association of melilitite-carbonatite rocks with hybrid arc-intraplate geochemical signatures has not been found elsewhere along the Apennines and remains unique to Vulture. Therefore, an exceptional magmatic association likely requires an unusual tectonic setting.

Peccerillo and Frezzotti (2015) suggested that the Vulture volcano is related to the occurrence of both a vertical and an along-strike tear faults in the Ionian-Adriatic slab. Such a structural setting may have started in the Middle Pleistocene, contemporaneously with the transition from a compressional to an extensional tectonic regime. The along-strike tear fault progressively propagated toward the south, as testified by the younger age of the tectonic inversion in the Gulf of Taranto (0.2–0.3 Ma; Pieri et al. 1997). This process brought to the formation of a narrow lithospheric slab that remained attached to the Ionian plate and retreated southeastward by rollback (see Chap. 7, Fig. 7.22). An important effect of such process was the suctioning of asthenospheric mantle material from the Apulia foreland, which migrated above the mantle wedge, flowing around the edge of the retreating Ionian slab. Such an allochthonous mantle was metasomatised by subduction fluids, ultimately generating a hybrid OIB-arc source. The Vulture magmas, as well as those of other volcanoes above the Ionian subduction zone, i.e. the Campania Province and Stromboli, were generated in such a hybrid mantle (De Astis et al. 2000, 2006; Peccerillo 2001). The stronger OIB-type signatures at Vulture relative to Campania and Stromboli could be simply related to its proximity to the slab tear zone, where geochemical effects of inflowing mantle were much stronger. Therefore, the magmatism of the Campania Province, Stromboli and Vulture would have similar geodynamic significance since they are all related to the inflow of intraplate mantle from the foreland and to the contamination by fluids released from the oceanic-type Ionian slab and associated sediments.

Partial melting beneath Mount Vulture could be related to fluid release from the sinking slab and/or mantle decompression melting in an extensional tectonics setting. However, these processes are expected to generate more widespread magmatism, and might not apply to a single isolated volcano. Therefore, a redox melting mechanism has been suggested by Peccerillo and Frezzotti (2015). Redox melting is a process by which melt is generated when a rock comes into contact with fluids characterised by contrasting oxidation state or also when rocks with distinct oxygen fugacities are brought close to each other by tectonic movements (e.g. Foley 2011). Redox melting by carbonation can occur in the presence of both CO_2-H_2O and is able to generate SiO_2-undersaturated magmas ranging from carbonatitic to melilititic, depending on the degree of partial melting (Foley et al. 2009). Dehydration reaction along a subducted slab, addition of carbonate sediments to the upper mantle and hydrothermal alteration of slab minerals are all able to induce heterogeneity of oxidation states in the upper mantle, thereby creating the conditions for redox melting at a local scale. Redox melting may be unlikely in subduction settings because of the low thermal gradients. However, in the case of Mount Vulture, tear-off of the slab, inflow of mantle and the consequent increase of temperatures may have favoured redox melting.

8.5 Summary and Conclusions

Mount Vulture is sited on the eastern side of the Apennine chain near to the Apulia foreland, away from the bulk of recent volcanism along the Tyrrhenian side of the Italian peninsula. The volcano consists of an association of silicate and carbonatic lavas and pyroclastic rocks, which is rarely, if ever, observed in orogenic areas. Silicate rocks are rich in alkalis and range from foidite and melilitite to phonolite. Incompatible element patterns show small negative anomalies of HFSE, a typical island-arc geochemical signature. However, there is also depletion in K and Rb relative to other LILE, which is typical of many OIB-type magmas. He-isotope ratios are much higher than other volcanoes in the Italian peninsula and resemble OIB-type rocks from Etna. Carbonatites have similar radiogenic isotope ratios as the silicate rocks supporting an origin by unmixing of a single carbonate-rich silicate liquid. Overall, geochemical data suggest a magma origin within an upper mantle characterised by OIB-type signatures, which was contaminated by fluids coming from an oceanic slab and associated sediments.

The geodynamic significance of Vulture volcano is still a matter of debate. A hypothesis is that volcanism developed above the intersection between a vertical and an along-strike tear zone of the Apulia-Ionian lithosphere. This brought to the formation of a detached narrow slab that retreated southeastward, sinking into the upper mantle. Slab retreat caused suctioning of asthenospheric mantle from the foreland above the sinking slab, where it underwent contamination by subduction fluids. The contribution by suctioned mantle and slab fluids imparted both OIB- and arc-type signatures to the Vulture magmas. A similar origin has been envisaged for the Campania Province and for Stromboli, although the role of OIB-type components is much less evident than at Vulture (Table 8.1).

Table 8.1 Major, trace element and isotopic compositions for representative rocks from the Mount Vulture volcano

Rock	Melilitite*	Haüynite	Basanite	Phonotephrite	Phonolite**	Carbonatite
–	1,2,3,5	3	3	4	2	4
SiO_2 %	38.48	41.57	45.1	51.5	55.75	16.3
TiO_2	2.08	0.91	1.07	1.15	0.32	0.46
Al_2O_3	13.67	18.54	15.27	19.1	22.33	5.41
Fe_2O_3	10.94	7.36	8.03	9.46	2.65	6.51
MnO	0.23	0.23	0.16	0.21	0.14	0.35
MgO	3.65	2.7	9.02	4.48	0.43	2.96
CaO	16.84	11.54	11.97	9.61	3.36	40.2
Na_2O	3.83	7.33	3.53	1.64	5.97	0.47
K_2O	5.00	5.14	2.88	2.23	6.74	0.17
P_2O_5	1.05	0.68	0.96	0.7	0.10	2.18
LOI	4.12	3.43	1.49	5.14	2.21	22.4
Sc ppm	13	5.2	28.1	13.7	–	7
V	273	199	194	200	62	316
Cr	7	9	483.3	15	–	7
Co	29	–	61	21.4	–	19
Ni	15	10	157	18	5	22
Rb	116	135	118	72	178	21.2
Sr	2497	3156	1794	1740	2434	10540
Y	71	51	37	46	28	87
Zr	548	513	339	411	467	210
Nb	126	155	51	82	142	332
Cs	7.08	22	9.9	3.8	–	5.3
Ba	2064	1872	1383	2050	1530	6451
La	217	327	165	223	274	604
Ce	426	554	291	400	413	1027
Pr	48	–	–	45.1	–	98.9
Nd	186	166	114.7	151	111	313
Sm	36.4	28.6	23.4	25.3	14.25	42
Eu	7.86	6.2	5.06	5.57	3.10	9.2
Gd	27.05	–	–	20.2	9.73	28.2
Tb	3.49	2.3	1.83	2.15	–	3.4
Dy	16.40	–	–	9.52	5.13	16.1
Ho	2.67	–	–	1.55	–	2.88
Er	5.90	–	–	4.06	2.27	6.8
Tm	0.74	–	–	0.45	–	1.03
Yb	4.295	4.3	2.48	3	1.97	5.6
Lu	0.49	0.54	0.3	0.42	0.28	0.76
Hf	11.3	6.89	6.46	8	–	–

(continued)

Table 8.1 (continued)

Rock	Melilitite*	Haüynite	Basanite	Phonotephrite	Phonolite**	Carbonatite
Ta	6.78	9.1	3.89	4.1	–	10.2
Pb	57.5	95	33	57.3	–	70
Th	51.8	86	39	58.4	106	85
U	12.38	21	10.5	8.44	–	56
$^{87}Sr/^{86}Sr$	0.70642	0.70574	0.70693	0.70595	0.705815	0.70615
$^{143}/^{144}Nd$	0.51260	0.51270	0.51269	0.51268	–	0.51258
$^{206}/^{204}Pb$	19.512	19.256	19.142	–	–	19.231
$^{207}Pb/^{204}Pb$	15.714	15.686	15.685	–	–	15.667
$^{208}Pb/^{204}Pb$	39.539	39.264	39.170	–	–	39.174

Data from: (1) De Fino et al. (1986); (2) Beccaluva et al. (2002); (3) De Astis et al. (2006); (4) D'Orazio et al. (2007); (5) Avanzinelli et al. (2008)
*Average of 5 analyses; **Average of 3 analyses

References

Avanzinelli R, Elliott T, Tommasini S, Conticelli S (2008) Constraints on the genesis of the potassium-rich Italian volcanics from U/Th disequilibrium. J Petrol 49:195–223

Beccaluva L, Coltorti M, Di Girolamo P, Melluso L, Milani L, Morra V, Siena F (2002) Petrogenesis and evolution of Mt. Vulture alkaline volcanism (Southern Italy). Mineral Petrol 74:277–297

Bell K, Castorina F, Lavecchia G, Rosatelli G, Stoppa F (2004) Is there a mantle plume below Italy? EOS 85:541–547

Bell K, Lavecchia G, Rosatelli G (2013) Cenozoic Italian magmatism - Isotope constraints for possible plume-related activity. J South Am Earth Sci 41:22–40

Bindi L, Cellai D, Melluso L, Conticelli S, Morra V, Menchetti S (1999) Crystal chemistry of clinopyroxene from alkaline undersaturated rocks of the Monte Vulture Volcano, Italy. Lithos 46:259–274

Boenzi F, La Volpe L, Rapisardi L (1987) Evoluzione geomorfologica del complesso vulcanico del Monte Vulture (Basilicata). Boll Soc Geol It 106:673–682

Bonini M, Sani F, Moratti G, Benvenuti MG (2011) Quaternary evolution of the Lucania Apennine thrust front area (Southern Italy), and its relations with the kinematics of the Adria Plate boundaries. J Geodyn 51:125–140

Caracausi A, Paternoster M, Nuccio PM (2015) Mantle CO_2 degassing at Mt. Vulture volcano (Italy): Relationship between CO_2 outgassing of volcanoes and the time of their last eruption. Earth Planet Sci Lett 411:268–280

Cavarretta G, Lombardi G (1990) Origin of sulfur in the quaternary perpotassic melts of Italy: evidence from haüyne sulfur isotope data. Chem Geol 82:15–20

Cello G, Guerra I, Tortorici L, Turco E, Scarpa R (1982) Geometry of the neotectonic stress field in Southern Italy: geological and seismological evidence. J Struct Geol 4:385–393

Conticelli S, Laurenzi MA, Giordano G, Mattei M, Avanzinelli R, Melluso L, Tommasini S, Boari E, Cifelli F, Perini G (2010) Leucite-bearing (kamafugitic/leucititic) and -free (lamproitic) ultrapotassic rocks and associated shoshonites from Italy: constraints on petrogenesis and geodynamics. In: Beltrando M, Peccerillo A, Mattei M, Conticelli S, Doglioni C (eds) J Virtual Explorer 36, pap er 20 doi:10.3809/jvirtex.2010.00251

D'Orazio M, Innocenti F, Tonarini S, Doglioni C (2007) Carbonatites in a subduction system: The Pleistocene alvikites from Mt. Vulture (southern Italy). Lithos 98:313–334

Dai L-Q, Zhao Z-F, Zheng Y-F (2014) Geochemical insights into the role of metasomatic hornblendite in generating alkali basalts. Geochem Geophys Geosyst 15. doi:10.1002/2014GC005486

De Astis G, Peccerillo A, Kempton PD, La Volpe L, Wu TW (2000) Transition from calc-alkaline to potassium-rich magmatism in subduction environments: geochemical and Sr, Nd, Pb isotopic constraints from the island of Vulcano (Aeolian arc). Contrib Mineral Petrol 139:684–703

De Astis G, Kempton PD, Peccerillo A, Wu TW (2006) Trace element and isotopic variations from Mt. Vulture to Campanian volcanoes: constraints for slab detachment and mantle inflow beneath southern Italy. Contrib Mineral Petrol 151:331–351

De Fino M, La Volpe L, Peccerillo A, Piccarreta G, Poli G (1986) Petrogenesis of Monte Vulture volcano, Italy: inferences from mineral chemistry, major and trace element data. Contrib Mineral Petrol 92:135–145

Del Ben A, Mocnik A, Volpi V, Karvelis P (2015) Old domains in the South Adria plate and their relationship with the West Hellenic front. J Geodyn 89:15–28

Doglioni C, Mongelli F, Pieri P (1994) The Puglia uplift (SE-Italy): an anomaly in the foreland of the Apenninic subduction due to buckling of a thick continental lithosphere. Tectonics 13:1309–1321 Mem Soc Geol It 52:457–468

Downes H (2001) Formation and modification of the shallow sub-continental lithospheric mantle: a review of geochemical evidence from ultramafic xenolith suites and tectonically emplaced ultramafic massifs of Western and Central Europe. J Petrol 42:233–250

Downes H, Kostoula T, Jones AP, Beard AD, Thirlwall MF, Bodinier JL (2002) Geochemistry and Sr-Nd isotopic compositions of mantle xenoliths from the Monte Vulture carbonatite-melilitite volcano, central southern Italy. Contrib Mineral Petrol 144:78–92

Foley SF (2011) A reappraisal of redox melting in the Earth's mantle as a function of tectonic setting and time. J Petrol 52:1363–1391

Foley SF, Yaxley GM, Rosenthal A, Buhre S, Rapp RP, Jacob DE (2009) The composition of near-solidus melts of peridotite in the presence of CO_2 and H_2O at 40 and 60 kbar. Lithos 112:274–283

Giannandrea P, La Volpe L, Principe C, Schiattarella M (2006) Unconformity-bounded stratigraphic units and evolutionary history of the middle Pleistocene Monte Vulture volcano, southern Apennines, Italy. Boll Soc Geol It 125:67–92

Jones AP, Kostoula T, Stoppa F, Wooley AR (2000) Petrography and mineral chemistry of mantle xenoliths in a carbonate-rich melilitic tuff from Vulture volcano, southern Italy. Mineral Mag 64:593–613

Klaver M, Djuly T, de Graaf S, Sakes S, Wijbrans J, Davies G, Vroon P (2015) Temporal and spatial variations in provenance of Eastern Mediterranean Sea sediments: implications for Aegean and Aeolian arc volcanism. Geochim Cosmochim Acta 153:149–168

La Volpe L, Patella D, Rapisardi L, Tramacere A (1984) The evolution of Monte Vulture volcano, Southern Italy: inferences from volcanological, geological and deep dipole electrical sounding data. J Volcanol Geoth Res 22:147–162

Marini L, Paiotti A, Principe C, Ferrara G, Cioni R (1994) Isotopic ratio and concentration of sulfur in the undersaturated alkaline magmas of Vulture volcano (Italy). Bull Volcanol 56:487–492

Martelli M, Nuccio PM, Stuart FM, Di Liberto V, Ellam RM (2008) Constraints on mantle source and interactions from He-Sr isotope variation in Italian Plio-Quaternary volcanism. Geochem Geophys Geosyst 9:Q02001. doi:10.1029/2007GC001730

Martin LHJ, Schmidt MW, Mattsson HB, Ulmer P, Hametner K, Günther D (2012) Element partitioning between immiscible carbonatite–kamafugite melts with application to the Italian ultrapotassic suite. Chem Geol 320–321:96–112

Melluso L, Morra V, Di Girolamo P (1996) The Mt. Vulture volcanic complex (Italy): evidence for distinct parental magmas and for residual melts with melilite. Mineral Petrol 56:225–250

Miller MS, Piana Agostinetti N (2011) Erosion of the continental lithosphere at the cusps of the Calabrian arc: evidence from S receiver functions analysis. Geophys Res Lett 38:L23301. doi:10.1029/2011GL049455

Mongelli G, Paternoster M, Rizzo G, Sansone MTC, Sinisi R (2013) Trace element geochemistry of the Mt Vulture carbonatites, southern Italy. Int Geol Rev 55:1541–1552

Nuccio PM, Caracausi A, Costa M (2014) Mantle-derived fluids discharged at the Bradanic foredeep/Apulian foreland boundary: The Maschito geothermal gas emissions (southern Italy). Mar Petrol Geol 55:309–314

Palladino G (2011) Tectonic and eustatic controls on Pliocene accommodation space along the front of the southern Apennine thrust-belt (Basilicata, southern Italy). Basin Res 23(5):591–614

Panina LI, Stoppa F (2009) Silicate-carbonate-salt liquid immiscibility and origin of the sodalite-hauyne rocks: study of melt inclusions in olivine foidite from Vulture volcano. S Italy Centr Eur J Geosci 1:377–392

Patacca E, Scandone P (2001) Late thrust propagation and sedimentary response in the thrust-belt-foredeep system of the southern Apennines (Pliocene-Pleistocene). In: Vai GB, Martini PI (eds) Anatomy of an Orogen. The Apennines and adjacent Mediterranean basins, Kluwer, Dordrecht, pp 401–440

Peccerillo A (2001) Geochemical similarities between Vesuvius, Phlegraean Fields and Stromboli volcanoes: petrogenetic, geodynamic and volcanological implications. Mineral Petrol 73:93–105

Peccerillo A, Frezzotti ML (2015) Magmatism, mantle evolution and geodynamics at the converging plate margins of Italy. J Geol Soc London 172:407–427

Pieri P, Vitale G, Beneduce P, Doglioni C, Gallicchio S, Giano SI, Loizzo R, Moretti M, Prosser G, Sabato L, Schiattarella M, Tramutoli M, Tropeano M (1997) Tettonica Quaternaria nell'area bradanico-ionica. Il Quaternario 10:535–542

Pilet S, Baker MB, Muentener O, Stolper E (2011) Monte carlo simulations of metasomatic enrichment in the lithosphere and implications for the source of alkaline basalts. J Petrol 52:1415−1442

Rosatelli G, Wall F, Stoppa F (2007) Calcio-carbonatite melts and metasomatism in the mantle beneath Mt. Vulture (Southern Italy). Lithos 99:229–248

Rosenbaum G, Gasparon M, Lucente FP, Peccerillo A, Miller MS (2008) Kinematics of slab tear faults during subduction segmentation and implications for Italian

magmatism. Tectonics 27, TC2008 doi:10.1029/2007TC002143
Solovova IP, Girnis AV, Kogarko LN, Konorikova NN, Stoppa F, Rosatelli G (2005) Compositions of magmas and carbonate-silicate liquid immiscibility in the Vulture alkaline igneous complex, Italy. Lithos 85:113–128
Stoppa F, Principe C (1997) Eruption style and petrology of a new carbonatitic suite from the Mt. Vulture Southern Italy: The Monticchio Lake formation. J Volcanol Geoth Res 78:251–265
Stoppa F, Woolley AR (1997) The Italian carbonatites: field occurrence, petrology and regional significance. Mineral Petrol 59:43–67
Stoppa F, Jones AP, Sharygin VV (2009) Nyerereite from carbonatite rocks at Vulture volcano: implications for mantle metasomatism and petrogenesis of alkali carbonate melts. Centr Eur J Geosci 1:131–151
Sun SS, McDonough WF (1989) Chemical and isotopic systematics of oceanic basalts: implications for mantle composition and processes. In: Saunders AD, Norry MJ (eds) Magmatism in ocean basins. Geol Soc London (Spec Publ 42):313–345
Villa IM, Buettner A (2009) Chronostratigraphy of Monte Vulture volcano (southern Italy): secondary mineral microtextures and $^{39}Ar/^{40}Ar$ systematics. Bull Volcanol 71:1195–1208
Washington HS (1906) The Roman Comagmatic Region. Carnegie Inst Washington Publ 57: 199 pp

9 The Aeolian Arc

Abstract

The Aeolian arc volcanic province consists of seven main islands and several seamounts in the Southern Tyrrhenian Sea. Rock compositions range from calcalkaline to shoshonitic and potassic alkaline, with an increase in potassium from the western to the central-eastern islands. $^{87}Sr/^{86}Sr$ increases and Nd–Pb–He isotope ratios decrease with potassium. Geochemical compositions of rocks from Stromboli resemble closely those of Campania volcanoes. Range of silica contents of single volcanoes reaches a maximum in the central islands of Vulcano and Lipari where abundant rhyolites have been erupted. These volcanoes were constructed along the Tindari-Letojanni strike-slip fault, a main tectonic feature that divides the Aeolian arc into two compositionally, volcanologically and structurally distinct sectors. Active volcanism is restricted to the central-eastern arc under which there is an active seismic Wadati-Benioff zone with depth of earthquake foci increasing northwestward. Aeolian arc magmas were generated in a heterogeneous upper mantle that was metasomatised by various types of fluids released from the Ionian subduction zone and associated sediments. The role of sediment increases significantly from the western islands of Alicudi and Filicudi to the eastern island of Stromboli. Pre-metasomatic mantle had OIB to MORB-type composition, with OIB components increasing at the margins of the arc. The rock compositional variation within single islands reflects both the characteristics of parental magmas and the shallow-level evolution history. In particular, the occurrence of abundant silicic volcanics in the central Aeolian islands results from extensive fractional crystallisation of magmas in large shallow reservoirs developed along pull-apart basins of the Tindari-Letojanni fault.

A. Peccerillo, *Cenozoic Volcanism in the Tyrrhenian Sea Region*,
Advances in Volcanology, DOI 10.1007/978-3-319-42491-0_9

Keywords
Lipari · Vulcano · Stromboli · Alicudi · Filicudi · Salina · Panarea · Calcalkaline rocks · Shoshonitic rocks · Potassic alkaline rocks · Wadati-Benioff zone · Active subduction · Ionian plate · Active volcanism · Tindari-Letojanni fault

9.1 Introduction

The Aeolian archipelago is an active volcanic arc in the southern Tyrrhenian Sea. It consists of several stratovolcanoes forming seven main islands and some seamounts, which develop around the Marsili basin (Romagnoli 2013; Romagnoli et al. 2013; Fig. 9.1). Volcanic activity exposed above the sea level took place entirely during the Quaternary, from about 250–300 ka to the present (Gillot 1987; Leocat 2011; Lucchi et al. 2013a). Seamounts have ages ranging from about 1.3 Ma to present.

Rock compositions range from mafic to silicic and show calcalkaline (CA), high-potassium calcalkaline (HKCA) and shoshonitic (SHO) affinity. Leucite-bearing alkaline rocks with a composition close to the Roman potassic series (KS) occur at Vulcano and Stromboli (Fig. 9.2). Arc tholeiites have been dredged along some seamounts (e.g. Beccaluva et al. 1982; Keller 1982; Francalanci et al. 2007).

There are important variations of structural, volcanic and magmatic features along the arc. These make the Aeolian arc an interesting example of a subduction-related magmatism, strongly correlated with and probably dependent on variable structural setting and nature of the subducted lithosphere along the arc. A summary of ages and petrological-volcanological characteristics of the Aeolian Islands is given in Table 9.1. Analyses of representative rocks are reported in Table 9.2. Seamounts are discussed in Chap. 12.

9.2 Regional Geology

The Aeolian arc has developed on the western margin of the Calabro-Peloritano basement. This is a fragment of the European plate, which was affected by metamorphism and tectonism during pre-Hercynian, Hercynian and Alpine orogenies, and migrated away from the Corsica-Sardinia block to its present position during the Miocene to Quaternary opening of the Tyrrhenian Sea. The Calabro-Peloritano basement runs from northern Calabria to northeastern Sicily, and is delimited by the Sangineto tectonic line in the north, and by the Tindari-Letojanni and the Taormina lines in the southwest (Bonardi et al. 2001, 2009; Gallais et al. 2013; Palano et al. 2015). The structure of the Calabro-Peloritano basement consists of a stack of various nappes composed of pre-Alpine metamorphic and granitoid rocks, often with Alpine metamorphic overprint, Mesozoic to Tertiary sedimentary rocks, ophiolitic sequences, and Quaternary sediments. Metamorphic rocks comprise a wide array of lithologies, including metavolcanics, phyllites, micaschists, gneiss, amphibolites, granulites, eclogites, marbles, and quartzites. Granitoid intrusions are Permo-Carboniferous calcalkaline granodiorites and tonalites, plus minor metaluminous to peraluminous granites and silicic dykes. The ophiolitic sequences are a typical association of metabasalts, radiolarites, limestones and pelites. Sedimentary rocks include evaporites, shales, sandstones, carbonates, and continental to shallow marine clastic

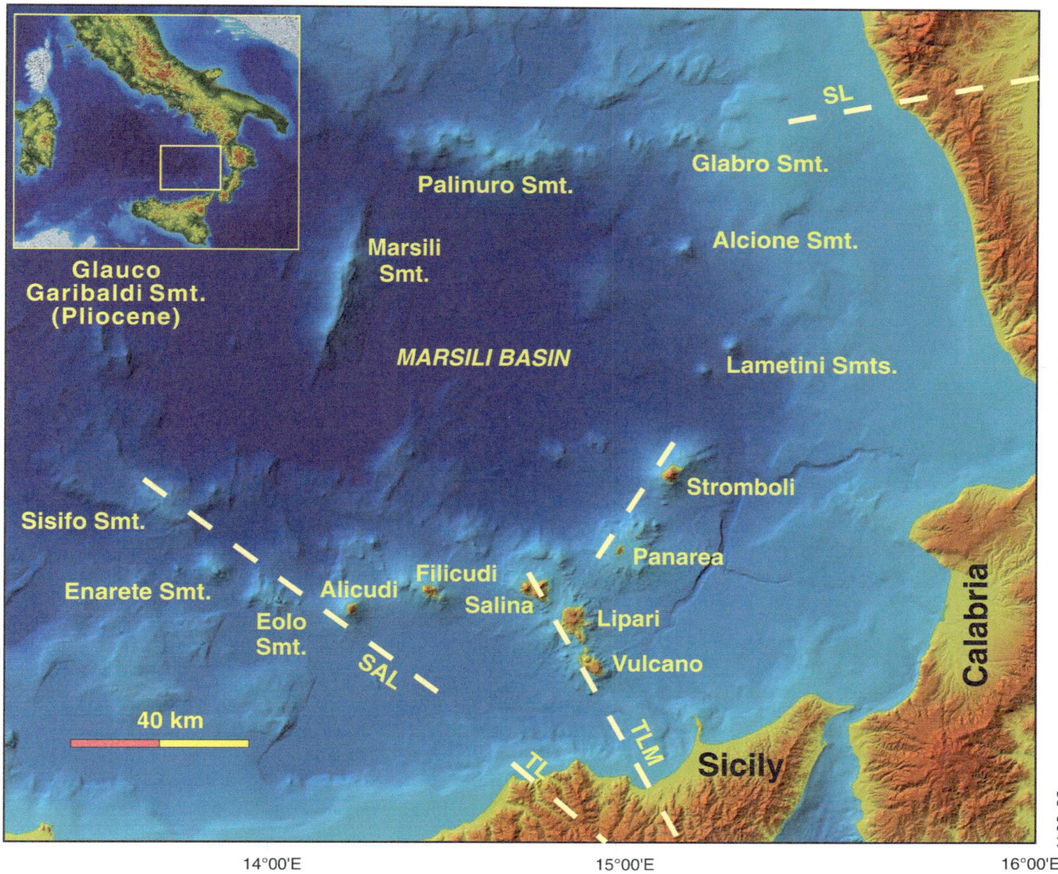

Fig. 9.1 Location of the Aeolian Islands and seamounts. *SAL* Sisifo-Alicudi line, *TLM* Tindari-Letojanni-Malta line, *TL* Taormina line, *SL* Sangineto line

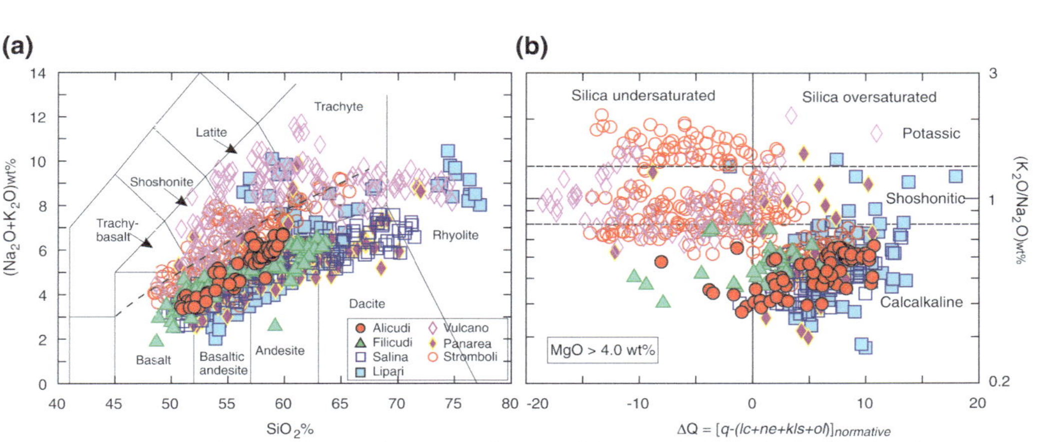

Fig. 9.2 **a** TAS classification diagram for the Aeolian arc volcanic rocks. The *dashed line* is the limit between the alkaline and subalkaline fields of Irvine and Baragar (1971). Data recalculated to 100 % on a LOI-free basis; **b** ΔQ versus K_2O/Na_2O classification diagram for mafic rocks ($MgO > 4.0$ wt%). ΔQ is the algebraic sum of normative quartz minus silica-undersaturated minerals (see Chap. 1)

Table 9.1 Summary of chronology, volcanology and petrology of the Aeolian arc volcanoes

Volcano	Age	Volcanology	Petrology and geochemistry
Alicudi	106 to 28 ka	– Composite volcano with summit calderas, made up of lava flows, domes and minor pyroclastics	– Calcalkaline basalts to andesites showing the most primitive compositions in the Aeolian arc ($^{87}Sr/^{86}Sr$ = 0.7035–0.7041; $^{3}He/^{4}He$: R/R_A = 6.5–7.1)
Filicudi	250 to 30 ka	– NW-SE trending alignment of composite volcanoes, monogenetic cones and domes. Dominant effusive and strombolian activity with a few phreatomagmatic, vulcanian and subplinian eruptions	– Basalt to dacite calcalkaline rocks evolved from mafic parents in various small-dimension magma chambers. Slightly higher Sr ($^{87}Sr/^{86}Sr$ = 0.7040–0.7047) and lower Nd isotopic compositions than Alicudi
Salina	250 to 16 ka	– Various partially coalescing and superimposed stratovolcanoes, e.g. Monte Fossa delle Felci and Monte dei Porri, consisting of dominant lavas. Final explosive rhyolitic activity	– Dominant calcalkaline mafic-intermediate rocks, with final rhyolite pumices. Similar radiogenic isotope signatures to Filicudi
Lipari	270 ka to 1220 AD	– Multicentre volcanic complex built up by partially superimposed composite lava and scoria cones, and by late acidic lava flows, domes and pumices	– Calcalkaline to high-K calcalkaline basaltic andesite and andesite, plus abundant rhyolites. Dubious presence of shoshonitic magmas intermingled with late erupted rhyolites. Variable radiogenic isotope compositions ($^{87}Sr/^{86}Sr$ = 0.7043 to 0.7067) indicating magma-crust interaction
Vulcano	130 ka to 1888–1890 AD	– Volcanic complex with two main polygenetic calderas, and several intra- and peri-caldera lava and pyroclastic cones and domes, includin the Lentia domes, the historical Vulcanello lava shoield and the active Cono della Fossa	– Early high-K calcalkaline and shoshonitic basalts to andesites (Primordial Vulcano), followed by latite to rhyolite (Lentia), undersaturated potassic alkaline shoshonite-latite (Vulcanello), and rhyolite pyroclastics and lavas (Cono della Fossa). High incompatible element abundances and moderately radiogenic isotopic compositions ($^{87}Sr/^{86}Sr$ = 0.7041–0.7059)
Panarea	155 to 8.7 ka	– Dome complex representing the subaerial summit of a large, mostly submerged dormant stratovolcano with a nearly flat top at about 100–150 m below sea level. Strong submarine fumarolic activity	– Dominant intermediate rocks, plus rhyolites and a few basaltic andesites with calcalkaline to shoshonitic affinities. Incompatible element contents increasing from calcalkaline to shoshonitic rocks. Radiogenic isotope ratios similar to Vulcano ($^{87}Sr/^{86}Sr$ = 0.7040 to 0.7057)
Stromboli	204 ka to present	– Active stratovolcano with summit calderas and flank collapses, and a satellite older neck (Strombolicchio), mostly consisting of lavas and scoriae	– Mafic to intermediate calcalkaline, shoshonitic and potassic alkaline rocks showing variable incompatible element contents and Sr isotope ratios ($^{87}Sr/^{86}Sr$ = 0.7050–0.7076) increasing from calcalkaline to potassic rocks. $^{3}He/^{4}He$ (R/R_A)=2.5-4.9
Aeolian Seamounts (Marsili, Sisifo, Enarete, Eolo, Lametini, Alcione, Palinuro, etc..)	1.3 to Present	– Volcanic cones at various depth, with evidence of central and flank collapses. Active release of hydrothermal fluids at several centres (e.g. Marsili, Palinuro, etc.)	– Calcalkaline and shoshonitic mafic to intermediate rocks similar to the Aeolian Islands. Arc tholeiites at Lametini and OIB-type basalt at Marsili

Table 9.2 Composition of selected Aeolian arc rocks. Data on the Aeolian seamounts are reported in Chap. 12

Volcano	Alicudi		Filicudi		Salina		
Rock type	Basalt	Andesite	Basalt	Dacite	Basalt	Andesite	Rhyolite
Data source	2,7,13	2,13	9,1,7	9	5	5	13
SiO_2 wt%	50.89	56.80	50.03	63.20	51.35	60.12	70.63
TiO_2	0.79	0.63	0.72	0.56	0.61	0.59	0.33
Al_2O_3	18.48	17.58	18.95	16.23	16.82	17.31	14.39
FeO_{total}	8.13	6.15	8.43	4.64	8.81	6.42	3.23
MnO	0.16	0.14	0.17	0.11	0.18	0.14	0.12
MgO	6.11	4.79	5.80	2.31	6.24	2.99	1.29
CaO	10.61	7.54	11.39	5.86	11.05	6.89	3.00
Na_2O	2.68	3.32	2.31	3.43	2.24	3.39	3.22
K_2O	1.06	2.08	1.10	2.89	1.09	1.97	3.57
P_2O_5	0.26	0.33	0.16	0.18	0.17	0.17	0.08
LOI	0.52	0.30	0.59	0.36	0.56	0.18	0.22
Sc ppm	32.9	27.3	30	17	–	–	16
V	–	–	–	–	313	204	77
Cr	172	124	68	12	54	6	18
Co	31	23	30	13	31	15	19
Ni	48	25	23	2	36	7	7
Rb	24	42	23	85	26	48	99
Sr	814	886	687	608	778	576	333
Y	14	25	17	15	16	21	15
Zr	84	107	75	123	60	90	136
Nb	6	13	7	10	3.14	5.35	19
Cs	0.61	0.80	0.26	–	0.99	0.97	2.6
Ba	347	787	352	646	351	496	1099
La	17	39	16	33	17	22	22
Ce	35	69	34	54	34	42	52
Nd	15	29	17	25	16	19	20
Sm	3.2	4.8	3.6	5.4	3.6	4.2	4.0
Eu	1.1	1.5	1.0	1.1	1.1	1.2	0.91
Tb	0.56	0.59	0.48	0.49	0.47	0.64	0.49
Yb	1.86	1.92	2.00	1.80	1.63	2.44	1.90
Lu	0.32	0.29	–	0.30	0.27	0.38	0.30
Hf	1.69	2.77	2.00	2.60	1.52	2.60	2.80
Ta	0.34	0.72	–	0.49	0.19	0.35	0.55
Pb	(5)	(4)	(6)	–	6.3	6.8	–
Th	2.93	8.7	3.0	8.7	3.4	5.1	9.0
U	3.71	2.84	1.02	2.80	1.08	1.97	2.60
$^{87}Sr/^{86}Sr$	0.70391	0.70343	0.70445	0.70436	0.70410	0.70439	0.70496
$^{143}Nd/^{144}Nd$	0.51281	0.51288	0.51273	0.51272	0.51279	0.51277	0.51272

(continued)

Table 9.2 (continued)

Volcano	Alicudi		Filicudi		Salina		
Rock type	Basalt	Andesite	Basalt	Dacite	Basalt	Andesite	Rhyolite
$^{206}Pb/^{204}Pb$	19.326	19.667	19.500	19.464	19.398	19.336	–
$^{207}Pb/^{204}Pb$	15.638	15.648	15.650	15.661	15.686	15.688	–
$^{208}Pb/^{204}Pb$	39.114	39.363	39.230	39.263	39.243	39.204	–
$^{176}Hf/^{177}Hf$	(0.283096)	–	(0.283074)	–	–	–	–

Volcano	Lipari				Vulcano			
Rock type	Basaltic andesite	Andesite	Cordierite-bearing dacite	Rhyolite	Shoshonitic basalt	Shoshonite	Trachyte	Rhyolite
Data source	1	12	12	8	4,7	3,8	3	8
SiO_2 wt%	53.76	60.53	64.47	73.35	48.10	53.35	60.28	72.76
TiO_2	0.61	0.56	0.67	0.10	0.82	0.67	0.49	0.13
Al_2O_3	16.27	19.36	16.43	12.70	12.75	16.36	16.36	13.24
FeO_{total}	8.26	4.39	6.66	2.00	11.12	8.13	5.81	2.62
MnO	0.16	0.09	0.12	0.11	0.22	0.16	0.11	0.11
MgO	6.25	0.86	1.36	0.34	8.34	3.61	1.72	0.34
CaO	9.86	5.59	4.19	0.92	12.69	7.25	4.52	0.99
Na_2O	2.14	3.34	2.2	4.02	2.20	4.17	4.20	4.25
K_2O	1.15	3.56	3.69	4.60	2.29	4.36	5.19	4.83
P_2O_5	0.17	0.2	0.21	0.02	0.24	0.36	0.32	0.04
LOI	1.03	1.52	–	1.67	0.03	1.14	0.59	0.68
Sc ppm	42	13	23	3	39.1	20.8	12.8	4
V	–	56	136	10	–	167	105	13
Cr	128	10	40	10	239	53	30	4
Co	32	7.5	15	2	22	28.18	20.7	2
Ni	33	13	40	3	50	21	12	2
Rb	27	125	163	267	53.7	153	190	280
Sr	580	674	403	52	1110	1212	887	81
Y	18	22.3	25.4	14	–	22	28	44
Zr	58	151	151	202	81	166	208	264
Nb	3.6	12	16	32	–	18	25	36
Cs	1.1	4.7	6.1	17	1.5	5.65	4.2	18.6
Ba	319	718	559	42	556	904	672	85
La	13	40.8	43.2	50	28	65	69.7	77
Ce	28	76.6	82.5	97	49	117	115	136
Nd	14.7	32.6	35.3	39	–	48.5	43	49
Sm	3.2	6.3	6.5	8.1	5.6	8.75	7.58	9.3
Eu	0.90	1.41	1.27	0.17	1.70	1.65	1.25	0.26
Tb	0.48	0.8	0.88	1.12	0.60	0.67	0.8	1.16
Yb	1.7	2.2	2.5	4.46	1.5	2.05	2.38	4.50
Lu	–	0.35	0.39	0.63	–	0.26	0.37	0.67

(continued)

Table 9.2 (continued)

Volcano	Lipari				Vulcano			
Rock type	Basaltic andesite	Andesite	Cordierite-bearing dacite	Rhyolite	Shoshonitic basalt	Shoshonite	Trachyte	Rhyolite
Hf	1.8	2.3	5	6.2	2.3	3.75	4.7	7.6
Ta	0.2	(0.6)	–	2.4	0.4	1.13	1.3	3.0
Pb	6.8	18	21	27	–	(20)	–	29
Th	3.4	11.1	13.2	46.7	6.1	21.8	31.6	58
U	0.91	3.68	3.77	13.5	1.9	6.7	8.5	16.6
$^{87}Sr/^{86}Sr$	0.70472	0.70431	0.705394	0.705123	0.70425	0.704578	0.704616	0.70527
$^{143}Nd/^{144}Nd$	0.51267	0.51283	0.512505	0.512554	0.51269	0.512671	0.512615	0.51257
$^{206}Pb/^{204}Pb$	19.44	19.633	19.264	19.345	19.528	19.412	19.375	19.379
$^{207}Pb/^{204}Pb$	15.66	15.655	15.675	15.678	15.656	15.676	15.658	15.694
$^{208}Pb/^{204}Pb$	39.26	39.421	39.187	39.257	39.314	39.291	39.221	39.321
$^{176}Hf/^{177}Hf$	–	–	–	–	0.282916	–	–	–

Volcano	Panarea			Stromboli			
Rock type	Basalt	Andesite	Rhyolite	HKCA basalt	CA Basaltic andesite	Shoshonitic basalt	KS Shoshonite
Data source	6	6	6	1,10	1	1,3	1,3
SiO_2 wt%	51.16	55.33	69.36	51.99	55.58	52.72	51.93
TiO_2	0.62	0.62	0.27	0.8	0.67	0.94	0.90
Al_2O_3	17.11	17.33	14.11	18.05	16.08	16.65	17.45
FeO	9.69	8.13	1.82	7.43	7.03	7.41	7.16
MnO	0.16	0.13	0.09	0.14	0.15	0.15	0.14
MgO	6.25	3.58	0.56	6.47	6.47	5.64	5.93
CaO	10.24	7.05	2.20	10.26	9.02	9.01	8.76
Na_2O	2.54	2.58	4.08	2.18	2.35	2.36	2.18
K_2O	0.90	1.52	4.46	1.83	1.63	3.61	4.63
P_2O_5	0.08	0.12	0.01	0.22	0.23	0.56	0.50
LOI	1.26	3.62	3.03	0.43	0.54	0.50	0.23
Sc	41	29	4	29	36	29	28
V	287	228	26	206	–	–	–
Cr	115	23	2.3	–	271	78	140
Co	33	21.5	2.5	26	33	30	30
Ni	27	13	1	51	53	36	34
Rb	23	53	151	64	40	113	152
Sr	566	441	289	514	540	708	773
Y	17	22	17	21	21	32	27
Zr	43	89	138	118	111	198	188
Nb	5	7	17	14	7	21	28
Cs	0.69	2.54	8.88	4.58	1.30	–	–
Ba	329	409	885	786	517	1538	1782

(continued)

Table 9.2 (continued)

Volcano	Panarea			Stromboli			
Rock type	Basalt	Andesite	Rhyolite	HKCA basalt	CA Basaltic andesite	Shoshonitic basalt	KS Shoshonite
La	12	20	47	25	23	50	56
Ce	23	38	77	50	47	103	99
Nd	12	19	24	24	22	42	45
Sm	2.7	3. 9	3.6	5.0	4.6	8.8	10.8
Eu	0.85	1.05	0.85	1.35	1.20	2.10	2.2
Tb	0.37	0.56	0.47	0.70	0.53	1.0	0.95
Yb	1.50	1.97	1.85	1.97	1.90	2.50	2.2
Lu	0.22	0.31	0.27	0.31	–	0.42	0.5
Hf	1.36	2.62	4.47	2.87	2.80	4.00	4.2
Ta	0.14	0.36	1.13	0.71	0.50	1.20	1.2
Pb	–	–	–	14.2	6.5	–	–
Th	2.5	6.5	30	12.5	8.2	18	18.1
U	1.3	4.0	9.3	3.4	1.9	–	–
$^{87}Sr/^{86}Sr$	0.70441	0.70491	0.70553	0.70625	0.70507	0.70683	0.70753
$^{143}Nd/^{144}Nd$	0.51284	0.51258	0.51259	0.512559	0.51261	0.51252	0.51250
$^{206}Pb/^{204}Pb$	19.43	19.233	19.186	19.000	19.17	19.002	18.975
$^{207}Pb/^{204}Pb$	15.71	15.680	15.682	15.680	15.66	15.658	15.654
$^{208}Pb/^{204}Pb$	39.36	39.142	39.163	39.091	39.08	39.041	39.007

Source of data: (1) Francalanci et al. (1989, 2013); (2) Peccerillo et al. (1993); (3) De Astis et al. (1997, 2000, 2103); (4) Del Moro et al. (1998); (5) Gertisser and Keller (2000); (6) Calanchi et al. (2002b); (7) Gasperini et al. (2002); (8) Gioncada et al. (2003); (9) Santo et al. (2004); (10) Tommasini et al. (2007); (11) Trua et al. 2011; (12) Forni et al. (2015); (13) Author's unpublished data. Numbers in parentheses indicate data on different, though compositionally similar samples from the same locality as those analysed for the other elements

sediments (Bonardi et al. 2001, 2009 and references therein).

The thickness of the crust is about 15–20 km beneath the Aeolian arc and increases in the Calabria peninsula and northern Sicily (about 25 km). In contrast, a typical oceanic thickness of about 10–12 km is found in the Marsili basin (e.g. Neri et al. 2002; Pontevivo and Panza 2006; Martinez-Arevalo et al. 2009). There is a belt of low-gravity anomalies east of the Calabria peninsula along the Ionian Sea coast, where values of 20–30 mGal have been measured. These contrast with 130–250 mGal of the Ionian Sea floor and 100 to 200 mGal of the southern Tyrrhenian Sea and Aeolian arc area. The belt of gravimetric lows offshore the Ionian Sea coast is likely related to the formation of an accretionary sedimentary wedge by the subducting Ionian crust against the Calabria peninsula (Catalano et al. 2001; Cella et al. 2006; Minelli and Faccenna 2010).

Based on rock compositions and volcanological characteristics, the Aeolian arc is divided into three main sectors. The western arc includes the islands of Alicudi, Filicudi and Salina, where magmas are typically calcalkaline with the dominance of mafic and intermediate rocks (Figs. 9.2 and 9.3). Ages range from about 250–16 ka (Keller 1980a; Gillot 1987; Leocat 2011); seismic activity is concentrated in the upper 20–30 km, with rare deep focus earthquakes (Orecchio et al. 2014). The central sector includes the islands of Lipari and Vulcano, which erupted repeatedly during historical times. Volcanism developed along the NNW-SSE striking Tindari-Letojanni

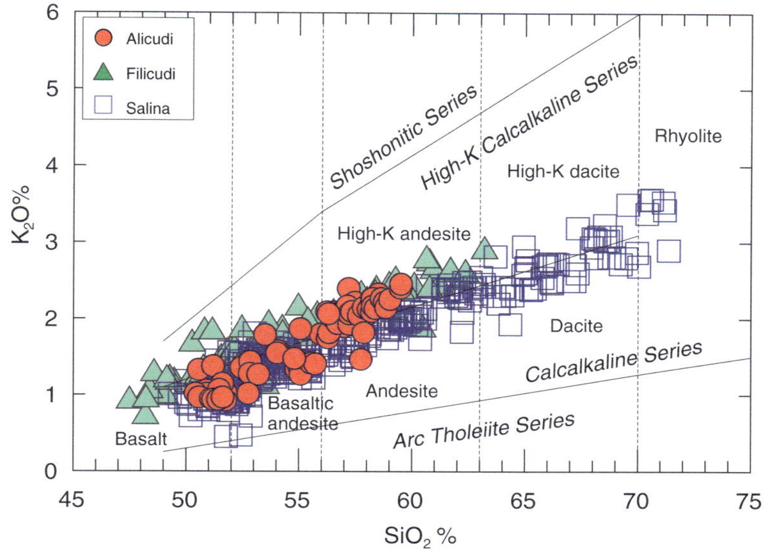

Fig. 9.3 K_2O versus SiO_2 classification diagram for the islands of Alicudi, Filicudi and Salina

tectonic line (Fig. 9.1), a dextral strike-slip fault running from Lipari to the mainland Sicily and to the Malta escarpment. Rock composition is much more variable than in the west, with the presence of mafic to silicic calcalkaline, shoshonitic and potassic alkaline products (Fig. 9.2). Seismicity is registered at various depths, with shallow earthquakes being concentrated along a NW-SE alignment (e.g. Orecchio et al. 2014). The eastern sector develops prevailing along NE-SW extensional faults and includes the active volcano of Stromboli and Panarea (Ventura 2013 and references therein). Measured ages are younger than 0.2 Ma. Rock compositions range from mafic to silicic with calcalkaline, shoshonitic to potassic alkaline affinities. Silicic rocks are restricted to Panarea, which, therefore, shows intermediate petrological characteristics between the central islands and Stromboli. The seismic activity is both at crustal and mantle levels, with earthquake foci defining a narrow and steep Wadati-Benioff plane dipping NNW and reaching a depth of more than 350 km (e.g. Neri et al. 2009; Ventura 2013 and references therein).

Active volcanism is restricted to the central and eastern sectors. At Lipari, the last eruption occurred about in 580 AD, whereas at Vulcano it dates to 1888–1890 AD (Mercalli 1907; Faraone 2002). Stromboli is characterised by a continuous mildly explosive strombolian activity. Fumaroles and hot springs characterise the submarine zone at Panarea (Chiodini et al. 2006; Capaccioni et al. 2007).

There are several seamounts located in the Aeolian arc region, which form a sort of ring around the Marsili backarc basin. Data on dredged and cored samples indicate ages from 1.3 to less than 0.1 Ma, and dominant calcalkaline to shoshonitic compositions, with a few arc tholeiitic and OIB-type rocks (see Chap. 12).

9.3 Alicudi

9.3.1 Volcanology and Stratigraphy

The Island of Alicudi rises about 2000 m above the sea floor and 675 m above sea level. The island was constructed from about 106–28 ka (Gillot 1987; Leocat 2011), during six epochs of dominantly effusive activity. The Scoglio Galera lavas and breccias, on the western margin of the island, are the oldest exposed rocks, whereas lava flows and domes at the summit crater and along the southwestern flank of the cone represent the youngest activity. Calcalkaline basalt and basaltic andesite lavas plus minor pyroclastics were emplaced during the first three epochs; andesite volcanism started during epoch 5 and become dominant during the latest eruptions (Montagnola synthem; Lucchi et al. 2013b).

9.3.2 Petrography and Mineral Chemistry

Rocks display a holocrystalline to hypocrystalline porphyritic texture, with phenocrysts of plagioclase ($\sim An_{80-65}$), olivine ($\sim Fo_{80-70}$), and diopside to augite clinopyroxene that are set in a groundmass consisting of the same phases plus Ti-magnetite and glass. Orthopyroxene ($\sim En_{73-67}$) and brown hornblende are present in basaltic andesites and andesites, where olivine is scarce and strongly resorbed; accessory phases include Fe–Ti oxides and apatite (Peccerillo and Wu 1992; Peccerillo et al. 1993).

A variety of metamorphic and magmatic xenoliths are found at Alicudi. Metamorphic xenoliths are more abundant in the mafic lavas than in the andesites, and are represented by predominant quartz-rich rocks displaying evidence of partial melting along grain boundaries, and by a few biotite gneiss and granulites (Peccerillo and Wu 1992; Peccerillo et al. 1993; Frezzotti et al. 2003). Igneous xenoliths include gabbros, diorites, and a few ultramafic nodules. Some ultramafic xenoliths are cumulate in origin, whereas others display a granoblastic texture with kinked olivine ($\sim Fo_{90}$), Cr-diopside, and accessory Cr-spinel, which probably denote an upper mantle origin (Peccerillo et al. 2004).

9.3.3 Petrology and Geochemistry

Alicudi rocks range in composition from calcalkaline basalt to high-K andesite (Fig. 9.3). Basalts exhibit the most primitive major and trace element compositions over the Aeolian arc (Table 9.2). Mg# (up to 0.72), Ni (up to 150 ppm), and Cr (up to 750 ppm) fall close to values of mantle equilibrated melts, although some of the highest values could partially result from olivine accumulation. There is an increase in K_2O, Na_2O, Al_2O_3, P_2O_5 and incompatible trace elements with increasing silica contents, which is mirrored by a decrease in MgO, FeO_t, CaO, and ferromagnesian trace elements (Fig. 9.4). Ratios of LILE/HFSE (e.g. Rb/Nb) are low relative to other western islands and do not show any significant variations with silica (Fig. 9.4g). REE are fractionated with flat HREE patterns (Fig. 9.5a). Incompatible element patterns of mafic rocks have negative anomalies of HFSE and positive spikes of Sr and Pb. High field-strength elements Ta and Nb show intermediate abundances between average OIB and N-type MORB (Fig. 9.5b).

The Alicudi rocks have the least radiogenic Sr- and the most radiogenic Nd-isotope compositions over the entire Aeolian arc ($^{87}Sr/^{86}Sr$ = 0.70343–0.70406; $^{143}Nd/^{144}Nd$ = 0.51280–0.51 290; Fig. 9.6a). However, an important feature is that the most primitive isotopic compositions (i.e. lowest $^{87}Sr/^{86}Sr$ ratios and highest $^{143}Nd/^{144}Nd$ ratios) are found in the evolved andesites rather than in the basaltic andesites and basalts (Fig. 9.4 h). Pb-isotope ratios are moderately radiogenic ($^{206}Pb/^{204}Pb$ ~ 19.19 to 19.67; $^{207}Pb/^{204}Pb$ ~ 15.62 to 15.67; $^{208}Pb/^{204}Pb$ ~ 39.07 to 39.36; Fig. 9.6b), and show a tendency to increase from basalts to andesites. Lower Pb-isotope ratios (e.g. $^{206}Pb/^{204}Pb$ ~ 18.10 to 19.29) were found in olivine-hosted melt inclusions by Rose-Koga et al. (2012). Sr–Nd–Pb isotopic disequilibrium between phenocrysts and groundmass has been observed (Peccerillo et al. 2004). Hf isotope ratio determined on a basaltic andesite sample yielded a value of $^{176}Hf/^{177}Hf$ = 0.283096 (Gasperini et al. 2002). He-isotope ratios for olivine and pyroxene phenocrysts have R/R_A = 6.5 to 7.1 (Di Liberto 2003; Martelli et al. 2008). $\delta^{11}B = -5.5$ was found for a basaltic andesite by Tonarini et al. (2001).

Whole-rock oxygen isotope ratios range from $\delta^{18}O_{SMOW}$ ~ +5.6 to +7.3 ‰ (Peccerillo et al. 1993), but values on separated clinopyroxenes show a more restricted range ($\delta^{18}O$ = +5.1 to +5.6). A rough positive correlation is observed between clinopyroxene $\delta^{18}O$‰ and whole rock $^{87}Sr/^{86}Sr$ ratios (Peccerillo et al. 2004).

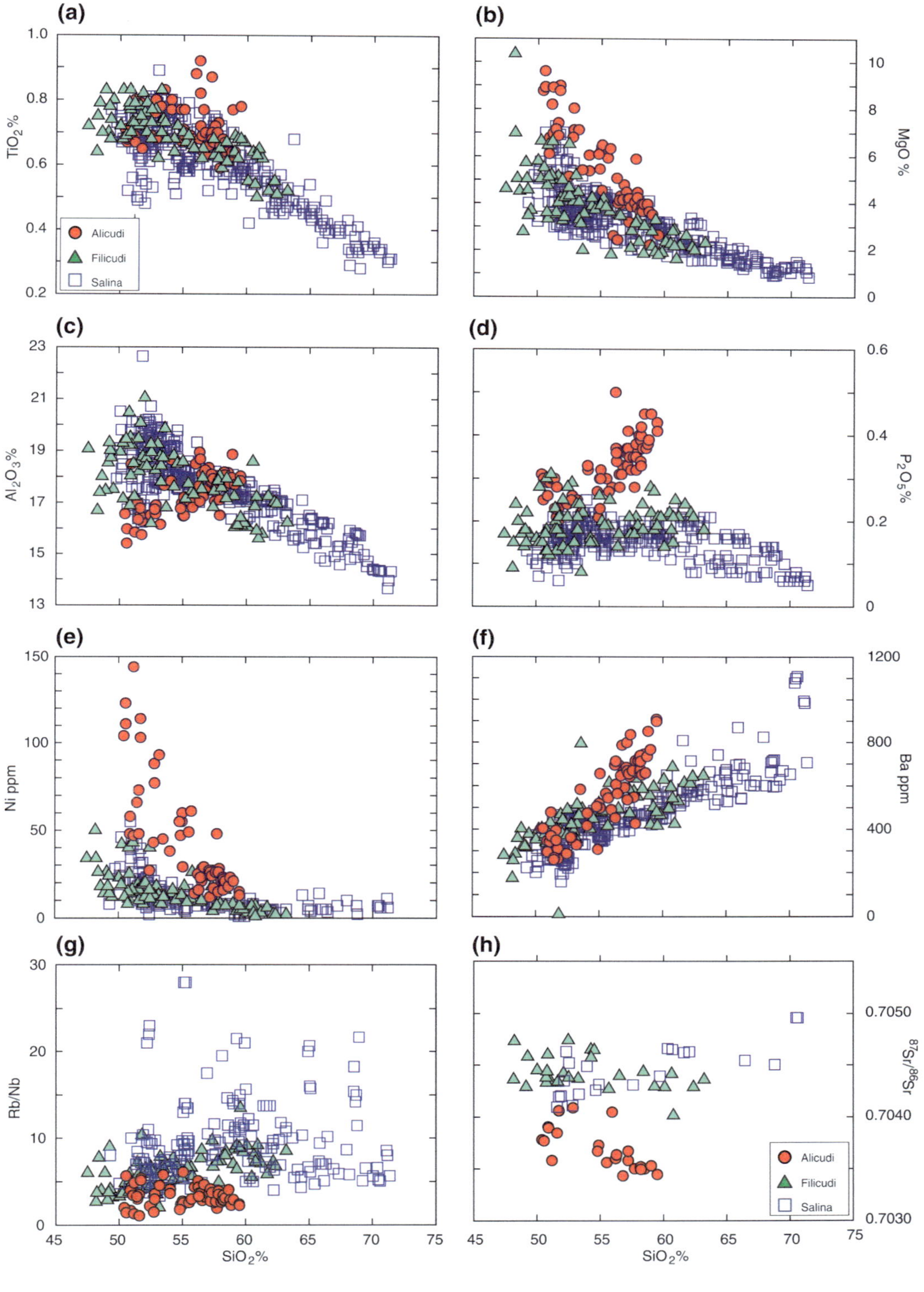

Fig. 9.4 Variation diagrams of selected major and trace elements and $^{87}Sr/^{86}Sr$ for the islands of Alicudi, Filicudi and Salina

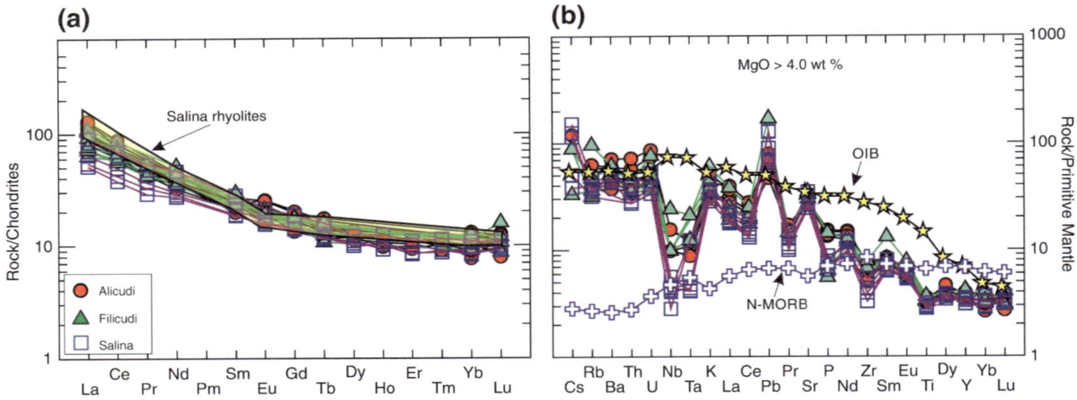

Fig. 9.5 **a** REE patterns for Alicudi, Filicudi and Salina rocks. Patterns of Salina rhyolites are reported as a *yellow* field; **b** Incompatible element patterns for mafic rocks. Average OIB and N-MORB compositions are also shown (Sun and McDonough 1989; Gale et al. 2013)

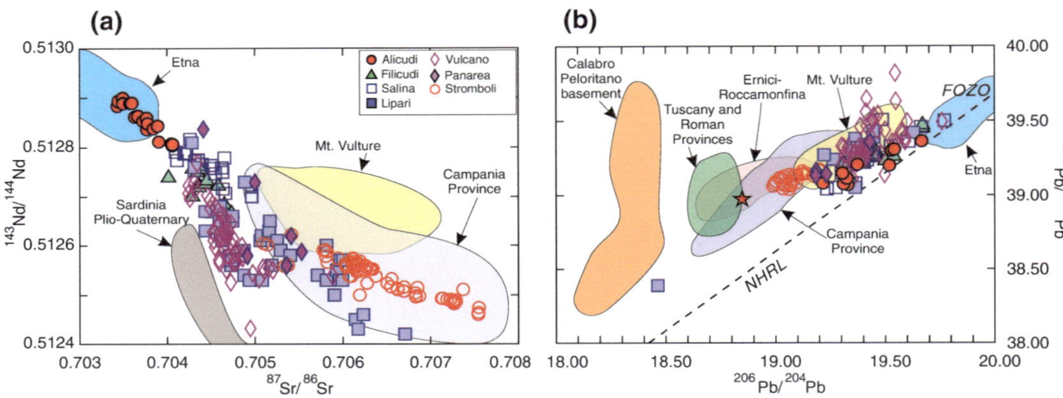

Fig. 9.6 Sr–Nd–Pb isotopic variations of the Aeolian arc rocks. The composition of mafic rocks of other Italian volcanic provinces and of the Calabro-Peloritano basement (Rottura et al. 1991) are also reported. NHRL is the Northern Hemisphere Reference Line of Hart (1984). The *red star* is the average sediment from the Ionian Sea (DSDP Site 374; Klaver et al. 2015)

9.3.4 Magma Evolution at Alicudi

Variations of several major and trace elements point to a derivation of the whole Alicudi series from a single magma by fractional crystallisation. However, variable isotopic signatures call for the involvement of some additional process of magma evolution. It has been suggested that the negative correlation between silica and Sr-isotope ratios at Alicudi (Fig. 9.4h) is the effect of a combined early fractional crystallisation process, followed by bulk assimilation of wall rocks. Fractional crystallisation of olivine, pyroxene and plagioclase produced a suite of basalt to andesite magmas, which underwent assimilation of wall rocks during their rise to the surface. The more fluid and hotter mafic magmas dissolved higher amounts of crustal rocks than the andesitic liquids, due to their higher temperature and to turbulent flow through volcanic conduits. This generated stronger isotopic modifications in the mafic than in the evolved magmas (Peccerillo and Wu 1992, Peccerillo et al. 1993, 2004).

A somewhat different scenario by Lucchi et al. (2013b) suggests a polybaric evolution and a migration of magma chambers from deep to shallow levels with time. The mafic volcanism of early activity was fed by a reservoir sited at about 20 km depth, near to the Moho (Nazzareni et al. 2001; Frezzotti et al. 2003). Here, assimilation of wall rocks generated relatively high Sr- and low-Nd isotope signatures in the magmas. However, intense mixing with newly injected mafic melts prevented evolution toward andesitic compositions. In contrast, the andesitic magmas of latest eruptive epochs were generated by fractional crystallisation of the same parent at shallow depths, where low temperature and high viscosity of magmas inhibited assimilation of wall rocks.

Whatever the case, the implication of data is that evolved andesites rather than mafic basalts better represent isotopic signatures of the Alicudi magma mantle source.

9.4 Filicudi

9.4.1 Volcanology and Stratigraphy

The Island of Filicudi and the islets of La Canna and Montenassari represent the emergent part of a NW-SE elongated volcanic complex formed by several partially-overlapping eruptive centres. The island rises about 1700 m above seafloor and reaches a maximum altitude of 774 m above sea level. The age of the exposed rocks is controversial, but a time interval between about 250 and 29 ka seems the most reliable on the basis of present knowledge (Leocat 2011; Lucchi et al. 2013c). Volcanic rocks consist of dominant lava flows and domes with minor pyroclastics, related to prevailingly effusive and strombolian activity, and some hydrovolcanic, vulcanian and subplinian eruptions.

The early activity took place in the western sector of the island with the formation of the Case Ficarisi and Fossa Felci stratocones. Successively, other centres become active in the central-southeastern (Chiumento stratocone and Monte Guardia scoria cone), and in the south-eastern sectors of Filicudi (Capo Graziano dome, Monte Torrione lavas, and Monte Montagnola dome). Finally, La Canna centre, presently represented by a towering neck offshore the western end of the island, was formed (Lucchi et al. 2013c).

9.4.2 Petrography and Mineral Chemistry

The Filicudi rocks range in composition from calcalkaline basalts to high-K andesites and dacites (Fig. 9.3). There is no evident variation of magma compositions with time, although the most evolved magmas were emplaced rather late in the history of the island. Rock textures are holocrystalline to hypocrystalline porphyritic, with phenocryst content ranging from 30 to 50 vol%. Phenocryst mineralogy of mafic rocks consists of zoned plagioclase ($\sim An_{95-50}$), diopside to augite clinopyroxene ($\sim Wo_{38-50}$, En_{38-50}, Fs_{5-20}), and minor olivine ($\sim Fo_{85-60}$) and orthopyroxene ($\sim En_{70-65}$) set in a groundmass made up of the same phases plus Fe–Ti oxides and glass (Santo 1998). Intermediate rocks have plagioclase as the dominant phase, plus minor augitic clinopyroxene and orthopyroxene; biotite and brown hornblende are present in some rocks. Ti-magnetite, ilmenite and apatite are found as accessory phases.

As at Alicudi, the Filicudi rocks contain both magmatic and metamorphic xenoliths. Igneous xenoliths consist of gabbros, granodiorites, and lavas. Metamorphic xenoliths, particularly abundant in rocks from La Canna neck, generally exhibit a granoblastic texture and have a quartz-rich composition with minor K-feldspar and plagioclase (Santo et al. 2004).

9.4.3 Petrology and Geochemistry

The Filicudi rocks show considerable scattering for many major and trace elements (Fig. 9.4). REE patterns are moderately fractionated, with almost flat HREE (Fig. 9.5a); mantle-normalised incompatible trace element patterns of mafic rocks are similar to Alicudi (Fig. 9.5b).

$^{87}Sr/^{86}Sr$ (~0.7040 to 0.7047) exhibits a poorly defined negative correlation with silica (Fig. 9.4h). Nd-isotope ratios range from 0.51267 to 0.51276; Pb-isotope ratios are poorly variable ($^{206}Pb/^{204}Pb$ ~ 19.31 to 19.67; $^{207}Pb/^{204}Pb$ ~ 15.64 to 15.69; $^{208}Pb/^{204}Pb$ ~ 39.11 to 39.47; Santo et al. 2004); Hf-isotope ratio for a basalt sample has a value of $^{176}Hf/^{177}Hf$ = 0.283096, close to the composition of Alicudi (Gasperini et al. 2002). He-isotope ratios have slightly lower values than at Alicudi (R/R_A = 5.1–6.7; Di Liberto 2003; Martelli et al. 2008). Oxygen isotopic compositions for clinopyroxene phenocrysts are moderately variable ($\delta^{18}O_{SMOW}$ = +5.37 ‰ to +6.20 ‰). The highest and lowest values have been respectively found for clinopyroxenes from basaltic and andesitic rocks (Santo and Peccerillo 2008).

9.4.4 Magma Evolution at Filicudi

Geochemical modelling by Santo et al. (2004) and Lucchi et al. (2013c) indicates that magma evolution at Filicudi cannot be explained by any specific process or even by a combination of evolutionary processes, starting from a single parent. Therefore, it has been suggested that distinct types of parental magmas underwent various degrees of fractional crystallisation, mixing, and assimilation, with variable roles of single processes at individual centres. Clinopyroxene crystal chemistry and fluid inclusion studies support polybaric magma ponding and crystallisation (Nazzareni et al. 2001; Frezzotti et al. 2003). This implies that many magma batches were injected at different times and evolved in small-size magma chambers beneath various volcanic centres. Such a model agrees with the particular structure of the Filicudi Island, which consists of an alignment of several diachronous magmatic centres. In general terms, basaltic melts underwent larger amounts of crustal assimilation than andesites and dacites, although such a process is less well defined than at Alicudi.

9.5 Salina

9.5.1 Volcanology and Stratigraphy

The Island of Salina (27 km^2) lies at the intersection between the Alicudi-Filicudi and the Vulcano-Lipari volcanic alignments. It represents the emergent part of an about 2000 m high structure, whose summit reaches 962 m above sea level. The island is characterised by the well-preserved twin cones of Monte Felci and Monte Porri (Fig. 9.7), which have been constructed upon the eroded volcanoes of Rivi, Capo and Corvo. The ages of the exposed rocks are controversial (Keller 1980a; Gillot 1987; De Rosa et al. 2003a), and an interval between about 250 and 15.6 ka has been recently accepted by Lucchi et al. (2013d), based on radiometric dating and field evidence. Early eruptions (about 244–226 ka) built up the Pizzo Corvo and Pizzo Capo volcanoes, respectively cropping out at the western and eastern ends of the island. After a long period of quiescence and erosion, the Monte Rivi and Monte Fossa delle Felci stratovolcanoes were constructed, from about 160–120 ka. Monte dei Porri andesite-dacite stratocone was built up between about 70 and 57 ka in the western sector of Salina. The final activity (about 30–15.6 ka) formed a few basaltic-andesite lava flows at Punta Perciato, and the rhyolitic pumice deposits emitted by two explosive eruptions of the Pollara crater, in the northwestern end of the island (Keller 1980a; Gillot 1987; Mazzuoli et al. 1995; Calanchi et al. 1993; Lucchi et al. 2013d).

9.5.2 Petrography and Mineral Chemistry

The Salina rocks range from calcalkaline basalt to rhyolite (Fig. 9.3). They show variably porphyritic textures with amounts of phenocrysts decreasing from lavas to scoriae and pumices. Basalts and basaltic andesites contain phenocrysts of zoned plagioclase (~An_{95-60}), olivine (~Fo_{75-60}), diopside to augite clinopyroxene, and orthopyroxene (~En_{70-60}) set in a matrix consisting of plagioclase, augite,

Fig. 9.7 View of Monte Porri cone, Salina

pigeonite and Ti-magnetite. Olivine is often surrounded by pyroxene coronas. Andesites are moderately porphyritic with phenocrysts of plagioclase, clinopyroxene, orthopyroxene, Ti-magnetite and rare strongly resorbed olivine. Amphibole with rims of opaque minerals is observed in some rocks. Phenocryst compositions are similar to those of mafic rocks. Groundmass is pilotaxitic to hyalopilitic and consists of plagioclase, clinopyroxene, orthopyroxene and magnetite; apatite is observed as an accessory phase. Dacites mainly occur as pyroclasts exhibiting less porphyritic textures and higher proportions of orthopyroxene and amphibole than andesites. Groundmass is pilotaxitic to vitrophyric and is composed of plagioclase, clinopyroxene, orthopyroxene, magnetite, and rare biotite. Quartz and alkali feldspar is observed in some samples. Rhyolite pumices occur at the Pollara centre. They contain phenocrysts of predominant plagioclase ($\sim An_{60-40}$), amphibole and biotite with minor clinopyroxene, orthopyroxene and rare xenocrystic olivine. Biotite is a major phenocryst phase in the latest erupted pumices (Keller 1980a; Gertisser and Keller 2000; Lucchi et al. 2013d; Nicotra et al. 2014).

A variety of xenoliths occurs in the Salina lavas and pyroclastics. Sedimentary xenoliths include silicoclastic sediments and marls; metamorphic rocks consist of gneisses, amphibolites, marbles, metagranitoids, contact metamorphic rock, and quartz-rich xenoliths whose texture, mineralogy and geochemistry suggest they are residues of crustal anatexis. Magmatic xenoliths consist of lava fragments, cumulate gabbros and diorites as well as rare ultramafic rocks (Zanon and Nikogosian 2004; Lucchi et al. 2013d).

9.5.3 Petrology and Geochemistry

Variation diagrams of major and trace elements versus SiO_2 exhibit negative correlations for Al_2O_3, FeO_{total}, CaO, MgO, ferromagnesian trace elements (Ni, Cr, Co, V), and Sr, and an opposite behaviour for LILE and HFSE (Fig. 9.4). LILE/HFSE (e.g. Rb/Nb) are scattered and overall higher than at Filicudi and Alicudi. REE patterns are moderately fractionated, with rhyolites exhibiting small negative Eu anomalies (Fig. 9.5a). Incompatible element patterns resemble those of Alicudi and Filicudi rocks, but there are higher positive spikes of Sr and Pb, and stronger depletion in HFSE, with Ta and Nb concentrations similar to N-MORB (Fig. 9.5b).

Sr-isotope ratios range between 0.7041 and 0.7050, and slightly increase with increasing silica (Fig. 9.4h). Nd-isotope ratios range from 0.51267 to 0.51282 (Fig. 9.6a). Pb isotopes are moderately radiogenic, with $^{206}Pb/^{204}Pb \sim 19.15$ to 19.77, $^{207}Pb/^{204}Pb \sim 15.60$ to 15.77, $^{208}Pb/^{204}Pb \sim 39.00$ to 39.55 (Fig. 9.6b). Oxygen isotope compositions of whole rocks range from $\delta^{18}O_{SMOW} = +6.4$ ‰ to +7.3 ‰ in the mafic volcanics and from $\delta^{18}O_{SMOW} = +7.0$ ‰ to +8.5 ‰ in the andesite-dacite field (Ellam et al. 1988; Ellam and Harmon 1990; Francalanci et al. 1993; Gertisser and Keller 2000; Lucchi et al. 2013d). $^3He/^4He$ on clinopyroxene and olivine ($R/R_A = 5.35$ and 4.50) are slightly lower than at Filicudi and Alicudi (Martelli et al. 2008).

9.5.4 Magma Evolution at Salina

Major and trace element variations indicate that fractional crystallisation of mafic phases and plagioclase was a first-order magma evolutionary process at Salina. However, variable Sr–Nd–O-isotope ratios indicate that crustal assimilation accompanied fractional crystallisation, although amount of assimilated crustal rocks is low relative to the mass of crystallised minerals (Gertisser and Keller 2000). Mixing between compositionally diverse magma batches was also an important evolution process, as testified, for instance, by mineral disequilibria (e.g. resorbed-reacted olivine crystals), occurrences of mafic enclaves in andesites, zoned pyroclastic deposits at Pollara (Calanchi et al. 1993), and variable compositions of melt inclusions in single crystals (Doherty et al. 2012). Therefore, the overall evolution history of Salina seems to be related to combined fractional crystallisation, mixing and assimilation processes, starting from mafic calcalkaline melts (Ellam and Harmon 1990; Gertisser and Keller 2000). Polybaric evolution processes are suggested by clinopyroxene crystal chemistry and fluid inclusion studies, with the older magmas hosted in deeper magma chambers than younger ones (Nazzareni et al. 2001; Zanon and Nikogossian 2004).

9.6 Lipari

9.6.1 Volcanology and Stratigraphy

Lipari covers an area of about 38 km^2 and is the largest of the Aeolian Islands (Fig. 9.8). It consists of several lava and pyroclastic centres, which represents the emergent portion of a large structure that includes the Island of Vulcano in the south, and extends northward up to the Secca del Capo seamount (Romagnoli 2013). These volcanoes are aligned in a NW-SE trend, along the continuation of the Tindari-Letojanni fault system (De Astis et al. 2003). Dated activity at Lipari ranges from 267 ka to late medieval times.

Detailed volcanological and stratigraphic investigations and geological mapping recognised various stages of activity (e.g. Pichler 1980; Crisci et al. 1991; Calanchi et al. 2002a; Forni et al. 2013; Fig. 9.8). According to Forni et al. (2013), the eruptive history of Lipari can be divided into nine epochs interrupted by periods of dormancy and accompanied by volcano-tectonic collapses.

The lowest exposed products (Paleo-Lipari) consist of basaltic andesitic lavas and pyroclastic rocks, cropping out along the western coast of the island. According to Pichler (1980) and Crisci et al. (1991) eruptions were partially submarine, a conclusion that is not shared by Forni et al. (2013). The Paleo-Lipari activity was followed by the construction of a number of basaltic andesite and andesite central and fissural lava and scoria cones with an age of 267–150 ka, still located along the western coast (e.g. Monte Mazzacaruso, Timpone Ospedale, etc.). Successively, there was a long period of dormancy, interrupted by the formation of the Monterosa scoria cones and of the Monte Sant'Angelo andesitic stratocone, respectively in the present eastern and central sectors of Lipari. The construction of this latter edifice occurred during various phases, lasted until about 81 ka. After a new period of quiescence, activity started in the south of Lipari with the formation of the rhyolitic lava domes and associated lava flows of Monte Falcone and Monte Perciato (43–40 ka), Monte Guardia, Punta San Giuseppe and Monte

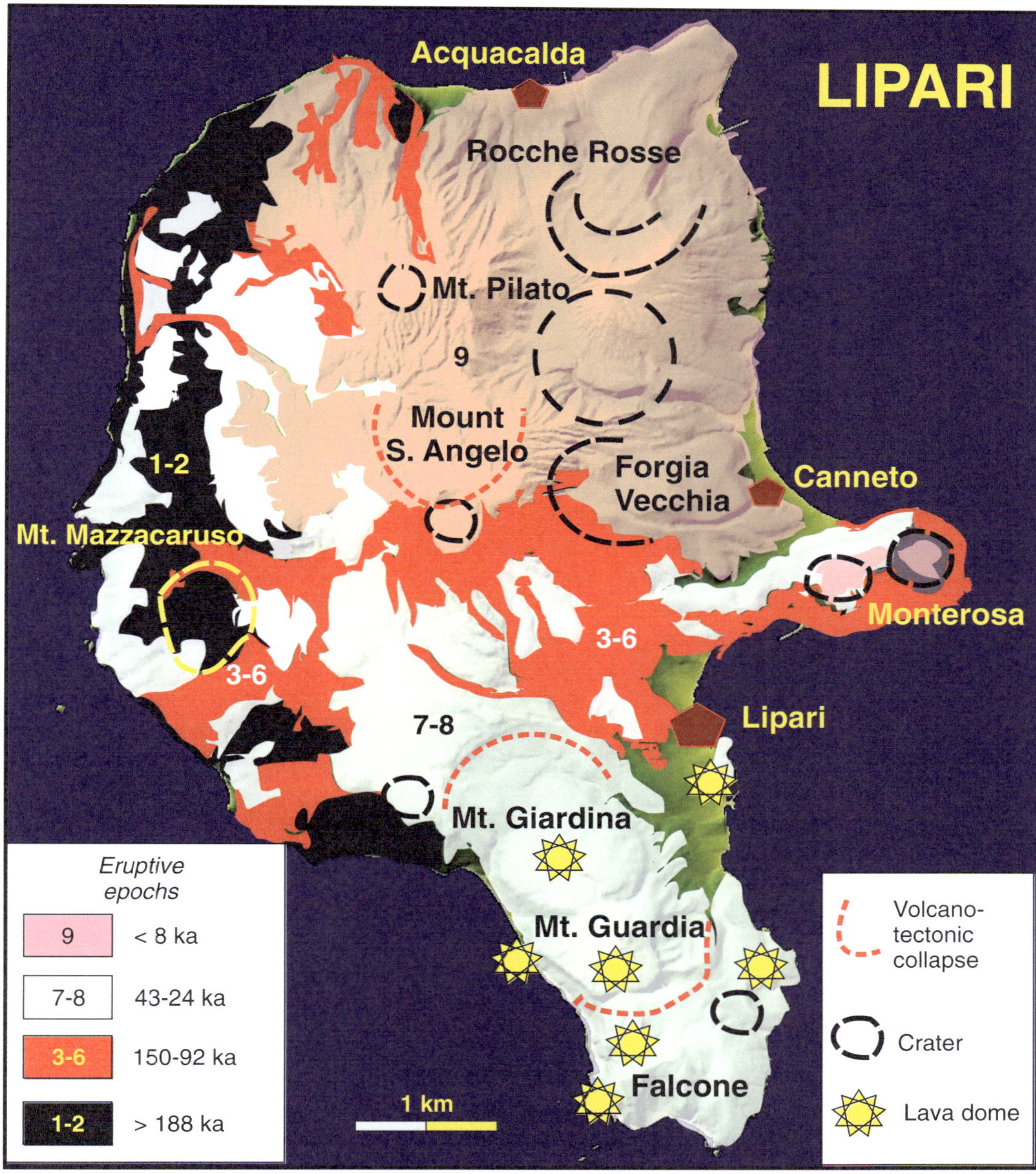

Fig. 9.8 Simplified geological map of Lipari Island. Modified after Forni et al. (2013, 2015)

Giardina (27–20 ka). The latest eruption took place in the north-eastern sector of the island with the formation of the lower Monte Pilato rhyolitic pumice cone (8.7–8.4 ka), the Forgia Vecchia lava, the upper Monte Pilato cone (AD 776) and the Rocche Rosse obsidian lava flow (1220 AD).

An important aspect of Lipari geology is the occurrence of pyroclastic deposits known as "Brown Tuffs", which also occur in the other Aeolian Islands and along the nearby coast of Sicily and Calabria (Crisci et al. 1983; Lucchi et al. 2008). These consist of brown-coloured, massive and weakly lithified latite to rhyolite pyroclastic layers, showing variable thickness, from a few centimetres to about twenty metres, depending on the paleotopography (Lucchi et al. 2008; Forni et al. 2013). Brown Tuffs originated from eruptive centres located between Lipari and Vulcano (Gioncada et al. 2003), or from the Fossa caldera at Vulcano (De Astis et al. 2013).

9.6.2 Petrography and Mineral Chemistry

The TAS and K_2O versus SiO_2 diagrams (Figs. 9.2 and 9.9) show that the Lipari volcanics range from calcalkaline basaltic andesite to rhyolite, with a compositional gap in the dacite field. Basaltic andesites show variable potassium contents, spanning the arc tholeiite and the high-K calcalkaline fields (Fig. 9.9). Shoshonitic magmas are only represented by latite enclaves intermingled with late emplaced rhyolites (Daví et al. 2009; Forni et al. 2015). Basaltic andesites and andesites have porphyritic textures with phenocrysts of zoned plagioclase ($\sim An_{90-60}$) and diopside to augite clinopyroxene set in a groundmass composed of the same phases, plus Fe–Ti oxides and glass (Pichler 1980; Forni et al. 2013, 2015). Orthopyroxene and pigeonite are abundant in the andesites. Hornblende and biotite are rare; corroded olivine ($\sim Fo_{90-65}$) occurs in some lavas. Abundant xenocrystic phases, including cordierite, garnet, andalusite, K-feldspar, sillimanite, corundum, and quartz, occur in some andesites, testifying to entrapment and fragmentation of basement rocks (Maccarrone 1963; Barker 1987; Di Martino et al. 2011). Well-known outcrops of cordierite-rich andesites occur in the southwestern sector of Mount Sant'Angelo stratocone, along the road to Terme San Calogero. Dacites are scarce at Lipari and some of them exhibit mingling textures suggesting an origin by interaction between rhyolitic and andesitic-latitic melts. The rhyolitic lava and pyroclastic rocks from the younger activity have variable textures. Pumice clasts from Monte Guardia contain K-feldspar, plagioclase ($\sim An_{25-20}$), hornblende, some biotite and accessory Ti-magnetite, apatite and zircon. These are closely associated and partially intermingled with grey high-K andesite to latite pumices containing clinopyroxene, plagioclase, minor olivine, Ti-magnetite and apatite (De Rosa et al. 2003b; Gioncada et al. 2003; Forni et al. 2015). Rhyolitic lavas have scarcely porphyritic textures with a few phenocrysts of K-feldspar ($\sim Or_{65-75}$), plagioclase ($\sim An_{25-20}$) and hornblende set in a glassy, sometimes eutaxitic matrix. Accessory phases include Ti-magnetite, zircon and apatite (Gioncada et al. 2003, 2005).

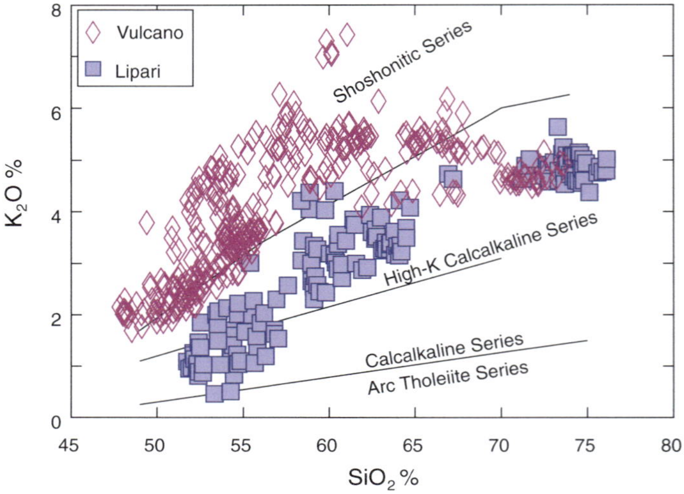

Fig. 9.9 K_2O versus SiO_2 classification diagram for the islands of Lipari and Vulcano

Xenoliths in the Lipari volcanics include granulites, metapelites, quartzites, gabbroic rocks, lava fragments, and mafic enclaves (e.g. Di Martino et al. 2010, 2011 and references therein). Metapelitic rocks are represented by xenoliths containing cordierite, garnet, andalusite, sillimanite, biotite, which are also found as single crystals derived from xenolith disaggregation. Mafic enclaves in the rhyolites show textural evidence of mingling with host magma when both were in a fluid state. These magmatic enclaves range from latite to andesite and trachyte and generally show porphyritic textures with phenocrysts of plagioclase, diopside to augite clinopyroxene and olivine, with composition up to Fo_{90} (Gioncada et al. 2005; Daví et al. 2009; Forni et al. 2015).

9.6.3 Petrology and Geochemistry

Major and trace elements are rather scattered in the Lipari rocks. As elsewhere in the Aeolian arc, Na_2O, K_2O and some trace elements (e.g. Rb, Th, Nb, Zr, etc.) are positively correlated with silica, whereas in FeO_{total}, MgO, CaO and ferromagnesian trace elements (Ni, Co, Cr etc.) show an opposite trend. Al_2O_3, TiO_2, P_2O_5 and several trace elements (e.g. La, Ba, etc.) increase from basaltic andesites to andesites and decrease in the rhyolites (Figs. 9.10 and 9.11). Sr is scattered in the basaltic andesites and andesites and decreases in the rhyolites (e.g. Crisci et al. 1991; Forni et al. 2013). Some inter-element variation diagrams show distinct, nearly parallel trends for mafic-intermediate rocks and for rhyolites, with a few samples plotting in the middle; some rhyolites overlap the acidic rocks from Vulcano (Fig. 9.11g and h). REE patterns are variably fractionated, with strong negative Eu anomalies in the rhyolites (Fig. 9.12a). Incompatible element patterns of mafic rocks show positive spikes of Sr and Pb and negative anomalies of HFSE, with Ta–Nb close to N-MORB abundances (Fig. 9.12b).

Radiogenic isotope ratios exhibit wide variations ($^{87}Sr/^{86}Sr$ = 0.7043 to 0.7067; $^{143}Nd/^{144}Nd$ = 0.51276 to 0.51242; $^{206}Pb/^{204}Pb$ = 18.46 to 19.63; $^{207}Pb/^{204}Pb$ = 15.61 to 15.71; $^{208}Pb/^{204}Pb$ = 39.05 to 39.43; Fig. 9.6). $^{87}Sr/^{86}Sr$ is scattered in the Lipari rocks (Fig. 9.11f), with the most radiogenic compositions being shown by cordierite-rich andesites (Esperança et al. 1992; Gioncada et al. 2003; Forni et al. 2013).

9.6.4 Magma Evolution at Lipari

The scattered distribution of several incompatible elements and radiogenic isotopes, the compositional gap in the dacite field, and the parallel trends observed on inter-element diagrams make the petrogenesis of Lipari rocks a challenging issue. Problems that have been long debated include the nature of primary mantle-derived melts, the role of various magma evolution processes, and the genesis of rhyolites.

According to Crisci et al. (1991), three distinct magmatic components occur at Lipari, two being generated in the mantle and one in the crust. The mantle-derived magmas had tholeiitic and K-rich composition, respectively, and were generated in a stratified source, characterised by decreasing degree of metasomatism with increasing depth. Early large degrees of partial melting of asthenospheric mantle generated tholeiitic magmas (Pasqual et al. 1998), whereas late melting of anomalous lithospheric mantle produced K-alkaline melts. Potassic mafic melts intruded into the volcanic system and mixed with tholeiitic magmas, causing a steep increase of potassium in the mafic rocks. Upward migration of mantle melting processes was related to an increase of isotherms, as a consequence of asthenospheric doming beneath the central sector of the Aeolian arc. Fractional crystallisation was the main evolution process, with separation of olivine and clinopyroxene in the mafic magmas, joined by plagioclase, Fe–Ti oxides and apatite in the intermediate compositions. Rhyolites could represent either the final products of evolution processes or independent melts, formed by crustal anatexis during the ascent of mafic magmas (Crisci et al. 1991).

According to Di Martino et al. (2011), partial melting in the lower continental crust and the generation of rhyolitic liquids have been main

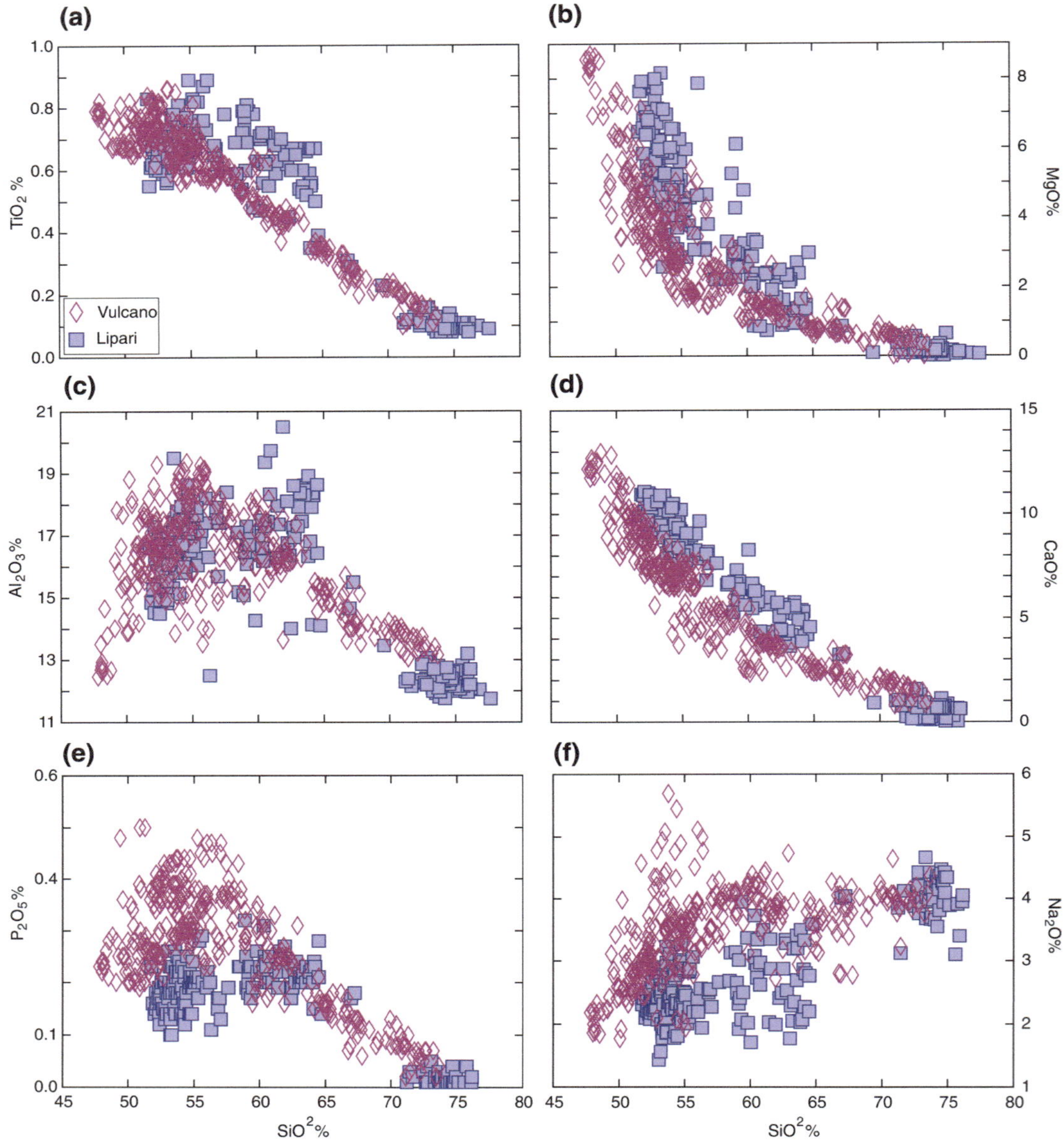

Fig. 9.10 Harker diagrams for the islands of Vulcano and Lipari

volcanological-petrogenetic events at Lipari. Crustal anatexis was triggered by intrusion of mafic magmas and intensive mixing took place between acid and mafic melts giving several contaminated magmas, including the cordierite-rich andesites. Occurrence of abundant rhyolitic magmas within the volcano plumbing system acted as a density filter and inhibited frequent mafic eruptions in the lat 100 ka.

Gioncada et al. (2003) noticed that isotopic compositions (especially Pb) of rhyolites are different from those of the Calabro-Peloritano basement and that there are two groups of acidic rocks with distinct Sr isotopic signatures but similar Nd isotopic ratios. Therefore, it was concluded that a generation of acid melts by crustal anatexis is unlikely, whereas an origin by fractional crystallisation plus crustal assimilation

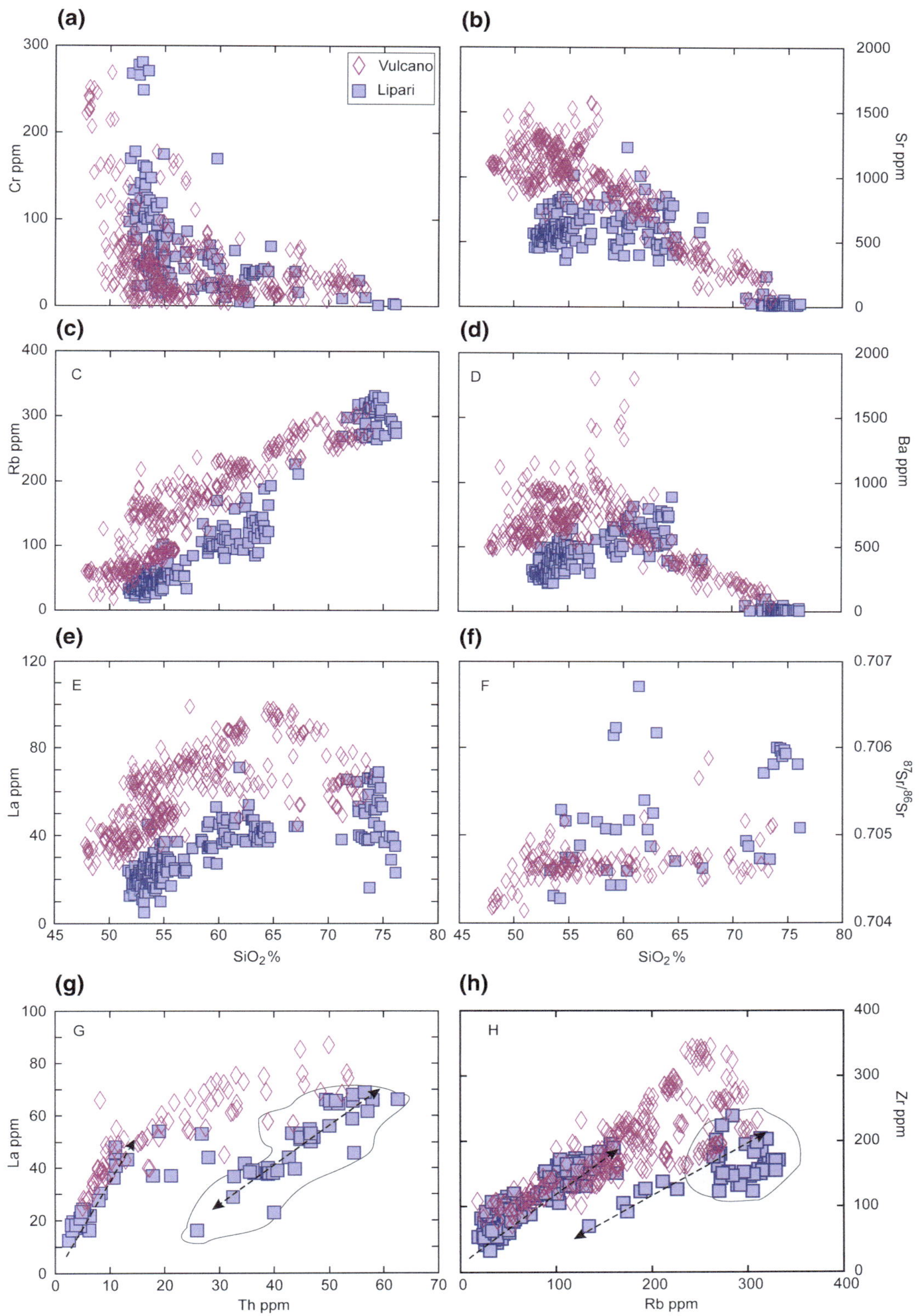
(a)
(b)
Vulcano
Lipari
Cr ppm
Sr ppm
(c)
(d)
C
D
Rb ppm
Ba ppm
(e)
(f)
E
F
La ppm
$^{87}Sr/^{86}Sr$
SiO_2 %
SiO_2 %
(g)
(h)
G
H
La ppm
Zr ppm
Th ppm
Rb ppm

◄**Fig. 9.11** Variation diagrams of selected trace elements and $^{87}Sr/^{86}Sr$ for the islands of Vulcano and Lipari. Parallel inter-element trends for the Lipari rocks are highlighted by *arrows* in panels G and H. *Grey* fields enclose rhyolitic rocks (SiO_2 > 70 wt%)

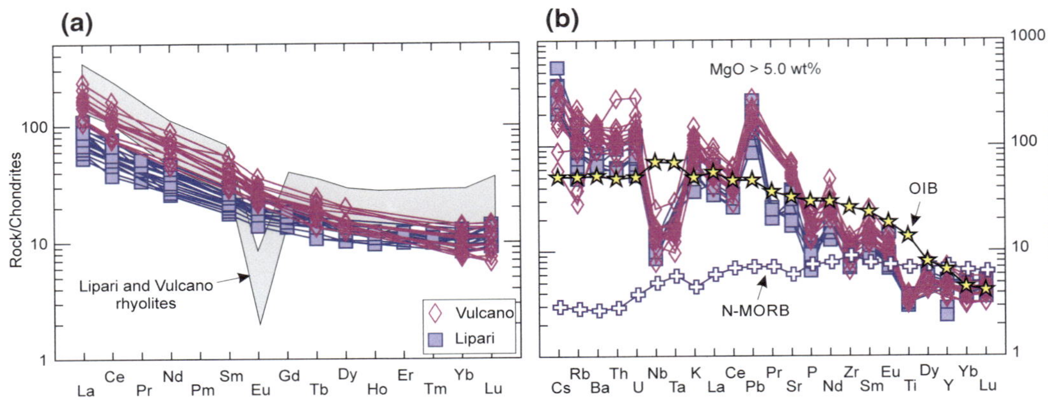

Fig. 9.12 REE (**a**) and incompatible element (**b**) patterns (restricted to mafic rocks) for the islands of Vulcano and Lipari

(AFC) and back-mixing with mafic-intermediate melts provides a better explanation for geochemical and isotopic data.

Forni et al. (2015) suggest that magma evolution at Lipari has been dominated by AFC processes of mafic calcalkaline magmas, with a variable participation of crustal material. Combined Sr–O isotope data suggest that mafic-intermediate magmas underwent heavy crustal contamination; in contrast, the role of the crust was much less significant in the formation of rhyolites. According to these authors, there is no convincing evidence for the occurrence of shoshonitic magmas at Lipari, and the K-rich enclaves occurring in some rhyolites actually represent remobilised cumulate material, rather than a shoshonitic magma. In this view, the shoshonitic affinity of enclaves would be the product of melting of potassic cumulate minerals, e.g. biotite. Whatever the case, it must be noted that the low concentrations of Ba in the rhyolites testify to an important role of alkali feldspar fractionation. Such a process typically occurs in the evolved shoshonitic magmas and is also observed at Vulcano, but not at Salina, where a derivation of rhyolites from calcalkaline parents is obvious. Therefore, heavy alkali feldspar separation implies the occurrence of evolved shoshonitic magmas at Lipari, at least during the latest stages of volcanism. Obviously, such a potassic character could have been imparted to evolved magmas by melting of cumulate phases.

Magma contamination and mixing were also important evolution processes, as mentioned earlier. Crustal contamination is obvious and particularly strong in the cordierite-rich andesites (e.g. Di Martino et al. 2011), but also affected other mafic-intermediate magmas (e.g. Esperança et al. 1992). Mixing is supported by a wealth of petrographic, mineralogical and geochemical evidence, best observed in silicic rocks. According to several authors (e.g. De Rosa and Sheridan 1983; De Rosa et al. 2003b; Gioncada et al. 2003), there was intensive interaction between rhyolitic and intermediate mafic melts during the late stage evolution of Lipari magmatism. Such a process could explain the double-trend distribution observed for some trace element variation diagrams (Fig. 9.11g and h). Such a feature easily develops during mixing of compositionally diverse batches of magmas, and is related to different mobility of elements (see Perugini et al. 2012; Morgavi et al. 2013). Therefore, a process of AFC generating silicic melts, plus back-mixing with mafic magmas could be an explanation for the complexity of rhyolite compositions.

Studies of magma evolution and geobarometric data on xenoliths provide constraints on physical condition of magma ponding and crystallisation, and, therefore, on the structure of the plumbing system at Lipari. According to Gioncada et al. (2003, 2005) a vertically zoned latitic to rhyolitic magmatic system fed by a deep mafic reservoir has been active during the last 40 ka. Di Martino et al. (2010) report on fluid inclusion geothermobarometric data supporting the occurrence of a large magma chamber at about 20 km depth, which fed shallow-level smaller reservoirs sited at various levels (14 to 5–3 km) beneath different centres. Overall, the two models seem to agree that deep processes drive magmatic and volcanic activity at Lipari, a conclusion also drawn for the nearby Island of Vulcano (Peccerillo et al. 2006).

9.7 Vulcano

9.7.1 Volcanology and Stratigraphy

The Island of Vulcano is exposed over an area of about 22 km^2, and represents the southern sector of a larger complex, which includes both Vulcano and Lipari (Gioncada et al. 2003). Rock compositions range from mafic to silicic, mostly with high-K calcalkaline to shoshonitic affinities, and minor potassic alkaline rocks (KS). Mafic shoshonitic and potassic volcanics are moderately oversaturated to undersaturated in silica (Fig. 9.2b). Ages of exposed rocks range from about 130 ka to present, the latest eruption being occurred at Cono della Fossa from August 1888 to March 1890.

Volcanological and stratigraphic studies (e.g. Keller 1980b; Sheridan et al. 1985; De Astis et al. 1997, 2013) documented a complex evolution during nine epochs of activity, characterised by effusive to strongly explosive eruptions, sometimes associated with caldera collapses, and by the emplacement of a wide variety of magmas (Fig. 9.13). The early activity built up an eroded shoshonite lava shield (Capo Secco) offshore the western coast of the island, and the high-K basaltic andesite stratocone of Primordial Vulcano (about 127–101 ka). After the collapse of the stratocone and the formation of Piano Caldera, volcanism was concentrated inside the caldera and at pericaldera centres (99.5–78 ka), successively extending northward, within the newly formed Fossa caldera (<70–42 ka). Shoshonitic basalts, shoshonites, and high-K basaltic andesites were erupted during this period. The following activity took place mainly along the western border of the island, with emplacement of latite to trachy-rhyolite lava domes, flows and pyroclastics at Lentia, shoshonitic basalt and shoshonite lavas and scoriae at Spiaggia Lunga, and shoshonite to trachyte scoriae and pumices at Quadrara (28–21 ka). The latest eruptions took place at the margins and inside the Fossa caldera. A new suite of latite-trachyte-rhyolite lava dome and flows at Lentia (about 15.5–8.5 ka), the high-K basaltic andesite and shoshonite lava and scoria cone of Monte Saraceno, and the historical Vulcanello shield and Cono della Fossa were formed during this period.

The Island of Vulcano is also probably the source of the widely distributed Brown Tuffs pyroclastic deposits whose proximal facies have been recognised in the Fossa caldera area. Three temporally distinct suites have been recognised, with ages of 70–56 ka (Lower Brown Tuffs), 56–27 ka (Intermediate Brown Tuffs) and 24–8 ka (Upper Brown Tuffs). Compositions of juvenile components mostly fall in the fields of latite and trachyte on the TAS diagram (Lucchi et al. 2008; Cicchino et al. 2011).

Vulcanello is a shoshonite (leucite bearing) to latite lava shield with three partially overlapping scoria cones, developed at the northern edge of Fossa caldera (Fig. 9.14). Based on geochronological data and on the occurrence of tephra layers from the 776 AD Pilato eruption (Lipari), De Astis et al. (2013) suggest that Vulcanello was formed about 2 Ka before present. This agrees with historical record by Strabo and Pliny, cited by Mercalli (1883) and Mercalli and Silvestri (1891), which report on eruptions during

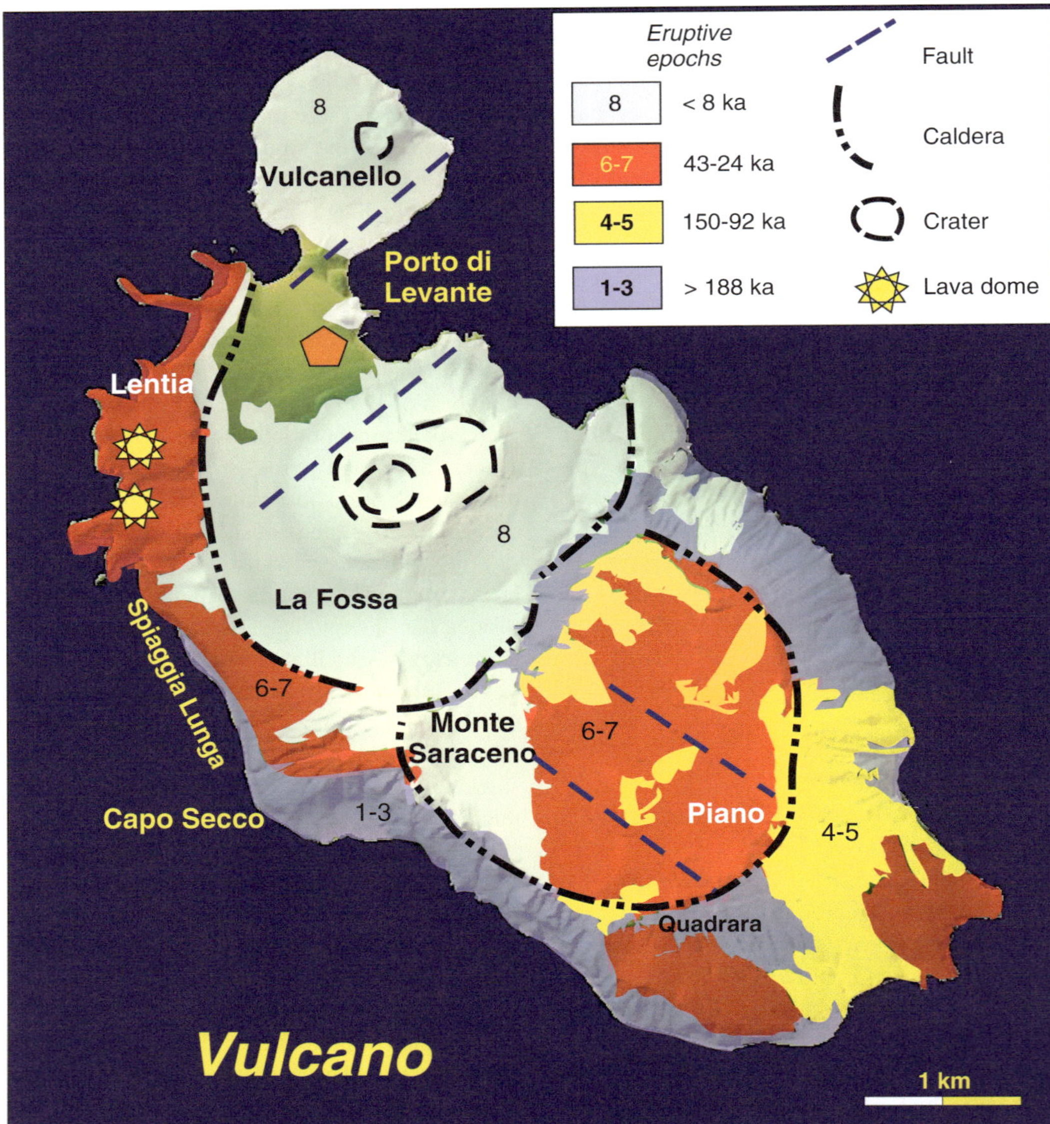

Fig. 9.13 Schematic geological map of Vulcano. Simplified after De Astis et al. (2013)

II–III century BC. In contrast, a two-stage construction during the eleventh-twelfth centuries and the seventeenth century has been suggested by other authors, mainly on the basis of magnetostratigraphic data (Tanguy et al. 2003; Arrighi et al. 2006; Fusillo et al. 2015). Sand accumulation in the isthmus area connected Vulcanello with the main island.

Cono della Fossa (La Fossa Cone) is a 391 m high cone that rises at the center of Fossa caldera; it was constructed during the last 2 ka by several pyroclastic and a few lava eruptions. The base of the sequence contains shoshonite to trachyte volcanics, whereas rhyolitic pyroclastics and obsidianaceous lava flows represent the latest emplaced products. According to historical chronicles summarised by Mercalli (1883), at least five eruptions took place from the XV to XVIII century at Cono della Fossa. These were characterised by emission of ashes, lapilli and

Fig. 9.14 The Island of Vulcano. View from Mt. Falcone, Lipari

bombs, and by emplacement in 1775 of an obsidian lava flow along the northwestern flank of the cone. The 1888–1890 eruption started on August 3rd, 1888 after a period of increased fumarolic activity and gas temperatures. The eruption was characterised by numerous "vulcanian" explosions that emitted a total of about one hundred thousand cubic metres of ashes, lapilli and bombs. The eruption ended on March 22nd, 1890 (Mercalli and Silvestri 1891).

At present, Vulcano is characterised by shallow seismicity and intense fumarolic activity, mostly occurring at the northern margin of Cono della Fossa crater, and near to Porto di Levante. Cono della Fossa has had several episodes of unrest during its recent history, with fumarole gas temperature increasing to more than 700 °C in 1993 (Chiodini et al. 1995). These anomalous values were accompanied by ground deformation, flank sliding, an upsurge of total gas flux, and by a large output of magmatic volatile components revealed by high F/Cl and $^3He/^4He$ ratios (Martini et al. 1986; Chiodini et al. 1995; Nuccio et al. 1999). Successively, gas temperatures decreased to around 250–300 °C in 2001–2002, to increase to about 400–450 °C in 2013, with continuous oscillations and sudden rise during main tectonic seismic events (Diliberto 2013; Madonia et al. 2013).

9.7.2 Petrography and Mineral Chemistry

The mafic rocks (high-K basalts, high-K andesites, shoshonitic basalts, shoshonites, according to classification of Peccerillo and Taylor 1976; Fig. 9.9), were mostly emplaced during the early stages of activity, and exhibit a variably porphyritic texture with phenocrysts of zoned plagioclase ($\sim An_{75-50}$), diopside to augite clinopyroxene, olivine (Fo_{91-45}), magnetite, and some biotite set in a microcrystalline to cryptocrystalline groundmass consisting of the same phases as the phenocrysts; microphenocrysts of leucite, often transformed to analcite, are observed in some rocks. Latites range from porphyritic to subaphyric, with phenocryst mineralogy consisting of augite to diopside clinopyroxene, plagioclase ($\sim An_{50-20}$), minor alkali feldspar (Or_{68-41}), biotite, and rare resorbed olivine ($\sim Fo_{75-45}$); leucite is also sometimes present as microphenocryst or in the groundmass; magnetite and apatite are main accessories. Trachytes contain phenocrysts of plagioclase often with sanidine rim, plus alkali feldspar ($\sim Or_{75-35}$), augite, some biotite, and sporadic xenocrysts of olivine set in hypocrystalline groundmass containing the same phases as the phenocrysts. Rhyolites are nearly aphyric to moderately porphyritic glassy rocks with phenocrysts of plagioclase ($\sim An_{55-20}$), K-feldspar (Or_{70-55}) plus some diopside, biotite and quartz. Crystals of amphibole and olivine (Ol_{70}) are rare; apatite and zircon occur as accessories (De Astis et al. 2013 with references).

9.7.3 Petrology and Geochemistry

The Vulcano rocks show scattered distribution of most major and trace elements in the mafic rocks, whereas more smooth linear trends are generally observed in the intermediate-silicic compositions (Figs. 9.10 and 9.11). The HKCA rocks of older activity (e.g. Primordial Vulcano) have lower incompatible element abundances than younger shoshonitic and KS products at a comparable degree of evolution (De Astis et al. 1997, 2013). Overall, incompatible elements exhibit stronger enrichments than at Lipari, at least for mafic rocks. Chondrite normalised REE patterns are fractionated and sometimes exhibit negative Eu anomalies, which increase considerably in the rhyolites (Fig. 9.12a). Mantle normalised incompatible element patterns of mafic rocks contain positive spikes of Pb and negative anomalies of HFSE, especially Ta and Nb whose abundance is close to N-MORB (Fig. 9.12b).

Sr-isotope ratios range from 0.7041 to 0.7059. The largest variations are observed in the older HKCA rocks, whose $^{87}Sr/^{86}Sr$ ratios (=0.7041 to 0.7052) show a negative correlation with MgO and a positive correlation with silica (De Astis et al. 1997). The rocks from Lentia ($^{87}Sr/^{86}Sr \sim 0.7046–0.7047$), Cono della Fossa and Vulcanello ($^{87}Sr/^{86}Sr \sim 0.7046–0.7048$) show moderate Sr-isotope variation, although a few rhyolites reach distinctively higher values ($^{87}Sr/^{86}Sr = 0.70565–0.70588$; De Astis et al. 1997; Del Moro et al. 1998). Overall, the most primitive HKCA rocks from Primordial Vulcano have slightly lower Sr isotopic ratios than the mafic SHO and KS rocks. Pb-isotope ratios ($^{206}Pb/^{204}Pb = 19.30–19.76$, $^{207}Pb/^{204}Pb = 15.66–18.86$, $^{208}Pb/^{204}Pb = 39.14–39.82$) show a rough negative correlation with Sr-isotope ratios. Hf-isotope composition of a high-K calcalkaline basalt ($^{176}Hf/^{177}Hf = 0.282916$; $\varepsilon Hf = 5.1$) has a lower value than Alicudi and Filicudi (Gasperini et al. 2002). He-isotopes measured on olivine and pyroxene phenocrysts yielded values of $R/R_A \sim 2.3$ to 4.9 (Di Liberto 2003; Martelli et al. 2008), lower than ratios measured in the fumaroles ($R/R_A \sim 5–6$; Tedesco et al. 1995; Tedesco and Nagao 1996). Oxygen isotope composition of whole rocks (Ellam and Harmon 1990) shows $\delta^{18}O_{SMOW} = +6.2$ ‰ to +8.3 ‰, whereas somewhat lower $\delta^{18}O_{SMOW} = 5.9–6.6$ ‰ has been measured in separated olivine and clinopyroxene (De Astis e al. 2013 and references therein).

9.7.4 Magma Evolution at Vulcano

Geochemical and petrological studies highlight complex magma evolution processes at Vulcano (De Astis et al. 1997, 2013; Del Moro et al. 1998). These took place at different depths, in partially interconnected magma chambers located at the mantle-crust boundary (17–21 km), within the felsic granulite upper crust (about 8–13 km), and at about 3–5 km along the contacts between the Calabro-Peloritano basement and the sedimentary-volcanic cover (Zanon et al. 2003; Peccerillo et al. 2006; De Astis et al. 2013).

The positive correlation of $^{87}Sr/^{86}Sr$ versus silica for Primordial Vulcano indicates strong interaction between mafic magmas and continental crust (De Astis et al. 1997). However, these rocks have a mafic composition despite the large amount of AFC that is needed to generate the observed Sr-isotope variations. This feature has been interpreted to reflect protracted mixing with mafic melts during AFC; such a process preserved the mafic composition of magmas, while Sr-isotopes ratios were increased significantly. The poorly variable isotopic composition of the Lentia rocks favours a derivation of rhyolites from mafic-intermediate parents by dominant fractional crystallisation. However, there are linear relationships for many inter-element variations, which suggest mixing between acid and mafic magmas at this centre (De Astis et al. 1997).

Finally, element variation at Cono della Fossa and Vulcanello is consistent with dominant fractional crystallization starting from shoshonitic mafic parents for both centres. The decrease in K_2O and some incompatible elements in the most evolved products indicates the separation of K-feldspar and of accessory phases such as zircon. Feldspar fractionation generated a drop in Sr and Ba and strong negative Eu anomalies in the rhyolites. The rhyolitic rocks from Cono della Fossa were also affected by mixing-mingling with mafic melts, as demonstrated by the occurrence of mafic enclaves with irregular crenulated margins and of xenocrystic olivine (Fo_{70-65}) and diopsidic clinopyroxene. The high $^{87}Sr/^{86}Sr$ ratios in some of the most acidic rocks reflect moderate crustal assimilation (Del Moro et al. 1998).

Overall, magmatism at Vulcano is characterised by an increase in the degree of rock evolution with time, accompanied by a less clear increase in potassium and incompatible elements in the mafic magmas. The time-related increase in the proportions of silicic versus mafic magmas has been attributed to an enhanced role of fractional crystallisation with respect to mafic magma input into the volcanic system. These effects are considered as related to an overall migration of magma chambers toward shallow levels (De Astis et al. 1997, 2013; Zanon et al. 2003; Peccerillo et al. 2006).

The variable enrichment in potassium and incompatible elements in the mafic melts has been attributed to the derivation of primary magmas from heterogeneous mantle sources. The early activity tapped less enriched mantle rocks than younger shoshonitic and KS volcanism. Magma generation at various levels in a stratified mantle would depend on isotherm increase, which triggered melting at progressively shallower depth with time (De Astis et al. 1997). However, there is an increase of La/Sm, at roughly constant Tb/Yb, from mafic high-K calcalkaline to KS rocks. Such a feature and the poorly variable radiogenic isotope signatures have been interpreted as evidence for different degrees of partial melting of rather homogeneous sources (Peccerillo et al. 2013).

9.8 Panarea

9.8.1 Volcanology and Stratigraphy

Panarea is the summit of a dissected volcanic structure that rises about 1700 m above the sea floor, with a nearly flat surface of about 50 km^2 at a depth of 100–150 m below sea level. The emergent portion of the volcano forms the main island of Panarea, the smaller island of Basiluzzo and a number of islets about 3 km east of Panarea (Dattilo, Panarelli, Lisca Bianca, Bottaro and Lisca Nera).

Exposed rocks consist mostly of intermediate-acidic lava domes, with minor lava flows and pyroclastics (Gabbianelli et al. 1990; Calanchi et al. 2002b; Dolfi et al. 2007; Lucchi et al. 2013e). Geological mapping, radiometric dating and correlation with the chronology of marine terraces indicate eruption ages between about 155 and 8.7 ka, with the bulk of the activity being concentrated between 155–149 ka and to 124–118 ka (Lucchi et al. 2013e and references therein). Early dome emplacement took place along the northern sector of Panarea, and successively shifted toward the south-southeast. At about 100 ka, a presently submerged volcanic centre located in the area of the islets east of Panarea emitted strombolian mafic scoriae that are found at several places on the main island, especially at Punta Falcone. The same provenance is also inferred for the much younger (24–8.7 ka) latite to trachyte pumice series (Drauto pumices). The rhyolitic dome of Basiluzzo, sited north of Panarea, has an age of about 54 ka. Several pyroclastic deposits of external origin are intercalated with local lavas and pyroclastics. These include ashes from the Campania Province, and from other Aeolian volcanoes, notably the Brown Tuffs tephra originated at Vulcano.

The Panarea complex is considered in a quiescent state. Intense submarine hydrothermal exhalative activity occurs in the islets area. Here, a major gas explosion occurred on the 3rd of November 2002 at a depth of 10–15 m, forming a crater about 15–20 m in diameter (e.g. Tassi et al. 2014 and references therein).

9.8.2 Petrography and Mineral Chemistry

The Panarea volcanics have mafic to felsic compositions (Fig. 9.15). Mafic rocks are represented by the Punta Falcone basaltic andesite strombolian scoriae and by a few lava lithics and enclaves from silicic rocks. Dacites and high-K dacites are the most common rocks, followed by andesites. Rhyolites make up the Basiluzzo dome and some pumice deposits. High-K calcalkaline andesites and dacites have been dredged from the submarine flanks of the volcano, offshore the Basiluzzo islet (Trua et al. 2004). Rocks with shoshonitic affinity are scarce and have been found for a few lithics, magmatic enclaves, lavas and pumices (Calanchi et al. 2002b; Cimarelli et al. 2008; Lucchi et al. 2013e).

The Panarea rocks are variably porphyritic, with zoned plagioclase ($\sim An_{90-25}$) as the dominant and ubiquitous phenocrystic phase. Diopside-augite clinopyroxene and orthopyroxene ($\sim En_{75-35}$) are also abundant in basaltic andesites; hornblende phenocrysts appear in the andesites and dacites; olivine (Fo_{78-76}) is found in small amounts in the basaltic andesites but also as a xenocryst (Fo_{74-72}) in some acidic rocks (Cimarelli et al. 2008; Doherty et al. 2015); biotite phenocrysts are present in the Basiluzzo rhyolite. Groundmass is hypocrystalline to holohyaline and contains plagioclase, opaque minerals, pyroxene and, sometimes, biotite. A glassy fluidal groundmass is present in the rhyolite lavas.

Many lavas and pyroclastic rocks at Panarea contain metamorphic and igneous xenoliths, sometimes reaching about 20 % of the total rock volume. Metamorphic xenoliths consist dominantly of quartz-rich rocks. Magmatic xenoliths include angular fragments of old lavas and mafic enclaves with crenulated margins, which reveal incorporation when they were still in a plastic state (Calanchi et al. 2002b).

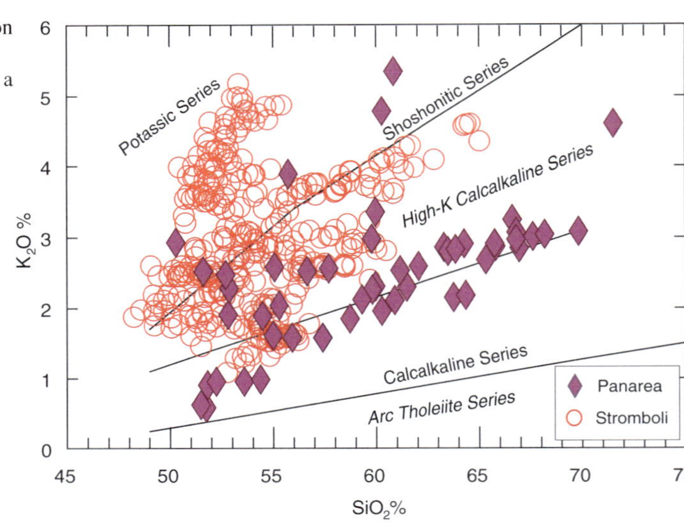

Fig. 9.15 K_2O versus SiO_2 classification diagram for Panarea and Stromboli. Panarea data are normalised to 100 % on a LOI-free basis

9.8.3 Petrology and Geochemistry

Geochemical variation diagrams highlight higher alkalies, P_2O_5, TiO_2, and incompatible trace elements for shoshonitic relative calcalkaline rocks (Figs. 9.16 and 9.17). In general, there is considerable data scattering, a feature that is drastically reduced if samples from single eruptive epochs are considered separately (Lucchi et al. 2013e).

REE of silicic rocks show small negative Eu anomalies (Fig. 9.18a), which are similar to Salina rhyolites and much smaller than at Lipari and Vulcano. Mantle normalised incompatible element patterns of mafic rocks have high LILE/HFSE ratios and a positive anomaly of Pb (Fig. 9.18b).

Sr-isotope composition ranges from 0.7040 to 0.7061 and increases from calcalkaline to shoshonitic rocks (Fig. 9.17f). Nd isotopic ratios

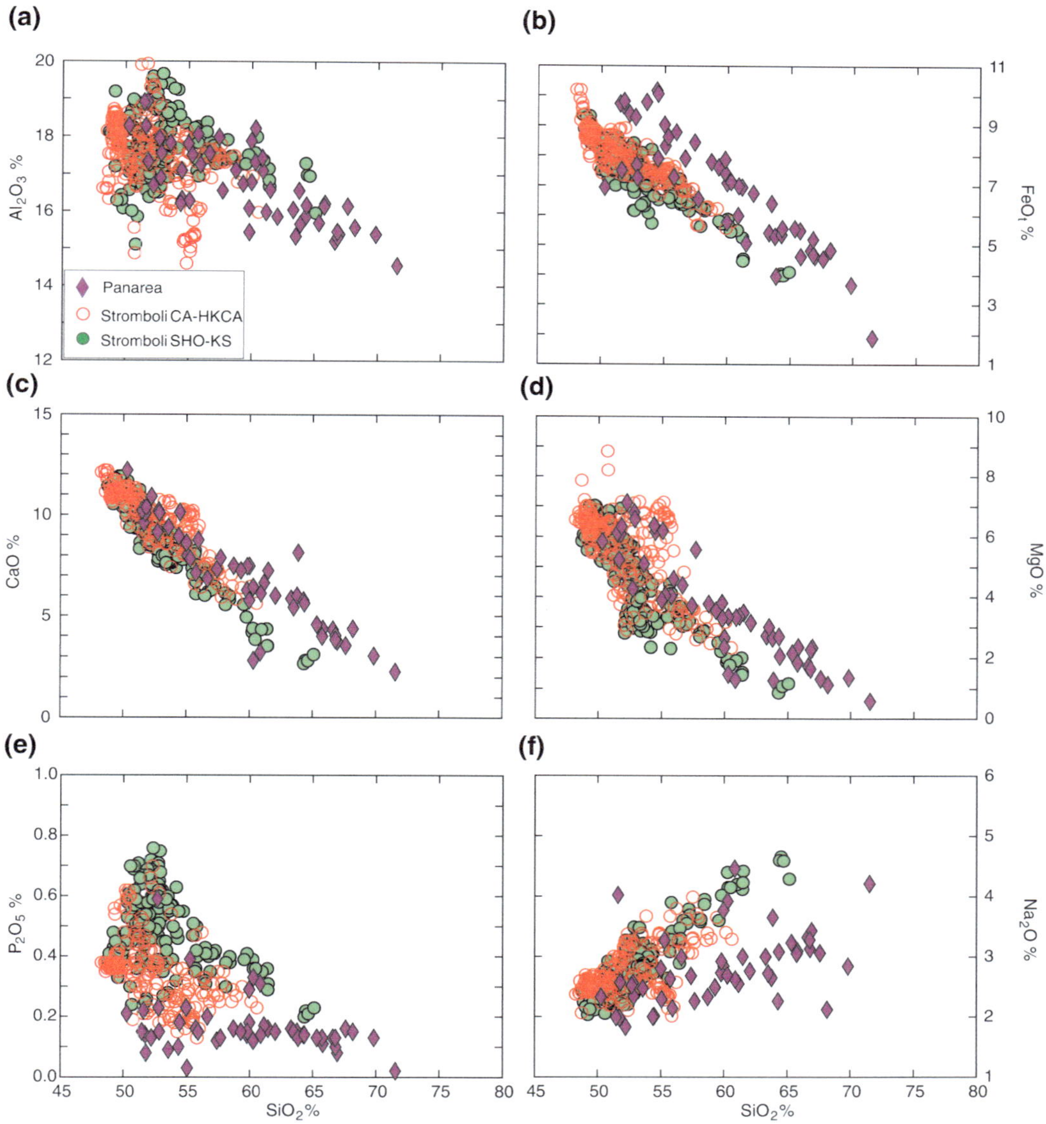

Fig. 9.16 Harker variation diagrams for the islands of Panarea and Stromboli. Panarea data are normalised to 100 % on a LOI-free basis

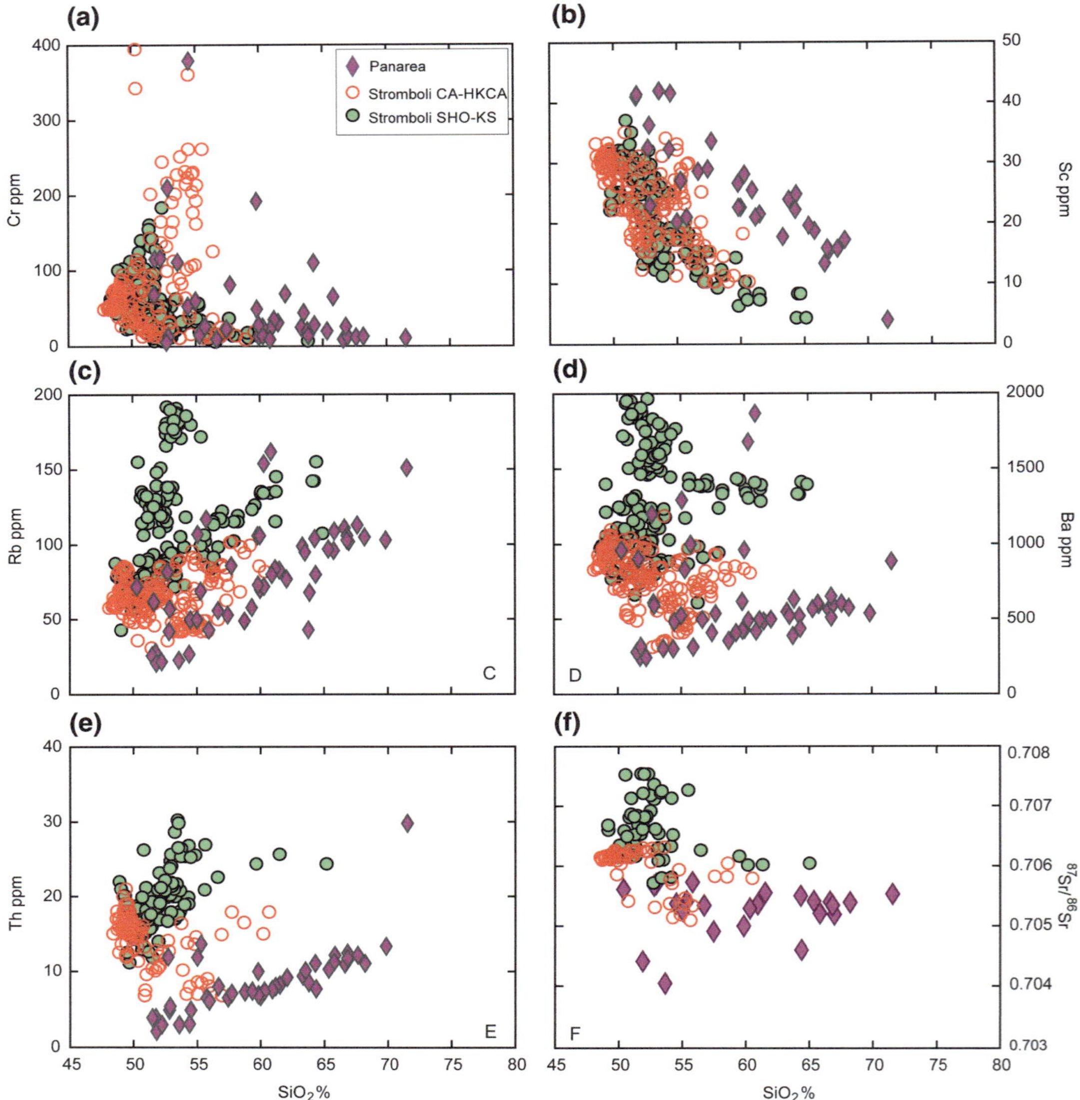

Fig. 9.17 Variation diagrams of selected trace elements and $^{87}Sr/^{86}Sr$ versus silica for the islands of Panarea and Stromboli

range from 0.51284 to 0.51257. Pb isotopic composition is available for a few samples and show a narrow range of values with $^{206}Pb/^{204}Pb$ = 19.18 to 19.43; $^{207}Pb/^{204}Pb$ = 15.67–15.71; $^{208}Pb/^{204}Pb$ = 39.13–39.36 (Calanchi et al. 2002b; Doherty et al. 2015).

9.8.4 Magma Evolution at Panarea

The very discontinuous and limited nature of the outcrops, which are mainly below sea level, makes it difficult to understand magma genesis and evolution at Panarea. Calanchi et al. (2002b)

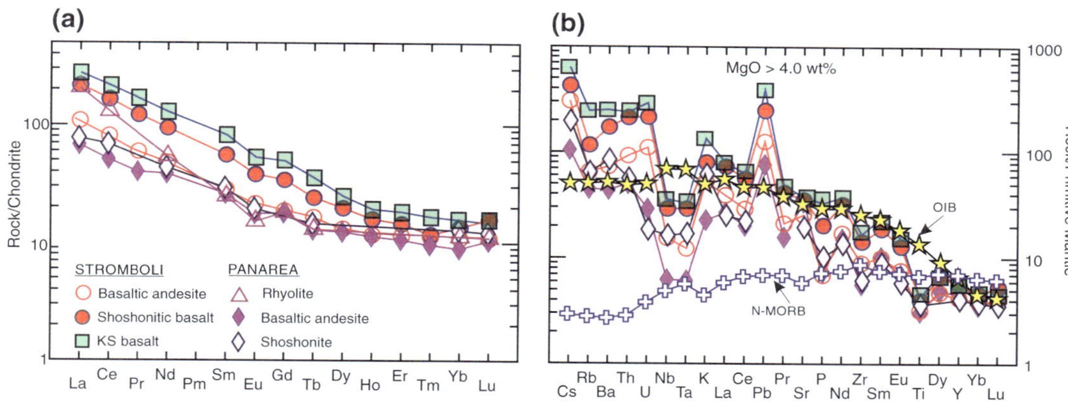

Fig. 9.18 REE (**a**) and incompatible element (**b**) patterns for selected rocks from Panarea and Stromboli

suggested that the variable incompatible element contents for calcalkaline to shoshonitic mafic rocks indicate the occurrence of distinct types of primary magmas. These underwent an evolution dominated by fractional crystallisation, magma mixing and moderate interaction with wall rocks in some cases, to produce intermediate and acidic products. Based on two-pyroxene geobarometry, extremely variable depths of crystallisation were suggested by Doherty et al. (2015), from the Moho to less than 1 km. Magma mixing is highlighted by within-rock compositional disequilibrium and by linear inter-element trends between mafic and silicic compositions defined by some samples (e.g. Fig. 9.17a). Rhyolites from Basiluzzo show geochemical and isotopic composition, which suggests a derivation from a shoshonitic mafic parent, probably represented by the mafic shoshonitic enclaves occurring in the rhyolite. However, the small Eu anomaly and relatively elevated Ba contents require that fractional crystallisation was less extreme and involved lower amounts of alkali-feldspar than at Lipari and Vulcano. This points to a calcalkaline rather than a shoshonitic parent for the Basiluzzo rhyolites.

The Panarea mafic calcalkaline rocks have Sr-, Nd-, and Pb-isotope compositions similar to those of the western-central Aeolian arc. In contrast, HKCA and shoshonitic rocks have higher Sr-isotope and lower Nd- and Pb-isotope ratios and fall midway between the western-central Aeolian arc and Stromboli. This is also observed for some trace element abundance and ratios (e.g. Ta, Nb, Ba/La, La/Ta, Ce/Sr; Calanchi et al. 2002b). If these characteristics are not related to intra-crustal magma evolutionary processes, it should be concluded that the mantle source of Panarea has transitional compositional features between the western-central Aeolian arc and Stromboli.

9.9 Stromboli

9.9.1 Volcanology and Stratigraphy

Stromboli is the summit of a NE-SW elongated cone sited at the eastern end of the Aeolian arc. The Strombolicchio islet is the remnant of a parasitic centre about 1.5 km NE of Stromboli. The volcanic edifice rises about 2000 m above the seafloor, reaching an altitude of 924 m above sea level. The outcropping rocks are younger than 85 ka at Stromboli, whereas an older K/Ar age of about 200 ka has been measured for Strombolicchio (Gillot 1987; Gillot and Keller 1993; Francalanci et al. 2013). Rock composition is mafic to intermediate, with calcalkaline (CA), high-K calcalkaline (HKCA), shoshonitic (SHO) and potassic alkaline (KS) affinities. Calcalkaline rocks are restricted to basaltic andesites; SHO and HKCA rocks range from basalt to high-K dacite and latite, which straddle the silica undersaturation line; KS are mafic and undersaturated in silica (Figs. 9.2b and 9.15).

Since ancient historical times, Stromboli has been characterised by continuous degassing and persistent strombolian explosions lasting a few seconds and emitting mafic HKCA to shoshonitic scoriae every few minutes (e.g. Rosi 1980; Ripepe et al. 2008; Rosi et al. 2013). Random larger explosions (paroxysm) and effusion of lava flows, with accompanying increased shallow seismicity (e.g. Scarfì et al. 2010), occur every few years, and about twenty events have been registered during the last hundred years. The paroxysm of September 11th, 1930 had dramatic effects on the resident population and inspired the drama movie "*Stromboli*" by Roberto Rossellini, one of the masterpieces of Italian neorealism (Rittmann 1931; Rosi et al. 2013 and references therein).

The active craters are located at an altitude of about 750 m, on a terrace sited at the top of the Sciara del Fuoco (Fig. 9.19). This is a large collapsed depression bounded by normal faults (Fig. 9.20), developed along the NW flank of the volcano and continuing under the sea level almost to the foot of the cone.

The basaltic andesites of the Strombolicchio neck are the oldest exposed rocks. At Stromboli, the outcropping volcanic sequence consists of mafic to intermediate calcalkaline and high-K calcalkaline lavas and minor pyroclastics (Paleostromboli I and Paleostromboli II stratovolcanoes; about 85–52 ka), overlain by HKCA and SHO mafic to intermediate products (Paleostromboli III; about 40–30 ka). Successively, the Vancori shoshonitic basalt to trachyte lava and scoria sequence was emplaced, between 26 and 13 ka. After a sector collapse that affected the entire northwestern sector of the stratovolcano, activity took place at the central crater area and at a number of eccentric vents mostly located in the western half of the island (Semaforo Labronzo, Timpone del Fuoco, etc.). Leucite-bearing KS lavas and scoriae erupted at these centres (Neostromboli series; e.g. Vezzoli et al. 2014). The youngest activity is restricted to the top of Sciara del Fuoco, with the exception of San Bartolo lava, which was outpoured from the northern flank of the Stromboli cone at about 2.4 ka (Keller et al. 1993).

Volcano-tectonic events had a major role in shaping the morphology and structure of Stromboli volcano. Francalanci et al. (2013) recognised five caldera and seven flank or sector collapses during the volcanic history of Stromboli. Some of these (e.g. between Vancori and Neostromboli) were associated with strong modification of magma compositions. Gravitational sliding is also a main morphogenetic factor, as well as a major source of hazard for the population at Stromboli and surrounding areas. The last dangerous event took place on December 30th 2002, when a large landslide along the northwestern flank of Stromboli generated two tsunamis, with waves up to 10 m high. These stroke San Vincenzo and Piscità villages and other Aeolian Islands in a few minutes, and reached the Calabria and Cilento coast, in the

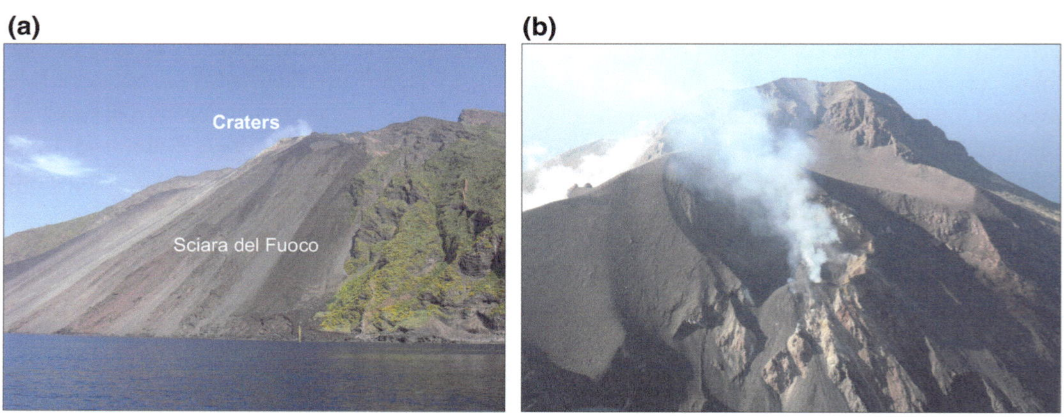

Fig. 9.19 **a** Sciara del Fuoco, Island of Stromboli; **b** Active crater area. Photos by F. Lucchi and G. De Astis

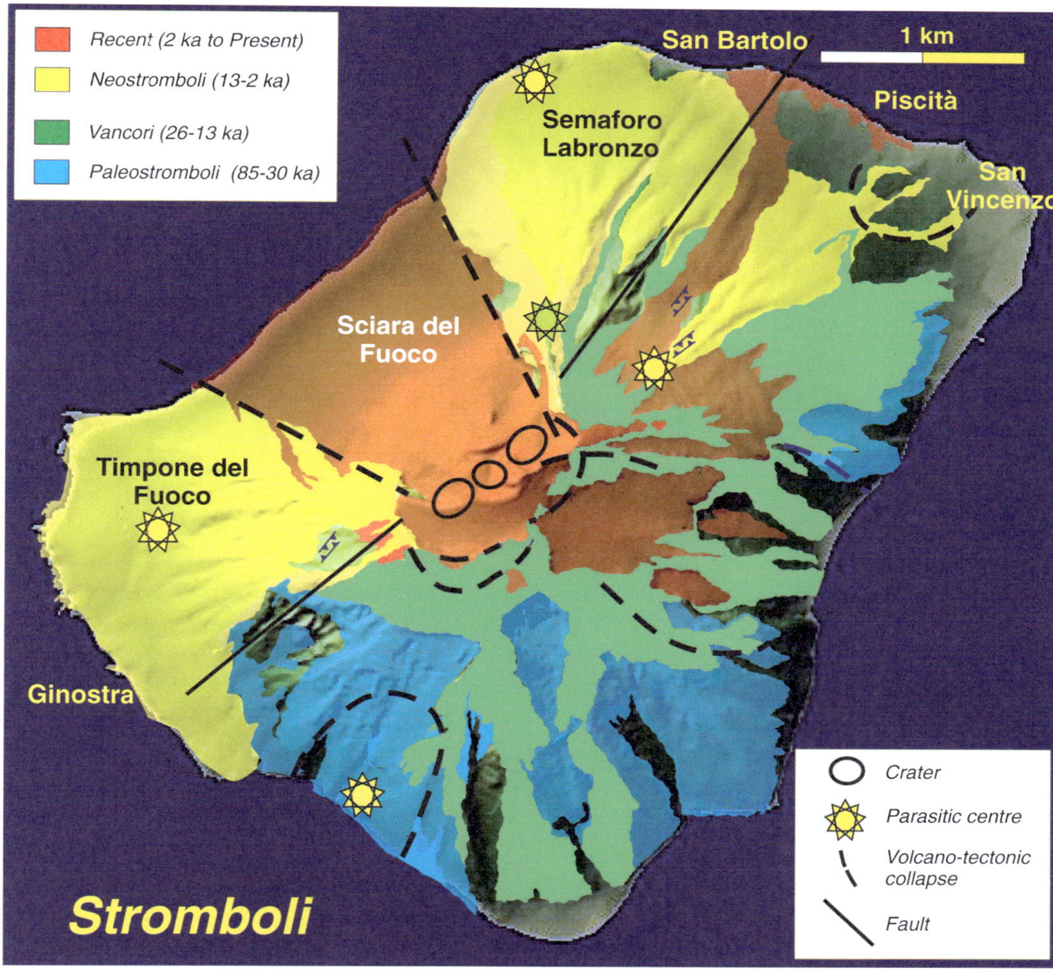

Fig. 9.20 Schematic geological map of Stromboli. Simplified after Francalanci et al. (2013)

southern Italian peninsula, after about one hour (Tinti et al. 2005). Because of wintertime, there were no people along the shoreline, which prevented catastrophic consequences of the tsunami.

The plumbing system of Stromboli probably has been characterised by multiple magma chambers, all through the history of the volcano (Francalanci et al. 2013). However, modification of magma compositions contemporaneously with tectono-volcanic collapses speaks for significant variations of the plumbing system at some stages of volcano evolution. The occurrence of largely prevailing mafic lava flows at Stromboli suggests that magma reservoirs have been mostly working as open systems, with frequent eruptions and continuous input of mafic melts from the source. An exception may be represented by the final stages of the Vancori period when closed-system evolution generated trachytic magma (e.g. Cortes et al. 2005). Depths of these reservoirs are controversial, and 4 and 11 km have been suggested by geochemical, petrological and fluid inclusion studies (Vaggelli et al. 2003; Pichavant et al. 2009; Francalanci et al. 2013). The present-day internal structure of the volcano is believed to consist of a complex system of magma bodies, occurring at different depth, from the uppermost mantle to a few kilometres below sea level.

Evidence for a deep reservoir sited at about 15–20 km is provided by both geothermobarometric data on ultramafic xenoliths (Laiolo and Cigolini 2006) and Vs seismic profiles (Martinez-Arevalo et al. 2009). These reservoirs are in a sort of steady-state regime, with continuous eruptions of strombolian scoriae and lavas, and uninterrupted ascent of mafic melts from a low-V_S layer detected at depths exceeding 32 km (Martinez-Arevalo et al. 2009).

9.9.2 Petrography and Mineral Chemistry

The Stromboli lavas display seriate porphyritic holocrystalline to hypocrystalline textures, with phenocryst contents varying from about 5–50 % of total rock volume (Francalanci et al. 1993, 2013; Bertagnini et al. 2003). Strongly zoned plagioclase ($\sim An_{90-45}$) is the dominant phenocryst in all the rocks, followed by diopside-augite. Olivine phenocrysts ($\sim Fo_{90-60}$), often corroded and altered to iddingsite, occur in the mafic rocks. Orthopyroxene is present as microphenocryst in the CA and HKCA rocks and in some latites. Leucite occurs as microphenocrysts and in the groundmass of some Neostromboli KS rocks. Rare biotite and amphibole are observed in andesites and latites; biotite is also found in the KS rocks as a groundmass phase. Small amounts of Ti-magnetite and ilmenite occur as microphenocryst and groundmass phases. Apatite is a common accessory mineral.

As all the Aeolian volcanoes, Stromboli erupted many types of magmatic and metamorphic xenoliths. Igneous lithologies consist of lava fragments, including some acidic types that are not found at the surface (Renzulli et al. 2001), gabbroic rocks, and some ultramafic xenoliths (dunite, wehrlite, clinopyroxenite), probably representing cumulates formed at a depth exceeding 20 km (Laiolo and Cigolini 2006). Metamorphic xenoliths include quartz-rich nodules and thermometamorphic rocks containing hercynite, sillimanite and corundum (Vaggelli et al. 2003; Del Moro et al. 2011).

9.9.3 Petrology and Geochemistry

CA and HKCA rocks have lower K_2O, P_2O_5 and incompatible trace element contents than the associated SHO and KS volcanics (Figs. 9.16 and 9.17). However, the mantle-normalised incompatible element patterns have similar shapes for all the mafic rocks, with positive spikes of Pb and negative anomalies of HFSE (Fig. 9.18b). A similar pattern is shown by the Mg-rich glass inclusions contained in olivine (Fo_{91-84}) from basaltic scoriae of the present-day activity (Bertagnini et al. 2003). Sr-isotope ratios are higher than for the other Aeolian Islands and increase from CA ($^{87}Sr/^{86}Sr \sim 0.7050$ to 0.7054) to KS rocks ($^{87}Sr/^{86}Sr \sim 0.7066$–0.7076), whereas $^{143}Nd/^{144}Nd$ ratios decrease from 0.51265 to 0.51243 (Fig. 9.6a; Ellam et al. 1988, 1989; Francalanci et al. 1993). Pb-isotope ratios (Fig. 9.6b) are lower than other Aeolian volcanoes ($^{206}Pb/^{204}Pb = 18.93$ to 19.20; $^{207}Pb/^{204}Pb = 15.65$ to 15.69; $^{208}Pb/^{204}Pb =$ 39.01 to 39.16). Slightly less radiogenic compositions (e.g. $^{208}Pb/^{204}Pb = 38.71$–38.81) have been found for olivine-hosted melt inclusions by Rose-Koga et al. (2012). Whole rock O-isotope ratios range from $\delta^{18}O = +6.1$ to +8.0 ‰, with a rough decrease from CA to KS rocks (Ellam and Harmon 1990). He-isotope ratios exhibit the lowest values in the Aeolian arc with $R/R_A = 2.5$–4.9, roughly decreasing from calcalkaline to potassic alkaline products (Di Liberto 2003; Martelli et al. 2008, 2014).

9.9.4 Magma Evolution at Stromboli

A large number of volcanological, petrological and geochemical studies have been aimed at understanding the origin and evolution of various magma types, and the relationships between magma compositional variations and volcano-tectonic events at Stromboli (e.g. Dupuy et al. 1981; Ellam et al. 1988; Luais 1988; Francalanci et al. 1989, 2013; Hornig-Kjarsgaard et al. 1993; Bertagnini et al. 2003; Vaggelli et al. 2003; Pichavant et al. 2009; Métrich et al. 2010; Pompilio et al. 2012; Bragagni et al. 2014).

Francalanci et al. (1989, 2013) noticed that the large trace element and radiogenic isotope variability of Stromboli is much reduced when rock samples are separated according to their stratigraphic position. This was interpreted as evidence that different batches of compositionally distinct mafic magmas (CA, HKCA, SHO, and KS) underwent various degrees of fractional crystallisation to generate distinct suites of derivative liquids with broadly homogeneous isotopic signatures. Such a process was particularly intensive during the upper Vancori stage of activity, when trachytic compositions were reached by fractional crystallisation of a shoshonitic basalt parent in a closed-system magma chamber (e.g. Cortes et al. 2005).

Mixing processes have been also working extensively all though the history of Stromboli. Evidence, however, is obvious only when magmas with contrasting compositions interacted within shallow reservoirs or conduits. A particularly striking case is observed at the transition from the trachytic Vancori stage to the Neostromboli KS mafic activity, where a well-defined progressive decrease in SiO_2 and increase of $^{87}Sr/^{86}Sr$ is observed along the stratigraphic sequence (Francalanci et al. 2013 and references therein).

However, the primary petrologic problem at Stromboli is the origin of the variable compositions of mafic magmas. Dupuy et al. (1981) suggested a generation in a heterogeneous mantle wedge affected by various degrees of subduction-related metasomatism, a view shared by several other authors (e.g. Ellam et al. 1988; Luais 1988). Francalanci et al. (1993) proposed that, in addition to source heterogeneity, a continuous process of AFC plus mixing of calcalkaline mafic magmas in a deep reservoir might generate a suite of mafic melts with variable enrichments in potassium, incompatible elements and radiogenic isotopes. Fractional crystallisation of these melts at shallow level would produce various compositionally distinct rocks suites.

Peccerillo (2001) pointed out that the Stromboli KS rocks have concentrations and ratios of incompatible elements (e.g. Ta, Zr, Th, Sr, Ba, REE, Th/La, Ba/Nb, Zr/Nb) and radiogenic isotope signatures that closely resemble those of Somma-Vesuvio, except for K and Rb, which are slightly lower at Stromboli (see Chap. 7). Moreover, the compositions of calcalkaline and shoshonitic rocks from Stromboli also exhibit close compositional affinities to the rocks from Procida. This close similarity argues against the idea that geochemical variation both at Stromboli and in Campania may derive from any kind of local process occurring inside the volcanic plumbing system, given the different type of basement rocks beneath various volcanoes. Therefore, compositionally similar mantle sources were invoked for the Stromboli and Campania magmas.

9.10 Aeolian Seamounts

Main seamounts in the Aeolian area are Marsili, Sisifo, Enarete, Eolo, Lametini, Alcione, Palinuro, and Glabro (Fig. 9.1). Description of these seamounts is given in Chap. 12. Other minor seamounts include Alicudi North, Banco di Filicudi and Filicudi North, respectively NW and north of the homonymous island, Secca del Capo, north of Lipari, and Castore and Polluce, west of Stromboli (Romagnoli et al. 2013; Passaro et al. 2009).

9.11 Petrogenesis of the Aeolian Arc Magmas

The compositional variability of the Aeolian magmas within the single islands and at the regional scale reflects both intra-crustal magma evolution processes, and the occurrence of different types of primary melts generated in heterogeneous mantle sources. The relative roles of these two factors have been amply debated (e.g. Ellam et al. 1988, 1989; Francalanci et al. 1993, 2004, 2007; De Astis et al. 2000, 2006; Tommasini et al. 2007; Peccerillo et al. 2013). A short summary of this discussion is given below.

9.11.1 Role of Magma Evolution Processes

Magma evolution in the Aeolian arc took place by fractional crystallisation, plus crustal assimilation and mixing, as for most if not all volcanoes on Earth. These processes have been working contemporaneously in most cases, but their relative roles have been extremely variable at various eruptive stages of single volcanic systems. Petrological, crystal chemistry and fluid inclusion studies demonstrate that magma evolution has been polybaric, and took place in magma reservoirs sited at the transition between lithologically different levels, especially at the Moho, at the transition between the lower and the upper crust, and along the interface between the substratum and the sedimentary-volcanic cover (Nazzareni et al. 2001; Zanon et al. 2003; Frezzotti et al. 2003, 2004; Peccerillo et al. 2006; Di Martino et al. 2010).

Magma evolution processes for single volcanoes have been discussed in the previous sections. In general terms, protracted ascent of mantle-derived mafic magmas and mixing with resident melts within deep reservoirs, concomitantly with some fractional crystallisation and assimilation, have been working extensively during the early stages of several Aeolian Islands. The continuous arrival of mafic melts inhibited magma evolution toward intermediate and acidic compositions. Therefore, thick suites of mafic lavas and scoriae make up the lowest exposed sections of several Aeolian volcanoes, especially Alicudi, Salina, Lipari and Vulcano. The occurrence of widespread mafic magmatism at Stromboli testifies that continuous mixing plus some fractional crystallisation has been going on during most of the evolution history of this volcano.

Fractional crystallisation is responsible for the origin of intermediate and acidic rocks. Most authors believe that such a process has been particularly extensive in the central islands of Lipari and Vulcano, where abundant rhyolitic magmas erupted. Generally, acidic magmas were emplaced during intermediate or latest stages of construction of Aeolian volcanoes, indicating a modification with time for the style of the leading evolution processes, from mixing to fractional crystallisation. Geobarometric data suggest that such a transition is related to a migration of magma chambers toward shallow depths, where lower temperature favour crystallisation and the arrival of fresh mafic magmas is less continuous, likely because of the increased distance from the mantle sources (e.g. Frezzotti et al. 2003, 2004; Peccerillo et al. 2006). Migration of magma chambers could be related to the modification of the structural setting of volcanoes. Barberi et al. (1994) report on geophysical and field evidence for extensive development of host and graben structures at Lipari and Vulcano, along the northern segment of the Tindari-Letojanni strike-slip fault. Such a geological setting favoured the formation of large shallow-level magma chambers, where extensive fractional crystallisation processes generated abundant silicic melts. Note, however, that other authors suggest the late occurrence of rhyolites at Lipari may reveal crustal melting by temperature increase generated by protracted injection of mafic melts (Crisci et al. 1991; Di Martino et al. 2011).

Finally, assimilation of crustal wall rocks is testified by plenty of lithological and geochemical evidence, including the widespread occurrence of crustal xenoliths and, in particular, of quartz-rich nodules that represent residues of partially melted upper crustal rocks. Stable and radiogenic isotope data testify to crustal contamination for both mafic and silicic magmas. At Alicudi, crustal assimilation in a deep reservoir took place contemporaneously with mixing and crystal fractionation, determining significant variations of Sr–Nd–Pb–O isotope ratios in the mafic melts. The same process probably occurred at Filicudi, and cannot be excluded for other islands, although clear isotopic evidence is lacking. These studies raise the problem if any of the mafic rocks remained unaffected by assimilation and may still be considered as representative of primary melts and their mantle sources.

Scrutiny of geochemical variation for some contaminated mafic rocks in the Aeolian arc can contribute to place some constraints on this problem. As discussed earlier, the Alicudi mafic magmas show radiogenic isotope signatures indicating significant crustal contamination. It has been calculated that at least 15–20 % crustal material with composition as the average Calabro-Peloritano metamorphic rocks was assimilated by the most contaminated basalts in this island (Peccerillo et al. 1993, 2004). However, several incompatible trace element ratios (e.g. Ba/La, Th/Ta, Rb/Nb, Zr/Nb etc.) do not show significant correlation with Sr isotope ratios or any evolution parameter. Therefore, the contaminated basalts show similar compositional range as the uncontaminated andesites (Fig. 9.4g). Similar results have also been obtained for a well-studied contaminated suite of mafic-intermediate magmas at Vulcano (De Astis et al. 1997, 2013). This suggests that many incompatible element ratios are basically unaffected by upper crust contamination in the Aeolian volcanoes, and can be used to infer compositions of primary melts (Peccerillo et al. 2013).

Oxygen versus Sr-isotope variations can add information on effects of magma contamination by the upper crust. A $\delta^{18}O$‰ and $^{87}Sr/^{86}Sr$ diagram for Aeolian rocks and separated phases (Ellam and Harmon 1990; Peccerillo et al. 2004, 2013; Santo and Peccerillo 2008; De Astis et al. 2013) shows that crustal assimilation has been working in most islands, producing a strong increase for $\delta^{18}O$‰ (vertical trends in Fig. 9.21). In contrast, variation of Sr-isotope ratios (and Nd–Pb) are moderate and much lower than those observed at regional scale. This leads to the conclusion that the first-order along-arc radiogenic isotope variations cannot be related to magma contamination but reflect primary magma (i.e. source) compositions (e.g. Ellam et al. 1989; Francalanci et al. 1993, 2004; De Astis et al. 2000, 2013; Peccerillo et al. 2013).

9.11.2 Genesis of Primary Magmas

There is an obvious consensus that the Aeolian arc magmas derived from primitive parents generated in a subduction-modified upper mantle wedge. However, several issues remain controversial. These include the composition of material released from the slab to the mantle wedge, mechanism of slab to mantle element transfer, the composition of pre-metasomatic mantle wedge, and the genetic relationships between CA-SHO-KS magmas.

As discussed earlier, regional radiogenic isotope variations cannot depend on shallow level processes; therefore, the interaction between the upper mantle and various amounts of crustal material is required by isotopic evidence. Isotopic and geochemical evidence, such as the eastward increase of Sr-isotope and decrease of Nd–Pb-isotope ratios (Fig. 9.22a and b), suggests an increasing contribution of upper crustal components (likely trench sediments) from the western-central islands to Stromboli.

It has been calculated that less than 1–2 % of upper crustal material with a composition as the Ionian Sea sediments (De Astis, unpublished data; Klaver et al. 2015) should be added to a pyrolite mantle to explain the Sr-isotope compositions of the western Aeolian arc and Marsili seamount. In contrast, higher amounts (4–10 %) should be added to the upper mantle beneath Stromboli to account for compositions of CA to KS mafic rocks (Peccerillo et al. 2013). This leads to the conclusion that fluids from an oceanic crust and from oceanic crust plus sediments were respectively responsible for mantle contamination beneath the western-central and the eastern sectors of the Aeolian arc (e.g. Tonarini et al. 2001; Tommasini et al. 2007; Klaver et al. 2015).

Along-arc variation of trace element ratios that are not modified by evolution processes provides constraints on the nature of mantle contaminant, mechanisms of element transfer, and on pre-metasomatic source compositions. Data reported in Fig. 9.22c and d show that there is an overall increase of Rb/Nb (and other LILE/HFSE ratios) from the margins to the centre of the arc, and an opposite tendency for

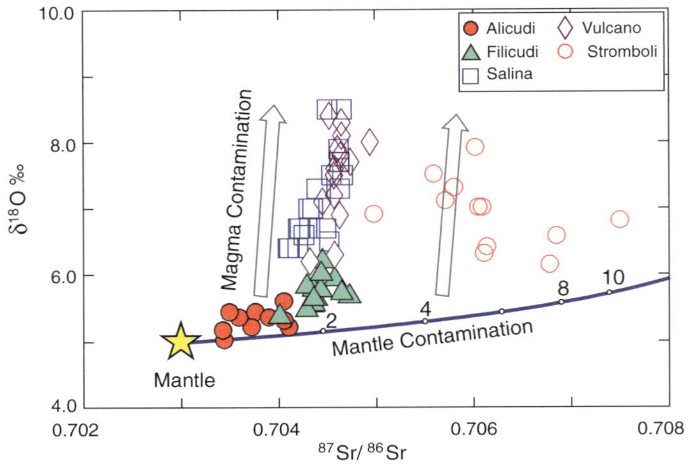

Fig. 9.21 $\delta^{18}O‰_{SMOW}$ versus $^{87}Sr/^{86}Sr$ for the Aeolian arc magmas. Vertical *arrows* are trends of magma contamination by average Calabro-Peloritano basement. The *curved blue line* is a mixing trend between the mantle and upper crust and mimics source contamination by subducted sediments. Numbers along the *line* indicate the amount of crustal material participating in the process

Nb/Zr. This is particularly evident if a homogeneous set of data from a single laboratory is used (see Peccerillo et al. 2013).

In general terms, there is a consensus that high LILE/HFSE ratios in the arc rocks reflect modification of the mantle sources by fluids migrating from the subduting slab (e.g. Kessel et al. 2005). Therefore, high LILE/HFSE ratios in the Aeolian arc are attributed to such a process. Variations along the arc may depend on different intensity of fluid modification or/and variable nature of fluids (i.e. aqueous fluids, supercritical fluids, and silicate melts), which may be variably enriched in LILE relative to HFSE. However, compositional variations may also depend on heterogeneous pre-metasomatic mantle sources.

Francalanci et al. (1993, 2007) and Tommasini et al. (2007) assume a homogeneous MORB-type pre-metasomatic mantle source along the arc and suggest that the range of regional geochemical and isotopic compositions can be explained as entirely related to fluids released from altered oceanic crust plus variable amounts of sediments. Other authors envisage a MORB- to OIB-type pre-metasomatic mantle wedge beneath the Aeolian arc (Ellam et al. 1988; Peccerillo et al. 2013 and references therein). According to this latter view, low LILE/HFSE ratios at Stromboli, Alicudi and Marsili would highlight a stronger role of OIB-type components for these volcanoes (Trua et al. 2011; Peccerillo et al. 2013). In contrast, a MORB-type mantle beneath the central arc would be responsible for development of high LILE/HFSE ratios in this sector.

HFSE are known to be less mobile than LILE during fluid transfer. HFSE/HFSE ratios vary considerably from MORB to OIB (Fig. 9.22d) and, therefore, may furnish reliable indications on nature of pre-metasomatic mantle rocks. Increase of Nb/Zr from the central to the external islands support a stronger role of OIB-type components in the latter.

The relative abundances of calcalkaline and shoshonitic magmatism in the various sectors of the Aeolian arc may be related to the nature and intensity of fluid flux from the slab. Aqueous fluids have strong effects on phase stability during mantle melting and, therefore, heavily control the petrological characteristics of magmas (e.g. Ulmer 2001; Grove et al. 2012). In particular, hydrated conditions lower the solidus temperature of peridotite, favouring large degrees of melting and the formation of calcalkaline with respect to potassic magmas (e.g. Wendlandt and Eggler 1980a, b). Accordingly, variable activity of hydrous fluids is

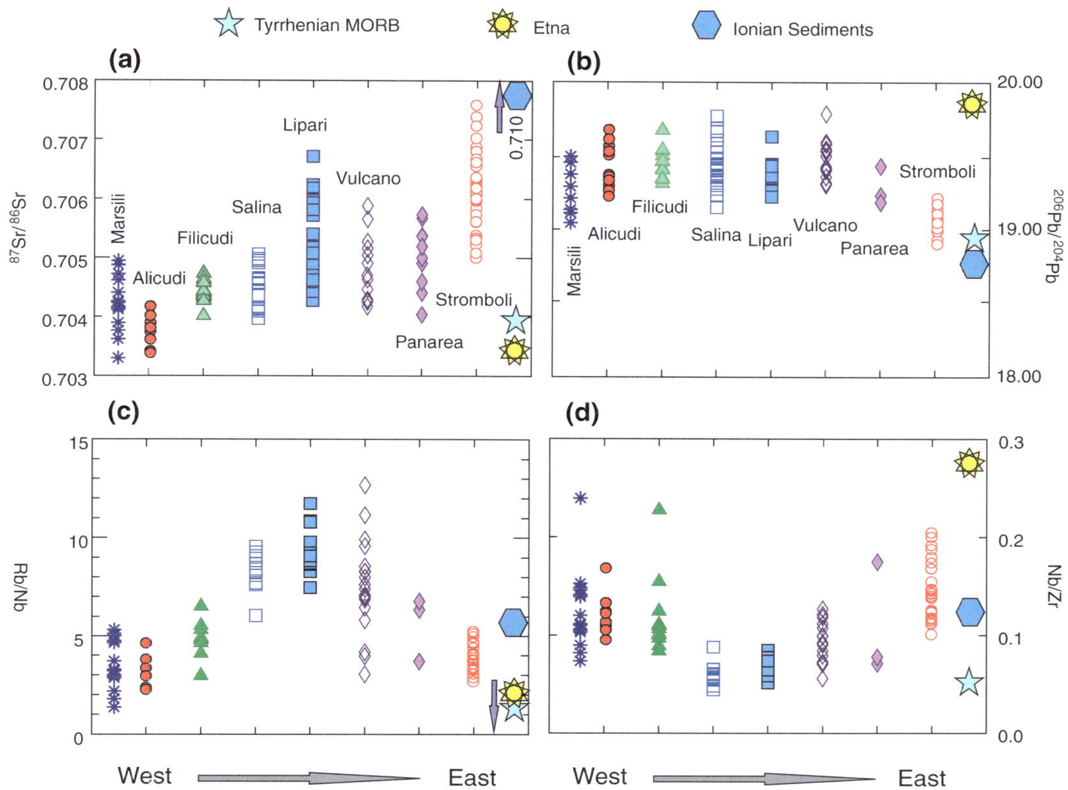

Fig. 9.22 Along-arc variations of $^{87}Sr/^{86}Sr$, $^{206}Pb/^{204}Pb$, and some key incompatible element ratios. Only mafic rocks from Peccerillo et al. (2013) are plotted for trace element diagrams, with the aim of avoiding interlaboratory bias. Compositions of Tyrrhenian MORB (ODP Site 655), Etna OIB-type rocks and Ionian Sea sediments (Klaver et al. 2015; Average of DSDP Site 274, and De Astis unpublished data) are also reported. For explanation, see text

a viable mechanism to explain the uneven distribution of magma types along the arc. This implies that water fluxes might be stronger in the western islands and Lipari where calcalkaline rocks dominate. In contrast, hydrous silicate melts or supercritical fluids from an oceanic slab plus variable amounts of sediments may have been released beneath Vulcano and Stromboli, where potassic and shoshonitic rocks dominate.

9.12 Geodynamic Implications

There are important along-arc structural and geophysical variations that parallel compositional modifications of magmas and make the Aeolian arc a stimulating area where magmatism and geodynamics are closely related. The Tindari-Letojanni strike-slip fault system represents a primary divide for both magmatism and structural setting (De Astis et al. 2003). Deep seismicity is mostly concentrated east of this fault system, continuing toward the NW, almost to the Campania volcanoes. Moreover, there is a strong variation of the attitude of the foreland monocline west and east of the Tindari-Letojanni fault, with the dip angle passing from 4 to 6° in northern Sicily to more than 20° offshore southern Calabria (Doglioni et al. 2001 and references therein).

Mantle tomographic images of the southeastern Tyrrhenian Sea (e.g. Faccenna et al. 2011; Orecchio et al. 2014) show the occurrence of a high-V_P slab beneath the eastern Aeolian arc associated with deep earthquake foci. In contrast, a vertical discontinuous aseismic high-V_P body is imaged at depth beneath the western arc, separated from the lithosphere by a low-V_P gap. A somewhat similar setting is observed northeast of the Aeolian arc where a low-V_P mantle window is detected at 100–200 km, between the lithosphere and a deep high-velocity slab (e.g. Chiarabba et al. 2008; Neri et al. 2009; Faccenna et al. 2011; Giacomuzzi et al. 2012). These data suggest that a vertically continuous active subducting slab only occurs beneath the eastern Aeolian arc. In contrast, a detached dehydrated fragment of oceanic crust is passively sinking into the upper mantle beneath the western arc (Peccerillo and Frezzotti 2015), and a slab detachment with formation of a slab window characterise the upper mantle northeast of the Aeolian arc.

Doglioni et al. (2001) suggested that subduction is active both west and east of the Tindari-Letojanni fault system, but subduction rate is higher in the east than in the west. Such a difference would be related to a variable nature of the undergoing crust, which would be of oceanic-type in Calabria and of continental-type in Sicily. The subduction of a high-density oceanic slab offshore Calabria would generate a steeper dip angle and a faster rate of slab rollback compared to northern Sicily. The lack of deep seismicity in the west is explained by the plastic behaviour of the easily fusible continental crust as opposed to the rigid characteristics of the oceanic Ionian plate. In such a model, the Tindari-Letojanni fault is viewed as a zone of accommodation of differential slab retreat rates (Doglioni et al. 2001). Gvirtzman and Nur (1999) proposed a somewhat different model, which hypothesises southeastward retreat of the Ionian plate, but no active subduction beneath the western Aeolian arc.

The occurrence of rocks with relatively primitive isotopic signatures (i.e. low Sr and high Nd-Pb isotope ratios) in the western Aeolian arc argues against the presence of a subducting continental-type slab in this sector. Moreover, the prevailing calcalkaline nature of the western arc suggests abundant hydrous fluid percolation through the mantle wedge. This is more easily accomplished by oceanic than continental crust. Therefore, petrological and geochemical data support the hypothesis of subduction of oceanic crust beneath the western arc and oceanic crust plus sediments in the eastern arc.

According to this model (Fig. 9.23), the Tindari-Letoianni fault system is a main structural divide, which separates the actively subducting Ionian oceanic slab + sediments from a western zone where an aseismic discontinuous slab is present. The deep slab segment in the western arc could represent a detached fragment of oceanic crust, which has already undergone dehydration at shallow depths inducing high degrees of mantle melting and formation of calcalkaline magmas, and is now sinking into the asthenosphere (Peccerillo and Frezzotti 2015).

This model also explains the occurrence of OIB-type mantle components revealed by incompatible element ratios at the external islands of Alicudi and Stromboli. De Astis et al. (2000) and Peccerillo (2001) suggest that OIB-type components were provided by inflow of intraplate-type mantle from the Ionian-African foreland around the margins of the slab (see also Miller and Piana Agostinetti 2011). Such a process modified the composition of mantle wedge at the margins of the arc, where OIB type peridotites from the foreland were contaminated by fluids released by an oceanic crust at Alicudi and from oceanic crust plus sediments at Stromboli. An inflow from the African foreland has been also suggested for Marsili (Trua et al. 2003, 2011).

9.13 Summary and Conclusions

The Aeolian Islands consist of several stratovolcanoes composed of calcalkaline to shoshonitic lavas and pyroclastics, and minor potassic alkaline rocks. There are strong variations of rock compositions both at local and regional scales. Regional variations consist of an increase in the amounts of potassic relative to calcalkaline volcanics from western to eastern islands, parallel to an increase of $^{87}Sr/^{86}Sr$ and a decrease of Nd, Pb and He isotope ratios. The eastern island of Stromboli resembles closely the volcanoes of

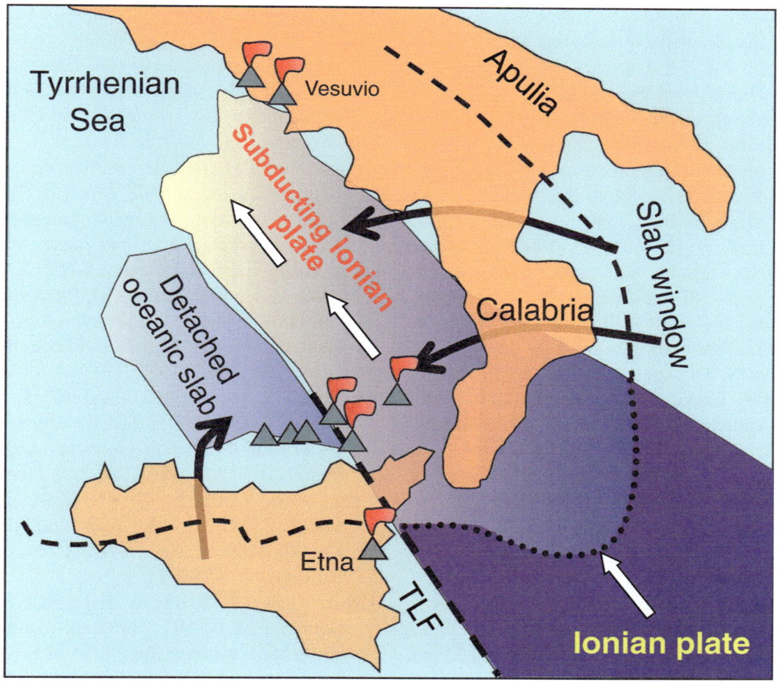

Fig. 9.23 Geodynamic model of the Aeolian arc. A narrow Ionian plate plus sediments are subducting beneath Calabria and the eastern Aeolian arc. A detached dehydrated oceanic slab is passively sinking into the mantle on the west of the Tintari-Letojanni-Malta fault (TLF). Asthenospheric mantle inflow from the foreland takes place around the margins of the slabs, providing OIB-type components to the source of magmas at the margins of the arc

the Campania Province. Range of silica contents in the single Aeolian volcanoes is also very variable and increases from the external to central islands.

Petrological and geochemical variations along the arc are considered to result from variable compositions of fluids released from the subduction zones and heterogeneous pre-metasomatic mantle sources. Aqueous fluids from an oceanic slab were responsible for metasomatism beneath the western-central islands, whereas supercritical fluids or silicate melts from an oceanic slab plus sediments contaminated the upper mantle beneath Stromboli. Variation of pre-metasomatic mantle composition, from MORB-type in the central arc to OIB-type in the external islands, is also considered as an important factor of compositional diversity along the arc.

The range of rock compositions occurring within single islands reflects both the characteristics of parental magmas and the shallow-level evolution history. In particular, the presence of abundant silicic rocks in the central Aeolian Islands likely reflects extensive fractional crystallisation suffered by magmas in large magma chambers developed along the horst-graben system of the Tindari-Letojanni strike-slip fault.

References

Arrighi S, Tanguy JC, Rosi M (2006) Eruptions of the last 2200 years at Vulcano and Vulcanello (Aeolian Islands, Italy) dated by high-accuracy archeomagnetism. Phys Earth Planet In 159:225–233

Barberi F, Gandino A, Gioncada A, La Torre P, Sbrana A, Zenucchini C (1994) The deep structure of the Aeolian arc (Filicudi-Panarea-Vulcano sector) in light of gravity, magnetic and volcanological data. J Volcanol Geoth Res 61:189–206

Barker DS (1987) Rhyolites contaminated with metapelite and gabbro, Lipari, Aeolian Islands, Italy: products of lower crustal fusion or of assimilation plus fractional crystallization? Contrib Mineral Petrol 97:460–472

Beccaluva L, Rossi PL, Serri G (1982) Neogene to recent volcanism of the Southern Tyrrhenian-Sicilian area: implications for the geodynamic evolution of the Calabrian Arc. Earth Evol Sci 3:222–238

Bertagnini A, Métrich N, Landi P, Rosi M (2003) Stromboli volcano (Aeolian Archipelago, Italy): an open window on the deep-feeding system of a steady state basaltic volcano. J Geophys Res 108(B7):2336. doi:10.1029/2002JB002146

Bonardi G, Cavazza W, Perrone V, Rossi S (2001) Calabria-Peloritani terraine and northern Ionian Sea. In: Vai GB, Martini PI (eds) Anatomy of an orogen. The Apennines and the Adjoining Mediterranean Basins. Kluwer, Dordercht, pp 287–306

Bonardi G, Ciarcia S, Di Nocera S, Matano F, Sgrossi Torre M (2009) Carta delle principali unità cinematiche dell'Appennino meridionale—Nota illustrativa. It J Geosci 128:47–60

Bragagni A, Avanzinelli R, Freymuth H, Francalanci L (2014) Recycling of crystal mush-derived melts and short magma residence times revealed by U-series disequilibria at Stromboli volcano. Earth Planet Sci Lett 404:206–219

Calanchi N, De Rosa R, Mazzuoli R, Rossi L, Santacroce R, Ventura G (1993) Silicic magma entering a basaltic magma chamber: eruptive dynamics and magma mixing an example from Salina (Aeolian Islands, Southern Thyrrenian Sea). Bull Volcanol 55:504–522

Calanchi N, Lucchi F, Pirazzoli PA, Romagnoli C, Tranne CA, Radtke U, Reyss JL, Rossi PL (2002a) Late Quaternary relative sea-level changes and vertical movements at Lipari (Aeolian Islands). J Quater Sci 17:459–467

Calanchi N, Peccerillo A, Tranne C, Lucchini F, Rossi PM, Kempton P, Barbieri M, Wu TW (2002b) Petrology and geochemistry of the Island of Panarea: implications for mantle evolution beneath the Aeolian island arc (Southern Tyrrhenian Sea, Italy). J Volcanol Geoth Res 115:367–395

Capaccioni B, Tassi F, Vaselli O, Tedesco D, Poreda R (2007) Submarine gas burst at Panarea Island (southern Italy) on 3 November 2002: a magmatic versus hydrothermal episode. J Geophys Res 112:B05201. doi:10.1029/2006JB004359

Catalano R, Doglioni C, Merlini S (2001) On the Mesozoic Ionian basin. Geophys J Int 144:49–64

Cella F, De Lorenzo S, Fedi M, Loddo M, Mongelli F, Rapolla A, Zito G (2006) Temperature and density of the Tyrrhenian lithosphere and slab and new interpretation of gravity field in the Tyrrhenian Basin. Tectonophysics 412:27–47

Chiarabba C, De Gori P, Speranza F (2008) The southern Tyrrhenian subduction zone: deep geometry, magmatism and Plio-Pleistocene evolution. Earth Planet Sci Lett 268:408–423

Chiodini G, Cioni R, Marini L, Panichi C (1995) Origin of the fumarolic fluids of Vulcano island, Italy and implications for volcanic surveillance. Bull Volcanol 57:99–110

Chiodini G, Caliro S, Caramanna G, Granieri D, Minopoli C, Moretti R (2006) Geochemistry of the submarine gaseous emissions of Panarea Aeolian Islands, Southern Italy: magmatic vs. hydrothermal origin and implications for volcanic surveillance. Pure appl Geophys 163:759–780

Cicchino AMP, Zanella E, De Astis G, Lanza R, Lucchi F, Tranne CA, Airoldi G, Mana S (2011) Rock magnetism and compositional investigation of Brown Tuffs deposits at Lipari and Vulcano (Aeolian Islands—Italy). J Volcanol Geoth Res 208:23–38

Cimarelli C, De Rita D, Dolfi D, Procesi M (2008) Coeval strombolian and vulcanian-type explosive eruptions at Panarea (Aeolian Islands, Southern Italy). J Volcanol Geoth Res 177:797–811

Cortes JA, Wilson M, Condliffe E, Francalanci L, Chertkoff DG (2005) The evolution of the magmatic system of Stromboli volcano during the Vancori period (26–13.8 ky). J Volcanol Geoth Res 147:1–38

Crisci GM, Delibrias G, De Rosa R, Mazzuoli R, Sheridan MF (1983) Age and petrology of the Late-Pleistocene Brown Tuffs on Lipari, Italy. Bull Volcanol 46:381–391

Crisci GM, De Rosa R, Esperança S, Mazzuoli R, Sonnino M (1991) Temporal evolution of a three component system: the Island of Lipari (Aeolian Arc, southern Italy). Bull Volcanol 53:207–221

Davì M, De Rosa R, Barca D (2009) A LA-ICP-MS study of minerals in the Rocche Rosse magmatic enclaves: evidence of a mafic input triggering the latest silicic eruption of Lipari Island (Aeolian Arc, Italy). J Volcanol Geoth Res 182:45–56

De Astis G, La Volpe L, Peccerillo A, Civetta L (1997) Volcanological and petrological evolution of Vulcano island (Aeolian arc, southern Tyrrhenian Sea). J Geophys Res 102:8021–8050

De Astis G, Peccerillo A, Kempton PD, La Volpe L, Wu TW (2000) Transition from calc-alkaline to potassium-rich magmatism in subduction environments: geochemical and Sr, Nd, Pb isotopic constraints from the island of Vulcano (Aeolian arc). Contrib Mineral Petrol 139:684–703

De Astis G, Ventura G, Vilardo G (2003) Geodynamic significance of the Aeolian volcanism (Southern Tyrrhenian Sea, Italy) in light of structural, seismological and geochemical data. Tectonics 22:1040–1057

De Astis G, Kempton PD, Peccerillo A, Wu TW (2006) Trace element and isotopic variations from Mt. Vulture to Campanian volcanoes: constraints for slab detachment and mantle inflow beneath southern Italy. Contrib Mineral Petrol 151:331–351

De Astis G, Lucchi F, Dellino P, La Volpe L, Tranne CA, Frezzotti ML, Peccerillo A (2013) Geology, volcanic history and petrology of Vulcano (central Aeolian archipelago). In: Lucchi F, Peccerillo A, Keller J, Tranne CA, Rossi PL (eds) The Aeolian Islands Volcanoes. Geol Soc London Mem 37:281–349

De Rosa R, Sheridan MF (1983) Evidence for magma mixing in the surge deposits of the Monte Guardia sequence, Lipari. J Volcanol Geoth Rese 17:313–328

De Rosa R, Gouillou H, Mazzuoli R, Ventura G (2003a) New unspiked K-Ar ages of volcanic rocks of the central and western sector of the Aeolian Islands: reconstruction of the volcanic stages. J Volcanol Geoth Res 120:161–178

De Rosa R, Donato P, Gioncada A, Masetti M, Santacroce R (2003b) The Monte Guardia eruption (Lipari, Aeolian Islands): an example of a reversely zoned magma mixing sequence. Bull Volcanol 65:530–543

Del Moro A, Gioncada A, Pinarelli L, Sbrana A, Joron JL (1998) Sr, Nd, Pb isotope evidence for open system evolution at Vulcano, Aeolian Arc, Italy. Lithos 43:81–106

Del Moro S, Renzulli A, Tribaudino M (2011) Pyrometamorphic processes at the magma-hydrothermal system interface of active volcanoes: evidence from buchite ejecta of Stromboli (Aeolian Islands, Italy). J Petrol 52:541–564

Di Liberto V (2003) Mantle-crust interaction under the Aeolian arc (Italy): inferences from He and Sr isotopes. PhD Thesis, University of Palermo, 73 pp

Di Martino C, Frezzotti ML, Lucchi F, Peccerillo A, Tranne CA, Diamond LW (2010) Magma storage and ascent at Lipari Island (Aeolian archipelago, Southern Italy) at 223–81 ka: the role of crustal processes and tectonic influence. Bull Volcanol 72:1061–1076

Di Martino C, Forni F, Frezzotti ML, Palmeri R, Webster JD, Ayuso RA, Lucchi F, Tranne CA (2011) Formation of cordierite-bearing lavas during anatexis in the lower crust beneath Lipari Island (Aeolian arc, Italy). Contrib Mineral Petrol 162:1011–1030

Diliberto IS (2013) Time series analysis of high temperature fumaroles monitored on the island of Vulcano (Aeolian Archipelago, Italy). J Volcanol Geoth Res 264:150–163

Doglioni C, Innocenti F, Mariotti G (2001) Why Mt. Etna? Terra Nova 13:25–31

Doherty AL, Bodnar RJ, De Vivo B, Bohrson WA, Belkin HE, Messina A, Tracy RJ (2012) Bulk rock composition and geochemistry of olivine-hosted melt inclusions in the Grey Porri Tuff and selected lavas of the Monte dei Porri volcano, Salina, Aeolian Islands, southern Italy. Centr Eur J Geosci 4:338–355

Doherty AL, Cannatelli C, Raia F, Belkin HE, Albanese S, Lima A, De Vivo B (2015) Geochemistry of selected lavas of the Panarea volcanic group, Aeolian Arc, Italy. Mineral Petrol 109:597–610

Dolfi D, De Rita D, Cimarelli C, Mollo S, Soligo M, Fabbri M (2007) Dome growth rates, eruption frequency and assessment of volcanic hazard: insights from new U/Th dating of the Panarea and Basiluzzo dome lavas and pyroclastics, Aeolian Islands, Italy. Quat Intern 162–163:182–194

Dupuy C, Dostal J, Girod M, Liotard JM (1981) Origin of volcanic rocks from Stromboli (Italy). J Volcanol Geoth Res 10:49–65

Ellam RM, Harmon RS (1990) Oxygen isotope constraints on the crustal contributions to the subduction-related magmatism of the Aeolian Islands, southern Italy. J Volcanol Geoth Res 44:105–122

Ellam RM, Menzies MA, Hawkesworth CJ, Leeman WP, Rosi M, Serri G (1988) The transition from calcalkaline to potassic orogenic magmatism in the Aeolian Islands, Southern Italy. Bull Volcanol 50:386–398

Ellam RM, Hawkesworth CJ, Menzies MA, Rogers NW (1989) The volcanism of southern Italy: role of subduction and the relationships between potassic and sodic alkaline magmatism. J Geophys Res 94:4589–4601

Esperança S, Crisci GM, De Rosa R, Mazzuoli R (1992) The role of the crust in the magmatic evolution of the Island of Lipari (Aeolian Islands, Italy). Contrib Mineral Petrol 112:450–562

Faccenna C, Molin P, Orecchio B, Olivetti V, Bellier O, Funiciello F, Minelli L, Piromallo C, Billi A (2011) Topography of the Calabria subduction zone (southern Italy): Clues for the origin of Mt. Etna. Tectonics 30, TC1003. doi:10.1029/2010TC002694

Faraone D (2002) I vulcani e l'uomo. Miti, Leggende e Storia. Liguori, Napoli, p 341

Forni F, Lucchi F, Peccerillo A, Tranne CA, Rossi PL, Frezzotti ML (2013) Stratigraphy and geological evolution of the Lipari volcanic complex (central Aeolian archipelago). In: Lucchi F, Peccerillo A, Keller J, Tranne CA, Rossi PL (eds) The Aeolian Islands Volcanoes. Geol Soc London Mem 37:213–279

Forni F, Ellis BS, Bachmann O, Lucchi F, Tranne CA, Agostini S, Dallai L (2015) Erupted cumulate fragments in rhyolites from Lipari (Aeolian Islands). Contrib Mineral Petrol 170:49

Francalanci L, Manetti P, Peccerillo A (1989) Volcanological and magmatological evolution of Stromboli volcano (Aeolian Islands): the roles of fractional crystallization, magma mixing, crustal contamination, and source heterogeneity. Bull Volcanol 51:355–378

Francalanci L, Taylor SR, McCulloch MT, Woodhead J (1993) Geochemical and isotopic variations in the calc-alkaline rocks of Aeolian Arc (Southern Italy): constraints on the magma genesis. Contrib Mineral Petrol 113:300–313

Francalanci L, Avanzinelli R, Petrone CM, Santo A (2004) Petrochemical and magmatological characteristics of the Aeolian Arc volcanoes, southern Tyrrhenian Sea, Italy: inferences on shallow level processes and magma source variations. Per Mineral 73:75–104

Francalanci L, Avanzinelli R, Tommasini S, Heumann A (2007) A west-east geochemical and isotopic traverse along the subaerial volcanism of the Aeolian Island arc, Southern Tyrrhenian Sea, Italy: Inferences on mantle source processes. In: Beccaluva L, Bianchini G, Wilson M (eds) Cenozoic Volcanism in the Mediterranean Area, Geol Soc Am, Spec Paper 418:235–263

Francalanci L, Lucchi F, Keller J, De Astis G, Tranne CA (2013) Eruptive, volcano-tectonic and magmatic history of the Stromboli volcano (north-eastern Aeolian archipelago). In: Lucchi F, Peccerillo A, Keller J, Tranne CA, Rossi PL (eds) The Aeolian Islands Volcanoes, Geol Soc London Mem 37:395–47

Frezzotti ML, Peccerillo A, Bonelli R (2003) Magma ascent rates and depths of crustal magma reservoirs beneath the Aeolian volcanic arc (Italy): inferences from fluid and melt inclusions in xenoliths. In: De Vivo B, Bodnar RJ (eds) Melt inclusions in volcanic systems: methods, applications and problems. Elsevier, Amsterdam, pp 185–205

Frezzotti ML, Zanon V, Peccerillo A, Nikogossian I (2004) Silica-rich melts in quartz xenoliths from Vulcano island and their bearing on processes of crustal anatexis and crust-magma interaction beneath the Aeolian Arc, southern Italy. J Petrol 45:3–26

Fusillo R, Di Traglia F, Gioncada A, Pistolesi M, Wallace PJ, Rosi M (2015) Deciphering post-caldera volcanism: insight into the Vulcanello (Island of Vulcano, Southern Italy) eruptive activity based on geological and petrological constraints. Bull Volcanol 77(9). doi:10.1007/s00445-015-0963-6

Gabbianelli G, Gillot P-Y, Lanzafame G, Romagnoli C, Rossi PL (1990) Tectonic and volcanic evolution of Panarea (Aeolian Islands, Italy). Mar Geol 92:313–326

Gale A, Dalton CA, Langmuir CH, Su Y, Schilling J-G (2013) The mean composition of ocean ridge basalts. Gechem Geophys Geosyst 14. doi:10.1029/2012GC004334

Gallais F, Graindorge D, Gutscher MA, Klaeschen D (2013) Propagation of a lithospheric tear fault (STEP) through the western boundary of the Calabrian accretionary wedge offshore eastern Sicily (Southern Italy). Tectonophysics 602:141–152

Gasperini D, Blichert Toft J, Bosch D, Del Moro A, Macera P, Albarède F (2002) Upwelling of deep mantle material through a plate window: evidence from the geochemistry of Italian basaltic volcanics. J Geophys Res 107(B12):2367. doi:10.1029/2001JB000418

Gertisser R, Keller J (2000) From basalt to dacite: origin and evolution of the calcalkaline series of Salina, Aeolian Arc, Italy. Contrib Mineral Petrol 139:607–626

Giacomuzzi G, Civalleri M, De Gori P, Chiarabba C (2012) A 3D Vs model of the upper mantle beneath Italy: insight on the geodynamics of central Mediterranean. Earth Planet Sci Lett 335–336:105–120

Gillot P-Y (1987) Histoire volcanique des Iles Eoliennes: arc insulaire ou complexe orogénique anulaire? Doc Trav IGAL 11:35–42

Gillot P-Y, Keller J (1993) Radiochronological dating of Stromboli. Acta Vulcanol 3:69–77

Gioncada A, Mazzuoli R, Bisson M, Pareschi MT (2003) Petrology of volcanic products younger than 42 ka on the Lipari-Vulcano complex (Aeolian Islands, Italy): an example of volcanism controlled by tectonics. J Volcanol Geoth Res 122:191–220

Gioncada A, Mazzuoli R, Milton AJ (2005) Magma mixing at Lipari (Aeolian Islands, Italy): insights from textural and compositional features of phenocrysts. J Volcanol Geoth Res 145:97–118

Grove TL, Till CB, Krawczynski MJ (2012) The role of H_2O in subduction zone magmatism. Annu Rev Earth Planet Sci 40:413–439

Gvirtzman Z, Nur A (1999) The formation of Mount Etna as the consequence of slab rollback. Nature 401:782–785

Hornig-Kjarsgaard I, Keller J, Koberski U, Stadlbauer E, Francalanci L, Lenhart R (1993) Geology, stratigraphy and volcanological evolution of the island of Stromboli, Aeolian arc, Italy. Acta Vulcanol 3:21–68

Irvine TN, Baragar WRA (1971) A guide to chemical classification of common volcanic rocks. Can J Earth Sci 8:523–548

Keller J (1980a) The Island of Salina. Rend Soc It Mineral Petrol 36:489–524

Keller J (1980b) The island of Vulcano. Rend Soc It Mineral Petrol 36:369–414

Keller J (1982) The Mediterranean island arc. In: Thorpe RS (ed) Andesites: Orogenic andesites and related rocks. Wiley, Chichester, pp 307–325

Keller J, Hornig-Kjarsgaard I, Koberski U, Stadlbauer E, Lenhart R (1993) Geological map of the island of Stromboli. Acta Vulcanol 3

Kessel R, Schmidt MW, Ulmer P, Pettke T (2005) Trace element signatures of subduction-zone fluids, melts and supercritical liquids at 120–180 km depth. Nature 437:724–727

Klaver M, Djuly T, de Graaf S, Sakes S, Wijbrans J, Davies G, Vroon P (2015) Temporal and spatial variations in provenance of Eastern Mediterranean Sea sediments: Implications for Aegean and Aeolian arc volcanism. Geochim Cosmochim Acta 153:149–168

Laiolo M, Cigolini C (2006) Mafic and ultramafic xenoliths in San Bartolo lava field: new insights on the ascent and storage of Stromboli magmas. Bull Volcanol 68:653–670

Leocat E (2011) Histoire éruptive des volcans du secteur occidental des Iles Eoliennes (Sud de la Mer Tyrrhénienne, Italie) et évolution temporelle du magmatisme. PhD Thesis, University Paris 11, Orsay, France

Luais B (1988) Mantle mixing and crustal contamination as the origin of the high-Sr radiogenic magmatism of Stromboli (Aeolian Arc). Earth Planet Sci Lett 88: 93–106

Lucchi F, Tranne CA, De Astis G, Keller J, Losito R, Morche W (2008) Stratigraphy and significance of Brown Tuffs on the Aeolian Islands (southern Italy). J Volcanol Geoth Res 177:49–70

Lucchi F, Peccerillo A, Keller J, Tranne CA, Rossi PL (eds) (2013a) The Aeolian Islands Volcanoes. Geol Soc London Mem 37:510

Lucchi F, Peccerillo A, Tranne CA, Rossi PM, Frezzotti ML, Donati C (2013b) Volcanism, calderas and magmas of the Alicudi composite volcano (western Aeolian archipelago). In: Lucchi F, Peccerillo A, Keller J, Tranne CA, Rossi PL (eds) The Aeolian Islands Volcanoes. Geol Soc London Mem 37:83–111

Lucchi F, Santo AP, Tranne CA, Peccerillo A, Keller J (2013c) Volcanism, magmatism, volcano-tectonics and sea-level fluctuations in the geological history of Filicudi (western Aeolian archipelago). In: Lucchi F, Peccerillo A, Keller J, Tranne CA, Rossi PL (eds) The Aeolian Islands Volcanoes. Geol Soc London Mem 37:113–153

Lucchi F, Gertisser R, Keller J, Forni F, De Astis G, Tranne CA (2013d) Eruptive history and magmatic evolution of the Island of Salina (central Aeolian archipelago). In: Lucchi F, Peccerillo A, Keller J, Tranne CA, Rossi PL (eds) The Aeolian Islands Volcanoes. Geol Soc London Mem 37:155–211

Lucchi F, Tranne CA, Peccerillo A, Keller J, Rossi PL (2013e) Geological history of the Panarea volcanic group (eastern Aeolian archipelago).). In: Lucchi F, Peccerillo A, Keller J, Tranne CA, Rossi PL (eds) The Aeolian Islands Volcanoes. Geol Soc London Mem 37:351–395

Maccarrone E (1963) Aspetti geochimico-petrografici di alcuni esemplari di andesite granato-cordieritifera dell'Isola di Lipari. Per Mineral 32:277–302

Madonia P, Cusano P, Diliberto IS, Cangemi M (2013) Thermal anomalies in fumaroles at Vulcano island (Italy) and their relationship with seismic activity. Phys Chem Earth 63:160–169

Martelli M, Nuccio PM, Stuart FM, Di Liberto V, Ellam RM (2008) Constraints on mantle source and interactions from He-Sr isotope variation in Italian Plio-Quaternary volcanism. Geochem Geophys Geosyst 9:Q02001. doi:10.1029/2007GC001730

Martelli M, Rizzo AL, Renzulli A, Ridolfi F, Arienzo I, Rosciglione A (2014) Noble-gas signature of magmas from a heterogeneous mantle wedge: The case of Stromboli volcano (Aeolian Islands, Italy). Chemical Geology 368:39–53

Martinez-Arevalo C, Musumeci C, Patanè D (2009) Evidence of a partial melt zone beneath Stromboli volcano (Italy) from inversion of teleseismic receiver functions. Terra Nova 21:386–392

Martini M, Legittimo PC, Piccardi G, Giannini L (1986) The fumaroles of Vulcano (Italy)—differences in chemical compositions produced by the surface environment. Geothermics 15:205–209

Mazzuoli R, Tortorici L, Ventura G (1995) Oblique rifting in Salina, Lipari and Vulcano Islands (Aeolian Islands, Southern Italy). Terra Nova 7:444–452

Mercalli G (1883) Vulcani e fenomeni vulcanici. In: Geologia d'Italia, Vallardi, Milano, vol 3, 374 pp

Mercalli G (1907) I vulcani attivi della Terra. Hoepli, Milano 421 pp

Mercalli G, Silvestri O (1891) L'eruzione dell'Isola di Vulcano incominciata il 3 agosto 1888 e terminata il 22 marzo 1890. Ann Uff Centr Meteor Geodin It 10:71–281

Métrich N, Bertagnini A, Di Muro A (2010) Conditions of magma storage, degassing and ascent at Stromboli: new insights into the volcano plumbing system with inferences on the eruptive dynamics. J Petrol 51:603–626

Miller MS, Piana Agostinetti N (2011) Erosion of the continental lithosphere at the cusps of the Calabrian arc: evidence from S receiver functions analysis. Geophys Res Lett 38:L23301. doi:10.1029/2011GL049455

Minelli L, Faccenna C (2010) Evolution of the Calabrian accretionary wedge (central Mediterranean). Tectonics 29, TC 4004 doi:10.1029/2009tc002562

Morgavi D, Perugini D, De Campos CP, Ertl-Ingrisch W, Lavallée Y, Morgan L, Dingwell DB (2013) Interactions between rhyolitic and basaltic melts unraveled by chaotic mixing experiments. Chem Geol 346: 199–212

Nazzareni S, Molin M, Peccerillo A, Zanazzi PF (2001) Volcanological implications of crystal chemical variations in clinopyroxenes from the Aeolian arc (Southern Tyrrhenian Sea, Italy). Bull Volcanol 63:73–82

Neri G, Barberi G, Orecchio B, Aloisi M (2002) Seismotomography of the crust in the transition zone between the southern Tyrrhenian and Sicilian tectonic domains. Geophys Res Lett 29–23:50–51

Neri G, Orecchio B, Totaro C, Falcone G, Presti D (2009) Subduction beneath southern Italy close the ending: results from seismic tomography. Seism Res Lett 80:63–70

Nicotra E, Viccaro M, De Rosa R, Sapienza M (2014) Volcanological evolution of the Rivi-Capo Volcanic Complex at Salina, Aeolian Islands: magma storage processes and ascent dynamics. Bull Volcanol 76:840

Nuccio PM, Paonita A, Sortino F (1999) Geochemical modeling of mixing between magmatic and hydrothermal gases: the case of Vulcano Island, Italy. Earth Planet Sci Lett 167:321–333

Orecchio B, Presti D, Totaro C, Neri G (2014) What earthquakes say concerning residual subduction and STEP dynamics in the Calabrian Arc region, south Italy. Geophys J Int 199:1929–1942

Palano M, Schiavone D, Loddo M, Neri M, Presti D, Quarto R, Totaro C, Neri G (2015) Active upper crust deformation pattern along the southern edge of the Tyrrhenian subduction zone (NE Sicily): insights from a multidisciplinary approach. Tectonophysics 657: 205–218

Pasqual D, Molin G, Zanazzi PF, Crisci GM (1998) Clinopyroxene from Lipari: Comparison with analogues from other Aeolian Islands, Italy. Can Mineral 36:97–105

Passaro S, Milano G, Sprovieri M, Tonielli R, Ruggieri S, De Martino G, Innangi S, Sammartino S, Marsella E (2009) Seamount bathymetries of the south-eastern Tyrrhenian Sea. In: Proceedings of GNGTS Conference 2009, Session 3.2, pp 656–660

Peccerillo A (2001) Geochemical similarities between Vesuvius, Phlegraean Fields and Stromboli volcanoes: petrogenetic, geodynamic and volcanological implications. Mineral Petrol 73:93–105

Peccerillo A, Frezzotti ML (2015) Magmatism, mantle evolution and geodynamics at the converging plate margins of Italy. J Geol Soc London 172:407–427

Peccerillo A, Taylor SR (1976) Geochemistry of Eocene calc-alkaline volcanic rocks of the Kastamonu area, northern Turkey. Contrib Mineral Petrol 58:63–81

Peccerillo A, Wu TW (1992) Evolution of calc-alkaline magmas in continental arc volcanoes: evidence from Alicudi, Aeolian Arc (Southern Tyrrhenian Sea, Italy). J Petrol 33:1295–1315

Peccerillo A, Kempton PD, Harmon RS, Wu TW, Santo AP, Boyce AJ, Tripodo A (1993) Petrological and geochemical characteristics of the Alicudi Volcano, Aeolian Islands, Italy: implications for magma genesis and evolution. Acta Vulcanol 3:235–249

Peccerillo A, Dallai L, Frezzotti ML, Kempton PD (2004) Sr–Nd–Pb–O isotopic evidence for decreasing crustal contamination with ongoing magma evolution at Alicudi volcano (Aeolian arc, Italy): implications for style of magma-crust interaction and for mantle source compositions. Lithos 78:217–233

Peccerillo A, Frezzotti ML, De Astis G, Ventura G (2006) Modeling the magma plumbing system of Vulcano (Aeolian Islands, Italy) by integrated fluid inclusion geo-barometry, petrology and geophysics. Geology 34:17–20

Peccerillo A, De Astis G, Faraone D, Forni F, Frezzotti ML (2013) Compositional variations of magmas in the Aeolian arc: implications for petrogenesis and geodynamics. In: Lucchi F, Peccerillo A, Keller J, Tranne CA, Rossi PL (eds) The Aeolian Islands Volcanoes. Geol Soc London Memoirs 37:491–510

Perugini D, De Campos CP, Ertel-Ingrisch W, Dingwell DB (2012) The space and time complexity of chaotic mixing of silicate melts: implications for igneous petrology. Lithos 155:326–340

Pichavant M, Di Carlo I, Le Gac Y, Rotolo SG, Scaillet B (2009) Experimental constraints on the deep magma feeding system at Stromboli volcano, Italy. J Petrol 50:601–624

Pichler H (1980) The Island of Lipari. Rend Soc It Mineral Petrol 36:415–440

Pompilio M, Bertagnini A, Métrich N (2012) Geochemical heterogeneities and dynamics of magmas within the plumbing system of a persistently active volcano: evidence from Stromboli. Bull Volcanol 74:881–894

Pontevivo A, Panza G (2006) The lithosphere-asthenosphere system in the Calabrian arc and surrounding seas, Southern Italy. Pure appl Geophys 163:1617–1659

Renzulli A, Serri G, Santi P, Mattioli M, Holm PM (2001) Origin of high-silica rhyolites at Stromboli volcano (Aeolian Islands, Italy) inferred from crustal xenoliths. Bull Volcanol 7:400–419

Ripepe M, Delle Donne D, Harris AJL, Marchetti E, Ulivieri G (2008) Dynamics of strombolian activity. In: Calvari S, Inguaggiato S, Puglisi G, Ripepe M, Rosi M (eds) The Stromboli Volcano: An Integrated Study of the 2002–2003 Eruption, Geophysical Monograph Series, vol 182. AGU, Washington, DC, pp 39–48

Rittmann A (1931) Der ausbruch des Stromboli am 11 September 1930. Zeitschr Vulkanol 14:47–77

Romagnoli C (2013) Characteristics and morphological evolution of the Aeolian volcanoes from the study of submarine portions. In: Lucchi F, Peccerillo A, Keller J, Tranne CA, Rossi PL (eds) The Aeolian Islands Volcanoes. Geol Soc London Mem 37:13–26

Romagnoli C, Casalbore D, Bortoluzzi G, Bosman A, Chiocci FL, D'Oriano F, Gamberi F, Ligi M, Marani M (2013) Bathy-morphological setting of the Aeolian Islands. In: Lucchi F, Peccerillo A, Keller J, Tranne CA, Rossi PL (eds) The Aeolian Islands Volcanoes. Geol Soc London Mem 37:27–36

Rose-Koga EF, Koga KT, Schiano P, Le Voyer M, Shimizu N, Whitehouse MJ, Clocchiatti R (2012) Mantle source heterogeneity for South Tyrrhenian magmas revealed by Pb isotopes and halogen contents of olivine-hosted melt inclusions. Chem Geol 334:266–279

Rosi M (1980) The Island of Stromboli. Rend Soc It Mineral Petrol 36:345–368

Rosi M, Pistolesi M, Bertagnini A, Landi P, Pompilio M, Di Roberto A (2013) Stromboli volcano, Aeolian Islands (Italy): present eruptive activity and hazards. In: Lucchi F, Peccerillo A, Keller J, Tranne CA, Rossi PL (eds) The Aeolian Islands Volcanoes. Geol Soc London Mem 37:473–490

Rottura A, Del Moro A, Pinarelli L, Petrini R, Peccerillo A, Caggianelli P, Bargossi GM, Piccarreta G (1991) Relationship between intermediate and acidic rocks in orogenic granitoid suites: petrological, geochemical and isotopic (Sr, Nd, Pb) data from Capo Vaticano (southern Calabria, Italy). Chem Geol 92:153–176

Santo AP (1998) Contribution of clinopyroxene EMP and SIMS data to the understanding of magmatic processes: an example from Filicudi Island (Aeolian Arc, Southern Tyrrhenian Sea). Neues Jb Miner Abh 173:207–231

Santo A, Peccerillo A (2008) Oxygen isotopic variations in the clinopyroxenes from the Filicudi volcanic rocks (Aeolian Islands, Italy): implications for open-system magma evolution. Open Miner J 2:1–12

Santo AP, Jacobsen SB, Baker J (2004) Evolution and genesis of calc-alkaline magmas at Filicudi volcano, Aeolian Arc (Southern Tyrrhenian Sea, Italy). Lithos 72:73–96

Scarfì L, Zuccarello L, Patanè D (2010) Magma dynamics of 2007 Stromboli effusive eruption as revealed by high precision location of seismic events. J Volcanol Geoth Res 198:405–415

Sheridan MF, Frazzetta G, La Volpe L (1985) Eruptive histories of Lipari and Vulcano, Italy, during the past 22,000 years. Geol Soc Am Spec Publ 212:29–34

Sun SS, McDonough WF (1989) Chemical and isotopic systematics of oceanic basalts: implications for mantle composition and processes. In: Saunders AD, Norry MJ (eds) Magmatism in ocean basins. Geol Soc London Spec Publ 42:313–345

Tanguy JC, Le Goff M, Principe C, Arrighi S, Chillemi V, Paiotti A, La Delfa S, Patanè G (2003) Archeomagnetic dating of Mediterranean volcanics of the last 2100 years: validity and limits. Earth Planet Sci Lett 211:111–124

Tassi F, Capaccioni B, Vaselli O (2014) Compositional spatial zonation and 2005–2013 temporal evolution of the hydrothermal-magmatic fluids from the submarine fumarolic field at Panarea Island (Aeolian Archipelago, southern Italy). J Volcanol Geoth Res 277:41–50

Tedesco D, Nagao K (1996) Radiogenic ^{4}He, ^{21}Ne and ^{40}Ar in fumarolic gases on Vulcano: implication for the presence of continental crust beneath the island. Earth Planet Sci Lett 144:517–528

Tedesco D, Miele G, Sano Y, Toutain JP (1995) Helium isotopic ratio in Vulcano island fumaroles: temporal variations in shallow level mixing and deep magmatic supply. J Volcanol Geoth Res 64:117–128

Tinti S, Manucci A, Pagnoni G, Armigliato A, Zaniboni F (2005) The 30 December 2002 landslide-induced tsunamis in Stromboli: sequence of the events reconstructed from the eyewitness account. Nat Hazards Earth Sys Sci 5:763–775

Tommasini S, Heumann A, Avanzinelli R, Francalanci L (2007) The fate of high-angle dipping slabs in the subductio factory: an integrated trace element and radiogenic isotope (U, Th, Sr, Nd, Pb) study of Stromboli Volcano, Aeolian Arc, Italy. J Petrol 48:2407–2430

Tonarini S, Leeman WP, Ferrara G (2001) Boron isotopic variations in lavas of the Aeolian volcanic arc, South Italy. J Volcanol Geoth Res 110:155–170

Trua T, Serri G, Marani MP (2003) Lateral flow of African mantle below the nearby Tyrrhenian plate: geochemical evidence. Terra Nova 15:433–440

Trua T, Serri G, Marani MP, Rossi PL, Gamberi F, Renzulli A (2004) Mantle domains beneath the southern Tyrrhenian: constraints from recent seafloor sampling and dynamic implications. Per Mineral 73:53–73

Trua T, Marani MP, Gamberi F (2011) Magmatic evidence for African mantle propagation into the southern Tyrrhenian backarc region. In: Beccaluva L, Bianchini G, Wilson M (eds) Volcanism and Evolution of the African Lithosphere. Geol Soc Am Spec Paper 478:307–331

Ulmer P (2001) Partial melting in the mantle wedge—the role of H_2O in the genesis of mantle-derived “arc-related” magmas. Phys Earth Planet In 127:215–232

Vaggelli G, Francalanci L, Ruggieri G, Testi S (2003) Persistent polybaric rests of calc-alkaline magmas at Stromboli volcano, Italy: pressure data from fluid inclusions in restitic quartzite nodules. Bull Volcanol 65:385–404

Ventura G (2013) Kinematics of the Aeolian volcanism (Southern Tyrrhenian Sea) from geophysical and geological data. In: Lucchi F, Peccerillo A, Keller J, Tranne CA, Rossi PL (eds) The Aeolian Islands Volcanoes. Geol Soc London Mem 37:3–11

Vezzoli L, Renzulli A, Menna M (2014) Growth after collapse: the volcanic and magmatic history of the Neostromboli lava cone (island of Stromboli, Italy). Bull Volcanol 76:821

Wendlandt RF, Eggler DH (1980a) The origins of potassic magmas: 1. Melting relations in the systems $KAlSiO_4$–Mg_2SiO_2 and $KAlSiO_4$–MgO–SiO_2–CO_2 to 30 kilobars. Am J Sci 280:385–420

Wendlandt RF, Eggler DH (1980b) The origins of potassic magmas: 2. Stability of phlogopite in natural spinel lherzolite an in the $KAlSiO_4$–MgO–SiO_2–H_2O–CO_2 at high pressures and high temperatures. Am J Sci 280:421–458

Zanon V, Nikogossian I (2004) Evidence of crustal melting events below the Island of Salina. Geol Mag 141:524–540

Zanon V, Frezzotti ML, Peccerillo A (2003) Magmatic feeding system and crustal magma accumulation beneath Vulcano Island (Italy): evidence from fluid inclusions in quartz xenoliths. J Geophys Res 108 (B6):2298. doi:10.1029/2002JB002140

The Sicily Province

10

Abstract

The Sicily Province consists of tholeiitic to Na-alkaline magmas that show OIB-type major and trace element features. Except for Pantelleria, compositions are prevailingly mafic. At Etna (0.5 Ma to Present), early-erupted tholeiitic basalts were followed by Na-alkaline hawaiites to trachytes; hawaiitic magmas are presently erupted from central and parasitic vents. Iblei magmatism (7–1.5 Ma) consists of an association of tholeiitic basalts-basaltic andesites and Na-alkaline basalts and nephelinites, some of which contain mantle xenoliths. Ustica (0.75–0.13 Ma) and Linosa (about 1–0.5 Ma) are built up by mafic rocks with minor benmoreite-trachyte. Pantelleria (320 to <10 ka) is characterised by a bimodal distribution of dominant peralkaline lavas and pyroclastics and minor basalts, with a large Daly gap for intermediate compositions. Seamounts in the Sicily Channel (Graham-Ferdinandea, Foerstner, etc.) are basaltic with variable enrichments in alkalies. Incompatible element contents of mafic rocks in the Sicily Province increase from tholeiites to alkali basalts and nephelinites. Sr–Nd isotopic compositions are poorly variable and resemble FOZO-type OIBs. Range of Pb isotopic ratios is larger and defines a trend between HIMU-FOZO and DMM mantle end-members. Compositions of Sicily mafic magmas are related to different degrees of partial melting of lithospheric to upper asthenospheric mantle sources. About 20% to 3% melting at increasing pressure is suggested to generate tholeiitic to nephelinitic magmas. The variable Pb isotopic signatures reveal mantle heterogeneities that result from interaction between lithosphere and enriched components of deep origin. These may be provided by an ascending plume head or, more probably, by asthenospheric fluids.

A. Peccerillo, *Cenozoic Volcanism in the Tyrrhenian Sea Region*,
Advances in Volcanology, DOI 10.1007/978-3-319-42491-0_10

Keywords
Sicily Magmatic Province · Etna · Iblei · Pantelleria · Linosa · Ustica · Sicily Channel seamounts · Ferdinandea Island · Tholeiites · Na-akaline rocks · Nephelinites · Peralkaline rhyolites · Mantle xenoliths · FOZO · DMM · HIMU

10.1 Introduction

The Sicily Province is made up by the volcanoes of Etna, Monti Iblei (Hyblean Mts.), Pantelleria, Linosa, the Sicily Channel seamounts, and the Island of Ustica and the Prometeo submarine lava field in the southern Tyrrhenian Sea (Fig. 10.1). The rocks have variable petrochemical affinity, from tholeiitic to Na-alkaline, all with typical intraplate OIB-type trace element signatures (e.g. low LILE/HFSE ratios). Ages range from about 7 Ma to present. A synthesis of ages and volcanological-petrological characteristics of Sicily volcanoes is given in Table 10.1. Geochemical analyses for representative rocks can be found in Table 10.2.

10.2 Regional Geology

Sicily is sited along the Africa-Europe convergent margins, between the Maghrebian chain and the Calabrian arc. The geology of the region is very complex and only a short summary is given here. Basically, four main lithological-structural domains have been recognised: (i) The foreland terrains of the Iblean-Pelagian block and the rifted Sicily Channel; (ii) Plio-Pleistocene foredeep sediments occurring along the northern border of the foreland and partially covered by the southern chain thrusts; (iii) a fold-thrust belt made up of E-SE verging Maghrebian-Sicilian units, which crop out over much of Sicily and extends westward to the northern Sicily Channel and Tunisia; (iv) the Calabro-Peloritani Units representing the southwestward continuation of the Calabrian arc (Fig. 10.1). Syn- to post-tectonic Middle Miocene to Pleistocene successions of clays, sandstone, and conglomerates, overlain by evaporites, limestones, calcarenite and marls are widespread. To the east, the Sicily tectonic units are sharply delimited by the Strait of Messina and by the NNW-SSE trending Tindari-Letojanni-Malta Escarpment fault, a major lithospheric boundary, resulting from a differential southeastward drift of the Ionian oceanic crust with respect to Sicily.

The complex puzzle of geologically and structurally diverse units in Sicily is the result of Africa-Europe convergence and collision. Contraction, nappe transport and block rotation started in the Lower Miocene and continued until about 1–0.8 Ma when subduction beneath Sicily came to an end (e.g. Catalano et al. 1996; Tortorici et al. 2001; Goes et al. 2004; Accaino et al. 2011). At present, deformation is generally small, and Sicily moves northward jointly with the African continent, at a rate of about 5 mm/year. Such movement is absorbed by an E-W running belt located south of the Island of Ustica, where several thrust-type moderate size earthquakes have been registered (e.g. Goes et al. 2004; Orecchio et al. 2014).

The foreland consists of a thick sequence of Triassic-Liassic platform carbonates, Jurassic-Eocene pelagic carbonates, and Tertiary clastic sediments. These rocks lie upon the African basement, which is generally believed as continental, although an oceanic nature is suggested by some authors (Manuella et al. 2015 with references). The undeformed foreland dips gently toward the north beneath the foredeep and the Maghrebian-Sicilian chain, reaching a depth of about 8–10 km at Etna (e.g. Musumeci et al. 2014; Carbone and Lentini 2015). A NW-SE oriented rift system (Malta, Linosa and Pantelleria grabens) occurs in the Sicily Channel (e.g. Boccaletti et al. 1984). The presence of such an extensional belt in a converging plate region is

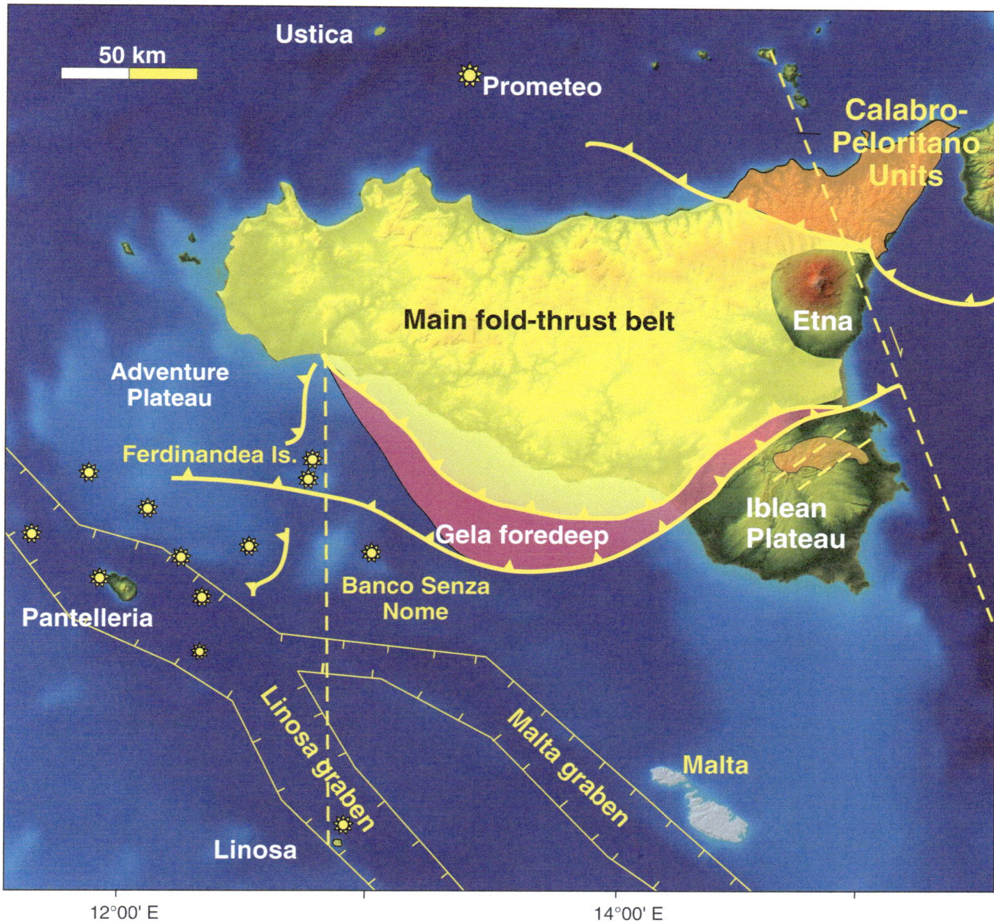

Fig. 10.1 Schematic structural map of the Sicily Province and location of main volcanic centres. *Stars* indicate seamounts. Modified after Catalano et al. (2010)

puzzling. According to Boccaletti et al. (1987), extension and localised volcanic activity developed in a different way than in typical continental rift systems, and were related to transtensional tectonic regime that brought to the development of large-scale pull apart basin in an area of continental collision. Such a tectonic regime has been related to the complex kinematics of distinct lithospheric blocks, which generates tensional or transtensional settings at the block boundaries, in an overall converging plate system (Mantovani et al. 2007, 2009; Viti et al. 2011). However, Corti et al. (2006) believe that the Sicily Channel grabens are a portion of a longer extensional belt running from Sardinia to Libya and Egypt.

The foredeep is a narrow weakly deformed depression (Gela basin) extending from the northwestern Iblean Plateau to offshore southwestern Sicily. Rocks consist of Plio-Pleistocene pelagic marly limestones and sandy clays, unconformably overlying Messinian evaporites.

The Maghrebian-Sicilian chain is a complex fold and thrust belt made up of sedimentary sequences originally deposited in the Tethyan ocean (Sicilide Units) and on the African continental margin (Maghrebian-Apennine Units). The Tethyan rocks consist of Upper

Table 10.1 Summary of geochronological, volcanological and petrological characteristics of the Sicily volcanoes

Volcano	Age	Volcanology and petrology	
Etna	About 0.5 Ma to present	– Several coalescing and superimposed stratovolcanoes (e.g. Timpa, Trifoglietto Cavigghiuni, Ellittico, Mongibello) mostly formed of lavas. The cone is spotted with numerous cinder cones, and is cut by three rift zones and by the Valle del Bove depression. Activity has been prevailingly effusive and strombolian with some vulcanian and plinian eruptions and related caldera collapses	– Early tholeiitic basalts followed by Na-alkaline rocks, mainly hawaiites. Benmoreite and trachyte magmas erupted at the end of Ellittico. Compositions characterised by typical intraplate geochemical and isotopic signatures (i.e. low LILE/HFSE) similar to FOZO-OIB rocks
Iblei	7 to 1.5 Ma	– Multicentre district with monogenetic cones, diatremes and lava flows formed during several phases of both submarine and subaerial activity. Abundant crustal and mantle xenoliths in the alkaline deposits around diatremes	– Na-alkaline hawaiite, basanite and nephelinite rock, with an Upper Pliocene-Pleistocene tholeiitic basalt to basaltic andesite phase
Pantelleria	320 to less than 10 ka	– Stratovolcano with central nested calderas made up of peralkaline rhyolitic (pantellerite) and trachytic ignimbrites and lava domes. Minor basaltic effusive and strombolian eruptions	– Bimodal distribution of dominant peralkaline trachyte-rhyolites and transitional to weakly Na-alkaline basalts
Linosa	1.06 to 0.53 Ma	– Several monogenetic hydrovolcanic, strombolian and effusive centres and two larger cones formed during three main stages of volcanic activity	– Dominant mildly alkaline basalts and hawaiites. Benmoreite-trachyte lithics in early-emplaced pyroclastic deposits
Sicily Channel seamounts	Miocene to Present	– Several seamounts (Graham, Banco Senza Nome, Foerstner, etc.) rising along NW-SE and N-S trending faults	– Basalt, hawaiite and basanite lavas and scoriae showing similar compositions as the equivalent rocks from Linosa and Pantelleria, but with some centres enriched in TiO_2 and P_2O_5
Ustica and Prometeo	750 to 130 ka	– Multicentre complex formed of submarine pillow lavas covered by subaerial lavas and pyroclastic rocks. Prometeo is a mugearite submarine lava field	– Dominant hawaiite and mugearite with minor benmoreite and trachyte lava flows and minor hydrovolcanic products and pumices. Trace element and isotope signatures similar to Etna

Jurassic-Oligocene basinal carbonates and sandy mudstones, and Upper Oligocene-Lower Miocene terrigenous turbiditic successions. The African sequences consist of Mesozoic-Lower Miocene deep-water carbonates and cherts, and Meso-Cenozoic shelf carbonates. The Maghrebian-Sicilian fold-thrust belt is cut to the east by the Tindari-Leojanni-Malta Escarpment fault and is intersected, to the west, by the Sicily Channel rift system (e.g. Corti et al. 2006).

The Peloritani Units crop out between the Taormina line and the Ionian Sea coast and represent the southwestern termination of the Hercynian-Alpine Calabrian arc (see Chap. 9). Outcropping rocks consist of Paleozoic-Mesozoic high-grade crystalline terrains and low-grade metapelites, Oligo-Miocene terrigenous deposits, and Late Miocene to Middle Pleistocene marine sediments (e.g. Palano et al. 2015).

Crustal thickness increases from about 20 km in the rifted foreland to 35 km in central Sicily, to decrease northward to about 20 km in the Ustica area (Boccaletti et al. 1984; Hirn et al. 1997; Nicolich 2001). The lithosphere is about 50–70 km thick with a maximum in the Etna and Iblei area (Calcagnile et al. 1982). Heath flow is about 50–70 mW/m^2 at the regional scale but is

Table 10.2 Compositions of Sicily volcanic rocks. Numbers in parentheses refer to data obtained on different, though compositionally similar samples

Volcano	Etna					Iblei	
Rock type	Tholeiitic basalt	Hawaiite	Mugearite	Benmoreite	Trachyte	Tholeiitic basalt	Nephelinite
Data source	9,1	12,13	5	5	5	7	7
SiO_2 wt%	49.18	47.55	52.74	57.66	61.02	50.03	38.31
TiO_2	1.47	1.76	1.23	1.18	1.29	1.57	1.89
Al_2O_3	17.20	18.16	17.44	18.36	16.94	16.91	11.15
FeO_{total}	9.58	10.18	7.44	5.27	5.32	10.60	10.77
MnO	0.15	0.18	0.13	0.16	0.17	0.16	0.21
MgO	8.73	4.97	5.20	2.04	1.88	7.38	14.84
CaO	9.27	10.5	7.76	4.12	3.43	8.56	13.69
Na_2O	3.20	3.52	4.28	6.64	5.95	2.89	2.88
K_2O	0.35	1.95	1.76	3.22	2.82	0.21	0.74
P_2O_5	0.38	0.45	0.50	0.50	0.45	0.32	2.18
LOI	0.39	0.32	1.00	0.72	0.52	1.04	2.77
Sc ppm	25	26	20	–	14	28	33
V	169	293	228	108	85	187	279
Cr	363	20	64	8	5	328	379
Co	45	36.7	32	14	13	51.5	58.8
Ni	144	20	40	3	3	234	311
Rb	5.5	53.1	36	60	62	6.4	16.1
Sr	455	1204	942	1041	857	319	1685
Y	22	29.1	17	26	30	32.7	41.1
Zr	99	216	162	333	408	118	297
Nb	17	52	53	94	102	15.8	186
Cs	0.17	1.0	0.30	1.75	2.56	0.24	0.39
Ba	142	659	644	1372	1335	247	1087
La	19.6	58	64	125	111	15.6	159
Ce	40	110	121	223	210	28.4	311
Nd	20	52.5	45	82	79	17	131
Sm	4.7	10.1	7.3	12.3	12.9	4.3	20.4
Eu	1.59	3.06	2.09	3.46	3.58	1.79	6.58
Tb	0.74	1.35	0.77	1.15	1.33	0.78	2.5
Yb	1.75	2.6	1.53	3.12	2.51	2.06	2.97
Lu	0.25	0.35	0.21	0.45	0.37	0.29	0.44
Hf	2.48	5	3.90	7.40	–	2.85	5.13
Ta	0.88	2.3	2.70	5.80	5.20	0.75	6.52
Pb	1.3	7	7.5	13.3	13.7	1.45	8.58
Th	2.22	8.8	11.8	19.4	16.3	1.0	11.1
U	0.49	2.47	3.50	4.90	3.50	0.35	4.18
$^{87}Sr/^{86}Sr$	0.703147	0.703615	0.703367	0.703409	–	0.70305	0.703223
$^{143}Nd/^{144}Nd$	0.512924	0.512852	0.512899	0.512887	–	0.51306	0.512962

(continued)

Table 10.2 (continued)

Volcano	Etna					Iblei	
Rock type	Tholeiitic basalt	Hawaiite	Mugearite	Benmoreite	Trachyte	Tholeiitic basalt	Nephelinite
$^{206}Pb/^{204}Pb$	(19.47)	19.730	–	–	–	19.18	19.79
$^{207}Pb/^{204}Pb$	(15.65)	15.640	–	–	–	15.61	15.62
$^{208}Pb/^{204}Pb$	(39.11)	39.490	–	–	–	38.82	39.37
$^{176}Hf/^{177}Hf$	–	0.282947	–	–	–	–	–

Volcano	Iblei	Linosa			Pantelleria
Rock type	Basanite	Hawaiite	Basalt	Trachyte	Basalt
Data source	7	11	14	11	6
SiO_2 wt%	43.24	49.46	44.97	60.82	46.44
TiO_2	3.15	2.15	2.21	0.91	3.42
Al_2O_3	14.02	17.1	15.33	20.11	14.13
FeO_{total}	11.74	10.59	11.94	1.84	13.25
MnO	0.19	0.18	0.21	0.03	0.18
MgO	10.14	5.32	11.87	0.38	6.45
CaO	10.05	8.7	8.83	2.7	10.41
Na_2O	2.89	3.9	3.02	6.7	3.55
K_2O	1.23	1.63	1.26	3.44	1.17
P_2O_5	0.99	0.62	0.38	0.21	1.28
LOI	1.99	0.05	–	1.75	-0.59
Sc ppm	33	24.77	–	9.57	22
V	279	181	201	38	283
Cr	256	98	455	5.8	158
Co	52	29	52	1	36
Ni	188	43	275	5	82
Rb	28	28	27	45.81	25
Sr	994	519.82	492	303	528
Y	37	29	21	30	33
Zr	369	24	228	378	220
Nb	97	55	49	96	46
Cs	0.44	0.31	0.3	0.75	–
Ba	420	328	376	1837	646
La	57	33	34	53	45
Ce	115	72	67	122	97
Nd	57	33	30	49	53
Sm	10.8	7.11	6.20	9.7	11.2
Eu	3.60	2.35	1.83	4.72	4.54
Tb	1.71	0.95	0.75	1.19	1.33
Yb	2.65	3.00	2.00	3.14	2.27
Lu	0.37	0.42	0.30	0.41	0.35
Hf	6.94	5.24	4.0	8.17	5.10

(continued)

Table 10.2 (continued)

Volcano	Iblei	Linosa			Pantelleria
Rock type	Basanite	Hawaiite	Basalt	Trachyte	Basalt
Ta	4.48	3.29	2.9	5.61	3.40
Pb	3.14	1.86	–	2.27	1.90
Th	5.62	3.85	4.10	7.9	4.32
U	1.6	1.14	1.15	4.33	1.20
$^{87}Sr/^{86}Sr$	0.703321	0.703041	0.703099	0.703111	0.703150
$^{143}Nd/^{144}Nd$	0.513061	0.512955	0.512910	0.512920	0.512960
$^{206}Pb/^{204}Pb$	19.70	–	19.330	–	19.910
$^{207}Pb/^{204}Pb$	15.64	–	15.630	–	15.674
$^{208}Pb/^{204}Pb$	39.29	–	38.980	–	39.520

Volcano	Pantelleria			Ustica		
Rock type	Hawaiite	Trachyte	Pantellerite	Hawaiite	Benmoreite	Trachyte
Data source	4	6	6	8,16	8,16	2
SiO_2 wt%	49.50	63.95	69.53	47.44	52.21	61.65
TiO_2	2.93	0.56	0.36	1.47	1.45	0.55
Al_2O_3	16.90	14.30	8.01	16.77	16.71	19.47
FeO_{total}	10.16	6.94	9.07	9.44	9.89	4.36
MnO	0.18	0.24	0.29	0.16	0.14	0.18
MgO	3.92	0.25	0.25	7.68	5.74	0.80
CaO	7.79	1.10	0.54	9.17	9.22	2.35
Na_2O	4.89	6.79	5.08	4.16	3.93	6.89
K_2O	1.80	5.08	4.33	1.40	0.97	3.29
P_2O_5	0.42	0.08	0.03	0.88	0.38	0.23
LOI	1.32	0.47	2.05	0.12	-0.25	1.69
Sc ppm	21	5.5	2.5	23	23	–
V	264	–	7	177	180	–
Cr	–	–	–	199	186	–
Ni	29	–	–	168	104	–
Rb	43	74	217	27	20	61
Sr	675	16	14	1050	496	333
Y	42	53	202	25	23	26
Zr	356	480	2363	185	120	320
Nb	79	98	425	50	32	82
Cs	0.17	–	–	0.3	–	–
Ba	686	413	30	723	454	–
La	75	67.6	288	85	40	97
Ce	145	125	520	142	67	–
Nd	63	58	187	50	26	37
Sm	11.8	12.51	40.86	7.9	4.9	5.5
Eu	3.72	3.0	4.89	2.56	1.74	–

(continued)

Table 10.2 (continued)

Volcano	Pantelleria			Ustica		
Rock type	Hawaiite	Trachyte	Pantellerite	Hawaiite	Benmoreite	Trachyte
Tb	1.55	–	–	0.93	0.80	–
Yb	2.86	4.82	17.9	2.22	2.10	1.9
Lu	0.43	0.87	3.04	0.32	0.29	–
Hf	6.74	–	–	3.7	3.2	5.3
Ta	4.34	–	–	3.03	1.79	3.6
Pb	6.8	4.6	13	12.1	11.6	–
Th	7.34	7	52	11.1	6.4	18.6
U	1.04	–	–	2.4	0.85	–
$^{87}Sr/^{86}Sr$	0.703888	0.70315	0.70316	0.70312	0.70322	0.70321
$^{143}Nd/^{144}Nd$	0.512895	0.51301	0.51301	0.51291	0.51294	–
$^{206}Pb/^{204}Pb$	19.086	19.481	19.612	(19.56)	(18.83)	–
$^{207}Pb/^{204}Pb$	15.672	15.591	15.641	(15.64)	(15.62)	–
$^{208}Pb/^{204}Pb$	39.160	39.020	39.170	(39.19)	(38.62)	–

Volcano	Tetide Seamount	Foerstner Seamount	Ferdinandea (Graham, Julia)	Banco Senza Nome (Nameless)	Pantelleria SE Seamount
Rock type	Tholeiitic basalt	Alkali-basalt	Hawaiite	Basalt	Basalt
Data source	3	15	10	10	10
SiO_2 wt%	47.02	44.2	49.1	45.72	48.61
TiO_2	2.43	4.34	2.74	3.6	1.65
Al_2O_3	15.09	14.1	17.06	13.49	16.52
FeO_{total}	11.37	15.6	11.13	13.26	8.38
MnO	0.17	0.24	0.15	0.19	0.17
MgO	7.62	5.55	4.47	8.2	7.09
CaO	9.10	9.6	8.4	10.99	13.8
Na_2O	3.22	3.43	4.64	1.68	2.95
K_2O	0.23	1.12	1.51	2.02	0.58
P_2O_5	0.38	1.73	0.79	0.85	0.25
LOI	2.77	0.55	2	4.31	7.05
V	–	278	81	391	285
Cr	–	14	102	234	191
Co	–	–	39	–	49
Ni	–	24	–	–	–
Rb	2	18	29	38	–
Sr	236	498	625	822	345
Y	28	37	31	27	14
Zr	142	197	300	346	89
Nb	12	47	53	67	–
Ba	69	479	583	563	59
La	6	39.7	58.8	50.2	5.7

(continued)

Table 10.2 (continued)

Volcano	Tetide Seamount	Foerstner Seamount	Ferdinandea (Graham, Julia)	Banco Senza Nome (Nameless)	Pantelleria SE Seamount
Rock type	Tholeiitic basalt	Alkali-basalt	Hawaiite	Basalt	Basalt
Ce	–	84.2	113.8	102	12.3
Nd	–	50.8	50.8	51.3	8.4
Sm	–	10.9	9.8	10.6	2.4
Eu	–	4.07	3.1	3.6	1.1
Tb	–	1.48	1.2	1.4	0.5
Yb	–	2.68	2.2	2.2	1.1
Lu	–	0.4	0.3	0.28	0.2
Hf	–	–	7	7.7	2.2
Ta	–	–	4.2	6.3	0.9
Pb	–	–	3.1	–	–
Th	–	–	7.6	6.0	1.0
U	–	–	1.8	1.6	0.5
$^{87}Sr/^{86}Sr$	–	–	0.703081	0.703316	0.70478
$^{143}Nd/^{144}Nd$	–	–	0.51303	0.513022	0.51299
$^{206}Pb/^{204}Pb$	–	–	19.681	19.603	19.153
$^{207}Pb/^{204}Pb$	–	–	15.664	15.694	15.697
$^{208}Pb/^{204}Pb$	–	–	39.415	39.361	38.949

Source of data: (1) Carter and Civetta (1977); (2) Cinque et al. (1988); (3) Calanchi et al. (1989); (4) Esperança and Crisci (1995); (5) D'Orazio et al. (1997); (6) Civetta et al. (1998); (7) Trua et al. (1998); (8) Trua et al. (2003); (9) Armienti et al. (2004); (10) Rotolo et al. (2006); (11) Di Bella et al. (2008); (12) Viccaro and Cristofolini (2008); (13) Viccaro et a. (2011); (14) Avanzinelli et al. (2014); (15) Kelly et al. (2014); (16) Author's unpublished data

higher than 80–100 mW/m^2 in the areas of recent magmatism (Della Vedova et al. 2001).

Volcanoes of the Sicily Province are located in a variety of structural settings and on different types of bedrocks. Mount Etna (about 0.5 Ma to present) occurs near to the Tindari-Letojanni-Malta Escarpment fault, at the junction of three major tectonic blocks (Mastrolembo Ventura et al. 2014). The bedrocks consist of Quaternary conglomerates, sands, and marly clays overlying Messinian evaporites, silicoclastic rocks and flysch sequences of the Sicilide tectonic units (e.g. Behncke 2001; Norini and Acocella 2011; Branca and Ferrara 2013).

The Iblei volcanoes (about 7–1.5 Ma) are sited on the Iblean Plateau, a region of the Pelagian Block that has been intermittently affected by volcanism since Triassic times. Miocene-Quaternary volcanism developed along NE-SW extensional faults cutting Oligocene to Miocene platform carbonates and foredeep sequences (e.g. Finetti et al. 2005; Catalano et al. 2010).

The volcanoes of the Sicily Channel (Linosa and Pantelleria and a number of seamounts; Miocene to present in age) are mainly located along extensional/transtensional NW-SE trending faults bordering the continental rift systems that affected the northern African lithosphere, or along N-S strike-slip faults (e.g. Calanchi et al. 1989).

The Ustica island (0.75–0.13 Ma) occurs in the southern Tyrrhenian Sea, west of the Aeolian volcanic arc, along the southern margin of the Tyrrhenian abyssal plain, probably at the intersection between E-W (Ustica-Aeolian line) and

NW-SE trending faults (e.g. Boccaletti et al. 1984; Gueguen et al. 2010). The Prometeo submarine lava field is located a few km southeast of Ustica (Trua et al. 2003).

10.3 Etna

Etna is an active volcano (Fig. 10.2) presently rising to about 3330 m above sea level, with an estimated rock volume of 350–500 km^3 covering an area of about 1260 km^2 (Tanguy et al. 1997; Branca and Ferrara 2013). Volcanic activity has been characterised by frequent effusive and explosive eruptions, which have been long observed and described since ancient Greek and Roman times (see Rodwell 1878; Rittmann 1961; Chester and Duncan 1985; Kilburn and McGuire 2001; Bonaccorso et al. 2004). The volcano has been constructed upon an uplifted substratum, at the intersection of various tectonic lineaments with NNW-SSE, NE-SW and E-W trends (e.g. Boccaletti et al. 1984; Chester and Dunkan 1985; Lanzafame and Bousquet 1997; Azzaro 1999; Mattia et al. 2007; Catalano et al. 2008; Azzaro et al. 2013). Etna shows an asymmetrical conical shape. Flanks inclination is low (5–10°) up to about 1800 m altitude, but increases sharply (about 25°) in the upper part of the volcano. Flanks are affected by strong instability, which has been attributed either to gravitational sliding over a weak clayish basement (Borgia et al. 1992) or/and to deformation related to magma intrusions at various depths (e.g. Chester and Duncan 1985; Norini and Acocella 2011; Bozzano et al. 2013 with references). The Valle del Bove is one of the main collapsed areas. It is a horseshoe shaped depression developed about 8–5 ka before present by sliding of the eastern flank of the volcano (Calvari et al. 1998). Other important volcano-tectonic features consist of fracture zones radiating from the central crater area and representing preferential sites of flank eruptions and formation of a large number of cinder cones (e.g. Azzaro 1999). The relations between these structures and the regional stress field are debated, although most authors agree that the stress regime on the eastern flank of Etna is linked to the Tindari-Letojanni-Malta Escarpment strike-slip fault system (e.g. Bousquet and Lanzafame 2004).

Etna is also well known for its intense release of volatiles, occurring both from the volcano vents and fractures and from the surrounding area (e.g. Caracausi et al. 2003; Giammanco et al. 2007). A total CO_2 emission of about 15–40 Mt/year has been estimated (Allard et al. 1991), which is about one tenth of the total output of carbon dioxide from Earth subaerial volcanoes (Gerlach 2011). The reason of such a massive gas release is debated. A contaminated mantle source or assimilation of the foreland carbonate rocks apparently occurring at a depth of about 10 km beneath the volcano, are possible explanation (Frezzotti et al. 2009; Chiodini et al. 2011; Paonita et al. 2012).

10.3.1 Volcanology and Stratigraphy

The history of Etna has been dominated by effusive and strombolian activity, with several high-energy vulcanian, subplinian and plinian eruptions (e.g. Chester and Duncan 1985; Branca and Del Carlo 2004; Branca et al. 2011). Four main evolution stages have been distinguished, starting from about 0.5 Ma to the present (Gillot et al. 1994; Branca et al. 2004, 2011; De Beni et al. 2011). The lowest exposed rocks consist of both submarine and subaerial tholeiitic basalts (Basal Tholeiites; Branca et al. 2011) that are observed as limited outcrops along the southwestern border of the volcano, and at Aci Castello along the Ionian coast (Tanguy 1978; Corsaro and Cristofolini 2000). Successively, eruptions of mafic lavas and minor pyrolastics built up volcanic suites progressively shifting from subalkaline to Na-alkaline (Timpe; 220–120 ky). Starting from about 110 ka up to 65 ky, volcanism migrated westward (Valle del Bove stage), forming several hawaiite to mugearite and benmoreite volcanoes (Trifoglietto, Giannicola, Cavigghiuni etc.). The latest activity (Stratovolcano suites; 60 ky to present), constructed the bulk of the present

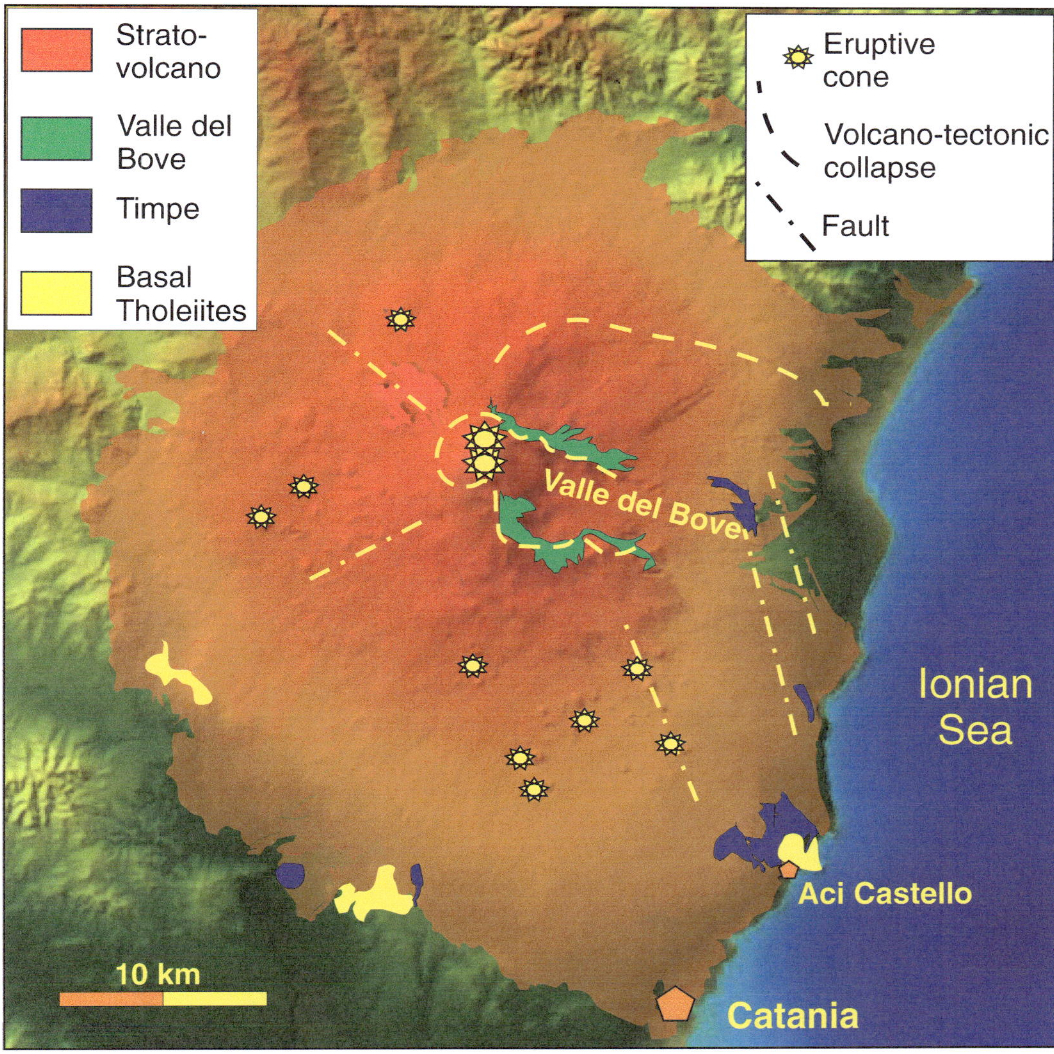

Fig. 10.2 Schematic geological map of Etna volcano. Simplified after Branca et al. (2011, 2015)

stratovolcano, by emission of lavas and pyroclastic rocks with hawaiite to benmoreite-trachyte compositions. This period has been subdivided into the Ellittico (60–15 ky) and the Mongibello stages (15 ky to present). The Ellittico activity started with hawaiite and mugearite lava effusions, and progressively shifted to plinian activity, which emplaced benmoreite-trachyte pumice fall and ignimbrite deposits and caused a caldera collapse. The post-Ellittico eruptions were dominated by basaltic lava flows and a few plinian to sub-plinian explosions, which filled the Ellittico caldera and built up the Mongibello volcano. The youngest activity is characterized by frequent eruptions of trachybasaltic lavas and pyroclastics, which show increased potassium contents with respect to older rocks. This is especially evident for magmas erupted after 1971 AD, but is also observed in some of older rocks, such as the prehistoric Mongibello lavas (Tanguy et al. 1997; Armienti et al. 2004; Métrich et al. 2004; Viccaro and Cristofolini 2008; Ferlito and Lanzafame 2010).

Fig. 10.3 Image of the 2002 flank eruption of Etna, with the degassing central crater in the background. Courtesy of Giovanni Marano and Luciano Giannini

Historical activity has been persistent and occurred both at the summit craters and along the flanks (e.g. Rodwell 1878; Bonaccorso et al. 2004). Some authors believe that both are fed by a single major central dike, whereas others suggest distinct magma chambers and dike systems, a hypothesis supported by emission of lavas with contrasting potassium contents at different vents during single eruptions (Ferlito et al. 2009). Activity at summit craters consists of quiet steam emission, interrupted by frequent strombolian to vulcanian and subplinian explosions and lava fountaining, sometimes accompanied by lava effusions, lasting from a few hours to months or sometimes years. As a consequence, the morphology of the summit area is continuously modifying and presently consists of a slightly N-S elongated circular platform generated by a small collapse, containing a major cone with two large summit craters (Bocca Nuova and Voragine) and two lateral cones (NE Crater, SE Crater). Flank eruptions have been prevailingly effusive and strombolian (Fig. 10.3), and often took place at low elevations (down to 300 m a.s.l.) close to inhabited areas, raising severe problems for civil defence. Among historical eruptions, worth of note are those occurred in 122 BC, 1614–1624, 1669, and 1991–1993 AD. The 122 BC plinian eruption is one of the most violent historical events and was associated with the formation of the Cratere del Piano summit collapse inside the Ellittico caldera (Coltelli et al. 1998). The 1614–1624 eruption is the longest recorded event. The 1669 eruption started on March 11 along a 12 km-long fissure on the southern flank of the volcano. By April 12 a lava flow, originating from the main cones of Monti Rossi at an altitude of 948 m, reached Catania, where it knocked down about 40 m of the city walls, destroyed part the town, and reached the sea on April 23. The eruption stopped on July 15 (Chester and Dunkan 1985; Faraone 2002). The 1991–1993 eruption started on December 14 on the eastern flank of the volcano, along a radial fracture extending between 2400 m above sea level and the central crater area, and lasted fifteen months, emitting voluminous lava flows that threatened the village of Zafferana Etnea (Calvari et al. 1994). This eruption is well known because of the successful lava flow diversion from its

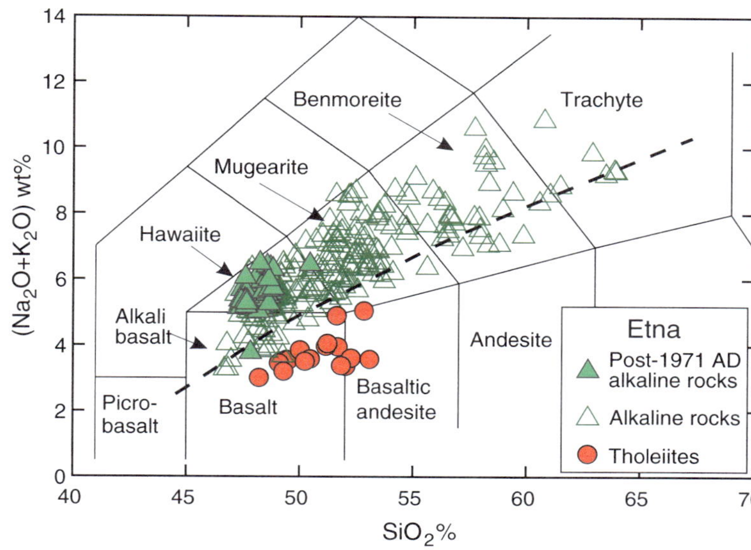

Fig. 10.4 TAS classification diagrams of Etna rocks. The dashed line is the boundary between subalkaline and alkaline fields of Irvine and Baragar (1971)

natural path to an artificial channel dug near to the vent (Barberi et al. 1993).

10.3.2 Petrography and Mineral Chemistry

The Etna rocks have tholeiitic to Na-alkaline affinity, with some products exhibiting a potassic tendency. Compositions range from basalt to hawaiite, mugearite, benmoreite, and trachyte on the TAS diagram (Fig. 10.4). Ol-hy normative tholeiites are the lowest exposed products. They show microcrystalline ophitic to porphyritic textures. Phenocrysts include plagioclase, clinopyroxene and olivine set in a groundmass formed of the same phases plus Fe–Ti oxides and glass. Olivine (about Fo_{86-60}) phenocrysts contain chromite and rare orthopyroxene inclusions and show decreasing forsterite contents from phenocryst core to rim and to the groundmass. Clinopyroxene ranges from diopside to augite and pigeonite. Plagioclase occurs as phenocrysts (mostly around An_{70-60}), whereas oligoclase to anorthoclase is found in the groundmass (Giacobbe 1993; Kamenetsky and Clocchiatti 1996; Corsaro and Cristofolini 1997; Tanguy et al. 1997; Armienti et al. 2004; Corsaro and Pompilio 2004).

Alkali basalts are aphyric to weakly porphyritic with phenocrysts of clinopyroxene, plagioclase and olivine (sometimes with chromite inclusions) set in a groundmass consisting of the same phases plus glass. Hawaiites (trachybasalts) are porphyritic with phenocrysts of plagioclase (An_{90-55}), olivine (up to Fo_{90}), clinopyroxene (Cr-diopside to augite), Fe–Ti oxides and sometimes kaersutite. Groundmass contains the same phases and, in some cases, nepheline, sodalite, and phlogopite. Hawaiitic lavas of the historic activity show strong variation in modal composition. Rocks older than the 1669 AD eruption are generally dominated by plagioclase phenocrysts with very few clinopyroxene and rare or absent olivine. Younger hawaiites have a typical phenocryst paragenesis consisting of plagioclase ($\sim An_{90-55}$), diopside-augite clinopyroxene, olivine (mostly Fo_{80-60}), and a few Fe–Ti oxides (e.g. Armienti et al. 1997; Kahl et al. 2015; Viccaro et al. 2015; Giacomoni et al. 2016). Locally, these rocks are referred to as *etnaites*. Kaersutite to Mg-hastingsite amphibole is found in some rocks (Viccaro and Cristofolini 2008). Mugearites and benmoreites are porphyritic with phenocrysts of dominant plagioclase (about An_{80-40}), augite clinopyroxene, olivine (Fo_{75-70}), and some amphibole and Ti-magnetite set in a groundmass containing the same minerals, plus some alkali feldspar, nepheline, sodalite, phlogopite, and apatite. Trachytes are aphyric to weakly porphyritic with microphenocrysts of plagioclase ($\sim An_{40}$), augitic clinopyroxene, Fe–Ti oxides, and some olivine ($\sim Fo_{75}$) and kaersutite enclosed in a groundmass containing alkali feldspar (anorthoclase to Na-sanidine), clinopyroxene, olivine, Ti-magnetite, biotite and sporadic sodalite and nepheline (Tanguy 1978; Kamenetsky and Clocchiatti 1996; Tanguy et al. 1997; Corsaro and Pompilio 2004).

10.3.3 Petrology and Geochemistry

Tholeiitic rocks display higher MgO and lower P_2O_5, alkali, and incompatible trace element abundances than alkaline rocks with the same silica content (Figs. 10.5 and 10.6). Among alkaline rocks, those erupted from 1971 AD to present are enriched in potassium, Rb, Cs and Cl, but abundances of other elements are within the range of the older rocks. Overall, Etna magmas contain high abundances of volatile components,

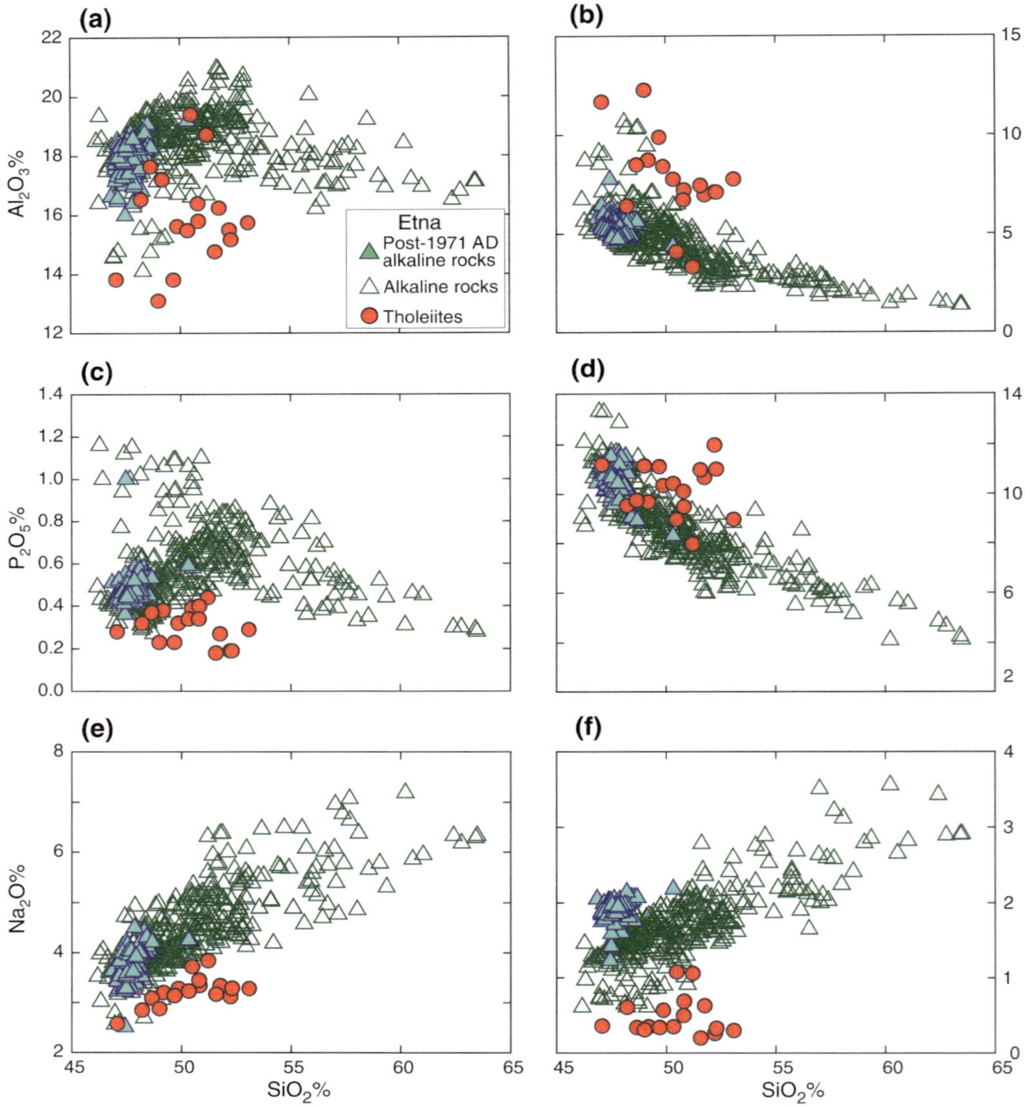

Fig. 10.5 Major element variation diagrams of Etna rocks

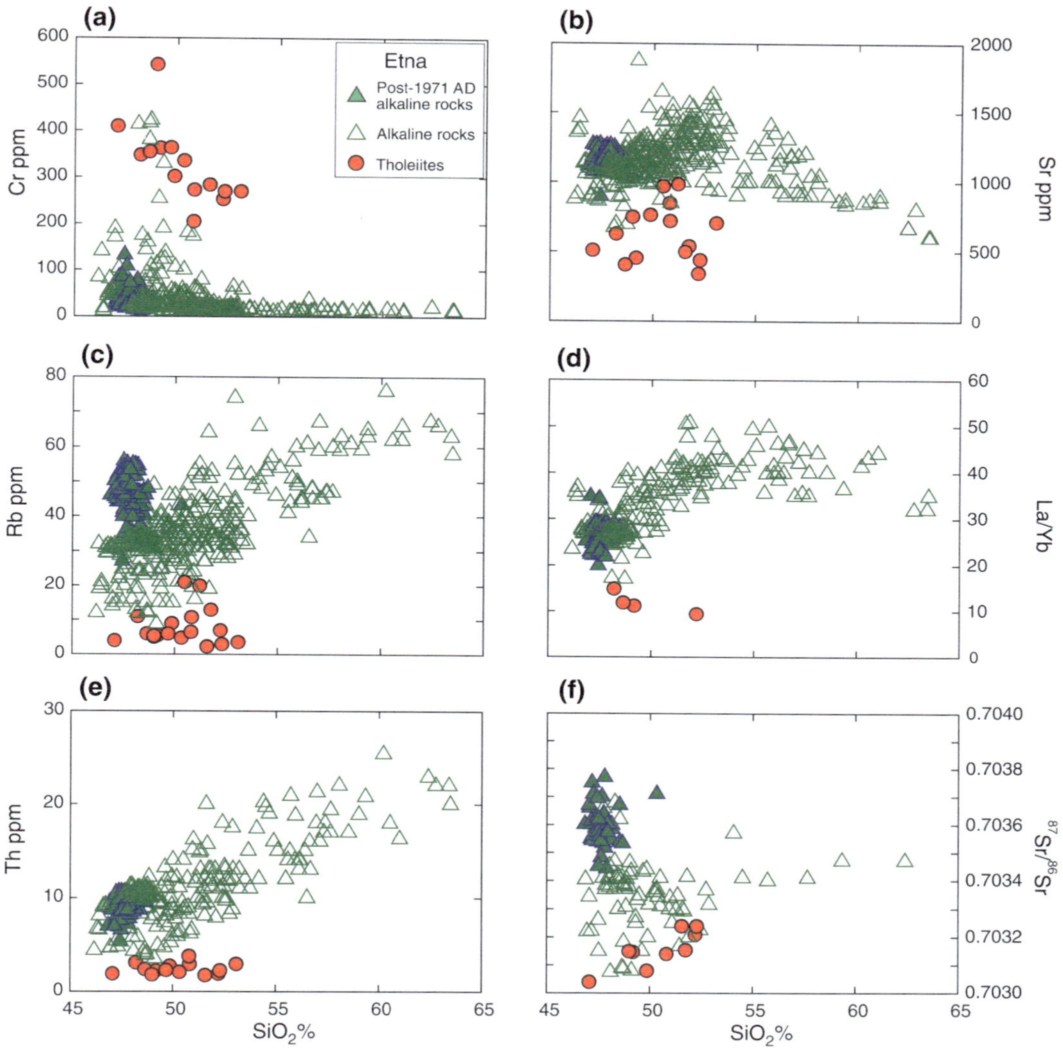

Fig. 10.6 Variation diagrams of selected trace elements and $^{87}Sr/^{86}Sr$ versus silica of Etna rocks

with olivine- and cpx-hosted glass inclusions showing H_2O up to 3–4 wt% and Cl = 0.13–0.18 wt% (Corsaro and Pompilio 2004; Armienti et al. 2004; Métrich et al. 2004; Spilliaert et al. 2006).

REE patterns are fractionated (Fig. 10.7a) for both light and heavy REE, with tholeiites showing lower La/Yb ratios than alkaline rocks. Incompatible element patterns normalised to primordial mantle compositions for mafic rocks are strikingly different from the Aeolian arc and central-southern Italy volcanics. Both tholeiitic and alkaline rocks exhibit an upward convexity, with negative spikes of K and Rb (Fig. 10.7b). Note, however, that alkaline lavas exhibit small negative anomalies for some HFSE (Zr, Ti) and positive spike of Pb, which are uncommon in typical intraplate Na-alkaline basalts (e.g. Sun and McDonough 1989; Wilson 1989; Willbold and Stracke 2006).

Sr and Nd isotope ratios ($^{87}Sr/^{86}Sr \sim 0.7031$ to 0.7040; $^{143}Nd/^{144}Nd \sim 0.51285$ to 0.51297) display moderate variation compared with other Tyrrhenian Sea volcanoes (Fig. 10.8a; e.g. Marty

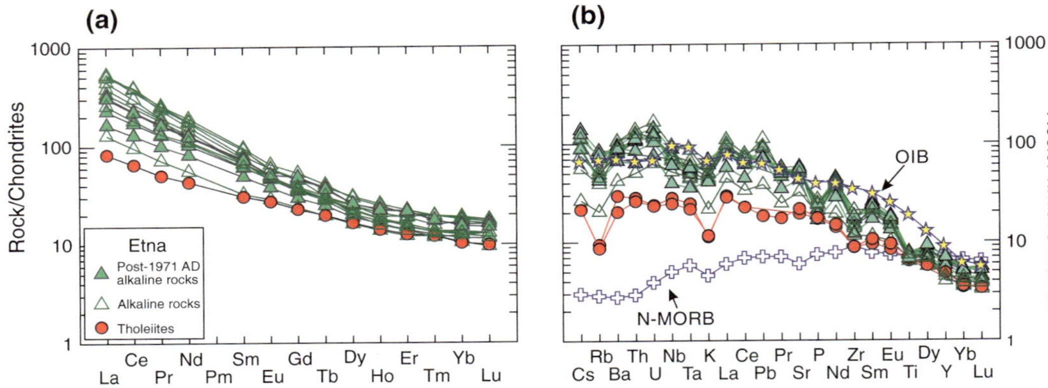

Fig. 10.7 **a** Chondrite normalised REE patterns of Etna volcanics; **b** Incompatible element patterns of mafic rocks. Average OIB and normal-type N-MORB are shown for comparison (Sun and McDonough 1989; Gale et al. 2013)

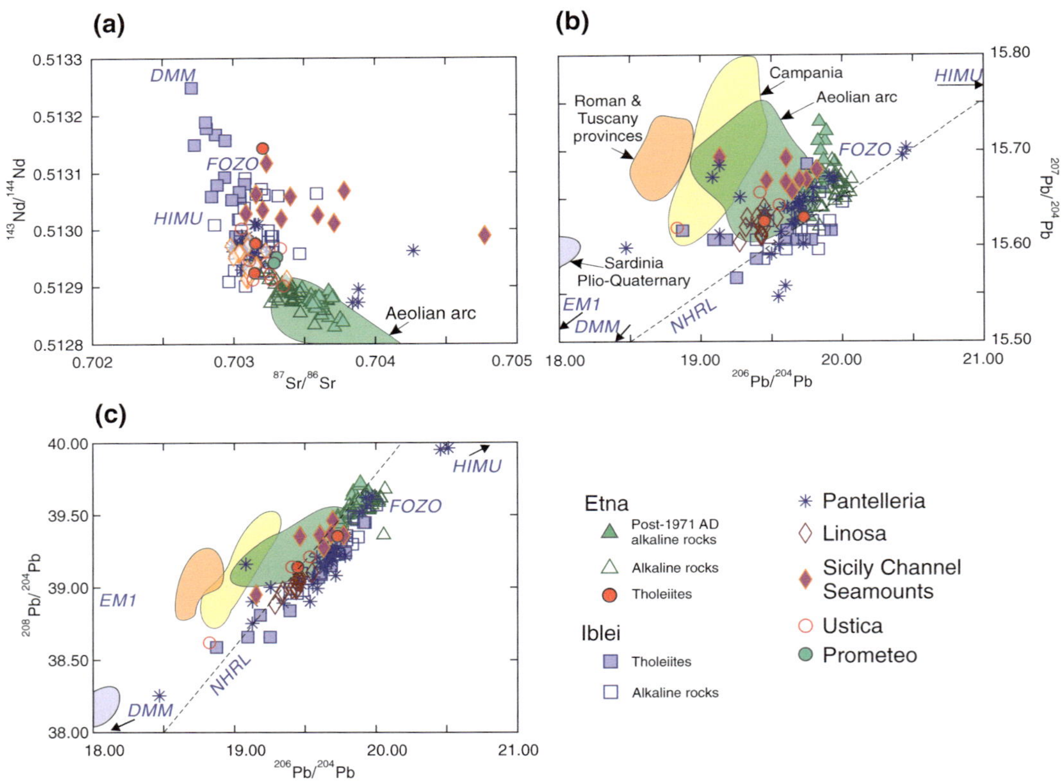

Fig. 10.8 Sr–Nd–Pb isotope plots of the Sicily Province. Compositions of other volcanic provinces and of HIMU, DMM, FOZO mantle end-members are also shown. The *dashed line* is the Northern Hemisphere Reference Line of Hart (1984)

et al. 1994; Giacobbe 1993; Tonarini et al. 1995, 2001; Armienti et al. 1996, 2004; Tanguy et al. 1997; Viccaro and Cristofolini 2008). Detailed studies have shown the occurrence of small but significant isotopic variations both among different eruptions and within single lava flows

(Armienti et al. 1984; Tonarini et al. 2001). Data on separated clinopyroxenes and surrounding matrix have often highlighted Sr isotopic disequilibrium, especially in the lavas erupted during the last 30–40 years (e.g. D'Orazio et al. 1997; Armienti et al. 2004). Pb-isotope ratios (Fig. 10.8b and c) show some of the most radiogenic compositions among recent Italian volcanic rocks (e.g. $^{206}Pb/^{204}Pb \sim 19.45$ to 20.06; $^{207}Pb/^{204}Pb \sim 15.62$ to 15.69; $^{208}Pb/^{204}Pb \sim 39.11$ to 39.68; Giacobbe 1993; Viccaro and Cristofolini 2008). Compositions of olivine-hosted melt inclusions yielded $^{206}Pb/^{204}Pb = 19.20–19.49$ and $^{208}Pb/^{204}Pb = 38.69–39.12$ for tholeiitic basalts of Aci Castello, and $^{206}Pb/^{204}Pb = 18.62–20.16$ and $^{208}Pb/^{204}Pb = 38.08–40.08$ for present-day lavas (Rose-Koga et al. 2012). These values fall within the field of FOZO mantle compositions, trending to HIMU. $^{176}Hf/^{177}Hf$ is around 0.28294–0.28299 (Gasperini et al. 2002; Viccaro et al. 2011).

Oxygen isotopic data on whole rocks show $\delta^{18}O_{SMOW}$ in the range +5.0 to +5.9 ‰ (Marty et al. 1994). Higher values of +6.1 to +7.1 ‰, increasing with decreasing MgO, have been measured on recent lavas by Viccaro and Cristofolini (2008) and Viccaro et al. (2008). Values of boron isotopic compositions range from $\delta^{11}B \sim -8$ to -3 ‰ (Tonarini et al. 2001).

Noble gas isotope compositions are within the range of MORB and OIB (Nakai et al. 1997). $^{3}He/^{4}He$ ratios measured on olivine phenocrysts show $R/R_A \sim 6.0–7.0$ (Marty et al. 1994; Nuccio et al. 2008; Coulson et al. 2011), whereas similar to slightly higher values have been found in the fumarolic gases (Nakai et al. 1997; Rizzo et al. 2006).

10.3.4 Evolution of Etna Magmas and Implications for the Plumbing System

Recurrent eruptions of mafic magma during much of the history of Etna suggest a continuous arrival of parental melts from the source, with moderate fractional crystallisation and intense mixing within an open-conduit plumbing system (e.g. Corsaro et al. 2013; Mollo et al. 2015; Giacomoni et al. 2016). However, explosive benmoreitic-trachytic plinian eruptions and associated caldera collapse at the end of the Ellittico activity suggest more protracted fractional crystallisation at this stage, likely in an obstructed-conduit system.

Petrological, geochemical and isotopic studies along with thermodynamic simulations suggest that fractional crystallisation at Etna was polybaric and involved separation of various proportions of olivine, plagioclase and clinopyroxene (e.g. Armienti et al. 2004). However, there is also clear evidence for other evolutionary processes, including magma mixing, element transfer by gaseous phases, and interaction between magma and wall rocks. Therefore, there is a general agreement for the occurrence of a multiple magma chamber plumbing system beneath the Etna volcano, although various models differ for several significant aspects.

According to Tanguy et al. (1997), a deep magma chamber sited at the mantle-crust boundary has been operating throughout the entire history of Etna volcano and is still active today. Occasionally, however, magmas ponded and fractionated in shallow reservoirs, giving evolved benmoreitic-trachytic products. Fractional crystallisation was dominated by separation of olivine and clinopyroxene at high pressure, whereas plagioclase was a main fractionating phase in the shallow reservoirs (Tanguy et al. 1997; Mollo et al. 2015). Polybaric (700 Ma to less than 100 MPa) magma crystallisation and mixing processes working for much of the Etna volcano history have been also suggested by Giacomoni et al. (2016) based on textural and geothermobarometric studies of clinopyroxenes. Assimilation of crustal rocks may also be an important factor of magma evolution, as testified by the occurrence of abundant crustal xenoliths in some lavas (Viccaro et al. 2008). Such a process is emphasized by chemical and isotopic disequilibria observed in several lavas, and may provide an explanation for the enrichment in K, Rb, Cs and radiogenic Sr in some rocks. However, a magma origin in a compositionally heterogeneous mantle source has been submitted as an alternative explanation for these features (Condomines et al. 1995; Tonarini et al. 2001; Armienti et al. 2004; Viccaro and Cristofolini 2008; Viccaro et al. 2011; Rizzo et al. 2013).

The evolution and dynamics of magmas erupted during the last decades received much attention, because of the obvious implication for forecasting future events and mitigating volcanic hazard. Geochemical data strongly support crystallisation and protracted mixing as main evolution mechanisms. These apparently took place along the entire length of the feeding system, within reservoirs sited at various depths beneath the volcano, from the Moho to 2–3 km beneath the central craters (e.g. Armienti et al. 2013). Mollo et al. (2015) suggest crystallisation for magma erupted between 2001 and 2012 at pressures ranging from 1.0 GPa and between 500 and 100 MPa. Corsaro et al. (2013) recognised intensive mixing between isotopically diverse magmas at a depth of about 6 km, followed by fractional crystallisation at about 2 km, for the 1995–2005 activity. Ferlito and Lanzafame (2010) envisage magma evolution and degassing in a complex system with a deeper reservoir sited at the Moho, which feeds two shallower reservoirs at about 10 and 5 km. Deep CO_2 and shallower H_2O degassing would transfer alkalies and chlorine to magmas residing in the uppermost section of the plumbing system, thus explaining enrichments in K and Rb of post-1971 rocks. According to Ferlito and Lanzafame (2010), one of the main evidence in favour of this hypothesis is furnished by the variable potassium contents of melt inclusions in olivine, which show decreasing enrichments with increasing depth of entrapment (Métrich et al. 2004; Spilliart et al. 2006; Ferlito and Lanzafame 2010). Such a hypothesis probably implies that wall rocks were the source of K- and Rb-rich fluids. Viccaro et al. (2015) suggest that large volumes of mafic magmas are intruded continuously from deep to intermediate and shallow levels. Increased degassing and soil uplift often precede magma eruptions, suggesting that gas transfer from the deep system into the uppermost 2–3 km of the plumbing system is the trigger of eruptions (e.g. Aiuppa et al. 2010). Magma ascent velocity has been estimated to range from a few cm to about 10 m/h by Armienti et al. (2013). Detailed mineral chemistry studies and thermodynamic modelling for the 1991–2008 AD activity, confirm the occurrence of various interconnected reservoirs beneath the Etna volcano, with at least five sites of magma ponding and crystallisation characterised by different confining pressure, redox conditions and H_2O content in the magmas. Sometimes, different reservoirs are contemporaneously tapped during a single eruption, thus explaining the variable compositions of lavas at different vents (e.g. Viccaro et al. 2008).

In summary, petrological and geochemical studies suggest a present-day volcano plumbing system consisting of vertical dikes that connect magma reservoirs sited at various depths. This model is supported by seismological studies that have identified several high-velocity bodies between 12 and 3–5 km, as well as low-P and S-wave volumes during eruptions, which indicate solidified gabbroic bodies and intrusions of volatile-rich magmas (e.g. Hirn et al. 1997; Patanè et al. 2003, 2006; Corsaro et al. 2014 with references). Magmas migrate almost continuously from deep to shallow levels, undergoing CO_2 degassing and shallow H_2O and SO_2 exsolution, decompression crystallisation and mixing. ^{226}Ra–^{230}Th and ^{228}Ra–^{210}Pb data suggest that the deep reservoir has been in a steady state for about 1550 years, before the injection of a new batch of magma that mixed with resident melts and fed the post-1971 activity (Condomines et al. 1995).

The magmas erupted at Etna also show significant long-term petrological-geochemical variations, especially the transition from subalkaline to Na-alkaline compositions during the early history of volcano (e.g. Clocchiatti et al. 1992: Corsaro and Cristofolini 1996; Kamenetshy and Clocchiatti 1996; Corsaro et al. 2013; Schiavi et al. 2015). These major variations are interpreted as related to the injection of compositionally distinct primary magmas into the volcano plumbing system (e.g. Tanguy et al. 1997;

Viccaro and Cristofolini 2008). The early erupted tholeiitic basalts show similar incompatible element patterns as the younger mafic alkaline rocks, suggesting different degrees of partial melting of broadly homogeneous source (e.g. Clocchiatti et al. 1998). However, the variable isotopic signatures are at odds with this conclusion and rather support compositional heterogeneities in the source.

The transition from subalkaline to alkaline compositions at Etna is accompanied by the manifestation of arc signatures in the rocks, such as small negative HFSE anomalies and positive spikes of Pb (Fig. 10.7b). These features are not found in typical OIB magmas, although do occur in many intraplate OIB-type rocks in the Mediterranean area (Lustrino and Wilson 2007). Therefore, a role for an arc component has been suggested for Etna magmatism; this may derive from the nearby Ionian subducting plate or, alternatively, from old recycling of upper crust into the mantle (e.g. Tonarini et al. 2001; Viccaro and Cristofolini 2008; Correale et al. 2014). Intriguingly, no convincing evidence for subduction-related geochemical signatures is found by Schiavi et al. (2015) in the melt inclusions entrapped in olivine and clinopyroxene phenocrysts of some recent eruptions. This supports a shallow-level origin (i.e. magma-crust interaction) for arc signatures in the Etna magmas.

10.4 Iblei

The Iblean region (southeastern Sicily) has been intermittently affected by volcanism from Triassic to Quaternary times. This review deals with Upper Miocene and Plio-Quaternary rocks, whereas older volcanism will be briefly addressed in Appendix 1.

The younger volcanic phases took place in the northern sector of the Iblean Plateau, along NE-SW extensional faults bordering graben zones (e.g. Grasso et al. 1983). The volcanic products cover an area of about 250 km^2 and have an estimated volume of about 30–40 km^3. They rest over a sedimentary substratum of Upper Cretaceous and Early Tertiary reef and marly limestones and Oligo-Miocene calcarenites and are interfingered with Mio-Pleistocene calcarenites, marls, clays and silicoclastic sediments (e.g. Schmincke et al. 1997). Activity was dominated by fissure-type submarine to subaerial eruptions.

10.4.1 Volcanology, Stratigraphy, and Petrography

Various pulses of subalkaline to strongly Na-alkaline volcanism characterise the recent volcanism at Iblei (Fig. 10.9). Miocene (Tortonian-Messinian) to Middle Pliocene volcanic cycles erupted submarine volcanoclastites from diatremic centres, pillow lavas and breccias, and minor subaerial lavas. Compositions are prevailingly Na-alkaline (alkalibasalt, basanite, nephelinite and ankaratrite[1]), possibly with a single tholeiitic basalt episode in the Tortonian (Schmincke et al. 1997; Di Grande et al. 2002; Behncke 2004). A wide variety of mantle and crustal xenoliths are found in the pyroclastic deposits of Miocene alkaline diatremes.

The Upper Pliocene to Lower Pleistocene activity erupted tholeiitic and Na-alkaline magmas (Fig. 10.10) in subaerial to shallow marine environments. Tholeiitic rocks are dominant and were emplaced over a very short time span, probably a few years (Behncke 2004). They range from slightly oversaturated (qz-tholeiites) to slightly undersaturated (ol-tholeiites) in silica (Beccaluva et al. 1998). Textures are aphyric to poorly porphyritic holocrystalline to hypocrystalline. Qz-normative rocks contain microphenocrysts of olivine ($\sim Fo_{81-76}$), orthopyroxene ($\sim En_{85}$), augite and plagioclase, with pigeonite in the groundmass. Ol-normative tholeiites contain olivine, augite clinopyroxene and plagioclase phenocrysts. Alkaline rocks have porphyritic textures with dominant MgO-rich olivine ($\sim Fo_{90-75}$), augite and plagioclase phenocrysts set in a groundmass containing the same

[1]Ankaratrite is a melanocratic variety of nephelinite, rich in olivine phenocrysts and/or xenocrysts.

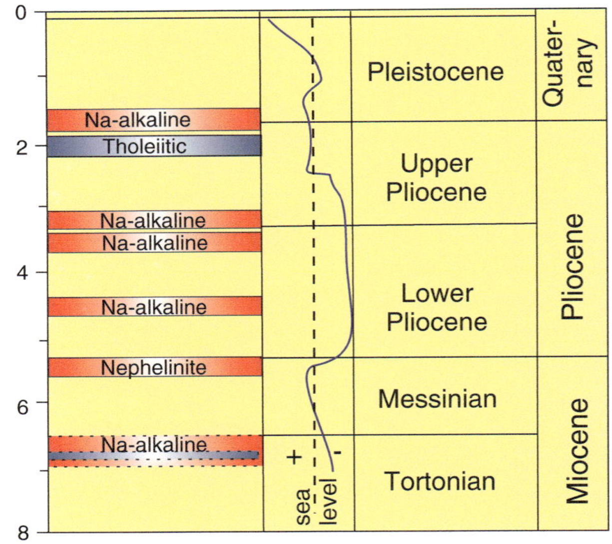

Fig. 10.9 Schematic stratigraphy of Miocene to Pleistocene volcanic activity at Iblei (Modified after Schmincke et al. 1997)

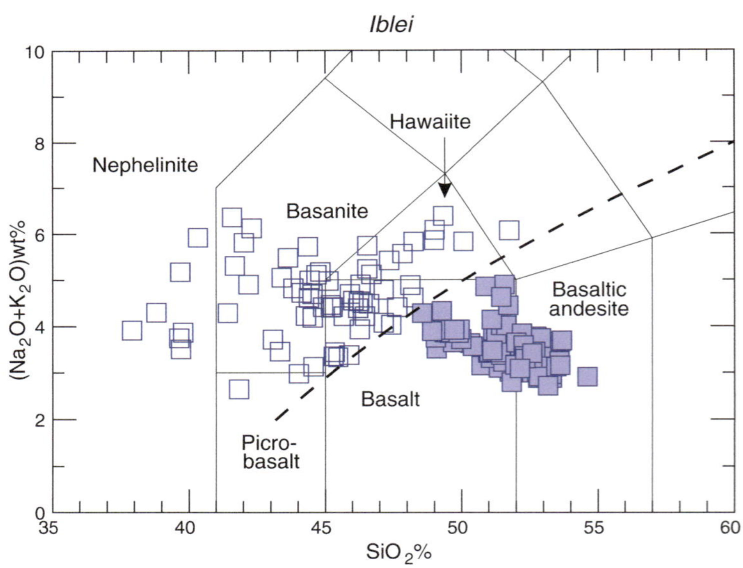

Fig. 10.10 TAS diagram for Iblean Upper Miocene-Quaternary volcanic rocks. Alkaline and subalkaline rocks are indicated with *open* and *full symbols*. Compositions recalculated to 100 % on a LOI-free basis

phases plus Fe–Ti oxides. Nepheline occurs in the groundmass of basanites and nephelinites. In the latter, haüyne, rare amphibole, and primary carbonate have also been found. Some basanites contain abundant phenocrysts and xenocrysts of olivine and clinopyroxene and have been classified as ankaratrites (Beccaluva et al. 1998; Trua et al. 1998; Bianchini et al. 1998, 1999).

Upper Miocene alkaline diatreme pyroclastic deposits contain abundant xenoliths, mainly of mantle origin. Ultramafic nodules consist of dominant spinel-harzburgites and lherzolites, plus minor websterites and clinopyroxenites. Crustal xenoliths are scarce and include gabbros, diorites, some mafic granulites and rare anorthosites (Sapienza and Scribano 2000; Scribano et al. 2006). A sample of a serpentinized and carbonated peridotite xenolith has been found to contain organic matter and nanodiamonds, which crystallised in an organic-water system during serpentinisation at relatively low pressure, in the stability field of graphite (Simakov et al. 2015). Small mantle and crustal xenoliths are also found in some Quaternary basanitic and nephelinitic lavas (e.g. Sapienza and Scribano 2000 and references therein).

10.4.2 Petrology and Geochemistry

Major and trace element data show an increase in SiO_2 and Al_2O_3, and a decrease in P_2O_5, MgO, CaO, and incompatible trace elements from alkaline to tholeiitic rocks (Fig. 10.11). In contrast, Ni and Cr do not show systematic variations and exhibit high concentrations in most samples. REE patterns are much more fractionated in the alkaline than in the tholeiitic rocks (Fig. 10.12a). Incompatible element patterns normalised to primordial mantle composition for the most mafic rocks (MgO > 5 wt%) have upward convex shapes and strong negative anomalies of K and Rb. Small negative spikes of Zr, Hf and Ti are observed in some rocks, especially in the alkaline volcanics (Fig. 10.12b).

$^{87}Sr/^{86}Sr$ ratios ($\sim$0.7027 to 0.7036) increase slightly from subalkaline to alkaline rocks (Figs. 10.8a and 10.11h); $^{143}Nd/^{144}Nd$ ratios (0.5129 to 0.5132) show the opposite trend. Pb isotopic ratios ($^{206}Pb/^{204}Pb$ $\sim$ 18.87 to 20.00; $^{207}Pb/^{204}Pb$ $\sim$ 15.57 to 15.69; $^{208}Pb/^{204}Pb$ $\sim$ 38.60 to 39.50) are more radiogenic in the alkaline than in the tholeiitic rocks (Fig. 10.8b and c), but with a large overlap of values (Beccaluva et al. 1998; Trua et al. 1998; Bianchini et al. 1999).

Much study has been dedicated to xenoliths from the Miocene diatreme deposits, with the aim of investigating the nature and composition of the crust and upper mantle. Sapienza et al. (2005) and Perinelli et al. (2008) found that xenoliths equilibrated at 870–1050 °C and 1.1–0.75 GPa, almost at the mantle-crust boundary. Tonarini et al. (1996) found that Sr-Nd isotopic signatures ($^{87}Sr/^{86}Sr$ = 0.70298–0.70334; $^{143}Nd/^{144}Nd$ = 0.51289–0.51299) overlap those of lavas, and suggested a metasomatic event of probable Permo-Triassic age. Bianchini et al. (2010) discovered that Sr–Nd isotopic signatures of peridotites are more primitive than those of pyroxenites, the latter falling in the compositional field of the Iblei magmas. Therefore, pyroxene-rich mantle rocks were envisaged as main source of Iblei magmas. $^3He/^4He$ ratios cluster around R/R_A $\sim$ 7.3, somewhat higher than other volcanoes in Sicily (Sapienza et al. 2005; Correale et al. 2012).

10.4.3 Evolution of Iblean Magmas

The high concentrations of MgO ($\sim$3 to 15 wt %), Ni ($\sim$100 to 300 ppm) and Cr ($\sim$100 to 500 ppm) of Iblean rocks suggest that there was little evolution during magma ascent to the surface, a conclusion supported by the monogenetic nature of volcanic centres and by the occurrence of mantle xenoliths and xenocrysts in alkaline rocks. The decrease in MgO from alkaline rocks to tholeiites is not related to evolutionary processes since Ni and Cr remain constant. Therefore, there is a general agreement that the Iblean subalkaline and alkaline rocks represent genetically distinct batches of magmas.

Geochemical and petrological investigations of mantle xenoliths and host rocks indicate an origin of Iblean parental magmas by different degrees of partial melting of metasomatised lherzolite and wherlite mantle sources. Beccaluva et al. (1998) suggested increasing degree of partial melting from alkaline to tholeiitic magmas at 30–90 km depth. An origin from enriched and heterogeneous sources is supported by geochemical data (e.g. Tonarini et al. 1996; Beccaluva et al. 1998; Trua et al. 1998; Bianchini et al. 1999; Sapienza et al. 2007).

There is debate on the nature (i.e. oceanic vs. continental) of the lithosphere where

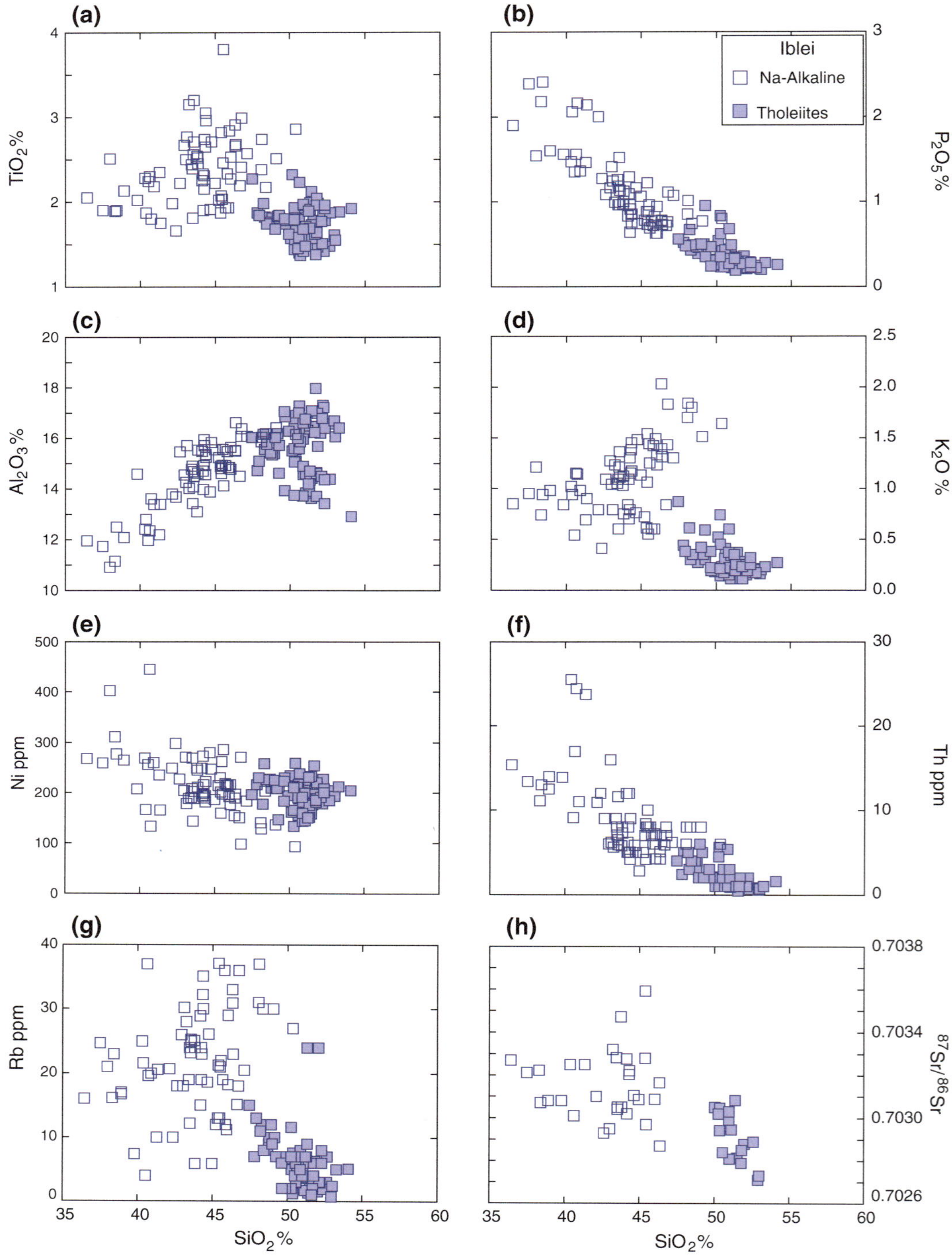

Fig. 10.11 Variation diagrams of selected major and trace elements against SiO_2 for Iblei Miocene to Quaternary volcanic rocks

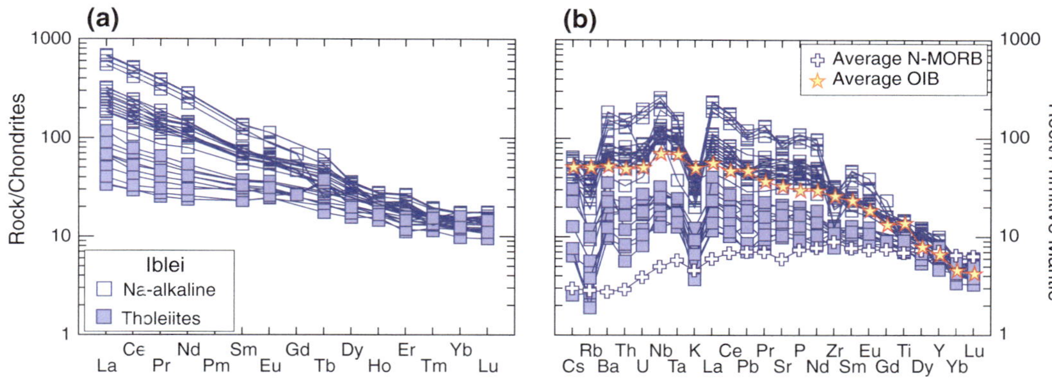

Fig. 10.12 **a** Chondrite normalised REE patterns of Iblean rocks; **b** Mantle normalised incompatible element patterns of mafic rocks (MgO > 5.0 wt%). Average OIB and normal-type N-MORB are shown for comparison (Sun and McDonough 1989; Gale et al. 2013)

metasomatism and melting took place. The scarcity of crustal xenoliths and the absence of typical continental lithologies (granites, felsic granulites, metasediments) suggest the occurrence of an oceanic-type lithosphere beneath Iblei (Trua et al. 1998; Scribano et al. 2006; Manuella et al. 2015). Such a hypothesis is challenged by Sapienza et al. (2007) who suggest the occurrence of a thinned African-type lithosphere, based on Hf–Os isotope signatures of zircons and sulfides from crustal and mantle xenoliths.

10.5 Pantelleria

The Island of Pantelleria is located in a NW-SE trending extensional zone (Pantelleria Rift) developed from Late Miocene to Early Pliocene in the Pelagian Block (e.g. Civile et al. 2010). The volcano rises about 1000 m above the sea floor and reaches an altitude of 836 m above sea level. The structural setting is dominated by NW-SE and NE-SW trending fractures (e.g. Catalano et al. 2009). A tensile fault system divides the island into two sectors, the northwestern sector containing most of the exposed basaltic rocks, and the southeastern sector that is constructed by silicic peralkaline volcanics (Fig. 10.13).

10.5.1 Volcanology and Stratigraphy

The Island of Pantelleria is a NW-SE elongated stratovolcano with polygenic calderas, built up by dominant peralkaline trachytes and rhyolites (pantellerites) and minor Na-transitional to mildly alkaline basalts. The lowest exposed rocks are 320 and 220 ka old, whereas the youngest dated activity on the island has an age of about 4 ka (e.g. Civetta et al. 1984, 1988, 1998; Mahood and Hildreth 1986; Scaillet et al. 2011). Volcanism in the area is still active, as testified by the submarine eruption occurred on 1891 a few km NW of Pantelleria (Foerstner volcano; Foerstner 1891; Washington 1909).

Most of the volcanic activity at Pantelleria was explosive and emitted silicic peralkaline pyroclastic products and some lavas. Basaltic lava flows and scoriae have been erupted episodically from 118 ka to less than 10 ka (Civetta et al. 1984, 1998; Mahood and Hildreth 1986). Several episodes of silicic pyroclastic eruptions have been recognised (e.g. Rotolo et al. 2013). The largest explosive events occurred at about 114 and 50 ka and generated caldera collapses in the southeastern sector of the island (Orsi and Sheridan 1984; Mahood and Hildreth 1986). The younger collapse is associated with the deposition of the so-called Green Tuff, a complex pyroclastic trachytic to pantelleritic

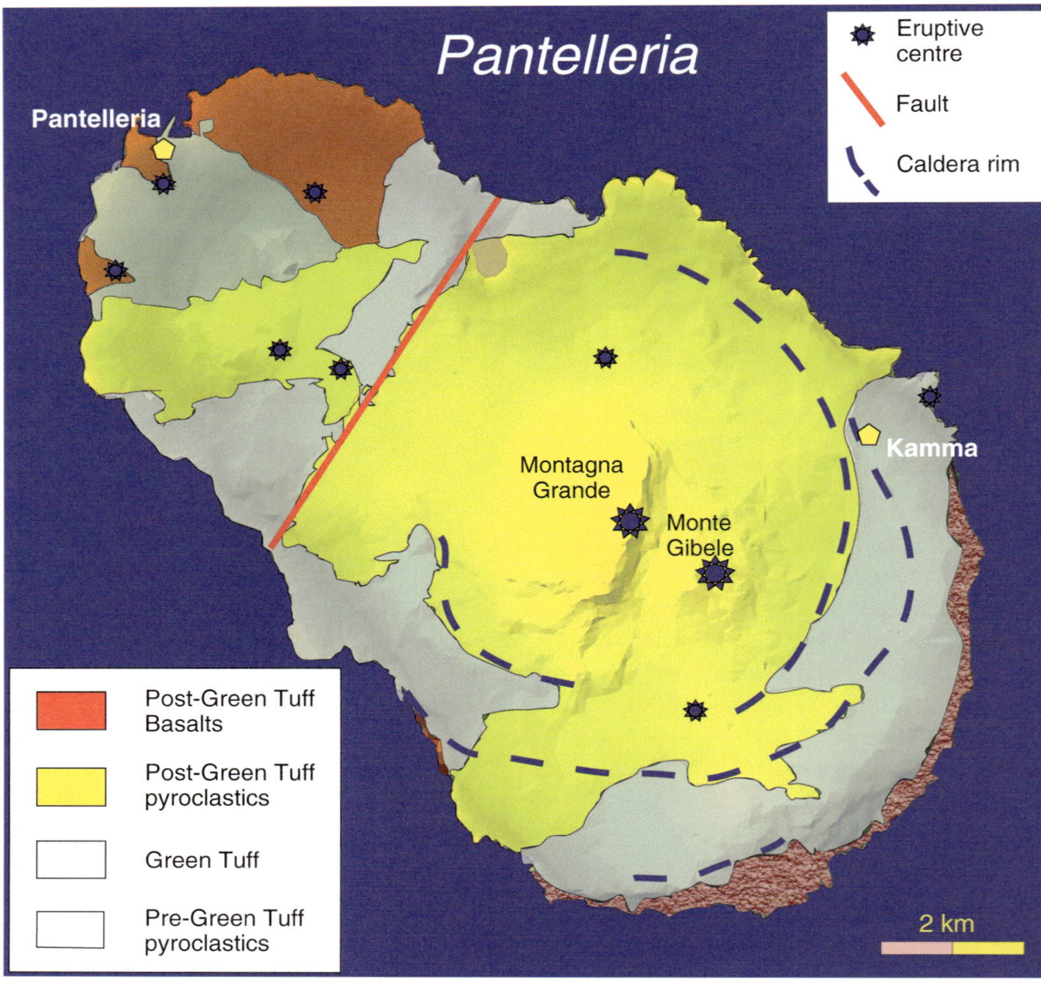

Fig. 10.13 Schematic geological map of the Island of Pantelleria. Simplified after Catalano et al. (2014) and references therein

deposit, formed of ignimbrite, fall and surge beds (Orsi and Sheridan 1984). Resurgence has taken place inside the caldera. Post-caldera silicic activity has occurred both inside and outside the caldera. Basaltic rocks crop out in the north-western sector of the island, and at some offshore submerged centres, such as Foerstner. At present, fumarolic and diffuse CO_2 emission and thermal water emanation are the only visible volcanic phenomena at Pantelleria (e.g. Granieri et al. 2014 with references).

10.5.2 Petrography and Mineral Chemistry

The mafic volcanic rocks from Pantelleria are classified as basalt and hawaiite, whereas the silicic products fall in the trachyte and rhyolite fields (Fig. 10.14a). Mafic rocks show transitional to weakly alkaline petrochemical affinity (Civetta et al. 1984; Mahood and Baker 1986). Transitional basalts are hyperstene-normative, whereas alkaline basalts contain small amounts of normative nepheline. The felsic volcanic rocks are generally peralkaline, and are classified as comendite, comenditic trachyte, pantelleritic trachyte and pantellerite, according to the Al_2O_3 versus FeO_{total} grid reported in Fig. 10.14b (Macdonald 1974; Le Maitre 2002).

Basalts and hawaiites are variably porphyritic, with 5–20 vol% of plagioclase, olivine, clinopyroxene and magnetite phenocrysts, set in a microcrystalline groundmass consisting of the same phases plus accessory apatite and sporadic analcite. Comendites are hypocrystalline porphyritic, with variable modal amounts of alkali feldspar, clinopyroxene, fayalitic olivine, Fe–Ti-oxides and rare aenigmatite phenocrysts enclosed in a glassy groundmass containing microlites of alkali feldspar, clinopyroxene, alkali-amphibole, and Fe–Ti oxides. Pantellerites are porphyritic, with phenocrysts of alkali feldspar, clinopyroxene, aenigmatite, olivine, ilmenite, and some corroded quartz set in a glassy, sometimes vesicular, groundmass containing scarce microlites of the same phases as the phenocrysts.

Clinopyroxene in the mafic rocks is generally zoned and ranges from diopside to augite. A few sodic augite, hedenbergite, and aegirinaugite are found in silicic rocks (e.g. White et al. 2009; Neave et al. 2012). Olivine shows a wide compositional variation both in the mafic ($\sim Fo_{86-45}$) and in the silicic rocks ($\sim Fo_{20-5}$). Plagioclase (about An_{80-40}) occurs in the mafic rocks both as a phenocryst and in the groundmass. Feldspar in the silicic rocks is mostly anorthoclase with Or_{30-40} (Mahood and Baker 1986; Civetta et al. 1998; Avanzinelli et al. 2004; White et al. 2009; Lanzo et al. 2013).

10.5.3 Petrology and Geochemistry

The Pantelleria rocks show a bimodal distribution into basalts and trachyte-rhyolites, with a "Daly gap" at intermediate compositions (Figs. 10.14 and 10.15). Such a feature is also observed for melt inclusions contained in the phenocrysts and for interstitial glasses (Neave et al. 2012). The few intermediate rocks mostly found as enclaves in silicic volcanics and as scanty low-volume lavas are considered hybrids between mafic and felsic magmas (e.g. Mahood and Baker 1986; Ferla and Meli 2006; Romengo et al. 2012).

There is an overall decrease in TiO_2, MgO, FeO_{total}, CaO, P_2O_5 and ferromagnesian trace elements with increasing silica, whereas K_2O and incompatible elements show the opposite trend. Al_2O_3 and Na_2O increase with silica in the mafic compositions and decrease in the silicic rocks, but with considerable scatter. Diagrams of

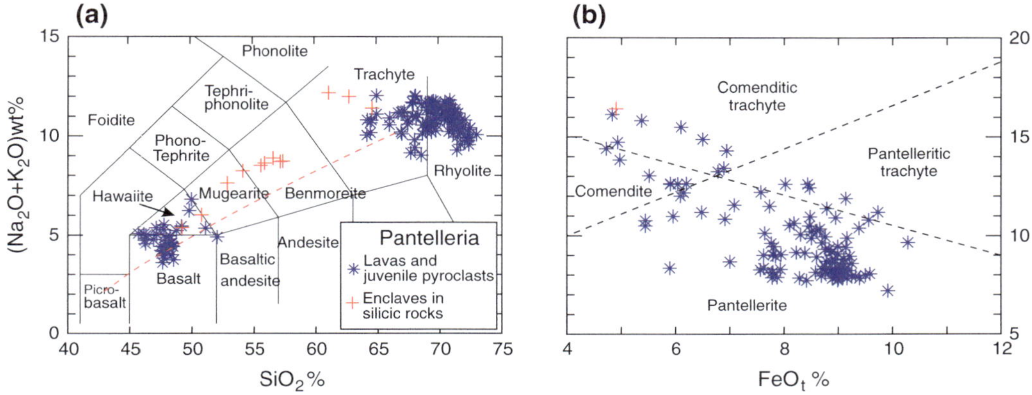

Fig. 10.14 **a** TAS classification diagram of Pantelleria volcanic rocks. The *dashed line* is the boundary between subalkaline and alkaline fields of Irvine and Baragar (1971); **b** Classification of Pantelleria peralkaline acid rocks ($SiO_2 > 63$ wt%; Macdonald 1974)

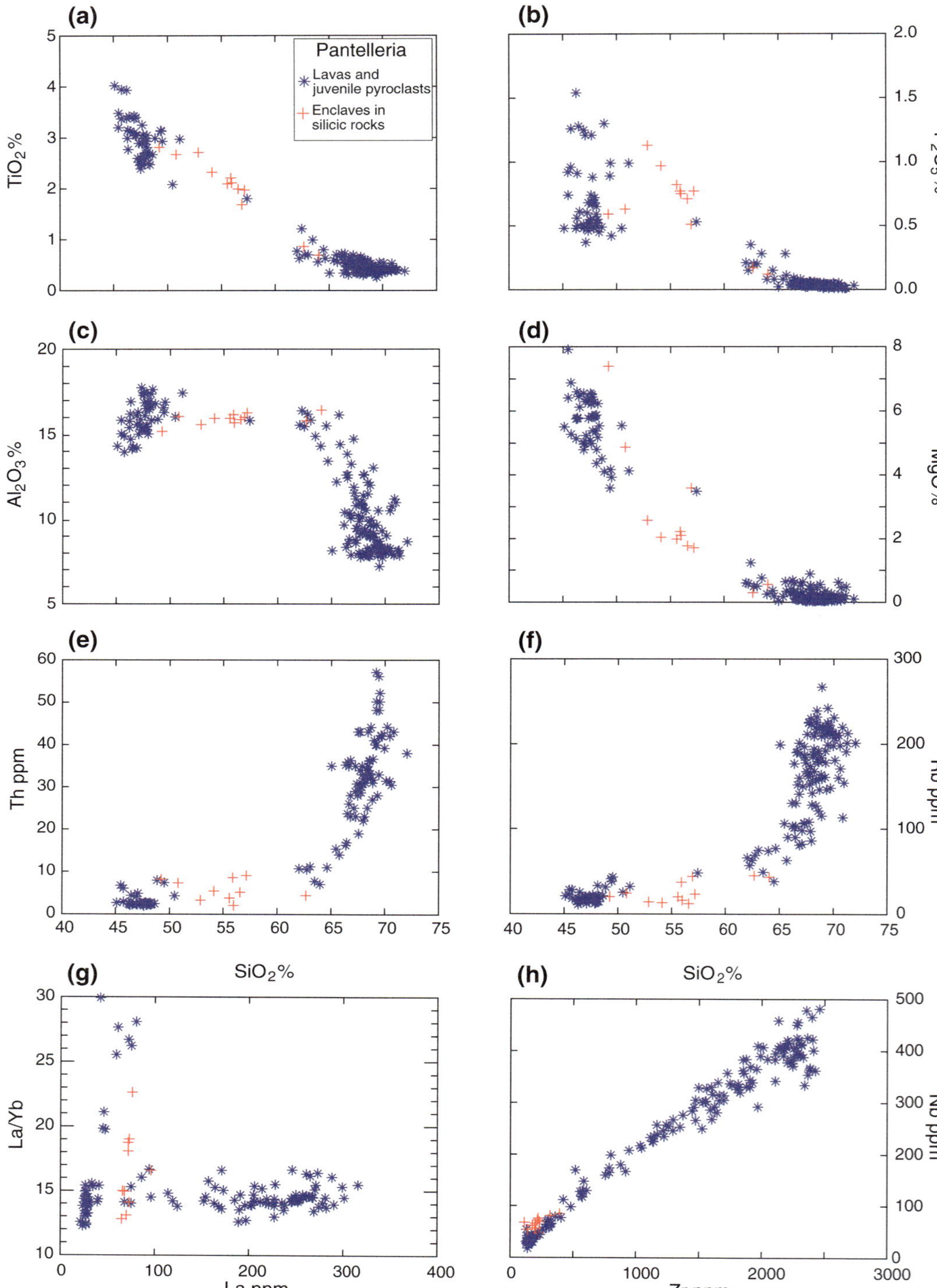

Fig. 10.15 Variation diagrams of selected major and trace elements for Pantelleria volcanic rocks

incompatible element pairs define very smooth positive trends (Fig. 10.15h). Enclaves occurring in the silicic rocks (indicated by crosses in the variation diagrams) define linear trends between mafic and silicic compositions, in most diagrams.

REE patterns for basalts show moderate fractionation for both light and heavy REE; some samples have a slightly positive Eu anomaly. Silicic rocks are more enriched in REE than basalts and display negative Eu anomalies (Fig. 10.16a). Extended incompatible element mantle-normalised plots of mafic rocks show an upward convex pattern with a maximum at Ta and Nb, a positive spike of Ba, and small negative anomalies of Sr and Zr (Fig. 10.16b). Some samples have marked Pb and U positive spikes, but all these data come from a single paper, and it is suspected they might represent analytical artefacts. Therefore, they have not been reported in Fig. 10.16b.

$^{87}Sr/^{86}Sr$ ratios mostly cluster around 0.7030–0.7032 in the mafic rocks, whereas in the silicic rocks they show comparable to higher values (Fig. 10.8a; Civetta et al. 1984, 1998; Esperança and Crisci 1995; Avanzinelli et al. 2014). The highest values of silicic rocks likely represent very scarce contamination by foreign material, which, however, had strong effects on Sr isotope ratios for rocks with very low Sr absolute abundances. Nd isotopic ratios show partially overlapping range of values in the mafic and felsic rocks ($^{143}Nd/^{144}Nd$ = 0.51287–0.51299). Pb isotopic ratios are very variable (Fig. 10.8b, c) encompassing the entire compositional range of Sicily volcanoes ($^{206}Pb/^{204}Pb$ ~ 18.45 to 20.45; $^{207}Pb/^{204}Pb$ ~ 15.55 to 15.71; $^{208}Pb/^{204}Pb$ ~ 38.25 to 40.10; Esperança and Crisci 1995; Civetta et al. 1998; Avanzinelli et al. 2014).

10.5.4 Evolution of Pantelleria Magmas

The Island of Pantelleria shows many characteristics that are commonly found in volcanoes from continental rift settings. These include the abundance of acid peralkaline rocks, the compositional gap between mafic and silicic volcanics, and the smooth inter-element trends. Therefore, the petrogenesis of Pantelleria magmas shares the same problems with peralkaline silicic rocks associated with transitional basalts in continental rift settings. In general terms, two main classes of hypotheses have been proposed for these magmas. One advocates fractional crystallisation starting from transitional basaltic parents, possibly with a small role for crustal contamination. The other suggests an origin of peralkaline silicic magmas by melting of underplated basaltic rocks. Basalt melting generated

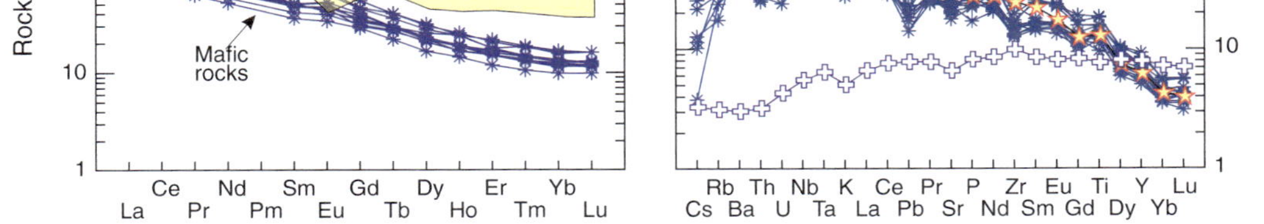

Fig. 10.16 **a** Chondrite normalised REE patterns of Pantelleria rocks; **b** Mantle normalised incompatible element patterns of mafic rocks. Average OIB and Normal-type MORB are shown for comparison (Sun and McDonough 1989; Gale et al. 2013)

trachytic magmas, which underwent fractional crystallisation plus gaseous transfer giving F, Cl, Na-rich pantelleritic compositions (e.g. Bailey and Macdonald 1975; Bohrson and Reid 1997; Peccerillo et al. 2003).

Thermodynamic, trace element and isotopic modelling for Panteleria magmas demonstrates that the silicic rocks can be derived from a weakly alkaline or transitional basaltic parent by 80–95 % fractional crystallisation at low pressure (e.g. Civetta et al. 1998; Ferla and Meli 2006; White et al. 2009). Separation of a solid made up of plagioclase, clinopyroxene and subordinate olivine, apatite, and Fe–Ti oxides drives mafic melt compositions to comendite-trachyte. Additional fractionation dominated by alkali feldspar plus minor amounts of other phases (clinopyroxene, ilmenite, aenigmatite) generates comenditic trachytes to pantelleritic rhyolites, at pressures around 50–150 MPa, and highly hydrated conditions (Di Carlo et al. 2010; Neave et al. 2012; Lanzo et al. 2013). Overall, the most evolved silicic magmas would result from some 95 % fractional crystallisation of transitional basalt. The "Daly gap" at Pantelleria would be the result of rapid differentiation in the intermediate compositions in a shallow-level density stratified magma chamber (e.g. White et al. 2009; Neave et al. 2012).

Experimental evidence shows that partial melting of basalt followed by fractional crystallisation can also give silicic peralkaline melts (e.g. Mahood and Baker 1986; Lowenstern and Mahood 1991). The occurrence of trachytic interstitial glasses derived by melting of hydrothermally-altered gabbroic xenoliths at Iblei, supports this possibility (Viccaro et al. 2009). However, smooth incompatible element variation trends along the entire erupted sequence observed at Pantelleria (as well as in many intra-continental volcanoes, such as in the Ethiopian Rift Valley; Peccerillo et al. 2003), support an origin of silicic magmas from mafic parents by fractional crystallisation rather than by gabbro melting plus fractional crystallisation.

As for the origin of mafic magmas, moderate fractional crystallization of primary mantle-derived melts is suggested to give the Pantelleria hawaiites (Neave et al. 2012). Primary melts would be formed by about 2 % melting of a primitive mantle source at depths of 60–100 km, mainly within the spinel-garnet transition zone. However, the Pantelleria mafic rocks show some strong compositional variations in the basalts (e.g. TiO_2 and P_2O_5 and Pb isotope ratios), which call for a more complex origin of primary melts (Civetta et al. 1998).

In summary, the origin and evolution of the Pantelleria magmas still show some puzzling aspects. For instance, mafic magmas have very variable REE fractionation, with La/Yb ranging from about 10–30 (Fig. 10.15g). Yet, these rocks show similar MgO, P_2O_5 and TiO_2, and radiogenic isotope signature, and come from both older and younger basaltic centres. Moreover, if the entire rock sequence is considered, various parallel or intersecting trends are observed in some variation diagrams (e.g. MgO vs. La/Yb, Cr vs. La). Some anomalous element enrichment is also observed in enclaves, which clearly derived from mixing processes (Ferla and Meli 2006). It is possible that some of these geochemical oddities may depend on analytical bias, as suggested by data of some basalts that show rather high MgO, Co, V, along with very low Cr and Ni, and high LREE and La/Yb . However, various trends remain when these data are expunged, suggesting they are real and may depend on mafic-felsic magma mixing. This process is briefly discussed in Chap. 9 for Lipari magmas, and is able to generate apparently erratic geochemical trends when elements with different diffusion coefficients are plotted (e.g. Morgavi et al. 2013).

10.6 Linosa

10.6.1 Volcanology and Stratigraphy

The Island of Linosa is the emergent summit of a large submarine cone located at the southwestern margin of the Linosa graben (Fig. 10.1). The exposed volcanism has been both explosive and effusive and developed during three main phases of activity from about 1.06–0.53 Ma (Di Paola 1973; Rossi et al. 1996). The lowest exposed deposits consist of scoria cones and tuff rings formed of basaltic to hawaiitic juvenile clasts and a few benmoreitic to trachytic lithic ejecta. The second phase gave hydromagmatic products, strombolian scoriae and some lavas with basalt to hawaiite compositions. Finally, the third phase erupted abundant basalt to hawaiite lavas, scoriae, and minor hydromagmatic products, which built up the main cones in the island.

10.6.2 Petrography and Mineral Chemistry

Linosa rocks range from mildly alkaline basalts to hawaiite, with a few benmoreite and trachyte lithic clasts (Fig. 10.17). Basalts show a phenocryst assemblage dominated by olivine ($\sim Fo_{85-80}$), variable amounts of plagioclase ($\sim An_{70-40}$), and minor diopside to augite clinopyroxene. Groundmass consists of the same phases plus opaque oxides and rare apatite and nepheline. Hawaiites contain phenocrysts of olivine ($\sim Fo_{80-70}$), diopside to augite clinopyroxene, plagioclase ($\sim An_{75-60}$), and sometimes amphibole, surrounded by a microcrystalline to hypocrystalline groundmass formed of the same minerals and opaque minerals. The benmoreite and trachyte lithics are porphyritic, with abundant plagioclase (An_{35-25}) and anorthoclase phenocrysts, minor clinopyroxene and kaersutitic amphibole, and some biotite and olivine set in a microcrystalline groundmass of plagioclase, clinopyroxene, and opaques. Accessory phases include apatite, magnetite, and perovskite (Rossi et al. 1996; Bindi et al. 2002; Di Bella 2007; Di Bella et al. 2008).

10.6.3 Petrology and Geochemistry

Major and trace elements (Fig. 10.18) show scattering in the mafic rocks and an overall increase in Al_2O_3, Na_2O, K_2O and incompatible trace elements (Rb, Th, Nb, Zr, REE) and a decrease in TiO_2, MgO, Cr and Ni from basalt to trachyte (Rossi et al. 1996; Bindi et al. 2002; Di Bella et al. 2008).

REE patterns are moderately fractionated, and some samples have a small positive Eu anomaly (Fig. 10.19a). Mantle normalised incompatible element diagrams show a bell-shaped pattern with a maximum at Nb–Ta and a slight negative anomaly of potassium (Fig. 10.19b).

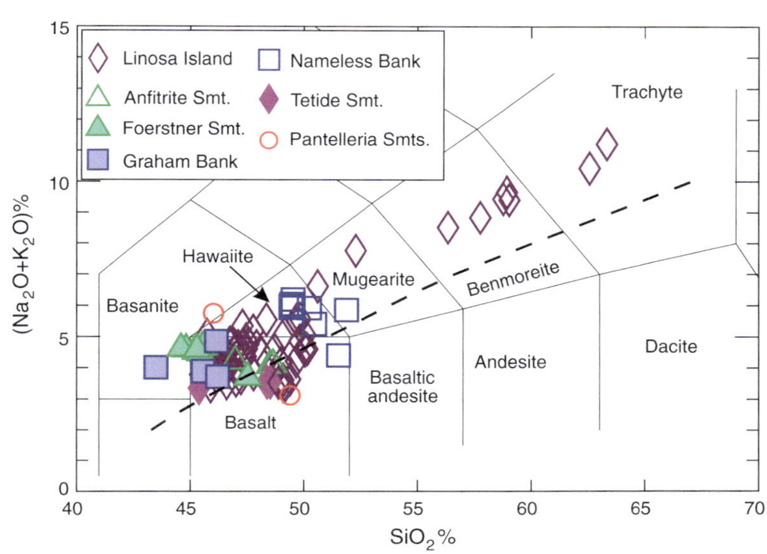

Fig. 10.17 TAS classification diagram for the Island of Linosa and the Sicily Channel seamounts. The *dashed line* is the boundary between subalkaline and alkaline fields of Irvine and Baragar (1971). Data have been recalculated to 100 % on a water-free basis

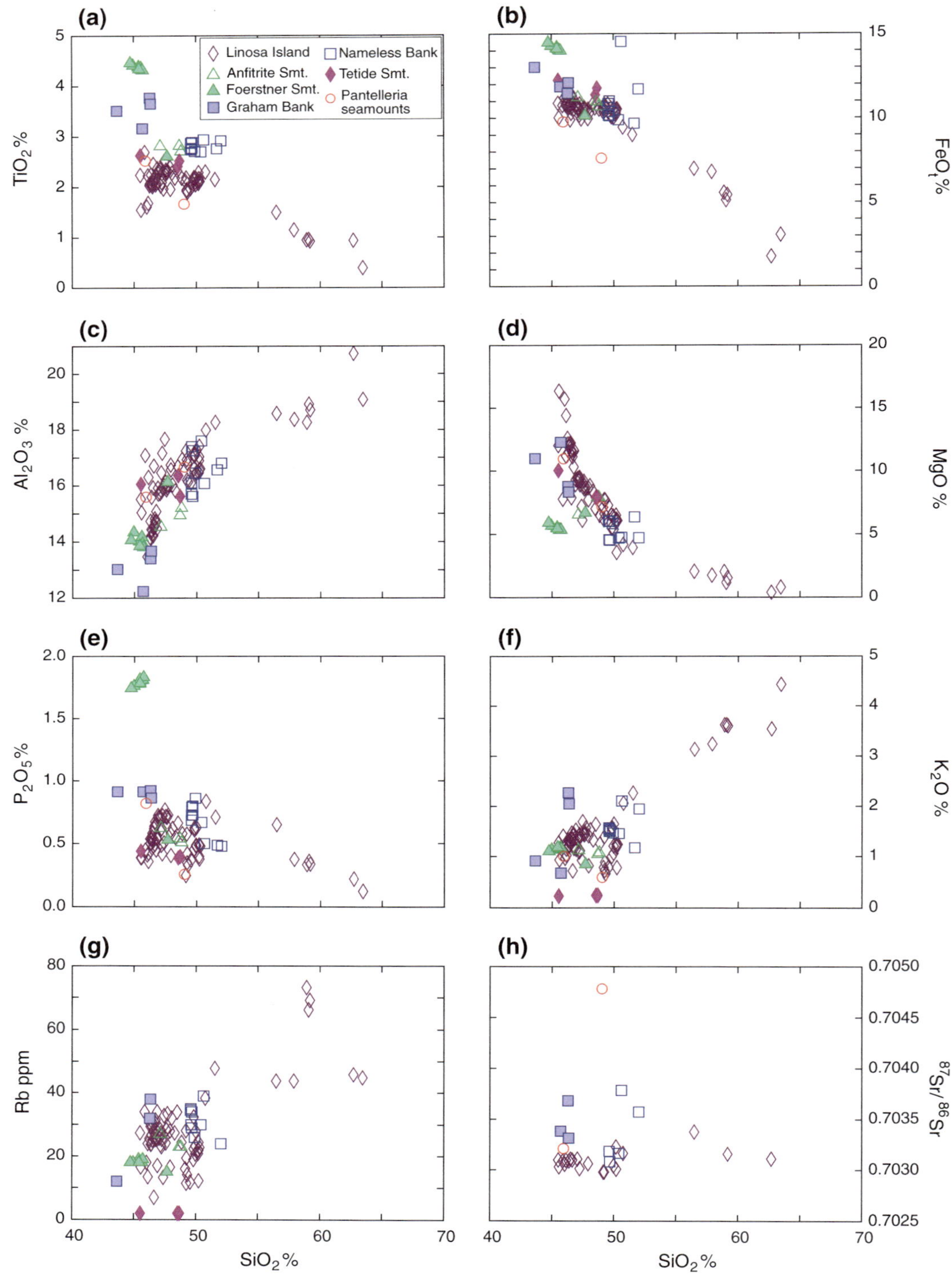

Fig. 10.18 Variation diagrams of selected major and trace elements, and Sr isotope ratios for Linosa Island and Sicily Channel seamounts. Major elements have been recalculated to 100 % on a water-free basis

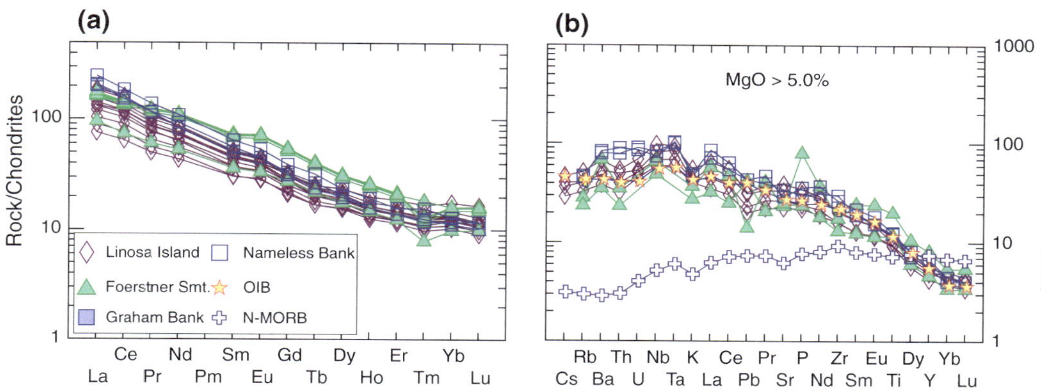

Fig. 10.19 **a** REE patterns of Linosa and Sicily Channel seamounts; **b** Mantle normalised incompatible element patterns of mafic rocks. Average N-MORB and OIB are also shown

Radiogenic isotope data (Figs. 10.8, 10.18 h) show small variations ($^{87}Sr/^{86}Sr$ = 0.70298–0.70338; $^{143}Nd/^{144}Nd$ = 0.51291–0.51298; $^{206}Pb/^{204}Pb$ ~ 19.28 to 19.54; $^{207}Pb/^{204}Pb$ ~ 15.60 to 15.64; $^{208}Pb/^{204}Pb$ ~ 38.89 to 39.11; Civetta et al. 1998; Di Bella et al. 2008; Avanzinelli et al. 2014).

10.6.4 Evolution of Linosa Magmas

Various lines of evidence point to polybaric evolution history for Linosa magmas. Bindi et al. (2002) investigated crystal chemistry of clinopyroxenes and found a decrease of cell volume, passing from early to late erupted basalts. This was interpreted as an evidence for increasing pressure of crystallisation with time. Therefore, it was suggested that early activity was fed by a shallow magma chamber where dominant fractional crystallisation of mafic magmas gave benmoreitic and trachytic melts. In contrast, the younger activity was fed by a deeper reservoir where fractional crystallisation was accompanied by continuous input of mafic melts from the source. This kept compositions in the mafic range, preventing evolution toward benmoreite-trachyte. Di Bella et al. (2008) found different variation trends of incompatible elements versus MgO and ferromagnesian trace elements, possibly reflecting the occurrence of slightly different parental magmas. There is also a slight but significant increase of Sr- and a decrease of Nd-isotope ratios from mafic to some of the evolved volcanics, which may reveal tiny crustal assimilation (Di Bella et al. 2008).

The Linosa mafic rocks have trace element and radiogenic isotope compositions that are not far from those of Pantelleria. Both plot along a trend between FOZO and DMM (Fig. 10.8), suggesting an origin in a mantle source that resulted from interaction between depleted and enriched components. Lower Pb isotope ratios at Linosa may suggest a more important role for the depleted component in this island.

10.7 Sicily Channel Seamounts

Numerous volcanic centres occur on the seafloor of the Sicily Channel (Fig. 10.1). Calanchi et al. (Calanchi et al. 1989) recognised at least ten recent submarine volcanoes of various dimensions, mostly located along NW-SE regional faults. Some of these seamounts erupted during historical times; others are covered by undisturbed Pliocene-Quaternary sediments and have been detected by seismic profiles and magnetic survey (e.g. Allan and Morelli 1971; Chiappini et al. 2000; Lodolo et al. 2012).

Tetide, *Anfitrite*, *Galatea,* and *Cimotoe* define a row of centres aligned in a NW-SE direction on the margin of the Banco Avventura (Adventure Plateau), a large erosional sedimentary plateau located off the south-western Sicily coast. The Tetide volcanic cone has a 3 km across base and rises from a seafloor about 70 m deep, reaching a depth of about 18 m below sea level. Anfitrite is about 1 km across and reaches a minimum depth of about 40 m. Galatea has a base diameter of 2 km, and the summit at about 74 m below sea level. Cimotoe consists of a series of conical peaks sited on the southeastern margin of Banco Avventura.

The *Graham Bank* is located east of Banco Avventura, about 50 km south of the Sicily coast. It is sited on a plateau about 200 m deep, shows a N-S elongated shape and a summit area formed by a 25–30 m deep oval-shaped terrace. A fractured neck of massive lavas rises from the terrace to a depth of about 9 m below sea level. Such a monolith represents the feeder neck of the 1831 eruption (Washington 1909; Mazzarella 2012) that formed the ephemeral Isola Ferdinandea (Graham or Julia, according to British and French scholars) rapidly dismantled by marine erosion (Fig. 10.20). Another volcanic cone located a few km to the south rises to a depth of 76 m below sea level. The alignment of these volcanoes highlights magma emplacement along a NNE-SSW trending fault system (Calò and Parisi 2014). Low-P velocity anomaly detected beneath this area (Calò et al. 2013) and high He-isotope signatures of gas emissions (Caracausi et al. 2005) clearly speak for active ascent of mantle material.

The *Banco Senza Nome* (*Nameless Bank*) is a pinnacle rising to about 80 m below sea level from a high probably made of sedimentary rocks. K/Ar isochrone dating on dredged samples gave an age of 9.5 Ma (Beccaluva et al. 1981).

The *Bannock* Seamount is located between the Island of Malta and the Pantelleria graben. It rises from a depth of about 1 km reaching 230 m below sea level. In spite of its conical shape, there are doubts about the volcanic nature of this seamount, since no magnetic anomalies have been detected (Lodolo et al. 2012).

Three seamounts (*Linosa 1, Linosa 2 and Linosa 3*) have been found north of Linosa island. Little is known about these cones and the volcanic nature of the northernmost centre (Linosa 3) is uncertain.

Foerstner volcano is sited, 5 km off the NW coast of Pantelleria. It erupted in October 1891 emitting basaltic scoriaceous bombs floating on the sea surface (lava balloons) (Foerstner 1891; Riccò 1892; Washington 1909; Conte et al. 2014; Kelly et al. 2014). The Foerstner volcano is a 90 m high cone rising from a 350 metres deep sea floor. Several other cones occur in the same zone (Conte et al. 2014).

Pantelleria East and *Pantelleria Southeast* seamounts are two of the many small submarine volcanoes sited east and southeast of Pantelleria, in the middle and on the shoulders of the Pantelleria Rift. Collected samples include lavas and glassy material (Rotolo et al. 2006).

Compositions of volcanic rocks from the Sicily Channel seamounts range from tholeiitic basalt to alkali basalt, hawaiite and basanite (Calanchi et al. 1989; Rotolo et al. 2006 and references therein; Fig. 10.17). Tholeiites have been found at Tetide and Pantelleria Southeast volcanoes. Textures are sparsely porphyritic, with phenocrysts of orthopyroxene rimmed by pigeonite, and olivine microphenocrysts set in a groundmass containing Ca-rich clinopyroxene, pigeonite, plagioclase, secondary calcite, and Fe-hydroxides. Transitional basalts are found at Anfitrite. These are porphyritic with phenocrysts of olivine, clinopyroxene and plagioclase set in a groundmass containing the same phases, along with Fe–Ti oxides, alkali feldspar and secondary minerals. Hawaiites from the Graham Bank (Rotolo et al. 2006) are porphyritic with plagioclase ($\sim An_{65-45}$), olivine ($\sim Fo_{70}$) and high-Ti clinopyroxene phenocrysts. Chemically, they show enrichment in TiO_2 and depletion in Al_2O_3 (Fig. 10.18). Alkali basalt to basanite from Banco Senza Nome contain plagioclase and clinopyroxene phenocrysts in a groundmass composed of the same phases, some alkali feldpspar, nepheline, accessory apatite and secondary minerals. Bombs erupted by the Foerstner volcano in October 1891 have alkali basalt to basanite compositions. Textures are vesicular hypocrystalline porphyritic with phenocrysts of

The Mirror
OF
LITERATURE, AMUSEMENT, AND INSTRUCTION.

No. 512.] SATURDAY, OCTOBER 8, 1831. [PRICE 2

NEW VOLCANIC ISLAND,

WHICH HAS JUST APPEARED OFF SICILY.

A hand invisible was rearing,
A new creation in the secret deep.—MONTGOMERY.

Fig. 10.20 The birth of a new island offshore Sicily announced by the Mirror of Literature, Amusement, and Instruction in the issue of October 8th, 1831 plagioclase and augite and microphenocrysts of olivine plus some apatite set in a highly glassy matrix (Washington 1909; Conte et al. 2014; Kelly et al. 2014). Chemical compositions are characterised by high enrichments in TiO_2 and P_2O_5.

Rocks from the Sicily Channel seamounts exhibit REE and incompatible element patterns very similar to average OIB and to mafic volcanics from Linosa and Pantelleria (Fig. 10.19). Available radiogenic isotope data show larger variations for Sr-isotope ($^{87}Sr/^{86}Sr = 0.7031–0.7048$) than for Nd-isotope ratios ($^{143}Nd/^{144}Nd = 0.5130–0.5131$), possible reflecting interaction with seawater. This also affected Pb isotope ratios (Fig. 10.8).

10.8 Ustica

The Island of Ustica is located in the southern Tyrrhenian Sea about 60 km north of the Sicily coast (Fig. 10.1). The volcano rises more than 2000 m above the sea floor and reaches an altitude of 244 m above sea level. The nearby submarine mugearite lava field of Prometeo shows close compositional affinity with Ustica (Trua et al. 2003). The island is strongly affected by tectonism and erosion. Faults are mostly normal, sometimes with a sinistral horizontal component, and show a prevailing ENE-WSW trend. Ustica shows evidence of at least five relative sea-level stands, with 350–80 ka old marine terraces located at elevations from 5 to 120 m above sea level (De Vita et al. 1998).

10.8.1 Volcanology and Stratigraphy

The exposed volcanism developed between 750 and 130 ka (De Vita et al. 1998). Activity occurred below sea level until about 500 ka, giving pillow lavas and pillow breccias that crop out along the southern border of the island. Successively, both effusive and explosive activity took place, the latter being related both to magmatic eruptions and to water-magma interaction. The best preserved hydrovolcanic centre is the Capo Falconiera tuff cone, sited on the eastern end of the island where steep erosion cliffs expose a section of the volcano. Rock compositions are predominantly mafic, with a single trachytic eruption and dome emplacement at about 424 ka (Romano and Sturiale 1971; Cinque et al. 1988; De Vita et al. 1998).

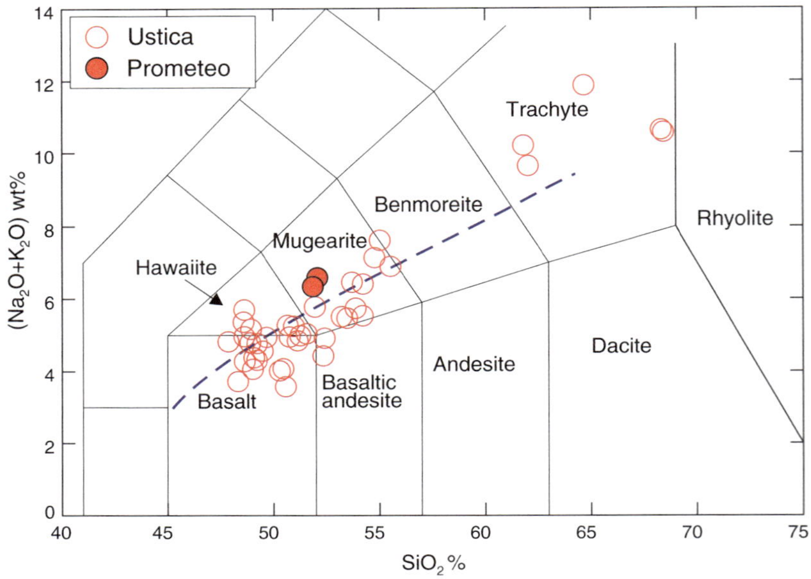

Fig. 10.21 TAS classification diagram of Ustica Island and Prometeo submarine lava field. The *dashed line* is the boundary between subalkaline and alkaline fields of Irvine and Baragar (1971). Compositions have been normalised to 100% on a LOI-free basis

10.8.2 Petrography and Mineral Chemistry

The Ustica rocks range from mildly Na-alkaline basalt to trachyte on the TAS classification diagram, with a small gap at intermediate compositions between about 55–60 wt% SiO_2 (Fig. 10.21). The mafic rocks straddle the boundary between the subalkaline and the alkaline fields, displaying a transitional to mildly alkaline affinity.

Textures of mafic rocks are variably porphyritic with phenocrysts of plagioclase, olivine and sporadic clinopyroxene. Fe–Ti oxides are generally found as microphenocrysts in the most differentiated rocks. Olivine ($\sim Fo_{85-75}$) is the most abundant mafic phase; clinopyroxene ranges from diopside to ferroan diopside- hedenbergite. Trachytic pumices are poorly porphyritic with phenocrysts of alkali feldspar, plagioclase, scarce biotite, and Fe–Ti-oxides set in a vesicular glassy matrix. Gabbroic, microsyenitic and ultramafic

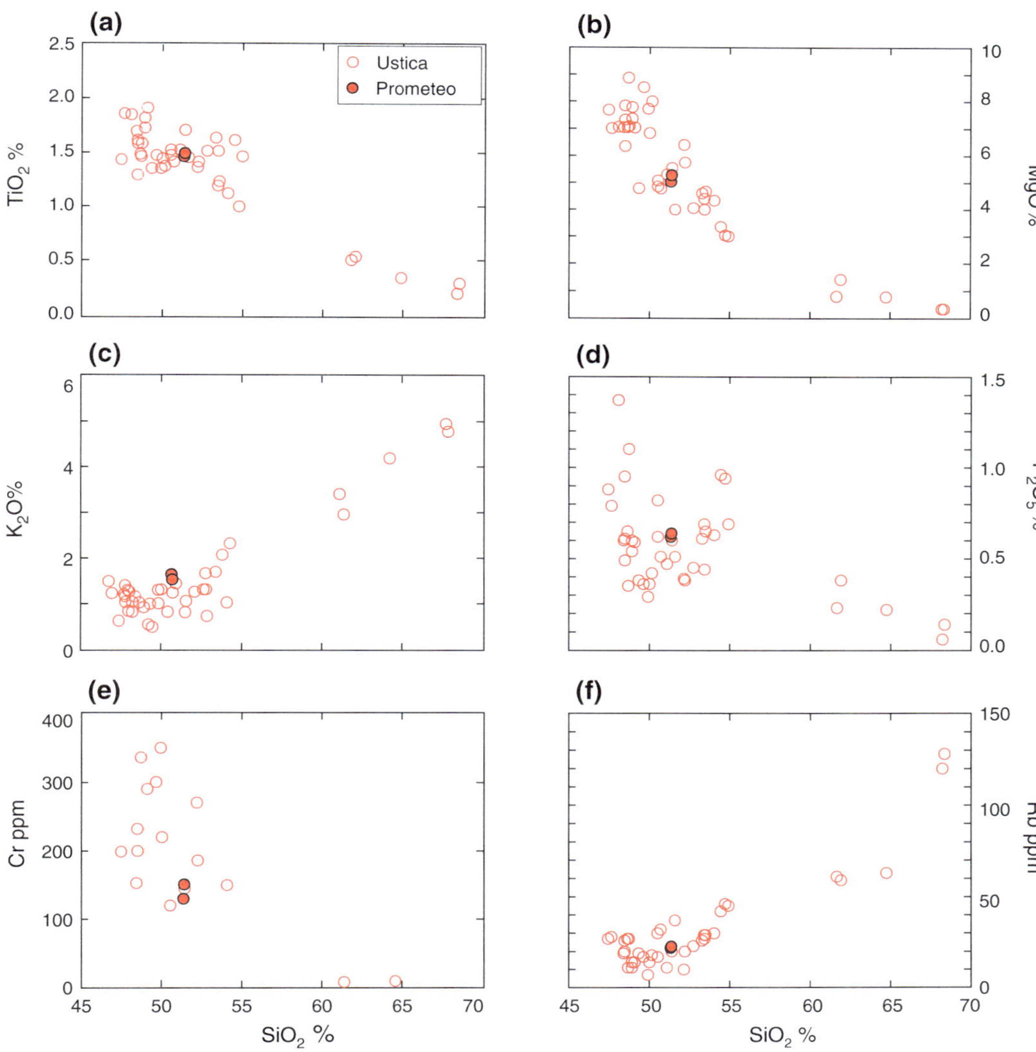

Fig. 10.22 Variation diagrams of selected major and trace elements against SiO_2 for volcanic rocks from Ustica Island and Prometeo lava field

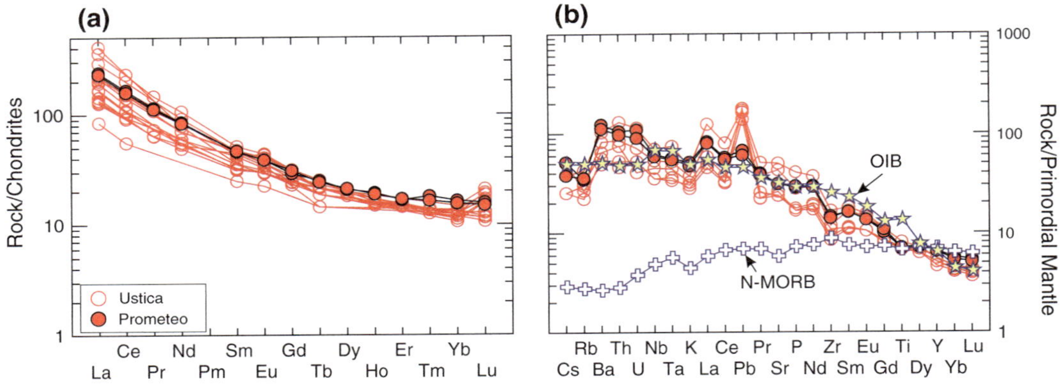

Fig. 10.23 **a** Chondrite normalised REE patterns of Ustica and Prometeo rocks; **b** Incompatible element patterns of mafic rocks (MgO > 4.0 %). Average N-MORB and OIB compositions are also reported

xenoliths found in some lavas represent cumulate rocks (Alletti et al. 2005).

10.8.3 Petrology and Geochemistry

Harker diagrams of Ustica rocks show a decrease in TiO_2, FeO_{total}, MgO, CaO, Cr, Ni and Sc, and an increase in K_2O and incompatible elements with increasing silica. P_2O_5 is scattered in the mafic rocks (Fig. 10.22). REE patterns exhibit variable degrees of fractionation, sometimes with a small positive anomaly of Eu (Fig. 10.23a). Mantle-normalised incompatible element plots of mafic rocks show an upward convex shape and negative anomalies of K, Zr, Hf and Ti (Fig. 10.23b).

The few available data on Sr-Nd–Pb isotope compositions (Fig. 10.8) show limited variations ($^{87}Sr/^{86}Sr$ = 0.70306 to 0.70343; $^{143}Nd/^{144}Nd$ = 0.51290 to 0.51300; $^{206}Pb/^{204}Pb$ = 18.84 to 19.56; $^{207}Pb/^{204}Pb$ = 15.62 to 15.66; $^{208}Pb/^{204}Pb$ = 38.61 to 39.18; Cinque et al. 1988; Trua et al. 2003; Author's unpublished data). Similar to more radiogenic Pb isotope compositions ($^{206}Pb/^{204}Pb$ = 19.03 and 19.96; $^{208}Pb/^{204}Pb$ = 38.64 and 40.53) have been found for two olivine-hosted melt inclusions by Rose-Koga et al. (2012). $^{3}He/^{4}He$ ratios have $R/R_A \sim 6.6$ (Martelli et al. 2008). Sr isotope ratios show a positive correlation with P_2O_5 and LILE/HFSE ratios in the mafic rocks (see Cinque et al. 1988), but do not define any correlation with silica or other common evolutionary geochemical parameters.

10.8.4 Evolution of Ustica Magmas

Based on major and trace element variations, it has been argued that fractional crystallisation was the dominant evolutionary mechanism for Ustica magmas (Cinque et al. 1988). Geobarometric studies of minerals from gabbroic xenoliths suggest a polybaric fractionation at pressure ranging from 1.0 GPa to 200 MPa (Alletti et al. 2005). Sr isotopic ratios suggest that assimilation of crustal rocks played a minor role during magma evolution. The variable contents of P_2O_5, incompatible elements and radiogenic Sr in the mafic rocks most likely reflect variations of primary melts.

In general terms, the mafic rocks from Ustica have similar compositions as the equivalent rocks from Etna A close match has been observed also for melt inclusions in olivine phenocrysts (Schiano et al. 2004; Rose-Koga et al. 2012). Therefore, a similar origin for the two alkaline volcanoes can be inferred.

10.9 Petrogenesis of the Sicily Province

The mafic rocks of the Sicily Province range from silica oversaturated quartz-tholeiitic to variably undersaturated Na-alkaline basalts and

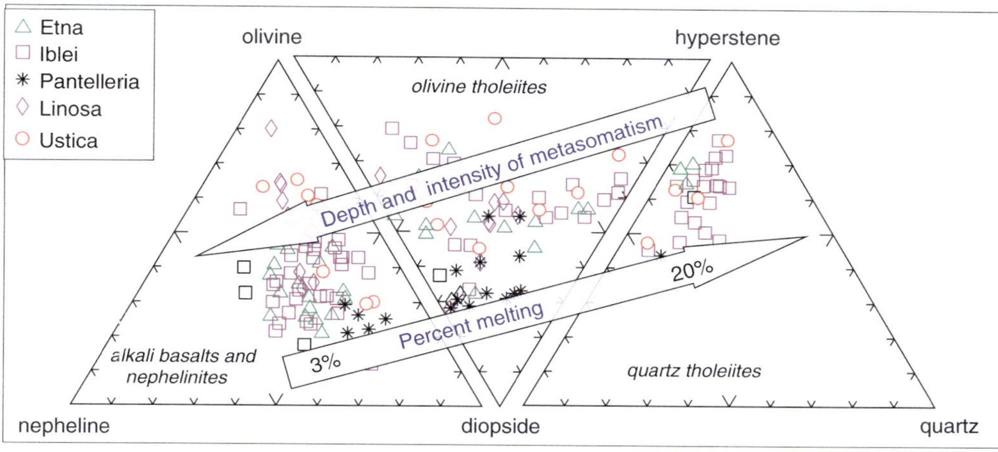

Fig. 10.24 Normative compositions of basaltic rocks (SiO_2 < 52 wt%; MgO > 4 wt%) from the Sicily Province. The *arrows* indicate variation of the depth of magma generation, degrees of source metasomatism, and degree of partial melting for nephelinitic to tholeiitic magmas

nephelinite (Fig. 10.24). There is a general agreement that the variable petrochemical affinity, degree of silica saturation and incompatible element enrichment of mafic rocks in Sicily are the integrated effects of several factors, including the degree of partial melting, the depth of magma generation and intensity of source enrichment (Fig. 10.24). Variable degrees of partial melting and depth of magma generation are supported by variation of La/Sm_N versus Tb/Yb_N of mafic rocks, which show a rough positive correlation at a regional scale (Fig. 10.25a). La/Sm essentially depends on degree of melting whereas high Tb/Yb is related to residual garnet, a phase which is stable at high mantle pressures. Geochemically heterogeneous sources are testified by the range of isotopic signatures found in the mafic rocks, both at local and regional scales (e.g. Tanguy 1978; Tanguy et al. 1997; Beccaluva et al. 1998; Trua et al. 1998).

Pb-isotope ratios of the Sicily Province plot along the Northern Hemisphere Reference Line (Hart 1984), between FOZO (Focal Zone) or HIMU (high-μ, i.e. $^{238}U/^{204}Pb$) and DMM (Depleted MORB Mantle) mantle end-members (Fig. 10.8b and c). There is a strong debate on

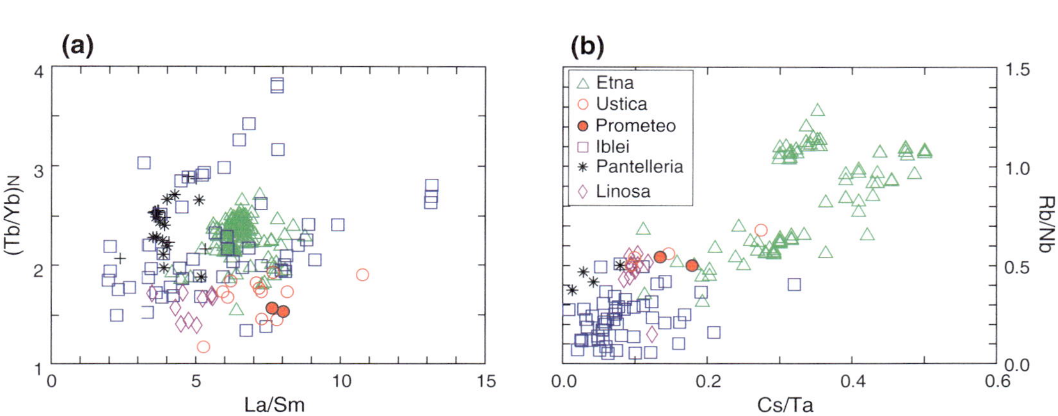

Fig. 10.25 **a** Tb/Yb normalised to chondritic values versus La/Sm variation of Sicily mafic rocks (MgO > 5.0 wt%); **b** Rb/Nb versus Cs/Ta variation of Sicily mafic rocks

the geodynamic significance of these compositions, both in the Sicily Province and at the global scale (see Foulger et al. 2005), as it will be briefly discussed later.

Major and trace element geochemistry of melt inclusions in olivine phenocrysts indicate that the primary alkaline and tholeiitic magmas at Etna are more mafic than the outcropping rocks. Melt inclusion studies also reveal variable compositions within single samples, indicating the coexistence of various types of primitive melts within the volcano feeding system at some stages. This leads to the conclusion that magmas erupted at the surface may actually result from mixing and homogenisation at shallow level (e.g. Kamenetsky and Clocchiatti 1996). Similar conclusion has been reached by Bryce and De Paolo (2004) on the basis of Pb isotope disequilibrium between phenocrysts and matrix.

Geochemical modelling has shown that magmas parental to the alkaline suite were generated by less than 5 % mantle melting, whereas primary tholeiitic magmas were formed by much more extensive melting of about 15–20 % (e.g. Corsaro and Cristofolini 1996; Tanguy et al. 1997). Thermodynamic modelling by Armienti et al. (2004) indicates that Etna parental magmas (both tholeiitic and alkaline) equilibrated at a pressure of 1.5–1.8 GPa. This is a minimum depth of partial melting, and other authors (Corsaro and Cristofolini 1996; Corsaro and Pompilio 2004) suggest an origin of alkaline magmas at somewhat higher pressure (1.5–3.0 GPa), within the stability field of garnet lherzolite. A role of residual garnet is supported by ratios of normalised Tb/Yb of mafic rocks, which are higher than one and increase from subalkaline to alkaline rocks (Fig. 10.25a).

Geochemical and isotopic investigations have shown significant time-related variation of magma composition at Etna, with an overall increase in radiogenic Sr, volatiles, and fluid-mobile elements (e.g. Rb, Cs, B, and K) and a decrease of Nd isotopic ratios from early tholeiites and alkaline rocks to recent activity, especially the post-1971 AD magmas. Secular compositional variations are generally attributed to mantle sources. Viccaro and Cristofolini (2008), Ferlito and Lanzafame 2010) and Viccaro et al. (2011) recall that enrichment in K and Rb are also observed in some of the older volcanic rocks, reaching the conclusion that these features reflect heterogeneous mantle sources containing variable amounts of hydrous phases such as phlogopite and amphibole. Mantle heterogeneity is attributed to ancient metasomatic events. In contrast, other authors suggest that enrichment in volatile-mobile elements could be related to input from the nearby Ionian subduction zone, which would also explains the arc signatures and chlorine isotope compositions observed both in the rocks and melt inclusions (e.g. Doglioni et al. 2001; Schiano et al. 2001; Tonarini et al. 2001; Rizzo et al. 2013). Such a hypothesis is supported by the overall higher LILE/HFSE ratios (e.g. Rb/Nb, Cs/Ta, K/Zr) of Etna and Ustica alkaline rocks relative to other Sicily volcanoes, which are away from the Ionian subduction zone and backarc areas (Fig. 10.25b). However, a role for wall rock assimilation in determining these variations cannot be excluded and is actually strongly supported by some melt inclusion studies recalled earlier (e.g. Ferlito and Lanzafame 2010).

The Island of Ustica and the nearby Prometeo lava field have been less extensively studied than Etna. However, the available data demonstrate close compositional affinities among these volcanoes. Such a similarity has been also observed for olivine-hosted melt inclusions, which are closer proxies of primary melts than magmas erupted at the surface. Overall, these data suggest a similar origin and mantle sources for Ustica, Prometeo and Etna volcanoes.

The Iblei Mio-Quaternary volcanoes show a wider variety of petrological and trace element compositions than Etna and Ustica. Beccaluva et al. (1998) used an integrated thermodynamic, geochemical and isotopic approach to conclude that tholeiitic to nephelinitic lavas were generated by variable degrees of lithospheric mantle melting (about 20 % for tholeiites and about 3 % for nephelinites) at pressures of 1.0–2.5 GPa passing from tholeiites to nephelinites. These represent lithospheric to upper asthenospheric depths in the Iblei area. Mantle source rocks were lherzolites containing some amphibole at shallow depth and amphibole plus phlogopite and carbonates at greater depth. Exotic mantle minerals were formed by reaction between depleted lithospheric rocks and percolating fluids or melts. Metasomatism was stronger at the base of the lithosphere with respect to the intermediate and upper lithospheric mantle.

Trua et al. (1998) also envisage an origin of Iblean magmas in a variably metasomatised upper mantle at the asthenosphere-lithosphere boundary. They noticed variable incompatible element ratios (e.g. Sm/Nd, Rb/Sr, Th/Pb, La/Yb, etc.) but not so variable Sr–Nd–Pb isotope composition in the Iblean magmas. It was suggested, therefore, that metasomatism was a relatively young process, possibly occurred during Triassic extensional events. Mantle underwent high elemental enrichment and fractionation, but there was insufficient time to develop high $^{87}Sr/^{86}Sr$ and low $^{143}Nd/^{144}Nd$ ratios.

The mafic rocks from Linosa are mildly enriched in alkalies and range from saturated to slightly undersaturated in silica. Their incompatible element abundances and ratios, and isotopic signatures are less variable than other rocks from the Sicily Province (Civetta et al. 1998; Bindi et al. 2002; Di Bella et al. 2008). This suggests derivation from a relatively homogeneous mantle source. Linosa basalts also show lower $(Tb/Yb)_N$ than other equivalent rocks from southeastern Sicily and Sicily Channel, supporting an origin by lower degree of melting and scarce role of residual garnet.

Basalts at Pantelleria show a much larger range of Pb isotope values than other Sicily volcanoes, and define a trend that extends from FOZO or HIMU toward DMM, along the NHRL (Fig. 10.8b and c). This has been interpreted as evidence that the source of the Pantelleria basalts is isotopically heterogeneous and basically consists of a mixture of HIMU- and DMM-type mantle material, possibly with some contribution by other enriched mantle components (e.g. Mahood and Baker 1986; Esperança and Crisci 1995; Civetta et al. 1998). According to Esperança and Crisci (1995), mantle heterogeneity results from the addition of asthenospheric melts to a thinned lithospheric mantle from at least Permo-Triassic times. In contrast, an origin from an asthenospheric source containing fragments of old recycled subducted eclogitic crust was suggested by Avanzinelli et al. (2014) on the basis of Sr–Nd–Pb and U-series isotope data. Magma interaction with the lithosphere and/or a role for other enriched mantle components was advocated as possible explanation for the distinct isotope signatures of some younger hawaiites (higher $^{87}Sr/^{86}Sr$ and lower $^{143}Nd/^{144}Nd$, $^{206}Pb/^{204}Pb$, $^{238}U/^{232}Th$, and $^{230}Th/^{232}Th$, but comparable $^{230}Th/^{238}U$).

In conclusion, the primary magmas of the Sicily Province were formed in heterogeneous mantle sources, consisting of a mixture of enriched and depleted components. The depleted end-member had a DMM-type signature, whereas an HIMU or FOZO type mantle is envisaged as enriched end-member. A role for other enriched mantle components (EM1, EM2) is also suggested by some authors (e.g. Beccaluva et al. 1998; Civetta et al. 1998; Viccaro et al. 2011; Avanzinelli et al. 2014; Schiavi et al. 2015). Melting occurred at different depths inside the lithospheric and upper asthenopsheric mantle, mostly within the stability field of garnet, as indicated by highly fractionated HREE observed in most samples.

10.10 Geodynamic Significance

The mafic magmas of the Sicily Province display typical intraplate trace element signatures, with low LILE/HFSE ratios, depletion in some LILE (e.g. K, Rb) and enrichments in Ta and Nb. Sr and Nd isotopic ratios show moderate variation when compared with orogenic rocks from central-southern Italy. This testifies to a negligible role of crustal components in the origin of Sicily magmas, compared to orogenic volcanic provinces. In contrast, Pb-isotopes are much more variable in Sicily (Fig. 10.8), requiring a contribution by various mantle components, essentially FOZO and DMM. There is a lively debate on the origin, physical nature and geodynamic significance of these mantle components, both in Sicily and at a global scale (e.g. Hofmann 1997; Meibom and Anderson 2003; Foulger et al. 2005; Foulger 2010).

DMM represents mantle material with low incompatible element contents, Rb/Sr, Th/Pb, U/Pb, and high Sm/Nd, a condition that was generated by extraction of basaltic magmas during previous melting events. In contrast, FOZO, HIMU, EM1, and EM2 compositions reflect enriched mantle reservoirs with variable trace element and radiogenic isotope ratios, basically derived by recycling of different types of crustal rocks. According to some authors, these enriched mantle rocks remained as isolated reservoirs in the deep mantle for a long time (of the orders of a few billion years) in order to develop peculiar isotopic signatures. These discrete deep mantle masses are emplaced as plumes into the shallow mantle (e.g. Courtillot et al. 2003; White 2010; Cagney et al. 2016).

Other authors are against the idea of more or less isolated regions of the mantle and argue that FOZO, HIMU, EM1 and EM2 may not represent discrete mantle reservoirs and do not imply the presence of active mantle plumes (e.g. Meibom and Anderson 2003). Therefore, an alternative model to the plume hypothesis is that the upper mantle is strongly heterogeneous and consists of more and less fusible portions, which show a wide range of trace element and isotopic compositions (e.g. Kellog et al. 2002; Meibom and Anderson 2003). The relative compositional homogeneity of MORBs, compared to OIBs, would be simply related to the higher degree of melting and to the larger region sampled during melting of the MORBs. These smooth out compositional heterogeneity and produce melts that show little variable geochemical and isotopic compositions (see discussion in Foulger et al. 2005).

The debate on global-scale mantle compositions is echoed by the discussion of the Sicily magmatism. The lithospheric mantle is generally envisaged as the source of the DMM component. As for the FOZO- or HIMU-type compositions, a deep mantle origin has been suggested by several authors (e.g. Condomines et al. 1982; Clocchiatti et al. 1998; Gasperini et al. 2002; Cadoux et al. 2007), a hypothesis apparently supported by a plume-like body imaged by P-wave tomography in the shallow mantle beneath Etna (Montelli et al. 2004). High He-isotope ratios ($R/R_A \sim 7.0$) of mantle xenoliths and lavas have been considered as evidence in favour of a plume origin for the enriched components in the Sicily Province. However, such a possibility is in contrast with potential temperatures of Iblean mantle, which are about 200 °C lower than in classical mantle plumes areas (Perinelli et al. 2008).

At Pantelleria, HIMU, DMM and possibly EM1 mantle components have been invoked to explain Pb-isotope variability. Civetta et al. (1998) proposed that DMM represents the uppermost portion of the asthenosphere, whereas the HIMU-type component derives from a deep mantle plume emplaced beneath the entire Sicily area. The plume could be either a single deep mantle body or may be linked to the larger homogeneous layer occurring at a continental scale beneath Europe (European Asthenospheric Reservoir, EAR), which is believed to represent an expanded plume head (Wilson and Downes 1991; Granet et al. 1995). This layer is believed to represent the source for many Miocene to Quaternary volcanoes in Europe, including the Eifel region (Germany), the Central French Massif and Central Spain (Lustrino and Wilson 2007). An origin by melting of mantle material ascending from the 650 km discontinuity is also suggested by Cadoux et al. (2007) for Etna and the entire central-southern Italy magmatism, although the post-emplacement history of deep mantle material was different for orogenic and anorogenic magmas.

In contrast, Esperança and Crisci (1995) and Trua et al. (1998) propose that asthenosphere-derived melts are the enriched components in Sicily, and that these were added to the lithospheric mantle during Permian-Triassic times. The young age of metasomatism is supported by Sr-Nd isotopic studies on the Iblei xenoliths, which give a pseudo-isochron of about 200 Ma (Tonarini et al. 1996). Avanzinelli et al. (2014) also exclude a role of deep mantle plumes for Pantelleria and Linosa magmas. Based on

U-series isotopes, these authors conclude that the enriched component of magmas is provided by fragments of eclogitic rocks mixed with depleted peridotite in the asthenosphere. Eclogites would represent ancient recycled oceanic crust.

10.11 Summary and Conclusions

The Sicily Province consists of tholeiitic to Na-alkaline magmas. Most rocks are mafic in composition, except at Pantelleria where peralkaline trachytes and rhyolites (pantellerites) dominate. A few trachytic rocks occur at Etna, Ustica and as lithic ejecta at Linosa.

Intermediate and acid rocks were formed by evolutionary processes dominated by fractional crystallisation of mafic parents. Evolution was extensively working at Pantelleria where huge amounts of peralkaline acid magmas were generated in a shallow magma reservoir, starting from mafic Na-transition parental melts. In contrast, at Etna, Ustica, and Linosa, fractional crystallisation played a minor role through most of the evolution history of these volcanoes, and prevailingly mafic magmas were erupted, likely by open-conduit feeding systems. Occasional eruptions of benmoreite-trachyte magmas testify to increased roles of fractional crystallisation during temporally-limited stages of evolution within obstructed-conduit plumbing systems. At Iblei, the dominance of mafic rocks with high ferromagnesian element contents attests rapid magma ascent to the surface, with little if any ponding and crystallisation within the crust.

Mafic rocks from the Sicily Province display variable abundances of incompatible elements, which increase from tholeiites to alkali basalts and nephelinites. However, all the rocks show low LILE/HFSE ratios, typical of intraplate basalts, in contrast with the high LILE/HFSE compositions of volcanism from the Aeolian arc and the Italian peninsula. They have a restricted range of Sr-Nd isotopic compositions, but variable Pb isotopic ratios, which fall along the NHRL between HIMU-FOZO and DMM mantle end-members. Such a contrasting behaviour of radiogenic isotopes indicates that, contrary to the orogenic volcanoes from central southern Italy, the role of continental crust in the origin of Sicily magmatism is negligible. In contrast, the interaction between different mantle reservoirs is testified by Pb isotope variations.

The variable petrogenetic affinity and incompatible element contents of the Sicily mafic magmas are mainly related to different degrees of partial melting at different depths within a heterogeneous mantle. Based on experimental and geochemical constraints, about 20–3 % melting is required to generate tholeiitic to nephelinitic magmas.

Isotopic compositions (especially Pb) reveal geochemically heterogeneous sources. The origin of this heterogeneity is debated and may derive either from shallow mantle processes (e.g. variable metasomatic modifications of the lithosphere by asthenospheric melts, or mixing of asthenosphere and old subducted oceanic crust) or from an interaction between a deep-mantle plume and asthenosphere-lithosphere material. Etna and Ustica show some trace element and isotopic characteristics (e.g. Rb/Nb, Ce/Pb, and boron isotopes) that recall volcanic-arc signatures, indicating a contribution by subduction-related components.

References

Accaino F, Catalano R, Di Marzo L et al (2011) A crustal seismic profile across Sicily. Tectonophysics 508:52–61

Aiuppa A, Cannata A, Cannavo F, Di Grazia G, Ferrari F, Giudice G, Gurrieri S, Liuzzo M, Mattia M, Montalto P, Patane D, Puglisi G (2010) Patterns in the recent 2007–2008 activity of Mount Etna volcano investigated by integrated geophysical and geochemical observations. Geochem Geophys Geosyst 11. doi:10.1029/2010gc003168

Allan TD, Morelli C (1971) A geophysical study of the Mediterranean Sea. Boll Geofis Teor Appl 13:100–142

Allard P, Carbonnelle J, Dajlevic D, Le Bronec J, Morel P, Robe MC, Maurenas JM, Faivre-Pierret R, Martin D, Sabroux JC, Zettwoog P (1991) Eruptive and diffuse emissions of CO_2 from Mount Etna. Nature 351:387–391

Alletti M, Pompilio M, Rotolo SG (2005) Mafic and ultramafic enclaves in Ustica Island lavas: inferences on composition of lower crust and deep magmatic processes. Lithos 84:151–167

Armienti P, Barberi F, Innocenti F, Pompilio M, Romano R, Villari L (1984) Compositional variation in the 1983 and other recent Etnean lavas: insights on the shallow feeding system. Bull Volcanol 47:995–1007

Armienti P, D'Orazio M, Innocenti F, Tonarini S, Villari L (1996) October 1995–February 1996 Mt. Etna explosive activity: trace element and isotopic constraints on the feeding system. Acta Vulcanol 8:1–6

Armienti P, Pareschi MT, Pompilio M (1997) Lava textures and time scales of magma storage at Mt. Etna (Italy). Acta Vulcanol 9:1–5

Armienti P, Tonarini S, D'Orazio M, Innocenti F (2004) Genesis and evolution of Mt. Etna alkaline lavas: petrological and Sr–Nd–B isotope constraints. Per Mineral 73:29–52

Armienti P, Perinelli C, Putirka KD (2013) A new model to estimate deep-level magma ascent rates, with applications to Mt. Etna (Sicily, Italy). J Petrol 54:795–813

Avanzinelli R, Bindi L, Menchetti S, Conticelli S (2004) Crystallisation and genesis of peralkaline magmas from Pantelleria Volcano, Italy: an integrated petrological and crystal-chemical study. Lithos 73:41–69

Avanzinelli R, Braschi E, Marchionni S, Bindi L (2014) Mantle melting in within-plate continental settings: Sr–Nd–Pb and U-series isotope constraints in alkali basalts from the Sicily Channel (Pantelleria and Linosa Islands, Southern Italy). Lithos 188:113–129

Azzaro R (1999) Earthquake surface faulting at Mount Etna volcano (Sicily) and implications for active tectonics. J Geodyn 28:193–213

Azzaro R, Bonforte A, Branca S, Guglielmino F (2013) Geometry and kinematics of the fault systems controlling the unstable flank of Etna volcano (Sicily). J Volcanol Geoth Res 251:5–15

Bailey DK, Macdonald R (1975) Fluorine and chlorine in peralkaline liquids and the need for magma generation in an open system. Mineral Mag 40:405–414

Barberi F, Carapezza ML, Valenza M, Villari L (1993) The control of lava flow during the 1991–1992 eruption of Mt. Etna. J Volcanol Geoth Res 56:1–34

Beccaluva L, Colantoni P, Di Girolamo P, Savelli C (1981) Upper-Miocene submarine volcanism in the Strait of Sicily (Banco Senza Nome). Bull Volcanol 44:573–581

Beccaluva L, Siena F, Coltorti M, Di Grande A, Lo Giudice A, Macciotta G, Tassinari R, Vaccaro C (1998) Nephelinitic to tholeiitic magma generation in a transtensional tectonic setting: an integrated model for the Iblean volcanism, Sicily. J Petrol 39:1547–1576

Behncke B (2001) Volcanism in the Southern Apennines and Sicily. In: Vai GB, Martini PI (eds) Anatomy of an Orogen. The Apennines and the adjacent Mediterranean basins. Kluwer, Dordrecht, pp 105–120

Behncke B (2004) Late Pliocene volcanic island growth and flood basalt-like lava emplacement in the Hyblean Mountains (SE Sicily). J Geophys Res 109:B09201. doi:10.1029/2003JB002937

Bianchini G, Clocchiatti R, Coltorti M, Joron JL, Vaccaro C (1998) Petrogenesis of mafic lavas from the northernmost sector of the Iblean district (Sicily). Eur J Mineral 10:301–315

Bianchini G, Bell K, Vaccaro C (1999) Mantle sources of the Cenozoic Iblean volcanism (SE Sicily, Italy): Sr–Nd–Pb isotopic constraints. Mineral Petrol 67: 213–221

Bianchini G, Clocchiatti R, Coltorti M, Joron JL, Vaccaro C (1998) Petrogenesis of mafic lavas from the northernmost sector of the Iblean district (Sicily). Eur J Mineral 10:301–315

Bianchini G, Yoshikawa M, Sapienza GT (2010) Comparative study of ultramafic xenoliths and associated lavas from South-Eastern Sicily: nature of the lithospheric mantle and insights on magma genesis. Mineral Petrol 98:111–121

Bindi L, Tasselli F, Olmi F, Peccerillo A, Menchetti S (2002) Crystal chemistry of clinopyroxenes from Linosa volcano: implications for modelling magmatic plumbing system. Mineral Mag 66:953–968

Boccaletti M, Nicolich R, Tortorici L (1984) The Calabrian arc and the Ionian Sea in the dynamic evolution of the central Mediterranean. Mar Geol 55:219–245

Boccaletti M, Cello G, Tortorici L (1987) Transtensional tectonics in the Sicily Channel. J Struct Geol 9:869–876

Bohrson WA, Reid MR (1997) Genesis of silicic peralkaline volcanic rocks in an ocean island setting by crustal melting and open-system processes: Socorro Island, Mexico. J Petrol 38:1137–1166

Bonaccorso A, Calvari S, Coltelli M, Del Negro C, Falsaperla S (eds) (2004) Mt. Etna: Volcano Laboratory. Am Geophys Un Mon 143, 384 pp

Borgia A, Ferrari L, Pasquarè G (1992) Role of gravitational spreading in the tectonic and volcanic evolution of Etna, Italy. Nature 357:231–235

Bousquet JC, Lanzafame G (2004) The tectonics and geodynamics of Mt. Etna: synthesis and interpretation of geological and geophysical data. In: Bonaccorso A, Calvari S, Coltelli M, Del Negro C, Falsaperla S (eds) Mt. Etna: Volcano Laboratory. Am Geophys Un Mon 143:29–47

Bozzano F, Gaeta M, Lenti L, Martino S, Paciello A, Palladino DM, Sottili G (2013) Modeling the effects of eruptive and seismic activities on flank instability at Mount Etna, Italy. J Geophys Res, Solid Earth 118:5252–5273

Branca S, Del Carlo P (2004) Eruptions of Mt. Etna during the past 3200 years: a revised compilation integrating historical and stratigraphic records. In: Bonaccorso A, Calvari S, Coltelli M, Del Negro C, Falsaperla S (eds) Mt. Etna: Volcano Laboratory. Am Geophys Un Mon 143:1–27

Branca S, Ferrara V (2013) The morphostructural setting of Mount Etna sedimentary basement (Italy): implications for the geometry and volume of the volcano and its flank instability. Tectonophysics 586:46–64
Branca S, Coltelli M, Groppelli G (2004) Geological evolution of Etna volcano. In: Bonaccorso A, Calvari S, Coltelli M, Del Negro C, Falsaperla S (eds) Mt. Etna: Volcano Laboratory. Am Geophys Un Mon 143:49–63
Branca S, Coltelli M, Groppelli G, Lentini F (2011) Geological map of Etna volcano, 1:50,000 scale. It J Geosci 130:265–291
Branca S, Coltelli M, Groppelli G (eds) (2015) Geological Map of Etna volcano. Mem Descr Carta Geol It 98, 126 p (http://www.isprambiente.gov.it/it/ispra)
Bryce JG, De Paolo DJ (2004) Pb heterogeneity in basaltic phenocrysts. Geochim Cosmochim Acta 68:4453–4468
Cadoux A, Blichert-Toft J, Pinti DL, Albarede F (2007) A unique lower mantle source for Southern Italy volcanics. Earth Planet Sci Lett 259:227–238
Cagney N, Crameri F, Newsome WH, Lithgow-Bertelloni C, Cotel A, Hart SR, Whitehead JA (2016) Constraining the source of mantle plumes. Earth Planet Sci Lett 435:55–63
Calanchi N, Colantoni P, Rossi PL, Saitta M, Serri G (1989) The Strait of Sicily continental rift systems: physiography and petrochemistry of the submarine volcanic centres. Mar Geol 87:55–83
Calcagnile G, D'Ingeo F, Farrugia P, Panza GF (1982) The lithosphere in the central-eastern Mediterranean area. Pure Appl Geophys 120:389–406
Calò M, Parisi L (2014) Evidences of a lithospheric fault zone in the Sicily Channel continental rift (southern Italy) from instrumental seismicity data. Geophys J Int 199:219–225
Calò M, Parisi L, Luzio D (2013) Lithospheric P- and S-wave velocity models of the Sicilian area using WAM tomography: procedure and assessments. Geophys J Int 195:625–649
Calvari S, Coltelli M, Neri M, Pompilio M, Scribano V (1994) The 1991–1993 Etna eruption: chronology and lava flow-field evolution. Acta Vulcanol 4:1–14
Calvari S, Tanner LH, Groppelli G (1998) Debris-avalanche deposits of the Milo Lahar sequence and the opening of the Valle del Bove on Etna volcano (Italy). J Volcanol Geotherm Res 87:193–209
Caracausi A, Favara R, Giammanco S, Italiano F, Paonita A, Pecoraino G, Rizzo A, Nuccio PM (2003) Mount Etna: Geochemical signals of magma ascent and unusually extensive plumbing system. Geophys Res Lett 30, 1057. doi:10.1029/2002gl015463
Caracausi A, Favara R, Italiano F, Nuccio PM, Paonita A, Rizzo A (2005) Active geodynamics of the central Mediterranean Sea: tensional tectonic evidences in western Sicily from mantle-derived helium. Geophys Res Lett 32:L04312. doi:10.1029/2004gl021608
Carbone S, Lentini F (2015) Inquadramento geologico-regionale e substrato dell'Etna. In: Branca S, Coltelli M, Groppelli G (eds) Carta Geologica del Vulcano Etna. Mem Descr Carta Geol It 98:9–29
Carter SR, Civetta L (1977) Genetic implications of the isotope and trace element variations in the eastern Sicilian volcanics. Earth Planet Sci Lett 36:168–180
Catalano R, Di Stefano P, Sulli A, Vitale FP (1996) Paleogeography and structure of the central Mediterranean: Sicily and its offshore area. Tectonophysics 260:291–323
Catalano S, De Guidi G, Romagnoli G, Torrisi S, Tortorici G, Tortorici L (2008) The migration of plate boundaries in SE Sicily: influence on the large-scale kinematic model of the African promontory in southern Italy. Tectonophysics 449:41–62
Catalano S, De Guidi G, Lanzafame G, Monaco C, Tortorici L (2009) Late Quaternary deformation on the island on Pantelleria: new constraints for the recent tectonic evolution of the Sicily Channel Rift (southern Italy). J Geodyn 48:75–82
Catalano S, Romagnoli G, Tortorici G (2010) Kinematics and dynamics of the Late Quaternary rift-flank deformation in the Hyblean Plateau (SE Sicily). Tectonophysics 486:1–14
Catalano S, Tortorici L, Viccaro M (2014) Regional tectonic control on large size explosive eruptions: insights into the Green Tuff ignimbrite unit of Pantelleria. J Geodyn 73:23–33
Chester DK, Duncan AN (1985) Mount Etna: The anathomy of a volcano. Stanford University Press, 412 pp
Chiappini M, Meloni A, Boschi E, Faggioni O, Beverini N, Carmisciano C, Marson I (2000) Onshore-offshore integrated shaded relief magnetic anomaly map at sea level of Italy and surrounding areas. Ann Geophys 43:983–989
Chiodini G, Caliro S, Aiuppa A, Avino R, Granieri D, Moretti R, Parello F (2011) First $^{13}C/^{12}C$ isotopic characterisation of volcanic plume CO_2. Bull Volcanol 73:531–542
Cinque A, Civetta L, Orsi G, Peccerillo A (1988) Geology and geochemistry of the island of Ustica (Southern Tyrrhenian Sea). Rend Soc It Mineral Petrol 43:987–1002
Civetta L, Cornette Y, Crisci G, Gillot P-Y, Orsi G, Requejo CS (1984) Geology, geochronology and chemical evolution of the island of Pantelleria. Geol Mag 121:541–562
Civetta L, Cornette Y, Gillot P-Y, Orsi G (1988) The eruptive history of Pantelleria (Sicily Channel) in the last 50 ka. Bull Volcan 50:47–57
Civetta L, D'Antonio M, Orsi G, Tilton GR (1998) The geochemistry of volcanic rocks from Pantelleria Island, Sicily Channel: petrogenesis and characteristics of the mantle source region. J Petrol 39:1453–1491
Civile D, Lodolo E, Accettella D, Geletti R, Ben-Avraham Z, Deponte M, Facchin L, Ramella R, Romeo R (2010) The Pantelleria graben (Sicily Channel, Central Mediterranean): an example of intraplate 'passive' rift. Tectonophysics 490:173–183

Clocchiatti R, Weisz J, Mosbah M, Tanguy JC (1992) Coexistence de «verres» alkalins et tholéiitiques saturés en CO_2 dans les olivines des hyaloclastites d'Aci Castello (Etna, Sicile, Italie). Arguments en faveur d'un manteau anormal et d'un réservoir profond. Acta Vulcanol 2:161–173

Clocchiatti R, Schiano P, Ottolini L, Bottazzi P (1998) Earlier alkaline and transitional pulsation of Mt Etna volcano. Earth Planet Sci Lett 163:399–407

Coltelli M, Del Carlo P, Vezzoli L (1998) The discovery of a Plinian basaltic eruption of Roman age at Etna volcano, Italy. Geology 26:1095–1098

Condomines M, Tanguy JC, Kieffer G, Allegre CJ (1982) Magmatic evolution of a volcano studied by $^{230}Th/^{238}U$ disequilibrium and trace elements systematics: The Etna case. Geochim Cosmochim Acta 46:1397–1416

Condomines M, Tanguy JC, Michaud V (1995) Magma dynamics at Mt Etna: constraints from U-Th-Ra-Pb radioactive disequilibria and Sr isotopes in historical lavas. Earth Planet Sci Lett 132:25–41

Conte AM, Martorelli E, Calarco M, Sposato A, Perinelli C, Coltelli M, Chiocci FL (2014) The 1891 submarine eruption offshore Pantelleria Island (Sicily Channel, Italy): Identification of the vent and characterization of products and eruptive style. Geochem Geophys Geosyst 15:2555–2574. doi:10.1002/2014gc005238

Correale A, Martelli M, Paonita A, Rizzo A, Brusca L, Scribano V (2012) New evidence of mantle heterogeneity beneath the Hyblean Plateau (southeast Sicily, Italy) as inferred from noble gases and geochemistry of ultramafic xenoliths. Lithos 132:70–81

Correale A, Paonita A, Martelli M, Rizzo A, Rotolo SG, Corsaro RA, Di Renzo V (2014) A two-component mantle source feeding Mt. Etna magmatism: Insights from the geochemistry of primitive magmas. Lithos 184:243–258

Corsaro RA, Cristofolini R (1996) Origin and differentiation of recent basaltic magmas from Mount Etna. Mineral Petrol 57:1–21

Corsaro RA, Cristofolini R (1997) Geology, geochemistry and mineral chemistry of tholeiitic to transitional Etnean magmas. Acta Vulcanol 9:55–66

Corsaro RA, Cristofolini R (2000) Subaqueous volcanism in the Etnaean area: evidence for hydromagmatic activity and regional uplift inferred from the Castle Rock of Aci Castello. J Volcanol Geoth Res 95:29–225

Corsaro RA, Pompilio M (2004) Dynamics of magmas at Mount Etna. In: Bonaccorso A, Calvari S, Coltelli M, Del Negro C, Falsaperla S (eds) Mt. Etna: Volcano Laboratory. Am Geophys Un Mon 143:91–110

Corsaro RA, Di Renzo V, Distefano S, Miraglia L, Civetta L (2013) Relationship between petrologic processes in the plumbing system of Mt. Etna and the dynamics of the eastern flank from 1995 to 2005. J Volcanol Geoth Res 251:75–89

Corsaro RA, Rotolo SG, Cocina O, Tumbarello G (2014) Cognate xenoliths in Mt. Etna lavas: witnesses of the high-velocity body beneath the volcano. Bull Volcanol 76. doi:10.1007/s00445-013-0772-8

Corti G, Cuffaro M, Doglioni C, Innocenti F, Manetti P (2006) Coexisting geodynamic processes in the Sicily Channel. In: Dilek Y and Pavlides S (eds) Postcollisional tectonics and magmatism in the Mediterranean region and Asia, Geol Soc Am Spec Paper 409:83–96

Coulson IM, Stuart FM, MacLean NJ (2011) Assessing the link between mantle source and sub-volcanic plumbing in the petrology of basalts from the 2001 and 2002/2003 eruptions of Mount Etna, Sicily: Evidence from geochemical and helium isotope data. Lithos 123:254–261

Courtillot V, Davaillie A, Besse J, Stock J (2003) Three distinct types of hotspots in the Earth's mantle. Earth Planet Sci Lett 205:295–308

De Beni E, Branca S, Coltelli M, Groppelli G, Wijbrans J (2011) $^{39}Ar/^{40}Ar$ isotopic dating of Etna volcanic succession. It J Geosci 130:292–305

De Vita S, Laurenzi MA, Orsi G, Voltaggio M (1998) Application of $^{40}Ar/^{39}Ar$ and ^{230}Th dating methods to the chronostratigraphy of Quaternary basaltic volcanic areas: the Ustica Island case history. Quat Intern 47 (48):117–127

Della Vedova B, Bellani S, Pelli G, Squarci P (2001) Deep temperatures and surface heath distribution. The Apennines and the adjacent Mediterranean basins. In: Vai GB, Martini PI (eds) Anatomy of an Orogen. Kluwer, Dordrecht, pp 65–76

Di Bella M (2007) Pleistocene magmatism at Linosa Island: 89 petrology, geochemistry, isotope geochemistry and relationships with Cretaceous magmatism from Capo Passero. PhD thesis, University of Messina, Italy

Di Bella M, Russo S, Petrelli M, Peccerillo A (2008) Origin and evolution of the Pleistocene magmatism of Linosa Island (Sicily Channel, Italy). Eur J Mineral 20:587–601

Di Carlo I, Rotolo SG, Scaillet B, Buccheri V, Pichavant M (2010) Phase equilibrium constraints on pre-eruptive conditions of recent felsic explosive volcanism at Pantelleria Island, Italy. J Petrol 51:2245–2276

Di Grande A, Mazzoleni P, Lo Giudice A, Beccaluva L, Macciotta G, Siena F (2002) Subaerial Plio-Pleistocene volcanism in the geo-petrographic and structural context of the north-central Iblean region (Sicily). Per Mineral 71:159–189

Di Paola GM (1973) The Island of Linosa (Sicily Channel). Bull Volcanol 37:149–174

Doglioni C, Innocenti F, Mariotti G (2001) Why Mt. Etna? Terra Nova 13:25–31

D'Orazio M, Tonarini S, Innocenti F, Pompilio M (1997) Northern Valle del Bove volcanic succession (Mt. Etna, Sicily): petrography, geochemistry and Sr–Nd isotope data. Acta Vulcanol 9:73–86

Esperança S, Crisci GM (1995) The island of Pantelleria: a case for the development of DMM-HIMU isotopic compositions in a long-lived extensional setting. Earth Planet Sci Lett 136:167–182

Faraone D (2002) I vulcani e l'uomo. Miti, Leggende e Storia. Liguori, Napoli, 341 pp

Ferla P, Meli C (2006) Evidence of magma mixing in the 'Daly gap' of alkaline suites: a case study from the enclaves of Pantelleria (Italy). J Petrol 47:1467–1507

Ferlito C, Lanzafame G (2010) The role of supercritical fluids in the potassium enrichment of magmas at Mount Etna volcano (Italy). Lithos 119:642–650

Ferlito C, Coltorti M, Cristofolini R, Giacomoni PP (2009) The contemporaneous emission of low-K and high-K trachybasalts and the role of the NE Rift during the 2002 eruptive event, Mt. Etna, Italy. Bull Volcanol 71:575–587

Finetti I, Lentini F, Carbone S, Del Ben A, Di Stefano A, Forlin E, Guarnieri P, Pipan M, Prizzon A (2005) Geological outline of Sicily and Lithospheric tectonodynamics of its Tyrrhenian margin from new CROP seismic data. In: Deep CROP (ed) Finetti IR. Elsevier, Seismic exploration of the Mediterranean region, pp 319–376

Foerstner H (1891) Das Gestein der 1891 bei Pantelleria entstandenen Vulcaninsel und seine Beziehungen zu den Jüngsten Eruptivoesteinen der Nachbarschaft. Tsch Min Petr Mitth 5:97–101

Foulger GR (2010) Plates versus A geological controversy. Wiley-Blackwell, Plumes, p 340

Foulger GR, Nathland JH, Presnall DC, Anderson DL (eds) (2005) Plates, plumes and paradigms. Geol Soc Am Spec Publ 388:881 pp

Frezzotti ML, Peccerillo A, Panza G (2009) Carbonate metasomatism and CO_2 lithosphere–asthenosphere degassing beneath the Western Mediterranean: An integrated model arising from petrological and geophysical data. Chem Geol 262:108–120

Gale A, Dalton CA, Langmuir CH, Su Y, Schilling J-G (2013) The mean composition of ocean ridge basalts. Gechem Geophys Geosyst 14. doi:10.1029/2012GC004334

Gasperini D, Blichert Toft J, Bosch D, Del Moro A, Macera P, Albarède F (2002) Upwelling of deep mantle material through a plate window: evidence from the geochemistry of Italian basaltic volcanics. J Geophys Res 107(B12):2367. doi:10.1029/2001JB000418

Gerlach T (2011) Volcanic versus anthropogenic carbon dioxide. EOS Trans Am Geophys Union 92:201–203

Giacobbe A (1993) An integrated petrologic, petrochemical and isotopic study of Mount Etna lavas: A 300,000 years of volcanic history. MA thesis, University of California at Santa Barbara

Giacomoni PP, Coltorti M, Bryce JP, Fahnestock MF, Guitreau M (2016) Mt. Etna plumbing system revealed by combined textural, compositional, and thermobarometric studies in clinopyroxenes. Contrib Mineral Petrol 171:34

Giammanco S, Parello F, Gambardella B, Schifano R, Pizzu S, Galante G (2007) Focused and diffuse effluxes of CO_2 from mud volcanoes and mofettes south of Mt. Etna (Italy). J Volcanol Geoth Res 165:46–63

Gillot P-Y, Kieffer G, Romano R (1994) The evolution of Mt. Etna in the light of potassium-argon dating. Acta Vulcanol 5:81–87

Goes S, Giardini D, Jenny S, Hollenstein C, Kahle H-G, Geiger A (2004) A recent tectonic reorganization in the south-central Mediterranean. Earth Planet Sci Lett 226:335–345

Granet M, Wilson M, Achauer U (1995) Imaging a mantle plume beneath the French Massif Central. Earth Planet Sci Lett 136:281–296

Granieri D, Chiodini G, Avino R, Caliro S (2014) Carbon dioxide emission and heat release estimation for Pantelleria Island (Sicily, Italy). J Volcanol Geoth Res 275:22–33

Grasso M, Lentini F, Nairn AEM, Vigliotti L (1983) A geological and paleomagnetic study of the Hyblean volcanic rocks, Sicily. Tectonophysics 98:271–295

Gueguen E, Tavarnelli E, Renda P, Tramutoli M (2010) The southern Tyrrhenian Sea margin: an example of lithospheric scale strike-slip duplex. It J Geosci 129:496–505

Hart SR (1984) A large-scale isotopic anomaly in the Southern Hemisphere mantle. Nature 309:753–757

Hirn A, Nicolich R, Gallart J et al (1997) Roots of Etna volcano in faults of great earthquakes. Earth Planet Sci Lett 148:171–191

Hofmann AW (1997) Mantle geochemistry: the message from oceanic volcanism. Nature 385:219–229

Irvine TN, Baragar WRA (1971) A guide to chemical classification of common volcanic rocks. Can J Earth Sci 8:523–548

Kahl M, Chakraborty S, Pompilio M, Costa F (2015) Constraints on the Nature and evolution of the magma plumbing system of Mt. Etna volcano (1991–2008) from a combined thermodynamic and kinetic modelling of the compositional record of minerals. J Petrol 56:2025–2068

Kamenetsky V, Clocchiatti R (1996) Primitive magmatism of Mt. Etna: insights from mineralogy and melt inclusions. Earth Planet Sci Lett 142:553–572

Kellogg JB, Jacobsen SB, O'Connel RJ (2002) Modeling the distribution of isotopic ratios in geochemical reservoirs. Earth Planet Sci Lett 204:183–202

Kelly JT, Carey S, Pistolesi M, Rosi M, Croff-Bell KL, Roman C, Marani M (2014) Exploration of the 1891 Foerstner submarine vent sit (Pantelleria, Italy): insights into the formation of basaltic balloons. Bull Volcanol 76:844

Kilburn C, McGuire B (2001) Italian volcanoes. Terra Publication, Edinburgh, UK 166 pp

Lanzafame G, Bousquet JC (1997) The Maltese escarpment and its extension from Etna to the Aeolian Islands (Sicily): importance and evolution of a lithospheric discontinuity. Acta Vulcanol 9:113–120

Lanzo G, Landi P, Rotolo SG (2013) Volatiles in pantellerite magmas: a case study of the Green Tuff

Plinian eruption (Island of Pantelleria, Italy). J Volcanol Geoth Res 262:153–163

Le Maitre RW (ed) (2002) A classification of igneous rocks and glossary of terms. Cambridge University Press, Cambridge 252 pp

Lodolo E, Civile D, Zanolla C, Geletti R (2012) Magnetic signature of the Sicily Channel volcanism. Mar Geophys Res 33:33–44

Lowenstern JB, Mahood GA (1991) New data on magmatic H_2O contents of pantellerites, with implications for petrogenesis and eruptive dynamics at Pantelleria. Bull Volcanol 54:78–83

Lustrino M, Wilson M (2007) The circum-Mediterranean anorogenic Cenozoic igneous province. Erath Sci Rev 81:1–65

Macdonald R (1974) Nomenclature and petrochemistry of the peralkaline oversaturated extrusive rocks. Bull Volcanol 38:498–516

Mahood GA, Baker D (1986) Experimental constraints on depths of fractionation of mildly alkalic basalts and associated felsic rocks: Pantelleria, Strait of Sicily. Contrib Mineral Petrol 93:251–264

Mahood GA, Hildreth W (1986) Geology of the peralkaline volcano at Pantelleria, Strait of Sicily. Bull Volcanol 48:143–172

Mantovani E, Viti M, Babbucci D, Tamburelli C (2007) Major evidence on the driving mechanism of the Tyrrhenian-Apennines trench-arc-back arc system from CROP seismic data. Boll Soc Geol It 126:459–471

Mantovani E, Babbucci D, Tamburelli C, Viti M (2009) A review on the driving mechanism of the Tyrrhenian-Apennines system: implications for the present seismotectonic setting in the Central-Northern Apennines. Tectonophysics 476:22–40

Manuella FC, Scribano V, Carbone S, Brancato A (2015) The Hyblean xenolith suite (Sicily): an unexpected legacy of the Ionian-Tethys realm. Int J Earth Sci 104:1317–1336

Martelli M, Nuccio PM, Stuart FM, Di Liberto V, Ellam RM (2008) Constraints on mantle source and interactions from He-Sr isotope variation in Italian Plio-Quaternary volcanism. Geochem Geophys Geosyst 9:Q02001. doi:10.1029/2007GC001730

Marty B, Trull T, Lussiez P, Basile I, Tanguy JC (1994) He, Ar, O, Sr and Nd isotope constraints on the origin and evolution of Mount Etna magmatism. Earth Planet Sci Lett 126:23–39

MastrolemboVentura B, Serpelloni E, Argnani A, Bonforte A, Bürgmann R, Anzidei A, Baldi P, Puglisi G (2014) Fast geodetic strain-rates in eastern Sicily (southern Italy): new insights into block tectonics and seismic potential in the area of the great 1693 earthquake. Earth Planet Sci Lett 404:77–88

Mattia M, Patane D, Aloisi M, Amore M (2007) Faulting on the western flank of Mt Etna and magma intrusions in the shallow crust. Terra Nova 19:89–94

Mazzarella S (2012) Dell'Isola Ferdinandea e di altre cose. Sellerio, Palermo 236 pp

Meibom A, Anderson DL (2003) The statistical upper mantle assemblage. Earth Planet Sci Lett 217:123–139

Métrich N, Allard P, Spilliaert N, Andronico D, Burton M (2004) 2001 flank eruption of the alkali- and volatile-rich primitive basalt responsible for Mount Etna's evolution in the last three decades. Earth Planet Sci Lett 228:1–17

Mollo S, Giacomoni PP, Coltorti M, Ferlito C, Iezzi G, Scarlato P (2015) Reconstruction of magmatic variables governing recent Etnean eruptions: constraints from mineral chemistry and P-T-fO_2-H_2O modeling. Lithos 212–215:311–320

Montelli R, Nolet G, Dahlen FA, Masters G, Engdahl ER, Hung S-H (2004) Finite-frequency tomography reveals a variety of plumes in the mantle. Science 303:338–343

Morgavi D, Perugini D, De Campos CP, Ertl-Ingrisch W, Lavallée Y, Morgan L, Dingwell DB (2013) Interactions between rhyolitic and basaltic melts unraveled by chaotic mixing experiments. Chem Geol 346:199–212

Musumeci C, Scarfi L, Palano M, Patane D (2014) Foreland segmentation along an active convergent margin: new constraints in southeastern Sicily (Italy) from seismic and geodetic observations. Tectonophysics 630:137–149

Nakai S, Wakita H, Nuccio M, Italiano F (1997) MORB-type neon in an enriched mantle beneath Etna, Sicily. Earth Planet Sci Lett 153:57–66

Neave DA, Fabbro G, Herd RA, Petrone CM, Edmonds M (2012) Melting, differentiation and degassing at the Pantelleria volcano, Italy. J Petrol 53:637–663

Nicolich R (2001) Deep seismic transects. In: Vai GB, Martini PI (eds) Anatomy of an Orogen. The Apennines and adjacent Mediterranean basins. Kluwer, Dordrecht, pp 47–52

Norini G, Acocella V (2011) Analogue modeling of flank instability at Mount Etna: understanding the driving factors. J Geophys Res B07206:116. doi:10.1029/2011jb008216

Nuccio PM, Paonita A, Rizzo A, Rosciglione A (2008) Elemental and isotope covariation of noble gases in mineral phases from Etnean volcanics erupted during 2001–2005, and genetic relation with peripheral gas discharges. Earth Planet Sci Lett 272:683–690

Orecchio B, Presti D, Totaro C, Neri G (2014) What earthquakes say concerning residual subduction and STEP dynamics in the Calabrian Arc region, south Italy. Geophys J Int 199:1929–1942

Orsi G, Sheridan MF (1984) The Green Tuff of Pantelleria: rheoignimbrite or rheomorphic fall? Bull Volcanol 47:611–626

Palano M, Schiavone D, Loddo M, Neri M, Presti D, Quarto R, Totaro C, Neri G (2015) Active upper crust deformation pattern along the southern edge of the Tyrrhenian subduction zone (NE Sicily): Insights from a multidisciplinary approach. Tectonophysics 657:205–218

Paonita A, Caracausi A, Iacono-Marziano G, Martelli M, Rizzo A (2012) Geochemical evidence for mixing between fluids exsolved at different depths in the magmatic system of Mt Etna (Italy). Geochim Cosmochim Acta 84:380–394

Patanè D, De Gori P, Chiarabba C, Bonaccorso A (2003) Magma ascent and the pressurization of Mount Etna's volcanic system. Science 299:2061–2063

Patanè D, Barberi G, Cocina O, De Gori P, Chiarabba C (2006) Time-resolved seismic tomography detects magma intrusions at Mount Etna. Science 313:821–823

Peccerillo A, Barberio MR, Yirgu G, Ayalew D, Barbieri M, Wu TW (2003) Relationships between mafic and peralkaline acid magmatism in continental rift settings: a petrological, geochemical and isotopic study of the Gedemsa volcano, Central Ethiopian Rift. J Petrol 44:2003–2032

Perinelli C, Sapienza GT, Armienti P, Morten L (2008) Metasomatism of the upper mantle beneath the Hyblean Plateau (Sicily): evidence from pyroxenes and glass in peridotite xenoliths. In: Coltorti M, Grégoire M (eds) Metasomatism in oceanic and continental lithospheric mantle, Geol Soc London, Spec Publ 293:197–221

Riccò A (1892) Terremoti, sollevamento ed eruzione sottomarina a Pantelleria nella seconda metà dell'ottobre 1891. Ann Uff Centr Meteorol, Geodin Rome 11:7–27

Rittmann A (1961) Volcanoes and their activity. Wiley, 305 pp

Rizzo A, Caracausi A, Favara R, Martelli M, Paonita A, Paternoster M, Nuccio PM, Rosciglione A (2006) New insights into magma dynamics during last two eruptions of Mount Etna as inferred by geochemical monitoring from 2002 to 2005. Geochem Geophys Geosyst 7. doi:10.1029/2005gc001175

Rizzo AL, Caracausi A, Liotta M, Paonita A, Barnes JD, Corsaro RA, Martelli M (2013) Chlorine isotope composition of volcanic gases and rocks at Mount Etna (Italy) and inferences on the local mantle source. Earth Planet Sci Lett 371:134–142

Rodwell GF (1878) Etna: A history of the mountain and of its eruptions. Kegan, London 218 pp

Romano R, Sturiale C (1971) L'isola di Ustica. Studio geovulcanologico e magmatologico. Riv Min Sic 127–129:1–61

Romengo N, Landi P, Rotolo SG (2012) Evidence of basaltic magma intrusions in a trachytic magma chamber at Pantelleria (Italy). Per Mineral 81:163–178

Rose-Koga EF, Koga KT, Schiano P, Le Voyer M, Shimizu N, Whitehouse MJ, Clocchiatti R (2012) Mantle source heterogeneity for South Tyrrhenian magmas revealed by Pb isotopes and halogen contents of olivine-hosted melt inclusions. Chem Geol 334:266–279

Rossi PL, Tranne CA, Calanchi N, Lanti E (1996) Geology, stratigraphy and volcanological evolution of the island of Linosa (Sicily Channel). Acta Vulcanol 8:73–90

Rotolo SG, Castorina F, Cellura D, Pompilio M (2006) Petrology and geochemistry of submarine volcanism in the Sicily Channel Rift. J Geol 114:355–365

Rotolo SG, Scaillet S, La Felice S, Vita-Scaillet G (2013) A revision of the structure and stratigraphy of pre-Green Tuff ignimbrites at Pantelleria (Strait of Sicily). J Volcanol Geoth Res 250:61–74

Sapienza G, Scribano V (2000) Distribution and representative whole-rock chemistry of deep-seated xenoliths from the Iblean Plateau, south-eastern Sicily, Italy. Per Mineral 69:185–204

Sapienza G, Hilton DR, Scribano V (2005) Helium isotopes in peridotite mineral phases from Hyblean Plateau xenoliths (south-eastern Sicily, Italy). Chem Geol 219:115–129

Sapienza GT, Griffin WL, O'Reilly SY, Morten L (2007) Crustal zircons and mantle sulfides: Archean to Triassic events in the lithosphere beneath south-eastern Sicily. Lithos 96:503–523

Scaillet S, Rotolo SG, La Felice S, Vita-Scaillet G (2011) High-resolution $^{40}Ar/^{39}Ar$ chronostratigraphy of the post-caldera (<20 ka) volcanic activity at Pantelleria, Sicily Strait. Earth Planet Sci Lett 309:280–290

Schiano P, Clocchiatti R, Ottolini L, Busà T (2001) Transition of Mount Etna lavas from a mantle-plume to an island-arc magmatic source. Nature 412:900–904

Schiano P, Clocchiatti IR, Ottolini L, Sbrana A (2004) The relationship between potassic, calcalkaline and Na-alkaline magmatism in South Italy volcanoes: a melt inclusion approach. Earth Planet Sci Lett 220:121–137

Schiavi F, Rosciglione A, Kitagawa H, Kobayashi K, Nakamura E, Nuccio PM, Ottolini L, Paonita A, Vannucci R (2015) Geochemical heterogeneities in magma beneath Mount Etna recorded by 2001–2006 melt inclusions. Geochem Geophys Geosyst 16:2109–2126. doi:10.1002/2015gc005786

Schmincke H-U, Behncke B, Grasso M, Raffi S (1997) Evolution of the northwestern Iblean Mountains, Sicily: uplift, Pliocene/Pleistocene sea-level changes, paleoenvironment, and volcanism. Geol Rund 86:637–669

Scribano V, Sapienza G, Braga R, Morten L (2006) Gabbroic xenoliths in tuff-breccia pipes from the Hyblean Plateau: insights into the nature and composition of the lower crust underneath South-eastern Sicily, Italy. Mineral Petrol 86:63–88

Simakov SK, Kouchi A, Mel'nik NN, Scribano V et al (2015) Nanodiamond finding in the hyblean shallow Mantle Xenoliths. Nature Sci Rep 7. doi:10.1038/srep10765

Spilliaert N, Allard P, Métrich N, Sobolev AV (2006) Melt inclusion record of the conditions of ascent, degassing, and extrusion of volatile-rich alkali basalt during the powerful 2002 flank eruption of Mount Etna (Italy). J Geophys Res, Solid Earth 111(B4) doi:10.1029/2005jb003934

Sun SS, McDonough WF (1989) Chemical and isotopic systematics of oceanic basalts: implications for mantle composition and processes. In: Saunders AD,

Norry MJ (eds) Magmatism in ocean basins. Geol Soc London Spec Publ 42:313–345

Tanguy JC (1978) Tholeiitic basalt magmatism of Mount Etna and its relations with the alkaline series. Contrib Mineral Petrol 66:51–67

Tanguy JC, Condomines M, Kieffer G (1997) Evolution of Mount Etna magma: constraints on the present feeding system and eruptive mechanism. J Volcanol Geoth Res 75:221–250

Tonarini S, Armienti P, D'Orazio M, Innocenti F, Pompilio M, Petrini R (1995) Geochemical and isotopic monitoring of Mt. Etna 1989–93 eruptive activity: bearing on the shallow feeding system. J Volcanol Geoth Res 64:95–115

Tonarini S, D'Orazio M, Armienti P, Innocenti F, Scribano V (1996) Geochemical features of eastern Sicily lithosphere as probed by Hyblean xenoliths and lavas. Eur J Mineral 8:1153–1173

Tonarini S, Armienti P, D'Orazio M, Innocenti F (2001) Subduction-like fluids in the genesis of Mt. Etna magmas: evidence from boron isotopes and fluid mobile elements. Earth Planet Sci Lett 192:471–483

Tortorici L, Monaco C, Bianca M (2001) Timing and modes of deformation in the western Sicilian thrust system, southern Italy. J Petrol Geol 24:191–211

Trua T, Esperança S, Mazzuoli R (1998) The evolution of the lithospheric mantle along the North African Plate: geochemical and isotopic evidence from the tholeiitic and alkaline volcanic rocks of the Hyblean Plateau, Italy. Contrib Mineral Petrol 131:307–322

Trua T, Serri G, Marani MP (2003) Lateral flow of African mantle below the nearby Tyrrhenian plate: geochemical evidence. Terra Nova 15:433–440

Viccaro M, Cristofolini R (2008) Nature of mantle heterogeneity and its role in the short-term geochemical and volcanological evolution of Mt. Etna (Italy). Lithos 105:272–288

Viccaro M, Ferlito C, Cristofolini R (2008) Complex evolution processes in the upper feeding system of Mt. Etna (Italy) as revealed by the geochemistry of recent lavas. Per Mineral 77:21–42

Viccaro M, Scribano V, Cristofolini R, Ottolini L, Manuella FC (2009) Primary origin of some trachytoid magmas: Inferences from naturally quenched glasses in hydrothermally metasomatized gabbroic xenoliths (Hyblean area, Sicily). Lithos 113:659–672

Viccaro M, Nicotra E, Millar IL, Cristofolini R (2011) The magma source at Mount Etna volcano: Perspectives from the Hf isotope composition of historic and recent lavas. Chem Geol 281:343–351

Viccaro M, Calcagno R, Garozzo I, Giuffrida M, Nicotra E (2015) Continuous magma recharge at Mt. Etna during the 2011–2013 period controls the style of volcanic activity and compositions of erupted lavas. Mineral Petrol 109:67–83

Viti M, Mantovani E, Babbucci D, Tamburelli C (2011) Plate kinematics and geodynamics in the Central Mediterranean. J Geodyn 51:190–204

Washington HS (1909) The submarine eruptions of 1831 and 1891 near Pantelleria. Am J Sci 27:131–150

White WM (2010) Oceanic island basalts and mantle plumes: the geochemical perspective. An Rev Earth Planet Sci 38:133–160

White JC, Parker DF, Ren M (2009) The origin of trachyte and pantellerite from Pantelleria, Italy: Insights from major element, trace element, and thermodynamic modelling. J Volcanol Geoth Res 179:33–55

Willbold M, Stracke A (2006) Trace element composition of mantle end-members: implications for recycling of oceanic and upper and lower continental crust. Geochem Geophys Geosyst 7:Q04004. doi:10.1029/2005GC001005

Wilson M (1989) Igneous petrogenesis. A global tectonic approach. Unwin Hyman, Boston 466 pp

Wilson M, Downes H (1991) Tertiary-Quaternary extension-related alkaline magmatism in Western and Central Europe. J Petrol 32:811–849

11 Sardinia

Abstract

Sardinia is the site of three stages of volcanic activity that show variable ages, geochemical-petrological characteristics and geodynamic significance. An early stage of Oligo-Miocene activity erupted dominant calcalkaline basalt to rhyolite lavas and pyroclastics. Similar magmatism also took place in the Corsica-Ligurian basin and in Provence, southeastern France. Compositions show typical arc-type (orogenic) characteristics and are related to a mantle wedge contaminated by fluids released by the west-dipping Neotethys ocean lithosphere. The following activity is Miocene-Quaternary in age and shows OIB-type anorogenic geochemical signatures. It took place during two stages. An early Upper Miocene to Lower Pliocene (about 12 and 4.5 Ma) activity emplaced low-volume Na-alkaline mafic to intermediate magma bodies. This was followed by a more voluminous Plio-Pleistocene stage that emplaced large amounts of tholeiitic to Na-alkaline volcanics in central-northern Sardinia. These two stages of volcanism share anorogenic trace element characteristics (i.e. low LILE/HFSE ratios) but show significant geochemical and isotopic diversities. The Mio-Pliocene anorogenic rocks are compositionally similar to other anorogenic centres in the Mediterranean region, and probably originated in the lower lithospheric or upper asthenospheric mantle passively ascending during the Mio-Pliocene extensional phases of the Sardinia block. The younger anorogenic stage shows unradiogenic Pb-isotope compositions similar to EM1-type mantle component, a unique case in Europe. The origin of such a component is controversial and could be related either to a deep plume or to an anomalous lithospheric mantle, which had undergone a Paleozoic or Pre-Cambrian contamination by acidic melts originated from delaminated lower continental crust.

A. Peccerillo, *Cenozoic Volcanism in the Tyrrhenian Sea Region*,
Advances in Volcanology, DOI 10.1007/978-3-319-42491-0_11

Keywords
Sardinia magmatism · Corsica-Ligurian basin · Provence · Orogenic magmatism · Anorogenic magmatism · EM1 · Mantle plume · Mantle metasomatism · Crustal delamination

11.1 Introduction

Sardinia has been the site of Cenozoic orogenic and anorogenic igneous activity, characterised by distinct ages, compositions, volcanological characteristics, petrogenesis and geodynamic significance. An Oligo-Miocene stage, with very tiny manifestation during Late Eocene, erupted dominant calcalkaline volcanics with arc-type (orogenic) petrological and geochemical characteristics. Contemporaneously, similar magmatism also occurred in other areas of the Western Mediterranean Sea, including the Valencia Trough, the Algeria basin (e.g. Martì et al. 1992), the Corsica-Ligurian basin, and Provence (e.g. Ottaviani-Spella et al. 2001; Réhault et al. 2012).

Starting from Middle Miocene until Pleistocene, there was a dramatic modification of magma compositions in Sardinia with eruption of tholeiitic to Na-alkaline products exhibiting OIB-type (anorogenic) geochemical signatures (e.g. Lustrino et al. 2004, 2007a, b; 2013). Such a younger volcanic activity is subdivided into two stages, showing distinct ages (Upper Miocene-Pliocene and Plio-Pleistocene, respectively), areal distribution, volume of erupted material, and radiogenic isotope signatures. Location of the main outcrops in Sardinia and position of coeval volcanoes occurring offshore western Corsica and in Provence are shown in Fig. 11.1. Chronology of volcanism in Sardinia is schematically shown in Fig. 11.2. Information on ages and petrological-volcanological characteristics is reported in Tables 11.1 and 11.2. Selected analyses are given in Tables 11.3 and 11.4.

11.2 Regional Geology

The Island of Sardinia, together with large part of Corsica (Variscan or Hercynian Corsica), is a continental-type microplate interposed between the Tyrrhenian and the Ligurian-Provençal extensional basins. Crustal thickness is approximately 28–30 km (Sartori et al. 2004; Grad et al. 2009) and decreases to 14 and 10 km in the Ligurian-Provençal and Tyrrhenian basins. Lithospheric thickness increases from 40 to 70 km from western to eastern Sardinia (Panza et al. 2007).

The island consists of a wide variety of Precambrian to Palaeozoic igneous and metamorphic rocks, overlain by Mesozoic carbonates, and Cenozoic sediments and volcanics. The oldest exposed rocks make up the Sardinia-Corsica Paleozoic basement, a segment of the European Variscan chain, which was separated from the southern European margin during Miocene. The Sardinia-Corsica basement is formed of a wide variety of Pre-Cambrian and Cambrian to Upper Paleozoic metamorphic lithologies, large volumes of calcalkaline granitoids (Sardinia-Corsica Batholith), and mafic to silicic igneous rocks.

Low-grade meta-sedimentary rocks crop out in the southwestern Sardinia, whereas medium-to high-grade metamorphic arc-type volcanics and sediments occur in the eastern sectors of the island. Hercynian intrusive rocks (about 308–275 Ma) are exposed over a large area of northeastern Sardinia and Corsica, as well in some zones of central and southern Sardinia (e.g. Carmignani et al. 2001). The Sardinia-Corsica batholith consists of an association of dominant monzogranites and granodiorites, minor

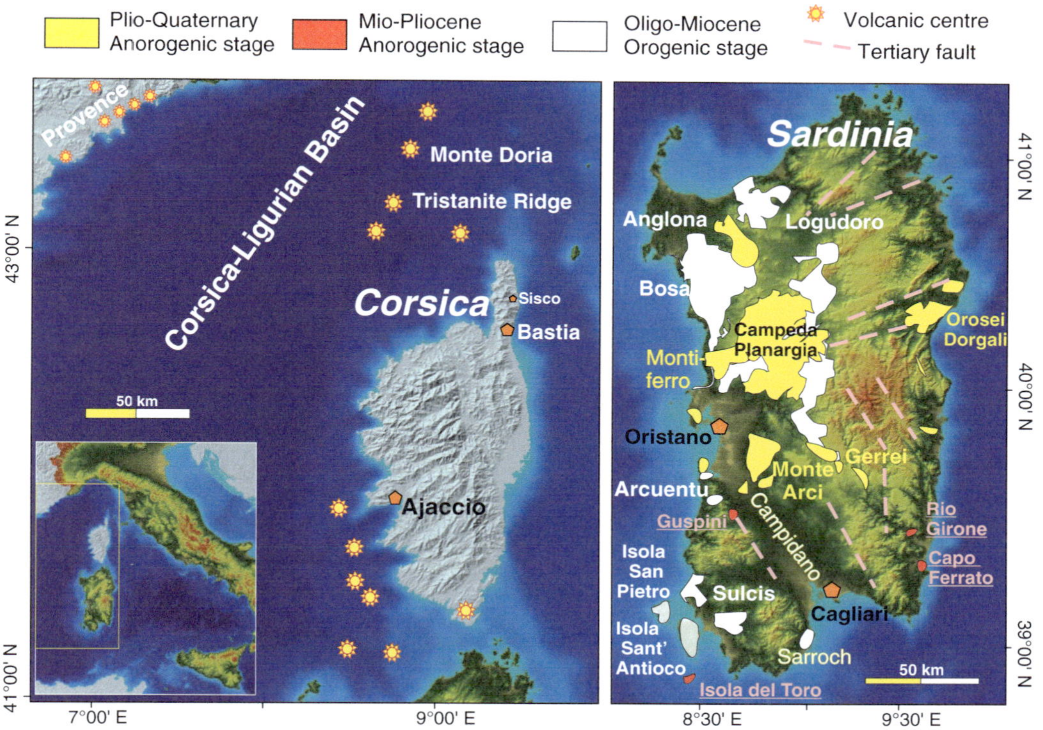

Fig. 11.1 **a** Distribution of Cenozoic magmatism in the Corsica-Ligurian basin, and Provence; **b** Distribution of Oligo-Miocene, Mio-Pliocene and Plio-Pleistocene volcanic stages in Sardinia

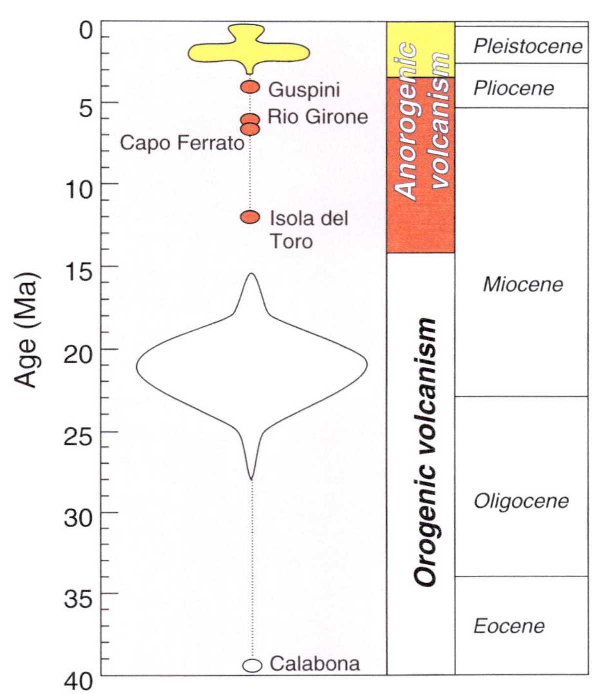

Fig. 11.2 Schematic chronogram of Cenozoic volcanism in Sardinia. Modified after Lustrino et al. (2009)

Table 11.1 Petrological characteristics and ages of the main centres of the Sardinia Oligo-Miocene orogenic volcanic stage

Oligo-Micene orogenic magmatic stage		
Main volcanic districts	Age (in Ma)	Volcanology and petrology
Bosa-Logudoro	~38–15	– Complex series of tholeiitic and calcalkaline mafic to rhyolitic pyroclastic deposits and lavas mostly erupted between 26 and 16 Ma
Anglona	~21–18	– Andesitic lava flows and domes plus ignimbrites with calcalkaline to shoshonitic affinities. Epithermal gold mineralisation at Osilo
Arcuentu	~30–17	– Basalt to andesite lava flows, domes, dikes, and pyroclastic deposits
Central Sardinia-Marmilla	~21–17	– Andesitic lava flows followed by ignimbrites and variably welded pyroclastic deposits. Precious metal mineralisations
Sarroch	~24–22	– Andesite breccia and conglomerates, plus minor calcalkaline basalts
Sulcis	~28–15	– Lower calcalkaline basalt to andesite lavas, followed by acidic lavas flows, domes and ignimbrites with calcalkaline to peralkaline affinity
Corsica-Ligurian basin- Provence	~43–7	– Dominant calcalkaline compositions with some shoshonites and minor anorogenic alkaline rocks (Toulon)

Table 11.2 Ages, petrological and volcanological characteristics of Miocene to Quaternary anorogenic magmatic stages in Sardinia

Miocene-Pleistocene anorogenic magmatism		
Main volcanic centres	Age (in Ma)	Volcanology and petrology
Miocene-Pliocene stage		
Isola del Toro	11.8	– Lava flows and dome made of mildly Na-alkaline benmoreite-trachyte
Capo Ferrato	6.6–5.0	– Lava flows, dykes and domes with a mugearitic to trachytic composition
Rio Girone and Guspini	6.4 and 4.4	– Small basanite and hawaiite necks containing ultramafic micronodules
Plio-Pleistocene stage		
Gerrei	3.8–2.1	– Hawaiite and basaltic andesite lavas containing mantle xenoliths
Monte Arci	3.8–2.6	– Volcanic ridge formed of dacite and rhyolite lavas plus minor basalts, trachybasalts and andesites
Orosei-Dorgali	3.6–2.0	– Subaerial and submarine lavas composed of hawaiites and mugearites, and minor basaltic andesites
Campeda-Planargia	3.1–2.0	– Lava plateaux composed of basaltic andesites, hawaiites and mugearites
Montiferro	3.9–1.6	– Multicentre volcanic complex formed of basanite and hawaiite to trachyte and phonolite rocks
Logudoro	3.1–0.1	– Cinder and spatter cones and associated lava fields formed predominantly of alkaline mafic rocks with Na to K affinity, and of minor tholeiites

tonalites, and a few gabbros and felsic rocks (Tommasini et al. 1995; Poli et al. 1989b). Compositions are mainly calcalkaline but shoshonitic volcanic and intrusive rocks are found in Corsica (Rossi et al. 2015). The Hercynian calcalkaline magmatism is followed by effusive and intrusive igneous activity, which emplaced tholeiitic and calcalkaline magmas,

Table 11.3 Representative compositions of Sardinia Oligo-Miocene orogenic rocks

Volcanic area	Logudoro	Oristano	Oristano	Sulcis	Arcuentu	Island of Sant'Antioco	Alghero	Anglona	Logudoro	Island of Sant'Antioco
Rock type	Basalt	Basalt	Basatic andesite	Basaltic andesite	Andesite	Andesite	Andesite	Dacite	Rhyolite	Comendite
Data source	5	1, 3, 5	5	5	2	4	5	5	5	5
SiO_2	46.36	48.33	53.72	54.25	56.47	58.93	63.00	68.28	74.51	75.57
TiO_2	1.12	0.74	0.95	1.01	0.81	0.68	0.43	0.47	0.18	0.20
Al_2O_3	18.31	15.38	16.61	16.83	17.35	17.99	19.54	14.87	12.87	10.85
Fe2O3	12.06	11.16	10.21	9.39	8.41	7.52	3.47	3.70	2.27	3.16
MnO	0.18	0.12	0.19	0.15	0.17	0.14	0.08	0.08	0.08	0.09
MgO	6.45	8.98	5.25	2.87	4.12	2.00	2.51	1.52	0.60	0.12
CaO	12.23	12.76	8.73	9.21	8.30	7.13	5.24	3.67	1.53	0.12
Na_2O	2.71	1.79	2.56	2.21	1.81	2.80	3.77	3.00	3.02	3.84
K_2O	0.29	0.51	1.52	1.18	1.69	1.57	1.67	4.20	4.89	4.88
P_2O_5	0.29	0.23	0.21	0.15	0.15	0.19	0.30	0.13	0.05	0.01
L.O.I.	–	–	1.48	–	0.62	1.41	–	2.56	3.45	–
Sc	53	–	32	32	34	–	12	10	3	3
V	442	311	246	249	210	117	52	53	–	–
Cr	37	345	–	–	76	6	7	50	4	2
Co	40	41	23	18	–	15	3	7	1	–
Ni	34	82	–	–	13	12	–	30	–	–
Rb	4	11	35	32	52	57	29	136	179	236
Sr	480	371	285	359	245	280	392	249	139	7
Y	20	16	26	21	28	24	15	24	24	61
Zr	45	34	82	107	143	121	132	216	165	735
Nb	3	2	7	7	8	7	10	15	11	105
Cs	0.92	0.20	2.30	0.60	–	–	1.20	5.00	6.60	4.20
Ba	86	51	195	400	552	539	423	572	486	22

(continued)

Table 11.3 (continued)

Volcanic area	Logudoro	Oristano	Oristano	Sulcis	Arcuentu	Island of Sant'Antioco	Alghero	Anglona	Logudoro	Island of Sant'Antioco
Rock type	Basalt	Basalt	Basatic andesite	Basaltic andesite	Andesite	Andesite	Andesite	Dacite	Rhyolite	Comendite
La	7.2	4.4	15	15	25	23	27	34	39	78
Ce	16.5	10.7	31	31	48	41	51	64	74	148
Pr	2.4	1.6	3.9	4.2	–	5.1	6.0	7.4	7.5	18.4
Nd	11.2	7.8	16.1	17.7	21.9	22.4	22.1	25.2	26.1	65.2
Sm	3.17	2.44	4.10	4.10	4.25	4.99	3.97	4.90	4.50	12.90
Eu	1.07	0.82	1.26	1.15	1.16	1.68	1.04	1.11	0.81	0.14
Gd	3.46	2.56	4.30	4.20	4.39	4.86	3.11	4.30	3.80	11.20
Tb	0.57	0.43	0.70	0.70	–	0.74	0.46	0.70	0.60	2.00
Dy	3.57	2.77	4.60	3.90	4.39	4.71	2.62	3.90	3.90	11.30
Ho	0.72	0.56	0.90	0.80	–	1.03	0.51	0.80	0.80	2.20
Er	2.01	1.51	2.80	2.30	2.50	2.99	1.46	2.40	2.50	6.60
Tm	0.28	0.23	0.42	0.35	–	0.44	0.22	0.38	0.40	1.04
Yb	1.87	1.60	2.80	2.30	2.42	3.01	1.57	2.50	2.80	6.90
Lu	0.29	0.25	0.43	0.37	0.40	0.45	0.25	0.40	0.45	1.12
Hf	1.41	1.01	2.50	2.60	3.50	4.31	3.56	5.60	4.80	16.00
Ta	0.21	0.13	0.40	0.40	0.54	0.87	0.76	1.00	1.00	7.50
Pb	2.8	1.9	8	8	20	–	10	58	22	31
Th	0.97	0.59	2.80	4.10	6.40	5.60	6.83	14.40	14.30	26.00
U	0.27	0.12	0.70	0.80	0.61	0.98	0.90	3.40	3.50	6.60
$^{87}Sr/^{86}Sr$	0.70435	0.70426	0.70611	0.70710	0.71106	0.70844	0.70685	0.70681	0.70676	0.71145
$^{143}Nd/^{144}Nd$	0.51258	0.51271	0.51251	0.51237	0.51220	–	0.51228	0.51242	0.51242	0.51261
$^{206}Pb/^{204}Pb$	18.682	18.707	18.563	18.473	–	–	18.651	18.677	18.683	18.864
$^{207}Pb/^{204}Pb$	15.621	15.644	15.606	15.604	–	–	15.640	15.638	15.627	15.663
$^{208}Pb/^{204}Pb$	38.690	38.663	38.471	38.521	–	–	38.871	38.743	38.721	38.942

Source of data: (1) Morra et al. (1994); (2) Downes et al. (2001); (3) Franciosi et al. (2003); (4) Conte et al. (2010); (5) Lustrino et al. (2013)

Table 11.4 Representative compositions of Sardinia Miocene-Quaternary rocks

Sample	Isola del Toro	Capo Ferrato	Rio Girone	Logudoro	Orosei Dorgali	Orosei Dorgali	Montiferro	Santu Lussurgiu	Mt. Arci	Mt. Arci
Rock type	Trachyte	Trachydacite	Trachybasalt	Trachybasalt	Basalt	Basaltic andesite	Phonolite	Trachybasalt	Basaltic trachyandesite	Rhyolite
Data source	6	3,6	3,6	2	4,6	4	5	3,6	6	1
SiO_2	66.94	61.43	45.78	49.70	48.00	52.74	61.08	50.10	50.59	74.71
TiO_2	0.35	0.89	3.13	2.07	1.76	1.86	0.46	2.40	2.38	0.11
Al_2O_3	15.06	16.81	15.21	16.28	15.66	16.85	19.86	14.18	16.34	14.06
Fe_2O_3	2.67	6.68	11.69	10.46	10.22	11.00	2.34	11.45	11.15	1.44
MnO	0.07	0.12	0.16	0.12	0.14	0.14	0.16	0.15	0.14	0.05
MgO	0.51	1.26	7.55	7.45	10.91	3.60	0.38	7.60	5.91	0.08
CaO	0.48	3.41	10.37	7.08	8.49	8.77	1.56	8.14	7.26	0.62
Na_2O	5.20	4.81	3.49	4.06	3.61	3.42	6.86	3.83	2.88	3.80
K_2O	6.83	4.22	2.23	2.31	0.82	1.31	7.21	1.54	2.79	5.05
P_2O_5	0.13	0.37	0.39	0.48	0.38	0.30	0.08	0.61	0.57	0.09
Sc	–	–	20	–	–	–	2	20	8	5
V	–	29	254	153	192	165	24	199	172	–
Cr	–	1	212	192	427	–	–	301	117	3
Co	–	8	44	41	52	39	–	49	39	–
Ni	–	6	126	122	298	168	3	196	157	5
Rb	245	99	49	52	21	21	164	60	50	227
Sr	50	321	970	743	690	536	57	821	883	25
Zr	899	508	223	200	159	134	966	201	253	99
Y	43	46	28	17	17	18	28	24	24	35
Nb	142	74	70	45	31	20	120	56	55	44
Cs	6.0		1.6	0.5	0.8	0.5	–	0.6	0.5	–
Ba	109	1022	528	854	648	507	–	1115	1150	131
La	101	67	47	38	26.5	20	136.3	42	48	20

(continued)

Table 11.4 (continued)

Sample	Isola del Toro	Capo Ferrato	Rio Girone	Logudoro	Orosei Dorgali	Orosei Dorgali	Montiferro	Santu Lussurgiu	Mt. Arci	Mt. Arci
Rock type	Trachyte	Trachydacite	Trachybasalt	Trachybasalt	Basalt	Basaltic andesite	Phonolite	Trachybasalt	Basaltic trachyandesite	Rhyolite
Ce	179	130	96	72	52.0	39	242.3	86	81	47
Pr	20.5	15.0	12.1	8.1	6.2	4.7	22.2	9.8	10.7	–
Nd	66.6	58.0	46.6	32.8	25.2	19.8	60.9	37.5	40.3	20.1
Sm	10.9	10.7	8.3	6.7	4.9	4.4	6.7	7.0	7.5	6.0
Eu	0.95	2.81	2.53	1.98	1.6	1.57	1.1	2.38	2.43	0.38
Gd	7.40	9.30	7.02	4.90	4.4	4.42	10.1	5.90	6.70	5.25
Tb	1.20	1.43	1.01	0.69	0.6	0.60	0.9	0.81	0.93	–
Dy	6.50	7.84	5.66	3.56	3.2	3.30	4.9	4.09	4.61	5.30
Ho	1.30	1.48	0.88	0.60	0.6	0.66	0.8	0.69	0.82	–
Er	3.70	4.00	2.26	1.36	1.5	1.59	2.6	1.71	2.00	2.61
Tm	0.58	0.64	0.35	0.22	0.2	0.23	0.4	0.23	0.24	–
Yb	4.10	3.97	1.97	1.15	1.2	1.36	2.8	1.39	1.50	2.32
Lu	0.67	0.61	0.30	0.19	0.2	0.22	0.4	0.20	0.20	–
Hf	15.7	11.8	4.8	4.3	3.5	3.3	22.0	5.7	6.2	4.2
Ta	11.9	5.5	4.5	3.1	2.1	1.4	8.3	3.9	3.7	3.0
Pb	11	14.9	5.9	7.0	3.4	2.9	19.3	5.8	9	16.0
Th	26.6	11.8	4.0	3.4	3.0	4.0		5.2	6.7	37.0
U	6.7	3.5	1.3	1.1	0.8	0.6	4.6	1.4	1.1	6.8
$^{87}Sr/^{86}Sr$	0.70547	0.70487	0.70401	0.70433	0.70442	0.70453	0.70739	0.70434	0.70428	0.71529
$^{143}Nd/^{144}Nd$	0.512745	0.512710	0.512850	0.512528	0.512571	0.512538	0.512445	0.512568	0.512632	0.512210
$^{206}Pb/^{204}Pb$	18.940	18.840	19.230	17.710	17.860	17.738	–	18.010	17.980	–
$^{207}Pb/^{204}Pb$	15.620	15.657	15.640	15.606	15.596	15.531	–	15.609	15.560	–
$^{208}Pb/^{204}Pb$	38.940	38.977	39.100	37.897	–	37.894	–	38.151	37.980	–

Source of data: (1) Montanini et al. (1994); (2) Gasperini et al. (2000); (3) Lustrino et al. (2000); (4) Lustrino et al. (2002); (5) Fedele et al. 2007; (6) Lustrino et al. (2013)

alkaline to peralkaline granites and rhyolites, and lamprophyric dykes (e.g. Cortesogno et al. 1998).

Mesozoic sequences consist of a wide variety of neritic and pelagic limestones, dolomitic limestones, and marls. Cenozoic rocks include orogenic and anorogenic volcanics, as well as continental and marine sedimentary successions concentrated along a belt crossing the island from its NW to SE corner.

Sardinia was constructed during a long and complex geodynamic evolutionary history. Hercynian (Variscan) orogeny is one of the most important geological events. It occurred during Carboniferous-Early Permian times by consumption of the Rheic Ocean floor and collision between the Gondwana and Laurussia plates (e.g. Tait et al. 1997; Carmignani et al. 2001), generating metamorphism, deformation and widespread late- to post-collisional magmatism (e.g. Carmignani et al. 1994a, 2001; Tommasini et al. 1995; Casini et al. 2015). Post-Hercynian geodynamic evolution was characterised by gravitational collapse of the previously thickened crust, and Middle Jurassic extension, with transgression and regression cycles and carbonate sedimentation over vast areas . A compressional tectonic phase took place in the Middle Cretaceous, likely related to early stages of Africa-Europe convergence. The Cenozoic geological history started with early compression, followed by Oligo-Miocene to Quaternary extension, contemporaneously with the counterclockwise rotation of Corsica-Sardinia block with respect to stable Europe, and the opening of the Balearic-Provençal-Ligurian and the Tyrrhenian Sea basins (e.g. Alvarez 1972; Carmignani et al. 1994b, 2001, 2004; Speranza et al. 2002; Gattacceca et al. 2007; Zattin et al. 2008; Carminati et al. 2012 with references). Marine to continental sedimentation took place during this period, with much of Plio-Quaternary sediments being concentrated within the Campidano Plain, an extensional system affecting the southwestern sector of the island, between the Gulf of Oristano and the Gulf of Cagliari (Fig. 11.1b).

Sardinia is the site of important mineral deposits exploited since pre-historical times. Pb–Zn and Au deposits are the most significant ones, but also barite, bauxite, kaoline, and bentonite are industrially significant resources. Mineralisation is related to various geological processes, including Paleozoic metamorphism and magmatism, Mesozoic sedimentation and supergene alteration and mobilization of pre-existing ore minerals, and Oligo-Miocene orogenic volcanism (e.g. Lattanzi 1999; Boni et al. 2002, 2009; Marcello et al. 2004; Palomba et al. 2006).

Most of the Corsica Island underwent a similar evolution as Sardinia (Variscan Corsica.). However, the northeastern sector (Alpine Corsica) was affected by Alpine metamorphism and deformation and represents the boundary between the Alpine chain and the Apennine system. The Oligo-Miocene orogenic magmatism discussed in this chapter is located west or inside the Variscan Corsica. In contrast, the Sisco lamproite discussed in Chap. 2, is located in the Alpine Corsica and, although showing orogenic geochemical signatures, is not related to Sardinia Oligo-Miocene stage but to the Tuscany Magmatic Province discussed in Chap. 2.

11.3 Oligo-Miocene Orogenic Volcanic Stage

Rocks belonging to the orogenic volcanic stage crop out mostly in western Sardinia along a N-S oriented Oligo-Miocene tectonic structure known as Fossa Sarda (e.g. Oggiano et al. 2009). The main volcanic districts occur, from north to the south, in the Bosa-Anglona area, Logudoro, Central Sardinia, Arcuentu, Sarroch and Sulcis (e.g. Conte 1997; Lustrino et al. 2004, 2013). The age of volcanism ranges between 38 and 15 Ma, but the maximum activity developed between about 24 and 16 Ma, contemporaneously with the main phases of Corsica-Sardinia block rotation (e.g. Gattacceca et al. 2007). The oldest dated rocks are found at Alghero (Calabona), northwestern Sardinia. The youngest activity is concentrated in the Sulcis regions, SW Sardinia (Lecca et al. 1997; Gattacceca et al. 2007; Lustrino et al. 2009; Oudet et al. 2010).

A summary of ages and volcanological-petrological characteristics of Oligo-Miocene orogenic magmatism is given in Table 11.1.

11.3.1 Volcanology and Stratigraphy

The Sardinia ororgenic volcanic stage shows a wide variety of products, including lava flows and domes, ignimbrites and pyroclastic fall and surge deposits, typical of the prevailing explosive activity of continental arc volcanoes. Eruptions took place during six main phases of activity, which alternately erupted mafic-intermediate and silicic products. The youngest activity (about 15 Ma) was concentrated in the Sulcis area, where rhyolitic lavas and pyroclastic rocks, and some peralkaline comendites were emplaced (e.g. Araña et al. 1974; Coulon 1977; Coulon et al. 1978; Lecca et al. 1997; Morra et al. 1994, 1997; Conte et al. 2010; Guarino et al. 2011).

The Sardinia Oligo-Miocene volcanic belt extends northward to southern Corsica, the Corsica-Ligurian basin, and Provence, south-eastern France (e.g. Rossi et al. 1998; Ottaviani-Spella et al. 2001; Rollet et al. 2002; Réhault et al. 2012). Ages of these volcanoes range from 41 to 7–6 Ma in Provence (Lustrino M, personal communication) and from about 19 to 8 Ma in the Corsica-Ligurian basin; an age of 43 Ma has been found for a single sample from Tristanite Ridge, Ligurian Sea (Réhault et al. 2012). As mentioned earlier, Oligo-Miocene calcalkaline to shoshonitic volcanism also took place elsewhere in the Western Mediterranean, e.g. in the Valencia and Algeria basins. Such volcanism is outside the subject of this book and will not be considered further.

11.3.2 Petrography and Mineral Chemistry

Compositions of orogenic rocks in Sardinia range from basalt to rhyolite according to the TAS diagram (Fig. 11.3a). K_2O versus SiO_2 classification diagram (Peccerillo and Taylor 1976) indicates a dominant calcalkaline and high-K calcalkaline affinity with a few intermediate to silicic shoshonitic rocks and some samples close to arc tholeiites (Fig. 11.3b).

The most mafic rocks (Arcuentu, Montresta in the Bosa-Logudoro area) are variably porphyritic with phenocrysts of olivine ($\sim Fo_{85-50}$), plagioclase, diopside to augite clinopyroxene, some orthopyroxene, and Fe–Ti oxides set in a groundmass containing both ortho- and clinopyroxene, some phlogopite, and accessory apatite. Olivine often shows reacted or corroded borders and contains Cr-spinel inclusions. Basaltic

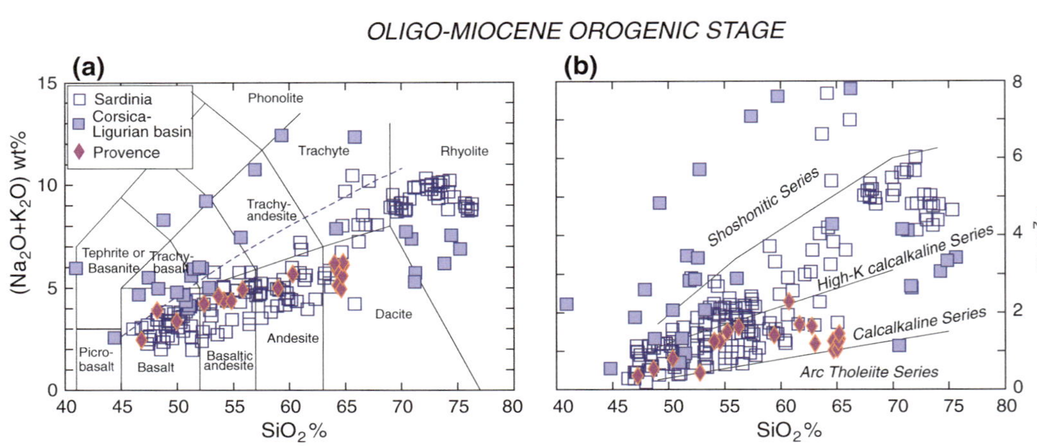

Fig. 11.3 **a** TAS classification diagram of Oligo-Miocene orogenic volcanic rocks from Sardinia, Corsica-Ligurian basin, and Provence; **b** K_2O versus SiO_2 classification diagram. Data recalculated to 100 % on a water-free basis for both diagrams

andesites and andesites are also generally porphyritic with phenocryst mineralogy dominated by compositionally variable plagioclase with minor augitic clinopyroxene, orthopyroxene, and Fe–Ti oxides; amphibole and brown mica are also observed; groundmass is composed of the same phases plus some glass. Dacites have similar composition to andesites but contain more biotite and hornblende. Rhyolites are variably porphyritic with minimum phenocryst contents in the ignimbrites, which are mostly vitreous. Main phenocrysts include plagioclase, alkali feldspar, brown mica, corroded quartz, amphibole, and opaque minerals. Comendites are variably porphyritic with phenocrysts of Na-sanidine, anortoclase, quartz, and minor arfvedsonite and aegirine microphenocrysts; aenigmatite is sometimes present (e.g. Conte 1993; Brotzu et al. 1997; Lecca et al. 1997; Morra et al. 1997; Mattioli et al. 2000; Conte et al. 2010; Guarino et al. 2011).

The rocks from the Corsica-Ligurian basin have variable potassium contents, mostly falling in the fields of calcalkaline and shoshonitic rocks; however, these rocks are all severely altered and classification based on mobile alkalies may be invalid. Rocks from Provence have less scattered compositions, mostly defining a classical basalt to dacite calcalkaline suite (Fig. 11.3). A few tholeiitic to weakly Na-alkaline rocks are also present both in the Corsica-Ligurian basin and Provence (e.g. Toulon). Rock textures are holocrystalline to hypocrystalline porphyritic with phenocrysts of plagioclase, diopside to augite clinopyroxene, and some olivine in the mafic rocks; amphibole, biotite, and some alkali feldspar occur in the more evolved samples (Rossi et al. 1998; Ivaldi et al. 2002; Réhault et al. 2012).

11.3.3 Petrology and Geochemistry

The Sardinia Oligo-Miocene rocks show scattered distribution for many major and trace element abundances and ratios in the mafic range (Fig. 11.4). FeO_{total} contents are somewhat higher than in the equivalent rocks from the Aeolian arc rocks (Conte et al. 2010). REE show variable fractionation and absolute abundances, with large range of values in the mafic rocks (Fig. 11.5a). Negative Eu anomalies are observed in silicic rocks, particularly in the Island of San Pietro comendites. Mantle normalised incompatible element patterns of mafic rocks show a large range of element abundance and fractionation, variable negative Ta and Nb anomalies, and little if any negative spike of Ti (Fig. 11.5b). HFSE show low values, close to depleted mid ocean ridge basalts (D-MORB). Initial $^{87}Sr/^{86}Sr$ ratios range from 0.7036 to 0.7127, and exhibit the usual negative correlation with Nd-isotopes; a few samples with very low Sr absolute abundance fall outside the main trend (Fig. 11.6a). $^{143}Nd/^{144}Nd$ ratios (0.51218–0.51274) decrease with silica in the mafic-intermediate compositions, and reverse their trend in the silicic rocks (Fig. 11.4h). Sr isotope ratios show a well-defined increase with silica in the mafic rocks (MgO > 4 %) and scattered decrease for other samples (not shown). Pb isotopic ratios are rather homogeneous ($^{206}Pb/^{204}Pb$ = 18.47–18.86; $^{207}Pb/^{204}Pb$ = 15.60–15.67; $^{208}Pb/^{204}Pb$ = 38.47–38.94), and significantly less radiogenic than the equivalent rocks from the Aeolian arc (Fig. 11.6b, c, d). Overall, Pb-isotope ratios plot above the Northern Hemisphere Reference Line (Hart 1984), shifted toward EM1 mantle compositions. Oxygen isotopic data for olivine and clinopyroxene from mafic rocks show $\delta^{18}O_{SMOW}$ = +5.41 ‰ to +7.4 ‰ (Downes et al. 2001; Lustrino et al. 2013). There is a well-defined negative correlation of $\delta^{18}O$ ‰ against whole rock $^{143}Nd/^{144}Nd$ ratio (Fig. 11.7a), and a positive trend versus SiO_2 and $^{87}Sr/^{86}Sr$ (not shown). Hf isotope ratios range from $^{176}Hf/^{177}Hf$ = 0.28278–0.28302 and are positively correlated with Nd isotopes. Os isotopes determined on a single basalt sample yielded $^{187}Os/^{188}Os$ = 0.1317 (Lustrino et al. 2013), a value lower than observed for other analysed orogenic volcanic rocks in the Tyrrhenian Sea area (Conticelli et al. 2007).

The available geochemical data on Corsica-Ligurian basin and Provence indicate variable magma compositions, especially for

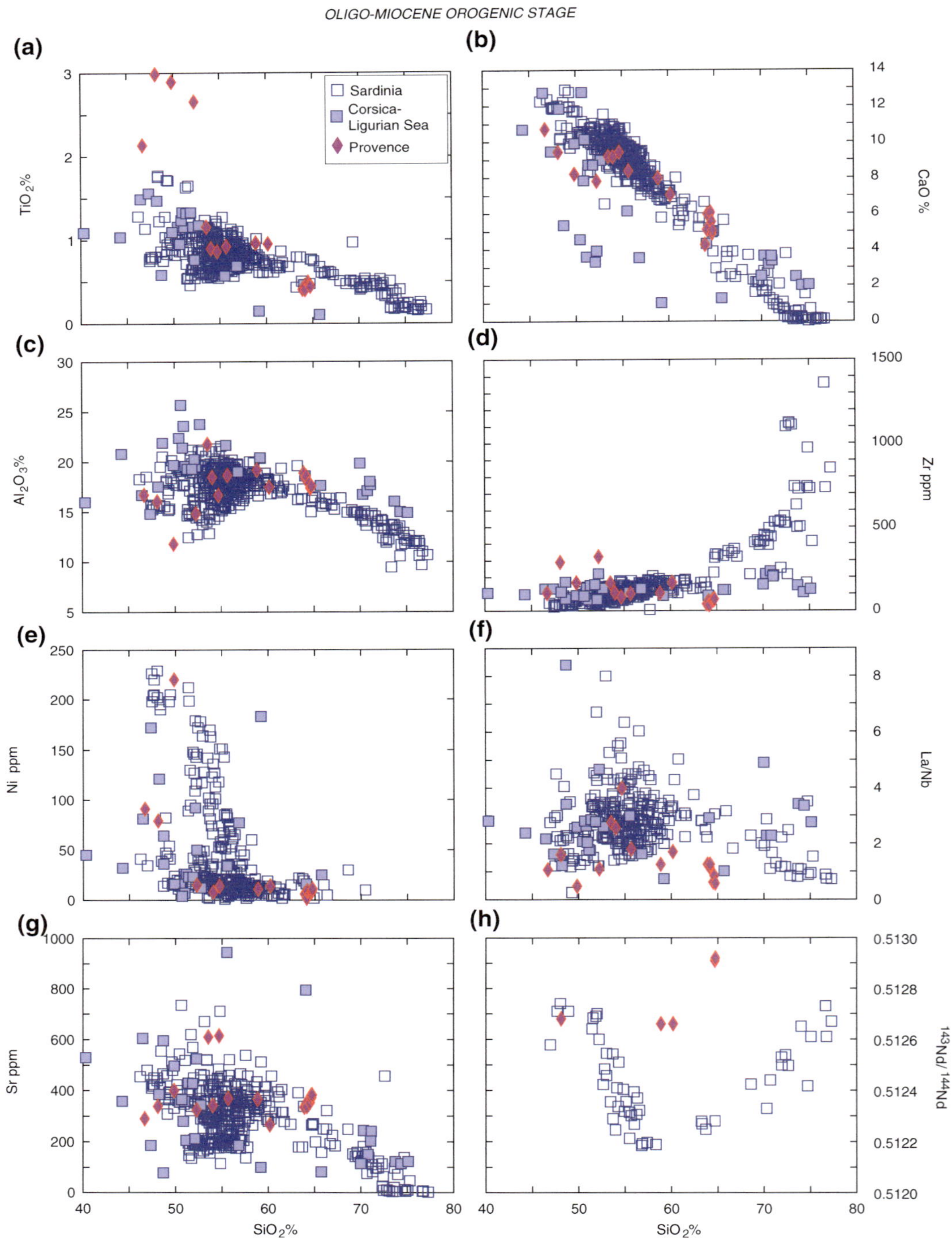

Fig. 11.4 Variation diagrams of selected major and trace elements, and Nd-isotope ratios for the Oligo-Miocene volcanic rocks from Sardinia, Corsica-Ligurian basin, and Provence. Major elements have been normalised to 100 % on a LOI-free basis

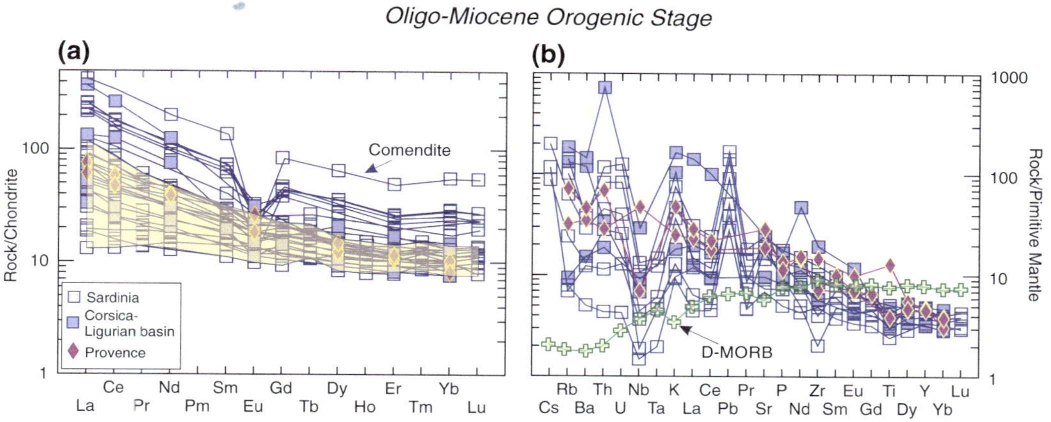

Fig. 11.5 a Chondrite-normalised REE patterns of selected rocks from the Oligo-Miocene volcanic belt of Sardinia, Corsica-Ligurian basin and Provence. The *yellow* field contains Sardinia mafic samples (MgO > 5 %). **b** Mantle-normalised incompatible elements patterns of mafic rocks. Composition of average Depleted D-MORB is from Gale et al. (2013)

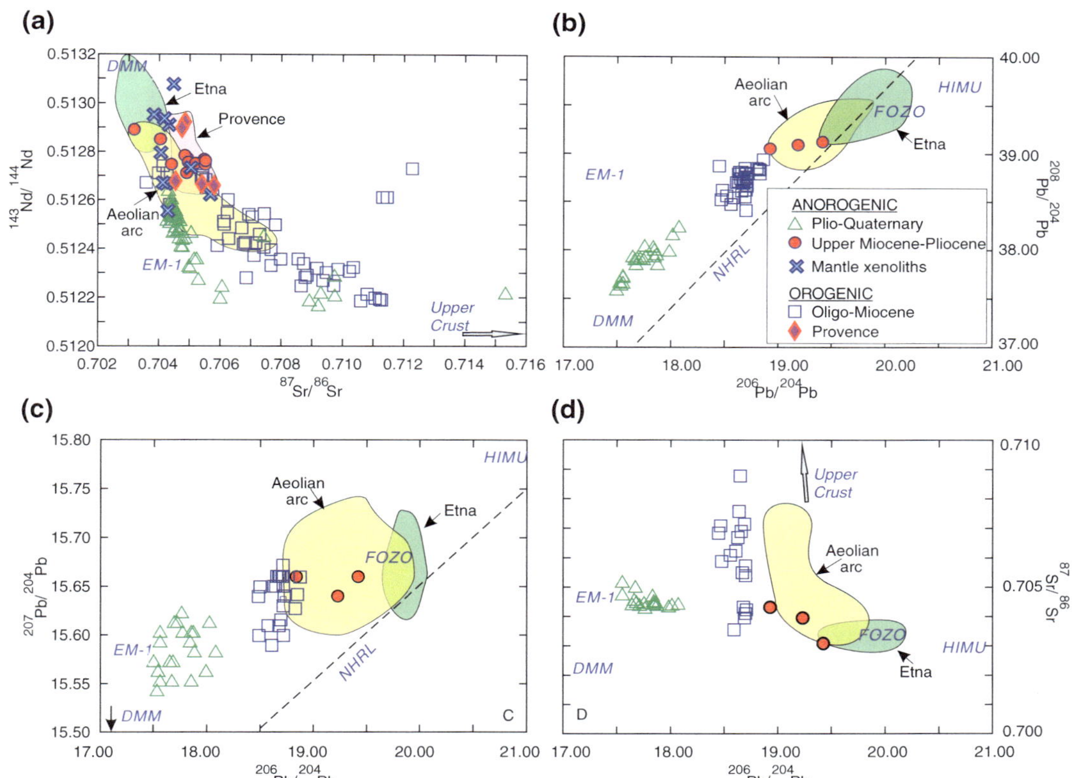

Fig. 11.6 Sr–Nd–Pb isotope variation for Cenozoic volcanic rocks from Sardinia. DMM (Depleted MORB Mantle), FOZO (FOcal ZOne), Enriched Mantle-1 (EM1) and HIMU (high-μ, $^{238}U/^{204}Pb_{t=0}$) mantle compositions are from Zindler and Hart (1986) and Stracke et al. (2005). NHRL is the Northern Hemisphere Reference Line of Hart (1984)

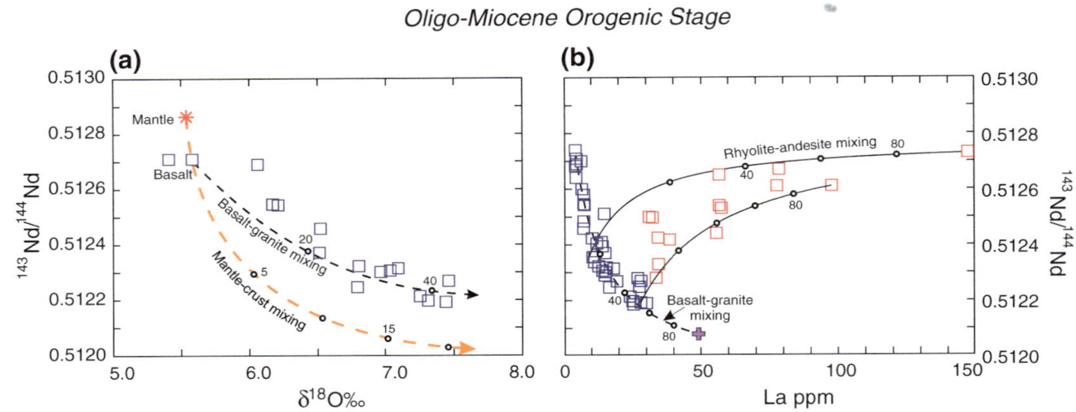

Fig. 11.7 **a** Variation of whole rock ^{143}Nd/^{144} versus δ^{18}O ‰ of olivine and clinopyroxene phenocrysts. Lines are mixing models between Hercynian granite and primitive calcalkaline basalt (magma contamination) and between mantle and granite (source contamination); **b** ^{143}Nd/^{144}Nd ratios versus La variation diagram of Sardinia Oligo-Miocene rocks. The *dashed line* is a mixing trend between basalt and average Hercynian granite (*full cross*; Di Vincenzo et al. 1996). *Full lines* are mixing models between rhyolitic and andesitic magmas. *Blue symbols* in both panels indicate basalt to andesite rocks (SiO_2 < 60 %); *red symbols* are silicic rocks (SiO_2 > 60 %). Numbers along mixing lines are amounts of one end-member. For further explanation, see text

seamounts. This could partially depend on secondary processes affecting submarine samples. Most rocks have clear arc geochemical signatures, and range from calcalkaline to shoshonitic. Others samples with subalkaline to transitional affinities have high $TiO_{2\%}$ and exhibit little or no HFSE negative anomalies (e.g. Cap Corse, Toulon), thus resembling the Sardinia anorogenic Miocene-Quaternary volcanic rocks. The Toulon outcrop has a small Nb positive spike and is also the youngest dated volcanic occurrence in Provence (7–6 Ma; Réhault et al. 2012). Sr–Nd isotope ratios show values of ^{87}Sr/^{86}Sr = 0.7045–0.7058, ^{143}Nd/^{144}Nd = 0.5129–0.50126 (Lustrino et al. 2011 and references therein).

11.3.4 Petrogenesis of Oligo-Miocene Orogenic Magmas

There is a general consensus that the origin of orogenic rocks from Sardinia is related to melting of a mantle wedge modified by subduction-related fluids or melts (e.g. Morra et al. 1994, 1997; Brotzu et al. 1997; Downes et al. 2001; Conte et al. 2010; Lustrino et al. 2013). However, there are many aspects of magma genesis and evolution that remain controversial. Negative trends of ^{143}Nd/^{144}Nd versus silica, incompatible elements, and δ^{18}O ‰ in the mafic-intermediate rocks (Figs. 11.4h, 11.7a) clearly indicate an interaction between mantle and crustal end-members. This can be accomplished either by addition of subducted sediments to the mantle wedge (source contamination) or/and by assimilation of Hercynian basement rocks by magmas (magma contamination) within the volcano feeding systems.

According to Downes et al. (2001), the range of isotopic compositions of mafic rocks reflects mantle source contamination by variable amounts (2–10 %) of subducted siliceous sediments. Franciosi et al. (2003) envisage a three-component mixing within the source, between a pre-metasomatic MORB-type mantle and two types of metasomatic fluids coming from an oceanic slab and from sediments. Isotopic heterogeneities at some centres (e.g. Arcuentu) were suggested to reveal a time-dependent modification of mantle source contamination. However, the observed isotopic variation may also reflect shallow-level magma contamination by wall rocks (e.g. Brotzu et al. 1997; Conte

1997; Morra et al. 1997). Simplified models of O-Nd isotope variations (Fig. 11.7a) show that source and magma contamination processes give qualitatively similar variation trends. However, the Sardinia samples seem to better fit the basalt-granite mixing model (i.e. magma contamination). This model shows that about 40 % bulk assimilation of average Hercynian granites (Di Vincenzo et al. 1996) can explain O–Nd isotope variation of Oligo-Miocene mafic-intermediate volcanics. Models of incompatible elements versus Nd-isotope ratios support this conclusion (Fig. 11.7b).

The origin of rhyolites and the reversed isotopic trend observed in silicic rocks are also intriguing. According to Lustrino et al. (2013), rhyolitic magmas were generated by anatexis of isotopically primitive (i.e. low Sr- and high Nd-isotope ratios) basaltic rocks emplaced early during the Oligo-Miocene magmatic stage. Successively, mixing between rhyolitic melts and contaminated (either in the source or/and within the volcano plumbing system) andesitic magmas generated the reverse trends observed in the silicic rocks. Trace element and isotope modelling supports the idea of andesite-rhyolite mixing (full lines in Fig. 11.7b). However, the rhyolitic magmas have exceedingly high concentrations of incompatible elements (Th, Zr, LREE) and very low abundances of compatible elements (Sr, Sc, Co, etc.). These compositions are difficult to obtain by partial melting of basaltic rocks, unless very low degrees of partial melting are assumed. In contrast, they can be easily modelled by about 60–70 % fractional crystallisation of basaltic parents.

The occurrence of peralkaline rocks in the Sulcis area suggests that a new type of magma with alkaline affinity had a role during the late history of Oligo-Miocene magmatism. Such a hypothesis is supported by data on melt inclusions from phenocrysts of Sulcis rocks (Lustrino et al. 2013). The low LILE/HFSE ratios in some rhyolites (Fig. 11.4f) point to the same conclusion. Note that the earliest manifestations of the Mio-Pliocene alkaline magmatic stage took place at Isola del Toro, a few km south of Sulcis coast, shortly after the end of Oligo-Miocene activity (Lustrino et al. 2007b).

Some of the most primitive Oligo-Miocene mafic rocks from Sardinia have high Nd- and low Sr-O-isotope ratios suggesting they are unaffected by shallow-level crustal contamination. Their calcalkaline character, along with variable but generally high LILE/HFSE ratios, clearly points to a subduction-related origin. Therefore, these magmas have been interpreted as representing poorly evolved melts formed in a mantle wedge contaminated by fluids released from the NW-directed subduction system of the Ionian Tethys ocean (e.g. Morra et al. 1997; Mattioli et al. 2000; Franciosi et al. 2003; Lustrino et al. 2009; Carminati et al. 2010; 2012). The low concentration of HFSE in these primitive magmas (Fig. 11.5b) supports a depleted MORB-type pre-metasomatic mantle source (Downes et al. 2001; Franciosi et al. 2003). The low Pb-isotope ratios relative to equivalent Aeolian arc rocks require a role for an unradiogenic end-member in the origin of orogenic magmas in Sardinia. This issue will be further discussed in the next paragraphs.

11.4 Miocene-Quaternary Anorogenic Volcanism

Volcanic activity with intraplate, anorogenic geochemical signatures started in Sardinia during Middle-Upper Miocene, marking one of the most abrupt modification of magma compositions in the Tyrrhenian Sea area. A similar type of magmatism also developed at a few other centres in the Ligurian Sea basin and southeastern France (e.g. Toulon; Réhault et al. 2012). This new outbreak of volcanism consists of two temporally, volumetrically, geographically and compositionally distinct stages. The first stage occurred between Tortonian and Upper Pliocene (11.8–4.4 Ma) and emplaced small volumes of magmas at a few centres scattered in southern Sardinia (Isola del Toro, Capo Ferrato, Guspini, Rio Girone; Fig. 11.1). The younger stage is Plio-Quaternary in age (3.9–0.1 Ma), and emitted

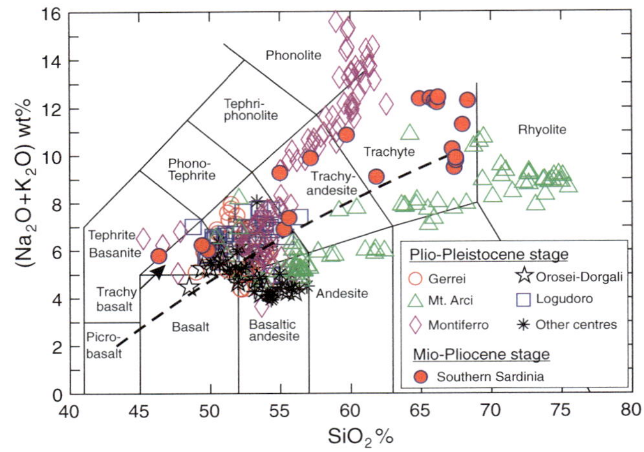

Fig. 11.8 TAS classification diagram (Le Maitre 2002) for Miocene-Quaternary volcanic rocks from Sardinia. Data are normalised to 100 % on a water-free basis

much larger volumes of magmas over an area of about 2200 km^2 in central-northern Sardinia. The two orogenic stages exhibit distinct petrological and geochemical characteristics (Figs. 11.8 and 11.9). In particular, the Mio-Pliocene rocks are more enriched in radiogenic Pb than the Miocene-Pliocene volcanics, and have been respectively referred to as Radiogenic-Pb Volcanics (RPV) and Unradiogenic-Pb Volcanics (UPV) by Lustrino et al. (2000, 2007a). Ages, volcanological and compositional features of the Miocene to Quaternary magmatic stages are summarised in Table 11.2.

11.4.1 Volcanology and Stratigraphy

The Mio-Pliocene volcanic stage consists of a few small outcrops, amounting some 1 % of anorogenic magmatism total volume. *Isola del Toro,* an about 300 m across island off the southwestern Sardinia coast, contains the oldest dated anorogenic rocks (benmoreite-trachyte; 11.8 Ma). *Capo Ferrato* (~6.6–5.0 Ma) is constructed over the Hercynian basement by mugearite to trachyte lava flows, dykes and dome. *Rio Girone* and *Guspini* (6.4 and 4.4 Ma) are two necks with a basanitic and hawaiitic composition, respectively (Lustrino et al. 2000, 2013 with references).

The Plio-Quaternary stage represents the bulk of anorogenic volcanism in Sardinia. It built up both small centres (Barisardo, San Pietro Baunei, Tharros, Gergei, Gesturi, Siddi, etc.) and large volcanic complexes and plateaux (Orosei-Dorgali, Monte Arci, Logudoro, Montiferro and Campeda-Planargia). *Barisardo* is a small volcano made up of hawaiite, mugearite, and a few basaltic andesite lava flows. A single basaltic lava flow approximately 5 km in length occurs at *S. Pietro Baunei*. *Capo Frasca* and *Tharros* are tiny outcrops of basaltic andesites (Lustrino et al. 2000). The *Gerrei* volcanic area (central Sardinia) is mostly formed of basaltic andesite and hawaiite lavas, sometimes containing ultramafic xenoliths (Rocco et al. 2012). *Monte Arci* is a 20 km long N-S trending volcanic ridge located on the eastern margin of the Campidano graben. It consists predominantly of dacitic to rhyolitic lava flows and minor basalts, trachybasalts and basaltic andesites (Cioni et al. 1982; Montanini et al. 1994). The *Orosei-Dorgali* volcanism consists of lava flows, sometimes with pillow structure, showing dominantly hawaiite and mugearite composition plus minor tholeiitic basaltic andesites, which spread over an area of about 150 km^2; activity took place along NE-SW trending faults cutting the crystalline basement and Mesozoic sedimentary rocks (Lustrino et al. 2002). The *Campeda-Planargia-Abbasanta-Paulilatino* basaltic plateaux are constructed by tholeiitic and Na-alkaline basaltic andesites, hawaiites, and mugearites, extending over an area of about 850 km^2 (Beccaluva et al.

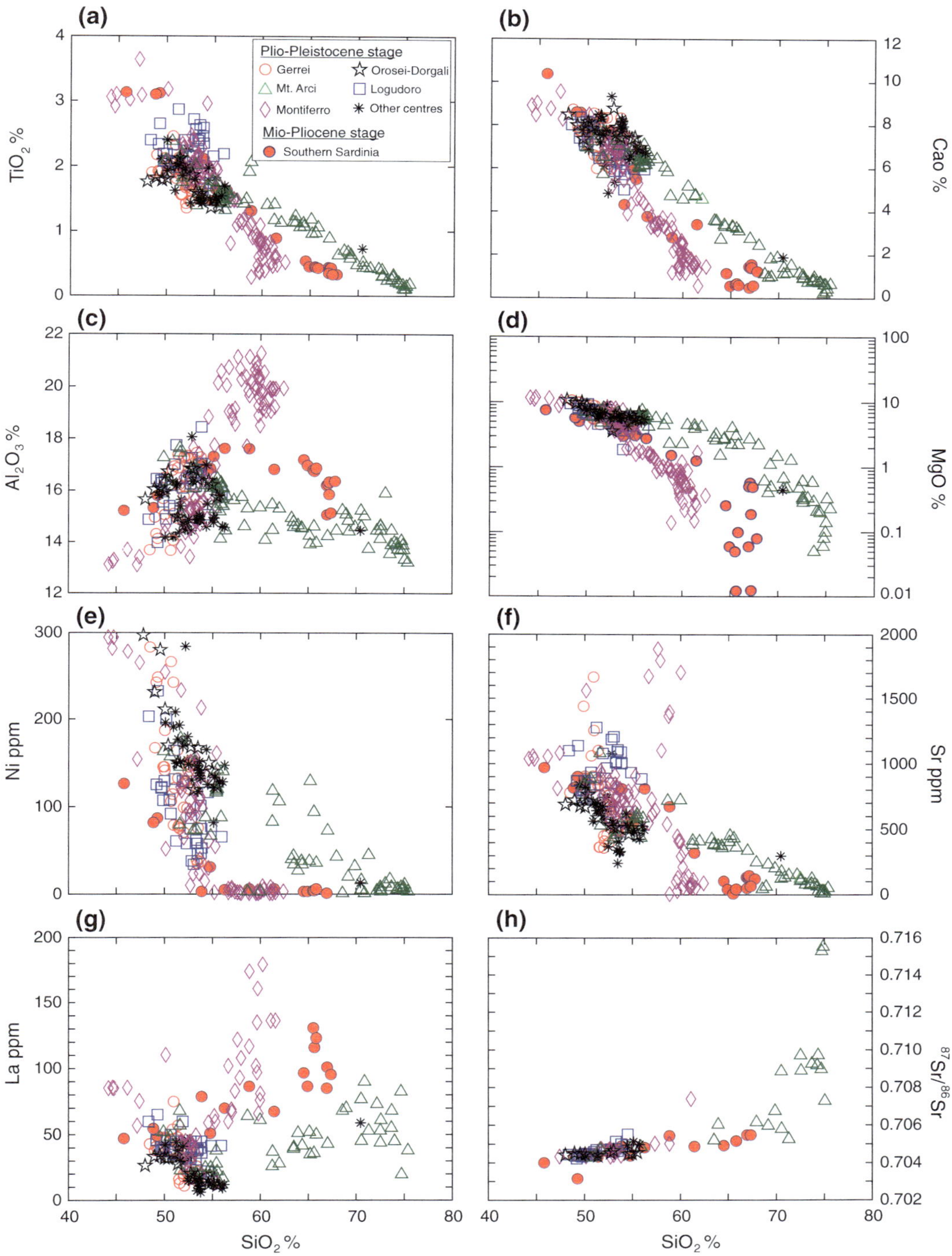

Fig. 11.9 Variation diagrams of selected major and trace elements, and Sr isotope ratio for anorogenic Miocene to Quaternary volcanic rocks from Sardinia

1977; Lustrino et al. 2004). The *Montiferro* complex consists of basanite, hawaiite, to trachyte-phonolite lava flows, dykes and domes cropping out over an area of about 400 km^2; volcanism is located in an uplifted zone of northern Sardinia at the intersection between NE-SW faults and the Campidano graben (Di Battistini et al. 1990; Lustrino et al. 2004; Fedele et al. 2007). The *Logudoro* volcanic field (3.2–0.1 Ma) represents the northernmost activity and contains the youngest rocks; eruptive centres developed along N-S and NE-SW faults, and erupted lavas, scoriae and minor pyroclastic deposits over an area of about 500 km^2; compositions are mafic and range from subalkaline to alkaline with sodic to mildly potassic affinity (Beccaluva et al. 1977; Savelli 1988; Gasperini et al. 2000).

11.4.2 Petrography and Mineral Chemistry

The rocks belonging to the Mio-Pliocene anorogenic volcanic stage range from trachybasalt to trachyte (Fig. 11.8). They have variably prophyritic textures. Plagioclase, clinopyroxene, and olivine are common phenocrysts of mafic rocks; plagioclase, alkali feldspars and clinopyroxene occur in the evolved rocks, which also contain kaersutitic amphibole, biotite and accessory apatite and titanite.

Plio-Quaternary rocks show variable compositions, from subalkaline to alkaline, and define two distinct evolution trends, toward phonolitic (Montiferro) and trachytic (Mt. Arci) compositions (Fig. 11.8). Mafic tholeiitic volcanics (mainly basaltic andesites) exhibit aphyric to porphyritic textures with olivine and plagioclase phenocrysts and microphenocrysts; orthopyroxene, augitic-pigeonitic clinopyroxene and Fe–Ti oxides crystallised later and are frequently confined to the matrix; the groundmass mostly contains the same phases and some glass (Beccaluva et al. 1975; Lustrino et al. 2000). Andesites, dacites and rhyolites from Monte Arci have porphyritic textures with variable absolute and relative abundances of plagioclase, orthopyroxene, clinopyroxene and opaques. Sanidine, biotite, and quartz phenocrysts occur in the rhyolites; the groundmass ranges from strongly to totally glassy, sometimes with perlitic texture; apatite and zircon are found as accessory phases (Cioni et al. 1982; Montanini et al. 1994). Basanites are porphyritic, with a phenocryst mineralogy characterised by variable amounts of olivine ($\sim Fo_{85-75}$), clinopyroxene (mostly Ti-augite), biotite, and Fe–Ti oxides; globular analcime occurs in the Montiferro basanites (Fedele et al. 2007); clinopyroxene and biotite sometimes occur as megacrysts. Hawaiites (trachybasalts) and mugearites (basaltic trachyandesites) range from subaphyritic to porphyritic with phenocrysts of olivine (up to Fo_{88}), labradoritic plagioclase ($\sim An_{60-50}$), diopside to augite clinopyroxene and rare amphibole set in a groundmass containing the same phases; sanidine is present in trachybasalts with potassic alkaline affinity. Trachytes contain abundant phenocrysts of sanidine plus minor clinopyroxene, biotite, plagioclase, and amphibole; orthopyroxene has been occasionally observed; zircon, apatite and opaque minerals are common accessories. The groundmass of trachytes is hypocrystalline to glassy.

Various types of xenoliths occur in the anorogenic volcanic rocks from Sardinia. These include high-pressure ultramafic nodules and a variety of crustal rocks. Ultramafic xenoliths (anhydrous spinel harzburgites and lherzolites, and minor websterites) have been found in alkaline basalts from Orosei-Dorgali, Montiferro, Logudoro, Rio Girone, and central Sardinia (e.g. Lustrino et al. 1999, 2004; Beccaluva et al. 2001; Rocco et al. 2012).

11.4.3 Petrology and Geochemistry

Volcanic rocks emplaced during the Mio-Pliocene stage show mildly alkaline character and define smooth trends on several major and trace element variation diagrams (Fig. 11.9). The Plio-Quaternary volcanics range from mafic to felsic and from subalkaline to alkaline. There is a scattering of composition in the mafic range, from which two main evolutionary series develop:

- A silica undersaturated Na-alkaline series at Montiferro, ranging from basanite-tephrite to phonolite, with some of the most evolved rocks reaching a peralkaline composition;
- A silica oversaturated subalkaline series (basalt-dacite-trachyte-rhyolite), represented by the large majority of Monte Arci rocks.

The undersaturated series exhibits more extreme enrichments in incompatible elements and a sharper decrease in ferromagnesian elements and Sr with increasing silica than the oversaturated series. The latter shows linear trends for several compatible elements, including Ni and Sr (Fig. 11.9e, f). REE show smooth and variably fractionated patterns, with negative Eu anomalies in the Monte Arci rhyolites (Fig. 11.10a). Incompatible element of mafic rocks exhibit indented mantle-normalised patterns, with some rocks exhibiting small positive spikes of Pb and negative anomalies of Nb and Ta (Fig. 11.10b); these rocks fall close or within the field of arc rocks on trace-element discriminant diagrams (Fig. 11.11). Sr-Nd-isotope compositions are highly variable with $^{87}Sr/^{86}Sr$ and $^{143}Nd/^{144}Nd$ ranging from about 0.7032 to 0.7155 and 0.51220 to 0.51289, respectively (Fig. 11.6a). The mafic rocks from the Mio-Pliocene southern occurrences plot on the mantle array, partially overlapping the fields of Etna and western Aeolian arc calcalkaline

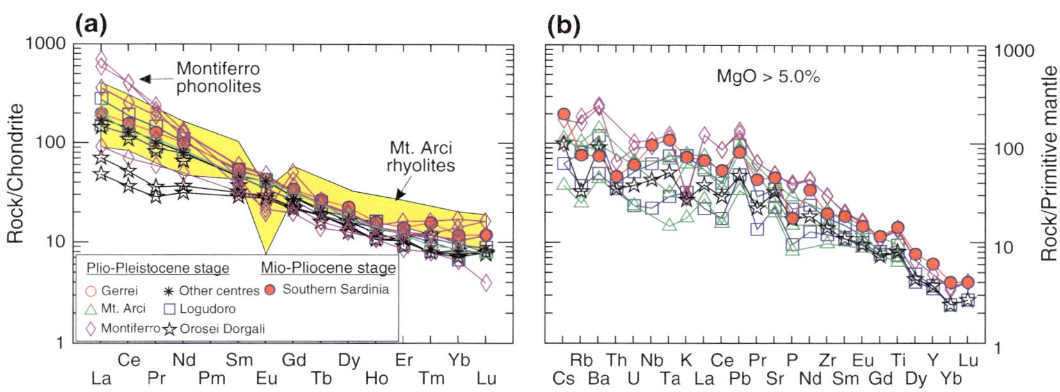

Fig. 11.10 **a** Chondrite normalised REE patterns of Miocene to Quaternary anorogenic volcanic rocks from Sardinia. The compositions of Mt. Arci rhyolites are reported as a yellow field; **b** Mantle-normalised incompatible element patterns of mafic rocks (MgO > 5 %)

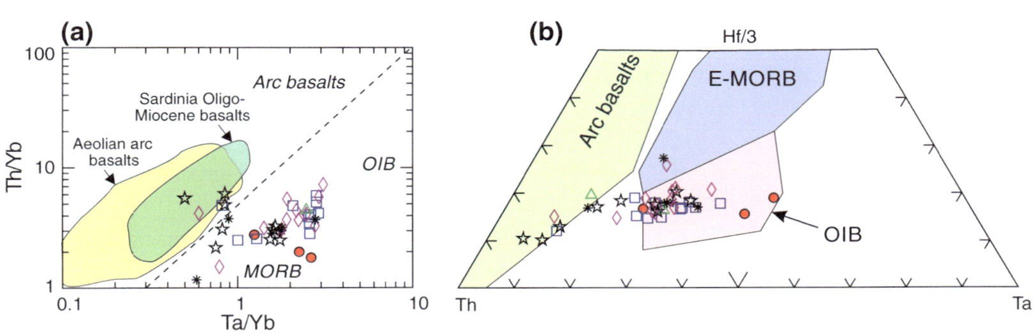

Fig. 11.11 Trace element discriminant diagrams for Miocene to Pleistocene mafic volcanic rocks from Sardinia. Symbols as in Fig. 11.10

basalts. The other rocks display a steeper trend, with the mafic rocks pointing toward the EM1 mantle end-member (Gasperini et al. 2000; Lustrino and Dallai 2003; Lustrino et al. 2013). There is an increase of Sr isotope ratios with silica for many volcanic centres, a feature that is particularly evident at Monte Arci (Fig. 11.9h). Pb-isotope ratios are also variable: the Mio-Pliocene volcanics have compositions close to FOZO ($^{206}Pb/^{204}Pb$ ~18.84 to 19.42; $^{207}Pb/^{204}Pb$ ~15.62 to 15.66; $^{208}Pb/^{204}Pb$ ~38.92 to 39.13), whereas the Plio-Quaternary rocks ($^{206}Pb/^{204}Pb$ ~17.37 to 18.07; $^{207}Pb/^{204}Pb$ ~15.53 to 15.62; $^{208}Pb/^{204}Pb$ ~37.44 to 38.23) have unradiogenic compositions, close to EM1 mantle end-member (Fig. 11.6). Hf isotopic ratios range from $^{176}Hf/^{177}Hf$ ~0.28258 to 0.28304 (Gasperini et al. 2000; Lustrino et al. 2013), with the highest values occurring in the Miocene-Pliocene centres. Os isotope ratios show $^{187}Os/^{188}Os = 0.1255–0.423$. Analyses of plagioclase, clinopyroxene and olivine gave $\delta^{18}O_{SMOW} = +6.46$ ‰ to +7.56 ‰ (Lustrino et al. 2013). Finally, Li isotope compositions determined for some Logudoro basalts yielded values of $\delta^{7}Li = +1.5$ to +3.6‰ relative to NIST L-SVEC standard (Kim et al. 2011).

Investigation on ultramafic xenoliths occurring at several Plio-Quaternary centres reveal variable REE fractionation and Sr-Nd isotope signatures for clinopyroxene; Sr–Nd isotope ratios ($^{87}Sr/^{86}Sr = 0.7038–0.7057$; $^{143}Nd/^{144}Nd = 0.51256–0.51295$) define two distinct trends that mimic those of anorogenic rocks (Fig. 11.6a). Overall, geochemical data have been interpreted to testify to processes of melt extraction, followed by metasomatic modification by alkaline melts or fluids (e.g. Lustrino et al. 1999; Beccaluva et al. 2001; Rocco et al. 2012).

11.4.4 Petrogenesis of Miocene-Quaternary Anorogenic Magmatism

Trends of major, trace elements and isotopic ratios versus silica indicate that different types of evolutionary processes occurred at the various anorogenic volcanic centres in Sardinia (e.g. Beccaluva et al. 1977). The Mio-Pliocene rocks (Isola del Toro, Capo Ferrato, Rio Girone, Guspini) show element variation consistent with a process of early fractionation of mafic phases, followed by separation of dominant alkali feldspar and some apatite during the advanced stages of fractionation. Increase of Sr isotope ratios with silica reveals crustal assimilation.

The Plio-Quaternary magmas underwent little evolution at most centres, except for Monte Arci and Montiferro. In the latter district, extensive fractional crystallisation acting on basanitic-tephritic parents generated a suite of alkaline daughter melts with phonolitic extreme compositions; separation of olivine, clinopyroxene and some Ti-magnetite dominated early stages of fractionation; late separation of alkali feldspar is testified by decrease in Sr and Ba in the phonolitic rocks. The Monte Arci subalkaline suite appears to have undergone both fractional crystallisation and interaction with the crust, as denoted by the increase of Sr-isotope ratios with silica and by the occurrence of partially digested crustal xenoliths in some lavas (Montanini et al. 1994). The low concentration of several incompatible elements such as LREE, Zr and Th in the acid rocks reveals an important role of fractionating accessory phases such as zircon and apatite. According to some authors, the most silica-rich rocks from Monte Arci were generated by crustal melting rather than by fractionation of mafic parents, whereas the intermediate magmas were derived from mixing between acid and mafic melts (Cioni et al. 1982; Montanini et al. 1994). Although a crustal anatectic origin for the most evolved rhyolites is questionable, the linear inter-element trends between mafic and silicic compositions militate in favour of mixing as a potentially important process of magma evolution.

Compositional variation among mafic rocks reflects contrasting compositions of primary melts. The radiogenic isotope diversity between the Mio-Pliocene and Plio-Quaternary volcanic stages is a first-order feature of anorogenic magmatism in Sardinia. As recalled earlier, the

Mio-Pliocene volcanic rocks have radiogenic isotope signatures not far from other anorogenic rocks from the Western Mediterranean area, including Etna and Iblei (Lustrino and Wilson 2007). Therefore, the Mio-Pliocene magmas likely derived from asthenospheric to lower lithospheric sources similar to other anorogenic rocks form the Tyrrhenian Sea area. However, the Sardinia Mio-Pliocene magmas have lower Pb isotope ratios than Etna and Iblei, requiring a role for a distinct unradiogenic-Pb component such as EM1.

Plio-Quaternary tholeiitic and alkaline rocks show distinct incompatible trace element abundances but overlapping radiogenic isotope compositions that have led to the conclusion that primary melts were generated by different degrees of partial melting of broadly homogeneous sources. Based on geochemical and phase equilibria constraints, about 3–6 % and 8–12 % melting at 1.0–2.0 GPa have been suggested for parental melts of alkaline and tholeiitic magmas, respectively (Lustrino et al. 2002, 2004; Beccaluva et al. 2010). These correspond to lithospheric to uppermost asthenospheric levels in Sardinia. The Plio-Quaternary rocks define a distinct Sr–Nd isotopic trend and have much lower Pb-isotope ratios than Mio-Pliocene rocks, approaching closely EM1 (Fig. 11.6). Such a composition is unique in Europe and there is debate concerning its origin both in Sardinia and elsewhere (e.g. Willbold and Stracke 2006 with references). Gasperini et al. (2000) noticed that the unradiogenic Pb-isotope signatures of Logudoro basalts are accompanied by high Ba/La, Ce/Pb, Eu/Eu* and Sr/Eu* values,[1] relative to oceanic basalts. In contrast, the Nb/U ratios are similar. These features were interpreted to indicate that the mantle source of the Logudoro magmas was an ancient basaltic plateau, which was recycled into the mantle over a long time and finally sampled by a plume. In particular, the excess Ba and Eu was considered to indicate a plagioclase component in the source, which was explained by assuming that the basaltic plateau material was composed of a high proportion of gabbros formed originally by plagioclase accumulation. Lustrino et al. (2004) argued that the volcanic rocks from north-central Sardinia do not show a significant correlation of Eu/Eu* versus Sr/Nd and Sr/Eu*, as expected for plagioclase-dominated compositions, and suggested that the unique EM1 signature of the Sardinian basalts could be related to local lithospheric mantle processes. It was postulated that strong syn-collisional crustal shortening and thickening during the Hercynian (or older) continental collision generated delamination and passive sinking of a dense, granulitic to eclogitic mafic lower crustal keel. This underwent partial melting, giving silicic liquids, which reacted with the surrounding mantle. The lower crust is depleted in U and Rb and enriched in Ba and Sr. Liquids generated by melting of delaminated lower crust inherited and transferred these characteristics to the lithospheric mantle. Viewed in this light, the EM1 composition in Sardinia would result from the ageing of a metasomatised lithosphere (Lustrino and Dallai 2003). Metasomatism should have occurred at least 1 Ga ago, in order to explain the low $^{206}Pb/^{204}Pb$ and $^{207}Pb/^{204}Pb$ of the anorogenic magmas from northern Sardinia (Lustrino et al. 2013). Finally, some Plio-Quaternary magmas also show arc-type trace element signatures that suggest a role of orogenic components (Fig. 11.11). A Miocene calc-alkaline heritage has been long recognised for the Monte Arci volcanism (Cioni et al. 1982). Rocks or fluids of the Oligo-Miocene orogenic magmatic stage, which were preserved within the lithosphere, may have imparted these arc-type signatures to younger magmas.

11.5 Geodynamic Settings of Orogenic and Anorogenic Magmas

The contrasting compositions of Cenozoic magmatism in Sardinia record variable types of geodynamic settings. The most widely accepted

[1]Eu* is the concentration of Eu estimated by interpolation between the nearby elements Sm and Gd on chondritic REE patterns.

hypothesis on the origin of the Oligo-Miocene orogenic volcanic stage is a genetic relationship with the Middle Eocene-Miocene NW-directed subduction of the Neotethys oceanic lithosphere (also named as Alpine Tethys, Ionian Tethys or Mesogean Ocean) beneath the Corsica-Sardinia sector of the southern European margin (e.g. Carminati et al. 2010 with references). However, only tiny magmatic manifestations took place during early subduction stages and the bulk of orogenic activity developed much later, at about 24–16 Ma. The climax of volcanic activity was coeval with the Fossa Sarda rifting in central Sardinia and with the maximum rotation rate of the Corsica-Sardina block with respect to the continental Europe (Gattacceca et al. 2007). According to Lustrino et al. (2013) the earliest episodes of orogenic activity are related to melting processes in the lower continental crust, whereas the bulk of orogenic magmatism originated within the mantle wedge, which was fluxed and contaminated by fluids released by the subducted Neotethys oceanic crust. The pre-metasomatic mantle source consisted of MORB-type rocks, but radiogenic Pb isotope signatures suggest a role of an EM1 component probably residing within the lithosphere. The role of such a component will become stronger during the following anorogenic magmatism.

The anorogenic magmatic stage started shortly after the end of the orogenic magmatism and of the Sardinia-Corsica rotation, contemporaneously with the onset of tectonic extension that ultimately brought to the formation of the Tyrrhenian Sea basin (Gattacceca et al. 2007). However, the bulk of anorogenic magmas was generated much later, during Upper Pliocene-Pleistocene times, probably triggered by ascending asthenosphere. The extensional tectonic regime responsible for mantle upwelling may be related to the opening of the backarc Tyrrhenian Sea basin, or also to a more extended rift system running from southeastern France, through Sardinia and the Sicily Channel, until Northern Africa (Corti et al. 2006).

As discussed earlier, the geodynamic significance of EM1-type Upper Pliocene-Pleistocene magmas is controversial. Gasperini et al. (2000) suggest an origin by melting of deep mantle plumes, a hypothesis challenged by Lustrino and coworkers (e.g. Lustrino et al. 2007a, 2013), who advocate a magma genesis within an old contaminated lithosphere. Such a hypothesis better explains the uniqueness of EM1-type composition in Sardinia. Its permanence within the lithosphere also accounts for the relatively unradiogenic Pb-isotope ratios observed in all the Sardinian Cenozoic volcanic rocks, including Oligo-Miocene calcalkaline magmas.

The geographic distribution of the two anorogenic stages is also worth of note. If not related to covering of tiny early outcrops by the abundant younger volcanics, such a distribution may reflect a heritage of old geological history of the island. The southern and northern Sardinia sectors represent two distinct pre-Carboniferous paleogeographic domain, which collided and welded during Hercynian orogeny (e.g. Carmignani et al. 2001b). Therefore, the different compositions of magmas erupted in distinct sectors could reflect different origin and evolution of Paleozoic blocks. This reinforces the idea of an important role of the lithosphere in the origin of anorogenic magmatism in Sardinia.

11.6 Summary and Conclusions

Sardinia is the site of extensive and compositionally variable Cenozoic volcanic activity. The first stage shows orogenic geochemical characteristics and developed in the Oligo-Miocene with a peak during Lower Miocene, contemporaneously with the maximum rotation rate of the Corsica-Sardinia block with respect to the European continent. Coeval orogenic-type magmatism also occurred in the Corsica-Ligurian basin and in Provence, southeastern France. Compositions are mostly calcalkaline with minor shoshonites and arc tholeiites. The orogenic magmatic stage is related to the subduction of the western margin of the Ionian Tethys ocean floor beneath the Sardinia-Corsica basement. Pb isotopic ratios of the most primitive uncontaminated basalts show significantly lower values than the equivalent rocks from the Aeolian arc, which are also related to subduction of the Ionian oceanic plate. This requires the involvement of an unradiogenic mantle component alike EM1, which will play a major role during the following anorogenic magmatic activity.

The following volcanism has anorogenic geochemical signatures, sharply contrasting with previous stage. Activity developed at several centres in Sardinia, but a very few occurrences have been found in the Ligurian Sea basin and in Provence. The anorogenic magmatism took place during two main stages. An early Upper Miocene to Lower Pliocene magmatic stage emplaced low-volume lavas and subvolcanic bodies in southern Sardinia between about 12 and 4.5 Ma; a younger much more voluminous stage occurred during Plio-Pleistocene time, between about 3.9 and 0.1 Ma in central-northern Sardinia. The two stages of anorogenic volcanism show significant geochemical and isotopic diversities, with the younger rocks exhibiting less radiogenic Pb isotopic signatures than the older southern Sardinian occurrences. These differences point to distinct types of mantle sources for primary magmas. The older anorogenic rocks from southern Sardinia are compositionally similar to other anorogenic centres in the Mediterranean region, and probably originated in the lower lithospheric or upper asthenospheric mantle passively ascending during extension. The younger anorogenic stage shows EM1-type radiogenic isotope composition that is unique in Europe. Its origin and composition are controversial and could be related either to melting of a mantle plume, or to a lithospheric mantle, which had been affected by a very old contamination event (Paleozoic or Pre-Cambrian) by acidic melts coming from delaminated lower crust. Some of the anorogenic magmas show relatively high LILE/HFSE ratios, which may be a heritage of the Oligo-Miocene magmatic stage.

References

Alvarez W (1972) Rotation of the Corsica-Sardinia microplate. Nature 235:103–105

Araña V, Barberi F, Santacroce R (1974) Some data on the comendite type-area of S. Pietro and S. Antioco islands, Sardinia. Bull Volcanol 38:725–736

Beccaluva L, Macciotta G, Venturelli G (1975) Dati geochimici e petrografici sulle vulcaniti Plio-Quaternarie della Sardegna centro-occidentale. Boll Soc Geol It 94:1437–1457

Beccaluva L, Deriu M, Macciotta G, Savelli C, Venturelli G (1977) Geochronology and magmatic character of the Pliocene-Pleistocene volcanism in Sardinia. Bull Volcanol 40:1–16

Beccaluva L, Bianchini G, Coltorti M, Perkins WT, Siena F, Vaccaro C, Wilson M (2001) Multistage evolution of the European lithospheric mantle: new evidence from Sardinian peridotite xenoliths. Contrib Mineral Petrol 142:284–297

Beccaluva L, Bianchini G, Bonadiman C, Natali C, Siena F (2010) Petrogenesis and geodynamic control of intraplate Cenozoic volcanism in Italy. In: Beltrando M, Peccerillo A, Mattei M, Conticelli S, Doglioni C (eds) The Geology of Italy: tectonics and life along plate margins, J Virtual Explorer 36, paper 19 doi: 10.3809/jvirtex.2010.00240

Boni M, Muchez P, Schneider J (2002) Permo-Mesozoic multiple fluid flow and ore deposits in Sardinia: a comparison with post-Variscan mineralization of Western Europe. In: Blundell DJ, Neubauer F, VonQuadt A (eds) Timing and location of major ore deposits in an evolving Orogen, vol 204, pp 199–211

Boni M, Balassone G, Fedele L, Mondillo N (2009) Post-Variscan hydrothermal activity and ore deposits in southern Sardinia (Italy): selected examples from Gerrei (Silius Vein System) and the Iglesiente district. Per Mineral 78:19–35

Brotzu P, Lonis R, Melluso L, Morbidelli L, Traversa G, Franciosi L (1997) Petrology and evolution of calcalkaline magmas from the Arcuentu volcanic complex (SW Sardinia, Italy). Per Mineral 66:151–184

Carmignani L, Carosi R, Di Pisa A, Gattiglio M, Musumeci C, Oggiano G, Pertusati PC (1994a) The Hercynian chain in Sardinia (Italy). Geodyn Acta 7:31–47

Carmignani L, Barca S, Disperati L, Fantozzi P, Funedda A, Oggiano G, Pasci S (1994b) Tertiary compression and extension in the Sardinian basement. Boll Geofis Teor Appl 36:5–62

Carmignani L, Oggiano G, Barca S, Conti P, Salvadori I, Eltrudis A, Funedda A, Pasci S (2001) Geologia della Sardegna. Note illustrative della Carta Geologica in scala 1:200.000. Mem Descr Carta Geol It 60:283 pp

Carmignani L, Funedda A, Oggiano G, Pasci S (2004) Tectono-sedimentary evolution of southwest Sardinia in the paleogene: Pyrenaic or Apenninic dynamic? Geodin Acta 17:275–287

Carminati E, Lustrino M, Cuffaro M, Doglioni C (2010) Tectonics, magmatism and geodynamics of Italy: what we know and what we imagine. In: Beltrando M, Peccerillo A, Mattei M, Conticelli S, Doglioni C

(eds) J Virtual Explorer 36, paper 9. doi:10.3809/jvirtex.2010.00226
Carminati E, Lustrino M, Doglioni C (2012) Geodynamic evolution of the central and western Mediterranean: tectonic vs. igneous constraints. Tectonophysics 579:173–192
Casini L, Cuccurru S, Puccini A, Oggiano G, Rossi P (2015) Evolution of the Corsica-Sardinia Batholith and late-orogenic shearing of the Variscides. Tectonophysics 646:65–78
Cioni R, Clocchiatti R, Di Paola GM, Santacroce R, Tonarini S (1982) Miocene calc-alkaline heritage in the Pliocene post-collisional volcanism of Monte Arci (Sardinia, Italy). J Volcanol Geoth Res 14:133–167
Conte AM (1993) High-temperature amphibole breakdown in calcalkaline subvolcanic rocks from Sarroch district (Sardinia, Italy); Implication for mixing processes. Neues Jb Miner Monat 1993:133–144
Conte AM (1997) Petrology and geochemistry of Tertiary calc-alkaline magmatic rocks from the Sarroch domain (Sardinia, Italy). Per Mineral 66:63–100
Conte AM, Palladino DM, Perinelli C, Argenti E (2010) Petrogenesis of the high-alumina basalt-andesite suite from Sant'Antioco Island, SW Sardinia, Italy. Per Mineral 79:27–55
Conticelli S, Carlson RW, Widom E, Serri G (2007) Chemical and isotopic composition (Os, Pb, Nd and Sr) of Neogene to Quaternary calc-alkalic, shoshonitic, and ultrapotassic mafic rocks from the Italian peninsula: inferences on the nature of their sources. In: Beccaluva L, Bianchini G, Wilson M (eds) Cenozoic volcanismin the Mediterranean area. Geol Soc Am Spec Paper 418:171–202. doi:10.1130/2007.2418(09)
Cortesogno L, Cassinis G, Dallagiovanna G, Gaggero L, Oggiano G, Ronchi A, Seno S, Vanossi M (1998) The Variscan post-collisional volcanism in Late Carboniferous Permian sequences of Ligurian Alps, Southern Alps and Sardinia (Italy): a synthesis. Lithos 45:305–328
Corti G, Cuffaro M, Doglioni C, Innocenti F, Manetti P (2006) Coexisting geodynamic processes in the Sicily Channel. In: Dilek Y and Pavlides S (eds) Postcollisional tectonics and magmatism in the Mediterranean region and Asia, Geol Soc Am Spec Paper 409: 83–96
Coulon C (1977) Le volcanisme calco-alcaline cénozoique de Sardaigne (Italie): petrography, géochimie et genèse des laves andesitiques et des ignimbrites. Signification geodinamique. PhD Thesis, Univer Marseille, 385 pp
Coulon C, Dostal J, Dupuy C (1978) Petrology and geochemistry of ignimbrites and associated lavas domes from N.W. Sardinia. Contrib Mineral Petrol 68:89–98
Di Battistini G, Montanini A, Zerbi M (1990) Geochemistry of volcanic rocks from southeastern Montiferro. Neues Jb Miner Abh 162:35–67
Di Vincenzo G, Andriessen PAM, Ghezzo C (1996) Evidence of two different components in a Hercynian peraluminous cordierite-bearing granite: the San Basilio intrusion (Central Sardinia, Italy). J Petrol 37:1175–1206
Downes H, Thirlwall MF, Trayhorn SC (2001) Miocene subduction-related magmatism in southern Sardinia: Sr-Nd and oxygen isotopic evidence for mantle source enrichment. J Volcanol Geoth Res 106:1–21
Fedele L, Lustrino M, Melluso L, Morra V, D'Amelio F (2007) The Pliocene Montiferro volcanic complex (central-western Sardinia, Italy): geochemical observations and petrological implications. Per Mineral 76:101–136
Franciosi L, Lustrino M, Melluso L, Morra V, D'Antonio M (2003) Geochemical characteristics and mantle sources of the Oligo-Miocene primitive basalts from Sardinia: the role of subduction components. Ofioliti 28:105–114
Gale A, Dalton CA, Langmuir CH, Su Y, Schilling J-G (2013) The mean composition of ocean ridge basalts. Gechem Geophys Geosyst 14. doi:10.1029/2012GC004334
Gasperini D, Blichert-Toft J, Bosch D, Del Moro A, Macera P, Télouk P, Albaréde F (2000) Evidence from Sardinian basalt geochemistry for recycling of plume heads into the Earth's mantle. Nature 408:701–704
Gattacceca J, Deino A, Rizzo R, Jones DS, Henry B, Beaudoin B, Vadeboin F (2007) Miocene rotation of Sardinia: new paleomagnetic and geochronological constraints and geodynamic implications. Earth Planet Sci Lett 258:359–377
Grad M, Tii Tiira et al (2009) The Moho depth map of the European Plate. Geophys J Int 176:279–292
Guarino V, Fedele L, Franciosi L, Lonis R, Lustrino M, Marrazzo M, Melluso L, Morra V, Rocco I, Ronga F (2011) Mineral compositions and magmatic evolution of the calcalkaline rocks of northwestern Sardinia, Italy. Per Mineral 80:517–545
Hart SR (1984) A large-scale isotopic anomaly in the Southern Hemisphere mantle. Nature 309:753–757
Ivaldi J-P, Bellon H, Guardia P, Mangan C, Müller C, Perez J-L, Terramorsi S (2002) Contexte lithostructural, âges ^{40}K-^{40}Ar et géochemie du volcanisme calco-alcalin tertiaire de Cap-d'Ail dans le tunnel ferroviaire de Monaco. C R Geosci 335:411–421
Kim T, Nakai S, Gasperini D (2011) Lithium abundance and isotope composition of Logudoro basalts, Sardinia: origin of light Li signature. Geochem J 45:323–340
Lattanzi P (1999) Epithermal precious metal deposits of Italy—an overview. Miner Deposita 34:630–638
Le Maitre RW (ed) (2002) A classification of igneous rocks and glossary of terms. Cambridge University Press, Cambridge, 252 pp
Lecca L, Lonis R, Luxoro S, Melis E, Secchi F, Brotzu P (1997) Oligo-Miocene volcanic sequences and rifting stages in Sardinia: a review. Per Mineral 66:7–61
Lustrino M, Dallai L (2003) On the origin of EM1 end-member. Neues Jb Miner Abh 179:85–100
Lustrino M, Wilson M (2007) The circum-Mediterranean anorogenic Cenozoic igneous province. Earth Sci Rev 81:1–65

Lustrino M, Melluso L, Morra V (1999) Origin of glass and its relationships with phlogopite in mantle xenoliths from central Sardinia (Italy). Per Mineral 68:13–42

Lustrino M, Melluso L, Morra V (2000) The role of lower continental crust and lithospheric mantle in the genesis of Plio-Pleistocene volcanic rocks from Sardinia (Italy). Earth Planet Sci Lett 180:259–270

Lustrino M, Melluso L, Morra V (2002) The transition from alkaline to tholeiitic magmas: a case study from the Orosei-Dorgali Pliocene volcanic district (NE Sardinia, Italy). Lithos 63:83–113

Lustrino M, Morra V, Melluso L, Brotzu P, D'Amelio F, Fedele L, Lonis R, Franciosi L, Petteruti Lieberknect AM (2004) The Cenozoic igneous activity in Sardinia. Per Mineral 73:105–134

Lustrino M, Melluso L, Morra V (2007a) The geochemical peculiarity of "Plio-Quaternary" volcanic rocks of Sardinia in the circum-Mediterranean area. In: Beccaluva L, Bianchini G, Wilson M (eds) Cenozoic Volcanism in the Mediterranean Area, Geol Soc Am Spec Paper 418:277301

Lustrino M, Morra V, Fedele L, Serracino M (2007b) The transition between 'orogenic' and 'anorogenic' magmatism in the western Mediterranean area: the Middle Miocene volcanic rocks of Isola del Toro (SW Sardinia, Italy). Terra Nova 19:148–159

Lustrino M, Morra V, Fedele L, Franciosi L (2009) Beginning of the Apennine subduction system in central western Mediterranean: constraints from Cenozoic "orogenic" magmatic activity of Sardinia, Italy. Tectonics 28 doi:10.1029/2008tc002419

Lustrino M, Duggen S, Rosenberg CL (2011) The Central-Western Mediterranean: anomalous igneous activity in an anomalous collisional tectonic setting. Earth Sci Rev 104:1–40

Lustrino M, Fedele L, Melluso L, Morra V, Ronga F, Geldmacher J, Duggen S, Agostini S, Cucciniello C, Franciosi L, Meisel T (2013) Origin and evolution of Cenozoic magmatism of Sardinia (Italy). A combined isotopic (Sr-Nd-Pb-O-Hf-Os) and petrological view. Lithos 180:138–158

Marcello A, Pretti S, Valera P, Agus M, Boni M, Fiori M (2004) Metallogeny in Sardinia (Italy): from the Cambrian to the Tertiary. Field Trip Guide Book 32nd Int Geol Congr, Florence, Italy

Martí J, Mitjavila J, Roca E, Aparicio A (1992) Cenozoic magmatism of the Valencia trough (western Mediterranean): relationship between structural evolution and volcanism. Tectonophysics 203:145–165

Mattioli M, Guerrera F, Tramontana M, Raffaelli G, D'Atri M (2000) High-Mg tertiary basalts in Southern Sardinia (Italy). Earth Planet Sci Lett 179:1–7

Montanini A, Barbieri M, Castorina F (1994) The role of fractional crystallization, crustal melting and magma mixing in the petrogenesis of rhyolites and mafic inclusion-bearing dacites from the Monte Arci volcanic complex (Sardinia, Italy). J Volcanol Geoth Res 61:95–120

Morra V, Secchi FA, Assorgia A (1994) Petrogenetic significance of peralkaline rocks from Cenozoic calc-alkaline volcanism from SW Sardinia, Italy. Chem Geol 118:109–142

Morra V, Secchi FA, Melluso L, Franciosi L (1997) High-Mg subduction-related Tertiary basalts in Sardinia, Italy. Lithos 40:69–91

Oggiano G, Funedda A, Carmignani L, Pasci S (2009) The Sardinia-Corsica microplate and its role in the Northern Apennine geodynamics: new insights from the Tertiary intraplate strike-slip tectonics of Sardinia. It J Geosc 128:527–541

Ottaviani Spella MM, Girard M, Rochette P, Cheilletz A, Thinon M (2001) Burdigalian volcanism in southern Corsica: petrology, K-Ar dating, palaeomagnetism. C R Acad Sci IIA 333:113–120

Oudet J, Munch P, Verati C, Ferrandini M, Melinte-Dobrinescu M, Gattacceca J, Cornee JJ, Oggiano G, Quillevere F, Borgomano J, Ferrandini J (2010) Integrated chronostratigraphy of an intra-arc basin: $^{40}Ar/^{39}Ar$ datings, micropalaeontology and magnetostratigraphy of the early Miocene Castelsardo basin (northern Sardinia, Italy). Palaeogeogr Palaeocl 295:293–306

Palomba M, Padalino G, Marchi M (2006) Industrial mineral occurrences associated with Cenozoic volcanic rocks of Sardinia (Italy): geological, mineralogical, geochemical features and genetic implications. Ore Geol Rev 29:118–145

Panza GF, Peccerillo A, Aoudia A, Farina B (2007) Geophysical and petrological modeling of the structure and composition of the crust and upper mantle in complex geodynamic settings: the Tyrrhenian Sea and surroundings. Earth Sci Rev 80:1–46

Peccerillo A, Taylor SR (1976) Geochemistry of Eocene calc-alkaline volcanic rocks of the Kastamonu area, northern Turkey. Contrib Mineral Petrol 58:63–81

Poli G, Ghezzo C, Conticelli S (1989) Geochemistry of granitic rocks from the Hercynian Sardinia-Corsica batholith: implication for magma genesis. Lithos 23:247–266

Réhault J-P, Honthaas C, Guennoc P, Bellon H, Ruffet G, Cotton J, Sosson M, Maury RC (2012) Offshore Oligo-Miocene volcanic fields within the Corsica-Liguria Basin: magmatic diversity and slab evolution in the western Mediterranean Sea. J Geodyn 58:73–95

Rocco I, Lustrino M, Morra V, Melluso L (2012) Petrological, geochemical and isotopic characteristics of the lithospheric mantle beneath Sardinia (Italy) as indicated by ultramafic xenoliths enclosed in alkaline lavas. Int J Earth Sci 101:1111–1125

Rollet N, Deverchère J, Beslier O, Guennoc P, Réhault J-P, Sosson M, Truffert C (2002) Back-arc extension, tectonic inheritance and volcanism in the Ligurian Ocean, Mediterranean Sea. Tectonics 21. doi:10.1029/2001tc900027

Rossi P, Guennoc P, Réhault J-P, Arnaud N, Bouchra J, Poupeau G, Tegyey Ferrandini J, Sosson M,

Beslier MO, Rollet N, Gloaguen R (1998) Importance du volcanisme calco-alcalin miocéne sur la marge sud-ouest de la Corse (campagne Marco). C R Acad Sci IIA 327:369–376

Rossi P, Cocherie A, Fanning CM (2015) Evidence in Variscan Corsica of a brief and voluminous Late Carboniferous to Early Permian volcanic-plutonic event contemporaneous with a high-temperature/low-pressure metamorphic peak in the lower crust. Bull Soc Geol Fr 186:171–192

Sartori R, Torelli L, Zitellini N, Carrara G, Magaldi M, Mussoni P (2004) Crustal features along a W-E Tyrrhenian transect from Sardinia to Campania margins (Central Mediterranean). Tectonophysics 383:171–192

Savelli C (1988) Late Oligocene to recent episodes of magmatism in and around the Tyrrhenian Sea: implications for the processes of opening in a young inter-arc basin of intra-orogenic (Mediterranean) type. Tectonophysics 146:163–181

Speranza F, Villa IM, Sagnotti L, Florindo F, Cosentino D, Cipollari P, Mattei M (2002) Age of the Corsica-Sardinia rotation and Liguro-Provençal basin spreading: new paleomagnetic and Ar/Ar evidence. Tectonophysics 347:231–251

Stracke A, Hofmann AW, Hart SR (2005) FOZO, HIMU, and the rest of the mantle zoo. Geochem Geophys Geosyst 6:Q05007

Tait JA, Bachtadse V, Franke W, Soffel HC (1997) Geodynamic evolution of the European Variscan Fold Belt: paleomagnetic and geological constraints. Geol Rund 86:585–598

Tommasini S, Poli G, Halliday AN (1995) The role of sediment subduction and crustal growth in Hercynian plutonism: isotopic and trace element evidence from the Sardinia-Corsica batholith. J Petrol 36:1305–1332

Willbold M, Stracke A (2006) Trace element composition of mantle end-members: implications for recycling of oceanic and upper and lower continental crust. Geochem Geophys Geosyst 7:Q04004. doi:10.1029/2005GC001005

Zattin M, Massari F, Dieni I (2008) Thermochronological evidence for Mesozoic-Tertiary tectonic evolution in the eastern Sardinia. Terra Nova 20:469–474

Zindler A, Hart SR (1986) Chemical geodynamics. Ann Rev Earth Planet Sci 14:493–571

12 Southern Tyrrhenian Sea

Abstract

The Southern Tyrrhenian seafloor (south of the 41° Parallel line) hosts a large number of Miocene to Quaternary volcanoes ranging in composition from MORB- to OIB- and arc-type. MORB-type rocks are concentrated in the Vavilov Basin, but their role is evident for other Tyrrhenian Sea volcanoes. OIB-type components similar to Etna and Ustica (FOZO) are better shown by the Prometeo lava field (SE of Ustica Island) and by a single sample from the Marsili Seamount. A distinct OIB-type composition, similar to Sardinia Plio-Quaternary magmas, is found for a few rocks collected offshore eastern Sardinia. Subduction-type rocks are widely distributed throughout the Tyrrhenian Sea basin, from the Upper Miocene potassic alkaline Cornacya volcano (offshore southeastern Sardinia) to the tholeiitic, calcalkaline and shoshonitic seamounts in the central Tyrrhenian Sea and the Aeolian area. The wide variety of rock types reveals a highly heterogeneous upper mantle in the Southern Tyrrhenian Sea. Such extreme heterogeneity is debated. Most authors agree that the Tyrrhenian Sea represents a back-arc basin where melting of various mantle rocks (ancient lithosphere, passively ascending asthenosphere, and mantle rocks inflown from the foreland) is responsible for the generation of MORB- and OIB-type magmas. Arc-type magmatism would be related to the release of fluids by the Ionian subduction systems that migrated south-eastward contemporaneously with the opening of the Southern Tyrrhenian Sea.

Keywords

Tyrrhenian seamounts · MORB · OIB · Arc-type rocks · Vavilov · Magnaghi · Marsili · Aeolian seamounts · Backarc basin · Ionian plate

A. Peccerillo, *Cenozoic Volcanism in the Tyrrhenian Sea Region*, Advances in Volcanology, DOI 10.1007/978-3-319-42491-0_12

12.1 Introduction

The Tyrrhenian Sea is an extensional basin formed from Miocene to Quaternary behind the eastward retreating Apennine-Maghrebian compression front. Conventionally, it is divided into a northern and a southern sector, separated by an important tectonic structure known as the 41° Parallel line. Magmatism within the Tyrrhenian Sea is characterized by a wide variety of rocks, ranging in age from approximately 12 Ma to the present. Igneous activity in the northern Tyrrhenian Sea has been dominated by granitoid magmatism and by minor mafic to felsic volcanism belonging to the Tuscany Magmatic Province (see Chap. 2). The southern Tyrrhenian Sea contains a wide variety of volcanic rocks spanning MORB-, OIB- and arc-type compositions. All these magmas are of ultimately mantle origin and their genesis and tectonic setting are discussed in this chapter.

Location of the main eruptive centres of the southern Tyrrhenian Sea is shown in Fig. 12.1. Information on age, petrological and volcanological characteristics is given in Table 12.1. Selected analyses are reported in Table 12.2. Many of the data discussed in this chapter are the results of two oceanographic research programs, the Deep Sea Drilling Project (DSDP) carried out in 1975, and the Ocean Drilling Program (ODP) in 1986. Reports of the ODP on DSDP can be found at the websites http://www-odp.tamu.edu/publications/107_SR/107TOC.HTM and http://www.deepseadrilling.org/42_1/dsdp_toc.htm.

12.2 Geological Setting

The Tyrrhenian Sea started to open in the Upper Miocene (about 13–11 Ma before present) and is still expanding in its southeastern sector (e.g. Sartori 2003; Sartori et al. 2004; Carminati et al.

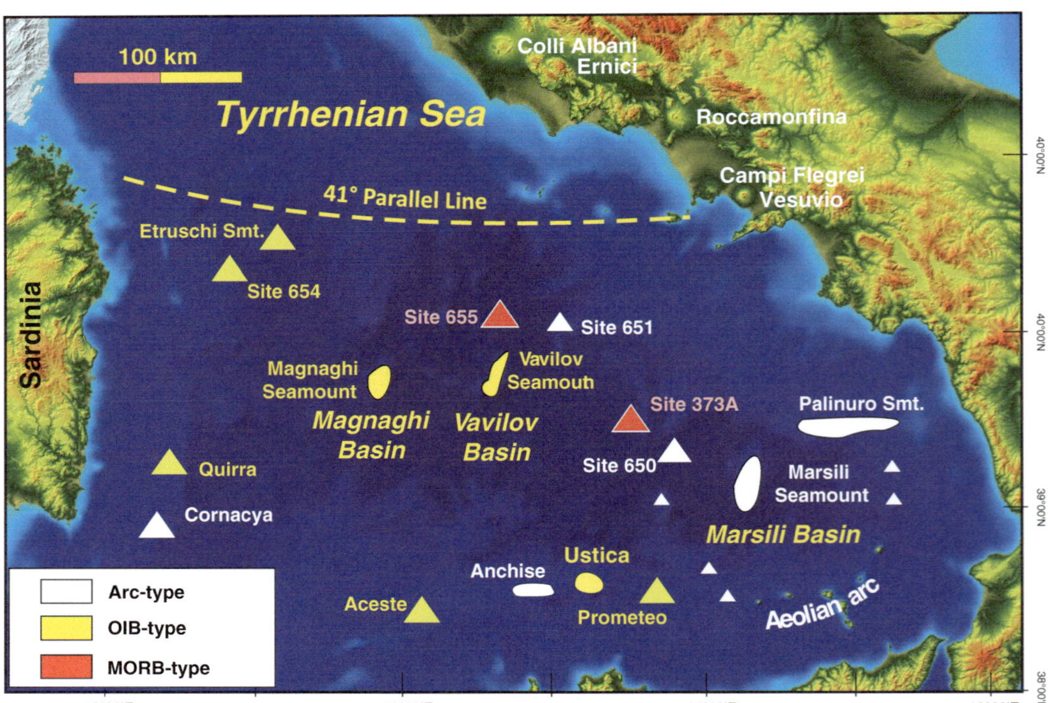

Fig. 12.1 Distribution of volcanism in the southern Tyrrhenian Sea. Detailed map of the Marsili Basin seamounts is reported in Fig. 12.8

Table 12.1 Petrological characteristics and ages of the main volcanic centres of the Tyrrhenian Sea floor

Magma types	Main volcanic centres	Age (in Ma)	Volcanology and petrology
MORB-type	DSDP Site 373A ODP Site 655	7.3– 4.3	– Abyssal tholeiitic basalt lavas from the Vavilov Basin seafloor
OIB-type	ODP Site 654	Plio-Quaternary	– Subalkaline basaltic andesite lavas similar to some Sardinia Plio-Quaternary volcanics
	South Etruschi Seamount	Plio-Quaternary	– Hawaiite lavas similar to moderately alkaline mafic rocks from Sardinia Plio-Quaternary volcanic stage
	Magnaghi Seamount	3.0–2.7	– N-S elongated, about 1400 m high cone containing Na-transitional basalts
	Vavilov Seamount	Late Pliocene to 0.1	– N-S elongated, about 2000 m high cone containing tholeiitic to Na-transitional basalts
	Aceste Seamount	Lower Pliocene	– E-W elongated volcano with hawaiitic to trachytic composition exhibiting intermediate signatures between OIB- and arc-type rocks
	Prometeo	Quaternary	– Mugearite lava field
ARC-type	Cornacya	12	– Altered shoshonitic lavas with lamproitic enclaves
	Anchise	5.2–3.6	– Mafic high-K calcalkaline to shoshonitic lavas
	ODP sites 650 and 651	3.0–2.6 1.9–1.7	– Basalt to andesite lavas with calcalkaline to HKCA composition
	Aeolian seamounts (Sisifo, Enarete, Eolo, Lametini, Alcione, Marsili, Palinuro, etc.)	1.3 to <0.1	– Arc tholeiites (Lametini and southern Marsili Basin) to calcalkaline and shoshonitic mafic to intermediate rocks

2010; Milia and Torrente 2014). The spreading rate has been much stronger in the south than in the north. This has resulted in a triangular shape for the basin and produced a major lithospheric structural discontinuity, known as the 41° Parallel line, roughly running from northern Sardinia to the Pontine Island and Campania volcanoes (e.g. Serri 1990; Bruno et al. 2000; Rosenbaum and Lister 2004a; Cocchi et al. 2015).

In the northern sector, extension of about 8–10 mm/year caused lithospheric boudinage (Gueguen et al. 1997, 1998) and led to the formation of small amounts of magmas, mostly belonging to the Tuscany Province (see Chap. 2). As a result, a few granitoid and lava bodies were emplaced in an area consisting of rifted continental crust with tilted blocks and half-graben structures (Savelli 2000; Cocchi et al. 2015). Moho is at a depth of about 20 km (e.g. Moeller et al. 2013). In the south, stronger lithospheric extension (up to about 5 cm/year) resulted in the diachronous opening of the Tortonian-Pliocene Vavilov (about 6 Ma) and the Plio-Quaternary Marsili basins (Kastens et al. 1988; Sartori 2003; Cocchi et al. 2009). There was a progressive modification of rifting direction in time and space, from an early E-W extension near to the Sardinia coast and in the Vavilov Basin, to a late NW-SE spreading in southeastern Tyrrhenian Sea (e.g. Rosenbaum and Lister 2004b; Milia et al. 2013).

Crustal thickness in the southern Tyrrhenian Sea decreases from approximately 20–25 km along its borders to less than 10 km in the Marsili Basin and a minimum of about 6 km beneath the Vavilov Basin (Sartori et al. 2004 with references). The shallow Moho clearly highlights the occurrence of oceanic-type crust in the central and southeastern Tyrrhenian Sea basins, a hypothesis corroborated by seismic and gravimetric evidence for a serpentinitic seafloor in the

Table 12.2 Representative compositions of Tyrrhenian Sea submarine volcanic rocks

Volcanic centre	ODP Site 655	DSDP Site 373	Magnaghi Seamount	Vavilov Seamount	Prometeo lava field	Cornacya
Rock type	MORB-type basalt	MORB-type basalt	OIB-type basalt	OIB-type basalt	Mugearite	Lamproite
Data source	3, 6	1	4	4	7	5
SiO_2 (wt%)	50.43	48.00	48.83	48.15	51.33	39.34
TiO_2	1.24	0.92	1.84	3.10	1.5	0.30
Al_2O_3	16.34	18.30	19.85	18.26	17.02	17.91
FeO_{total}	8.19	7.32	9.21	11.30	8.35	2.08
MnO	0.13	0.10	0.16	0.17	0.17	–
MgO	7.33	7.66	3.95	3.82	5.03	10.15
CaO	9.83	10.3	10.30	9.57	8.09	0.98
Na_2O	3.40	3.16	3.70	3.44	4.94	4.08
K_2O	0.33	0.33	0.51	0.56	1.55	4.32
P_2O_5	0.17	0.12	0.35	0.36	0.62	0.11
LOI	1.70	3.39	0.30	0.02	0.69	20.24
Sc (ppm)	28	29	28	28	17	2.3
V	203	–	232	346	152	13
Cr	222	227	82	33	130	1.1
Co	32	30	–	–	–	–
Ni	101	–	31	33	76	67
Rb	3.3	1	7	10	22	188
Sr	180	232	305	343	653	156
Y	22	23	42	42	29.8	11
Zr	88	77	228	228	164	237
Nb	6	3.1	15	21	44	19
Ba	45	59	95	100	863	554
La	5.5	6.3	14	18	57	41
Ce	14	15	36	43	102	92
Nd	9.8	–	–	–	40	26
Sm	3.05	1.9	6.17	7.17	7.1	3.8
Eu	1.05	0.9	2.03	2.47	2.26	0.65
Tb	0.56	0.63	1.18	1.23	0.94	0.38
Yb	2.37	2.2	3.63	3.31	2.79	1.36
Lu	0.36	0.29	0.54	0.5	0.4	0.18
Hf	2.25	1.6	5.3	5.4	3.6	6.47
Ta	0.32	–	–	–	2.29	1.51
Pb	1.6	2.2	–	–	4.8	31
Th	0.75	0.42	1.5	1.8	9	25
U	0.2	0.24	–	–	2.3	7.8
$^{87}Sr/^{86}Sr$	0.70375	–	–	–	0.70329	0.70742

(continued)

Table 12.2 (continued)

Volcanic centre	ODP Site 655	DSDP Site 373	Magnaghi Seamount	Vavilov Seamount	Prometeo lava field	Cornacya
Rock type	MORB-type basalt	MORB-type basalt	OIB-type basalt	OIB-type basalt	Mugearite	Lamproite
$^{143}Nd/^{144}Nd$	0.51311	–	–	–	0.51295	0.51251
$^{206}Pb/^{204}Pb$	18.698	–	–	–	–	–
$^{207}Pb/^{204}Pb$	15.573	–	–	–	–	–
$^{208}Pb/^{204}Pb$	38.666	–	–	–	–	–
$^{176}Hf/^{177}Hf$	0.28317	–	–	–	–	–

Volcanic centre	ODP Site 651	Anchise Seamount	Enarete Seamount	Eolo Seamount	Palinuro Seamount	Marsili Seamount	
Rock type	Calc-alkaline basalt	Andesite	Shoshonitic basalt	Latite	Andesite	Calcalkaline basalt	OIB-type basalt
Data source	3,6	2	7	7	7	7	7
SiO_2 (wt%)	52.75	51.50	48.48	59.62	54.83	49.94	49.38
TiO_2	1.32	1.28	0.74	0.60	0.67	0.9	1.13
Al_2O_3	15.39	17.52	19.98	18.92	15.55	16.71	18.41
FeO_{total}	7.68	7.78	7.80	4.56	7.32	8.11	8.3
MnO	0.12	0.11	0.13	0.03	0.15	0.13	0.14
MgO	7.10	4.12	4.31	0.93	6.15	7.44	6.53
CaO	7.06	9.93	8.76	2.82	9.64	11.53	9.63
Na_2O	3.56	3.30	2.66	3.92	2.12	3.19	3.7
K_2O	1.68	1.80	2.32	5.64	1.64	0.78	1.21
P_2O_5	0.20	0.29	0.46	0.20	0.18	0.18	0.68
LOI	2.29	1.69	3.57	2.48	1.19	0.65	1.16
Sc (ppm)	–	–	25	14	34	–	–
V	209	–	253	149	228	196	232
Cr	173	–	69	3.3	126	122	105
Co	35	–	–	–	–		
Ni	125	–	34	5.7	33	97	116
Rb	18	40	37	184	49	9	19
Sr	225	457	723	425	444	450	945
Y	23	30	17.9	25.2	23.5	20.1	25
Zr	84	119	114	242	116	97	128
Nb	5.1	17	12.4	19.2	12.5	8.5	31.5
Cs	1.0	–	0.34	1.57	3.46	0.37	0.51
Ba	131	–	556	584	558	167	657
La	8.6	–	36	46	33	15.1	56
Ce	18.9	–	61	81	65	30.7	92
Nd	12.1	–	30	30	29	15.5	41
Sm	2.9	–	5.7	5.4	6	3.5	6.5
Eu	1.05	–	1.68	1.44	1.34	1.08	2.14

(continued)

Table 12.2 (continued)

Volcanic centre	ODP Site 651	Anchise Seamount	Enarete Seamount	Eolo Seamount	Palinuro Seamount	Marsili Seamount	
Rock type	Calc-alkaline basalt	Andesite	Shoshonitic basalt	Latite	Andesite	Calcalkaline basalt	OIB-type basalt
Tb	0.57	–	0.65	0.7	0.76	0.58	0.82
Yb	2.20	–	1.77	2.84	2.06	1.84	2.31
Lu	0.33	–	0.25	0.41	0.3	0.28	0.33
Hf	2.1	–	3.2	6	3.1	2.2	3
Ta	0.41	–	0.66	1.16	0.7	0.5	1.98
Pb	2.7	–	14.8	47	12.1	3.3	5.1
Th	1.76	–	12.9	21.8	10.3	3.89	9
U	0.56	–	0.7	6.5	2.9	0.99	2.6
$^{87}Sr/^{86}Sr$	0.70438	–	–	–	–	0.70365	0.703222
$^{143}Nd/^{144}Nd$	0.51294	–	–	–	–	0.512858	0.512921
$^{206}Pb/^{204}Pb$	18.834	–	–	–	–	19.129	19.554
$^{207}Pb/^{204}Pb$	15.644	–	–	–	–	15.653	15.634
$^{208}Pb/^{204}Pb$	38.967	–	–	–	–	39.059	39.177
$^{176}Hf/^{177}Hf$	0.28307	–	–	–	–	–	–

Source of data: (1) Dietrich et al. (1977); (2) Calanchi et al. (1984); (3) Beccaluva et al. (1990); (4) Serri (1990); (5) Mascle et al. (2001); (6) Gasperini et al. (2002); (7) Trua et al. (2003, 2004, 2011)

D'Ancona Ridge, between the Vavilov and Magnaghi seamounts (e.g. Cocchi et al. 2009; Prada et al. 2014). The lithosphere thickness decreases from about 50 to 70 km along the borders to 20 km beneath the central Tyrrhenian Sea. The Tyrrhenian basis is characterised by strong regional positive Bouguer gravity anomaly with maximum values of 250 Mgal in the Vavilov and Marsili basins, where high heat flow values of about 140 mW/m^2 have been measured (e.g. Della Vedova et al. 2001; Zito et al. 2003; Cella et al. 2006). These features are consistent with the occurrence of high-density and hot mantle material at shallow depth (Panza 1984; Doglioni et al. 2004 and references therein). Magnetic anomalies reveal the many volcanic bodies occurring within the Tyrrhenian Sea (e.g. Cella et al. 1998; De Ritis et al. 2010; Florio et al. 2011).

The uppermost mantle beneath the central and southern Tyrrhenian Sea consists of relatively low S-wave velocity rocks, which reach a minimum Vs $\sim$ 3.0–3.3 km/s beneath the Vavilov Basin (Panza et al. 2003, 2004, 2007), suggesting the occurrence of liquid material. A layer with high seismic wave velocity is detected at a depth of about 650 km and has been interpreted as the accumulation of subducted slab material (e.g. Lucente et al. 1999). Singularly, little attention has been dedicated to the petrologic significance of this layer as a possible source of some of the geochemical components recognised in the Tyrrhenian Sea magmas (e.g. Cadoux et al. 2007).

Various rock types make up the Tyrrhenian Sea floor. Rocks near to the Sardina-Corsica coast consist of Plio-Quaternary clastic layers overlying Messinian evaporitic and Middle Miocene sediments. In the Vavilov and Marsili basins, Plio-Pleistocene hemipelagic sediments and some turbidities rest over a basaltic crust. In the area between these two basins (Issel Swell), a veneer of hemipelagic deposits cover a substratum of shallow- and deep-water carbonates, plus siliciclastic and metamorphic rocks (Borsetti et al. 1990; Kastens and Mascle 1990; Sartori 2003; Sartori et al. 2004; Doglioni et al. 2004; Gamberi and Marani 2004).

The morphology of the Tyrrhenian Sea floor is complex. The northern sector is characterised by a basin-and-range seafloor topography resulting from tilting of crustal blocks. Similar topography and structures are found along the margins of the southern Tyrrhenian Sea, e.g. near to Sardinia.

The Cornaglia Terrace (2500–3000 m deep) marks the transition from the Sardinia tilted blocks to the Tyrrhenian abyssal plains, whose center are occupied by Magnaghi and Vavilov seamounts, and by the D'Ancona serpentinitic ridge (Sartori 2001). The Vavilov Basin is a 3500 m deep triangular depression delimited to the west by a NE-SW oriented scarp (Selli line) along which old Alpine "Corsica-type" units and Miocene rocks occur. De Marchi and Flavio Gioia seamounts, sited in the western and eastern sector of the abyssal plain, are tilted crustal blocks. The Gortani Ridge is a 40 km long volcanic structure northwest of the Vavilov Seamount (Marani and Gamberi 2004).

The Marsili Basin is about 3500 m deep and contains the Marsili Seamount at its center and several other seamounts and the Aeolian islands along the border (see Fig. 12.8). The Marsili Seamount is a huge NE-SW elongated volcanic ridge rising about 3000 m above the seafloor. The Palinuro-Glabro volcanic lineament marks the northeastern margin of the Marsili Basin. Sisifo, Enarete, Eolo seamounts occur southwest of Marsili; the Lametini and Alcione seamounts are sited on the east (Marani and Gamberi 2004; Romagnoli 2013; Romagnoli et al. 2013).

12.3 Igneous Activity

Igneous activity in the southern Tyrrhenian Sea emplaced a wide variety of magmas. Volcanic centres exposed above sea level (Aeolian arc, Ustica and Pontine Islands) have been already discussed in the previous chapters. Here, the submarine volcanism in and around the Vavilov and Marsili basins will be focused.

Many rocks samples from the Tyrrhenian Sea floor exhibit severe alteration with high LOI values, which modified significantly the concentrations of several major and trace elements, especially mobile elements such as alkalis. Therefore, immobile trace element criteria have been adopted to classify these rocks (Barberi et al. 1978; Drietrich et al. 1977, 1978; Beccaluva et al. 1982, 1990). Here, only samples with LOI < 4.0 wt% will be plotted in the classification and variation diagrams that employ mobile elements.

TAS classification diagram shows that most samples have basaltic to basaltic andesite compositions (Fig. 12.2). Based on immobile-element discriminant diagrams (e.g. Th/Yb vs. Ta/Yb; Th–Hf/3-Ta), three broad groups of rocks are recognised spanning MORB, OIB and island-arc geochemical signatures (Fig. 12.3).

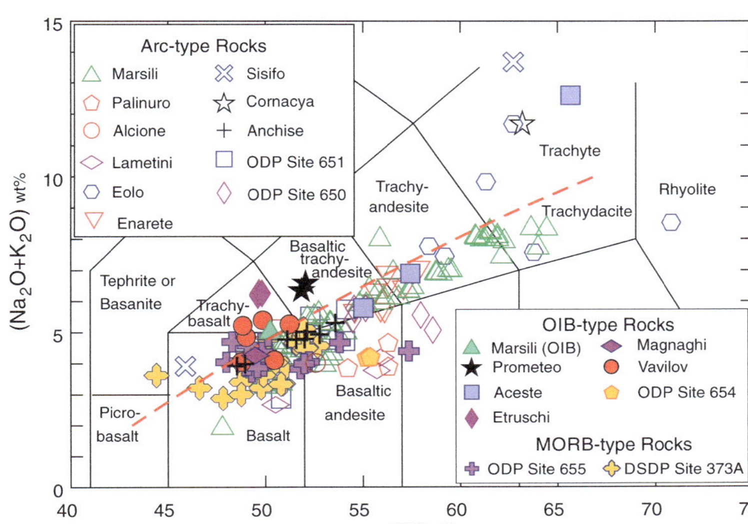

Fig. 12.2 TAS diagram for submarine volcanic rocks from the Tyrrhenian Sea. Compositions have been recalculated to 100 % on a water-free basis. Only samples with Loss on Ignition (LOI) lower than 4.0 wt% have been plotted. The *thick red dashed line* is the divide between subalkaline and alkaline rocks of Irvine and Baragar (1971). *Full symbols*, *open symbols*, and *full crosses* are used to indicate OIB-, arc- and MORB-type rocks, respectively

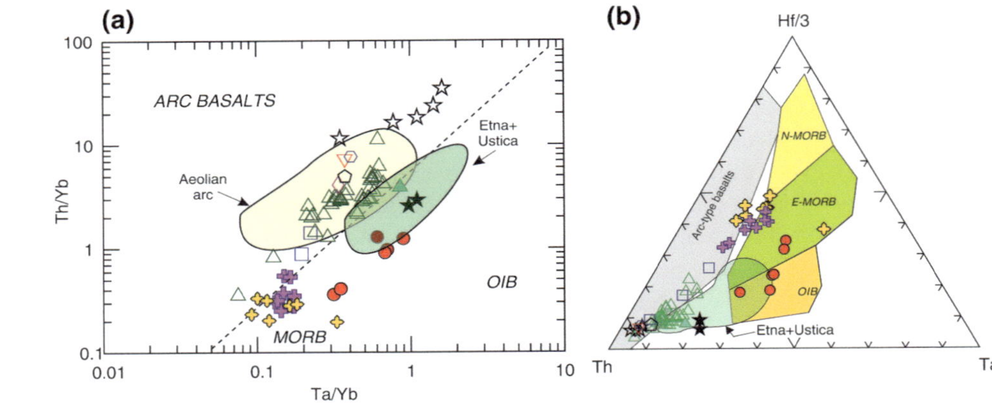

Fig. 12.3 Discriminant element plots for mafic rocks from the Tyrrhenian Sea seamounts (Pearce 1982; Wood 1980). Symbols as in Fig. 12.2

The regional distribution of these magma types is schematically shown in Fig. 12.1.

Incompatible element normalised to primitive mantle composition (Fig. 12.4) show that MORB-type rocks have poorly fractionated patterns and element abundances slightly higher than average Enriched-type MORB (Gale et al. 2013). Some samples show transitional characteristics towards arc-type rocks on immobile discriminant diagrams (Fig. 12.3). OIB-type rocks are slightly more enriched in incompatible elements, close to average OIB (Sun and Mcdonough 1989). Positive spikes of Pb observed in some OIB and MORB rocks are rather unusual, and could depend on secondary processes (e.g. Hoefig et al. 2014). Arc-type rocks show strong element fractionation with negative anomalies of Ta and Nb, and positive spikes of Pb, as observed in the rocks from the Aeolian islands.

12.3.1 MORB-Type Rocks

Rocks with a composition similar to the Mid-Ocean Ridge Basalts have been recovered in the Vavilov Basin at the ODP Site 655 and DSDP Site 373A (Fig. 12.1).

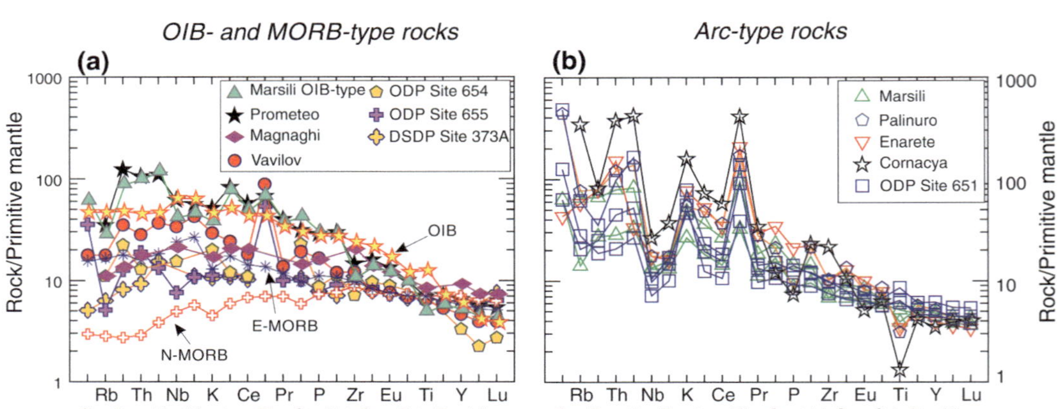

Fig. 12.4 Incompatible element patterns for selected submarine mafic volcanics (MgO > 4.0 %) from the southern Tyrrhenian Sea

1. ***ODP Site 655***. Drilling along the NNE-SSW trending Gortani Ridge, near to the western edge of the Vavilov Basin, encountered basaltic pillow lavas underlying 3.4–3.6 Ma old sediments. $^{40}Ar/^{39}Ar$ dating yielded an age of about 4.3 Ma for these lavas (Feraud 1990). Rock textures vary from aphyric to porphyritic with phenocrysts and glomerophyric aggregates of olivine (Fo_{87-82}), plagioclase (An_{85-60}) and some diopside to augitic clinopyroxene set in a groundmass containing the same phases, Fe–Ti oxides, and altered glass. Variable amounts of secondary carbonates and zeolites occur as vein and vesicle fillings, and in the groundmass (Beccaluva et al. 1990; Bertrand et al. 1990). Major and trace element composition is characterised by low silica (SiO_2 ~ 48–53 wt%, LOI-free basis), moderate enrichment in TiO_2 (~1.2–1.7 wt%), MgO (~4–7 wt %), Ni (~70–110 ppm) and Cr (~120–220 ppm), and low concentrations of incompatible elements (Fig. 12.5). Isotope ratios (Fig. 12.6) show relatively unradiogenic Sr and Pb, and radiogenic Nd and Hf signatures ($^{87}Sr/^{86}Sr = 0.7037$ to 0.7045; $^{143}Nd/^{144}Nd$ ~ 0.5131–0.5133; $^{206}Pb/^{204}Pb$ ~ 18.70; $^{207}Pb/^{204}Pb$ ~ 15.57; $^{208}Pb/^{204}Pb$ ~ 38.68; $^{176}Hf/^{177}Hf$ ~ 0.28311–0.28321; Beccaluva et al. 1990; Gasperini et al. 2002).
2. ***DSDP Site 373A***. Drilling at this site penetrated through a thin Plio-Quaternary sedimentary veneer before encountering a thick pile of basaltic rocks with K/Ar ages of 3.5–7.3 Ma (Barberi et al. 1978; Kreuzer et al. 1978; Savelli and Lipparini 1978). Volcanic samples show aphyric to porphyritic textures with phenocrysts of plagioclase (about An_{80}) and a few altered olivine set in a strongly altered groundmass containing diopside to augite clinopyroxene, opaques, and various secondary minerals such as chlorite, amphibole, smectite, and carbonates (Barberi et al. 1978; Dietrich et al. 1978; Kreuzer et al. 1978). Based on the abundance of TiO_2, two groups of lavas have been distinguished (Dietrich et al. 1977; Barberi et al. 1978). The high-Ti group ($TiO_2 > 1.4$ wt%) has higher K, P, Zr, Hf and REE, and lower MgO and Cr than the low-Ti group ($TiO_2 < 1.1$). Sr isotope ratios are poorly variable with $^{87}Sr/^{86}Sr$ ~ 0.703 (Barberi et al. 1978).

Overall, MORB-type rocks from the Vavilov Basin have lower ferromagnesian element contents than mantle equilibrated magmas (Trua et al. 2011), suggesting variable degrees of fractional crystallisation, mainly with removal of olivine. Incompatible element abundances are relatively low (Fig. 12.4a). REE patterns (not shown) are poorly fractionated, and exhibit positive spikes of Eu for some samples (Dietrich et al. 1977). Immobile incompatible element contents and ratios mostly resemble enriched E-MORB, whereas LILE concentration is somewhat lower (Fig. 12.4a). $^{87}Sr/^{86}Sr$ and $^{208}Pb/^{204}Pb$ ratios are higher than typical depleted-MORB values (Beccaluva et al. 1990; Gasperini et al. 2002).

12.3.2 OIB-Type Rocks

Rocks akin to the Ocean Island Basalts have been recovered from the Vavilov, Managhi, Quirra, Etruschi and Aceste seamounts, and from the Prometeo lava field (Fig. 12.1).

1. ***ODP Site 654***. A block of lava (possibly a re-sedimented boulder) was encountered by drilling within the Upper Pliocene volcano-sedimentary sequence overlying a tilted block of the eastern Sardinian margin (Beccaluva et al. 1990). Texture is aphyric to poorly porphyritic with a few microphenocrysts of olivine (about Fo_{83}) and labradorite plagioclase set in groundmass containing clinopyroxene, glass, and secondary minerals. Some chromite is included in the olivine. Composition is basaltic andesitic, according to TAS diagram. Trace elements are characterised by relatively high Ni and Cr (about 170–180 and 250–270 ppm, respectively), and modest enrichment in several incompatible elements (Figs. 12.4a, 12.5).

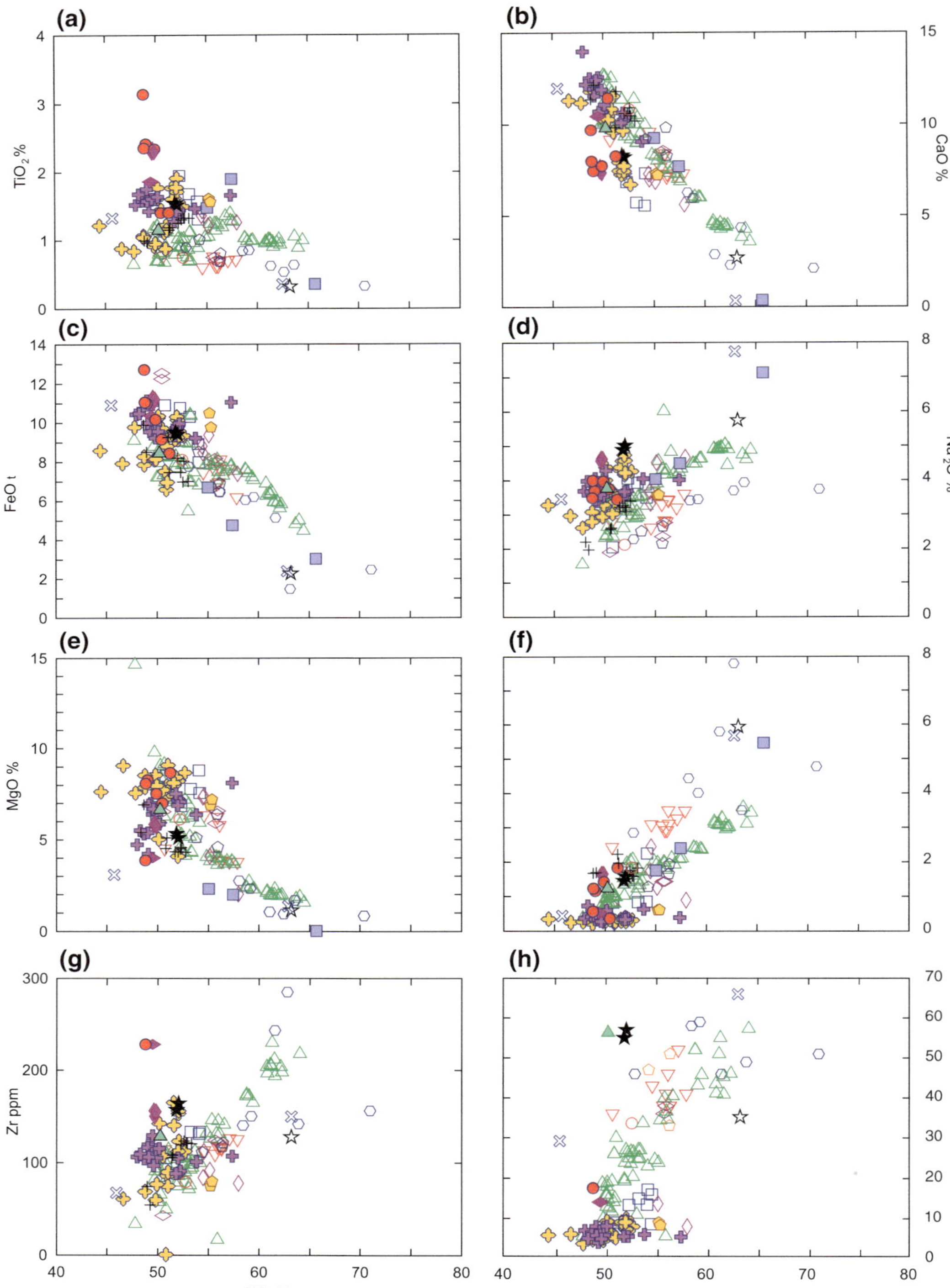

Fig. 12.5 Variation diagrams of selected major and trace elements for submarine volcanoes from the southern Tyrrhenian Sea. Major element abundances have been normalized to 100 % on a LOI-free basis. Only rocks with LOI < 4 % have ben plotted. Symbols as in Fig. 12.2

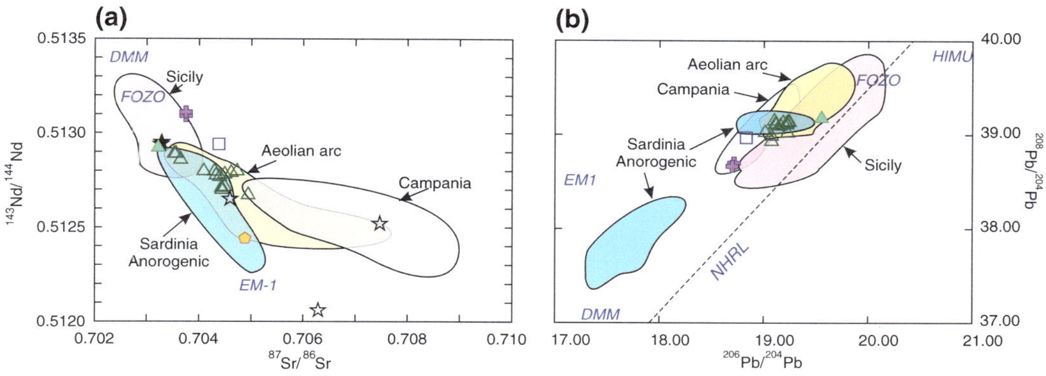

Fig. 12.6 Sr–Nd–Pb isotope diagrams for the Tyrrhenian Sea submarine volcanoes. Symbols as in Fig. 12.2

$^{87}Sr/^{86}Sr$ ratio is 0.70488 and $^{143}Nd/^{144}Nd$ is 0.51244 (Beccaluva et al. 1990; Fig. 12.6). Overall, this rock has a composition close to continental tholeiites occurring in Sardinia, and can be considered as belonging to the Plio-Quaternary magmatic stage from this island.

2. ***Etruschi***. Basaltic lapilli have been recovered at the northern margin of the Cornaglia Terrace (40° 45.2′N, 11° 05.6′E; Core RC 9-195). This site has been indicated as the Etruschi Seamount (e.g. Serri 1990), although the Etruschi Ridge is sited much more to the north, off the southeastern coast of Corsica. The rocks are mostly glassy with a few small grains of olivine and acicular plagioclase crystals. Compositions are undersaturated hawaiitic, similar to the analogous rocks from Sardinia (Keller 1981).
3. ***Magnaghi Seamount***. Magnaghi (3.0–2.7 Ma; Argnani and Savelli 1999) is a 25 km long NNE-SSW elongated volcanic edifice rising to about 1470 m above its surroundings. The very few available data indicate a basaltic composition with mildly Na-alkaline affinity (e.g. Serri 1990). Incompatible elements show moderate enrichments and a smooth mantle-normalised pattern (Fig. 12.4a).
4. ***Vavilov Seamount***. The Vavilov Seamount is a NNE-SSW elongated volcano, about 40 km long and 10 km large, which rises about 2800 m above the abyssal plain, reaching 730 m below sea level. Activity apparently started in the Late Pliocene and ages younger than 0.4–0.1 Ma have been suggested for the summit area by Savelli (1988). The recovered samples are variably altered and show transitional to mildly Na-alkaline basaltic compositions (e.g. Robin et al. 1987; Kastens and Mascle 1990; Serri 1990; Trua et al. 2011). Incompatible elements are moderately enriched and mantle-normalised patterns show a slight upward convexity (Fig. 12.4a). Positive anomalies of Pb probably result from secondary processes on samples with extremely high LOI values (Trua et al. 2011). Pb isotope ratios of melt inclusions from olivine phenocrysts are $^{206}Pb/^{204}Pb = 18.96–19.02$ and $^{208}Pb/^{204}Pb = 39.40–39.56$ (Rose-Koga et al. 2012).
4. ***Aceste and Quirra seamounts.*** The Aceste Seamount and the volcanic rocks collected near Quirra Seamount are Pliocene in age and show Na-alkaline affinity. Aceste rocks range from hawaiite to trachyte, and reveal somewhat intermediate compositions between OIB and arc-type rocks (Beccaluva et al. 1984). Textures are slightly porphyritic to microgranular, with microphenocrysts of plagioclase and clinopyroxene in the mafic rocks, and sanidine in the trachytes. The groundmass consists of the same minerals plus opaques, glass, and secondary phases. Alkali-olivine basalts have been found south of the Quirra Seamount (Colantoni et al. 1981).

5. ***Prometeo***. This lava field is located southeast of the Island of Ustica. The recovered lavas are mugearites with trace element composition similar to Ustica. Sr–Nd isotope ratios are $^{87}Sr/^{86}Sr$ ~ 0–7033 and $^{143}Nd/^{144}Nd$ ~ 0.5129 (Trua et al. 2004).
6. ***Marsili Seamount***. Marsili is mainly made up of arc-type rocks. However, a single sample analysed by Trua et al. (2011) has a clear OIB-type affinity and resembles closely the benmoreite lavas from Prometeo (Fig. 12.4a).

12.3.3 Arc-Type Rocks

Rocks with petrological-geochemical arc affinity (high LILE/HFSE) are widespread in the southern Tyrrhenian Sea (Fig. 12.1). According to several authors volcanic centres are arranged along arcuate structures becoming younger from west to the east. One of these arcs (Central Tyrrhenian Arc) has a Pliocene age (5–2 Ma) and would be formed by the Anchise Seamount in the south, the Glauco (Garibaldi) Seamount in the center, and by the Pontine Islands and the buried calcalkaline rocks of the Campanian Plain in the north (e.g. Savelli 1988; Locardi 1993; Argnani and Savelli 1999). K_2O versus SiO_2 diagram for the arc-type rocks from the southern Tyrrhenian seamounts is reported in Fig. 12.7.

1. ***Cornacya Seamount***. This is an about 12 Ma old volcano located off the southern Sardinia coast (Mascle et al. 2001). Recovered samples consist of strongly altered lavas that contain enclaves of mica-rich lamprophyres. Lava textures are porphyritic with ubiquitous phenocrysts of zoned plagioclase and biotite plus some amphibole and clinopyroxene that are surrounded by a glass-rich matrix containing minor Na-rich plagioclase, anorthoclase, biotite, and secondary products. Accessory minerals include Fe–Ti oxides, apatite, and zircon. The lamprophyric enclaves are porphyritic with phenocrysts of altered olivine, amphibole, and phlogopite that are set in a fine-grained groundmass of the same phases plus plagioclase. All the Cornacya samples have high to extreme LOI (up to about 23 wt%), which indicates severe secondary alteration. Nevertheless, mineral chemical data and immobile element distributions indicate a shoshonitic to ultrapotassic lamproitic affinity for lavas and enclaves, respectively. These rocks have fractionated REE patterns (not shown) generally with a small negative Eu anomaly. Patterns of mantle-normalised incompatible elements are strongly fractionated, with positive spikes of Th, U and Pb, and negative anomalies of Ba, Sr, Ti and HFSE (Fig. 12.4b). Sr- and Nd-isotope ratios are variable ($^{87}Sr/^{86}Sr$ =

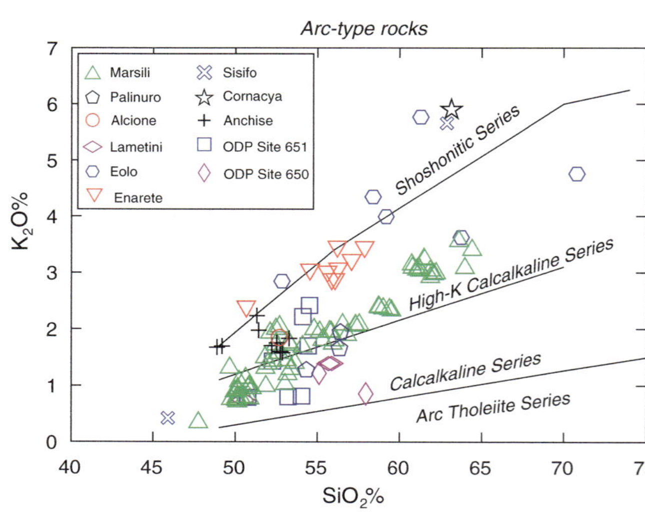

Fig. 12.7 K_2O vs. SiO_2 diagram for arc-type volcanics from the southern Tyrrhenian seamounts. Compositions have been recalculated to 100 % on a water-free basis. Only rocks with LOI < 4.0 wt% have been plotted

$0.70465–0.70742$; $^{143}Nd/^{144}Nd = 0.51206–0.51265$; Fig. 12.6a). Overall, the Cornacya rocks show petrological and trace element compositions that resemble closely the lamproites and shoshonites from Tuscany, although Sr–Nd isotope compositions differ significantly (see Chap. 2)

2. ***Anchise Seamount***. This volcano is a W-E elongated irregular cone situated west of the Island of Ustica. It consists of several coalescing eruptive centres, which rise from a depth of 1150–532 m below sea level (Calanchi et al. 1984). K/Ar ages on whole rocks yielded values of 5.2–3.6 Ma (Savelli 1988). Lavas from Anchise have a vesicular porphyritic texture, with phenocrysts of plagioclase, clinopyroxene and minor olivine set in a groundmass of the same phases plus Fe–Ti oxides, sanidine, minor biotite, and glass. Secondary phases include calcite, Fe-hydroxides and clay minerals. Mafic high-K calcalkaline compositions are indicated by the K_2O versus SiO_2 classification diagram (Fig. 12.7).
3. ***ODP Site 651***. Drilling in the northern Vavilov Basin revealed the occurrence of two thick basaltic lava series separated by a doleritic sill. Serpentinised peridotites were found beneath the lava units (Bonatti et al. 1990). The igneous sequence is covered by Plio-Pleistocene sediments. $^{40}Ar/^{39}Ar$ dating of three basalt samples yielded ages of approximately 3.0–2.6 Ma (Feraud 1990). Rock textures are microcrystalline to poorly porphyritic with microphenocrysts of strongly altered olivine ($Fo_{89–81}$), zoned plagioclase ($An_{72–50}$) and augite set in a groundmass containing the same phases plus opaque minerals and variably altered glass (Bertrand et al. 1990). Biotite and amphibole have been observed in some rocks. Clay minerals, zeolites, Fe-hydroxides, and carbonates are common secondary phases. The occurrence of chlorite, talc and minor actinolite in some samples indicates low-grade metamorphism (Beccaluva et al. 1990). Data on the least altered rocks reveal the occurrence of two types of magmas with different alkali contents, decreasing from older to younger flows. Incompatible element patterns of both groups are fractionated with negative anomalies of Nb and Ta, a positive spike of Pb and no depletion in Ti (Fig. 12.4b). Isotopic compositions show $^{87}Sr/^{86}Sr \sim 0.7044–0.7073$; $^{143}Nd/^{144}Nd = 0.51294$; $^{176}Hf/^{177}Hf = 0.28297–0.28307$; $^{206}Pb/^{204}Pb = 18.83–18.92$; $^{207}Pb/^{204}Pb = 15.64–15.68$; $^{208}Pb/^{204}Pb = 38.97–39.18$; $^{176}Hf/^{177}Hf = 0.28297–0.28307$ (Beccaluva et al. 1990; Gasperini et al. 2002).
4. ***ODP Site 650***. This site on the western margin of the Marsili Basin is characterised by the occurrence of basaltic rocks beneath an approximately 600 m thick pile of sediments, whose base has a magnetostratigraphic and biostratigraphic age of 1.9–1.7 Ma. Recovered volcanic rocks are vesicular basalts, displaying a texture that is characterised by a network of skeletal plagioclase and altered olivine and some clinopyroxene surrounded by altered glass mesostasis. Identification of the petrogenetic affinity of these rocks is hampered by severe alteration. However, a composition not far from some Stromboli calcalkaline basaltic andesites has been recognised by Beccaluva et al. (1990) on the basis of immobile element ratios.
6. ***Aeolian seamounts***. Several seamounts occur in the Aeolian arc area (e.g. Passaro et al. 2009), inside and at the margins of the Marsili Basin (Fig. 12.8). Many of these seamounts have exhalative activity with associated hydrothermal Fe–Mn–Pb–Zn mineral deposits at some centres (e.g. Marani et al. 1997; Dekov and Savelli 2004).

Volcanic rocks recovered during several research cruises generally show porphyritic textures, with phenocrysts of plagioclase, clinopyroxene, and Ti–Fe oxides set in a matrix made of the same phases plus glass. Olivine is present in the mafic rocks from Marsili, Eolo, Palinuro and along the southern margin of the Marsili Basin. Orthopyroxene appears in the intermediate calcalkaline, HKCA, and shoshonitic rocks, whereas biotite and hornblende generally occur

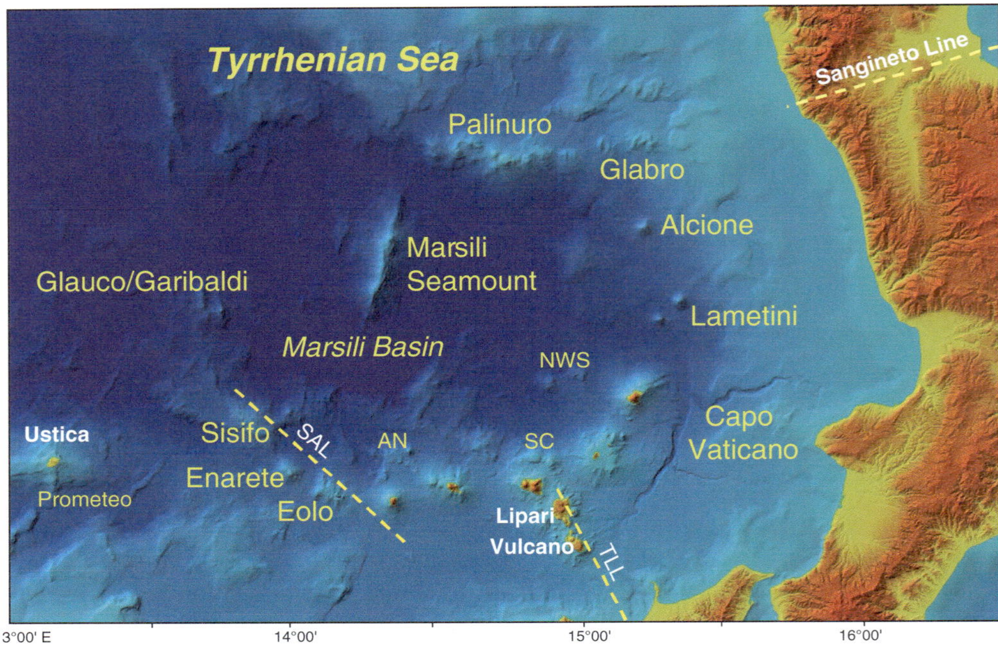

Fig. 12.8 The Marsili Basin and associated seamounts. *AN* Alicudi North Seamount; *SC* Secca del Capo; *NWS* North-West Stromboli Seamount; *SAL* Sisifo-Alicudi line; *TLL* Tindari-Letojanni line

in the evolved rocks (Trua et al. 2004). Incompatible trace elements display variable concentrations, with an increase from tholeiitic to shoshonitic rocks.

Glauco or Garibaldi Seamount rises about 1600 m above the seafloor, on the western margin of the Marsili Basin. The few altered dredged samples have a broadly calcalkaline mafic composition (Beccaluva et al. 1981).

Sisifo is a 7 km long volcano developed along a dextral strike-slip fault known as the Sisifo-Alicudi line (De Astis et al. 2003). It contains the oldest dated rocks among the Aeolian seamounts (1.3–0.9 Ma). Dredged samples have calcalkaline basalt to trachyte compositions (Beccaluva et al. 1985).

Enarete has NW-SE elongated subconical shape, rising from 1650 m deep sea floor to about 300 m below sea level. Rocks are 0.78–0.67 Ma old and consist of shoshonitic basalts and shoshonites.

Eolo is a NW-SE elongated volcano with collapsed flanks and a flat top; hydrothermal activity has been detected about 600 m below sea level (Dekov and Savelli 2004). Compositions of recovered rocks are mafic to felsic, with shoshonitic to high-K calcalkaline affinity. Ages are 0.85 and 0.66 Ma.

Lametini seamounts are located about 30 km northeast of Stromboli. They are two symmetrical cones (North Lametino and South Lametino) rising about 800–900 m from a 2000 m deep sea floor. The absence of He isotope anomalies in the water column suggests these seamounts are extinct (Lupton et al. 2011). Rocks with island arc tholeiitic compositions have been recovered from Lametini.

Alcione Seamount is a cone with two peaks separated by a fault scarp. The volcano rises about 1600 m above its surroundings, with peaks at 950 and 850 m below sea level. A high-K calcalkaline basalt composition has been found for a dredged sample. $^3He/^4He$ ratio of seawater over the summit area is near the regional background, suggesting scarce or absent hydrothermal activity (Lupton et al. 2011).

Palinuro (0.8–0.3 Ma) is a W-E oriented, 55 km long and 25 large ridge formed of several partially overlapping cones. It rises from a seafloor about 1880–3200 deep, reaching 84 m below sea level. Volcanism develops along an E-W strike-slip fault (Del Ben et al. 2008), which likely represents the western extension of the Sangineto line (Passaro et al. 2009, 2010 and references therein). Samples of high-alumina basalt, basaltic andesite, and high-K andesite have been dredged from the summit of Palinuro.

Glabro Seamount represents the eastern continuation of the Palinuro volcanic ridge. It consists of two NW-SE aligned peaks with the base at 1200 m and the summit at 826 and 900 m below sea level.

Capo Vaticano is a recently discovered seamount, although a submarine peak was indicated by old maps in the area. It consists of NE trending ridge, sited about 12 km offshore the Capo Vaticano promontory, Calabria. Paleomagnetic data suggest an age of about 1.8–1.9 Ma. The top of the seamount, at about 70 m b.s.l., is the site of exhalative activity with emission of fluids rich in CO_2, CH_4, and ^3He (Loreto et al. 2015).

Marsili is the largest and best studied Aeolian seamount. It is a 70 km long and 30 km wide NE-SW elongated volcanic ridge, consisting of several coalescing cones sited at the center of the Marsili Basin. Marsili rises 3200 m above the surrounding seafloor, reaching about 500 m below sea level. The Marsili Basin is an extensional structure with about 10 km thick oceanic-type crust, and a 30 km thick lithosphere (e.g. Pontevivo and Panza 2006). According to magnetic data, the opening of the Marsili Basin started about 1.8 Ma before present, whereas the Marsili Seamount was constructed between about 1.07 Ma and the present (Cocchi et al. 2009). Recent studies suggest that the Marsili Seamount is still active (e.g. Iezzi et al. 2014). Analysed rocks include lavas and pyroclastics, which show dominant calcalkaline basalt compositions, followed by basaltic andesites and high-K calcalkaline andesites. An OIB-type sample has been also recovered. Basalts have vesicular, variably porphyritic textures. Main phenocrysts are plagiocase and olivine, plus clinopyroxene in some rocks and rare biotite. Groundmass range from microcrystalline to glassy and contain the same phases as the phenocrysts plus Fe–Ti oxides (Trua et al. 2002). Plagioclase is zoned ($\sim An_{93-65}$); olivine ($\sim Fo_{92-75}$) contains Cr-spinel inclusions; clinopyroxene has a diopside composition. Basaltic andesites have plagioclase, clinopyroxene, orthopyroxene, olivine ($\sim Fo_{78-74}$) plus some amphibole microcrysts. Andesites are moderately porphyritic with phenocrysts of plagioclase (An_{90-42}), Mg-rich augitic clinopyroxene, orthopyroxene ($\sim En_{75-70}$) and opaques, set in a microcrystalline to glassy groundmass. Apatite is present as an accessory phase (Trua et al. 2002, 2014). The tephra consists of ash particles with low vesicularity and contain plagioclase, clinopyroxene, and olivine plus orthopyroxene, amphibole and accessory minerals (Iezzi et al. 2014).

The Marsili rocks show smooth trends for most elements on the Harker diagrams (Figs. 12.5). Basalts are variably enriched in incompatible elements, with pyroxene-phyric types showing the highest concentrations (Trua et al. 2002, 2004 and references therein). Incompatible element patterns are fractionated, with negative anomalies of HFSE, and positive spikes of Pb and Sr (Fig. 12.4b). LILE/HFSE ratios (e.g. Rb/Nb) are lower than the central Aeolian Islands and close to the Alicudi and Stromboli values (see Chap. 9, Fig. 9.22). Very small HFSE negative anomalies are observed in the pattern of the OIB-type basalt, which resembles closely the Prometeo mugearite (Fig. 12.4a). Sr- and Nd-isotope ratios are similar to the western Aeolian islands ($^{87}Sr/^{86}Sr \sim 0.7032$–$0.7049$; $^{143}Nd/^{144}Nd \sim 0.5127$–$0.5129$; $^{206}Pb/^{204}Pb = 19.07$–$19.55$, $^{207}Pb/^{204}Pb = 15.64$–$15.68$, and $^{208}Pb/^{204}Pb = 38.93$–$39.18$; Trua et al. 2011). Analyses of two olivine-hosted melt inclusions by Rose-Koga et al. (2012) gave $^{206}Pb/^{204}Pb = 18.71$–$19.51$ and $^{208}Pb/^{204}Pb = 38.55$–$40.22$.

The bulk of geochemical data suggests that various mafic magmas erupted at Marsili. Most of them have clear island-arc geochemical signatures, but one sample shows OIB-type trace element pattern. Arc-type basalts underwent differentiation in the lower crust with intensive mixing and fractional crystallisation of olivine, clinopyroxene and plagioclase, plus amphibole in the intermediate compositions (Trua et al. 2014). The occurrence of both arc- and OIB-type basalts has been interpreted as an evidence for heterogeneous mantle sources beneath the seamount (Trua et al. 2011).

12.4 Petrogenesis of the Tyrrhenian Sea Magmatism

The occurrence of MORB-, OIB- and arc-type volcanic rocks in the southern Tyrrhenian Sea calls for a variety of mantle sources and melting processes. MORB-type rocks occurring in the Vavilov Basin display the lowest incompatible element concentrations, but similar incompatible element ratios to the OIB rocks of the Magnaghi and Vavilov seamounts (Figs. 12.4 and 12.9). This suggests an origin in the same source, with a larger degree of partial melting for MORB- than for OIB-type rocks. Seismic studies reveal the occurrence of a low S-wave velocity zone beneath the Vavilov Basin (Panza et al. 2007) at a depth about 30–35 km; this layer may represent the source of MORB-type magmas in the Tyrrhenian Sea. Such a depth of origin fits experimental data that suggest an origin of MORB magmas at the transition between the plagioclase- and the spinel-lherzolite stability fields, possibly with some contribution by small melts fractions from the deeper mantle (Presnall et al. 2002). The Tyrrhenian Sea MORB-type rocks, however, have different radiogenic isotope ratios than typical mid-ocean basalts, and plot above the Northern Hemisphere Reference Line, near to the OIB-type rocks of the Sardinia Mio-Pliocene activity (Fig. 12.6b). The enriched nature for Tyrrhenian MORB rocks is not surprising, considering that they have been emplaced in a much more complex setting than the mid-ocean ridges, in an area where a variety of mantle sources and several enrichment processes have been demonstrated by geochemical data. Moreover, it has to be recalled that a layer at a depth of 650 km extends beneath the Tyrrhenian Sea area. This layer is believed to result from accumulation of subducted slab and may represent the source of some components for magmas erupted in the area. Such a possibility, however, has received little attention and remains largely speculative.

OIB-type rocks have variable incompatible trace element abundances and ratios. Some samples (e.g. Vavilov Seamount) have similar trace element patterns as the associated MORB-type rocks, supporting a genesis by different degrees of partial melting in a common source. Other OIB-type rocks from the western Tyrrhenian Sea floor (Quirra and ODP Site 654) have compositions, and most likely a genesis, similar to the volcanic rocks from the Plio-Quaternary volcanic stage of Sardinia (see Chap. 11). Finally, the Na-alkaline rocks from the Prometeo lava field and an OIB-type sample from Marsili resemble closely the Etna and Ustica lavas. As discussed in Chap. 10, Etna and Ustica magmas have been generated in a FOZO-like upper mantle, which underwent very modest contamination by arc-type components (e.g. Beccaluva et al. 1982; Cinque et al. 1988; Schiano et al. 2001; Marani and Trua 2002; Trua et al. 2003). Therefore, a similar origin can be envisaged for Prometeo and the OIB-type basalt from Marsili. The occurrence of FOZO-type components in the southern Tyrrhenian Sea has been explained as an effect of mantle migration from the foreland, by flowing around the edges of the Ionian subducting plate or across tear faults of the slab (e.g. Trua et al. 2004, 2011; De Astis et al. 2006; Rosenbaum et al. 2008).

As for the arc-type rocks, the Cornacya volcanics and enclosed lamprophiric xenoliths resemble the shoshonitic and lamproitic rocks from the Tuscany Province, although Sr- and Nd-isotope ratios are less extreme at Cornacya, suggesting a lesser degree of source contamination for the latter. It has been hypothesised that Tuscany potassic magmas derive from lithospheric mantle sources that had been contaminated by upper crustal material during Alpine subduction processes (see Chap. 2). A similar origin can be envisaged for Cornacya rocks,

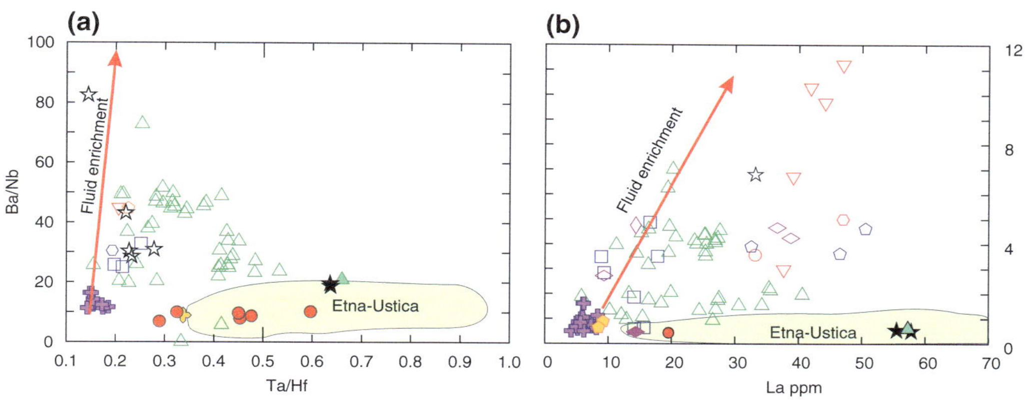

Fig. 12.9 Ba/Nb versus Ta/Hf and Rb/Nb versus La diagrams for submarine mafic volcanics from the southern Tyrrhenian Sea. *Red arrows* indicate subduction-related fluid enrichments. Symbols as in Fig. 12.2

implying that a Tuscany-type contaminated lithosphere extends to zones offshore the western coast of Sardinia.

The other arc-type rocks exhibit variable element enrichments and incompatible element ratios. Assuming these features represent primary magmatic compositions, the bulk of geochemical data supports the hypothesis of magma origin from mantle sources that were variably contaminated by subduction-related fluids. The ratios of some incompatible elements, which are not modified by subduction processes (e.g. Zr/Nb, Nb/Yb, Ta/Hf; Pearce and Parkinson 1993; Pearce and Peate 1995) generally show MORB-type values, suggesting depleted pre-metasomatic magma sources. However, variation diagrams of LILE/HFSE versus HFSE/HFSE and REE suggest a somewhat more complex magmatological framework (Fig. 12.9). Basically, the Tyrrhenian Sea arc-type magmas plot inside the space defined by two extreme trends, both starting from MORB-type basalts. One trend is characterised by variable LREE (and other LILE) concentrations and Ta/Hf ratios (and other HFSE/HFSE) at nearly constant LILE/HFSE (e.g. Ba/Nb, Rb/Nb) values. Rocks from Prometeo, Magnaghi, Vavilov, plus Etna, Ustica and the Marsili OIB-type basalt, plot along this trend. Another extreme trend has MORB-type LREE abundance and HFSE/HFSE, but shows extremely variable LILE/HFSE ratios. Such a trend clearly reflects enrichments by subduction-related fluids. The most obvious explanation for these variations is that three types of end-member compositions played variable roles in the origin of arc-related rocks in the Tyrrhenian Sea. One component has a depleted MORB-type trace element signature, another one has an enriched OIB-FOZO type composition similar to Etna, and a third component is a high LILE/HFSE typical of subduction-related fluids (Ellam et al. 1988; Trua et al. 2004). From a physical point of view, such a geochemical pattern can be interpreted as indicating a heterogeneous pre-metasomatic mantle consisting of variable proportions of FOZO- and MORB-type components, which was contaminated by different amounts of subduction-related fluids.

12.5 Geodynamic Setting

The compositional and temporal evolution of magmatism through the Tyrrhenian Sea is better explained by a model that envisages: (1) eastward and south-eastward discontinuous migration of the Ionian-Adriatic subducting plate from Oligo-Miocene to Present, (2) asymmetric back-arc spreading, (3) passive ascent of the asthenospheric mantle in backarc position, (4) mantle inflow from the foreland, and (5) contamination of mantle rocks by various amounts and types of subduction-related fluids.

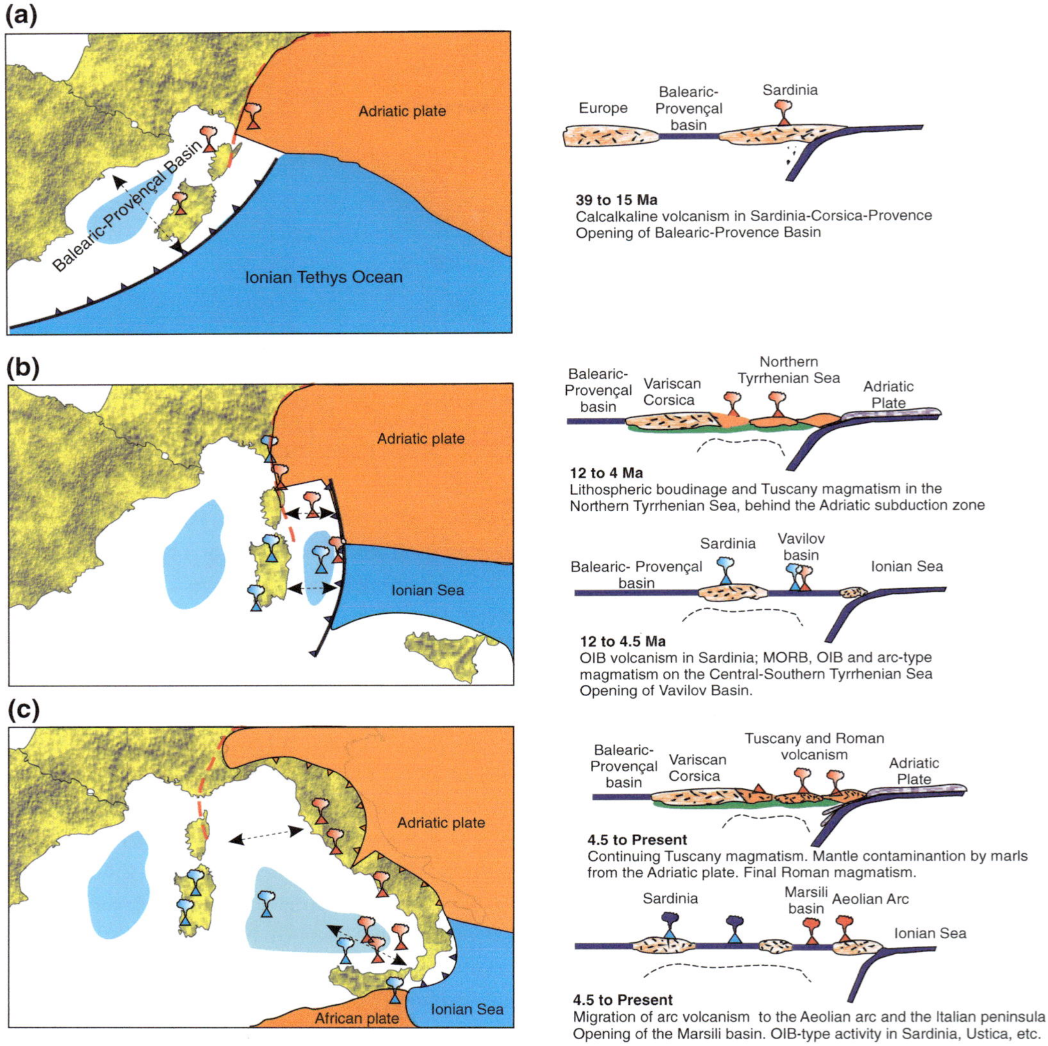

Fig. 12.10 Schematic evolution of the Tyrrhenian Sea and Sardinia from Eocene to present (simplified after Faccenna et al. 2010). **a** Eocene to Miocene westward subduction of the Ionian Tethys ocean floor beneath the southern European margin determines backarc spreading and calcalkaline magmatism in Corsica-Sardinia; **b** Subduction of the Ionian oceanic slab in the south and of delaminated Adriatic continental lithosphere in the north generates differential spreading in the Tyrrhenian basin and eastward migration of orogenic magmatism in Tuscany and central-southern Tyrrhenian Sea; **c** Subduction of the delaminated Adriatic slab is associated with Tuscany and Roman magmatism; subduction of residual Ionian plate triggers opening of the Marsili Basin and volcanism in the southern Tyrrhenian Sea. *Double arrows* indicate the direction of extension

The idea of the Tyrrhenian Sea as a backarc basin behind the eastward migrating Apennine subduction zone dates back to the early years of plate tectonics theory and was first discussed in a number of seminal papers by Boccaletti and Guazzone (1974) and Boccaletti et al. (1971, 1984). The following research clarified fundamental aspect of these process (e.g. Carminati et al. 2010 with references), but the basic ideas envisioned by early authors are still valid today.

Figure 12.10 is a synthesis of the most widely accepted ideas on the evolution of the Tyrrhenian Sea and origin of magmatism, simplified after several authors (e.g. Carminati et al. 2010; Faccenna et al. 2010 and references therein). According to this view, both oceanic and continental lithosphere has been subducted from the Eocene to Present. In the northern sector (Tuscany-Roman area), an early stage of oceanic crust consumption was followed by subduction of the Adriatic continental plate. In contrast, an oceanic-type lithosphere has been persistently subducting in the southern Tyrrhenian Sea (Carminati et al. 2005). After an early stage of oceanic subduction and orogenic volcanism in Corsica-Sardinia (Fig. 12.10a), there was slab rollback and eastward migration of the compression front, accompanied by E-W back-arc extension and orogenic magmatism migrating in the same direction (Fig. 12.10b). Starting from the end of Pliocene, there was a change in stretching direction in the Southern Tyrrhenian Sea, which became NW-SE oriented and caused the opening of the Marsili Basin (e.g. Sartori 2003); in contrast, there was still a roughly E-W directed extension in the northern sector (Fig. 12.10c). This modification of back-arc extension was related to the different resistance opposed to slab retreat by the Adriatic continental foreland with respect to the thinner Ionian oceanic lithosphere. The occurrence of ultrapotassic rocks in central Italy is also related to modifications in the nature of the undergoing slab, since the continental Adriatic slab brought large amounts of upper crustal material (marls) into the source of central Italy magmas (e.g. as Peccerillo et al. 1988; Serri et al. 1993). In contrast, the dominant calcalkaline to moderately potassic magmatism in the Aeolian arc and Campania area is related to a lower degree of mantle contamination by the Ionian oceanic crust.

Such a model explains the wide variety of anorogenic magma types in the Tyrrhenian Sea and Sardinia by assuming melting of three possible different sources, represented by (i) passively uprising asthenospheric rocks, and/or (ii) ancient contaminated lithosphere, and/or (iii) mantle material suctioned from the African plate around the edges of the Ionian subduction plate during slab roll-back (e.g. Lustrino et al. 2000; Trua et al. 2002; Faccenna et al. 2010).

Alternative hypotheses about the geodynamic evolution of the Tyrrhenian Sea advocate either convective movements of asthenospheric material or an active upwelling of deep-mantle plumes. Locardi and Nicolich (1988) and Locardi (1993) suggested that both the orogenic and anorogenic magmas are the result of uprising and melting of a large asthenospheric body, which had been softened by fluids of deep origin. Bell et al. (2004, 2013) advocate the expansion of a large mantle plume head centred in the Eastern Atlantic Ocean (Duggen et al. 1995). Fluids enriched in K-Ca-CO_2-H_2O released by the plume, modified upper mantle rocks whose melting gave the wide variety of magmas occurring in the area. However, a role for active mantle ascent has been questioned based, among others, on both the scarcity of volcanism on the Tyrrhenian Sea floor and on the very sharp subsidence of the Tyrrhenian Sea, which has reached a depth of more than 3600 m in a time interval of a few million years (Peccerillo and Lustrino 2005). The ascent of mantle plumes not rooted in the lower mantle and possibly triggered by the subducted slab itself, can well play a role in the origin of magmatism, especially for alkaline volcanoes in backarc areas (Faccenna et al. 2010). Further discussion on these issues can be found in Chap. 13.

12.6 Summary and Conclusions

A wide variety of orogenic and anorogenic magmas erupted in the Tyrrhenian Sea area from Miocene to Quaternary. Such a variability reveals a complex composition for the upper mantle. Geochemical data show that mantle sources of magmas are mixtures of MORB-, and OIB-type components, plus subduction-related fluids. MORB-type compositions are concentrated beneath the Vavilov Basin, but their role is evident at several other sectors of the Tyrrhenian Sea. An OIB-type component similar to Etna and

Ustica volcanoes (FOZO) is better shown by Prometeo lavas and by a single sample from Marsili Seamount. A distinct OIB-type composition, similar to Sardinia Plio-Quaternary volcanics, is shown by a few rocks collected offshore eastern Sardinia. Subduction-type components are widely distributed throughout the Tyrrhenian Sea basin, from the Upper Miocene Cornacya volcano (offshore southeastern Sardinia) to the central Tyrrhenian seafloor and the Aeolian area.

The reasons for this upper mantle heterogeneity in the Tyrrhenian Sea area are debated. Most authors agree that the Tyrrhenian Sea represents a back-arc basin where melting of various mantle rocks (ancient lithosphere, passively ascending asthenosphere, and mantle rocks inflown from the foreland) is responsible for the generation of MORB- and OIB-type magmas. Arc-type magmatism would be related to the release of fluids from subduction systems (both oceanic and continental) that contaminated various types of pre-metasomatic mantle rocks.

References

Argnani A, Savelli C (1999) Cenozoic volcanism and tectonics in the southern Tyrrhenian Sea: space-time distribution and geodynamic significance. Geodynamics 27:409–432

Barberi F, Bizouard H, Capaldi G, Ferrara G, Gasparini P, Innocenti F, Joron JL, Lambert B, Treuil M, Allegre C (1978) Age and nature of basalts from the Tyrrhenian Abyssal Plain. In: Hsu K, Montadert L et al (eds) Init Rep Deep Sea Drilling Project 42:509–514

Beccaluva L, Colantoni P, Di Girolamo P, Savelli C (1981) Upper-Miocene submarine volcanism in the Strait of Sicily (Banco Senza Nome). Bull Volcanol 44:573–581

Beccaluva L, Rossi PL, Serri G (1982) Neogene to recent volcanism of the Southern Tyrrhenian-Sicilian area: implications for the geodynamic evolution of the Calabrian Arc. Earth Evol Sci 3:222–238

Beccaluva L, Morlotti E, Torelli L (1984) Notes of the geology of the Elimi Chain area (southwestern margin of the Tyrrhenian Sea). Mem Soc Geol It 27:213–232

Beccaluva L, Gabbianelli G, Lucchini F, Rossi PL, Savelli C (1985) Petrology and K/Ar ages of volcanics dredged from the Eolian seamounts: implications for geodynamic evolution of the southern Tyrrhenian basin. Earth Planet Sci Lett 74:187–208

Beccaluva L, Bonatti E, Dupuy C, Ferrara G, Innocenti F, Lucchini F, Macera P, Petrini R, Rossi PL, Serri G, Seyler M, Siena F (1990) Geochemistry and mineralogy of the volcanic rocks from ODP sites 650, 651, 655, and 654 in the Tyrrhenian Sea. In: Kastens KA, Mascle J et al (eds) Proc Ocean Drilling Program, Sci Res 107:49–74

Bell K, Castorina F, Lavecchia G, Rosatelli G, Stoppa F (2004) Is there a mantle plume below Italy? EOS 85:541–547

Bell K, Lavecchia G, Rosatelli G (2013) Cenozoic Italian magmatism—isotope constraints for possible plume-related activity. J South Am Earth Sci 41:22–40

Bertrand H, Boivin P, Robin C (1990) Petrology and geochemistry of basalts from the Vavilov Basin (Tyrrhenian Sea), Ocean drilling program LEG 107, Holes 651A and 655B. In: Kastens KA, Mascle J et al (eds) Proc Ocean Drilling Program, Sci Res 107:75–89

Boccaletti M, Guazzone G (1974) Remnant arcs and marginal basins in the Cainozoic development of the Mediterranean. Nature 252:18–21

Boccaletti M, Elter P, Guazzone G (1971) Plate tectonics model for the development of Western Alps and Northern Apennines. Nature 234:108–111

Boccaletti M, Nicolich R, Tortorici L (1984) The Calabrian arc and the Ionian Sea in the dynamic evolution of the central Mediterranean. Mar Geol 55:219–245

Bonatti E, Seyler M, Channell J, Geraudeau J, Mascle G (1990) Peridotites drilled from the Tyrrhenian Sea, ODP LEG 107. In: Kastens KA, Mascle J et al (eds) Proc Ocean Drilling Program, Sci Res 107:37–47

Borsetti AM, Curzi PV, Landuzzi V, Mutti M, Ricci Lucchi F, Sartori R, Tomadin L, Zuffa GG (1990) Messinian and pre-Messinian sediments from ODP leg 107 sites 652 and 654 in the Tyrrhenian Sea: sedimentologic and petrographic study and possible comparisons with Italian sequences. In: Kastens KA, Mascle J et al (eds) Proc Ocean Drilling Program, Sci Res 107:169–186

Bruno PP, Di Fiore V, Ventura G (2000) Seismic study of the "41st Parallel" fault system offshore the Campanian-Latial continental margin, Italy. Tectonophysics 324:37–55

Cadoux A, Blichert-Toft J, Pinti DL, Albarede F (2007) A unique lower mantle source for Southern Italy volcanics. Earth Planet Sci Lett 259:227–238

Calanchi N, Colantoni P, Gabbianelli G, Rossi PL, Serri G (1984) Physiography of Anchise Seamount and of the submarine part of Ustica (south Tyrrhenian): petrochemistry of dredged volcanic rocks and geochemical characteristics of their mantle sources. Mineral Petrogr Acta 28:141–215

Carminati E, Negredo AM, Valera JL, Doglioni C (2005) Subduction-related intermediate-depth and deep seismicity in Italy: insights from thermal and rheological modelling. Phys Earth Planet Inter 149:65–79

Carminati E, Lustrino M, Cuffaro M, Doglioni C (2010) Tectonics, magmatism and geodynamics of Italy: what we know and what we imagine. In: Beltrando M, Peccerillo A, Mattei M, Conticelli S, Doglioni C (eds) J Virtual Explorer 36, paper 9. doi:10.3809/jvirtex.2010.00226

Cella F, Fedi M, Glorio G, Rapolla A (1998) Boundaries of magnetic anomaly sources in the Tyrrhenian region. Ann Geofis 41:433–447

Cella F, De Lorenzo S, Fedi M, Loddo M, Mongelli F, Rapolla A, Zito G (2006) Temperature and density of the Tyrrhenian lithosphere and slab and new interpretation of gravity field in the Tyrrhenian Basin. Tectonophysics 412:27–47

Cinque A, Civetta L, Orsi G, Peccerillo A (1988) Geology and geochemistry of the island of Ustica (Southern Tyrrhenian Sea). Rend Soc It Mineral Petrol 43:987–1002

Cocchi L, Tontini FC, Muccini F, Marani MP, Bortoluzzi G, Carmisciano C (2009) Chronology of the transition from a spreading ridge to an accretional seamount in the Marsili backarc basin (Tyrrhenian Sea). Terra Nova 21:369–374

Cocchi L, Masetti G, Muccini F, Carmisciano C (2015) Geophysical mapping of Vercelli Seamount: implications for Miocene evolution of the Tyrrhenian back arc basin. Geosci Front 2015:1–15

Colantoni P, Fabbri A, Gallignani P, Sartori R (1981) Lithologic and stratigraphic map of the Italian seas. Litografia Artistica Cartografica, Firenze

De Astis G, Ventura G, Vilardo G (2003) Geodynamic significance of the Aeolian volcanism (Southern Tyrrhenian Sea, Italy) in light of structural, seismological and geochemical data. Tectonics 22:1040–1057

De Astis G, Kempton PD, Peccerillo A, Wu TW (2006) Trace element and isotopic variations from Mt. Vulture to Campanian volcanoes: constraints for slab detachment and mantle inflow beneath southern Italy. Contrib Mineral Petrol 151:331–351

De Ritis R, Ventura G, Chiappini M, Carluccio R, von Frese R (2010) Regional magnetic and gravity anomaly correlations of the Southern Tyrrhenian Sea. Phys Earth Planet Inter 181:27–41

Dekov VM, Savelli C (2004) Hydrothermal activity in the SE Tyrrhenian Sea: an overview of 30 years of research. Mar Geol 204:161–185

Del Ben A, Barnaba C, Taboga A (2008) Strike-slip systems as the main tectonic features in the Plio-Quaternary kinematics of the Calabrian Arc. Mar Geophys Res 29:1–12

Della Vedova B, Bellani S, Pelli G, Squarci P (2001) Deep temperatures and surface heath distribution. In: Vai GB, Martini PI (eds) Anatomy of an Orogen. The Apennines and the adjacent Mediterranean basins. Kluwer, Dordrecht, pp 65–76

Dietrich V, Emmermann R, Keller J, Puchelt H (1977) Tholeiitic basalts from the Tyrrhenian Sea floor. Earth Planet Sci Lett 36:285–296

Dietrich V, Emmermann R, Puchelt H, Keller J (1978) Oceanic Basalts from the Tyrrhenian Basin, DSDP Leg 42A, Hole 373A. K, Montadert L et al (eds) Init Rep Deep Sea Drilling Project 42:515–530

Doglioni C, Innocenti F, Morellato C, Procaccianti D, Scrocca D (2004) On the Tyrrhenian Sea opening. Mem Descr Carta Geol It 64:147–164

Duggen S, Garbe-Schoenberg D, Hoernle K, Zhang YS, Graham D (1995) Seismic and geochemical evidence for large-scale mantle upwelling beneath the eastern Atlantic and western and central Europe. Nature 374:34–112

Ellam RM, Menzies MA, Hawkesworth CJ, Leeman WP, Rosi M, Serri G (1988) The transition from calcalkaline to potassic orogenic magmatism in the Aeolian Islands, Southern Italy. Bull Volcanol 50: 386–398

Faccenna C, Becker TW, Lallemand S, Lagabrielle Y, Funiciello F, Piromallo C (2010) Subduction-triggered magmatic pulses: a new class of plumes? Earth Planet Sci Lett 299:54–68

Feraud G (1990) ^{39}Ar-^{40}Ar analysis on basaltic lava series of Vavilov Basin, Tyrhenian Sea (Ocean Drilling Program, LEG 107, Holes 655B and 651A). In: Kastens KA, Mascle J et al (eds) Proc Ocean Drilling Programme, Sci Res 107:93–97

Florio G, Fedi M, Cella F (2011) Insights on the spreading of the Tyrrhenian Sea from the magnetic anomaly pattern. Terra Nova 23:127–133

Gale A, Dalton CA, Langmuir CH, Su Y, Schilling J-G (2013) The mean composition of ocean ridge basalts. Gechem Geophys Geosyst 14. doi:10.1029/2012GC004334

Gamberi F, Marani MP (2004) Deep-sea depositional systems of the Tyrrhenian basin. Mem Descr Carta Geol It 44:127–146

Gasperini D, Blichert Toft J, Bosch D, Del Moro A, Macera P, Albarède F (2002) Upwelling of deep mantle material through a plate window: evidence from the geochemistry of Italian basaltic volcanics. J Geophys Res 107(B12):2367. doi:10.1029/2001JB000418

Gueguen E, Doglioni C, Fernandez M (1997) Lithospheric boudinage in the Western Mediterranean back-arc basin. Terra Nova 9:184–187

Gueguen E, Doglioni C, Fernandez M (1998) On the post-25 Ma geodynamic evolution of the western Mediterranean. Tectonophysics 298:259–269

Hoefig TW, Geldmacher J, Hoernle K, Hauff F (2014) From the lavas to the gabbros: 1.25 km of geochemical characterization of upper oceanic crust at ODP/IODP Site 1256, Eastern Equatorial Pacific. Lithos 210–211:289–312

Iezzi G, Caso C, Ventura G, Vallefuoco M, Cavallo A, Behrens H, Mollo S, Paltrinieri D, Signanini P, Vetere F (2014) First documented deep submarine explosive eruptions at the Marsili Seamount (Tyrrhenian Sea, Italy): a case of historical volcanism in the Mediterranean Sea. Gondwana Res 25:764–774

Irvine TN, Baragar WRA (1971) A guide to chemical classification of common volcanic rocks. Can J Earth Sci 8:523–548

Kastens K, Mascle J (1990) The geological evolution of the Tyrrhenian Sea: an introduction to the scientific results of the ODP LEG 107. In: Kastens KA, Mascle J, et al. (eds) Proc Ocean Drilling Program, Sci Res 107:3–26

Kastens KA, Mascle J et al (1988) ODP Leg 107 in the Tyrrhenian Sea: insights into passive margin and back-arc basin evolution. Bull Geol Soc Am 100:1140–1156

Keller J (1981) Alkalibasalts from the Tyrrhenian Sea Basin: magmatic and geodynamic significance. Bull Volcanol 44:327–337

Kreuzer H, Mohr M, Wendt I (1978) Potassium-argon age determination of basalt samples from LEG 42A, Hole 373A, Core 7. In: Hsu K, Montadert L et al (eds) Init Rep Deep Sea Drilling Project 42:531–537

Locardi E (1993) Dynamics of deep structures in the Tyrrhenian-Apennines area and its relation to neotectonics. Il Quaternario 6:59–66

Locardi E, Nicholich R (1988) Geodinamica del Tirreno e dell'Appennino centro-meridionale: la nuova carta della Moho. Mem Soc Geol It 41:121–140

Loreto MF, Pepe F, De Ritis R, Ventura G et al (2015) Geophysical investigation of Pleistocene volcanism and tectonics offshore Capo Vaticano (Calabria, southeastern Tyrrhenian Sea). J Geodyn 90:71–86

Lucente FP, Chiarabba C, Cimini GB, Giardini D (1999) Tomographic constraints on the geodynamic evolution of the Italian region. J Geophys Res 104:20307–20327

Lupton J, De Ronde C, Sprovieri M, Baker ET, Bruno PP, Italiano F, Walker S, Faure K, Leybourne M, Britten K, Greene R (2011) Active hydrothermal discharge on the submarine Aeolian Arc. J Geophys Res 116:B02102. doi:10.1029/2010JB007738

Lustrino M, Melluso L, Morra V (2000) The role of lower continental crust and lithospheric mantle in the genesis of Plio-Pleistocene volcanic rocks from Sardinia (Italy). Earth Planet Sci Lett 180:259–270

Marani MP, Gamberi F (2004) Distribution and nature of submarine volcanic landforms in the Tyrrhenian Sea: the arc vs. the back-arc. Mem Descr Carta Geol It 44:109–126

Marani MP, Trua T (2002) Thermal constriction and slab tearing at the origin of a superinflated spreading ridge: Marsili volcano (Tyrrhenian Sea). J Geophys Res 107 (B2):2188. doi:10.1029/2001JB000285

Marani MP, Gamberi F, Savelli C (1997) Shallow-water polymetallic sulfide deposits in the Aeolian island arc. Geology 25:815–818

Mascle GH, Tricart P, Torelli L, Bouillin J-P, Rolfo F, Lapierre H, Monié P, Depardon S, Mascle J, Peis D (2001) Evolution of the Sardinia channel (western Mediterranean): new constraints from a diving survey on Cornacya seamount off SE Sardinia. Mar Geol 179:179–202

Milia A, Torrente MM (2014) Early-stage rifting of the Southern Tyrrhenian region: the Calabria-Sardinia brekup. J Geodyn 81:17–29

Milia A, Torrente MM, Massa B, Iannace P (2013) Progressive changes in rifting directions in the Campania margin (Italy): new constrains for the Tyrrhenian Sea opening. Glob Planet Change 109:3–17

Moeller S, Grevemeyer I, Ranero CR, Berndt C, Klaeschen D, Sallares V, Zitellini N, De Franco R (2013) Early-stage rifting of the northern Tyrrhenian Sea Basin: results from a combined wide-angle and multichannel seismic study. Geochem Geophys Geosyst 14:3032–3052. doi:10.1002/ggge.20180

Panza GF (1984) Structure of the lithosphere-asthenosphere system in the Mediterranean region. Ann Geophys 2:137–138

Panza GF, Pontevivo A, Chimera G, Raykova R, Aoudia A (2003) The lithosphere-asthenosphere: Italy and surroundings. Episodes 26:169–174

Panza GF, Pontevivo A, Sarao' A, Aoudia A, Peccerillo A (2004) Structure of the lithosphere-asthenosphere and volcanism in the Tyrrhenian Sea and surroundings. Mem Descr Carta Geol It 64:29–56

Panza GF, Peccerillo A, Aoudia A, Farina B (2007) Geophysical and petrological modeling of the structure and composition of the crust and upper mantle in complex geodynamic settings: the Tyrrhenian Sea and surroundings. Earth Sci Rev 80:1–46

Passaro S, Milano G, Sprovieri M, Tonielli R, Ruggieri S, De Martino G, Innangi S, Sammartino S, Marsella E (2009) Seamount bathymetries of the south-eastern Tyrrhenian Sea. Proc GNGTS Conf 2009, Session 3.2, pp 656–660

Passaro S, Milano G, D'Isanto C, Ruggieri S, Tonielli R, Bruno P, Sprovieri M, Marsella E (2010) DTM-based morphometry of the Palinuro seamount (Italy, Eastern Tyrrhenian Sea): geomorphological and volcanological implication. Geomorphology 115:129–140

Pearce JA (1982) Trace element characteristics of lavas from destructive plate boundaries. In: Thorpe RS (ed) Andesites: orogenic andesites and related rocks. Wiley, Chichester, pp 525–548

Pearce JA, Parkinson IJ (1993) Trace element models for mantle melting: application to volcanic arc petrogenesis. Geol Soc London Spec Publ 76:373–403

Pearce JA, Peate DW (1995) Tectonic implications of the composition of volcanic arc magmas. Ann Rev Earth Planet Sci 23:251–285

Peccerillo A, Lustrino M (2005) Compositional variations of the Plio-Quaternary magmatism in the circum-Tyrrhenian area: deep- vs. shallow-mantle processes. In: Foulger GR, Nathland JH, Presnall DC, Anderson DL (eds) Plates, plumes and paradigms. Geol Soc Am Spec Publ 388:421–434

Peccerillo A, Poli G, Serri G (1988) Petrogenesis of orenditic and kamafugitic rocks from Central Italy. Can Mineral 26:45–65

Pontevivo A, Panza G (2006) The lithosphere-asthenosphere system in the Calabrian arc and surrounding seas, Southern Italy. Pure appl Geophys 163:1617–1659

Prada M, Sallares V, Ranero CR, Vendrell MG, Grevemeyer I, Zitellini N, De Franco R (2014) Seismic structure of the Central Tyrrhenian basin: geophysical constraints on the nature of the main crustal domains. J Geophys Res 119:52–70

Presnall DC, Gudmundur H, Gudfinnsson H, Walter MJ (2002) Generation of mid-ocean redge basalts at pressures from 1 to 7 GPa. Geochim Cosmochim Acta 66:2073–2090

Robin C, Colantoni P, Gennesseaux M, Réhault JP (1987) Vavilov seamount: a mildly alkaline quaternary volcano in the Tyrrhenian Basin. Mar Geol 78:125–136

Romagnoli C (2013) Characteristics and morphological evolution of the Aeolian volcanoes from the study of submarine portions. In: Lucchi F, Peccerillo A, Keller J, Tranne CA, Rossi PL (eds) The Aeolian Islands volcanoes. Geol Soc London Mem 37:13–26

Romagnoli C, Casalbore D, Bortoluzzi G, Bosman A, Chiocci FL, D'Oriano F, Gamberi F, Ligi M, Marani M (2013) Bathy-morphological setting of the Aeolian Islands. In: Lucchi F, Peccerillo A, Keller J, Tranne CA, Rossi PL (eds) The Aeolian Islands volcanoes. Geol Soc London Mem 37:27–36

Rose-Koga EF, Koga KT, Schiano P, Le Voyer M, Shimizu N, Whitehouse MJ, Clocchiatti R (2012) Mantle source heterogeneity for South Tyrrhenian magmas revealed by Pb isotopes and halogen contents of olivine-hosted melt inclusions. Chem Geol 334:266–279

Rosenbaum G, Lister GS (2004a) Neogene and Quaternary rollback evolution of the Tyrrhenian Sea, the Apennines and the Sicilian Maghrebides. Tectonics, 23, TC1013. doi:10.1029/2003TC001518

Rosenbaum G, Lister GS (2004b) Formation of arcuate orogenic belts in the western Mediterranean region. In: Sussman AJ, Weil AB (eds) Orogenic curvature: integrating paleomagnetic and structural analyses. Geol Soc Am Spec Paper 383:41–56

Rosenbaum G, Gasparon M, Lucente FP, Peccerillo A, Miller MS (2008) Kinematics of slab tear faults during subduction segmentation and implications for Italian magmatism. Tectonics 27, TC2008. doi:10.1029/2007TC002143

Sartori R (2001) Corsica-Sardinia block and the Tyrrhenian Sea. In: Vai GB, Martini PI (eds) Anatomy of an Orogen. The Apennines and the adjacent Mediterranean basins. Kluwer, Dordrecht, pp 367–374

Sartori R (2003) The Thyrrhenian backarc basin and subduction of the Ionian lithosphere. Episodes 26:217–221

Sartori R, Torelli L, Zitellini N, Carrara G, Magaldi M, Mussoni P (2004) Crustal features along a W-E Tyrrhenian transect from Sardinia to Campania margins (Central Mediterranean). Tectonophysics 383:171–192

Savelli C (1988) Late Oligocene to recent episodes of magmatism in and around the Tyrrhenian Sea: implications for the processes of opening in a young inter-arc basin of intra-orogenic (Mediterranean) type. Tectonophysics 146:163–181

Savelli C (2000) Subduction-related episodes of K-alkaline magmatism (15–0.1 Ma) and geodynamic implications in the north Tyrrhenian, central Italy region: a review. J Geodyn 30:575–591

Savelli C, Lipparini E (1978) K/Ar determinations on basalt rocks from Hole 373A. In: Hsu K, Montaderd L et al (eds) Init Rep Deep Sea Drilling Project 81:537–538

Schiano P, Clocchiatti R, Ottolini L, Busà T (2001) Transition of Mount Etna lavas from a mantle-plume to an island-arc magmatic source. Nature 412:900–904

Serri G (1990) Neogene-Quaternary magmatism of the Tyrrhenian region: characterization of the magma sources and geodynamic implications. Mem Geol Soc It 41:219–242

Serri G, Innocenti F, Manetti P (1993) Geochemical and petrological evidence of the subduction of delaminated Adriatic continental lithosphere in the genesis of the Neogene-Quaternary magmatism of central Italy. Tectonophysics 223:117–147

Sun SS, McDonough WF (1989) Chemical and isotopic systematics of oceanic basalts: implications for mantle composition and processes. In: Saunders AD, Norry MJ (eds) Magmatism in ocean basins. Geol Soc London Spec Publ 42:313–345

Trua T, Serri G, Marani M, Renzulli A, Gamberi F (2002) Volcanological and petrological evolution of Marsili seamounts (southern Tyrrhenian Sea). J Volcanol Geoth Res 114:441–464

Trua T, Serri G, Marani MP (2003) Lateral flow of African mantle below the nearby Tyrrhenian plate: geochemical evidence. Terra Nova 15:433–440

Trua T, Serri G, Marani MP, Rossi PL, Gamberi F, Renzulli A (2004) Mantle domains beneath the southern Tyrrhenian: constraints from recent seafloor sampling and dynamic implications. Per Mineral 73:53–73

Trua T, Marani MP, Gamberi F (2011) Magmatic evidence for African mantle propagation into the southern Tyrrhenian backarc region. In: Beccaluva L, Bianchini G, Wilson M (eds) Volcanism and evolution of the African lithosphere. Geol Soc Am Spec Paper 478:307–331

Trua T, Marani M, Barca D (2014) Lower crustal differentiation processes beneath a back-arc spreading ridge (Marsili seamount, Southern Tyrrhenian Sea). Lithos 190:349–362

Wood DA (1980) The application of Th-Hf-Ta diagram to problems of tectonomagmatic classification and to establish the nature f crustal contaminant of basaltic lavas of the British Tertiary volcanic province. Earth Planet Sci Lett 50:11–30

Zito G, Mongelli F, De Lorenzo S, Doglioni C (2003) Heat flow and geodynamics in the Tyrrhenian Sea. Terra Nova 15:425–443

Magmatism and Geodynamics in the Tyrrhenian Sea Region

13

Abstract

The Tyrrhenian Sea region is the site of widespread Cenozoic igneous activity that shows extreme compositional variations in space and time, highlighting complex evolution history for magmas and their sources. Based on ages and volcanological-petrological-geochemical characteristics of mafic rocks, several magmatic provinces are distinguished, which are related to compositionally distinct sectors of the upper mantle beneath the Tyrrhenian Sea region. Rocks with MORB- and OIB-type geochemical signatures (i.e. low LILE/HFSE ratios, and EM1, FOZO, DMM isotopic compositions) originated from mantle rocks that were not affected by young subduction events. FOZO- and DMM-type magmas are best represented by the Sicily volcanics and may derive from an ascending plume head or from a metasomatised lithosphere-asthenosphere permeated by deep asthenospheric fluids. EM1-type rocks crop out in Sardinia and are unique in Europe. They could derive either from a mantle plume or from an ancient metasomatised lithosphere. The petrogenesis of arc-type magmas is related to contamination of various mantle rocks (MORB to OIB-type) by subducted upper crustal components. Four main types of subduction-related mantle metasomatic events are recognised in the Tyrrhenian Sea area, with a role of upper crustal components (sediments) increasing from Sardinia and the southern Tyrrhenian Sea to Central Italy. Metasomatism beneath Tuscany was Alpine in age, whereas mantle contamination beneath the Roman Province was provided by Miocene-Quaternary subduction of the Adriatic continental plate beneath the Northern Apennines. In contrast, subduction of the Ionian oceanic plate plus some sediments generated metasomatism and magmatism in Campania and the central-southern Tyrrhenian Sea.

A. Peccerillo, *Cenozoic Volcanism in the Tyrrhenian Sea Region*,
Advances in Volcanology, DOI 10.1007/978-3-319-42491-0_13

Keywords
Tyrrhenian Sea · Magmatism · Geodynamics · MORB · OIB · Arc-type magmas · Mantle metasomatism · Subduction · Sediment subduction · Adriatic plate · Ionian plate

13.1 Introduction

The magmatic provinces discussed in the previous chapters exhibit exceedingly large compositional variations that almost entirely cover the range of magmatic rocks occurring worldwide. Except for a few crustal anatectic rocks, mostly occurring in Tuscany, the bulk of magmatism is of ultimate mantle origin. Many of the observed compositional characteristics are related to magma evolution processes, which, however, are unable to explain the first-order geochemical and petrological variations, especially for mafic rocks. It has been, therefore, unanimously concluded that the wide variety of magmas occurring in the Tyrrhenian Sea region reveals extremely heterogeneous mantle sources. These are the result of a long and complex evolution history suffered by the Tyrrhenian Sea and Western Mediterranean, from Paleozoic until the present day.

In this chapter, the main structural, petrological-geochemical and petrogenetic aspects of individual magmatic provinces in the Tyrrhenian Sea region will be first summarised. Attention will be focused on those issues that, in this author's opinion, are crucial for a better understanding of the relationships between magmatism and geodynamics. Successively, the most popular hypotheses on the geodynamic evolution of the Tyrrhenian Sea area will be briefly examined with the aim of exploring their ability to explain the first-order magmatological features of the Tyrrhenian Sea region.

13.2 Compositional and Structural Characteristics of Volcanism in the Tyrrhenian Sea Region

Petrological and geochemical data highlight the existence of two broad groups of magmas in the Tyrrhenian Sea region. One group shows geochemical characteristics typical of rocks occurring along converging plate margins (i.e. high LILE/HFSE ratios and radiogenic isotope signatures intermediate between mantle and crust) and has been defined as "orogenic". The other group has compositions falling within the range of magmas emplaced in intraplate settings or along passive plate margins (i.e. low LILE/HFSE and radiogenic isotope ratios close to MORB and OIB), and has been referred to as "anorogenic".

The rocks with orogenic geochemical signatures range in age from Eocene to the present, and are found in Sardinia-Ligurian Sea-Provence, at various sites in the Tyrrhenian Sea basin, and, most notably, in the Italian peninsula and the Aeolian arc. These magmas show a clear decrease in age from west to the east, along with and an overall increase in potassium from the dominant calcalkaline rocks of Sardinia-Ligurian Sea to the abundant shoshonitic and ultrapotassic magmatism of central-southern Italy.

Magmas with anorogenic compositions are Upper Miocene to Quaternary in age and occur in Sicily, the Sicily Channel, Sardinia, and at various centres in the Tyrrhenian Sea basin. A few occurrences have been also found in Provence (Toulon) and the Ligurian Sea. There is no clear age polarity for anorogenic rocks, but, in single volcanic districts, most notably in Sardinia, the anorogenic rocks are younger than the spatially associated orogenic products.

Radiogenic isotope ratios exhibit large variations in the mafic rocks. In general terms, Sr–Nd–Hf–He isotope ratios are rather homogeneous in the anorogenic rocks from Sicily, but different values are shown by Plio-Quaternary rocks from Sardinia. Much larger range of values is observed in the orogenic rocks (e.g. Sano et al. 1989; Gasperini et al. 2002; Martelli et al. 2008; Alagna et al. 2010 with references). Pb-isotope ratios define a somewhat different picture, showing strong variations both for orogenic and anorogenic rocks (Fig. 13.2).

It has been demonstrated that isotopic heterogeneity of volcanism in the Tyrrhenian Sea region reveals interaction among compositionally distinct mantle end-members, and between these and the upper crust. A two-end-member mixing between mantle (FOZO) and upper continental crustal components has been suggested for orogenic magmas by various authors (e.g. Hawkesworth and Vollmer 1979; Gasperini et al. 2002; Cadoux et al. 2007 and many others). However, data on mafic rock indicate distinct trends for Sr–Nd isotopes, suggesting that various mantle and crustal end-members were involved in the mixing (Fig. 13.2a). Trace element data militate in favour of this hypothesis, as discussed in the previous chapters.

Additional complexities are revealed by Pb-Sr isotope plot, in which various trends are obvious (Fig. 13.2e). These highlight interaction among different types of mantle end-members (FOZO, DMM, and EM1) and no role for the crust in the anorogenic magmas. In contrast, trends connecting mantle and upper crust compositions are observed for orogenic magmas, highlighting important roles for both these Earth reservoirs. Notably, some orogenic volcanoes (e.g. Procida and Ventotene, in the Pontine and Campania provinces) divert from the main mantle-crust mixing trend, pointing toward EM1-type compositions.

In summary, the broad groups of orogenic and anorogenic magmas display strong internal compositional variations. Based on petrological and geochemical data, several distinct magmatic provinces have been recognised (Fig. 13.1; see also Fig. 1.8). Some of these provinces are bounded by important tectonic lines and are characterised by a specific structure of the crust-upper mantle system (e.g. Moho depth, mechanical characteristics of the upper mantle, seismicity, etc.; Peccerillo and Panza 1999; Peccerillo and Frezzotti 2015). Therefore, structural and geophysical evidence supports the subdivision of magmatism inferred from petrological-geochemical data.

The *Tuscany Magmatic Province* consists of an association of mafic to felsic intrusive rocks and lavas and a few pyroclastics. Ages (14.5–0.2 Ma) decrease easward from the Alpine Corsica (Sisco) to the Tuscany archipelago and mainland, parallel with the migration of backarc extension and of the foredeep-thrust belt system of Northern Apennines (e.g. Boccaletti et al. 1990a). Magmatism developed in a zone of thinned continental crust, with Moho reaching a minimum depth of about 20 km in southern Tuscany (Fig. 13.1). The upper mantle is characterised by the occurrence of a rigid vertical layer, which cuts the asthenosphere (e.g. Mele and Sandvol 2003; Panza et al. 2003). According to Benoit et al. (2011) this slab reaches a depth of about 300 km and extends no further south than about 43°N, with no deep continuity with the Central Apennine. Remnants of an older slab have been imaged in the western sector of the Tuscany Province (e.g. Finetti et al. 2001). In spite of their small volumes, mafic magmas show extremely variable petrological-geochemical compositions, from calcalkaline to potassic and ultrapotassic lamproitic. Lamproites are particularly depleted in Na, Ca, Al, Sc, revealing a cpx-poor harzburgitic mantle source. The incompatible trace element patterns of Tuscany mafic rocks (normalised to primitive mantle) resemble closely those of some upper crustal rocks, such as granites, pelites or their metamorphic equivalents; radiogenic isotope signatures are closer to upper continental crust than to typical mantle values (e.g. $^{87}Sr/^{86}Sr \sim 0.709$–0.717; $^{143}Nd/^{144}Nd \sim 0.5121$–0.5230). Overall these data indicate that mafic calcalkaline to

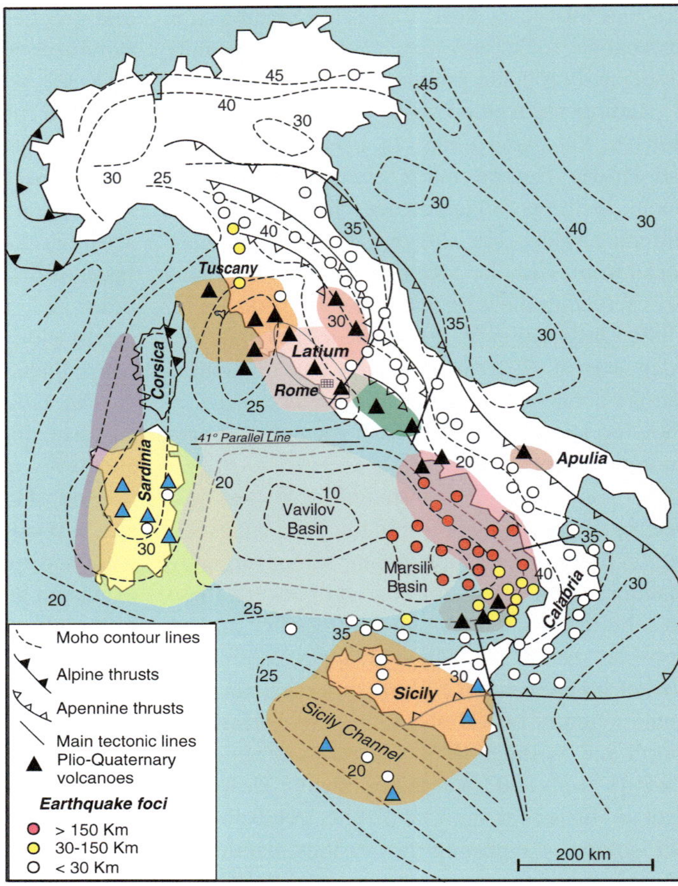

Fig. 13.1 Sketch map showing Moho depth, distribution of earthquakes hypocenters and main structural features in the Tyrrhenian Sea region. Simplified after Boccaletti et al. (1984) Locardi and Nicolich (1988), Mele et al. (2006), Di Stefano et al. (2011), and Doglioni and Carminati (2012). Distribution of magmatic provinces is shown by shaded areas

lamproitic magmas of the Tuscany Province were generated in a depleted lherzolitic to harzburgitic upper mantle (likely asthenospheric to lithospheric rocks) that was contaminated by siliceous upper crustal materials such as metapelites. A group of intermediate calcalkaline rocks shows adakite-type enrichments in some incompatible elements (Ba, Sr, and LREE), but their origin needs further investigation to be clarified. The Tuscany mafic rocks resemble closely the Oligocene calcalkaline to lamproitic magmas from the Western Alps, indicating a common origin in depleted mantle source contaminated by siliceous sediments, with little role for OIB-type components (see Appendix 1).

Felsic rocks in Tuscany (granodiorites, monzogranites, rhyolites, etc.) are polygenetic. A few of them were originated by crustal anatexis, as a consequence of temperature increase during mafic magma emplacement. Others are hybrids between crustal anatectic and mantle-derived magmas (Poli and Peccerillo 2016).

The *Roman Province* consists of large volcanic complexes and stratovolcanoes, which erupted about 900 km^3 of prevailingly pyroclastic rocks and minor lavas. Volcanism developed over a relatively short period of time (mostly between 0.8 Ma and about 20 ka) along NW–SE trending normal faults parallel to the Tyrrhenian Sea border, in a zone situated west of the so-called Olevano-Antrodoco (or Ancona-Anzio) tectonic line. The Roman volcanoes are partially superimposed upon the Tuscany Magmatic Province, and there is evidence for hybridisation between magmas of these two provinces. Compositionally, the Roman rocks

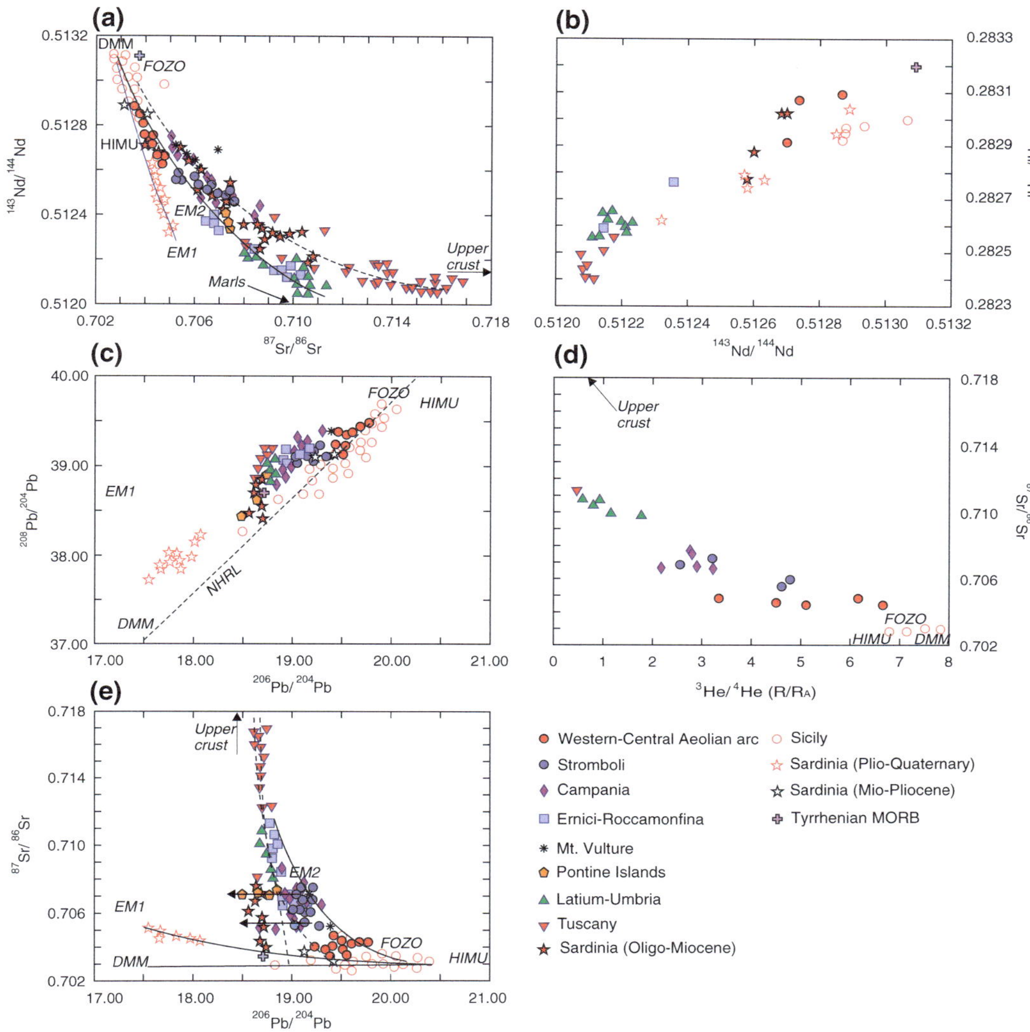

Fig. 13.2 Sr–Nd–Pb–Hf–He isotope variation in mafic rocks (MgO > 5 %) from the Tyrrhenian Sea region. Only representative samples covering the entire compositional field of various provinces have been plotted for clarity. *Lines* in panels A and E indicate mixing trends between different mantle and crustal components. See text for explanation

show variable enrichments in potassium and incompatible elements, ranging from potassic series (KS) to ultrapotassic series (HKS). Radiogenic isotope ratios show intermediate values between the mantle and the upper continental crust (e.g. ^{87}Sr/^{86}Sr is typically around 0.709–0.711), but are closer to mantle compositions than potassic and ultrapotassic rocks from Tuscany (Fig. 13.2). Geochemical and isotopic data suggest that the Roman magmas were generated by melting of anomalous mantle sources that were enriched by the addition of upper crustal material, with a broadly carbonated pelite (marly) composition (Peccerillo et al. 1988; Grassi et al. 2012). The mantle contamination beneath the Roman Province probably took place more recently than in Tuscany, as supported by the younger age of Roman volcanism. Low concentrations of HFSE in the most mafic rocks denote a depleted MORB-type composition for pre-metasomatic mantle sources of Roman magmas. This does not conflict with radiogenic isotope data, since the Roman rocks can be envisaged as plotting on a mixing trend between DMM and marls (Fig. 13.2a). Experimental petrology studies indicate pressures exceeding 2–3 GPa for the origin of silica undersaturated ultrapotassic melts, corresponding to asthenospheric depths in the area (Mueller and Panza 1986).

The *Intra-Apennine Province* (0.6–0.25 Ma) consists of a few monogenetic centres that exhibit similar trace element and radiogenic isotopic signatures to those of the Roman Province, suggesting a similar type of source metasomatism. The lower Al_2O_3 and Na_2O combined with higher CaO, K_2O/Na_2O and degree of silica undersaturation of Intra-Apennine rocks suggest a magma origin at higher pressure, by melting of mantle sources that had a different pre-metasomatic modal mineralogy (i.e. harzburgite instead of lherzolite) than the Roman Province. Some intra-Apennine pyroclastic rocks consist of a mixture of silicate and carbonate material and have been suggested to represent carbonatitic magmas, a hypothesis which has been argued to conflict with many geochemical and isotopic data, at least for some outcrops (Peccerillo 1998, 2004).

The *Ernici-Roccamonfina Province* (0.6–0.15 Ma) has very peculiar volcanological, petrological and geochemical characteristics, given by small volumes of erupted magmas and very large compositional variations (calcalkaline-shoshonitic to ultrapotassic) for mafic rocks. The Ernici-Roccamonfina Province is delimited by the Ancona-Anzio line in the north and by the Ortona-Roccamonfina tectonic line in the south. Tomographic images highlight a discontinuity (slab window) in the high-velocity vertical layer that characterises the Northern and Southern Apennines (e.g. Giacomuzzi et al. 2012; Rosenbaum and Piana Agostinelli 2015). Early erupted ultrapotassic magmas exhibit incompatible element contents and radiogenic isotope ratios similar to the Roman volcanoes, whereas late calcalkaline-shoshonitc rocks resemble the Campania Province. Therefore, the Ernici-Roccamonfina Province represents a sort of transition zone, in which there were two mantle metasomatic processes showing Roman-type and Campania-type compositions, respectively. Timing of magma emplacement suggests an earlier occurrence for the Roman-type than for Campania-type metasomatism. Low abundances of HFSE in the mafic rocks speak for a depleted MORB-type composition of pre-metasomatic mantle sources.

The *Pontine Islands* (4.2–0.1 Ma) probably represent the northern end of orogenic Plio-Pleistocene volcanic arcs developed across the Tyrrhenian Sea (Argnani and Savelli 1999). Sr–Nd–Pb isotopic ratios of mafic rocks plot along the general crust-mantle mixing trend defined by the bulk of orogenic igneous rocks. However, some mafic samples from the Island of Ventotene show a clear trend toward lower Pb-isotope compositions, suggesting a role for a distinct component with low Pb-isotope ratios such as the EM1-type Plio-Quaternary rocks from Sardinia (Fig. 13.2e).

The *Campania Province and Stromboli*. The Campania Province (0.3 Ma to Present) is formed of shoshonitic to ultrapotassic rocks; thick calcalkaline (2 Ma) lava sequences have been drilled beneath the Campanian Plain. Stromboli (0.2 Ma to present) is sited in the eastern Aeolian arc and is formed of mafic-intermediate rocks with calcalkaline to slightly undersaturated potassic alkaline affinity. There is overwhelming geochemical and radiogenic isotope evidence demonstrating that compositions of Stromboli and Campania mafic rock resemble closely each other. This has led to the conclusion that the Campania volcanism is not the southern manifestation of the Roman Province, as stated by early authors, but rather represents the northern end of the eastern Aeolian arc magmatism (Peccerillo 2001). Additional important evidence in favour of this hypothesis has been recently provided by experimental petrology studies, which demonstrate that the leucite-bearing rocks from Vesuvio may be derived from a potassic trachybasalt parental magma by fractional crystallisation and assimilation of carbonate wall rocks (Pichavant et al.

2014). The implication of experimental data is that primary ultrapotassic magmas as those erupted in the Roman Province are absent in Campania. Geochemical and petrological data are paralleled by seismic evidence showing that Stromboli and Campania volcanoes are sited above or close to a narrow zone of deep-focus earthquakes, which extends from Calabria to the southern limb of the Campania Province. In spite of petrological and geochemical evidence, several authors still consider the Campania volcanism as a district of the Roman Province. This disagreement is not simply a problem of classification of volcanic rocks but deals with the much more important issue of the structural significance of the discontinuity between the northern-central and the southern Apennine arcs, a divide marked, among others, by the large slab window beneath the Ernici-Roccamonfina area (e.g. Rosenbaum and Piana Agostinetti 2015). This slab windows separates two orogenic systems that have been evolving independently from Upper Pliocene to present, as a result of northward migration of the Adriatic plate and a shifting of the Ionian plate toward the southeast (e.g. Malusà et al. 2015, 2016). Radiogenic isotope ratios of Campania-Stromboli volcanoes fall between FOZO mantle compositions and upper crustal values. HFSE concentrations are higher and LILE/HFSE ratios are lower than for Roman and many other Italian "orogenic" volcanoes, recalling OIB-type compositions. Collectively, these features have been interpreted as evidence for a FOZO-type mantle source that was contaminated by metasomatic material characterised by much fainter crustal signatures than for Central Italy volcanoes (Serri 1990; Beccaluva et al. 1991).

Mount Vulture (0.75–0.14 Ma) shows unique petrological-geochemical characteristics and structural setting among Italian volcanoes. Compositions (mostly basanite and trachyphonolite with minor foidite, tephrite, phonotephrite and melilitite) are rich in both Na_2O and K_2O and contain haüyne as the main foid. Carbonatitic magmas erupted during the final stages of volcanic activity. Composition of silicate rocks shows the coexistence of arc-type and OIB-type geochemical signatures. Such a feature is mirrored by the peculiar structural position of Vulture volcano, at the limit between the Southern Apennine chain and Apulia, a sector of Adriatic foreland. Petrological-geochemical data suggest a similar origin as Stromboli and Campania volcanoes, but with a stronger role of FOZO-type components (e.g. Peccerillo and Frezzotti 2015).

The *Western and Central Aeolian arc* (>0.3 to 1888–1890 AD) is formed by calcalkaline, high-K calcalkaline, and shoshonitic rocks, with some slightly undersaturated potassic alkaline products at Vulcano. Volcanism developed west and along the Tindari-Letojanni fault over an about 20 km thick crust, likely represented by the Calabro-Peloritano basement. A vertical rigid body has been imaged within the asthenosphere beneath the western Aeolian arc, but deep seismicity is scanty. Sr isotopic ratios of mafic rocks ($^{87}Sr/^{86}Sr \sim 0.7035$–$0.7045$) include some of the lowest values among the orogenic volcanoes in the Tyrrhenian area. LILE/HFSE ratios are high and decrease slightly from the central arc to the external island of Alicudi. Petrological, geochemical and isotopic data suggest that the western-central Aeolian magmas were generated in a depleted MORB-type mantle that was contaminated by LILE-rich fluids characterised by relatively unradiogenic Sr-isotope signatures. This suggests a very limited role for upper crustal rocks in the origin of the western-central Aeolian magmas. Lower LILE/HFSE ratios at the westernmost margin of the arc are attributed to a role of OIB-type components.

The *Sicily Province* (Upper Miocene to Present) includes a large number of volcanoes (Etna, Iblei, Ustica and Sicily Channel islands and seamounts) with tholeiitic to Na-alkaline and nephelinite compositions. Volcanism developed in different structural settings. Etna is sited close to the Tindari-Letojanni-Malta Escarpment lineament, which defines the western border of the Ionian oceanic crust in this area; the Iblei volcanoes developed along a NE–SW fault zone of the Pelagian block, a promontory of the African continent; Pantelleria, Linosa and several nearby seamounts are sited along NW to SE trending grabens of the northern Africa foreland; Ustica occurs in the southern Tyrrhenian Sea, west of the Aeolian arc, basically in a backarc position. Trace element signatures exhibit low LILE/HFSE ratios, typical of OIB- and MORB-type magmas. Sr–Nd isotope ratios are poorly variable and close to FOZO mantle composition. In contrast, Pb-isotope ratios define a larger range of values that plot between FOZO and DMM. The most popular petrogenetic hypotheses suggest an origin of Sicily primary magmas by variable degrees of partial melting of heterogeneously enriched mantle rocks within the asthenosphere and/or at the lithosphere-asthenosphere boundary. Etna, Ustica, and the nearby Prometeo submarine lava field show some arc-type signatures, such as high contents of volatile components and relatively low TiO_2, which suggest a role of arc components in their origin.

Sardinia has been the site of both orogenic and anorogenic volcanism, which developed from Eocene to Upper Pleistocene during three temporally distinct stages of activity. The Oligo-Miocene stage erupted orogenic, dominantly cal-calkaline magmas, whose parental liquids originated in mantle sources modified by fluids rich in LILE and moderately enriched in radiogenic Sr. The participation of a component with low Pb-isotope ratios, probably residing within the lithosphere, had also a role in determining the composition of orogenic magmas. The subsequent Miocene-Quaternary magmatism has anorogenic geochemical features, utterly contrasting with the previous activity. It consists of two volumetrically and compositionally distinct stages, which respectively occurred in southern and central-northern Sardinia during Mio-Pliocene and Upper Pliocene-Quaternary times. The Plio-Quaternary volcanic stage erupted magmas with strikingly low $^{206}Pb/^{204}Pb$ and $^{207}Pb/^{204}Pb$ ratios similar to EM1 mantle composition, a unique case in Europe. The Mio-Pliocene volcanic stage has intermediate Pb-isotope compositions between Plio-Quaternary magmas and FOZO (Fig. 13.3e). The origin and geodynamic significance of EM1 component are controversial. However, its geographically restricted occurrence points to an origin within the local lithospheric mantle (Lustrino et al. 2013). Mantle xenoliths occurring in the Plio-Quaternary volcanics have Sr–Nd isotopic ratios falling within the field of the host rocks, thus supporting a role of the lithosphere in the origin of anorogenic magmatism. However, an extensive Pb-isotope study of xenoliths would be necessary to eventually strength this hypothesis.

Tyrrhenian Sea. A wide variety of arc-, MORB- and OIB-type magmas have been emplaced in the Tyrrhenian Sea basin. Among

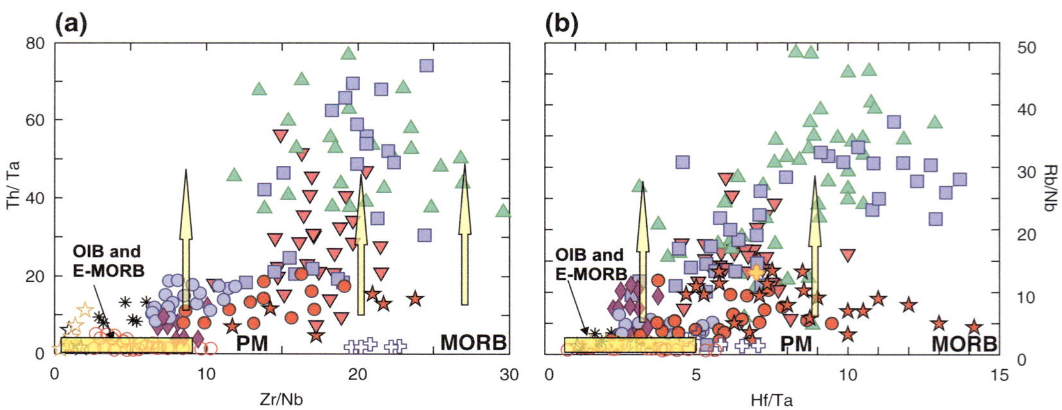

Fig. 13.3 HFSE/HFSE (Zr/Nb, Hf/Ta) and LILE/HFSE (Th/Ta, Rb/Nb) variations in the mafic rocks (MgO > 5 wt%) from the Tyrrhenian Sea region. *Arrows* indicate the effects of subduction-related metasomatism starting from mantle compositions with OIB, Enriched E-MORB, Primitive Mantle (PM), and Normal MORB compositions. Symbols as in Fig. 13.2

arc-type rocks, the Cornacya volcanics (12 Ma) occurring off southeastern Sardinia, has similar composition to mafic potassic rocks from the Tuscany Province. Other arc-type magmas are scattered throughout the Tyrrhenian Sea floor, showing a decrease in age from west to the east. MORB- and OIB-type rocks do not show clear age polarity but they are younger than the associated arc-type rocks in the single sectors. A variety of mantle sources, showing MORB to EM1 and FOZO compositions, locally contaminated by compositionally variable fluids, gave the wide range of magmas erupted in the Tyrrhenian Sea basin.

13.3 Geodynamic Evolution of the Tyrrhenian Sea Region: A Short Summary

The geodynamic evolution history of the Tyrrhenian Sea region has been characterised by a complex succession of rifting, magmatism and continental collision from Palaeozoic to the present (e.g. Stampfli and Borel 2004). The Upper Cretaceous to Quaternary geodynamic evolution is crucial for the development of the complex association of igneous rocks occurring in the Tyrrhenian Sea region. However, magmas may preserve the record of older geodynamic events in their geochemical-isotopic composition.

The Hercynian (Variscan) orogenesis is a first-order yet ancient geodynamic process. It occurred during Carboniferous to Permian times by consumption of a ca. 3000-km-wide oceanic basin, separating Gondwana and Laurussia (central Europe). Plate convergence and continent-continent collision generated metamorphism, deformation and orogenic magmatism. The latter is recorded by the intrusive calcalkaline to shoshonitic complexes in several places around the Tyrrhenian Sea (e.g. western Corsica, Sardinia, Calabria, and the Alps; Lustrino 2000; Carmignani et al. 2001 and references therein). Hercynian orogenesis was followed by rifting, extensive mantle melting, extraction of basaltic magmas and the formation of the so-called Ligurian-Piemontese oceanic basin during Middle-Late Jurassic break-up of Pangea.

Beginning in the early Cretaceous ($\sim$120 Ma), a new compressive phase took place, marking the onset of Alpine orogenic cycle. The Ligurian-Piemontese ocean basin was consumed by convergence between the European and Adria-Africa plates. Continental collision gave rise to the Alpine chain (e.g. Boccaletti et al. 1971; 1990b, c; Dal Piaz 2010; Turco et al. 2012; Malusà et al. 2015).

The Oligocene to Middle Miocene evolution of the western Mediterranean was dominated by the opening of the Ligurian-Provençal, Algerian, and Valencia basins, accompanied by separation of the Corsica-Sardinia block from the southern European continent as the former rotated counter-clockwise some 40–50° (e.g. Alvarez 1972; Rosenbaum and Lister 2004b; Carminati et al. 2010). Contemporaneously, calcalkaline magmatism took place in the Balearic-Ligurian-Provençal area and in Sardinia, where it extensively developed between about 32 and 15 Ma (e.g. Carminati et al. 2010, 2012; Malusà et al 2015 with references).

The Tyrrhenian Sea opened from Middle Miocene to Quaternary (about 13 to 0 Ma; e.g. Carminati et al. 2010), with an eastward migration of rifting and magmatism (e.g. Savelli and Gasparotto 1994). At the same time, compression occurred in the Apennine front, with a progressive migration of the thrust belt-foredeep system towards the present-day Padanian-Adriatic-Ionian foreland (Keller et al. 1994; Turco and Zuppetta 1998; Doglioni et al. 1999; Sartori 2003). In other words, there was a compression wave migrating eastward, followed by extension and orogenic magmatism shifting in the same direction (e.g. Boccaletti et al. 1990a; Brunet et al. 2000; Barchi et al. 2001; Rosenbaum and Lister 2004a; Barchi 2010). Crustal blocks originally close to Corsica-Sardinia, such as Calabria, were drifted eastwards to their present position during the Tyrrhenian Sea opening (Doglioni et al. 1998; Malusà et al. 2015). Contemporaneously, there was rifting and Miocene-Quaternary anorogenic magmatism in Sardinia, in the

Tyrrhenian Sea (e.g. Vavilov, Magnaghi, Ustica, etc.) and along the northern margin of the Africa foreland (Iblei, Sicily Channel, Etna).

Formation of the Tyrrhenian Sea and the counter-clockwise rotation of the Apennines were associated with longitudinal stretching of the chain and the formation of several arc sectors separated by important transverse tectonic lines (e.g. Locardi 1988; Rosenbaum and Lister 2004b). The main tectonic lineaments crossing the Apennine-Maghrebian chain include the Olevano-Antrodoco (Ancona-Anzio), Ortona-Roccamonfina, and the Tindari-Letojanni fault systems (e.g. Locardi 1988; Turco and Zuppetta 1998). These structures separate arc sectors characterised by different drifting velocity and variable styles of tectonic evolution. A main structural discontinuity developed in the Central Apennine area, resulting from the different evolution undergone by the northern and southern Apennine arcs and the opposed northward and south-eastward drift of the Adriatic and Ionian sectors of the foreland.

13.4 Relationship between Petrogenesis and Geodynamics

The geodynamic and magmatic features of the Tyrrhenian Sea region outlined above represent factual evidence that, in principle, can be framed in a variety of evolutionary scenarios. In broad terms, three extreme classes of hypotheses have been proposed to explain the Cenozoic geodynamic evolution of the Tyrrhenian Sea region, and its relationship with magmatism:

1. Impinging upon the lithosphere of upwelling deep-rooted mantle plume;
2. Lithospheric stretching, passive upwelling of asthenosphere and variable degrees of decompression melting at various mantle levels in an intra-continental rift setting;
3. Subduction of compositionally variable lithospheric segments between the converging Africa and European plates, heterogeneous contamination of sub-arc mantle, and generation of variable types of orogenic magmas; anorogenic magmas would be formed by passive mantle upwelling in the backarc areas or along the foreland margins.

13.4.1 Plume-Related Hypotheses

There are several versions of the plume-related hypothesis for the origin of Tyrrhenian Sea magmatism. Some suggest that a deep-rooted plume head was emplaced earlier or contemporaneously with subduction, contributing significantly to upper mantle heterogeneity and compositionally variability of magmas. Other more extreme theories invoke plume emplacement as the only driving mechanism of volcanism and geodynamics in the Tyrrhenian Sea region.

A plume-related origin for the Italian Plio-Quaternary volcanism was first advocated by Vollmer (1976). He suggested that smooth hyperbolic trends of Sr versus Pb isotopes along the Italian peninsula were generated by mixing between magmas from a mantle plume and the upper crust. The decrease in age of the volcanism from Tuscany to southern Italy was suggested to represent the trace of a fixed hotspot beneath the rotating Italian peninsula.

Locardi and Nicolich (1988) suggested upwelling of soft, low-density mantle rocks as an active mechanism of formation of the Tyrrhenian basin and associated magmatism. However, the plume-like body in this model does not come from the deep mantle but rather from the asthenosphere; the decrease in density necessary to generate mantle upwelling is ascribed to fluids ascending from the deep mantle. This body was hypothesised to be mobile and to move eastward, thus explaining the migration of extension and magmatism with time.

According to Bell et al. (2004, 2013), none of the mantle end-members recognised in the Tyrrhenian Sea area, in particular FOZO and EM1, are known to be associated with present-day subduction-related magmatism, suggesting they may represent deep mantle plumes.

Consequently, magmatism in the Tyrrhenian area is postulated as genetically related to a plume body extending from France and the Western Alps to Corsica-Sardinia and southern Italy. This model suggests that the plume has been intermittently active from Late Triassic-Early Jurassic to present. Its emplacement produced the opening of the Alpine Tethys and of the Oligocene to Quaternary western-central Mediterranean Sea basins, at a later stage. Stretching and extension of the Mediterranean lithosphere were caused by the progressive eastward growth of the plume head. K, Ca, CO_2 and H_2O released by the plume, prompted metasomatic modification of the uppermost mantle and the origin of the wide variety of magmas encountered in the Tyrrhenian Sea region. The trend of radiogenic isotope ratios between mantle and crustal compositions (Fig. 13.2) is interpreted as a two end-member mixing between FOZO-type plume material and crustal-like upper mantle rocks occurring beneath the Alps and Tuscany. Such a contaminated mantle, in turn, could be related to ancient metasomatism by upper crustal material (Bell et al. 2004, 2013).

Several objections have been raised to this model, as amply discussed by Peccerillo and Lustrino (2005). Here, it has to be reminded that the highly radiogenic Sr end-member occurring is Tuscany and in the Western Alps (the so-called ITEM, Italian Enriched Mantle, of Bell and coworkers) is not restricted to Italy, but is widely distributed along the entire Alpine-Himalayan orogenic belt. This speaks for diffuse mantle contamination by upper crustal rocks during Alpine-Himalayan subduction (e.g. Peccerillo and Martinotti 2006; Gao et al. 2007; Tommasini et al. 2011; Prelevic et al. 2008; Conticelli et al. 2009). Lack of consideration for trace element geochemistry is a much more serious drawback of the plume hypothesis proposed by Bell et al. (2013). An obvious consequence of such an oversight is that no difference is made between rocks exhibiting arc- and intraplate-type trace element signatures, thus removing the need of explaining geochemically contrasting groups of magmas. Hence, the association of leucitites, kamafugites, and carbonatites from Italy is considered as petrogenetically and geodynamically akin to the analogous rocks association occurring in intraplate settings, such as East Africa. Yet, carbonatites, melilitites, and K-alkaline rocks from the Tyrrhenian Sea region have strong arc-type trace element signatures, which are not found in Africa and other classical intraplate occurrences, as pointed out repeatedly during the last few decades (e.g. Peccerillo 1985; Peccerillo and Lustrino 2005; see also Chap. 3, Fig. 3.4b).

Less extreme hypotheses accept subduction as a first-order geodynamic and petrogenetic process, but also advocate an important role for material ascended from deep mantle layers. Ayuso et al. (1998) invoked the presence of a plume beneath Campania to explain high contents of HFSE and low LILE/HFSE of Somma-Vesuvio. Gasperini et al. (2002) suggested that mantle materials containing both enriched and depleted components ascended vertically from depth and contributed substantially to the variability of magmatism in the Tyrrhenian Sea region. The orogenic magmas would reflect contamination of these mantle rocks by subducted upper crustal components dominated by pelagic sediments. According to this hypothesis, emplacement of deep mantle into the source of orogenic magmas was favoured by the slab window that opened beneath southern Italy as a result of differential rates of rollback between the northern and southern Apennines subduction systems. Mantle rocks ascending from the 670 km discontinuity were also suggested by Cadoux et al. (2007) as a unique pre-metasomatic mantle composition for all the Italian magmas. Piromallo et al. (2008) propose that the common FOZO mantle component in the Mediterranean Sea area, as well as in the rest of Europe, reveals the occurrence of a large plume head, originally emplaced beneath the Central Atlantic Ocean and successively expanded to Europe. Such a process occurred during Late Cretaceous-Paleocene time when the Euro-Mediterranean region was located southwest of its present position, close to the plume hotspot location. Cenozoic rifting, back-arc spreading, and strike-slip faulting favoured upwelling and decompression melting of the plume head, thus giving anorogenic magmas with homogeneous isotopic signatures at the continental scale. FOZO-type rocks also emplaced above subducting zones, which have been active in the Tyrrhenian Sea region during the Cenozoic. Here, they underwent contamination by slab-derived fluids, acquiring high LILE contents and crustal-like isotopic signatures that were inherited by orogenic magmas.

A homogeneous plume head could represent an acceptable explanation for the uniform isotopic compositions of the bulk of anorogenic magmatism in Europe. However, mantle compositions in the Tyrrhenian Sea are very heterogeneous, and enriched and depleted sources have been highlighted by geochemical data for both anorogenic and orogenic magmas (Fig. 13.2). As recalled earlier in this chapter, orogenic magmas generally had depleted pre-metasomatic sources, with the exception of the Stromboli- Vulture-Campania area where a FOZO-type mantle has been recognised.

13.4.2 Passive-Rifting-Related Hypotheses

The hypothesis that circum-Tyrrhenian magmatism is related to intracontinental rifting basically stems from the occurrence of abundant K-alkaline and ultra-alkaline magmatism, which is believed to be typical of rift zones. According to Lavecchia and Stoppa (1990, 1996) the magmatism in Italy, the Apennine chain, and the Tyrrhenian Sea basin are related to asymmetric passive rifting, stretching of the lithosphere, upwelling and decompression melting of the asthenosphere. Stretching is related to external forces acting laterally along the plates, and there is no need for active mantle upwelling or deep mantle plumes. Evidence in favour of this hypothesis is the low amount of N–S shortening (about 100 km) compared to a large W–E extension in the western Mediterranean area (some 800 km), during the last 23 Ma (Gueguen et al. 1998 and references therein). This requires that the compressive movements between Africa and Europe were minor compared to W–E extensional movements over this time period.

According to passive rift hypotheses, the variable composition of the magmatism in the Tyrrhenian Sea area is the result of melting mantle sources of differing composition. Magmatism in Sicily, Sardinia, and the western Tyrrhenian Sea would be derived from mantle material unaffected by metasomatism (Lavecchia and Stoppa 1990, 1996). Melting of amphibole-bearing peridotite at pressure within the stability field of spinel would produce Na-alkaline rocks; increasing degrees of partial melting at lower pressure would be responsible for tholeiitic magma genesis. In contrast, lower degrees of partial melting in a metasomatised phlogopite-bearing mantle at a pressure exceeding about 2.5 GPa would generate potassic alkaline magmas. The metasomatism responsible for anomalies in the source of potassic magmas would not be related to subduction processes, but to upwelling of deep mantle fluids with anomalous geochemical and isotopic compositions. Magmas from non-metasomatic sources would be generated at shallower levels than those from metasomatic ones. Melting processes for all types of magmas would occur in the thermal boundary layer at the base of the lithosphere. Therefore, the asymmetric distribution of Na-alkaline and K-alkaline magmatism in the Tyrrhenian Sea would be related to the variable thickness of the lithosphere, which increases eastward (Lavecchia and Stoppa 1990, 1996).

The rift-related models outlined above leave a large number of problems unresolved, especially with regard to the origin of the metasomatising agents of potassic and other orogenic magma sources. The contention that the unusual compositions of potassic magmas depend on deep metasomatic fluids shifts the problem to the lower mantle rather than solving it.

Other authors invoke stretching of the lithosphere and passive rifting in the Tyrrhenian Sea region, but also accept subduction as a main factor of orogenic magmatism. Boccaletti et al. (1984) suggest that tensional fissures and extensional basins in the western Mediterranean area may have developed as a consequence of the non-parallel convergence of rigid African and European plates. Mantovani and coworkers (e.g. Mantovani et al. 2009 and references therein) propose that the Tyrrhenian Sea opening and rotation of southern Apennines are the effects of block extrusion processes during the oblique collision between Africa. Magmatism would occur in extensional or transtensional areas and orogenic and anorogenic compositions would depend on the nature of mantle sources, which change from the foreland to backarc basins and subduction zones.

13.4.3 Subduction-Related Hypotheses

According to these models, most of the Cenozoic magmatism in the Tyrrhenian Sea region is related to lithospheric subduction processes resulting from convergence between Africa and Europe. Rocks with island-arc geochemical signatures would be directly related to subduction, being generated by melting of mantle sources metasomatised by fluids (aqueous fluids, supercritical fluids, silicate or carbonate melts) released by the subducted lithosphere. Anorogenic magmatism occurring in back-arc areas (i.e. Tyrrhenian Sea and Sardinia) would be related to focused mantle upwelling induced by slab trajectories and fragmentation (Faccenna et al. 2010). In contrast, magmas from Sicily would be generated by passive asthenospheric upwelling and decompression melting in zones of extensional or transtensional tectonic regime along the margins of the foreland (e.g. Boccaletti et al. 1990b, c; Viti et al. 2011).

The large petrological-geochemical variation of orogenic rocks is explained by assuming that distinct amounts and types of contaminants (aqueous fluids, silicate melts, etc. from oceanic slab and sediments) released from different types of subducted rocks affected heterogeneous mantle wedge along the Apennine-Maghrebian subduction system. Variability of contaminants would be essentially related to lateral heterogeneous nature of the foreland. Mantle heterogeneity would be related to several factors, including vertical variability along lithosphere-asthenosphere sections, mantle migration from the foreland, and mantle upwelling from asthenosphere. Therefore, trends highlighted by radiogenic isotopes (Fig. 13.2) would not result from single two-end-member mixing between FOZO-type mantle and upper crustal material, as suggested by several authors. Rather, they are the result of multiple mantle contamination events affecting heterogeneous pre-metasomatic mantle rocks and involving distinct contaminants at different places along the Apennine subduction zone (Peccerillo 1999).

Plots of HFSE/HFSE versus LILE/HFSE ratios best highlight such a complex process. HFSE are relatively immobile during subduction-related metasomatism, whereas LILE migrate easily from slab to the mantle wedge (e.g. Kessel et al. 2005). As a consequence, HFSE/HFSE ratios of mantle wedge are less dramatically modified than LILE/HFSE ratios by subduction-related metasomatism, although residual accessory phases in the slab can modify significantly this behaviour (Hermann and Rubatto 2009). Plots of mafic rocks from the Tyrrhenian Sea region show that anorogenic magmas have variable HFSE/HFSE reflecting MORB to OIB-type sources, but LILE/HFSE ratios are invariably low (Fig. 13.3). In contrast, orogenic magmas show wide ranges for both ratios. These relationships are consistent with the hypothesis of variable intensities of subduction-related enrichments (indicated by arrows) over OIB- to MORB-type pre-metasomatic mantle rocks. En passing, such a complexity is not observed in igneous rocks along the Alps for which only depleted mantle rocks were apparently affected by subduction-related metasomatism (see Appendix 1).

Several scenarios for the subduction-related origin of Tyrrhenian Sea have been proposed, starting from the pioneering papers by Boccaletti and coworkers (Boccaletti et al. 1971; Boccaletti and Guazzone 1974). Based on a wealth of literature (e.g. Doglioni et al. 1999; Gvirtzman and Nur 1999; Faccenna et al. 2007, 2010; Carminati et al. 2010, 2012; Malusà et al. 2015, and many others), the following main steps can be recognised in the Cenozoic geodynamic evolution and

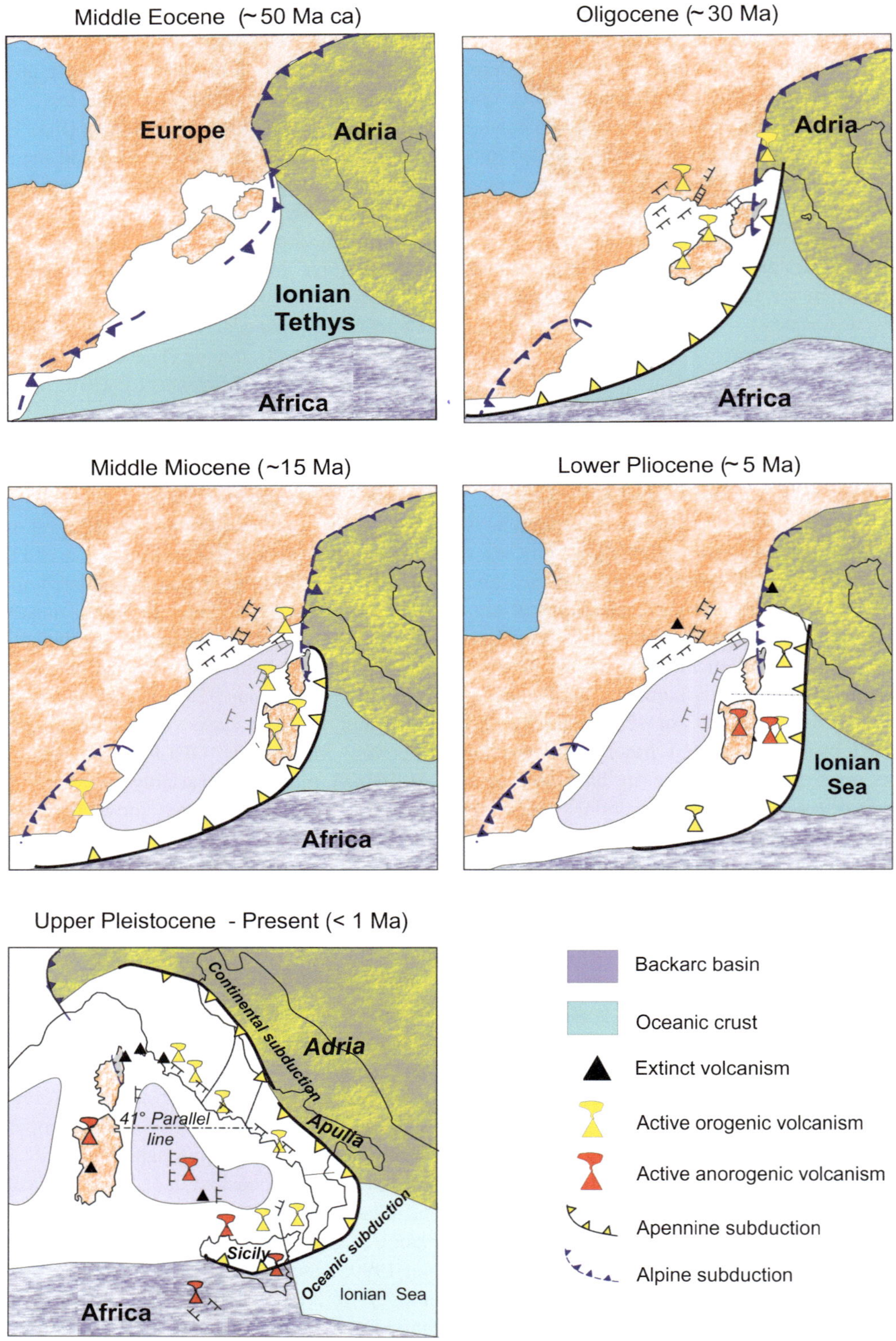

Fig. 13.4 Schematic geodynamic evolution model of the Western Mediterranean and the Tyrrhenian Sea region from Oligocene to Present. For explanation, see text

igneous petrogenesis of the Tyrrhenian Sea region (Fig. 13.4):

Step 1.

Late Cretaceous to Eocene subduction of the European plate beneath the northern border of the African-Adria-Neotethys plate (Alpine subduction stage) delivers silicic upper crustal rocks (e.g. pelites) to the mantle wedge beneath the Alpine belt. Bulk sediments and slices of oceanic crust mix with front-arc lithospheric mantle (probably consisting of residual harzburgites) to form sediment-blueschist-harzburgite mélange bodies. These do not undergo melting, probably because of the low ambient temperature, and are preserved as discrete megaliths above the slab [see Chap. 2, Fig. 2.14; Poli and Peccerillo 2016)]. A similar scenario also applies to the Western Alps and the Betic Cordillera where an association of mafic rocks similar to Tuscany is observed (see Appendix 1; Peccerillo and Martinotti 2006).

Step 2.

Paleogene subduction of Adriatic-Ionian lithosphere takes place beneath Europe, initially along the Sardinia transect. Fluids released by the slab prompt mantle wedge metasomatism and orogenic magmatism in Sardinia, Corsica-Ligurian Basin and Provence (about 43–15 Ma). Low Pb-isotope ratios of Sardinia rocks relative to other calcalkaline volcanics in the Tyrrhenian Sea area (Fig. 13.2) denote the involvement of an EM1-type component, probably residing within the local lithosphere.

Step 3.

Oceanic crust of the Ionian Tethys basin along the northern sector of the Apennines is consumed, with Adria reaching the Alpine wedge of Corsica by the end of Oligocene (Malusà et al. 2016). The subsequent subduction in the area is, therefore, accomplished by immersion of delaminated Adriatic continental plate, an idea first suggested by Boccaletti et al. (1980) and successively adopted by Serri et al. (1993). In contrast, there is still consumption of the oceanic Ionian oceanic floor along the Central-Southern Apennine arc. Therefore, Upper Oligocene-Middle Miocene marks the onset of two contrasting types of mantle wedge contamination along the Apennine system. In the north, carbonated pelites (i.e. marls) from the Adriatic lithosphere are added the mantle wedge until Quaternary times, where they mix with the resident mantle to form marl-peridotite mélange bodies (see Chap. 4, Fig. 4.21; Peccerillo and Frezzotti 2015). This long-lived process delivers large amounts of upper crustal material to the mantle wedge, which will generate abundant ultrapotassic magmatism at a later stage. Note that such a marly metasomatic agent is distinct from the one that affected the Tuscany Province. In the central-southern Apennine area, subduction of the oceanic-type Ionian slab releases fluids or melts that cause calcalkaline magmatism.

Step 4.

Continuing west-directed Mio-Pliocene subduction of the delaminated Adriatic plate in the north-central Apennines and of the oceanic Ionian crust in the south determines the asymmetric opening of the Tyrrhenian Sea backarc basin. Orogenic magmatism migrates eastward behind the retreating Adriatic-Ionian subduction zone (e.g. Locardi and Nicolich 1988; Argnani and Savelli 1999). In the north, backarc extension and asthenospheric ascent prompt melting of the lithosphere contaminated during the Alpine subduction stage, generating magmatism in Tuscany (Poli and Peccerillo 2016; see Fig. 2.14). Extension is less intense than in the south, favouring ponding of mafic melts within the crust, anatexis, and the generation of acid magmatism. Noteworthy, Adria still moves northward relative to Europe, whereas the Ionian plate shifts southeastward (e.g. Malusà et al. 2015), leading to a breakup of the plate and to the opening of a wide slab gap beneath Central Apennines. As a consequence, mantle contamination in this region will result from fluid migration from both the subducting plates of the Northern and Southern Apennines. Anorogenic magmatism with MORB- to OIB-type compositions takes place in the central-southern Tyrrhenian area and Sardinia. OIB-type magmas also erupt in the Sicily Channel and at Iblei, as an effect of

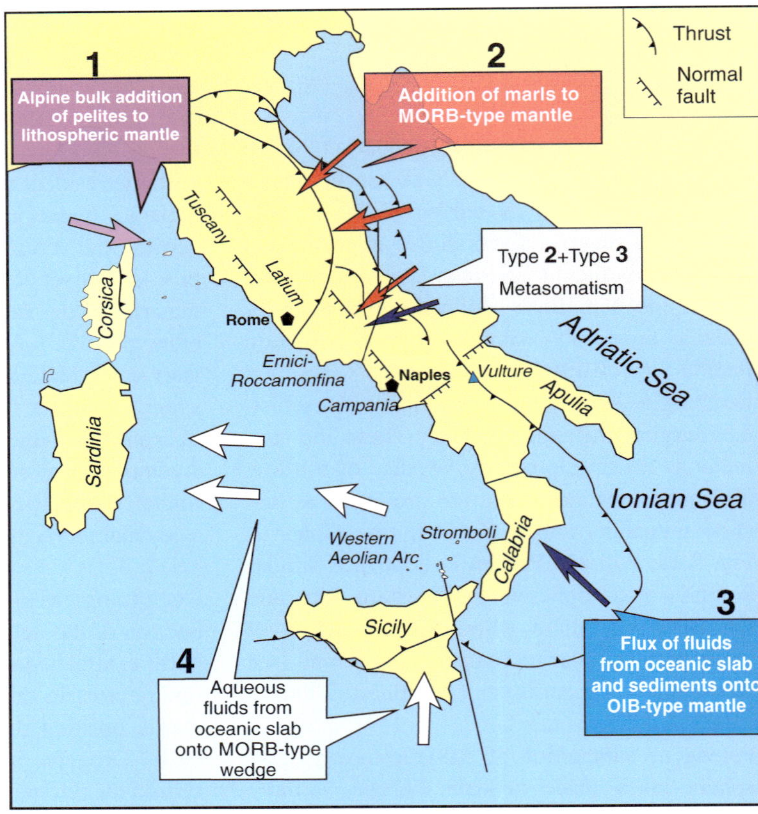

Fig. 13.5 Schematic overview of mantle metasomatic processes for various sectors of the orogenic magmatism in the Tyrrhenian Sea area. For explanation, see text

transtensional tectonics related to lateral block expulsion (e.g. Boccaletti et al. 1987; Viti et al. 2011) or to regional rifting affecting a long belt from Central Europe to North Africa (Corti et al. 2006).

Step 5.

Collision of Apulia foreland against the Southern Apennines produces slab breakoff and southeastward sinking of the residual Ionian crust during Upper Pliocene-Pleistocene (see Chap. 7, Fig. 7.22). An along-strike slab tear is formed in the Ionian slab, which favours inflow of plastic asthenosphere from the foreland. South-eastward shift of subduction is associated with further migration of orogenic volcanism to its present position in the Aeolian area and with the opening of the Marsili backarc basin. Mantle melting is related to aqueous fluids released by the oceanic Ionian slab plus sediments in the eastern Aeolian arc, Campania Province and Vulture. Inflow of asthenosphere from Ionian-Africa foreland provides FOZO-type geochemical components that are observed in these volcanoes (Peccerillo and Frezzotti 2015). In contrast, metasomatism by aqueous fluids from an oceanic slab with little or no participation of sediments takes place in the western Aeolian arc; a similar process previously occurred in Sardinia and through the southern Tyrrhenian Sea.

In the Northern Apennines, abundant ultrapotassic and potassic volcanism takes place in the Roman Province, with emplacement of huge amounts of magmas over a relatively short period of time. Such volcanism is probably related to slowdown of active subduction, which favours temperature increase and melting of marl-mantle mélange bodies. Magmatism also continues in southern Tuscany, where there is geochemical evidence of hybridism with Roman-type magmas. In the Ernici-Roccamonfina area, small volumes of calcalkaline to ultrapotassic magmas are erupted in a segment of Central Apennines where the absence of a dehydrating subducting lithosphere is suggested by geophysical evidence. The variable geochemical compositions of magmas in this area reflect contamination by compositionally different fluids released by the Northern and Southern Apennine subducting slab sectors.

13.5 Conclusions

Cenozoic magmatism in the Tyrrhenian Sea area shows strong compositional complexities, which have made these volcanoes much studied but not yet fully understood geological subjects. Various hypotheses have been proposed to explain the compositional variations of orogenic and anorogenic magmatic suites occurring in this region. The anorogenic magmas result from melting of mantle sources that have not suffered contamination by young subduction processes. Their petrological and geochemical signatures can be explained by assuming different degrees of partial melting of heterogeneous upper asthenospheric to lower lithospheric mantle sources. These are made up of rocks that show distinct compositions, plotting between FOZO, DMM and EM1. FOZO-type rocks are best represented by Etna and other magmas of the Sicily Province, but are widespread in the Mediterranean Sea and Europe. Its origin is controversial and most authors suggest a derivation from deep mantle material ascended from the asthenosphere or from deeper levels. The EM1 component is even more mysterious, although its restricted occurrence to Sardinia points to an origin in the local lithosphere.

The petrogenesis of orogenic magmas is much more complex and an origin from various types of mantle rocks contaminated by compositionally different subducted crustal materials is the most likely petrogenetic scenario. An synopsis of such a complex picture is given in Fig. 13.5.

According to this view, four main types of subduction-related mantle metasomatic events took place in the Tyrrhenian Sea area. The first contamination process (*Metasomatism-1*) took place during the Alpine subduction of the European plate beneath the northern African margin, and affected the lithospheric mantle sources of the Tuscany Magmatic Province. Melting in Tuscany, however, occurred much later during the Miocene to present opening of the Northern Tyrrhenian Sea.

A separate episode of metasomatism (*Metasomatism-2*) took place from Miocene to present beneath the Roman Province. It was accomplished by introduction of marly sediments into a mantle wedge characterised by depleted MORB-type composition. A similar type of metasomatism also affected the Intra-Apennine Magmatic Province.

Metasomatism-3 took place beneath the Campania, Vulture and Stromboli volcanoes. The metasomatic agent was represented by fluids released from the Ionian oceanic slab and associated sediments. The pre-metasomatic mantle consisted of enriched OIB-type rocks, which probably were suctioned from the foreland during the south-eastward rollback of the Ionian slab.

The upper mantle beneath the western Aeolian arc and many other volcanoes in Sardinia and through the Tyrrhenian Sea basin was contaminated by fluids from the oceanic Ionian slab, with little or no participation of sediments (*Metasomatism-4*). The composition of pre-metasomatic mantle in Sardinia was intermediate between FOZO and EM1, whereas MORB-type mantle rocks occurred beneath the western Aeolian arc.

The upper mantle in the Ernici-Roccamonfina area consisted of a depleted MORB-type peridotite and was affected by two metasomatic events with Roman type and Campania-type compositions, respectively (*Metasomatism-2* plus *Metasomatism-3*). This explains the occurrence in a very limited area of a wide variety of orogenic volcanics spanning the Roman and Campanian compositions.

In conclusion, the orogenic magmatism results from various mixing processes in which several mantle and crustal components interacted above subduction zones. These generated various trends of radiogenic isotope variation, which merged and partially superimposed on each other to generate an apparently continuous single trend, which brought several authors to suggest a simple two-end-member mixing process for the entire orogenic magmatism.

References

Alagna K, Peccerillo A, Martin S, Donati D (2010) Tertiary to present evolution of orogenic magmatism in Italy. J Virtual Explorer 36, paper 18 doi: 10.3809/jvirtex.2010.00233
Alvarez W (1972) Rotation of the Corsica-Sardinia microplate. Nature 235:103–105
Argnani A, Savelli C (1999) Cenozoic volcanism and tectonics in the southern Tyrrhenian Sea: space-time distribution and geodynamic significance. Geodynamics 27:409–432
Ayuso RA, De Vivo B, Rolandi G, Seal RR II, Paone A (1998) Geochemical and isotopic (Nd–Pb–Sr–O) variations bearing on the genesis of volcanic rocks from Vesuvius, Italy. J Volcanol Geoth Res 82:53–78
Barchi M (2010) The Neogene-Quaternary evolution of the Northern Apennines: crustal structure, style of deformation and seismicity. In: Beltrando M, Peccerillo A, Mattei M, Conticelli S, Doglioni C (eds) J Virtual Explorer 36, paper 11 doi: 10.3809/jvirtex.2010.00220
Barchi M, Landuzzi A, Minelli G, Pialli G (2001) Outhern Northern Apennines. In: Vai GB, Martini PI (eds) Anatomy of an Orogen. The Apennines and the adjacent Mediterranean basins. Kluwer, Dordrecht, pp 215–254
Beccaluva L, Di Girolamo P, Serri G (1991) Petrogenesis and tectonic setting of the Roman volcanic province, Italy. Lithos 26:191–221
Bell K, Castorina F, Lavecchia G, Rosatelli G, Stoppa F (2004) Is there a mantle plume below Italy? EOS 85:541–547
Bell K, Lavecchia G, Rosatelli G (2013) Cenozoic Italian magmatism—Isotope constraints for possible plume-related activity. J South Am Earth Sci 41:22–40
Benoit MH, Torpey M, Liszewski K, Levin V, Park J (2011) P and S wave upper mantle seismic velocity structure beneath the northern Apennines: new evidence for the end of subduction. Geochem Geophys Geosyst 12:Q06004. doi:10.1029/2010GC003428
Boccaletti M, Guazzone G (1974) Remnant arcs and marginal basins in the Cainozoic development of the Mediterranean. Nature 252:18–21
Boccaletti M, Elter P, Guazzone G (1971) Plate tectonics model for the development of Western Alps and Northern Apennines. Nature 234:108–111
Boccaletti M, Coli M, Decandia FA, Giannini E, Lazzarotto A (1980) Evoluzione dell'Appennino settentrionale secondo un nuovo modello strutturale. Mem Soc Geol It 21:31–49
Boccaletti M, Nicolich R, Tortorici L (1984) The Calabrian arc and the Ionian Sea in the dynamic evolution of the central Mediterranean. Mar Geol 55:219–245
Boccaletti M, Cello G, Tortorici L (1987) Transtensional tectonics in the Sicily channel. J Struct Geol 9:869–876
Boccaletti M, Calamita F, Deiana G, Gelati R, Massari F, Moratti G, Ricci Lucchi F (1990a) Migrating foredeep-thrust belt systems in the northern Apennines and southern Alps. Palaeogeogr Palaeocl 77:3–14
Boccaletti M, Ciaranfi N, Cosentino D, Deiana G, Gelati R, Lentini F, Massari F, Moratti G, Pescatore T, Ricci Lucchi F, Tortorici L (1990b) Palinspastic restoration and paleogeographic reconstruction of the peri-Tyrrhenian area during the Neogene. Palaeogeogr Palaeocl 77:41–50
Boccaletti M, Nicolich R, Tortorici L (1990c) New data and hypothesis on the development of the Tyrrhenian Basin. Palaeogeogr Palaeocl 77:15–40
Brunet C, Monié P, Jolivet L, Cadet JP (2000) Migration of compression and extension in the Tyrrhenian Sea, insights from 40Ar/39Ar ages on micas along a transect from Corsica to Tuscany. Tectonophysics 321:127–155
Cadoux A, Blichert-Toft J, Pinti DL, Albarede F (2007) A unique lower mantle source for Southern Italy volcanics. Earth Planet Sci Lett 259:227–238
Carmignani L, Oggiano G, Barca S, Conti P, Salvadori I, Eltrudis A, Funedda A, Pasci S (2001) Geologia della Sardegna. Note illustrative della Carta Geologica in scala 1:200,000. Mem Descr Carta Geol It 60:283 pp
Carminati E, Lustrino M, Cuffaro M, Doglioni C (2010) Tectonics, magmatism and geodynamics of Italy: what we know and what we imagine. In: Beltrando M, Peccerillo A, Mattei M, Conticelli S, Doglioni C (eds) J virtual explorer 36, paper 9 doi: 10.3809/jvirtex.2010.00226
Carminati E, Lustrino M, Doglioni C (2012) Geodynamic evolution of the central and western Mediterranean: tectonic vs. igneous constraints. Tectonophysics 579:173–192
Conticelli S, Guarnieri L, Farinelli A, Mattei M, Avanzinelli R, Bianchini G, Boari E, Tommasini S, Tiepolo M, Prelevic D, Venturelli G (2009) Trace elements and Sr-Nd-Pb isotopes of K-rich, shoshonitic, and calc-alkaline magmatism of the Western Mediterranean Region: Genesis of ultrapotassic to calc-alkaline magmatic associations in a post-collisional geodynamic setting. Lithos 107:68–92
Corti G, Cuffaro M, Doglioni C, Innocenti F, Manetti P (2006) Coexisting geodynamic processes in the Sicily channel. In: Dilek Y and Pavlides S (eds) Postcollisional tectonics and magmatism in the Mediterranean region and Asia, Geol Soc Am Spec Paper 409:83–96
Dal Piaz GV (2010) The Italian Alps: a journey across two centuries of Alpine geology. In: Beltrando M, Peccerillo A, Mattei M, Conticelli S, Doglioni C

(eds) J virtual explorer 36, paper 8 doi: 10.3809/jvirtex.2010.00234
Di Stefano R, Bianchi I, Ciaccio MG, Carrara C, Kissling E (2011) Three-dimensional Moho topography in Italy: new constraints from receiver functions and controlled source seismology. Geochem Geophys Geosyst 12:Q09006. doi:10.1029/2011GC003649
Doglioni C, Carminati E (2012) Alps vs. Apennines: the paradigm of a tectonically asymmetric Earth. Earth Sci Rev 112:67–96
Doglioni C, Mongelli F, Pialli G (1998) Boudinage of the Alpine belt in the Apenninic back-arc. Mem Soc Geol It 52:457–468
Doglioni C, Harabaglia P, Merlini S, Mongelli F, Peccerillo A, Piromallo C (1999) Orogens and slabs vs. their direction of subduction. Earth Sci Rev 45:167–208
Faccenna C, Funiciello F, Civetta L, D'Antonio M, Moroni M, Piromallo C (2007) Slab disruption, mantle circulation, and the opening of the Tyrrhenian basins. In: Beccaluva L, Bianchini G, Wilson M (eds) Cenozoic volcanism in the Mediterranean area, Geol Soc Am Spec Paper 418:153–169
Faccenna C, Becker TW, Lallemand S, Lagabrielle Y, Funiciello F, Piromallo C (2010) Subduction-triggered magmatic pulses: a new class of plumes? Earth Planet Sci Lett 299:54–68
Finetti IR, Boccaletti M, Bonini M, Del Ben A, Geletti R, Pipan M, Sani F (2001) Crustal section based on CROP seismic data across the North Tyrrhenian-Northern Apennines-Adriatic Sea. Tectonophysics 43:135–163
Gao Y, Hou Z, Kamber BS, Wei R, Meng X, Zhao R (2007) Lamproitic rocks from a continental collision zone: evidence for recycling of subducted Tethyan oceanic sediments in the mantle beneath Southern Tibet. J Petrol 48:729–752
Gasperini D, Blichert Toft J, Bosch D, Del Moro A, Macera P, Albarède F (2002) Upwelling of deep mantle material through a plate window: evidence from the geochemistry of Italian basaltic volcanics. J Geophys Res 107(B12):2367. doi:10.1029/2001JB000418
Giacomuzzi G, Civalleri M, De Gori P, Chiarabba C (2012) A 3D Vs model of the upper mantle beneath Italy: insight on the geodynamics of central Mediterranean. Earth Planet Sci Lett 335–336:105–120
Grassi D, Schmidt MW, Gunther D (2012) Element partitioning during carbonated pelite melting at 8, 13 and 22 GPa and the sediment signature in the EM mantle components. Earth Planet Sci Lett 327–328:84–96
Gueguen E, Doglioni C, Fernandez M (1998) On the post-25 Ma geodynamic evolution of the western Mediterranean. Tectonophysics 298:259–269
Gvirtzman Z, Nur A (1999) The formation of Mount Etna as the consequence of slab rollback. Nature 401:782–785
Hawkesworth CJ, Vollmer R (1979) Crustal contamination vs. enriched mantle: $^{143}Nd/^{144}Nd$ and $^{87}Sr/^{86}Sr$ evidence from the Italian volcanics. Contrib Min Pet 69:151–165
Hermann J, Rubatto D (2009) Accessory phase control on trace element signature of sediment melt in subduction zones. Chem Geol 265:512–526
Keller JVA, Minelli G, Pialli G (1994) Anatomy of late orogenic extension: the northern Apennines case. Tectonophysics 238:275–294
Kessel R, Schmidt MW, Ulmer P, Pettke T (2005) Trace element signatures of subduction-zone fluids, melts and supercritical liquids at 120–180 km depth. Nature 437:724–727
Lavecchia G, Stoppa F (1990) The Tyrrhenian zone: a case of lithosphere extension control of intra-continental magmatism. Earth Planet Sci Lett 99:336–350
Lavecchia G, Stoppa F (1996) The tectonic significance of Italian magmatism: an alternative view to the popular interpretation. Terra Nova 8:435–446
Locardi E (1988) The origin of the Apenninic arc. Tectonophysics 146:105–123
Locardi E, Nicholich R (1988) Geodinamica del Tirreno e dell'Appennino centro-meridionale: la nuova carta della Moho. Mem Soc Geol It 41:121–140
Lustrino M (2000) Volcanic activity during the Neogene to present evolution of the western Mediterranean area: a review. Ofioliti 25:87–101
Lustrino M, Fedele L, Melluso L, Morra V, Ronga F, Geldmacher J, Duggen S, Agostini S, Cucciniello C, Franciosi L, Meisel T (2013) Origin and evolution of Cenozoic magmatism of Sardinia (Italy). A combined isotopic (Sr–Nd–Pb–O–Hf–Os) and petrological view. Lithos 180:138–158
Malusà MG, Faccenna C, Baldwin SL, Fitzgerald PG, Rossetti F, Balestrieri ML, Danisik M, Ellero A, Ottria G, Piromallo C (2015) Contrasting styles of (U) HP rock exhumation along the Cenozoic Adria-Europe plate boundary (Western Alps, Calabria, Corsica). Geochem Geophys Geosyst 16 doi:10.1002/2015GC005767
Malusà MG, Danisik M, Kuhlemann J (2016) Tracking the Adriatic-slab travel beneath the Tethyan margin of Corsica-Sardinia by low-temperature thermochronometry. Gondwana Res 31:135–149
Mantovani E, Babbucci D, Tamburelli C, Viti M (2009) A review on the driving mechanism of the Tyrrhenian-Apennines system: implications for the present seismotectonic setting in the Central-Northern Apennines. Tectonophysics 476:22–40
Martelli M, Nuccio PM, Stuart FM, Di Liberto V, Ellam RM (2008) Constraints on mantle source and interactions from He-Sr isotope variation in Italian Plio-Quaternary volcanism. Geochem Geophys Geosyst 9:Q02001. doi:10.1029/2007GC001730
Mele G, Sandvol E (2003) Deep crustal roots beneath the northern Apennines inferred from teleseismic receiver functions. Earth Planet Sci Lett 211:69–78
Mele G, Sandvol E, Cavinato GP (2006) Evidence of crustal thickening beneath the central Apennines

(Italy) from teleseismic receiver functions. Earth Planet Sci Lett 249:425–435

Mueller S, Panza GF (1986) Evidence of a deep-reaching lithospheric root under the Alpine Arc. In: Wezel FC (ed), The origin of arcs, vol 21. Elsevier, pp 93–113

Panza GF, Pontevivo A, Chimera G, Raykova R, Aoudia A (2003) The lithosphere-asthenosphere: Italy and surroundings. Episodes 26:169–174

Peccerillo A (1985) Roman Comagmatic Province (Central Italy): evidence for subduction-related magma genesis. Geology 13:103–106

Peccerillo A (1998) Relationships between ultrapotassic and carbonate-rich volcanic rocks in central Italy: petrogenetic implications and geodynamic significance. Lithos 43:267–279

Peccerillo A (1999) Multiple mantle metasomatism in central-southern Italy: geochemical effects, timing and geodynamic implications. Geology 27:315–318

Peccerillo A (2001) Geochemical similarities between Vesuvius, Phlegraean fields and Stromboli volcanoes: petrogenetic, geodynamic and volcanological implications. Min Pet 73:93–105

Peccerillo A (2004) Carbonate-rich pyroclastic rocks from central Apennines: carbonatites or carbonated rocks? A commentary. Per Min 73:165–175

Peccerillo A, Frezzotti ML (2015) Magmatism, mantle evolution and geodynamics at the converging plate margins of Italy. J Geol Soc Lond 172:407–427

Peccerillo A, Lustrino M (2005) Compositional variations of the Plio-quaternary magmatism in the circum-Tyrrhenian area: deep- vs. shallow-mantle processes. In: Foulger GR, Nathland JH, Presnall DC, Anderson DL (eds) Plates, plumes and paradigms. Geol Soc Am Spec Publ 388:421–434

Peccerillo A, Martinotti G (2006) The Western Mediterranean lamproitic magmatism: origin and geodynamic significance. Terra Nova 18:109–117

Peccerillo A, Panza GF (1999) Upper mantle domains beneath central-southern Italy: petrological, geochemical and geophysical constraints. Pure Appl Geophys 156:421–443

Peccerillo A, Poli G, Serri G (1988) Petrogenesis of orenditic and kamafugitic rocks from Central Italy. Can Min 26:45–65

Pichavant M, Scaillet B, Pommier A, Iacono-Marziano G, Cioni R (2014) Nature and evolution of primitive Vesuvius magmas: an experimental study. J Pet 55:2281–2309

Piromallo C, Gasperini D, Macera P, Faccenna C (2008) A late cretaceous contamination episode of the European-Mediterranean mantle. Earth Planet Sci Lett 268:15–27

Poli G, Peccerillo A (2016) The upper Miocene magmatism of the Elba Island: compositional characteristics, petrogenesis and implications for the origin of the Tuscany Magmatic Province. Min Pet (in press)

Prelevic D, Foley SF, Romer R, Conticelli S (2008) Mediterranean Tertiary lamproites derived from multiple source components in postcollisional geodynamics. Geochim Cosmochim Acta 72:2125–2156

Rosenbaum G, Lister GS (2004a) Neogene and quaternary rollback evolution of the Tyrrhenian Sea, the Apennines and the Sicilian Maghrebides. Tectonics, 23, TC1013 doi:10.1029/2003TC001518

Rosenbaum G, Lister GS (2004b) Formation of arcuate orogenic belts in the western Mediterranean region. In: Sussman AJ, Weil AB (eds) Orogenic curvature: integrating paleomagnetic and structural analyses. Geol Soc Am Spec Paper 383:41–56

Rosenbaum G, Piana Agostinetti N (2015) Crustal and upper mantle responses to lithospheric segmentation in the northern Apennines. Tectonics 34 doi:10.1002/2013TC003498

Sano Y, Wakita H, Italiano F, Nuccio M (1989) Helium isotopes and tectonics in Italy. Geophys Res Lett 16:511–514

Sartori R (2003) The Thyrrhenian backarc basin and subduction of the Ionian lithosphere. Episodes 26:217–221

Savelli C, Gasparotto G (1994) Calc-alkaline magmatism and rifting of the deep-water volcano of Marsili (Aeolian back-arc, Tyrrhenian Sea). Mar Geol 119:137–157

Serri G (1990) Neogene-Quaternary magmatism of the Tyrrhenian region: characterization of the magma sources and geodynamic implications. Mem Geol Soc It 41:219–242

Serri G, Innocenti F, Manetti P (1993) Geochemical and petrological evidence of the subduction of delaminated adriatic continental lithosphere in the genesis of the Neogene-Quaternary magmatism of central Italy. Tectonophysics 223:117–147

Stampfli GM, Borel G (2004) The TRANSMED Transects in space and time: Constraints on the paleotectonic evolution of the Mediterranean Domain. In: Cavazza W, Roure F, Spakman W, Stampfli GM, Ziegler PA (eds) (2004) The TRANSMED Atlas (with CD-ROM). The Mediterranean region from crust to mantle. Springer, p 53–80

Tommasini S, Avanzinelli R, Conticelli S (2011) The Th/La and Sm/La conundrum of the Tethyan realm lamproites. Earth Planet Sci Lett 301:469–478

Turco E, Zuppetta A (1998) A kinematic model for the Plio-Quaternary evolution of the Tyrrhenian-Apenninic system: implications for rifting processes and volcanism. J Volcanol Geoth Res 82:1–18

Turco E, Macchiavelli C, Mazzoli S, Schettino A, Pierantoni PP (2012) Kinematic evolution of Alpine Corsica in the framework of Mediterranean mountain belts. Tectonophysics 579:193–206

Viti M, Mantovani E, Babbucci D, Tamburelli C (2011) Plate kinematics and geodynamics in the Central Mediterranean. J Geodyn 51:190–204

Vollmer R (1976) Rb–Sr and U–Th–Pb systematics of alkaline rocks: the alkaline rocks from Italy. Geochim Cosmochim Acta 40:283–295

Erratum to: Cenozoic Volcanism in the Tyrrhenian Sea Region

Erratum to:
A. Peccerillo, *Cenozoic Volcanism in the Tyrrhenian Sea Region,* Advances in Volcanology, DOI 10.1007/978-3-319-42491-0

The original version of the book was inadvertently published with incorrect appendices in the back matter. The appendices have been revised. The erratum book has been updated with the changes.

The updated original online version for this book can be found at DOI 10.1007/978-3-319-42491-0

A. Peccerillo, *Cenozoic Volcanism in the Tyrrhenian Sea Region*,
Advances in Volcanology, DOI 10.1007/978-3-319-42491-0_14

Appendix 1: The Alpine Magmatic Stage

Introduction

The volcanic activity described in the previous chapters is entirely linked to the formation of the Tyrrhenian Sea basin and of the Apennine-Maghrebian chain. As discussed in Chap. 13, the geodynamic evolution of this region is the result of the convergence between Europe and Africa, of the subduction of various segments of the African lithosphere (Ionian Tethys, Adria) beneath the southern European margin, and of the Eocene-Oligocene to present eastward migration of the compression front and backarc spreading Beltrando et al. (2010). Magmatism took place above subduction zones, in backarc settings and along the border of the foreland, and has been collectively indicated as the Apennine magmatic stage (see Chap. 1).

Slightly earlier and partially coeval with Apennine magmatism, a distinct stage of igneous activity developed contemporaneously with the formation of the Alps, connected with Africa-Europe convergence/collision and the subduction of European plate beneath Northern Adria-Africa. Such igneous activity developed essentially along the Alps with some occurrences in the Adria-Africa plate, and is referred to as the Alpine stage magmatism. Ages range from Upper Cretaceous to Miocene. Rocks compositions show many characteristics in common with the Apennine stage, but there are also significant differences whose scrutiny may help significantly to achieve a better understanding of magmatism and geodynamics along the Africa-Europe converging margins.

Location and Ages

The Alpine-stage igneous activity includes both intrusive and volcanic rocks. Location and ages of the main outcrops along the Alps and the nearby Veneto area (Veneto Volcanic Province) are shown in Fig. A1.1. An andesitic volcano drilled beneath the Po Plain, near Mortara, about 40 south-west of Milan, also belongs to the Alpine stage (Mattioli et al. 2002). Other occurrences in the Italian peninsula and south-eastern Sicily (Pietre Nere, La Queglia, Capo Passero) are indicated in Chap. 1, Fig. 1.1.

Igneous rocks along the Alps consist of several plutonic bodies, dykes, and a few volcanics making up the so-called Periadriatic Magmatic Province (Alagna et al. 2010; Lustrino et al. 2011 and references therein). Fragments of Alpine volcanics are also found in sandstone and conglomerates along the Alps (e.g. Taveyanne sandstone and Val d'Agola formations; e.g. Boyet et al. 2001; Martin and Macera 2014 and references therein) and in the Apennines (Aveto-Petrignacola formation; Mattioli et al. 2012). Igneous activity in the Western Alps is represented by dioritic to granitic intrusions, such as Traversella (30 Ma), Valle del Cervo (31 Ma), Miagliano (32–29 Ma) and associated dikes, by a few lava and pyroclastic rocks (Biella zone), and by a number of calcalkaline to potassic

A. Peccerillo, *Cenozoic Volcanism in the Tyrrhenian Sea Region*,
Advances in Volcanology, DOI 10.1007/978-3-319-42491-0

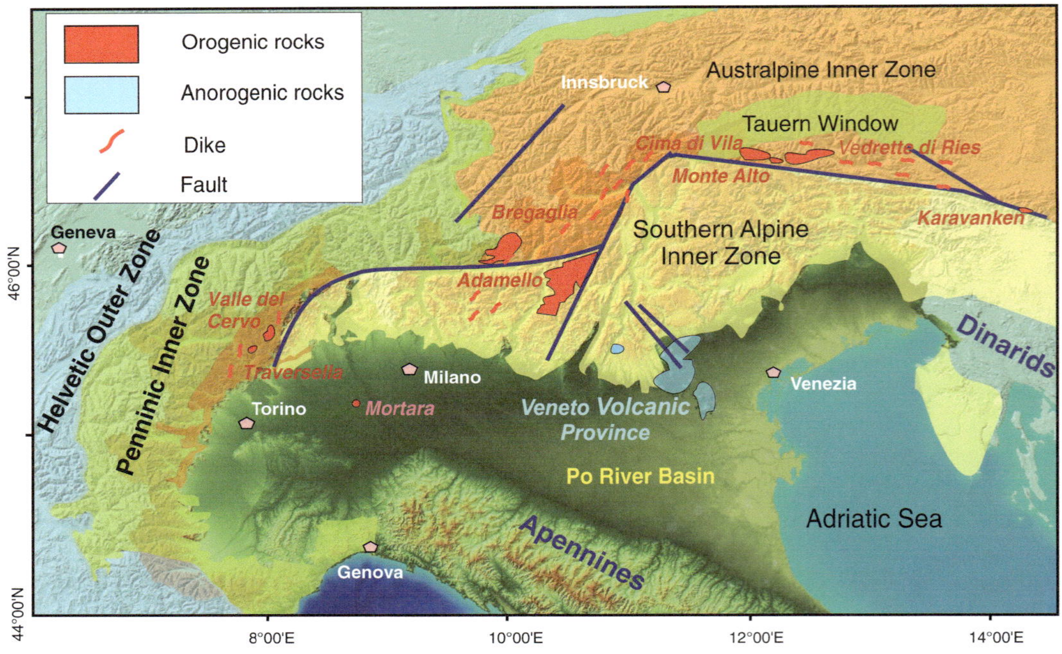

Fig. A1.1 Location of main outcrops of Alpine-stage igneous rocks in the Alps area

alkaline lamprophyric dikes. Central and Eastern Alps contain several single and composite plutons (Bregaglia-Bergel, Novate, Adamello, Rensen, Monte Alto, Vedrette di Ries-Rieserferner, Karavanken, Pohorje) and numerous dikes with dominant tonalite and granodiorite compositions plus minor gabbro, diorite, and granite. Ages are about 33–26 Ma at Bregaglia, 43–29 Ma at Adamello, 29–24 at Rensen and Monte Alto, 30–28 Ma at Karavanken, and 19 Ma at Pohorje in eastern Slovenia.

The Mortara volcano has been drilled at about 5500 m beneath the Plio-Quaternary sediments of the Po Plain. Compositions of sampled rocks are andesitic-dacitic (Mattioli et al. 2002).

The Veneto Volcanic Province (about 60–25 Ma) contains less than 20 km^3 of prevailingly mafic lavas, pyroclastics and subvolcanic rocks, sometimes containing mantle xenoliths (Macera et al. 2003, 2008; Beccaluva et al. 2010).

La Queglia outcrop (Abruzzi region) is represented by lamprophyric dykes intruded into Eocene limestones (Bianchini et al. 2008; Avanzinelli et al. 2012).

Pietre Nere outcrops (65–58 Ma) consist of two distinct hypabyssal bodies with melasyenite to melgabbro compositions (De Astis et al. 2006 and references therein).

The southeastern Sicily outcrops of Capo Passero, Portopalo and Pachino are made of submarine to subaerial basaltic lavas, volcanoclastic material and some dikes showing a K/Ar age of 84–70 Ma, with the most reliable age of about 71 Ma (Avanzinelli et al. 2012 with references).

Petrology and Geochemistry

Classification diagrams for the Alpine-stage rocks are reported in Fig. A1.2. Compositions range from mafic to felsic, from subalkaline to alkaline, and from oversaturated to undersaturated in silica. Strongly alkaline tephrite to phonolite suites, such as those erupted by the Roman and Plio-Quaternary Sardinia volcanoes, are absent. Rocks cropping out along the Alps show arc-type trace element characteristics, whereas Veneto, La Queglia, Pietre Nere and southeastern Sicily fall

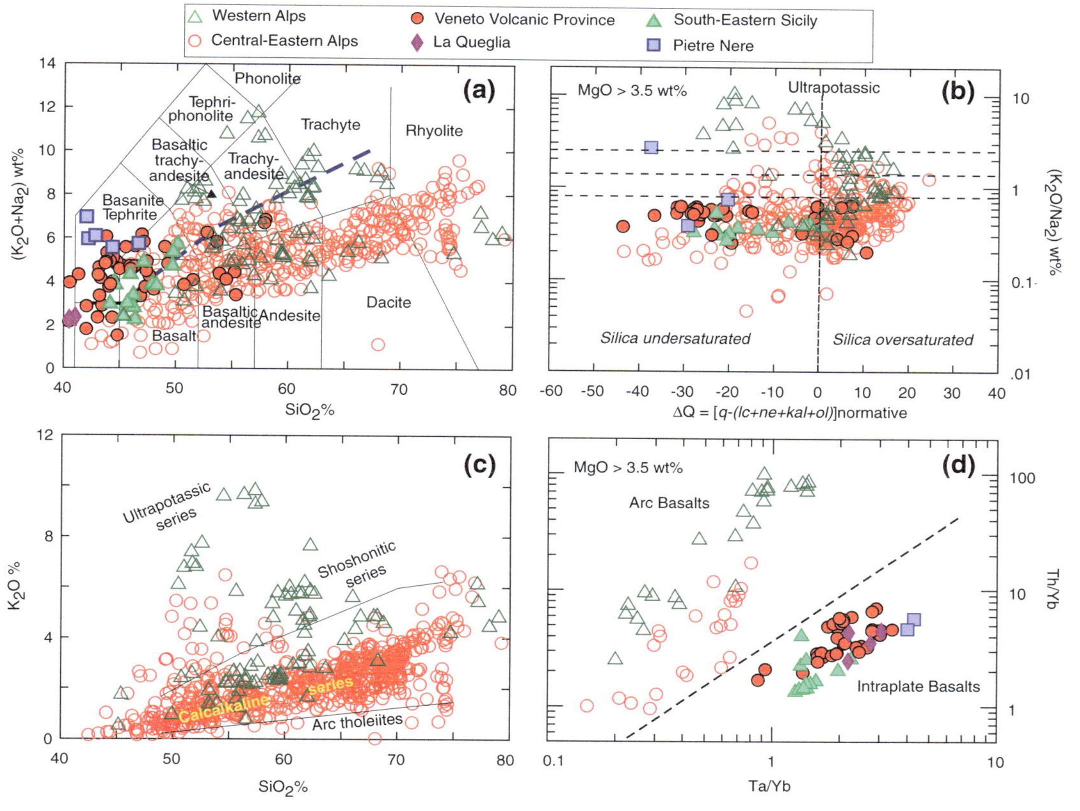

Fig. A1.2 **a** TAS classification diagram for the Alpine stage igneous rocks. Note that the TAS nomenclature only applies to volcanic and hypabyssal rocks; **b** ΔQ versus K_2O/Na_2O diagram for mafic rocks (MgO > 3.5%); **c** Potassium silica diagram for orogenic rocks; **d** Th/Yb versus Ta/Yb discriminant diagram for mafic rocks

in the field of intraplate "anorogenic" rocks (Fig. A1.2d),

Magmatism from the Western Alps shows a very wide range of compositions, from basalt (gabbro) to rhyolite (granite) and trachyte (syenite), with some high-potassium rocks falling within the field of tephriphonolites in the TAS diagram. Mafic rocks are oversaturated to moderately undersaturated in silica, with a calcalkaline to ultrapotassic affinity (Fig. A1.2b, c). The latter are classified as lamproites (Conticelli et al. 2009 and references therein). In the Central-Eastern Alps, there is a decrease in potassium and calcalkaline suites dominate over minor shoshonitic and scanty ultrapotassic rocks

The Veneto Volcanic Province ranges from subalkaline basalt and basaltic andesite to nephelinite-basanite, trachybasalt and trachyandesite (Fig. A1.2a). La Queglia dikes are ultrabasic in compositions. Pietre Nere intrusions fall in the field of basanites, whereas southeastern Sicily volcanics are tholeiitic to transitional basalts.

Incompatible element patterns of orogenic rocks (Fig. A1.3a) show negative anomalies of HFSE, positive spikes of Pb, and variable LILE concentrations, with maximum enrichments in the ultrapotassic lamproitic dikes from the Western Alps. These latter rocks show mantle normalised patterns that resemble very closely those of lamproites from Tuscany (see Chap. 2, Fig. 2.9a). Rocks from Veneto, La Queglia, Pietre Nere and SE Sicily have rather smooth upward convex patterns, sometimes with a small negative anomaly of potassium (Fig. A1.3b). These patterns resemble closely average OIB (Sun and McDonough 1989).

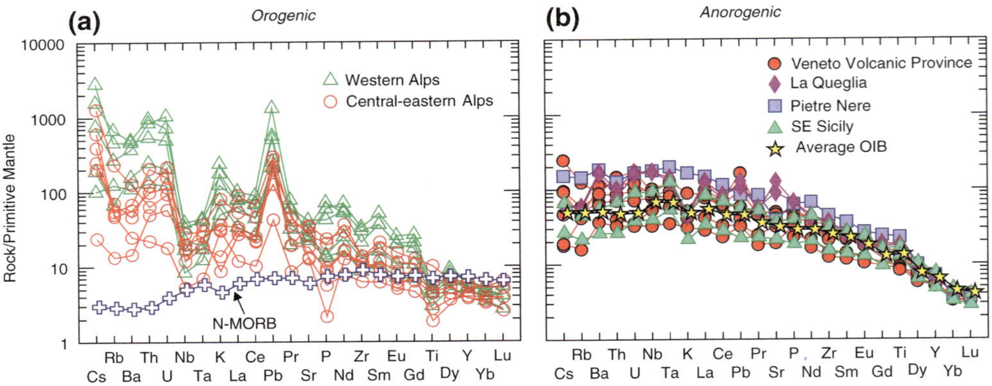

Fig. A1.3 a Patterns of incompatible elements normalised to primitive mantle for Alpine mafic orogenic rocks. The pattern of N-MORB is shown for comparison (Gale et al. 2013); **b** Patterns of anorogenic rocks. Average OIB composition is shown for comparison (Sun and McDonough 1989)

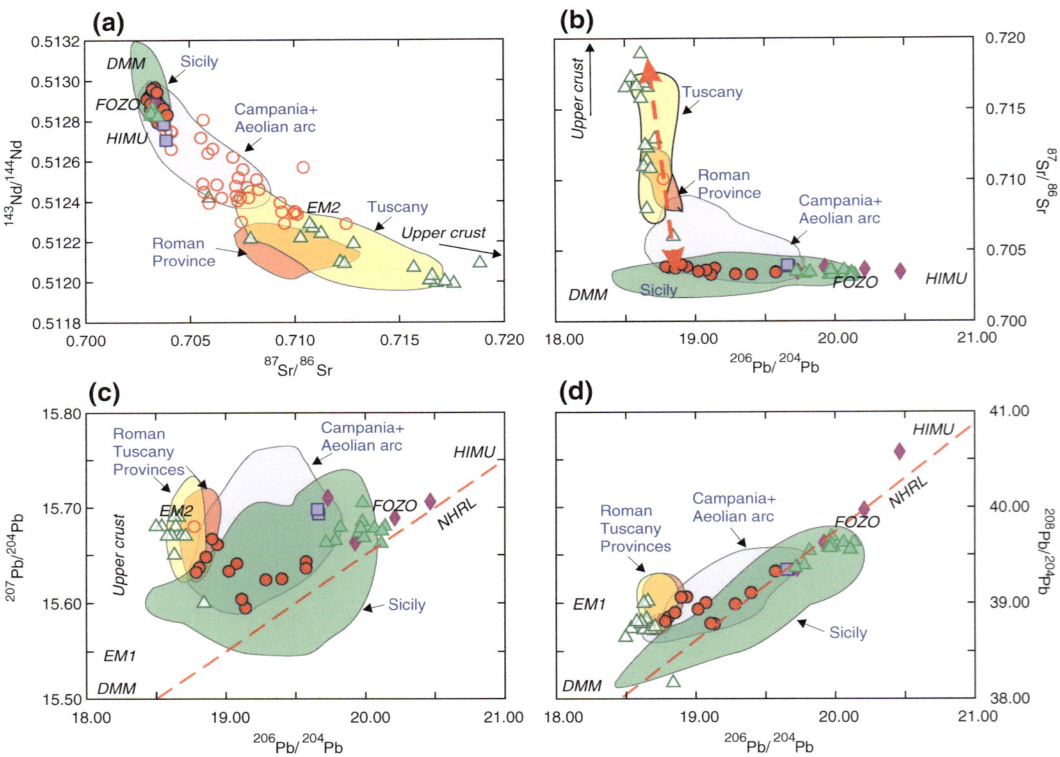

Fig. A1.4 Radiogenic isotope variation in the mafic rocks of the Alpine magmatic stage. The fields of mafic rocks of some Apennine-stage magmatic provinces are reported for comparison. Compositions of some mantle end-members (DMM, HIMU, FOZO, EM1, EM2) and the North Hemisphere Reference Line (NHRL) are also shown

The radiogenic isotope compositions of anorogenic rocks plot in the space between FOZO-HIMU and DMM-type mantle end-members (Fig. A1.4). In contrast, orogenic rocks have intermediate radiogenic isotope ratios between the upper continental crust and DMM-type mantle compositions, a feature better shown by Sr-Pb-isotope data (Fig. A1.4b). Notably, the available data do not show any intermediate isotopic ratio between the upper crust and HIMU-FOZO. Note, finally, that the Western Alps calcalkaline to ultrapotassic rocks basically overlap the mafic rocks of the Tuscany Province.

Petrogenesis and geodynamics

The petrogenesis and geodynamic setting of the Alpine magmatism are extremely complex topics whose discussion is far beyond the scopes of this book. The petrological and geochemical characteristics summarised above, however, furnish some clues that may be worth of further attention and could teach us much about igneous petrogenesis for both the Alpine and the Apennine magmatic stages.

The Alpine rocks show several characteristics in common with Apennine stage igneous activity. These include:

(a) The occurrence of both orogenic and anorogenic rock suites, suggesting two geodynamic distinct settings of magmatism, as for the Apennnine stage;
(b) The radiogenic isotope ratios of anorogenic rocks plotting between the HIMU-FOZO and DMM, indicating the interaction between depleted and enriched mantle compositions. Similar array is observed in the Sicily Magmatic Province (Chap. 8);
(c) The intermediate isotopic compositions between mantle and the upper continental crust for orogenic rocks, which indicate crust-mantle mixing, as envisaged for the Apennine magmatic stage;
(d) The close similarity between Tuscany and Western Alps mafic calcalkaline to lamproitic rocks, which calls for similar mantle sources and petrogenetic processes.

On the other hand, the Alpine and Apennine magmatic stages also show important differences. In particular:

(a) Strongly silica undersaturated ultrapotassic suites, such as those abundantly occurring in the Roman Province, are lacking in the Alpine magmatic stage;
(b) The isotopic signatures of Alpine anorogenic rocks mostly overlap the compositions of the Sicily Province. However, La Queglia rocks are closer to typical HIMU basalts than observed for any Apennine-stage volcano. Moreover, EM1-type compositions are lacking in the Alpine rocks;
(c) The mantle end-member of the mantle-crust mixing trend of Alpine orogenic rocks is substantially different from that envisaged for the Apennine stage. In the latter, the mantle end-member has high Pb-isotope ratios similar to FOZO (e.g. Etna). In contrast, the mantle end-member of the Alpine stage orogenic suite has a clear unradiogenic-Pb isotopic signature, which resembles closer DMM than FOZO or HIMU (Fig. A1.4b).

Overall, these contrasting features encourage some comments on key aspects of magmatism and geodynamics in the Alps-Apennine orogenic belts, which, although speculative, can furnish interesting hints on mantle contamination, magma origin and geodynamic settings.

1. According to Macera et al. (2003, 2008) the range of isotopic compositions in the Veneto Volcanic Province results from mixing between a depleted component similar to DMM and an enriched end-member with HIMU-type signatures. According to these authors, the depleted component originated within the lithospheric mantle, whereas deeper rocks were the source of HIMU-FOZO-type magmas. The latter would derive from a deep mantle plume head, which ascended during Late Cretaceous beneath the eastern Atlantic Ocean and successively expanded to Europe (Piromallo et al. 2008; see Chap. 13).
2. Whatever the origin of the HIMU-FOZO component, it is obvious that it had little or no role in the origin of orogenic magmatism

along the Alps, a conclusion required by the absence of intermediate isotopic compositions between FOZO-HIMU and the upper crust. In contrast, these compositions are abundantly present among Apennine stage orogenic rocks.

3. If DMM components come from the lithospheric mantle (Macera et al. 2003, 2008), the conclusion is to be drawn that the whole orogenic magmatism along the Alps was generated within a shallow contaminated mantle. Such a shallow source could be related to the dipping angle of the subducted European slab beneath the Adria plate margin, which, according to some authors (e.g. Doglioni et al. 1999), was flatter than in the Apennine system. Therefore, the asthenosphere could have not been contaminated by slab fluids and, therefore, was unable to melt and contribute to the Alpine orogenic magmatism.
4. The slab inclination and trajectory could also provide an explanation for the absence of strongly undersaturated orogenic magmas in the Alps. As discussed in Chaps. 3 and 4, the origin of strongly undersaturated ultrapotassic magmas requires contamination and melting of asthenospheric mantle sources, at pressures exceeding 2–3 GPa. A relatively flat subduction would be unable to reach and contaminate deep asthenospheric layers, which seem to be the source of strongly undersaturated orogenic magmas.
5. The absence of strongly undersaturated ultrapotassic rocks in the Alps could also depend on the nature of the subducting slab. As discussed earlier in this book, the geochemical signatures of ultrapotassic Roman-type and kamafugitic magmas in Central Italy call for mantle contamination by marly sediments. These were provided by the subducted Adriatic plate, which contains abundant carbonate rocks. In contrast, it can be hypothesised that contamination of mantle rocks by marly sediments had little or no role in the Alpine subduction system, basically because of the compositional diversity of the European plate in respect to Adria.
6. The common origin of orogenic magmatism in the Western Alps and Tuscany supports the hypothesis that the lithospheric contamination in both provinces took place during a single geodynamic event, i.e. during the east-directed subduction of the European plate beneath the northern African margin. Contamination generated a lithospheric mantle that was characterised by upper crustal trace element and isotopic signatures (the so-called CLIMA, Crustal-like Lithospheric Mantle; Peccerillo and Martinotti 2006). The post-contamination geodynamic evolution of Western Alps diverged dramatically from that of Tuscany. In the Western Alps, there was continental collision, followed by an Oligocene extensional phase that induced melting of CLIMA and the generation of calcalkaline to ultrapotassic rocks. In Tuscany, there was an inversion of the subdution direction, which generated Middle Miocene to Pleistocene backarc dismembering and melting of CLIMA with the generation of calcalkaline to lamproitic magmas compositionally similar to Western Alps (see Chaps. 2 and 13).
7. An additional important implication of the similar radiogenic isotope signatures of Western Alps and Tuscany rocks is that the Tuscany magmatism is not a member of a continuous mixing suite between FOZO and upper crustal material, as it has been suggested by some authors (e.g. Bell et al. 2013). Rather, rocks from both Tuscany and Western Alps represent depleted mantle sources that were contaminated by upper crustal material, without any role of FOZO or other OIB-type components. This conclusion fits well the hypothesis discussed in Chap. 13 that the orogenic magmatism of the Apennine stage is the result of multiple metasomatic events affecting compositional distinct mantle sources rather than the consequence of two-end-member mixing between mantle and upper crust.

Appendix 2: Classification of K-rich Rocks

Classification and Nomenclature

Rocks enriched in potassium and containing abundant potassium minerals (e.g. leucite, phlogopite, K-feldspar) are found in several tectonic environments, including continental cratons, active subduction zones and post-collisional settings (e.g. Müller and Groves 2000; Gupta and Fife 2003; Gupta 2015). They are limited in abundance over a global scale but are widespread in Italy (Peccerillo 1992, 2003). Potassium-rich rocks cover a very wide range of petrological and geochemical compositions, from alkaline to transitional, from strongly undersaturated to oversaturated in silica, from ultrabasic (SiO_2 < 45.0 wt%) and basic (SiO_2 = 45.0–52.0 wt%) to intermediate (SiO_2 = 52.0–60.0 wt%). Mineralogical composition is even more variable, being dependent not only on the chemical composition of magmas but also on the pressure-temperature conditions of crystallisation (Yoder 1986). Consequently, rocks with similar major and trace element compositions sometimes display different mineralogical compositions (*heteromorphism*). As a consequence, nomenclature based on modal mineralogy uses a plethora of names to designate these rocks (wyomingite, vesbite, verite, coppaellite, and many others; see Yoder 1986).

In order to overcome this problem, it is appropriate to define and classify potassic rocks on the basis of chemical parameters. *Potassic* or *potassium-rich* are very general terms to indicate rocks enriched in potassium relative to sodium. Limits of potassium enrichment, however, are extremely flexible, and change from one author to author and/or with the particular group of rocks under consideration. K_2O concentration of about 2 wt% and K_2O/Na_2O > 1%, at MgO > 3.5–4.0 wt%, are used by many authors as lower limits of potassic rock compositions. For instance, Middlemost (1986) proposed $0.5 < K_2O/Na_2O < 2$ and $K_2O/Na_2O > 2$ to define "potassic" and "high-potassic" groups, respectively, at $SiO_2 \leq 53.5$ wt %. Restriction to mafic (or basic) compositions is necessary to exclude those rocks, such as some dacites and rhyolites, whose potassium enrichment is related to magma evolution rather than to a potassium-rich nature of parental melts.

In this book, *potassic rocks* are defined as a very wide group of rocks that have K_2O higher than about 2 wt % and K_2O/Na_2O (expressed as wt%) ranging from 0.8 to 2.5 at a MgO > 4 wt % and SiO_2 < 55–57 wt %. A *Shoshonitic group* having K_2O/Na_2O = 0.8–1.5 is distinguished among potassic rocks. *Ultrapotassic rocks* have K_2O > 3 wt % and K_2O/Na_2O > 2.5 at a MgO > 4 wt % or SiO_2 < 55 wt% (see Chap. 1, Fig. 1.3). As for any classification scheme, the criteria outlined above are questionable and not applicable on a global scale. However, in this author

A. Peccerillo, *Cenozoic Volcanism in the Tyrrhenian Sea Region*,
Advances in Volcanology, DOI 10.1007/978-3-319-42491-0

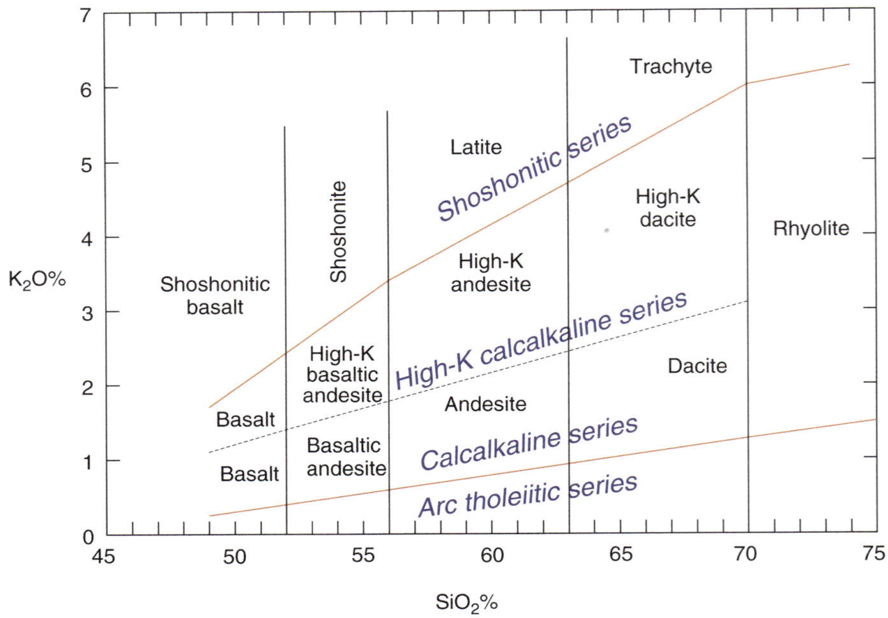

Fig. A2.1 K_2O versus SiO_2 classification grid for arc rocks. Modified from Peccerillo and Taylor (1976)

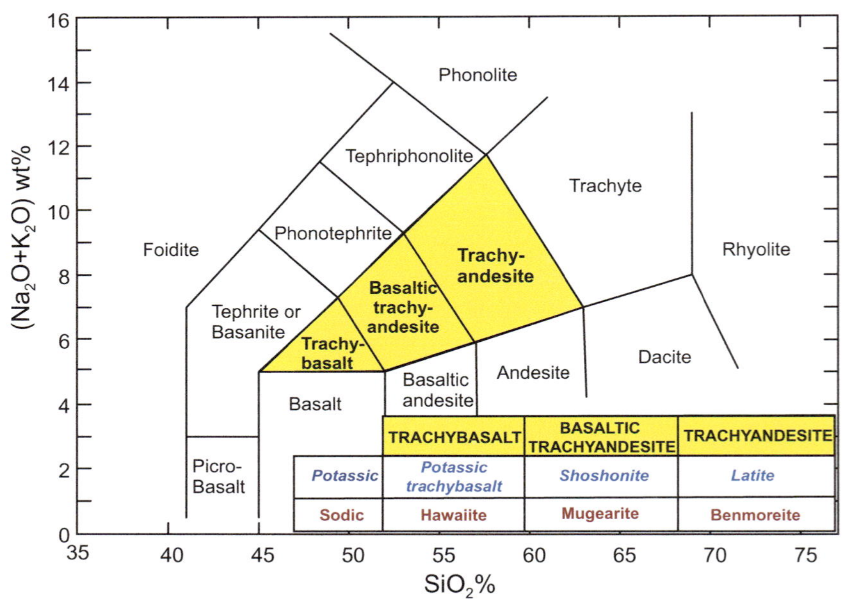

Fig. A2.2 Total alkali versus silica (TAS) classification diagram of volcanic rocks (modified from Le Maitre 2002). Rocks falling in the shaded area may be subdivided as shown in the inset table, on the basis of their sodic (Na_2O-2.0 $\geq$ K_2O) or potassic (Na_2O-2.0 $\leq$ K_2O) affinity opinion, they are well suitable to classify rocks in the Tyrrhenian Sea region.

The K_2O versus SiO_2 and the TAS diagrams of Peccerillo and Taylor (1976; Fig. A2.1) and Le Maitre (2002; Fig. A2.2) are commonly used to classify potassium-rick rocks, and these schemes are followed, as close as possible, in this book.

However, the extreme compositional variability, which pertains to many other major petrological parameters in addition to silica and alkalis, makes K_2O versus SiO_2 and TAS diagrams insufficient to express petrochemical affinities and variability of potassium-rich magmas. Therefore, some additional information has to be given in order to better explain the nomenclature used through the book.

Following several authors (e.g. Morrison 1980; Foley et al. 1987) the potassium-rich rocks have been divided into several suites: 1—Shoshonitic series; 2—Potassic series (KS); 3—Roman-type high potassic series (HKS); 4—Lamproites; 5—Kamafugites.

Shoshonitic series are widespread in the Tyrrhenian Sea orogenic volcanoes, especially in the central-eastern Aeolian arc and the Italian peninsula. The mafic rocks of this suite are slightly oversaturated to undersaturated in silica and have K_2O around 1.5–2.5 wt % (Peccerillo and Taylor 1976) and K_2O/Na_2O around unity. The evolved rocks show decreasing FeO_{total}, MgO and CaO with increasing silica. Overall, they fall in the shoshonitic basalt (K-trachybasalt), shoshonite, latite, trachyte fields on the K_2O versus SiO_2 and TAS diagrams (Figs. A2.1 and A2.2). Some K-rich alkaline rhyolites may also belong to the shoshonitic suite. High Al_2O_3 and CaO in the mafic-intermediate magmas favour crystallisation of plagioclase as a main phenocryst and groundmass mineral.

Potassic series (KS) has similar composition to shoshonitic suites, but slightly higher $K_2O\%$ (about 2.5–3.0 in the mafic range) and K_2O/Na_2O ($\sim$1.5–2.5). Mafic KS rocks are saturated to slightly undersaturated in silica, with the possible occurrence of leucite, especially in the groundmass. The entire rock suite includes shoshonitic basalt or K-trachybasalts, shoshonite, latite and trachyte, similar to the shoshonitic suite. Potassic rocks show the same occurrence as shoshonitic rocks in the Tyrrhenian Sea.

Roman-type high-potassium series (HKS) are ultrapotassic ($K_2O > 3$ wt %, $K_2O/Na_2O > 2.5$ in the mafic compositions) undersaturated in silica, and rich in Al_2O_3 ($\sim$12–20 wt % ca). The mafic types have high CaO, typically around 10–13 wt % and Na_2O around 2–3 wt % (Fig. A2.3). Rocks of this series fall in the foidite, basanite-tephrite, phonotephrite, tephriphonolite and phonolite fields in the TAS diagram. Leucite and plagioclase, along with clinopyroxene are common and sometimes abundant phenocryst phases of these rocks.

Lamproites are ultrapotassic, slightly undersaturated to oversaturated in silica, and typically have low Al_2O_3 (<10–11 wt %), Na_2O (<1.5–2 wt%) and CaO (<6–7 wt %; Fig. A2.3). SiO_2 is variable (< about 45% to 60 wt%) in the Cenozoic lamproites from the Tyrrhenian Sea area, but Tuscany occurrences have SiO_2 $\sim$55–60% and are named high-silica lamproites. Mg# is high to very high in most of these lamproites, sometimes reaching 75–80. This suggests that silica variations do not depend on magma evolution processes, but likely represent primary features of mantle-equilibrated melts. The mineralogy of lamproites is plagioclase-free and consists of highly magnesian olivine, Al-poor diopside, phlogopite, sanidine, K-richterite, and several uncommon phases such as priderite (K–Ba–Fe^{3+}-Ti oxide) and rutile.

Kamafugites owe their name to the katungite-mafurite-ugandite series of eastern Africa (Sahama 1974). They are found in the Intra-Apennine Province, with some occurrences at Colli Albani and Ernici. Kamafugitic rocks are ultrapotassic and share with lamproites low Al_2O_3 and Na_2O and high MgO abundances (Fig. A2.3).

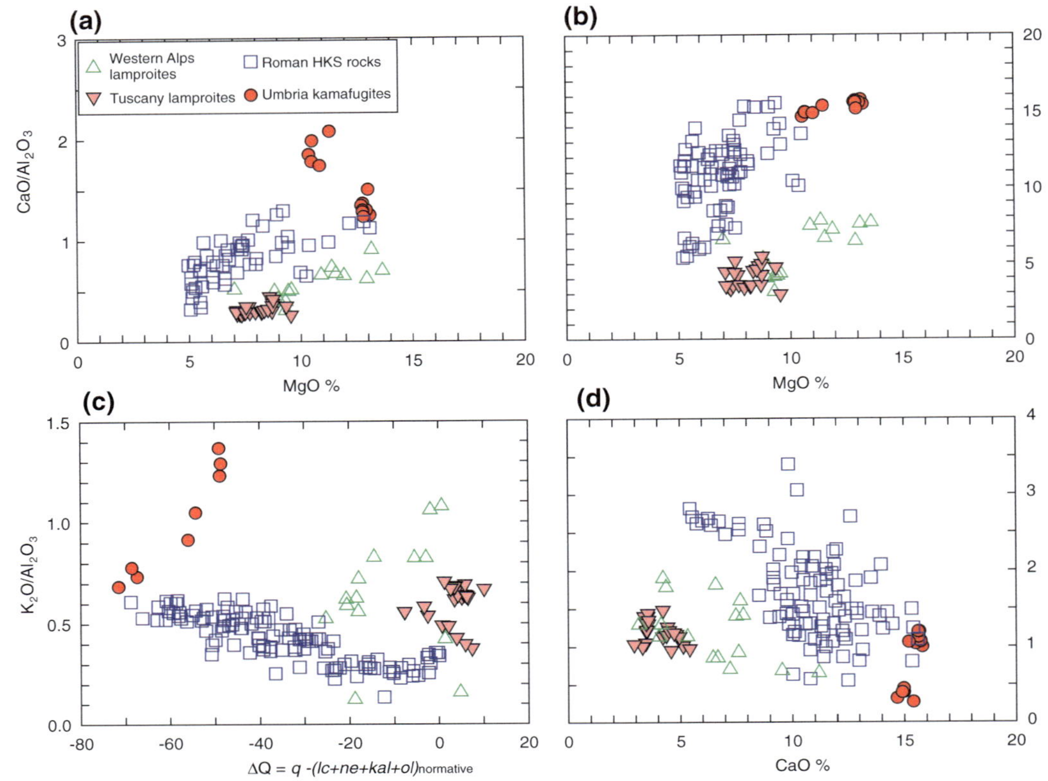

Fig. A2.3 Selected major element plots of ultrapotassic rocks from Central Italy and the Western Alps

However, they are rich in CaO (up to 18 wt %) and strongly undersaturated in silica. Typical minerals include melilite, leucite, kalsilite, Mg-rich olivine, diopside, monticellite, phlogopite, and perovskite. Most kamafugites fall in the foidite field in the TAS diagram. Based on modal mineralogy, kamafugitic rocks are classified as olivine kalsilite melilitites.

References

Alagna K, Peccerillo A, Martin S, Donati D (2010) Tertiary to present evolution of orogenic magmatism in Italy. J Virtual Explorer 36, paper 18. doi:10.3809/jvirtex.2010.00233

Avanzinelli R, Sapienza GT, Conticelli S (2012) The Cretaceous to Paleogene within-plate magmatism of Pachino-Capo Passero (southeastern Sicily) and Adria (La Queglia and Pietre Nere, southern Italy): geochemical and isotopic evidence against a plume-related origin of circum-Mediterranean magmas. Eur J Mineral 24:73–96

Beccaluva L, Bianchini G, Bonadiman C, Natali C, Siena F (2010) Petrogenesis and geodynamic control of intraplate Cenozoic volcanism in Italy. In: Beltrando M, Peccerillo A, Mattei M, Conticelli S, Doglioni C, The Geology of Italy: tectonics and life along plate margins, J Virtual Explorer 36, paper 19 doi: 10.3809/jvirtex.2010.00240

Beltrando M, Peccerillo A, Mattei M, Conticelli S, Doglioni C (eds) (2010) The geology of Italy: tectonics and life along plate margins. J Virtual Explorer 36

Bianchini G, Beccaluva L, Siena F (2008) Post-collisional and intraplate Cenozoic volcanism in the rifted Apennines/Adriatic domain. Lithos 101:125–140

Boyet M, Lapierre H, Tardy M, Bosch D, Maury R (2001) Nature des sources des composants andesitiques des Gres du Champsaur et des Gres de Taveyannaz; implications dans l'evolution des Alpes occidentales au Paleogene. Bull Soc Geol Fr 172:487–501

Conticelli S, Guarnieri L, Farinelli A, Mattei M, Avanzinelli R, Bianchini G, Boari E, Tommasini S, Tiepolo M, Prelevic D, Venturelli G (2009) Trace elements and Sr-Nd-Pb isotopes of K-rich, shoshonitic, and calc-alkaline magmatism of the Western Mediterranean Region: genesis of ultrapotassic to calc-alkaline magmatic associations in a post-collisional geodynamic setting. Lithos 107:68–92

De Astis G, Kempton PD, Peccerillo A, Wu TW (2006) Trace element and isotopic variations from Mt. Vulture to Campanian volcanoes: constraints for slab detachment and mantle inflow beneath southern Italy. Contrib Mineral Petrol 151:331–351

Doglioni C, Harabaglia P, Merlini S, Mongelli F, Peccerillo A, Piromallo C (1999) Orogens and slabs vs. their direction of subduction. Earth Sci Rev 45:167–208

Foley SF, Venturelli G, Green DH, Toscani L (1987) The ultrapotassic rocks: characteristics classification, and constraints for petrogenetic models. Earth Sci Rev 24:81–134

Gupta AK (2015) Origin of potassium-rich silica-deficient igneous rocks. Springer, New Delhi, 536 pp

Gupta AK, Fyfe WS (2003) The young potassic rocks. ANE Books, New Delhi, 370 pp

Le Maitre RW (ed) (2002) A classification of igneous rocks and glossary of terms. Cambridge University Press, Cambridge, 252 pp

Lustrino M, Duggen S, Rosenberg CL (2011) The Central-Western Mediterranean: anomalous igneous activity in an anomalous collisional tectonic setting. Earth Sci Rev 104:1–40

Macera P, Gasperini D, Piromallo C, Blichert-Toft J, Bosch D, Del Moro A, Martin S (2003) Geodynamic implications of deep mantle upwelling in the source of Tertiary volcanics from the Veneto region (South-Eastern Alps). J Geodyn 36:563–590

Macera P, Gasperini D, Ranalli G, Mahatsente R (2008) Slab detachment and mantle plume upwelling in subduction zones: an example from the Italian South-Eastern Alps. J Geodyn 45:32–48

Martin S, Macera P (2014) Tertiary volcanism in the Italian Alps (Giudicarie fault zone, NE Italy): insight for double alpine magmatic arc. It J Geosci 133:63–84

Mattioli M, Di Battistini G, Zanzucchi G (2002) Geochemical features of the tertiary buried Mortara volcanic body (Northern Apennines, Italy). Boll Soc Geol It, Spec 1:239–249

Mattioli M, Lustrino M, Ronca S, Bianchini G (2012) Alpine subduction imprint in Apennine volcaniclastic rocks. Geochemical–petrographic constraints and geodynamic implications from Early Oligocene Aveto-Petrignacola Formation (N Italy). Lithos 134–135:210–220

Middlemost EAK (1975) The basalt clan. Earth Sci Rev 11:337–364

Morrison GW (1980) Characteristics and tectonic setting of the shoshonite rock association. Lithos 13:97–108

A. Peccerillo, *Cenozoic Volcanism in the Tyrrhenian Sea Region*,
Advances in Volcanology, DOI 10.1007/978-3-319-42491-0

Müller D, Groves DI (2000) Potassic igneous rocks and associated gold-copper mineralizations. Springer, Berlin, pp 252

Peccerillo A (1992) Potassic and ultrapotassic magmatism: compositional characteristics, genesis and geologic significance. Episodes 15:243–251

Peccerillo A (2003) Plio-Quaternary magmatism in Italy. Episodes 26:222–226

Peccerillo A, Martinotti G (2006) The Western Mediterranean lamproitic magmatism: origin and geodynamic significance. Terra Nova, 18:109–117

Peccerillo A, Taylor SR (1976) Geochemistry of Eocene calc-alkaline volcanic rocks of the Kastamonu area, northern Turkey. Contrib Mineral Petrol 58:63–81

Piromallo C, Gasperini D, Macera P, Faccenna C (2008) A late Cretaceous contamination episode of the European-Mediterranean mantle. Earth Planet Sci Lett 268:15–27

Sahama TG (1974) Potassium-rich alkaline rocks. In: Sorensen H (ed) The alkaline rocks. Wiley, London, pp 94–109

Sun SS, McDonough WF (1989) Chemical and isotopic systematics of oceanic basalts: implications for mantle composition and processes. In: Saunders AD, Norry MJ (eds) Magmatism in ocean basins. Geol Soc London Spec Publ 42:313–345

Yoder HS (1986) Potassium-rich rocks: phase analysis and heteromorphic relations. J Petrol 27:1215–1228

Index

A. Peccerillo, *Cenozoic Volcanism in the Tyrrhenian Sea Region*, Advances in Volcanology, DOI 10.1007/978-3-319-42491-0

MIX
Papier aus verantwortungsvollen Quellen
Paper from responsible sources
FSC® C105338

If you have any concerns about our products,
you can contact us on
ProductSafety@springernature.com

In case Publisher is established outside the EU,
the EU authorized representative is:
Springer Nature Customer Service Center GmbH
Europaplatz 3, 69115 Heidelberg, Germany

Printed by Libri Plureos GmbH
in Hamburg, Germany